Epidemiology of Brain and Spinal Tumors

Epidemiology of Brain and Spinal Tumors

Jahangir Moini, MD, MPH
Professor of Science and Health (Retired)
Eastern Florida State College,
Palm Bay, FL, United States

Nicholas G. Avgeropoulos, MD
Director of the Brain and Spine Tumor Center and
Medical Director of Neuro-Oncology
Orlando Health / UF Health Center

Mohtashem Samsam, MD, PhD
Director of Research and Professor
Advent Health University,
Orlando, FL, United States

Academic Press is an imprint of Elsevier
125 London Wall, London EC2Y 5AS, United Kingdom
525 B Street, Suite 1650, San Diego, CA 92101, United States
50 Hampshire Street, 5th Floor, Cambridge, MA 02139, United States
The Boulevard, Langford Lane, Kidlington, Oxford OX5 1GB, United Kingdom

Copyright © 2021 Elsevier Inc. All rights reserved.

No part of this publication may be reproduced or transmitted in any form or by any means, electronic or mechanical, including photocopying, recording, or any information storage and retrieval system, without permission in writing from the publisher. Details on how to seek permission, further information about the Publisher's permissions policies and our arrangements with organizations such as the Copyright Clearance Center and the Copyright Licensing Agency, can be found at our website: www.elsevier.com/permissions.

This book and the individual contributions contained in it are protected under copyright by the Publisher (other than as may be noted herein).

Notices

Knowledge and best practice in this field are constantly changing. As new research and experience broaden our understanding, changes in research methods, professional practices, or medical treatment may become necessary.

Practitioners and researchers must always rely on their own experience and knowledge in evaluating and using any information, methods, compounds, or experiments described herein. In using such information or methods they should be mindful of their own safety and the safety of others, including parties for whom they have a professional responsibility.

To the fullest extent of the law, neither the Publisher nor the authors, contributors, or editors, assume any liability for any injury and/or damage to persons or property as a matter of products liability, negligence or otherwise, or from any use or operation of any methods, products, instructions, or ideas contained in the material herein.

British Library Cataloguing-in-Publication Data
A catalogue record for this book is available from the British Library

Library of Congress Cataloging-in-Publication Data
A catalog record for this book is available from the Library of Congress

ISBN: 978-0-12-821736-8

For Information on all Academic Press publications
visit our website at https://www.elsevier.com/books-and-journals

Publisher: Nikki Levy
Acquisitions Editor: Natalie Farra
Editorial Project Manager: Hilary Carr
Production Project Manager: Selvaraj Raviraj
Cover Designer: Mark Rogers

Typeset by MPS Limited, Chennai, India

Dedication

Dr. Moini: To the memory of my parents, who taught me the value of perseverance and hard work.

To my wife Hengameh, my daughters Mahkameh and Morvarid, and also my precious granddaughters Laila and Anabelle, thanks for your understanding and tremendous support.

Dr. Avgeropoulos: I would like to dedicate this effort to the man I admire most, his namesake, and my namesake who followed after that. May we all relentlessly seek to separate what we believe from what we know; in doing so, may we ever implement the knowledge that brings forth change for good.

Dr. Samsam: I dedicate this book to my wife and my precious daughter.

Contents

About the authors	xiii
Preface	xv
Acknowledgments	xvii

Part I
Structure, function, and development

1. Anatomy and physiology — 3

- Major subdivisions of the nervous system — 3
- The brain — 3
  - Cerebral hemispheres — 5
  - Diencephalon — 9
  - Brain stem — 12
- Blood supply of the brain — 13
- Functional brain systems — 13
  - Limbic system — 14
  - Reticular formation — 19
- Higher mental functions — 19
  - Language — 20
  - Memory — 21
  - Sleep and sleep-wake cycles — 22
- Meninges — 23
  - Cerebrospinal fluid — 23
- Spinal cord and reflex center — 25
  - Structure of the spinal cord — 26
  - Neuronal pathways — 27
- Peripheral nervous system — 29
  - Sensory receptors — 30
  - Cranial nerves — 31
  - Spinal nerves — 33
- Autonomic nervous system — 33
  - Sympathetic division — 33
  - Parasympathetic division — 37
- Key terms — 38
- Further reading — 40

2. Cytology of the nervous system — 41

- Glial cells — 41
  - Glial cells of the central nervous system — 41
  - Glial cells of the peripheral nervous system — 43
- Neurons — 44
  - Functional characteristics — 44
  - Structure — 45
  - Classification — 47
- Membrane potential — 49
  - Resting membrane potential — 49
  - Membrane channel effects on resting membrane potential — 50
  - Graded potentials — 51
  - Action potentials — 52
  - Generation of action potentials — 53
  - Action potential propagation — 54
  - Threshold and the all-or-none principle — 55
  - Axon diameter and propagation speed — 55
- Synapses — 56
  - Structure — 56
  - Classification — 56
  - Function of electrical synapses — 57
  - Function of chemical synapses — 57
- Neuronal integration of stimuli — 59
  - Postsynaptic potentials — 59
  - Presynaptic regulation — 60
  - Rate of action potential generation — 60
- Nerve cell degeneration and regeneration — 61
- Key terms — 62
- Further reading — 63

3. Embryology — 65

- Neural tube — 65
- Neural crest — 69
- Cranial placodes — 70
- Medulla spinalis — 71
- Myelencephalon — 71
- Metencephalon — 75
- Mesencephalon — 75
- Diencephalon — 75
- Telencephalon — 75
- Central nervous system malformation — 77
- Key terms — 78
- Further reading — 79

Part II
Epidemiology

4. Global epidemiology of brain and spinal tumors (in comparison with other cancers) — 83

- World population — 83
- Aging of the population — 83
- Life expectancy — 85
- Distribution of brain and spinal tumors by age, gender, and race — 87
- Attributable deaths — 91
- Global prevalence — 92
- Disability adjusted life years — 93
- Quality adjusted life years — 94
- Global mortality from brain and spinal tumors — 94
- Burden of brain and spinal tumors in the United States — 95
- Mortality from brain and spinal tumors in the United States — 96
- Comparison of brain and spinal tumors with other cancers — 97
- Key terms — 100
- Further reading — 100

5. Prevalence and incidence in epidemiology — 103

- Prevalence — 103
- Incidence — 104
- Incidence rate — 104
- Cumulative incidence — 105
- Measures of disease frequency — 106
- Relationship between prevalence and incidence — 106
- Prevalence of brain and spinal tumors — 106
- Key terms — 108
- Further reading — 108

Part III
Classifications of tumors

6. Astrocytoma — 111

- Overview — 111
- Astrocytoma classifications — 111
  - Pilocytic astrocytoma — 112
  - Subependymal giant cell astrocytoma — 115
  - Anaplastic astrocytoma — 121
- Clinical cases — 123
- Key terms — 126
- Further reading — 126

7. Glioblastoma — 129

- Glioblastoma — 129
  - Isocitrate dehydrogenase-wild-type — 139
  - Isocitrate dehydrogenase-mutant — 141
  - Gliosarcoma — 141
- Clinical cases — 141
- Key terms — 144
- Further reading — 144

8. Oligodendroglioma — 147

- Oligodendroglioma — 147
  - IDH-mutant and 1p/19q-codeleted oligodendroglioma — 147
  - NOS oligodendroglioma — 151
- Anaplastic oligodendroglioma (IDH-mutant and 1p/19q-codeleted) — 153
- NOS anaplastic oligodendroglioma — 156
- Clinical cases — 157
- Key terms — 158
- Further reading — 159

9. Ependymal tumors — 161

- Classic ependymoma — 161
- Subependymoma — 165
- Myxopapillary ependymoma — 168
- Papillary ependymoma — 171
- Clear cell ependymoma — 171
- Tanycytic ependymoma — 171
- RELA fusion-positive ependymoma — 171
- Anaplastic ependymoma — 172
- Ependymoblastoma — 173
- Clinical cases — 174
- Key terms — 176
- Further reading — 176

10. Schwannoma — 179

- Schwannoma — 179
  - Cellular schwannoma — 184
  - Plexiform schwannoma — 185
  - Melanotic schwannoma — 188
- Schwannomatosis — 190
- Clinical cases — 192
- Key terms — 195
- Further reading — 195

11. Neurofibroma 197

Neurofibroma	197
Atypical neurofibroma	200
Neurofibromatosis type 1	202
Neurofibromatosis type 2	207
Clinical cases	211
Key terms	213
Further reading	213

12. Meningioma 215

Meningioma	215
Meningothelial meningioma	219
Fibrous meningioma	220
Psammomatous meningioma	220
Transitional meningioma	220
Metaplastic meningioma	221
Angiomatous meningioma	221
Microcystic meningioma	222
Secretory meningioma	222
Lymphoplasmacyte-rich meningioma	222
Chordoid meningioma	222
Atypical meningioma	223
Clear cell meningioma	225
Anaplastic meningioma	225
Rhabdoid meningioma	227
Papillary meningioma	227
Clinical cases	227
Key terms	230
Further reading	231

13. Choroid plexus tumors 233

Choroid plexus papilloma	233
Atypical choroid plexus papilloma	236
Choroid plexus carcinoma	238
Clinical cases	239
Key terms	241
Further reading	241

14. Mesenchymal tumors 243

Overview	243
Hemangioblastoma	243
Hemangiopericytoma	246
Hemangioma	249
Epithelioid hemangioendothelioma	252
Angiosarcoma	253
Chondrosarcoma	255
Ewing sarcoma	256
Fibrosarcoma	257
Kaposi sarcoma	258
Leiomyosarcoma	260
Liposarcoma	261
Osteosarcoma	262
Rhabdomyosarcoma	264
Lipoma	266
Angiolipoma	268
Desmoid-type fibromatosis	269
Inflammatory myofibroblastic tumor	270
Myofibroblastoma	271
Benign fibrous histiocytoma	272
Malignant fibrous histiocytoma	273
Leiomyoma	275
Rhabdomyoma	275
Chondroma	276
Osteochondroma	278
Osteoma	279
Hibernoma	280
Clinical cases	280
Key terms	283
Further reading	283

15. Pituitary tumors 285

Overview	285
Pituitary adenoma	285
Somatotroph adenoma	291
Prolactinoma	295
Thyrotropin-secreting tumors	297
Adrenocorticotropic hormone–secreting tumors	299
Craniopharyngioma	302
Nonfunctioning pituitary tumors	305
Pituitary carcinoma	307
Pituicytoma	312
Spindle cell oncocytoma	314
Secondary tumors	317
Various parasellar masses	318
Clinical cases	319
Key terms	321
Further reading	322

16. Pineal parenchymal tumors 323

Overview	323
Pineocytoma	323
Pineal parenchymal tumor with intermediate differentiation	326
Papillary tumor of the pineal region	328
Pineoblastoma	330
Clinical cases	333
Key terms	335
Further reading	335

17. Melanocytic tumors — 337

- Overview — 337
- Meningeal melanoma — 337
- Meningeal melanocytoma — 339
- Meningeal melanomatosis — 340
- Meningeal melanocytosis — 342
- Clinical cases — 344
- Key terms — 346
- Further reading — 346

18. Primary lymphoma of the brain — 349

- Overview — 349
- Diffuse large B-cell lymphoma of the central nervous system — 349
- Intravascular large B-cell lymphoma — 353
- Low-grade B-cell lymphomas — 355
- Immunodeficiency-associated central nervous system lymphomas — 356
- T-cell and NK/T-cell lymphomas — 357
- MALT lymphoma of the dura — 359
- Anaplastic large cell lymphoma — 360
- Clinical cases — 361
- Key terms — 365
- References — 365

19. Li–Fraumeni syndrome — 367

- Overview — 367
- Li–Fraumeni syndrome — 367
- Clinical cases — 373
- Key terms — 376
- Further reading — 376

20. Turcot syndrome — 379

- Overview — 379
- Turcot syndrome — 379
  - Brain tumor-polyposis syndrome 1 — 379
  - Brain tumor-polyposis syndrome 2 — 381
- Clinical cases — 384
- Key terms — 386
- Further reading — 386

21. Von Hippel–Lindau disease — 389

- Von Hippel–Lindau disease — 389
- Clinical cases — 394
- Key terms — 396
- Further reading — 396

22. Cowden syndrome — 399

- Cowden syndrome — 399
- Clinical cases — 403
- Key terms — 405
- Further reading — 405

23. Tuberous sclerosis — 407

- Tuberous sclerosis — 407
- Clinical cases — 413
- Key terms — 416
- Further reading — 416

24. Brain tumors in children — 417

- Overview — 417
- Germ cell tumors — 417
  - Germinoma — 421
  - Teratoma — 423
  - Yolk sac tumor — 426
  - Embryonal carcinoma — 428
  - Choriocarcinoma — 429
  - Mixed germ cell tumors — 431
  - Medulloblastoma — 432
- Clinical cases — 433
- Key terms — 435
- Further reading — 436

25. Peripheral nerve sheath tumors — 439

- Hybrid nerve sheath tumors — 439
- Malignant peripheral nerve sheath tumors — 442
  - Malignant peripheral nerve sheath tumor with divergent differentiation — 444
  - Epithelioid malignant peripheral nerve sheath tumor — 446
  - Malignant peripheral nerve sheath tumor with perineurial differentiation — 449
- Perineurioma — 450
- Clinical cases — 451
- Key terms — 453
- Further reading — 454

26. Metastatic tumors — 455

- Metastatic tumors — 455
- Clinical cases — 461
- Key terms — 463
- Further reading — 463

27. Spinal cord tumors 465

Spinal cord tumors 465
   Extradural tumors 469
   Intradural tumors 471
Clinical cases 474
Key terms 476
Further reading 476

28. Rare brain tumors 479

Overview 479
Rare brain tumors 479
   Atypical teratoid-rhabdoid tumor 479
   Diffuse midline gliomas 483
   Gliomatosis cerebri 487
   Pleomorphic xanthoastrocytoma 488
   Anaplastic pleomorphic xanthoastrocytoma 491
   Primitive neuro-ectodermal tumors 493
Clinical cases 495
Key terms 497
Further reading 497

Glossary 499
Index 519

About the authors

Dr. **Jahangir Moini** was an assistant professor at Tehran University, Medical School, Department of Epidemiology and Preventive Medicine, Iran for 9 years. For 18 years, he was the Director of Epidemiology for the Brevard County Health Department, United States. For 15 years, he was the Director of Science and Health for Everest University in Melbourne, FL, United States. He was also a Professor of Science and Health at Everest for a total of 24 years. For 6 years, he was a Professor of Science and Health at Eastern Florida State College, United States, but is now retired. Dr. Moini has been actively teaching for 39 years and has been an international author of 45 books for 20 years. His *Anatomy and Physiology for Health Professionals* has been translated into the Japanese and Korean languages.

Dr. **Nicholas G. Avgeropoulos** is the Director of the Brain and Spine Tumor Center at Orlando Health/UF Health Cancer Center, United States, where he also serves as the Medical Director of Neuro-oncology. He is board-certified in Neurology and Medical Neuro-oncology and has been in clinical practice for over 20 years, with a particular focus on clinical research. His current research affiliation is with the Brain Tumor Trials Collaborative, a National Cancer Institute centered consortium. Dr. Avgeropoulos has received fellowship training in neurovirology/neuroimmunology at the Medical University of South Carolina, in neuropathology at Yale New Haven Hospital and neuro-oncology at Massachusetts General Hospital.

Dr. **Mohtashem Samsam** is a Director of Research and Professor at Advent Health University, Orlando, FL, United States, and a former Professor of Medicine and a faculty at the Burnett School of Biomedical Sciences and College of Medicine at the University of Central Florida. Dr. Samsam studied medicine in the English language program of Albert Szent-Györgyi Medical University, in Szeged, Hungary (1991–96) and received his PhD from the Department of Cell Biology and Pathology, Faculty of Medicine, University of Salamanca, Spain, in 2002. He completed his postdoc studies at Wuerzburg University, Germany (1999–2002).

Preface

The authors welcome you to this first edition of *Epidemiology of Brain and Spinal Tumors*. Our goals are to discuss the global epidemiology of these tumors, with a detailed focus on the clinical considerations. The book will meet the needs of readers by providing concrete information about structures and functions, cytology, epidemiology (including molecular epidemiology), etiology, risk factors, pathology, clinical manifestations, diagnosis, treatment, and prognosis. It has been written to provide logical, step-by-step information on the brain and spinal tumors. The chapters are formatted to include clinical considerations and references. "Point to Remember" boxes feature certain conditions related to various tumors. Tables summarize information into easy-to-remember topics throughout the chapters. Clinical cases at the end of the chapters help to give real-life examples of brain and spinal tumors. The book is organized into three parts that contain a total of 28 chapters. It is designed for graduate and postgraduate students including medical students; residents in internal medicine, neurology, and oncology; nurse practitioners; and physician assistants.

Acknowledgments

The authors appreciate the contributions of everyone who assisted in the creation of this book, especially Cathleen Sether—Publishing Director; Nikki Levy—Publisher; Natalie Farra—Senior Acquisitions Editor, Neuroscience; Hilary Carr—Content Manager; Selvaraj Raviraj—Project Manager; and Greg Vadimsky. The authors also wish to acknowledge Dr. Morvarid Moini for contributing the necessary artwork.

Part I

Structure, function, and development

Chapter 1

Anatomy and physiology

Chapter Outline

Major subdivisions of the nervous system	3	Cerebrospinal fluid	23
The brain	3	**Spinal cord and reflex center**	25
Cerebral hemispheres	5	Structure of the spinal cord	26
Diencephalon	9	Neuronal pathways	27
Brain stem	12	**Peripheral nervous system**	29
Blood supply of the brain	13	Sensory receptors	30
Functional brain systems	13	Cranial nerves	31
Limbic system	14	Spinal nerves	33
Reticular formation	19	**Autonomic nervous system**	33
Higher mental functions	19	Sympathetic division	33
Language	20	Parasympathetic division	37
Memory	21	**Key terms**	38
Sleep and sleep-wake cycles	22	**Further reading**	40
Meninges	23		

Major subdivisions of the nervous system

The central nervous system (CNS) consists of the brain and spinal cord. It works with the peripheral nervous system (PNS) to regulate body functions. The PNS includes nervous tissue that is outside of the CNS and includes the cranial and spinal nerves (sensory and motor). The subdivisions of the nervous system are shown in Fig. 1.1. The motor nerves are subdivided into somatic and autonomic nerves. The autonomic nervous system (ANS) is subdivided into the sympathetic nervous system and parasympathetic nervous system. The PNS is mostly made up of axons of the sensory and motor neurons, which pass between the CNS and the rest of the body. It contains the neurons innervating the secretory glands, cardiac muscle, and smooth muscle. Primarily, the ANS controls the internal environment of the body. Neurons in the gastrointestinal tract form the **enteric nervous system (ENS)** and sustain local reflex activity independent of the CNS.

The brain

The human brain coordinates everything that we see, hear, evaluate, and distinguish. Nearly 97% of the body's nervous tissue is contained within the adult human brain. The four visible brain regions include the cerebral hemispheres, diencephalon, brain stem, and cerebellum. The major regions of the adult brain are shown in Fig. 1.2. There is a unique distribution of gray and white matter throughout the brain. The gray matter is made up of short, nonmyelinated neurons and neuron cell bodies. The white matter consists of mostly myelinated and some nonmyelinated axons. Think of nerves as electrical wires where the bare copper is covered by a plastic coating to decrease waste of energy and to increase electrical conduct. Nerves are no different, and axons are similar to the bare copper, while the myelin sheath is similar to the plastic coating. In both cases, this increases the electrical conduction. The *cortex* is the gray colored outer layer of the brain, which is involved in higher mental and physical functions. The cortex is connected to the spinal cord thorough a network of tracts. These tracts send signals to or from the cortex in a fraction of a second, and therefore, these tracts all consist of myelinated axons that conduct electricity (action potential) much faster than the unmyelinated axons. Since myelin protein is white in color, areas of the brain and spinal cord that mostly consist of these myelinated axons are also white in color and are called the *white matter*. The brain weighs approximately 1600 grams (3.5 pounds)

4 PART | I Structure, function, and development

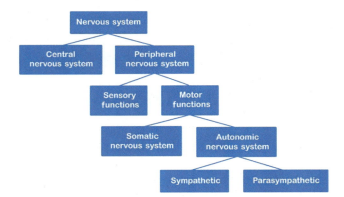

FIGURE 1.1 Subdivisions of the nervous system.

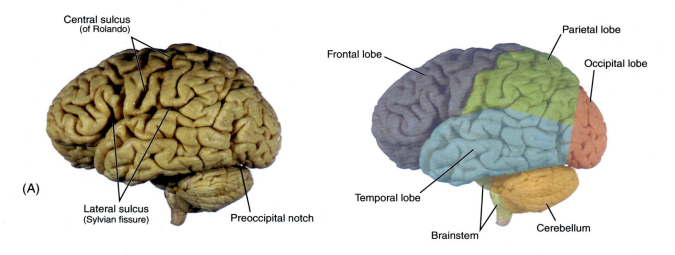

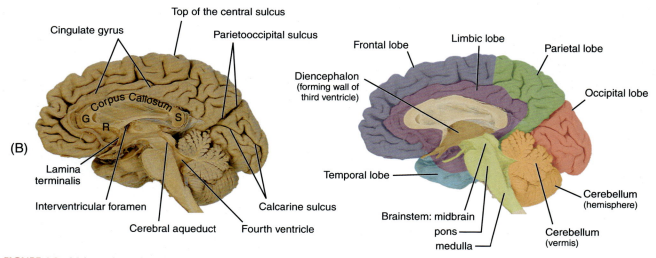

FIGURE 1.2 Major regions of the adult brain as seen in a lateral view (A) and in a medial view of a hemisected brain (B). G, genu; R, rostrum; S, splenium of the corpus callosum. *Dissection courtesy: Grant Dahmer, Department of Cell Biology and Anatomy, University of Arizona College of Medicine.*

in men, and 1450 grams (3.2 pounds) in women. The difference in brain sizes is proportional to body size, not intelligence.

While the spinal cord has the same basic pattern, extra gray matter consists of cell bodies and nuclei located at the center. The white matter and tracts surround this H-shaped gray matter area. The cerebral hemispheres and cerebellum also have an outer layer of gray matter cortex. The most **rostral** region of the CNS is the cerebral cortex, while the most **caudal** region is the brain stem.

Please note that there are other deep areas in the brain that are mostly consistent of nuclei and gray matter, such as the thalamus, caudate, putamen, and globus pallidus in the basal ganglia area.

Cerebral hemispheres

The two large masses of the cerebrum, the **cerebral hemispheres**, are nearly identical and side-by-side (Fig. 1.3). A wide, flat, heavily myelinated axon bundle called the corpus callosum connects them. They are separated by the **falx cerebri**, which is a layer of dura mater. Their surfaces are marked by **gyri**, which are the many convolutions and ridges that are separated by grooves. Basically any groove that is shallow to slightly deep is called a **sulcus**, while each extremely deep groove is called a **fissure**. Similar in all human brains, these elevations and depressions are very complex. The **longitudinal fissure** separates the right and left cerebral hemispheres. A **transverse fissure** separates the cerebrum and cerebellum. Sulci divide each hemisphere into five *lobes*.

Lobes

The lobes of the cerebral hemispheres have been named from the skull bones they underlie, and include the following:

- **Frontal lobe**—forms the anterior portion of each cerebral hemisphere. It is bordered posteriorly by a **central sulcus** (*Rolandic fissure*). This passes outward from the longitudinal fissure at a 90-degree angle. The frontal lobe is

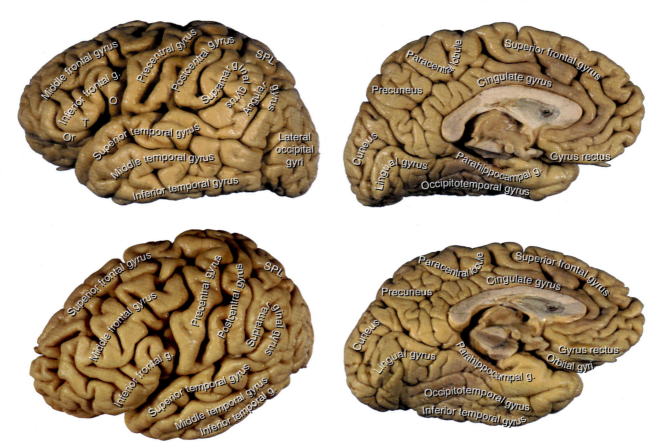

FIGURE 1.3 Cerebral hemispheres.

inferiorly bordered by the **lateral sulcus** (Sylvian fissure), which emerges from the underside of the brain, along the sides.
- **Parietal lobe**—posterior to the frontal lobe and separated from this lobe via the central sulcus.
- **Temporal lobe**—inferior to the frontal and parietal lobes and separated from them via the lateral sulcus.
- **Occipital lobe**—forms the posterior portion of each cerebral hemisphere. It is separated from the cerebellum by an extension of dura mater (the **tentorium cerebelli**). No distinct boundaries are present between the occipital, parietal, and temporal lobes.
- **Insula**—also called the *island of Reil*, deep with the lateral sulcus. Its name refers to it being covered by parts of the frontal, parietal, and temporal lobes. The insula is separated from other brain lobes by a **circular sulcus**.

The various lobes of the brain are shown in Fig. 1.4.

Cerebral cortex

The **cerebral cortex** is the brain layer covering the cerebral hemispheres. It accounts for about 40% of total brain mass, containing $14 - 16$ billion **neurons**. The **neocortex**, with six layers, makes up about 90% of the cerebral cortex. Its neuronal layers have different thicknesses between the various parts of the cerebrum. The layers have different compositions of cells, connections of synapses, neuron size, and locations where axons terminate. All axons that leave the cortex and enter the white matter arise from layers III, V, and VI. Layer V is thickest in motor areas. Layer IV is thickest in sensory areas.

Certain areas of the cerebral cortex have less layers. The type of cortex appearing in earlier vertebrates was called the **paleocortex**, with one to five layers. In humans, this was limited to just part of the insula, and some areas to the temporal lobe, which processed the sense of smell. A three-layered **archicortex** evolved in the hippocampus, the temporal lobe's memory-forming center. The neocortex was the final type of cortex to evolve.

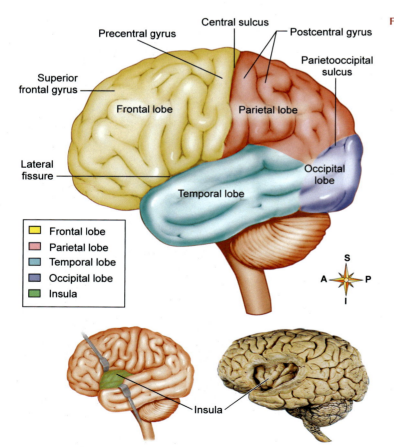

FIGURE 1.4 The various lobes of the brain.

Motor areas

The **primary motor cortex** is located in the surface of the precentral gyrus. Its neurons have **pyramidal cells** and control voluntary movements by regulating somatic motor (SM) neurons of the brain stem and spinal cord. When a specific motor neuron is stimulated, a contraction is generated in a certain skeletal muscle.

Sensory areas

At each sensory area of the cerebral cortex, data are reported in the pattern of neuron activity present. The **primary somatosensory cortex** is within the surface of the postcentral gyrus. Its neurons receive generalized somatic sensory (SS) information from pain, pressure, temperature, touch, and vibration receptors. We are aware of these sensations when nuclei of the thalamus relay information to the primary somatosensory cortex. Other areas of the cerebral cortex receive the sensations of sight, smell, sound, and taste (see Fig. 1.5).

Within the occipital lobe, the **visual cortex** receives visual information. In the temporal lobe, the **auditory cortex** receives information about hearing, while the **olfactory cortex** receives information about smell. In the anterior insula and adjacent frontal lobe areas, the **gustatory cortex** receives taste information from receptors of the pharynx and tongue. Therefore, any seizures initiated from the insula can cause gustatory hallucinations or abnormal tastes in the mouth.

Association areas

Association areas connect with the sensory and motor areas of the cerebral cortex, interpreting incoming data or coordinating motor responses. *Sensory association areas* monitor and interpret sensory information. They include the *somatosensory association cortex, visual association area,* and *premotor cortex*. In the primary somatosensory cortex, these activities are monitored by the **somatosensory association cortex**, which allows recognition of light touch sensations, such as an insect landing on the skin.

The special senses of hearing, sight, and smell have individual association areas. The **visual association area** monitors and interprets activity in the visual cortex. It can process letters seen by the eyes into recognizable words. If this area becomes damaged, an individual could still see printed letters as symbols, but there would be no meaning obtained from them. The **auditory association area** monitors activities in the auditory cortex, where word recognition occurs. These association areas are generally involved in a higher level of function and more complex tasks. If the primary visual cortex is responsible for recognizing and visualizing simple objects, the association cortex forms more complex pictures.

The **premotor cortex**, or *SM association area*, coordinates learned movements. The primary motor cortex does nothing by itself, and its neurons must be stimulated by other cerebral neurons. A voluntary movement happens when the premotor cortex sends instructions to the primary motor cortex. The proper pattern of stimulation is stored within the premotor cortex when the actions are repeated. The movement then becomes easier and smoother, since the *pattern* is triggered instead of individual neurons being controlled. The frontal eye field of the premotor cortex controls

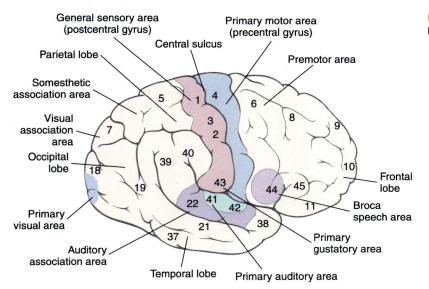

FIGURE 1.5 Sensory and motor areas of the cerebral cortex.

"learned" eye movements. If damaged, the individual can understand written letters and words, but cannot read since the eyes are unable to follow the lines that are printed.

Integrative centers and higher mental functions

Information from association areas is received by **integrative centers**, which control complicated motor activities and analyze information. From sensory association areas, the **prefrontal cortex** integrates information, performing intellectual functions, including prediction of possible outcomes of various responses. The integrative centers are found in each cerebral hemisphere's lobes and cortical areas. They handle mathematical computations, speech, understanding of spatial relationships, and writing. For these functions, the integrative centers are mostly restricted to one hemisphere. **Hemispheric lateralization** refers to the various functions between the cerebral hemispheres. Related regions on the opposite hemisphere are active, with less well-defined functions. Yet, the cerebral hemispheres appear to be almost the same. A left-side weakness in the setting of stroke lateralizes the stroke lesion to the right side of the brain.

Cerebral white matter

The cerebrum mostly contains white matter, and its axons are either classified as *association fibers, commissural fibers*, or *projection fibers*. **Association fibers**, in one cerebral hemisphere, interconnect areas of the cerebral cortex. The shorter **arcuate fibers** curve similar to arcs, passing from one gyrus to another. Longer fibers form bundles known as *fasciculi*. **Longitudinal fasciculi** connect the frontal lobe to the other lobes within the same cerebral hemisphere.

- **Commissural fibers**—interconnect between both hemispheres to allow communication. Their bands link the hemispheres and include the **corpus callosum** and **anterior commissure**. There are over 200 million axons in the corpus callosum (the largest commissure), which can carry about 4 billion impulses per second.
- **Projection fibers**—link the cerebral cortex to the diencephalon, brain stem, cerebellum, and spinal cord. They pass through the diencephalon, where ascending axons linking sensory areas pass near descending axons from the motor areas which run horizontally. A dissected brain shows similarities of ascending and descending fibers. The **internal capsule** is the total collection of projection fibers.

Basal nuclei

The **basal nuclei** consist of gray matter in both brain hemispheres, deep to the floor of the lateral ventricle (see Fig. 1.6). The nuclei are embedded in the cerebrum's white matter. Surrounding the basal nuclei are radiating projection

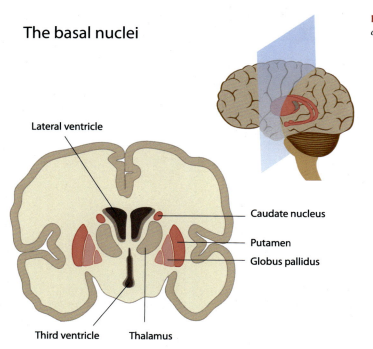

FIGURE 1.6 Basal nuclei. *Source: https://www.shutterstock.com/collections/281545102-7d165da1.*

fibers and commissural fibers. The nuclei have been considered part of an extensive functional group called the *basal ganglia*. This group included the basal nuclei and related motor nuclei in the diencephalon and midbrain. Since the term *ganglia* mostly concern the PNS, the term *basal nuclei* is preferred over the term *basal ganglia*. There is a physical continuity above the orbital surface of the brain's frontal lobe.

The **caudate nucleus** has a large *head* and a thin, curved *tail* following the lateral ventricle's curve. It appears to be shaped like a "comma", with a "C"-shape, similar to that of the lateral ventricles. The head, deep in the frontal lobe, is anterior to the **lentiform nucleus**, a lens-shaped structure made up of a lateral **putamen** and a medial **globus pallidus**. The head of the caudate nucleus merges with the *nucleus accumbens*, which then merges with the anterior putamen. The tail of the caudate nucleus is in the temporal lobe. The caudate nucleus and putamen have similar histology. They are sometimes referred to as the *corpus striatum* or simply *striatum*, which means "striated body." It has a striped internal capsule, as the fibers pass through these nuclei. The lentiform (or *lenticular*) nucleus is lateral and slightly anterior to the thalamus, but separated from the thalamus and most of the head of the caudate nucleus by a thick, fibrous internal capsule. It contains most fibers interconnecting the cerebral cortex, thalamus, basal nuclei, and brain stem. The best way to locate the head of the caudate is to locate the frontal horn of the lateral ventricle. The adjacent structure is the head of the caudate.

The basal nuclei also have subthalamic nuclei in the lateral floor of the diencephalon, as the **substantia nigra** of the midbrain. A thin layer of gray matter called the **claustrum** lies near putamen. Part of the limbic system, the **amygdaloid body**, lies anterior to the tail of the caudate nucleus, and inferior to the lentiform nucleus. The amygdaloid body (*amygdala*) has an "almond" shape. The basal nuclei begin forming during week five of gestation. They become prominent in weeks 6 − 7 of gestation. The basal nuclei, including the caudate nucleus, putamen, and globus pallidus, have a total volume of about 8 cm^3. The basal nuclei receive most of their blood from small, perforating branches of the middle cerebral artery (MCA) called the *lenticulostriate arteries*. The thalamus receives blood from the tiny perforating branches of the posterior cerebral artery (PCA), also called lenticulostriate arteries.

Diencephalon

The **diencephalon** is part of the *prosencephalon*, or *forebrain*. It develops from the primary cerebral vesicle, but differentiates, forming the rostral *telencephalon* and the caudal diencephalon. From the side of the telencephalon, the cerebral hemisphere forms, containing the lateral ventricles. These, plus the third ventricle, are able to communicate through the foramen of Monro, or *interventricular foramen*. The diencephalon is largely related to the structures developing lateral to the third ventricle. Superiorly, the lateral diencephalon walls form the *epithalamus*, and centrally, the *thalamus*. Inferiorly, they form the *subthalamus* and *hypothalamus*. Structures on either side of the third ventricle make up the thalamus (see Fig. 1.7).

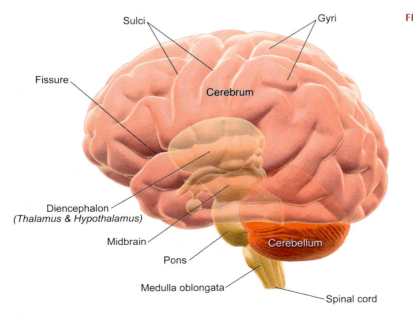

FIGURE 1.7 The subportions of the thalamus.

Most sensory, motor, and limbic pathways stop in the diencephalon (thalamus). Most limbic and motor pathways also are related to telencephalic structures. Nearly all connections between the cerebral cortex and subcortical structures, but primarily the diencephalon, course through the internal capsule. The only area of the diencephalon visible on an intact human brain is the inferior surface of the hypothalamus, which includes the mammillary bodies and infundibulum.

Thalamus

The **thalamus** is the last relay point for ascending sensory information projected to the primary sensory cortex. It filters information, amplifies signals, and passes on desired sensory information. The thalamus also regulates activities of the basal nuclei and cerebral cortex, relaying information. The thalamus ventral anterior/ventral lateral (VA/VL) nucleus is the last part of the dopamine pathway. The thalamus has bilateral, egg-shaped nuclei located on either side of the third ventricle. The thalamus is about 4 cm in length and is deep in the brain, making up 80% of the diencephalon.

Inside, the major portion of the thalamus contains anterior, medial, and lateral nuclear subportions. A vertical, "Y"-shaped sheet of white matter known as the internal medullary lamina creates the subdivisions. The thalamic nuclei are basically named from their locations. Each nucleus has functional specialties, projecting fibers into and receiving fibers from specific regions of the cerebral cortex. The thalamus receives afferent impulses from the body, sorting and filtering information. Impulses related to similar functions are relayed in groups through the internal capsule and reach the correct areas of the sensory cortex as well as specific cortical association areas. Afferent impulses are determined to be pleasant or unpleasant. Localization of specific stimuli, and their discrimination, occurs in the cerebral cortex. Nearly all inputs rising to the cerebral cortex move through the thalamic nuclei. They include inputs assisting regulation of emotion and visceral functions from the hypothalamus, through the anterior nuclei. Also included are instructions helping with direction of activity of motor cortices, from the cerebellum and basal nuclei. Inputs for memory or sensory integration are sent to various association cortices. The thalamus mediates sensation, cortex arousal, motor activities, learning, and memory. Understanding the thalamus has aided in treatment of many disorders, including epilepsy, pain, Parkinson's disease, and psychiatric disorders.

Hypothalamus

The **hypothalamus** is below the thalamus, and is above the brain stem. It creates the inferolateral walls of the third ventricle. While merging into the midbrain inferiorly, the hypothalamus extends from the optic chiasma, where the optic nerves cross, to the posterior margin of the **mammillary bodies**. Each of these bodies has three to four *mammillary nuclei*, primarily relaying signals from the limbic system to the thalamus. The mammillary bodies are paired and round. They bulge ventrally from the hypothalamus, relaying information in the olfactory pathways. The **infundibulum** is a stalk of tissue between the mammillary bodies and optic chiasma and connects the **pituitary gland** to the base of the hypothalamus. The hypothalamus is only 4 cm^2 of neural tissue, making up 0.3% of the entire brain. There are many important nuclei in the hypothalamus, much like the thalamus. The tiny hypothalamus is the primary visceral control center, extremely important for homeostasis. It influences most body tissues. The homeostatic activities of the hypothalamus are listed in Table 1.1.

Hypothalamic lesions are linked with unusual endocrine disorders, and emotional, metabolic, motor, and visceral disturbances. The hypothalamus normally controls the endocrine system. It acts through parvocellular neurosecretory projections to the median eminence and through the ANS. The hypothalamus controls endocrine output of the anterior pituitary and peripheral endocrine organs. **Vasopressin** and **oxytocin** are the hormones mostly involved in control of osmotic homeostasis and reproductive functions, respectively. The hypothalamus influences secretions from the thyroid gland, suprarenal cortex, gonads, and mammary glands and processes of growth and metabolic homeostasis. It affects the parasympathetic and sympathetic divisions of the ANS. Parasympathetic effects mostly occur when the anterior hypothalamus is stimulated. Sympathetic effects are more related to the posterior hypothalamus. When the anterior hypothalamus and paraventricular nucleus are stimulated, there may be decreases in blood pressure (BP) and heart rate.

Damage to the anterior hypothalamus may cause uncontrolled increases in body temperature. Projections to the ventromedial hypothalamus help regulate food intake. If damaged, there may be uncontrollable eating and obesity. When the posterior hypothalamus is stimulated, there is sympathetic arousal, **piloerection**, vasoconstriction, increased metabolic heat production, and shivering. Control of shivering occurs in the dorsomedial posterior hypothalamus. Many different areas of the hypothalamus, when stimulated, alter cardiac output, heart rate, peripheral resistance, vasomotor tone, differential blood flow, respiration depth and frequency, GI motility and secretion, erection, and ejaculation.

TABLE 1.1 Homeostatic activities of the hypothalamus.

ANS control	Controls activity in brain stem and spinal cord	Blood pressure, heartbeat, digestive motility, pupil size, many visceral activities
Physical responses to emotions	Within the limbic system, it has nuclei for perceiving pleasure, fear, anger, sex drive, other biological rhythms	Initiates most physical expressions of emotion: hypertension, pallor, sweating, pounding heart, dry mouth
Body temperature	Acts as thermostat, neurons monitor blood temperature and receiving input from thermoreceptors	Initiates sweating or shivering, to maintain constant temperature
Food intake	Responds to changing levels of glucose, amino acids, or hormones	Controls hunger and satiety
Water balance and thirst	Body fluids overconcentration activates osmoreceptors; they excite nuclei, triggering release of antidiuretic hormone (ADH) from posterior pituitary	ADH causes kidneys to retain water; thirst center is stimulated, increasing fluid intake
Sleep-wake cycles	Works with other brain areas to regulate sleep	Suprachiasmatic nucleus sets timing of sleep cycle, via visual responses to daylight and darkness
Endocrine functions	Various hormones control secretion of anterior pituitary gland hormones	Supraoptic and paraventricular nuclei produce ADH and oxytocin
Memory	Via mammillary nuclei in signal pathway from hippocampus to the thalamus	Important in memory; lesions cause memory deficits

Epithalamus

The **epithalamus** is the most distal part of the diencephalon, forming the roof of the third ventricle. The **pineal gland** extends from its posterior border, being externally visible. The caudal border of the epithalamus forms from the **posterior commissure**. The epithalamus is also made up of anterior and posterior *paraventricular nuclei*, medial and lateral *habenular nuclei*, and the *stria medullaris thalami*.

> **Point to remember**
> The pituitary gland greatly regulates body metabolism, affecting appetite, weight gains and weight loss. It also regulates behaviors, moods, many mental and emotional health issues, thirst, urinary output, and endocrine gland functions.

Pineal gland

The pineal gland (also called the *epiphysis cerebri* or *pineal body*) is shaped like a pinecone. The gland is small, reddish-gray in color, and located in a depression between the superior colliculi. It is inferior to the **splenium** of the corpus callosum. The pineal gland is about 8 mm in length, and its base is directed anteriorly, attached via a peduncle. There is a rich blood supply, as pineal arteries branch from the medial posterior choroidal arteries, which also branch, from the PCA. The veins open into the internal cerebral veins, great cerebral vein, or both. The pineal gland excretes **melatonin**, an antioxidant and sleep-inducing hormone, in response to varied circadian rhythms sensed by the hypothalamus. The pineal gland and hypothalamic nuclei help regulate the sleep — wake cycle. The precursor to melatonin is **serotonin**, synthesized from tryptophan by pinealocytes, then secreted into the fenestrated capillaries. The pineal gland has major regulatory actions.

> **Point to remember**
> The body makes less melatonin over time, so melatonin supplementation may help the elderly to sleep better. Melatonin levels begin to rise in the blood at about 9:00 p.m., reducing alertness. Levels remain in the blood for about 12 h, until the morning light causes them to reduce to the barely detectable daytime levels.

Brain stem

The brain stem is between the cerebrum and spinal cord and consists of the midbrain, pons, and medulla (see Fig. 1.7). Pathways for fiber tracts run between higher and lower neural areas. Brain stem nuclei are associated with 10 of the cranial nerve pairs and greatly involved with innervation of the head. It collectively accounts for only 2.5% of total brain weight.

Midbrain

The **midbrain**, located at the top of the brain stem, lies below the *diencephalon*.

A hollow *cerebral aqueduct* runs through it, as well as the entire brain stem. The aqueduct connects the third and fourth ventricles. **Superior colliculi** act as visual reflex centers, coordinating eye movements when following moving objects, even without conscious visualization. **Inferior colliculi** act as auditory relays from hearing receptors to the sensory cortex. They help react reflexively to sounds, such as in the *startle reflex*. A way to remember this is to note that the eyes are superior to the ears. The oculomotor nerve (CN III) exits the midbrain ventrally in the middle. Cranial nerves II and IV exit the lateral midbrain. Cranial nerve IV is the only cranial nerve that exiting the brain stem dorsally.

Pons

The **pons** is the bulged area of the brain stem, separated from the cerebellum dorsally by the fourth ventricle. It mostly contains conduction tracts in two directions, and links the cerebellum with the midbrain, diencephalon, cerebrum, and spinal cord. Deep projection fibers of the pons run longitudinally, between higher brain centers and the spinal cord. The superficial ventral fibers are transverse as well as dorsal. The *trigeminal*, *abducens*, and *facial* nerves have tissue from the pontine nuclei. Other nuclei are part of the reticular formation, and some aid the medulla oblongata in control of normal breathing rhythm.

Medulla oblongata

The **medulla oblongata**, or *medulla*, is the inferior part of the brain stem. It is cone-shaped, blending with the spinal cord near the foramen magnum of the skull. When viewed sectionally, the inferior part of the medulla looks very much like that of the spinal cord, with more complex organization of gray and white matter. As the central canal of the cord continues into the medulla it widens to form the fourth ventricle's cavity. The medulla and pons help form the ventral wall of the fourth ventricle. Here the canal looks nearly identical to that of the spinal cord. There are three groups of nuclei in the medulla. The first includes nuclei and processing centers controlling visceral functions. The coordination of complex autonomic reflexes is within the medulla. The second group has sensory and motor nuclei of the CNS. The third group has relay stations conducting communications between the brain and spinal cord, via ascending or descending tracts through the medulla. These groups are further explained below.

Cerebellum

The cerebellum makes up just 10% of the brain mass. It has a very large surface, similar to the cerebral cortex, containing over 50% of all brain neurons—about 100 billion in total. The cerebellum, dentate gyrus of the hippocampus, dorsal cochlear nucleus, olfactory bulb, and cerebral cortex have granule cells, which are the most abundant nuerons in the entire brain. The most distinctive neurons are the large, round **Purkinje cells**, which are in a single line, and the thick, dendritic planes that are parallel to each other. Axons travel to the deep nuclei, to synapse on output neurons that send fibers to the brain stem. The Purkinje cells are the primary output neurons of the cerebellar cortex. The cerebellum looks much like a cauliflower, with lobes having rough correspondences to its separate functional areas (see Fig. 1.7).

The cerebellum is dorsal to the pons, medulla oblongata, and fourth ventricle. It protrudes underneath the occipital lobes of the cerebral hemispheres and is separated from the hemispheres by a transverse cerebral fissure, or tentorium cerebelli. The cerebellum is bilaterally symmetrical and is the largest portion of the hindbrain. The two **cerebellar hemispheres** are about the size of apples, medially connected by the **vermis**, which resembles a worm. It is extremely convoluted on its service, with fine gyri called **folia**, resembling leaves with a transverse orientation. The folia are less prominent than the cerebral cortex folds.

The cerebellum's outer cortex, like the cerebrum, is thin and made up of gray matter. It has internal white matter, and small, deep, paired masses of gray matter. The cortex has three layers: *molecular* (basket, Golgi, and stellate cells);

Purkinje (Purkinje cell bodies); and granular. The medial areas of the cerebellum generally regulate motor activities of the trunk and axial muscles. Intermediate portions of each hemisphere regulate distal areas of the limbs, and skilled movements. The lateral parts of each hemisphere manage information from association areas of the cerebral cortex. They may assist in planning movements, but not in executing movements. **Flocculonodular lobes** receive input from the equilibrium centers of the inner ears, to adjust posture and maintain balance.

Though the cerebellum is much smaller than the cerebrum, if unfolded, it would be about 50% of the size of the cerebral cortex.

> **Point to remember**
> The cerebellum was long considered to control balance and all motor functions. Today, we know that it is important role for other cognitive functions, including attention, focusing and image capturing. Breathing, sleeping, BP, and heart rate are also controlled by the cerebellum, since it sends information to the spinal cord and nearby brain sections as well.

Blood supply of the brain

Two pairs of arteries supply the brain, plus much of the spinal cord: the **internal carotid arteries** and the **vertebral arteries**. The internal carotid arteries, which are anterior circulation, supply the anterior brain areas, most of the telencephalon, and much of the diencephalon. The vertebral system, or posterior circulation, supplies the posterior brain areas. The vertebral arteries supply the brain stem, cerebellum, and sections of the diencephalon, spinal cord, plus occipital and temporal lobes of the brain. The carotid arteries and vertebral arteries are classified amongst the *superficial arteries* of the head and neck (see Fig. 1.8).

On each side of the neck, one internal carotid artery ascends. Each continues superiorly, along the optic chiasm, then splits into the **MCA** and the **anterior cerebral artery**. It joins the *PCA*, part of the vertebral artery system. The posterior communicating artery connects the anterior circulation to the posterior circulation. The MCA is large, running laterally into the lateral sulcus. It divides into many branches supplying the insula, exits from the lateral sulcus, and spreads to supply most of the lateral surface of the cerebral hemisphere. The vertebral arteries arise from the subclavian arteries, at the root of the neck, forming the **basilar artery**. The cerebral arteries are classified amongst the *deep arteries* of the head and neck (see Fig. 1.9).

The **cerebral arterial circle** is polygon-shaped, and is also called the **circle of Willis**. In this circle, the PCA connects to the internal carotid artery via the posterior communicating artery. The anterior cerebral, internal carotid, and posterior cerebral arteries of both sides are interconnected here. The anterior communicating artery is a short anastomosis between the two anterior cerebral arteries. Arterial pressure in the internal carotid arteries is similar to the pressure in the posterior cerebral arteries. Normally, only small amounts of blood flow through the posterior communicating arteries.

The brain's primary venous drainage occurs via a system of cerebral veins. They empty into the venous dural sinuses and then into the **internal jugular veins**. The cerebral veins are divided into **superficial veins** and **deep veins**. Each of these types of veins is organized into systems. All cerebral veins lack valves. Superficial veins usually lie on the surfaces of the cerebral hemispheres (see Fig. 1.10). Most empty into the superior sagittal sinus. Deep veins primarily drain internal structures, and eventually empty into the straight sinus (see Fig. 1.11). The **basal vein** empties into the straight sinus. Unlike cerebral arteries, the cerebral veins are connected by many anastomoses, both within one group of veins, and between superficial and deep vein groups.

Blood circulating in the brain is collected in large, thin-walled **venous dural sinuses**. These blood-filled spaces lie between the dura mater layers. The dura mater has a periosteal layer lying against the cranial bone, and a meningeal layer lying against the brain. In some areas, there is a space between these layers, accommodating a sinus that collects blood. The **superior sagittal sinus** overlies the brain's longitudinal fissure. It begins anteriorly near the **crista galli** and extends posteriorly. The **inferior sagittal sinus** within the inferior margin of the falx cerebri arches over the corpus callosum, lies deep in the longitudinal fissure. It joins the *great cerebral vein* posteriorly, forming the **straight sinus**.

Functional brain systems

The functional brain systems are neuronal networks that are not localized to any certain brain regions, but interact over larger distances. Examples include the *limbic system* and *reticular formation*. The limbic system is positioned on the medial aspect of each cerebral hemisphere and the diencephalon. The reticular formation extends through the cores of the medulla oblongata, pons, and midbrain.

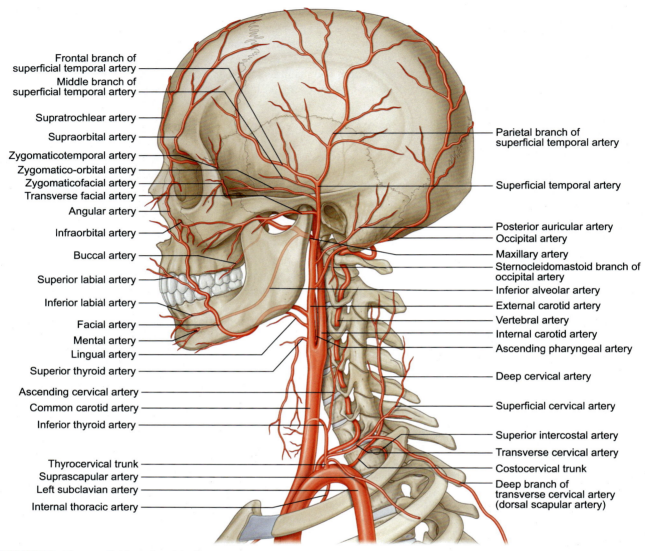

FIGURE 1.8   The superficial arteries of the head and neck.

Limbic system

The **limbic system** is a complex neural network (see Fig. 1.12). It handles emotions, homeostasis, memory, motivations, unconscious drives, and olfaction. Its complexity makes the study of this system clinically difficult. Improvements in behavioral studies, deep-brain stimulation, functional magnetic resonance imaging (MRI), and perfusion have allowed for better understanding of the limbic system. The **limbic lobe** contains the structures of this system, which include:

- *Amygdala (amygdaloid nuclear complex)*—fear and emotion center;
- *Various hypothalamic nuclei*—homeostasis, hunger, satiety, sleep onset, and thermoregulation;
- *Olfactory cortex*—sense of smell;
- *Septal nuclei*—below the rostrum of the corpus callosum, they are essential in generating the theta rhythm of the hippocampus, and play a role in reward and reinforcement, along with the nucleus accumbens;
- *Nucleus accumbens*—a region in the basal forebrain, rostral to the preoptic area of the hypothalamus, playing an important role in processing rewarding stimuli and reinforcing stimuli; these include exercise, sex, and drugs;
- *Hippocampal formation*—mainly involved in memory, and believed to play a role in spatial navigation and control of attention;

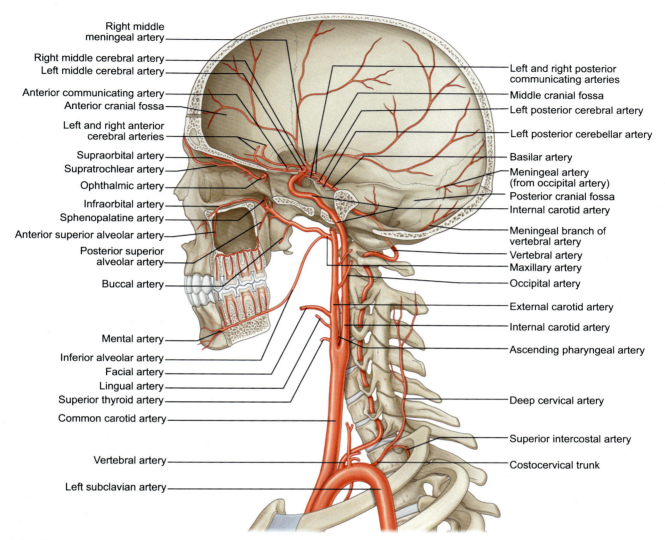

FIGURE 1.9 The deep arteries of the head and neck.

- *Cingulate cortex*—in the medial aspect of the cerebral cortex, involved with emotion formation and processing, learning, and memory;
- *Areas of the basal ganglia*—at the base of the forebrain and top of the midbrain, involved in reward learning, cognition, and frontal lobe functioning;
- *Ventral tegmental area*—close to the midline, on the floor of the midbrain, involved in drug and natural reward circuitry in the brain;
- *Limbic midbrain areas*—including the periaqueductal gray matter, play critical roles in autonomic function, motivated behavior, and behavioral responses to threatening stimuli.

The *limbic brain* includes these structures and their projections, which reach the forebrain, midbrain, lower brain stem, and the *spinal cord limbic systems*. The spinal cord limbic systems are reached primarily via the fornix, stria terminalis, **ventral amygdalofugal pathway**, and **mammillothalamic tract**.

Functions of the limbic system

The cingulate gyri appear to be involved in exploratory behaviors, memory processing, and visually focused attention. The limbic system is believed to be more effective in the nondominant brain hemisphere. The cingulate gyri function in both cognition

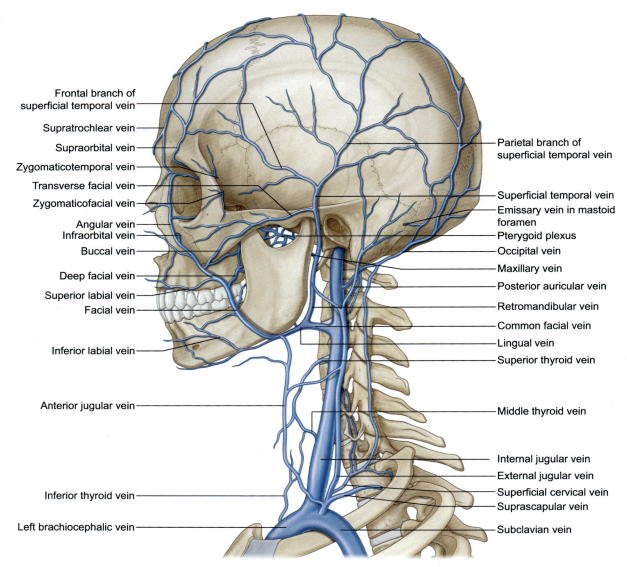

FIGURE 1.10 The superficial veins of the head and neck.

and emotions. Norepinephrine is present in its highest amounts in the hypothalamus, and then in the medial limbic system areas. About 70% of the norepinephrine is concentrated in axon terminals arising in the medulla and locus ceruleus of the rostral pons. A large amount of serotonin is within axons of other ascending fibers, especially those starting in the reticular formation of the midbrain, and ending in the amygdala, septal nuclei, and lateral areas of the limbic lobe.

Neuronal axons of the ventral tegmental parts of the midbrain have large amounts of dopamine. This may indicate why a severe depressive reaction can be initiated by electrical stimulation of the substantia nigra via an electrode that had been placed before, to treat Parkinson's disease. The term *limbic system* is actually a simple description, since its parts have widely different connections with the neocortex and central nuclei. The neurotransmitters also differ, as do the effects of the parts when they are damaged. Limbic system lesions usually do alter emotions.

Hippocampal formation

The **hippocampal formation** includes the **dentate gyrus**, **hippocampus**, and **subicular complex** (see Fig. 1.13). The subicular complex includes the **subiculum**, **presubiculum**, and **parasubiculum**. The neocortex of the parahippocampal gyrus passes medially from the collateral sulcus, and joins the transitional **juxtallocortex** of the subiculum. This

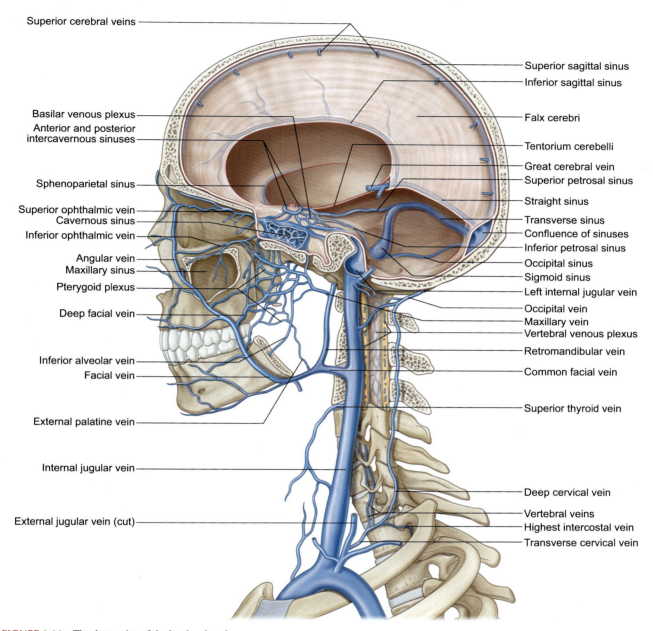

FIGURE 1.11 The deep veins of the head and neck.

structure is curved superomedially to the inferior surface of the dentate gyrus, and then curves laterally to the laminae of the hippocampus. The curve is continued superiorly and then medially, above the dentate gyrus. It ends while pointing to the center of the superior surface of the dentate gyrus. Three pathways in the hippocampal formation are believed to utilize glutamate, aspartate, or both as the major excitatory neurotransmitter.

A circuit within in the limbic system that was first described by the anatomist named James Papez back in 1937, was initially believed to play a large role in emotions. Today, the *Papez circuit* is known to play a major role in *memory* formation and processing, and has many other functions. The circuit is very complex, with structures include the following:

- Hippocampus;
- Mammillary body;
- Anterior nucleus of the thalamus;
- Cingulate gyrus.

18 PART | I Structure, function, and development

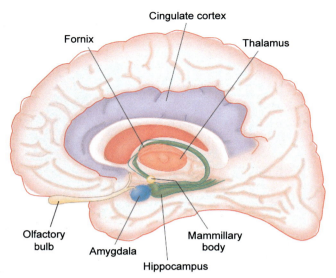

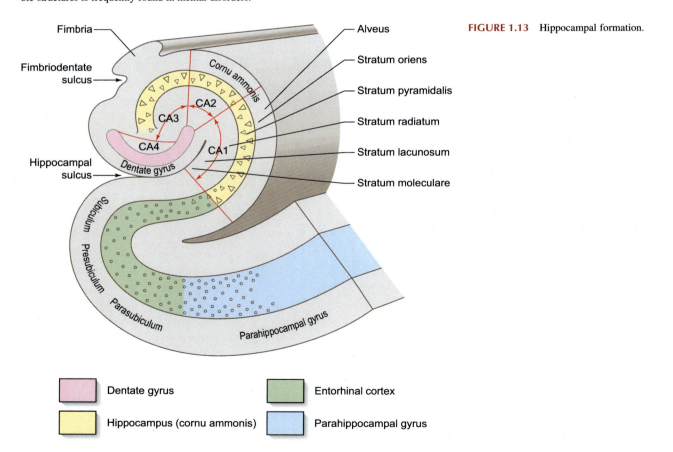

FIGURE 1.12 The Limbic System. Structures of the limbic system play important roles in emotion, learning, and memory. Pathophysiology in limbic structures is frequently found in mental disorders.

FIGURE 1.13 Hippocampal formation.

The Papez circuit (see Fig. 1.14) begins at the hippocampus, connecting to the mammillary body via the *fornix*. The connection between the mammillary body and anterior nucleus of the thalamus is through mammillothalamic fibers. The connection between the anterior nucleus and the **cingulate gyrus** is via the *cingulate bundle, entorhinal cortex*, and subiculum, then back to the hippocampal formation to complete the circuit. This circuit may be the most well described circuit out of the many circuits of the limbic system.

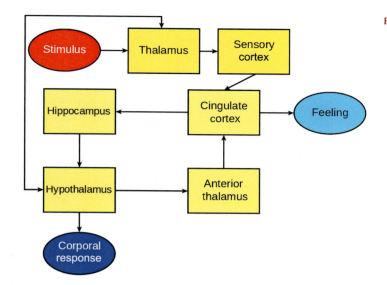

FIGURE 1.14 The Papez circuit.

Reticular formation

The *reticular formation* consists of loose groupings of neurons in the white matter (see Fig. 1.15). Reticular neurons have long axonal connections, and individually, project to the hypothalamus, thalamus, cerebral cortex, cerebellum, and spinal cord. The neurons are specialized for regulating brain arousal. Some form the **reticular activating system**, sending continued impulses to the cerebral cortex, affecting consciousness and cortex excitability. This system also filters sensory input, allowing just the most important data to reach consciousness. It is inhibited by the hypothalamic sleep centers, and depressed by alcohol and various drugs. The reticular formation also has motor nuclei, projecting to motor neurons in the spinal cord via the *reticulospinal tracts*, which help control skeletal muscles as coarse limb movements are made. Cardiac, respiratory, and vasomotor centers of the medulla are reticular motor nuclei, and regulate visceral motor (VM) functions.

Higher mental functions

Cerebral functions concern higher brain activities. The cerebrum interprets impulses from the sense organs and initiates voluntary muscular movements. It also stores information as memories and uses reasoning to retrieve this information. The cerebrum is responsible for intelligence and as personality. Both cerebral hemispheres are involved in basic functions, including the receipt and analyses of sensory impulses, memory storage, and control of skeletal muscles on opposing sides of the body. One side is usually the *dominant hemisphere*. For most people, the left hemisphere is dominant. Its functions include language, reading, speech, and writing. The left hemisphere also dominates for complicated intellectual functions requiring analytical, computational, and verbal skills. A 2014 French study revealed some interesting statistics concerning hemisphere dominance:

- A dominant left hemisphere was present in 88% of right-handed people and in 78% of left-handed people.
- There was no clearly dominant hemisphere present in 12% of right-handed people and 15% of left-handed people.
- A dominant right hemisphere was present only in left-handers, at a rate of 7%.

In humans, the nondominant hemisphere is specialized for nonverbal functions, yet still handles basic functions. Nonverbal tasks include motor skills that require orientation of the body in its environment, visual experiences, and understanding and interpreting musical patterns. The nondominant hemisphere also provides thought processes involved in emotions and intuition. The area of the nondominant hemisphere related to the motor speech area does not really control speech, but affects emotional factors used when speaking.

Nerve fibers of the corpus callosum connect to the cerebral hemispheres. The fibers can transfer sensory information reaching the nondominant hemisphere to the general area of the dominant hemisphere, where interpretation of information happens. This is where information can be used to make decisions. The left cerebral motor areas control the muscles of the body's right side, but it is not known why this crossover of activities exists. One area of the cerebral cortex

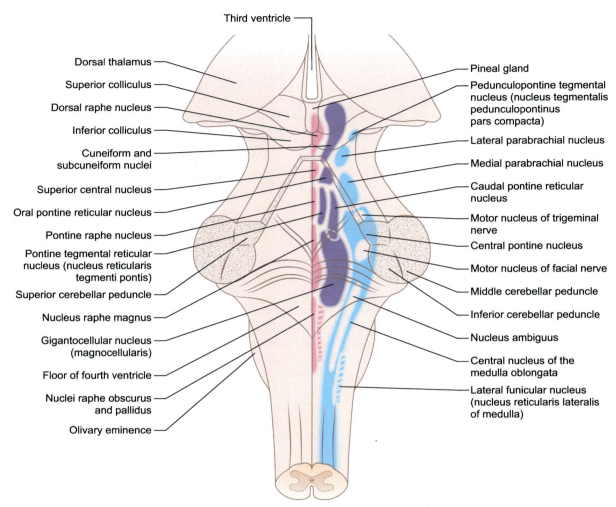

FIGURE 1.15 Reticular formation.

often has functions that overlap other areas. Consciousness and other cortical functions are not only controlled by one area. The functions of each lobe or area of the cerebral cortex are further summarized in Table 1.2. It is important to note that in the cerebellum, the left half controls balance in the *ipsilateral (same side)* of the body, and the right half controls balance in the right side of the body—which is different than how the cerebral cortex operates.

Language

Language processing is complicated, occurring in both cerebral hemispheres, yet varies between individuals. Mostly related to the left hemisphere are *Wernicke's area* and *Broca's area*. **Wernicke's area** is close to the auditory cortex, located in the superior temporal gyrus. This area is associated with language comprehension, receiving data from sensory association areas. It is important for personality, controlling sensory information and access to auditory and visual memories.

Broca's area, the *motor speech area*, is near the motor cortex and used in speech production (see Fig. 1.5). It is located in the inferior frontal gyrus, regulating breathing patterns while speaking and vocalizations used in normal speech. Broca's area coordinates activates of the respiration muscles, larynx, pharynx, and muscles of the cheeks, jaws, lips, and tongue. If the area is damaged, sounds can be made, but words cannot be formed. The *receptive speech area* is another name for the *auditory association area*. It uses feedback to adjust motor commands from the motor speech area. Many speech-related problems occur because of damage to a specific sensory area. Some patients have difficulty

TABLE 1.2 The functions of the cerebral cortex.

Lobe and areas	Functions
FRONTAL LOBE Primary motor cortex	Voluntary control of skeletal muscles
PARIETAL LOBE Primary somatosensory cortex	Conscious perception of pain, pressure, taste, temperature, touch, and vibration; Control of visuospatial tasks
OCCIPITAL LOBE Visual cortex	Conscious perception of visual stimuli
TEMPORAL LOBE Auditory cortex, olfactory cortex	Conscious perception of auditory and olfactory stimuli, as well as memory
ALL LOBES Motor, sensory, association, and somatosensory areas	Sensory integration and processing; motor processing and initiation

speaking but understand the correct words to use. Others can speak consistently yet use many incorrect words. While Broca's and Wernicke's areas are the main language centers, in order to read, speak, and write, other areas of the brain are needed as well.

In most people, the left cerebral hemisphere contains the specialized language areas. The premotor cortex, for hand movements, is larger on the left side for people who are right-handed than for those who are left-handed. The left hemisphere is crucial for logical decision making, mathematical calculations, and other analytical functions. The right cerebral hemisphere processes sensory information, helping the body relate to its environment. It helps identify familiar sights, smells, tastes, or touches, and is the dominant hemisphere for face recognition or understanding three-dimensional images. The right hemisphere analyzes the emotions of people we are speaking to. When the right hemisphere is damaged, a person may be unable to add emotional inflections to their speech. This is also called *prosody* and can occur with nondominant frontal lobe lesions. To assess language, patients are tested for fluency, repeating ability, and comprehension, as well as reading and writing skills.

Only about 10% of the human population is left-handed. The right hemisphere's primary motor cortex usually controls motor function for the dominant left hand. However, centers for analytical speech and function are in the left hemisphere, just as they are in right-handed people. Interestingly, a very high percentage of artists and musicians are left-handed. Also, when a person favors one hand over the other, connections with the opposite side of the brain strengthen.

Memory

To remember school studies, phone numbers, addresses, tastes, and many other types of information, humans require memories. These are stored groups of information collected through a lifetime. Different types of memories include:

- **Fact memories**—specific information. such as colors or odors;
- **Skill memories**—learned motor behaviors, such as how to use a key, or how to mix ingredients when cooking. Skill memories become unconsciously controlled over time. Those related to eating are stored in specific parts of the brain stem. Complex skill memories, such as playing a guitar, need integration of motor patterns in the basal nuclei, cerebral cortex, and cerebellum.

The two classes of memories are further explained as follows:

- **Short-term memories**—information can be immediately recalled but only for a short time. These involve small pieces of information, such as a name or address. Repeating the information reinforces the original short-term memory, which can then be converted to a long-term memory.
- **Long-term memories**—last longer, even throughout life. Conversion between short-term and long-term memories is called **memory consolidation**. There are two subtypes: *secondary memories*, which fade over time, and may need a great effort to recall; and *tertiary memories*, which remain throughout life, such as your own name and appearance.

> **Point to remember**
> The ability to retain memories begins just 20 weeks after conception. The brain's ability to store information is almost limitless. Sleep is significant to memory, helping retrieve and store long-term memories. Memory loss experienced by older people is not due to aging—it is because they have fewer engaging activities than earlier in life. Lack of "brain exercise" results in loss of memory. Thinking about past experiences creates stronger connections between active neurons.

Brain regions used in memory

Memory consolidation involves the *amygdaloid body* or *amygdala*, and the *hippocampus*, both part of the limbic system. With damage to the hippocampus, there is an inability to covert short-term memories to new long-term memories. Existing long-term memories still remain intact. Tracts leading from the amygdaloid body to the hypothalamus may link memories to certain emotions. The **nucleus basalis** plays a role in storage and retrieval of memories. It is connected via tracts with the hippocampus, amygdaloid body, and cerebral cortex. If damaged, there is altered intellectual function, emotional states, and memory.

The cerebral cortex stores most long-term memories. Specific association areas handle conscious sensory and motor memories. Visual memories are stored in the visual association area. The premotor cortex stores memories of voluntary motor activity. Words, voices, and faces are remembered via special areas of the occipital and temporal lobes. Specific memories usually involve activity of just one neuron. In an area of the temporal lobe, an individual neuron responds to a specific word while ignoring others. The neuron may be activated by a correct combination of sensory stimuli that relates to a certain individual.

Many different locations in the brain collect information on a certain subject. A long-time friend's face is remembered in the visual association area, while his voice is remembered in the auditory association area. The speech center remembers his name, and the frontal lobes remember his house, hobbies, likes, and dislikes. Other areas store related information. When an area is damaged, the memory of the old friend is diminished in some way.

Sleep and sleep-wake cycles

We progress through our sleep stages, which are usually followed by a short interval of **rapid eye movement (REM) sleep**. This occurs in cycles of five to six times every night. Each person needs different amounts of sleep, usually from 6 to 10 h per day. Infants sleep for a longer part of each day. With aging, total sleep time and deep sleep usually decrease, with more sleep interruptions occurring. Elderly people may not even enter stages three or four of **nonrapid eye movement (NREM) sleep**. These changes may explain increasing **Ehlers—Danlos syndrome (EDS)** and fatigue with aging, but are of unclear clinical significance.

Types of sleep

There are two *types* or *states* of sleep, each with physiological characteristics. Nonrapid eye movement sleep makes up 75% − 80% of total sleep time in adults. Its four stages have an increased depth of sleep. There are slow, rolling eye movements, characterizing quiet wakefulness and early *stage one sleep*. These are not present in deeper sleep stages, and muscle activity decreases. Stages three and four are called *deep sleep* since the arousal threshold is high. These stages are perceived by many as "high-quality" sleep. Rapid eye movement sleep follows every cycle of NREM sleep. There is low-voltage, quick activity when measured on an EEG, and postural muscle **atonia**. Large fluctuations of respiration rate and depth occur, with most dreams happening at this time.

Regulation of sleep

Alternating sleep-wake cycles show the natural 24-h *circadian rhythm*. The sleep cycle is regulated by the hypothalamus and its *suprachiasmatic nucleus*. This acts as a biological clock, regulating its sleep-inducing center, the *preoptic nucleus*. Inhibition of the reticular activating system in the brain stem results in the preoptic nucleus causing the cerebral cortex to enter "sleep mode." As the arousal system is switched off, dreaming and other sleep stages are regulated. Just before awakening, the hypothalamus releases *orexins*—peptides that cause arousal. Neurons in the reticular formation, firing at their highest rates, arouse the cerebral cortex.

Importance of Sleep

The reasons concerning slow-wave NREM stages 3 and 4 and REM sleep restoring the body are based on these stages being when most neural activities slow down to basal levels. With sleep deprivation, a person spends longer periods of time than normal in slow-wave sleep during the next sleep period. Consistent REM sleep deprivation causes depression, moodiness, and symptoms similar to personality disorders. Good REM sleep allows the brain to review daily events and address emotions while dreaming. It can eliminate synaptic connections that are not overly important, helping us to forget noncritical information.

Meninges

The nervous tissue of the brain is separated from the general blood circulation, via a type of *biochemical isolation*, called the *blood − brain barrier*. Layers of the **cranial meninges** are continuous with the *spinal meninges*. They include the *dura mater*, *arachnoid mater*, and *pia mater*. Except for only a few areas, there is normally no space on each side of the cranial dura. One side is attached to the skull, while the other is attached to the arachnoid mater. However, there are two **potential spaces** associated with the dura mater: the **epidural space** and the **subdural space**.

Dura mater and dural folds

Outer and inner fiber layers make up the **dura mater**, which means "tough matter," the strongest of the meninges. A dura layer surrounds the brain, consisting of two layers of fibrous connective tissue: the *meningeal layer* and *periosteal layer*. The outer layer is connected to the periosteum of the bones of the cranium. The deeper meningeal layer forms the true external brain covering. It continues caudally, in the vertebral canal, as the *spinal dura mater*. This means there is no superficial epidural space above the dura mater, but in the spinal cord, there is such a superficial epidural space. The outer *periosteal cranial dura* and inner *meningeal cranial dura* are fused together.

Arachnoid mater

The **arachnoid mater** gets its name because it resembles a spider web. It is a thin, avascular membrane with a cell layer attached to the dural border cell layer of the dura mater. It has no natural space. The **subarachnoid space** is between the arachnoid mater and pia mater. The subarachnoid space is filled with cerebrospinal fluid (CSF). It contains the largest blood vessels serving the brain. Since the arachnoid mater is thin and elastic, its blood vessels do not have significant protection. The arachnoid granulations absorb CSF into the venous blood of the sinus.

Pia mater

The **pia mater** covers all external surfaces of the CNS. The term "pia mater" means "tender matter." It is made up of delicate connective tissue, with many tiny blood vessels. The pia mater is the only layer clinging tightly to the brain and follows all of its convolutions. Cerebral arteries and veins travel in the subarachnoid space, enveloped by pia mater, which is also found near the cerebral blood vessel branches where they penetrate the brain surface, and reach its internal structures.

Cerebrospinal fluid

CSF is mostly produced in the **choroid plexus**, between each ventricle (see Fig. 1.16). The plexus is present in all four brain ventricles, containing strands of convoluted, vascular membranes. A long, continuous band of the plexus, like the C-shaped course of each lateral ventricle, extends from nearly the tip of the inferior horn. It continues through the body of the ventricle, reaching the interventricular foramen. There is no choroid plexus in the anterior or posterior horn. The total volume of CSF in adults is between 140 and 270 mL, yet about 600 − 700 mL are produced every day. The CSF has many important functions as it surrounds and coats exposed CNS surfaces. The functions include:

- *Brain support*—with the brain basically floating in CSF, it is suspended in the cranium. When supported by CSF, the human brain weighs about 1.8 ounces (50 g). This is much different from its weight outside of the body, which is 3.09 pounds (1400 g).
- *Cushioning of neural structures*—the CSF cushions the brain and spinal cord against physical trauma.
- *Transportation of chemical messengers, nutrients, and wastes*—ependymal cell linings are freely permeable, except where CSF is produced. There is constant chemical information exchange between the CSF and interstitial fluid that surrounds the CNS neurons and neuroglia.

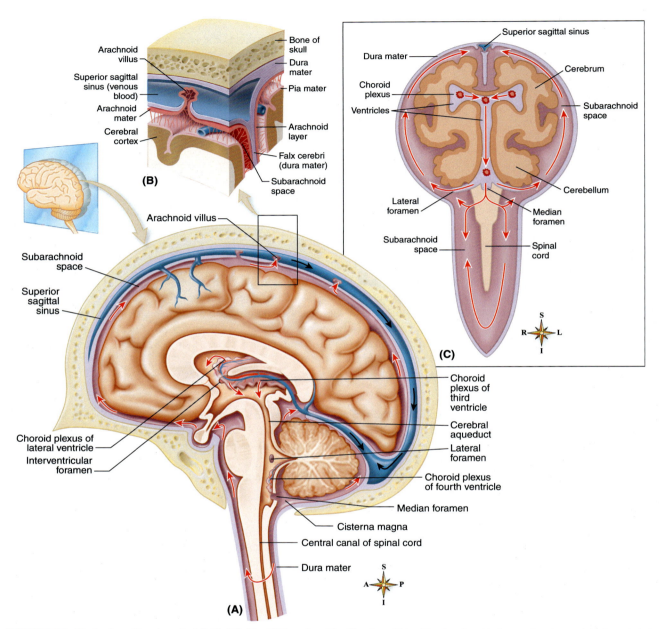

FIGURE 1.16 Production of cerebrospinal fluid. (A), The CSF produced by filtration of blood by the choroid plexus of each ventricle flows inferiorly through the lateral ventricles, interventricular foramen, third ventricle, cerebral aqueduct, fourth ventricle, and subarachnoid space and to the blood. (B), Inset showing arachnoid villus, where CSF is reabsorbed into the blood of the superior sagittal sinus. (C), Simplified diagram showing flow of CSF.

Point to remember

CSF is normally produced and drained continuously. If a tumor or other obstruction stops this from occurring, CSF accumulates, placing pressure upon the brain, in a condition known as *hydrocephalus*. In newborns, hydrocephalus causes the head to enlarge since the skull bones have not yet fused. If this condition occurs in adults, the rigidity of the skull means that brain damage is likely, due to compression of blood vessels and soft nervous tissue. Hydrocephalus is treated by inserting a shunt into the brain ventricles, to drain excess CSF into the abdominal cavity.

Spinal cord and reflex center

The **spinal cord** is enclosed in the vertebral column. It extends from the skull's **foramen magnum** to the first or second lumbar vertebra, and is slightly inferior to the ribs. The spinal cord is a shiny white structure and provides two directions of conduction: descending from the brain, and ascending to the brain. The cord is protected by the vertebral bones, meninges, and CSF. The dural and arachnoid membranes extend inferiorly to the level of *sacral spinal nerve 2* (S_2), which is far lower than the end of the spinal cord. The cord usually ends between *lumbar nerve 1* (L_1) and *lumbar nerve 2* (L_2). Therefore, the subarachnoid in the meningeal sac, inferior to this point, is the ideal location for removal of CSF, since there is no spinal cord at this level, preventing any damage. The procedure is called a **lumbar puncture**. The spinal cord terminates inferiorly in cone-shaped, tapered **conus medullaris**. A fibrous extension, covered by pia mater, is called the **filum terminale**, which extends inferiorly from the conus medullaris to the coccyx. Here, the filum terminale anchors the spinal cord, for vertical support. The spinal cord is also bound to the dura mater meninges over its length by the **denticulate ligaments**. These are saw-toothed, shelf-like structures of pia mater that provide horizontal support to the cord. For most of its length, the spinal cord is as wide as an adult human thumb, with enlargements where the nerves arise that serve the arms and legs. The upper ones are called **cervical enlargements**, while the lower ones are called *lumbar enlargements* (see Fig. 1.17). The cervical enlargements usually extend from *cervical spinal nerve 5* (C_5) to *thoracic spinal nerve 1* (T_1). The lumbar enlargements are also called *lumbosacral enlargements*, and usually extend from L_2 to *sacral spinal nerve 3* (S_3).

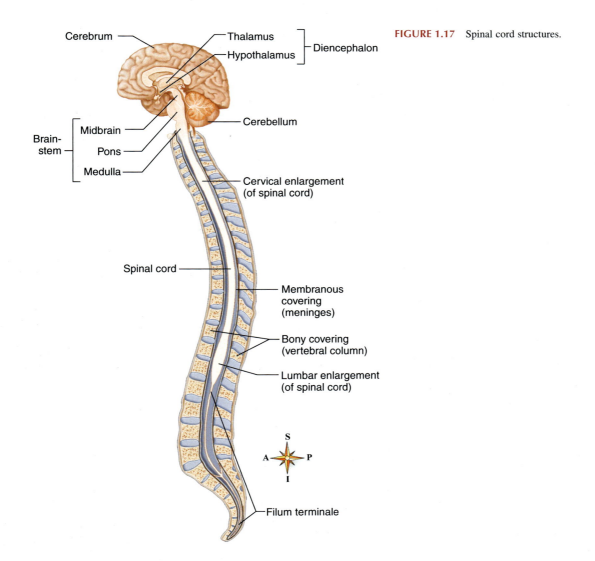

FIGURE 1.17 Spinal cord structures.

> **Point to remember**
> With the meninges extending beyond the spinal cord, there is a good area to perform a lower lumbar puncture without damaging the cord. The procedure allows CSF to be withdrawn from the subarachnoid space, in the lumbar region of the vertebral column. A needle is inserted just above or below the fourth lumbar vertebra. The cord ends about 1 in. above this level. The fourth lumbar vertebra is located easily, lying on a line with the iliac crest. The patient is placed on one side, with the knees and chest drawn together in the fetal position, arching the back and separating the vertebrae. The needle is inserted once the patient's head is pointed towards the belly bottom, in a 45-degree angle. When it enters the CSF, the thin nerve roots roll off the needle's tip, allowing CSF to be collected with no nerve tissue damage. Correct positioning is the key to success. CSF is tested for bacteria, blood cells, or other abnormalities indicating injury or an infection such as meningitis. A *manometer* may be attached to the needle. It is a sensor that determines CSF pressure in the subarachnoid space. Lumbar puncture is also used to inject diagnostic and therapeutic agents such as radiopaque dyes or chemotherapy agents into the subarachnoid space.

Structure of the spinal cord

From front to back, the spinal cord is slightly flat. There are two grooves that mark its surface: the wide **anterior median fissure** and the narrow **posterior median sulcus**. The grooves follow the entire length of the cord and partially divide it into two halves. The two sides of the cord are able to communicate via a thin band of neural tissue near the central canal. *The* **posterior intermediate sulcus** is positioned at the cervical and upper thoracic levels. A **glial septum** projects from this sulcus, partially subdividing each **posterior funiculus**.

The gray matter lies inside the spinal cord, while the white matter is outside. The **central canal** is filled with CSF and runs through the entire cord. If cross-sectioned, the spinal cord's gray matter appears like a letter "H" or is sometimes described as "butterfly-like." It has mirrored-image lateral gray masses connected by the **gray commissure**, which encloses the central canal. Through this commissure, the central canal carries CSF throughout the cord. The two **posterior horns** are projections of gray matter. There are also similar **anterior horns** resembling the letter "H". Viewed in three dimensions, the horns form columns of gray matter through the entire spinal cord.

Neurons with cell bodies in the cord's gray matter are *multipolar*, while the posterior horns are totally made up of **interneurons**. The anterior horns have mostly cell bodies of **SM neurons**, and some interneurons. The motor neurons send axons out to the skeletal muscles and emerge from a poorly defined **anterolateral sulcus**. These rootlets fuse to become the **anterior roots** of the spinal cord. Size of anterior gray matter, at any spinal cord level, reveals how much skeletal muscle is innervated at that level. This means the anterior horns are largest in the cervical and lumbar regions that innervate the limbs, since they innervate muscles of the upper and lower extremities. This is why there are enlargements of the spinal cord in these areas. The lateral horns primarily contain cell bodies of the sympathetic division of autonomic motor neurons serving visceral organs. Motor axons leave the cord through *anterior roots*, which have both somatic and autonomic efferent fibers of the PNS.

From peripheral sensory receptors, afferent fibers form *posterior roots*, which spread out as **posterior rootlets** before entering the spinal cord. The rootlets enter the cord in the shallow, longitudinal **posterolateral sulcus**. Cell bodies of associated sensory neurons are within an enlarged region of the posterior root, the **posterior root ganglion**. Once entering the cord, axons of these neurons take a many routes. Some enter the posterior white matter directly, to synapse at higher spinal cord or brain levels. Others synapse with interneurons in the posterior horns of the spinal gray matter, at the level of entry. Posterior and anterior roots are very short, fused laterally, and form the **spinal nerves** (containing motor and sensory nerves). Spinal gray matter is subdivided based on neuron involvement in innervation of the somatic and visceral regions. Spinal gray matter has four divisions or zones: the **SS division**, **visceral sensory (VS) division**, **visceral motor (VM) division**, and **SM division**.

White matter

The **white matter** of the spinal cord consists mostly of myelinated nerve fibers, and to a lesser degree, nonmyelinated nerve fibers. These allow communication between different areas of the cord, and between the cord and the brain. The fibers run in three different directions, as follows:

- **Ascending** (sensory)—running upwards to higher centers;
- **Descending** (motor)—running downwards to the spinal cord from the brain, or within the cord, to lower levels;
- **Transverse** (commissural)—these run across the midline, from one side of the spinal cord to the other.

The myelinated ascending and descending tracts make up most of the white matter of the spinal cord. On each side of the spinal cord, the white matter is divided into three *white columns*, called **funiculi**, named by their position, as follows:

- **Dorsal (posterior) funiculi**;
- **Lateral funiculi**;
- **Ventral (anterior) funiculi**.

Each funiculus has several fiber tracts, consisting of axons with similar functions and destinations. The names of the spinal tracts usually identify their origins and destinations.

Neuronal pathways

There are three general types of nerve fibers in the white matter, which include:

- *Long ascending fibers*—projecting to the thalamus, cerebellum, cortex and various brain stem nuclei;
- *Long descending fibers*—projecting from the cerebral cortex or from several brain stem nuclei, towards the spinal gray matter;
- *Shorter propriospinal fibers*—interconnecting different levels of the cord, and include fibers used in coordinating flexor reflexes.

Fibers with similar connections are often banded together, forming various spinal cord tracts. Propriospinal fibers are usually in a thin shell surrounding the gray matter. This **propriospinal tract** is also called the *fasciculus proprius*. Many ascending and descending tracts are primarily described by where they originate and terminate. Some have unknown functions. The appearance of the fibers is not as simple as how they are depicted in books. Each primary afferent may be part of one or more reflex arcs, or one or more ascending tracts. Most sensory information reaches the thalamus and cerebellum by several routes, so other tracts often compensate for loss of a single tract.

Ascending pathways

Ascending tracts are found in all three of the funiculi. Data reaching the thalamus is relayed to the cerebral cortex, and consciously perceived. Information reaching the cerebellum helps regulate movements, unconsciously. Each *posterior column* contains all parts of a posterior funiculus, except for its area of interaction with the propriospinal tract. The posterior columns are primarily made up of ascending collaterals of large, myelinated primary afferents. They carry data from various mechanoreceptors. Large numbers of second-order and unmyelinated fibers are also present. This is the primary pathway in which data reaches the cerebral cortex, from low-threshold receptors of the skin, joints, and muscles.

Cell bodies of spinal primary afferent fibers are within ipsilateral dorsal root ganglia. The fibers are of many different diameters and amounts of myelination. Where each posterior rootlet enters the spinal cord, the fibers are separated into a **medial division** and a **lateral division**. The medial division has heavily myelinated, large-diameter afferents. The lateral division has smaller diameter afferents with fine myelination or a lack of myelination. The medial division fibers enter the posterior column, and ascend to the brain stem. Many types are collateral to the deeper laminae of gray matter in the spine, reaching the caudal medulla before their final synapses. Each posterior column caudal to the T_6 level is an undivided bundle called a **fasciculus gracilis**. Rostral to T_6, fibers may leave this structure, but not many fibers are added. Afferents that enter rostral to T_6 form another bundle, the slightly triangular **fasciculus cuneatus**, lateral to the fasciculus gracilis. The two fasciculi are partially separated by a glial partition, extending inward. The partition is called the **posterior intermediate septum**.

> **Point to remember**
> As sensory nerve fibers reach the spinal cord, they are bundled according to function. They are called *nerve tracts* or *fasciculi* and are found in the spinal white matter. The name of each ascending nerve tract is based on its origin and termination.

Descending pathways

Descending pathways are primarily found in the lateral and anterior funiculi. They influence activities of **lower motor neurons**. Alpha and gamma motor neurons of the anterior horn are largely controlled by the supraspinal centers. These centers are partially found in the brain stem. The main descending outflow is from the cerebral cortex, especially from the precentral gyrus. These structures comprise the **corticospinal system**. The **lateral corticospinal tract** is also called the *pyramidal tract* since it continues through a pyramid of the medulla. The tract is large, and has a "crossed" structure. About 85% of the fibers from the contralateral pyramid are present, crossing in the pyramidal decussation. The lateral corticospinal tract is within the lateral funiculus of the spinal cord. It is medial to the posterior spinocerebellar tract, and fibers originate in the precentral gyrus and nearby areas of the cerebral cortex. They descend throughout the internal capsule, cerebral peduncle, basal pons, and medullary pyramid. The fibers decussate at the spinomedullary junction, and end in the anterior horn or intermediate gray matter. While terminating on the motor neurons of the anterior horn, they are also present in large numbers on smaller interneurons that synapse on these motor neurons. Interruption of lower motor neurons that supply a muscle causes **flaccid paralysis** and eventual muscle atrophy.

Upper motor neurons are those with axons descending from the cerebral cortex or brain stem, ending on the lower motor neurons directly, or via an interneuron. With corticospinal damage, an upper motor neuron lesion will have very different effects than a lower motor neuron lesion. These effects are compared in Table 1.3. The muscles involved usually have hyperactive reflexes, are **hypertonic**, and have increased resting tension. There is **paresis**, which is paralysis or weakness of fine voluntary and involuntary movements. These symptoms are described as **spastic paralysis**. Many pathologic reflexes are from upper motor neuron lesions. *Babinski's sign* involves dorsiflexion of the big toe and fanning of the other toes after the sole of the foot is firmly stroked.

About 15% of fibers in each pyramid do not cross in the pyramidal decussation, but reach the anterior funiculus. This is located near the anterior median fissure. When they reach it, the fibers are described as the **anterior corticospinal** tract. The fibers end on motor neurons or interneurons in medial areas of the anterior horn or intermediate gray matter. They can affect activity of motor neurons for the axial muscles. Many fibers cross in the anterior white commissure before synapsing. Most of the fibers end in the cervical and thoracic portions, probably helping to control of the neck and shoulder muscles. When damaged, there is no obvious weakness, which may be because of bilateral fiber distribution from the contralateral tract. Therefore, the *pyramidal tract* refers to the lateral as well as anterior corticospinal tracts.

Reflex activity

Reflexes make up many types of control systems in the body. They may be *inborn (intrinsic)* or *learned (acquired)*. Inborn reflexes are rapid, predictable motor responses to stimuli. They are unconscious, unlearned, and involuntary. These reflexes help avoid pain, maintain posture, and control visceral activities. Contacting a hot surface causes an instantaneous inborn spinal reflex, and the hand from the surface. Similar reflexes continue without awareness, such as visceral reflexes regulated by subconscious lower CNS regions—mostly the brain and spinal cord.

A *learned* reflex is developed through practice and repetition, an example of which is the activities required to drive an automobile. The process is mostly automatic, and only develops after a lot of time is used to master the related skills. The difference between inborn and learned reflexes is unclear. Most inborn reflexes are modified by learning and conscious effort. Pain signals identified by the spinal cord's interneurons are quickly transmitted to the brain. The individual becomes aware of pain and its cause. The **withdrawal reflex** is serial processing regulated by the spinal cord. Pain awareness is based on simultaneous and parallel processes of the sensory stimuli.

Components of a reflex arc

Reflexes occur over highly specific neural paths. *Reflex arcs* consist of five components, as follows:

- **Receptor**—the location of stimulus action;
- **Sensory neuron**—transmitting afferent impulses to the CNS;

TABLE 1.3 Upper and lower motor neuron damage and related effects.

Effect	Damage to upper motor neurons	Damage to lower motor neurons
Atrophy	Slight	Severe
Muscle tone	Increased	Decreased
Strength	Decreased	Decreased
Stretch reflexes	Increased	Decreased
Other signs	Pathologic reflexes such as Babinski's sing; clonus	Fibrillations; fasciculations

- **Integration center**—in a simple reflex arc, this can be a single synapse, between sensory and motor neurons, such as the **monosynaptic reflex**; in a more complex reflex arc, many synapses and chains of interneurons are involved, such as the **polysynaptic reflex**;
- **Motor neuron**—carries efferent impulses from an integration center to an effector organ;
- **Effector**—a gland cell or muscle fiber that responds, via contraction or secretion, to efferent impulses.

Functional classes of reflexes include **somatic reflexes** and **autonomic (visceral) reflexes**. Somatic reflexes activate skeletal muscle. Autonomic reflexes activate smooth or cardiac muscle, or glands, which are described as *visceral effectors*.

Spinal reflexes

The spinal cord is involved in sensory processing, motor outflow, and reflexes. **Spinal reflexes** are motor outputs caused by certain afferent inputs, often involving neural circuitry found only in the spinal cord. Except for axon reflexes, all reflex pathways involve at one or more receptor structures, and an associated afferent neuron. The cell body of an afferent neuron must lie in a posterior root ganglion or another sensory ganglion. Reflex pathways also involve efferent neurons, which have cell bodies in the CNS. Except for a **stretch reflex**, all reflexes involve one or more interneurons. Reflexes can be simple to very complex.

Stretch reflexes can be used for clinical testing, such as tapping the patellar tendon to stimulate the **knee-jerk reflex**, which slightly stretches the quadriceps muscle. Testing many stretch reflexes helps determine integrity of peripheral nerves and predictable spinal cord areas (Table 1.4). Since stretch reflexes are often initiated by tapping a tendon, they are commonly called **deep tendon reflexes**. The actual responsible receptors in muscles are attached to the tapped tendons. Stretch reflexes may be important for ongoing, automatic corrections of movements and postures, though other reflexes may be more important. While standing, there is a slight swaying of the body. Some muscles are stretched, causing a reflex contraction that helps return the body to the desired position.

> **Point to remember**
>
> With **peripheral nerve damage** or ventral horn injury of a certain body area, stretch reflexes are usually hypoactive or absent. Reflexes are absent in people with chronic diabetes mellitus or neurosyphilis, and in comatose individuals. They are hyperactive when corticospinal tract lesions reduce inhibitory effect of the brain upon the spinal cord, as in stroke patients.

Peripheral nervous system

The **PNS** allows the CNS to send commands to voluntary muscles and other effectors. This results in needed body movements. The PNS utilizes white nerves throughout most body areas, allowing the CNS to receive information and complete functions. The PNS includes all neural structures outside the brain and spinal cord, including sensory receptors, peripheral nerves and their ganglia, and efferent motor endings.

TABLE 1.4 Common clinically-tested deep tendon reflexes.

Reflex	Main spinal cord segment	Affected muscle	Peripheral nerve
Biceps	C5	Biceps brachii	Musculocutaneous
Brachioradialis	C6	Brachioradialis	Radial
Triceps	C7	Triceps brachii	Radial
Knee-jerk (patellar)	L4	Quadriceps femoris	Femoral
Ankle-jerk (Achilles)	S1	Gastrocnemius, soleus	Tibial

Sensory receptors

Sensory receptors have specialized functions, and respond to environmental changes in stimuli. Usually, activation of these receptors by stimuli causes graded potentials triggering nerve impulses along the afferent PNS fibers reaching the CNS. Awareness of stimulus is called *sensation*, while interpretation of sensation is called *perception*. **Chemoreceptors** respond to chemicals in a solution, such as molecules that are tasted or smelled, or any changes in chemistries of blood or interstitial fluid. **Mechanoreceptors** react to mechanical forces, including BP, other types of pressure, touch, stretching, and vibrations. **Nociceptors** respond to damaging stimuli, resulting in pain, which include extreme cold or heat, excessive pressure, and inflammatory chemicals. Their signals stimulate subtypes of chemoreceptors, mechanoreceptors, and thermoreceptors. **Photoreceptors** react to light, including the receptors in the retinas of the eyes. **Thermoreceptors** respond to temperature changes.

The three classifications of sensory receptors are based on their location. **Enteroceptors** respond to stimuli from outside the body and are usually near or upon the body surface. They react to pressure, touch, pain, and temperature in the skin, and also include most receptors of the special senses (hearing, vision, equilibrium, smell, and taste). **Interoceptors** respond to stimuli inside the body, including from blood vessels and internal viscera. They are also called *visceroceptors*, monitoring chemical changes, temperature, and tissue stretching. They may cause discomfort, pain, hunger, or thirst, but are usually not perceived. **Proprioceptors** also respond to internal stimuli, with more restricted locations. They are within connective tissue coverings of bones and muscles, and skeletal muscles, joints, ligaments, and tendons. They inform the brain of movements by monitoring stretching that occurs in the organs in which they are contained.

Most sensory receptors are within the **general senses**. They are modified dendritic endings of sensory neurons, throughout the body, and monitor most general sensory information. Receptors for the *special senses* are in the complex *sense organs*. In the eyes, there are sensory neurons and nonneural cells forming the lens and supporting wall. There are also simple general sense receptors, involved in *tactile sensation*, mixing pressure, touch, stretching, and vibrations. They are also involved in temperature, pain, and proprioceptor "muscle sensing." One type can respond to different stimuli. Different types can respond to similar stimuli.

General sensory receptors are *nonencapsulated (free)* or *encapsulated*. **Nonencapsulated nerve endings** are most common in the epithelia and connective tissues, responding mostly to temperature and painful stimuli, but also to tissue movement caused by pressure. Nerve endings responding to cold temperature are in the superficial dermis, while those responding to hot temperatures are deeper in the dermis. **Tactile (Merkel) discs** are in the deepest epidermis, acting as light touch receptors. **Hair follicle receptors** wrap around hair follicles, and are light touch receptors detecting hair bending.

All **encapsulated nerve endings** have one or more fiber terminals of sensory neurons in a connective tissue capsule. Most are mechanoreceptors, with variances in distribution, shape, and size. **Tactile corpuscles** (Meissner's corpuscles) are just below the epidermis and are used for touch discrimination (see Fig. 1.18). **Lamellar corpuscles** (Pacinian corpuscles) are deep in the dermis and subcutaneous tissue, stimulated by deep pressure, and respond only when pressure is first applied (see Fig. 1.19). **Bulbous corpuscles** (Ruffini endings) in the dermis, subcutaneous tissue, and joint

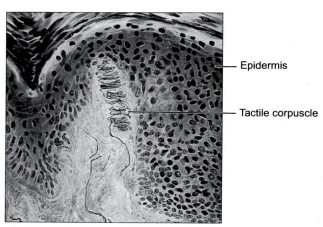

FIGURE 1.18 Tactile (Meissner's) corpuscles.

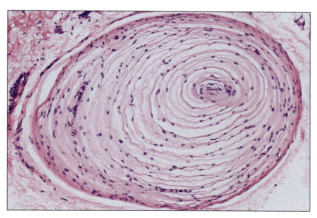

FIGURE 1.19 Lamellar (Pacinian) corpuscles.

capsules react to deep, continuous pressure. **Muscle spindles** in the perimysium of each skeletal muscle detect muscle stretching, initiating a reflex that resists the stretch. **Tendon organs** are near the junctions of skeletal muscles and tendons, respond to stretching of tendon fibers, then reacting with a reflex that causes relaxation of the contracting muscle. **Joint kinesthetic receptors** monitor stretch in articular capsules around synovial joints, providing data on joint position and motion.

Cranial nerves

The **cranial nerves** are part of the PNS and are directly connected to the brain, in 12 pairs. The cranial nerves are named in relation to function or distribution. Each nerve is numbered by its position along the brain's longitudinal axis, beginning at the cerebrum. Each cranial nerve is abbreviated as "CN," followed by its appropriate Roman numeral. For example, "CN I" refers to cranial nerve I, the olfactory nerve. When the full name of each CN is used, then just its Roman numeral is required, such as *olfactory nerve (I)*.

Every cranial nerve is attached to the brain near its associated nuclei, whether sensory, motor, or both. Motor nuclei information is similarly processed, as input is received from higher brain centers, or from nuclei along the brain stem. The cranial nerves are classified as follows:

- **Mostly sensory**—carrying SS information (pressure, touch, temperature, vibration, pain) or special sensory (sight, hearing, balance, smell);
- **Motor**—dominated by axons of SM neurons;
- **Mixed (sensory and motor)**—mixtures of sensory and motor fibers.

Cranial nerves have these primary functions, and also secondary functions.

Cranial nerves III, VII, IX, and X are also part of the parasympathetic autonomic system. The cranial nerves connect to the under portion of the brain, mostly on the brain stem (see Fig. 1.20). They pass through tiny holes called *foramina* in the cranial cavity of the skull, which allows them to extend between the brain and the peripheral connections. Sensory fibers of the cranial nerves are *proprioceptive*.

The classifications of the 12 pairs of cranial nerves are listed in Table 1.5.

Sensory and motor nuclei of the cranial nerves

In the medulla, there are sensory and motor nuclei related to cranial nerves VIII, IX, X, XI, and XII. These nerves send motor commands to muscles of the pharynx, neck, and back. They also send motor commands to visceral organs in the thoracic and peritoneal cavities. Cranial nerve VIII carries sensory information, from receptors in the internal ear, to the vestibular and cochlear nuclei, which extend from the pons into the medulla. The vestibular nuclei affect balance, while the cochlear nuclei affect hearing.

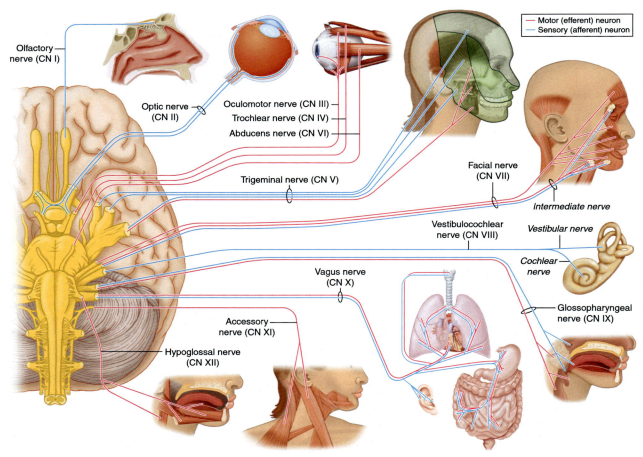

FIGURE 1.20 Cranial nerves.

TABLE 1.5 Classifications of the cranial nerves.

Name and number	Function
Olfactory (I)	Special sensory (smell)
Optic (II)	Special sensory (vision)
Oculomotor (III)	Motor (eye movements)
Trochlear (IV)	Motor (eye movements)
Trigeminal (V)	Mixed (sensory and motor, to the face)
Abducens (VI)	Motor (eye movements)
Facial (VII)	Mixed (sensory and motor, to the face)
Vestibulocochlear (VIII)	Special sensory (balance and equilibrium via the vestibular nerve, and hearing via the cochlear nerve)
Glossopharyngeal (IX)	Mixed (sensory and motor, to the head and neck)
Vagus (X)	Mixed (sensory and motor, widely distributed in the thorax and abdomen)
Accessory (XI)	Motor (to muscles of the neck and upper back)
Hypoglossal (XII)	Motor (tongue movements)

> **Point to remember**
> Damage to the brain stem can be fatal, since breathing, swallowing, or other basic motor functions may be lost. Most unconscious physiologic functions such as breathing, heartbeat, and digestion depend on a normal brain stem. The brain stem structures continually function, without need for input from the rest of the brain. People with brain damage can still have almost normal spontaneous body functions (normal breathing, heart rate, and BP) as long as the medulla and other brain stem areas are not harmed. A patient may be considered brain dead, with no possibility of recovery, when the brain stem is irreversibly damaged. Opiates and alcohol may cause dysfunction in different parts of the brain stem. In overdoses, this may be fatal if the brain stem cannot function normally. Opiates and alcohol suppress the respiratory center in the medulla, which can cause breathing cessation and death.

Spinal nerves

There are 31 pairs of *spinal nerves* are attached to the spinal cord by paired roots. They are part of the PNS. Each segment of the cord is designated by the paired spinal nerves arising from it. For example, the first lumbar spinal cord segment (L_1) is where the first lumbar nerves (lumbar nerve L_1) emerge from the cord. However, the spinal cord is continuous over its length, with only small changes in its internal structure. Each nerve exits the vertebral column, passing superiorly to its related vertebra, via the intervertebral foramen, except for the eighth cervical nerve, which exits the spine below the C7 vertebrae. Each nerve then connects to the area of the body that it supplies. The 31 segments of the spinal cord are as follows:

- **Cervical**—8 segments;
- **Thoracic**—12 segments;
- **Lumbar**—5 segments;
- **Sacral**—5 segments;
- **Coccygeal**—1 segment.

The spinal cord does not reach the end of the vertebral column. It ends at the level of L1-L2 in adults, and its segments are superior to where their related spinal nerves emerge, via the intervertebral foramina. Before reaching their intervertebral foramina, the lumbar and sacral spinal nerve roots are angled sharply downward. They continue inferiorly through the vertebral canal for a large distance. At the inferior end of the vertebral canal, there are nerve roots that form the **cauda equina**, named because it looks like horse's tail (Fig. 1.21). This structure forms due to fetal growth of the vertebral column is faster than that of the spinal cord, leading to extension of the nerves beyond the end of spinal cord at the L1 − L2 level. This results in the lower spinal nerve roots to point toward their exit points, inferiorly through the vertebral canal. The cauda equina fills the **lumbar cistern**, an area from L_1 or L_2 to the end of the dural sheath at S_2.

> **Point to remember**
> The first seven cervical nerves leave the vertebral canal *above* their corresponding vertebrae. The first cervical nerve emerges between the *occiput* and *atlas*, which is the first cervical vertebra. However, since there are only seven cervical vertebrae, the eighth cervical nerve emerges between the seventh cervical and first thoracic vertebrae. Each subsequent nerve leaves *below* its corresponding vertebra.

Autonomic nervous system

The ANS is the visceral part of the nervous system. It has neurons in the CNS and PNS. Its target tissues include the cardiac muscle, smooth muscle of the viscera and blood vessels, and glands. The ANS maintains homeostasis, which is a constant internal environment. Since it controls involuntary effectors, the ANS is the subconscious regulator of body functions under normal circumstances. It uses efferent and afferent pathways, and various neurons of the brain and spinal cord. The ANS allows for interactions with the external environment. A threat can immediately increase the heart rate, making it possible to run away for safety. Parts of the CNS working with the ANS include *nuclei* for the cranial nerves III, VII, X and IX and the ventrolateral medulla, anterior cingulate gyrus, insula, amygdala, and hypothalamus. The PNS includes some areas of the ANS, basically the parasympathetic and sympathetic nerves. One example is the vagus nerve (CN X).

Sympathetic division

The sympathetic nervous system, or thoracolumbar division (T1 to L2) of the ANS, forms from preganglionic cell bodies in the intermediolateral cell columns of the 12 thoracic spinal cord segments, and with the upper two lumbar

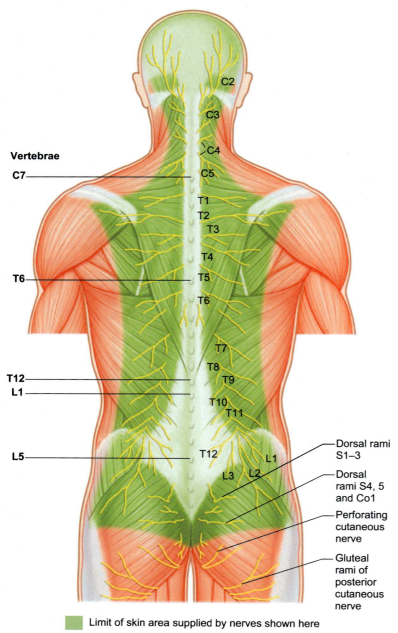

FIGURE 1.21 Cauda equina.

segments (Fig. 1.22). It consists of preganglionic sympathetic fibers, the adrenal medulla, and postganglionic sympathetic fibers. Preganglionic sympathetic neurons are located in the lateral horn of the thoracic spinal cord.

Preganglionic sympathetic fibers

Most preganglionic fibers in the lateral horn of the thoracic spinal cord are myelinated. The fibers follow the ventral roots, forming the **white communicating** rami of the thoracic and lumbar nerves (Fig. 1.23). Through these rami, the fibers reach the ganglia of sympathetic chains or trunks, which appear like strings of beads along the spinal cord on each side. These **trunk ganglia** are located upon lateral sides of the bodies of thoracic and lumbar vertebrae. Entering the ganglia, the fibers may synapse with ganglion cells. They may also continue up or down to the sympathetic trunk, synapsing with ganglion cells at different levels. They may pass through the trunk ganglia, out to one of the collateral (intermediary) sympathetic ganglia. These

Anatomy and physiology **Chapter | 1** 35

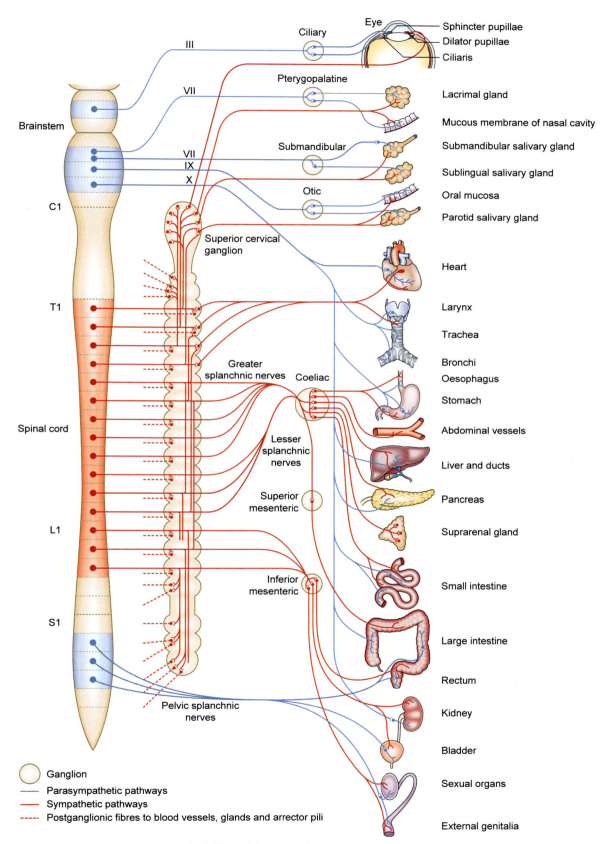

FIGURE 1.22 Sympathetic and parasympathetic divisions of the autonomic nervous system.

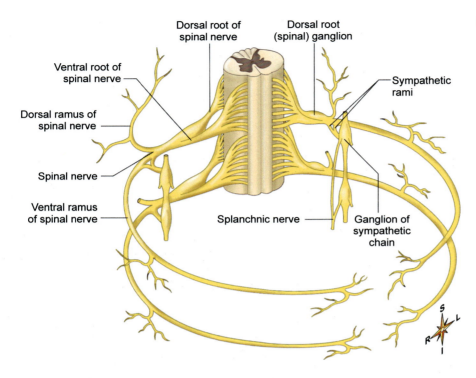

FIGURE 1.23 White communicating rami.

include the **celiac ganglia** and **mesenteric ganglia**. Normally, there are 22 sympathetic ganglia on either side of the vertebral column: cervical (3), thoracic (11), lumbar (4), and sacral (4).

The **splanchnic nerves** emerge from the lower seven thoracic spinal cord segments, passing through the trunk ganglia to the celiac ganglia and **superior mesenteric ganglia**. Synaptic connections in these areas occur with ganglion cells having postganglionic axons passing to the abdominal viscera, through the celiac plexus. Splanchnic nerves from the lowest thoracic and upper lumbar regions send fibers to synaptic stations in the **inferior mesenteric ganglion**, and to small ganglia related to the **hypogastric plexus**. Through this plexus, postsynaptic fibers are distributed to viscera of the lower abdomen and pelvis. The lumbar splanchnic sympathetic nerves are part of the superior and inferior hypogastric plexuses. This is part of the innervation of the bladder neck, ductus deferens, prostate, and other structures. If there is nerve damaged, sexual dysfunction may occur.

> **Point to remember**
> By tracing the axon in a sympathetic chain ganglion, the preganglionic fiber can branch along three paths. It may synapse with a sympathetic postganglionic neuron. From preganglionic fibers, there are ascending or descending branches through the sympathetic trunk. These synapse with postganglionic neurons in other chain ganglia. They may pass through one or more ganglia, with no synapses.

> **Point to remember**
> Preganglionic neurons passing through chain ganglia with no synapses continue through the splanchnic nerves to other sympathetic ganglia. These *collateral (prevertebral) ganglia* are paired sympathetic ganglia near to the spinal cord. They are named for nearby blood vessels, such as the *celiac ganglion (or solar plexus)*, which lies next to the celiac artery.

Adrenal medulla

Preganglionic sympathetic axons of the splanchnic nerves also project to the adrenal glands. They synapse on chromaffin cells of the adrenal medulla. The **adrenal chromaffin cells** receive direct synaptic input from preganglionic sympathetic axons. These cells form from the neural crest, being considered to be *modified postganglionic cells* without axons.

Postganglionic sympathetic fibers

Postganglionic sympathetic fibers are mostly unmyelinated (and gray in color), with their dendrites and cell bodies in sympathetic chain ganglia. They form the **gray communicating rami**. The fibers may follow the spinal nerve for a long distance, or go directly to their target tissues. The gray communicating rami join each spinal nerve. They distribute vasomotor, **pilomotor**, and sweat gland innervation to all somatic areas. Branches of the **superior cervical sympathetic ganglion** enter the sympathetic carotid plexuses, to surround the internal and external carotid arteries, for distribution of sympathetic fibers to the head.

The postganglionic sympathetic axons project to the lacrimal and salivary glands, muscles raising the eyelids, muscles dilating the pupils, facial sweat glands, cranial sweat glands, and blood vessels. From the three paired cervical sympathetic ganglia, the superior **cardiac nerves** pass to the **cardiac plexus**, at the heart's base. They distribute cardioaccelerator fibers to the myocardium. From the upper five thoracic ganglia, vasomotor branches continue to the thoracic aorta and posterior *pulmonary plexus*. Though this plexus, dilator fibers pass to the bronchi.

> **Point to remember**
>
> In the sympathetic division, preganglionic neurons are relatively short, while postganglionic neurons are long. An axon of one sympathetic preganglionic neurons synapses with many postganglionic neurons. These often terminate in organs that are far apart. This means that sympathetic responses are usually widespread and involve many different organs.

> **Point to remember**
>
> The sympathetic division is mostly an "emergency" system. Physical or psychological stress increases sympathetic outflow greatly. Responses make the body ready to use large amounts of energy (the "fight-or-flight reaction"). There is increased heart rate, strength of cardiac muscle contraction, dilation of coronary vessels, dilation of skeletal muscle blood vessels, dilation of respiratory airways, and increased sweating.

Parasympathetic division

The parasympathetic (craniosacral) division of the ANS arises from preganglionic cell bodies in the brain stem's gray matter and the S2 to S4 segments of the sacral spinal cord. Involved cell bodies include the medial portion of the **oculomotor nucleus**, the **Edinger − Westphal nucleus**, and the **salivatory nuclei**. Most preganglionic fibers from S2, S3, and S4 continue uninterrupted from their central origin in the cord. They reach the wall of the viscus that they supply, or the location where they synapse with terminal ganglion cells. These cells are related to the **plexus of Meissner** and **plexus of Auerbach**, in the wall of the intestinal tract. Since parasympathetic postganglionic neurons are near to tissues they supply, their axons are relatively short. Parasympathetic distribution is only to visceral structures (See Fig. 1.22).

There are four cranial nerves conveying preganglionic parasympathetic fibers have visceral efferent functions. The *oculomotor nerve* (cranial nerve III), *facial nerve* (CN VII), and *glossopharyngeal nerve* (CN IX) distribute parasympathetic or visceral efferent fibers to the head. Parasympathetic axons of these nerves synapse with postganglionic neurons in the ciliary, sphenopalatine, submaxillary, and otic ganglia respectively. The *vagus nerve* (CN X) distributes autonomic fibers to thoracic and abdominal viscera, through the *prevertebral plexuses*. The **pelvic nerve**, or **nervus erigentes**, distributes parasympathetic fibers to most the large intestine, and to the pelvic viscera and genitals, through the *hypogastric plexus*.

Table 1.6 compares the pathways of the sympathetic and parasympathetic nervous systems.

> **Point to remember**
>
> Over 75% of all parasympathetic preganglionic fibers travel with the vagus nerve for at least 12 in., prior to synapsing with postganglionic fibers in the terminal ganglia, near effectors in the chest and abdomen.

TABLE 1.6 Sympathetic and parasympathetic neurons and pathways.

Neurons	Sympathetic pathways	Parasympathetic pathways
Preganglionic		
Dendrites and cell bodies	In lateral gray columns of the thoracic cord; in the first 2–3 lumbar segments of the cord	In brain stem nuclei, and in lateral gray columns of the sacral cord
Axons	In anterior spinal nerve roots, to the thoracic and first four lumbar spinal nerves, to and through white rami; terminating in sympathetic ganglia at various levels or through sympathetic ganglia, to and through splanchnic nerves, ending in collateral ganglia	From brain stem nuclei—through CN III to the ciliary ganglion; from pons nuclei through CN VIII—to a sphenopalatine or submaxillary ganglion; from nuclei in the medulla—through CN IX to the otic ganglion, or through CN X and XI to cardiac and celiac ganglia, respectively
Distribution	Short fibers—from CNS to a ganglion	Long fibers—from CNS to a ganglion
Neurotransmitter	Acetylcholine	Acetylcholine
Ganglia	Sympathetic chain ganglia in 22 pairs; collateral (celiac, superior, inferior mesenteric) ganglia	Terminal ganglia (in or near an effector)
Postganglionic		
Dendrites, cell bodies	In the sympathetic and collateral ganglia	In the parasympathetic ganglia (ciliary, sphenopalatine, submaxillary, otic, cardiac, celiac)—located in or near visceral effector organs
Receptors	Cholinergic (nicotinic)	Cholinergic (nicotinic)
Axons	In autonomic nerves, the plexus innervating thoracic, abdominal viscera and blood vessels in these cavities; in gray rami to spinal nerves, to smooth muscles of skin, blood vessels, hair follicles, and to sweat glands	In short nerves—to visceral effector organs
Distribution	Long fibers from ganglion—to widespread effectors	Short fibers from ganglion—to one effector
Neurotransmitter	Norepinephrine (most); acetylcholine (less)	Acetylcholine

Key terms

Adrenal chromaffin cells
Amygdaloid body
Anterior cerebral artery
Anterior commissure
Anterior corticospinal tract
Anterior horns
Anterior median fissure
Anterior roots
Anterolateral sulcus
Arachnoid mater
Archicortex
Arcuate fibers
Association areas
Association fibers
Atonia
Auditory association area
Auditory cortex
Autonomic (visceral) reflexes
Basal nuclei
Basal vein

Basilar artery
Broca's area
Bulbous corpuscles
Cardiac nerves
Cardiac plexus
Cauda equina
Caudal
Caudate nucleus
Celiac ganglia
Celiac plexus
Central canal
Central sulcus
Cerebellar hemispheres
Cerebral arterial circle
Cerebral cortex
Cerebral hemispheres
Cerebrospinal fluid
Cervical enlargements
Chemoreceptors
Choroid plexus

Cingulate gyrus
Circle of Willis
Circular sulcus
Claustrum
Conus medullaris
Corpus callosum
Corticospinal system
Cranial meninges
Cranial nerves
Crista galli
Deep tendon reflexes
Deep veins
Dentate gyrus
Denticulate ligaments
Diencephalon
Dura mater
Edinger-Westphal nucleus
Ehlers-Danlos syndrome (EDS)
Encapsulated nerve endings
Enteric nervous system (ENS)

Enteroceptors
Epidural space
Epithalamus
Falx cerebri
Fasciculus cuneatus
Fasciculus gracilis
Filum terminale
Fissure
Flaccid paralysis
Flocculonodular lobes
Folia
Foramen magnum
Funiculi
General senses
Glial septum
Globus pallidus
Granule cells
Gray commissure
Gray communicating rami
Gustatory cortex
Gyri
Hair follicle receptors
Hemispheric lateralization
Hippocampal formation
Hippocampus
Hypertonic
Hypogastric plexus
Hypothalamus
Inferior colliculi
Inferior mesenteric ganglion
Inferior sagittal sinus
Infundibulum
Integrative centers
Internal capsule
Internal carotid arteries
Internal jugular veins
Interneurons
Interoceptors
Joint kinesthetic receptors
Juxtallocortex
Knee-jerk reflex
Lamellar corpuscles
Lateral corticospinal tract
Lateral division
Lateral sulcus
Lentiform nucleus
Limbic lobe
Limbic system
Longitudinal fasciculi
Longitudinal fissure
Lower motor neurons
Lumbar cistern
Lumbar enlargements
Lumbar puncture
Mammillary bodies
Mammillothalamic tract
Mechanoreceptors
Medial division
Medulla oblongata
Melatonin
Memory consolidation
Mesenteric ganglia
Midbrain
Middle cerebral artery
Monosynaptic reflex
Muscle spindles
Neocortex
Neurons
Nociceptors
Nonencapsulated nerve endings
Nonrapid eye movement (NREM) sleep
Nucleus basalis
Oculomotor nucleus
Olfactory cortex
Oxytocin
Paleocortex
Parasubiculum
Paresis
Pelvic nerve
Peripheral nervous system
Photoreceptors
Pia mater
Piloerection
Pilomotor
Pineal gland
Pituitary gland
Plexus of Auerbach
Plexus of Meissner
Polysynaptic reflex
Pons
Posterior cerebral artery
Posterior commissure
Posterior funiculus
Posterior horns
Posterior intermediate septum
Posterior intermediate sulcus
Posterior median sulcus
Posterior root ganglion
Posterior roots
Posterolateral sulcus
Potential spaces
Prefrontal cortex
Premotor cortex
Presubiculum
Prevertebral plexuses
Primary motor cortex
Primary somatosensory cortex
Proprioceptors
Propriospinal tract
Pulmonary plexus
Purkinje cells
Putamen
Pyramidal cells
Rapid eye movement (REM) sleep
Reticular activating system
Reticular formation
Rostral
Salivatory nuclei
Sense organs
Sensory receptors
Serotonin
Somatic motor neurons
Somatic motor (SM) division
Somatic reflexes
Somatic sensory (SS) division
Somatosensory association cortex
Spastic paralysis
Special senses
Spinal cord
Spinal nerves
Spinal reflexes
Splanchnic nerves
Splenium
Straight sinus
Stretch reflex
Subarachnoid space
Subdural space
Subicular complex
Subiculum
Substantia nigra
Sulcus
Superficial veins
Superior cervical sympathetic ganglion
Superior colliculi
Superior mesenteric ganglia
Superior sagittal sinus
Tactile corpuscles
Tactile (Merkel) disks
Tendon organs
Tentorium cerebelli
Thalamus
Thermoreceptors
Transverse fissure
Trunk ganglia
Upper motor neurons
Vasopressin

Venous dural sinuses
Ventral amygdalofugal pathway
Vermis
Vertebral arteries
Visceral motor (VM) division

Visceral sensory (VS) division
Viscus
Visual association area
Visual cortex
Wernicke's area

White communicating rami
White matter
Withdrawal reflex

Further reading

Avilable at https://radiopaedia.org/articles/blood-supply-of-the-meninges.
Avilable at https://radiopaedia.org/articles/dural-venous-sinuses.
Avilable at https://radiopaedia.org/articles/innervation-of-the-meninges.
Agyen-Mensah, K. (2017). *Imaging appearances, diagnosis & treatment of atypical brain abscesses: Imaging appearances, a guide to making prompt diagnosis and recommended treatment of atypical brain abscesses*. Lap Lambert Academic Publishing.
Anderson, M. W., & Fox, M. G. (2016). *Sectional anatomy by MRI and CT (4th Ed.)*. Elsevier.
Berkowitz, A. (2016). *Lange clinical neurology and neuroanatomy: A localization-based approach*. McGraw-Hill Education/Medical.
Blumenfeld, H. (2010). *Neuroanatomy through clinical cases (2nd Ed.)*. Sinauer Associates/Oxford University Press.
Broussard, D. M. (2013). *The Cerebellum: Learning Movement, Language, and Social Skills*. Wiley-Blackwell.
Browning, W. (2018). *The veins of the brain and its envelopes*. Sagwan Press.
Bujis, R. M., & Swaab, D. F. (2013). *Autonomic nervous system (Handbook of clinical neurology, 117)*. Elsevier.
Cardinali, D. P. (2017). *Autonomic nervous system: Basic and clinical aspects*. Springer.
Carter, R. (2014). *The human brain book: An illustrated guide to its structure, function, and disorders*. DK Publishing.
Cramer, G. D., & Darby, S. A. (2013). *Clinical anatomy of the spine, spinal cord, and ANS (3rd Ed.)*. Mosby.
DeArmond, S. J., & Fusco, M. M. (1989). *Structure of the human brain: A photographic atlas (3rd Ed.)*. Oxford University Press.
Eroschenko, V. P. (2017). *Atlas of histology with functional correlations (13th Ed.)*. LWW.
Filley, C. (2012). *The behavioral neurology of white matter (2nd Ed.)*. Oxford University Press.
Gabella, G. (2012). *Structure of the autonomic nervous system*. Springer.
Geary, R. T. (2014). *The limbic system: Anatomy, functions and disorders*. Nova Biomedical.
Hugdahl, K., Westerhausen, R., Sun, T., Gannon, P. J., et al. (2010). *The two halves of the brain: Information processing in the cerebral hemispheres*. The MIT Press.
Irani, D. N. (2008). *Cerebrospinal fluid in clinical practice*. Saunders.
Isaacson, R. L. (2011). *The limbic system*. Springer.
Ito, M. (2011). *The cerebellum: Brain for an implicit self*. FT Press Science.
Iwase, S., Hayano, J., & Orimo, S. (2016). *Clinical assessment of the autonomic nervous system*. Springer.
Jirillo, E., Magrone, T., et al. (2018). *Immunity to helminths and novel therapeutic approaches (Immune response to parasitic infections)*. Bentham Science Publishers.
Joseph, R. G. (2017). *Limbic system: Amygdala, hypothalamus, septal nuclei, cingulate, hippocampus: Emotion, memory, language, development, evolution, love, attachment, ... aggression, dreams, hallucinations, amnesia*. Science Publishers.
Kiernan, J., & Rajakumar, R. (2013). *Barr's the human nervous system: An anatomical viewpoint (10th Ed.)*. LWW.
Kirshblum, S., & Lin, V. W. (2018). *Spinal cord medicine (3rd Ed.)*. DemosMedical.
Lalonde, R. (2017). *The brainstem and behavior*. Nova Science Publishers, Inc.
Mai, J. K., & Paxinos, G. (2011). *The human nervous system (3rd Ed.)*. Academic Press.
Marieb, E. N., & Hoehn, K. (2018). *Human anatomy & physiology* (11th Ed.). Pearson.
McCance, K. L., & Huether, S. E. (2018). *Pathophysiology: The biologic basis for disease in adults and children* (8th Ed.). Mosby.
Nolte, J. (2013). *The human brain in photographs and diagrams* (4th Edition). Saunders.
Pandya, D., & Seltzer, B. (2015). *Cerebral cortex: Architecture, connections, and the dual origin concept*. Oxford University Press.
Robertson, D., Biaggioni, I., Burnstock, G., Low, P. A., & Paton, J. F. R. (2011). *Primer on the autonomic nervous system (3rd Ed.)*. Academic Press.
Rolls, E. T. (2016). *Cerebral cortex: Principles of operation*. Oxford University Press.
Ruiz, A., & Fleming, D. (2013). *Encephalitis, encephalomyelitis and encephalopathies: Symptoms, causes, and potential complications*. Nova Biomedical.
Saladin, K. S. (2017). *Anatomy & physiology: The unity of form and function (8th Ed.)*. McGraw-Hill Education.
Sengul, G., & Watson, C. (2012). *Atlas of the spinal cord*. Academic Press.
Shier, D. N., & Butler, J. L. (2015). *Hole's human anatomy & physiology (14th Ed.)*. McGraw-Hill Education.
Smythies, J. R., Edelstein, L., & Ramachandran, V. S. (2013). *The claustrum: structural, functional, and clinical neuroscience*. Academic Press.
Soper, H. V., Comstock, T., et al. (2017). *Understanding the frontal lobe of the brain: Fractioning the prefrontal lobes and the associated executive functions* (Vol. 11). Fielding University Press.
Spetzler, R. F., Kalani, M. Y. S., Nakaji, P., & Yagmurlu, K. (2017). *Color atlas of brainstem surgery*. Thieme.
Vanderah, T., & Gould, D. J. (2015). *Nolte's the human brain: An Introduction to its functional anatomy*. Elsevier.

Chapter 2

Cytology of the nervous system

Chapter Outline

Glial cells	41
Glial cells of the central nervous system	41
Glial cells of the peripheral nervous system	43
Neurons	44
Functional characteristics	44
Structure	45
Classification	47
Membrane potential	49
Resting membrane potential	49
Membrane channel effects on resting membrane potential	50
Graded potentials	51
Action potentials	52
Generation of action potentials	53
Action potential propagation	54
Threshold and the all-or-none principle	55
Axon diameter and propagation speed	55
Synapses	56
Structure	56
Classification	56
Function of electrical synapses	57
Function of chemical synapses	57
Neuronal integration of stimuli	59
Postsynaptic potentials	59
Presynaptic regulation	60
Rate of action potential generation	60
Nerve cell degeneration and regeneration	61
Key terms	62
Further reading	63

Glial cells

Glial cells separate and protect neurons, support **nervous tissue**, and aid in regulation of interstitial fluid composition. The various types of **neuroglia** have unique functions. Most (about 90%) of the nervous system is made up of neuroglia. There are many more neuroglia than neurons. In the central nervous system (CNS), there are four types of neuroglia, but in the peripheral nervous system (PNS), there are only two types. *Macroglia* include astrocytes and oligodendrocytes, which are derived from the ectoderm. In some circumstances, they are able to regenerate. Table 2.1 summarizes various types of glial cells and their functions.

Glial cells of the central nervous system

The CNS contains four types of glial cells (neuroglia): *microglia, astrocytes, oligodendrocytes,* and *ependymal cells*. Also, the PNS contains two types of glial cells: *satellite cells* and *Schwann cells*. Fig. 2.1 shows all six types of glial cells.

Astrocytes

Astrocytes are the most common and largest neuroglia in the CNS. They are named because of their star-like shape and have many thin cytoplasmic processes. These processes end in *vascular feet* connecting them to outside areas of nearby capillary walls. Astrocytes contain microfilaments that extend across their widths and across their processes. They are connected by **tight junctions** and have the following functions:

- *Maintaining the blood—brain barrier (BBB) maintenance*. The **BBB** is a selectively semipermeable membrane between the CNS and the capillary blood flow. It allows certain substances to pass through, but blocks others, and is created by astrocytes, isolating the CNS from the overall circulation. Astrocytes release chemicals that help maintain permeability of capillary endothelial cells.
- *Building a framework for the CNS*. Astrocyte cytoskeletons help provide a three-dimensional structural framework for CNS neurons.

TABLE 2.1 Types of glial cells.			
Central nervous system	1. Astrocytes (macroglia)		These cells have many radiating processes. They maintain the blood – brain barrier; provide structural support; regulate amounts of dissolved gases, ions, and nutrients; absorb and recycle neurotransmitters; and form scar tissue after injuries occur. There are two large classes: *protoplasmic* and *fibrous* astrocytes. The protoplasmic astrocytes are thinner, with many branched processes, and found in the gray matter. Fibrous astrocytes have more fibers, with little branching of their processes, though the processes contain glial fibrils.
	2. Oligodendrocytes (macroglia)		These cells have a central body and processes that wrap around their neuron processes, and myelinate CNS axons while providing framework.
	3. Ependymal cells		Forming sheets with the motile cilia, these cells line ventricles of the brain and central spinal cord canal. They produce and propel cerebrospinal fluid.
	4. Microglia (microglial cells)		Small, stationary phagocytes that remove debris, pathogens, and wastes from cells. They can enlarge and move, due to stimulation, and also eliminate synapses (this is known as synaptic pruning). The microglia conduct immune surveillance in the CNS.
Peripheral nervous system	1. Schwann cells		Entirely wrapping around neuron processes, these cells have an outer portion (neurilemma). They provide myelination; surround all axons; and assist in repairs after injuries.
	2. Satellite cells		These cells cover surfaces of nerve cell bodies in sensory, sympathetic, and parasympathetic ganglia. In neurons around ganglia, they control levels of neurotransmitters, oxygen, carbon dioxide, and nutrients. They also surround neuron cell bodies.

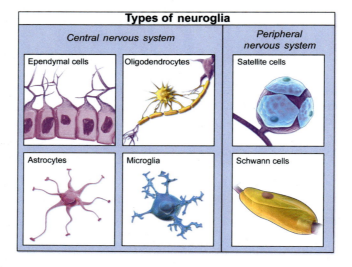

FIGURE 2.1 Types of glial cells.

- *Repairing nervous tissue damage.* When damaged, CNS tissue usually does not regain its normal functions. Astrocytes structurally repair damaged areas, stabilizing tissues and preventing any more injury from occurring.
- *Controlling the interstitial environment.* Astrocytes regulate interstitial fluid by:
  - Balancing concentrations of sodium and potassium ions, and carbon dioxide;
  - Quickly moving dissolved gases, ions, and nutrients between capillaries and neurons;
  - Controlling blood flow volume through capillaries;
  - Absorbing and recycling neurotransmitters;
  - Releasing chemicals that increase or reduce movement of information over axon terminals.

Astrocytes form a covering on all of the CNS, proliferating to repair damaged neural tissues. These *reactive astrocytes* are larger in size and are more easily stained. They can be histologically identified since they contain a characteristic protein called *glial fibrillary acidic protein (GFAP)*. The chronic proliferation of astrocytes leads to *gliosis* (glial scarring), which may or may not be beneficial or harmful to the nervous system.

Oligodendrocytes and myelination

Oligodendrocytes also have thin cytoplasmic extensions. The cell bodies are smaller than those of astrocytes and have fewer processes. These processes have unknown functions and are attached to neuron surfaces. Processes ending on *axon* surfaces are more widely understood. They insulate oligodendrocytes from the extracellular fluid (ECF) and surround many CNS **axons**. Maturing plasma membranes near oligodendrocyte processes form large pad-like structures, as the cytoplasm becomes thinner. The flattened mass wraps around the axolemma to form concentric plasma membrane layers. The wrapping is *myelin*, formed from membranes. It provides electrical insulation, increasing speeds of electrical *action potentials* as they move along the axon. Oligodendrocytes form a *myelin sheath* over an axon's length, and the axon is referred to as *myelinated*. Larger areas of the axon wrapped in myelin are *internodes*, about 1−2 mm in length. Separating nearby internodes are tiny gaps of a few micrometers in length, called the *nodes*, or **nodes of Ranvier**. Collateral axon branches begin at these nodes.

Myelinated axons have a shiny white appearance, mostly due to lipids in the myelin. Areas that are mostly made up of myelinated axons form the *white matter* of the CNS. *Unmyelinated axons* do not have total coverings of oligodendrocytes. They are common where collaterals and short axons synapse with neuron cell bodies. Regions having **dendrites**, neuron cell bodies, and unmyelinated axons have a gray color, and form the *gray matter* of the CNS. Oligodendrocytes have structural importance, joining clusters of axons.

Ependymal cells

The longitudinal axis of the CNS has a centralized passageway filled with supportive *cerebrospinal fluid (CSF)*. The fluid surrounds the brain and spinal cord, which actually "float" within it. The spinal cord's inner passageway is the *central canal*. In areas of the brain, this canal forms enlarged cavities called *ventricles*. The canal and ventricles are lined with **ependymal cells** aiding with production and control of CSF. Ependymal cells have *cilia* assisting with CSF circulation in the brain ventricles.

Microglia

Microglia are the least common and smallest of the CNS glial cells. They are phagocytic macrophages with thin, finely branched processes, and move through nervous tissue to engulf cellular debris, pathogens, and wastes. Though some microglia are always present in the brain, an injury or infection brings others into the brain from the blood vessels. Microglia originate early in embryonic development, from the mesodermal cell layer. They move into the CNS as it is forming.

Extracellular space

The fluid-filled extracellular space lies between various cells of the CNS and, usually, makes up about 20% of the total volume of the brain and spinal cord. Since transmembrane gradients of ions such as potassium and sodium are required for proper electrical signaling, regulation of these ion levels, known as *ionic homeostasis*, is important, and partially conducted by the astrocytes. In the CNS, capillaries are totally mixed with glial or neural processes. The capillary endothelial cells of the brain, as compared to those of other organs, form *tight junctions*. These are impermeable to diffusion and create the *blood-brain barrier*, which protects the brain's extracellular space from direct contact with the intravascular compartment.

Glial cells of the peripheral nervous system

Cell bodies of the PNS are clustered together, forming ganglia. Neuroglial processes insulate neuronal cell bodies and most PNS axons from their surroundings. The two types of PNS neuroglia are Schwann cells and satellite cells (see Fig. 2.1).

Schwann cells

Schwann cells, also called *neurolemmocytes*, have two structural types. They may form a thick myelin sheath, or an indented plasma membrane that folds around peripheral axons in the PNS. Where a Schwann cell covers an axon, the outer cell surface is called the **neurilemma**. This shields the axon from interstitial fluids. *Nodes* are gaps between Schwann cells, which are basically the oligodendrocytes of the PNS. Only peripheral nerves have a neurilemma as well as the myelin sheath. The term neurilemma is also spelled *neurolemma*. Myelinating Schwann cells wrap a single axon, while a CNS oligodendrocyte may wrap several axons. Nonmyelinating Schwann cells may surround areas of several unmyelinated axons. A series of Schwann cells encloses an axon along its entire length.

Satellite cells

Satellite cells surround the neuron cell bodies in the ganglia. They control the interstitial fluid around the neurons, similar to the function of the astrocytes of the CNS.

> **Point to remember**
> When there are improper amounts of glial cells, health can be affected. While these cells may divide quickly, causing tumors, neurons do not divide. Damage to the spinal cord may destroy neuroglia. Axons can no longer produce myelin. Due to overgrowth of neuroglia, scars form, and recovering lost function takes a longer time.

Neural responses to injuries

Neurons have only limited responses to injury. In most cases, more significant repair occurs in PNS neurons than in those of the CNS. Nissl bodies scatter and increased protein synthesis result in the nucleus moving away from its centralized location. If it starts to function normally again, the neuron will return to its original appearance. Recovery is based on events inside axons. Crushing injuries cause a local decrease in blood flow and oxygen. The affected axolemma becomes unexcitable. If pressure is relieved in under 72 h, the neuron can sometimes recover. More long-term or severe pressure causes similar effects to those that occur when an axon is severed.

Schwann cells of the PNS aid in repairing damaged nerves. With axonal damage, a degeneration process begins in axons distal to the injury area. This is called **Wallerian degeneration**. Macrophages arrive to clear any debris. Schwann cells proliferate, forming a cellular cord resembling the original axon path. The neuronal axon grows into the area of injury. Schwann cells wrap around the axon. Any axon growing along the correct cord of Schwann cells may reform normal synaptic contacts eventually. If this does not occur, or the axon stops growing, normal function will not return. Most often, the growing axon arrives at the correct location if the edges of the original cut nerve bundle are still intact. After injury to an axon, the neuronal cell body enters a phase called the **axon reaction**, also known as *chromatolysis*. Axons generally regenerate at about 1 mm per day. CNS regeneration is more difficult. This is because:

1. There is much more axon involvement;
2. Axon growth across the injury is prevented by scar tissue that is related to astrocytes;
3. Axon growth is blocked by chemicals released from astrocytes.

Neurons

The *neuron* is the basic structural and functional unit of the nervous system. They are usually large, highly specialized cells that can provide electrical impulses along their axons. Neurons have different structures than glial cells. They also can be varied in size and complexity. Motor neurons are usually larger than sensory neurons. Any nerve cells that have long processes, such as the dorsal root ganglion cells, are larger than nerve cells that have short processes. Certain neurons project outward from the cerebral cortex to the lower spinal cord, which can be more than four feet long in adults. Other neurons have extremely short processes and only reach between cells within the cerebral cortex. These small **interneurons** have short axons that end locally.

Functional characteristics

Neurons usually survive for much longer than glial cells and have a high metabolic rate. Their excitable plasma membranes are similar to those of skeletal muscle cells. Energy in the neuron is needed in large amounts due to generation and **propagation** of action potentials. Significant energy is used to synthesize and secrete chemical compounds needed to transfer information between neurons. Mitochondria produce much energy in active neurons. Most neurons do not have centrioles, which are the organelles that function in mitosis. Centrioles help organize microtubules that transport chromosomes, along with the cytoskeleton. The lack of centrioles in CNS neurons prevents them from dividing. They are therefore not replaced if lost to injury or disease.

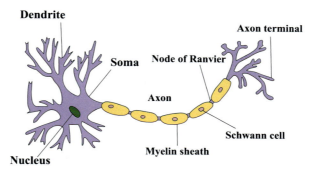

FIGURE 2.2 Structure of a neuron. Source: *Quasar Jarosz CC BY SA 3.0, via Wikimedia Commons.*

Structure

Neurons are primarily located in the CNS, with different shapes (see Fig. 2.2). Each neuron has four basic regions:

- A large *cell body*;
- Short, branched *dendrites*;
- One long *axon*;
- Terminal axon branches called *telodendria*.

The cell body

The *cell body*, (also called the *soma* or **perikaryon**), has a large, round nucleus plus an obvious *nucleolus*. The nucleus is usually centered within the cell body, and contains genes. The perikaryon contains the cytoplasm surrounding the nucleus, which is also surrounded by a plasma membrane. Mature neurons do not duplicate chromosomes, and function solely in gene expression. The chromosomes are uncoiled, and are not compacted. The nucleolus helps with synthesis of ribosomal ribonucleic acid (rRNA), and with the assembling of ribosome subunits. The size of the nucleolus is large because of high amounts of protein synthesis. The *nuclear envelope* is a specialized region of rough endoplasmic reticulum (RER), in the cytoplasm. The two-layered envelope has small **nuclear pores** and allows materials to diffuse in and out of the nucleus. Ribosomal subunits pass through the nuclear pores into the cytoplasm with little difficulty.

The cytoplasm has granular and agranular forms of endoplasmic reticulum. The cytoplasm also includes: the Golgi apparatus (or Golgi complex), Nissl bodies (or *Nissl substance*), microfilaments, microtubules, mitochondria, centrioles, lysosomes, lipofuscin, melanin, glycogen, and lipids. The **Golgi apparatus** is a network of irregular threads around the nucleus, with flat cisternae and small vesicles of smooth endoplasmic reticulum. Protein from the Nissl bodies is moved inside the Golgi complex via transport vesicles, and stored there. Carbohydrates may be added, forming *glycoproteins*. The Golgi complex aids in producing lysosomes and in synthesis of cell membranes. This synthesis is highly important in formation of synaptic vesicles at axon terminals. The Nissl bodies synthesize proteins, giving a gray color to areas containing neuron cell bodies—the *gray matter*. Nissl bodies are located throughout the cytoplasm, except for those close to the axon, which is known as the **axon hillock** (meaning "little hill").

The *mitochondria* are found throughout the cell body, dendrites, and axons. Their inner membranes have folds (*cristae*) that project into their centers. The double-membraned mitochondria are important in energy production. The multiple mitochondria, ribosomes, and membranes cause the perikaryon to have a rough, grainy appearance. Mitochondria generate adenosine triphosphate (ATP) for the energy requirements of neurons, and ribosomes and RER synthesize proteins. Certain parts of the perikaryon have clusters of RER and free ribosomes. These areas stain that darkly with **cresyl violet** are the actual *Nissl bodies*. In the perikaryon, the cytoskeleton has **neurofilaments** and **neurotubules** (*microtubules*) that are similar to intermediate filaments and microtubules of different cells. The main component of the cytoskeleton is formed by neurofilaments. **Neurofibrils** are bundles of neurofilaments that extend into dendrites and the axon, providing internal support. The large amounts of neurofibrils run parallel to each other. The perikaryon has organelles that provide energy and synthesize organic materials, mostly the chemical *neurotransmitters* required for cell-to-cell communication.

Actin **microfilaments** below the plasma membrane of neurons form a dense network. Along with microtubules, the microfilaments help form new cell processes and remove older ones. The **microtubules** extend through the cell bodies and their processes. In an axon, microtubules are parallel to each other. The proximal end of each microtubule points at the cell body, while the distal end points away from it. *Lysosomes* are vesicles that are connected to the plasma

TABLE 2.2 Structures and functions of nerve cell bodies.

Structures	Functions
Nucleus	Controls cellular activities
Perikaryon (cytoplasm)	Surrounds the nucleus and occupies the entire cell body
Ribosomes	Protein synthesis
Nissl bodies	Protein synthesis
Golgi complex	Adds carbohydrates to protein; forms products that will be carried to nerve terminals; forms cell membranes
Mitochondria	Produce chemical energy
Neurofibrils	Determine shapes of neuron
Microfilaments	Help form and retract cell processes; aid in cellular transport
Microtubules	Aid in cellular transport
Lysosomes	Digest lipids, melanin, and pigment
Centrioles	Aid in cell division and microtubule maintenance
Lipofuscin	A harmless byproduct of metabolism
Melanin	Related to formation of *dopa*, the precursor of *dopamine*

membrane. They have hydrolytic enzymes and form by budding off the Golgi complex. Lysosomes digest lipids, myelin, and pigments. **Centrioles** are small and paired with immature, dividing nerve cells. A pigment material, **lipofuscin**, has yellowish-brown granules in the cytoplasm. It accumulates with aging, but is a harmless metabolic byproduct. *Melanin granules* are in the cell cytoplasm as well as in the substantia nigra of the midbrain. They are related to the neurons' ability to synthesize catecholamine. *Dopamine* is the neurotransmitter of these neurons. Table 2.2 summarizes the structures and functions of nerve cell bodies.

Nerve cell bodies within the cerebral and cerebellar cortices are aggregated and form layers known as *laminas*. In the spinal cord, brain stem, and cerebrum, nerve cell bodies form *nuclei*, which are compacted groupings. Each nucleus has *projection neurons*, with axons carrying impulses to other areas of the nervous system, plus *interneurons* that act as short relays inside the nucleus. Within the PNS, these groupings of nerve cell bodies are called *ganglia*. Nerve cell groups are connected by pathways created by bundled axons, which are sometimes well defined and called *tracts* or *fasciculi*. However, sometimes there are no discrete axon bundles. Groupings of tracts inside the spinal cord are called *columns* or *funiculi*. In the brain, some tracts are called **lemnisci**. In certain brain regions, axons are found amongst dendrites, and are not bundled, making their pathways difficult to visualize. Such networks are referred to as the *neuropil*.

Neuron processes

Around the cell body, there are processes that include axons, dendrites, and **telodendria**. An axon is one long cytoplasmic process, up to 1 m in length. The axon extending from the lumbar spine to the big toe is up to four feet in length. Axons are among the longest cells, and long axons are also called *nerve fibers*. There is only one axon per neuron. Nerves are made up of several axons, which can propagate electrical impulses known as *action potentials*, which move away from the cell body. The **axoplasm** contains enzymes, lysosomes, mitochondria, neurofibrils, neurotubules, and tiny vesicles. The **axolemma** surrounds the axoplasm. In the CNS, an axolemma may be exposed to interstitial fluid or may be covered by neuroglial cellular processes. The *initial segment* of an axon of many neurons joins its cell body at the thick, cone-shaped *axon hillock*. The axolemma of the initial segment has a large density of sodium channels, which allow it to act as a *trigger zone*, where action potentials are generated. The action potentials travel along the axon to invade terminal axonal branches, triggering synaptic activity, which affects other neurons. The initial segment lacks **Nissl substance**. The axon narrows to form a thin process of the same diameter over the rest of its length. Some neurons have extremely short axons, or totally absent. Other neurons have axons making up nearly the entire length of the neuron. Axons may be branched along their lengths, producing side branches called **collaterals** that allow a neuron to share information with several other cells.

Dendrites are thin, branched extensions important for intercellular communication. The sensitive dendrites usually have many branches, some of which feature thin, lengthened, spiked projections called *dendritic spines*, which function in synapses. The receptive part of a neuron is called the **dendritic zone**. The CNS neurons receive information from neurons located mostly at the dendritic spines, which can make up 80% − 90% of the neuron's total surface area. Dendrites of motor neurons are short and tapered. There are usually hundreds of stick-like dendrites near the cell body. Almost all organelles in the cell body are also present in dendrites, the main *input regions*, creating an enlarged surface area receiving signals from other neurons. In many brain regions, thinner dendrites are specialized to collect information. Dendrites bring incoming messages to cell bodies as short-length electrical signals called *graded potentials*. The primary axon trunk and collaterals end in thin extensions called telodendria, also called *terminal branches*. Neurons may have over 10,000 terminal branches. These are also known as *terminal arborizations*. The end point of an axon is its *terminus*. Telodendria end at knob-like **axon terminals** (*synaptic terminals* or *arborizations*), assisting in communication with other cells. A **synapse** is the point where a neuron communicates with other cells, which can be other neurons.

Axonal (axoplasmic) transport is movement of materials between cell bodies and axon terminals, traveling along axons via neurotubules in the axoplasm. They are "pulled" by molecular the motor proteins **kinesin** and *dynein*, using ATP. Some materials move slowly, at a few millimeters per day. This is called "slow stream" transport. Some vesicles move faster, traveling as a "fast stream", up to 1000 mm per day. **Axonal transport** occurs in both directions at the same time. Flow of materials from a cell body to an axon terminal is carried by *kinesin*, and is known as **anterograde flow** or *anterograde transport*. Simultaneously, substances are transported by *dynein* from the axon terminal toward the cell body. This is **retrograde flow** or *retrograde transport*. Materials flow in both directions along the axon via anterograde flow and retrograde flow. An action potential can only travel away from the cell body, in one direction. Debris or substances in the axon terminal are delivered to the cell body quickly by retrograde flow. When in the cell body, these may change cellular activity by activating or inactivating specific genes.

> **Point to remember**
> When neurons are deprived of oxygen, their nuclei shrink, and the neurons change shape and disintegrate. *Hypoxemia, ischemia*, or toxins may cause an oxygen deficiency. Toxins can block aerobic respiration, preventing neurons from using oxygen.

Classification

Neurons are classified by their structures and functions. Understanding the classifications helps to remember the important concepts of neurons.

Structural classification

Neurons are structurally classified by the relationships between dendrites, cell body, and axons. They may be *anaxonic, bipolar, unipolar*, or *multipolar* (see Fig. 2.3):

- *Anaxonic neurons:* found in the brain and special sense organs. They are small in size, with many dendrites, and no obvious axons, even under a microscope. Their functions are not completely understood.
- *Multipolar neurons:* the most common CNS neurons, they have two or more dendrites and one axon. All motor neurons that control skeletal muscles are multipolar. The longest axons are those that carry motor commands from the spinal cord to the small muscles of the toes.
- *Bipolar neurons:* relatively small, uncommon neurons that are only located in the special sense organs. They have one dendrite and one axon, with the cell body between them, and transmit data about sight, smell, or hearing.
- *Unipolar neurons:* are also called *pseudounipolar neurons*, with dendrites and axons that are continuous with each other, with the cell body located on one side. The neuronal base lies where dendrites meet. The remaining process is considered to be an axon, carrying action potentials. Most sensory neurons of the PNS are unipolar, having long axons of 1 m or more, ending at synapses in the CNS. The longest of these axons carry impulses from the toes to the spinal cord.

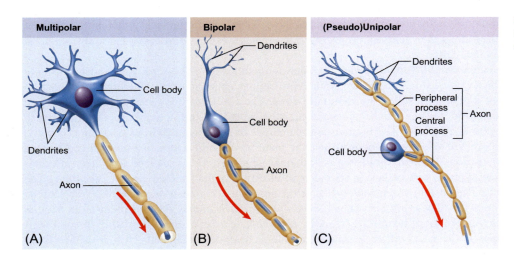

FIGURE 2.3 Types of neurons. (A) Multipolar neuron. (B) Bipolar neuron. (C) (Pseudo)Unipolar neuron.

Functional classification

Neurons are also classified by functionality, as:

- *Sensory neurons*;
- *Motor neurons*;
- *Interneurons*.

The classifications are based on directions that nerve impulses travel to and from the CNS.

Sensory neurons

Sensory neurons form the *afferent* division of the PNS, and transmit impulses from sensory receptors in the skin or internal organs to the CNS. A **ganglion** consists of neuron cell bodies within in the PNS. Cell bodies in the brain and spinal cord form *nuclei*, including the *trigeminal nuclei*. Sensory neurons are unipolar, and almost all of their cell bodies are in the peripheral *sensory ganglia*. Processes of sensory neurons are *afferent fibers*, and extend between sensory receptors and the CNS. Information travels from sensory receptors to the brain or spinal cord. Approximately 10 million sensory neurons exist in the body. Each collects information about the external and internal environments. *Somatic sensory neurons* monitor the external environment. **Visceral sensory neurons** monitor the internal environment and organ systems. Within the peripheral nerves, somatic fibers innervate skin, muscle, joints and the walls of the body. Visceral fibers innervate blood vessels and internal organs. *Sensory fibers* are *afferent*, while *motor fibers* are *efferent*. Sensory receptors are of three groups:

- *Interoceptors:* monitor the cardiovascular, digestive, reproductive, respiratory, and urinary systems. They signal contraction or distention of visceral structures. General visceral afferent fibers bring interoceptive data from the visceral organ receptors.
- *Exteroceptors:* handle pressure, temperature, or touch information, and the senses of equilibrium (balance), hearing, sight, smell, and taste.
- *Proprioceptors:* monitor skeletal muscle and joint movement, and position.

Also, somatic afferent fibers carry information from the proprioceptors and exteroceptors.

Motor neurons

Motor neurons form the *efferent* division of the PNS. There are about 500,000 of them, carrying information from the CNS to peripheral effectors in the peripheral tissues and organ systems. *Efferent fibers* are the axons of motor neurons, and carry data away from the CNS. The two primary efferent systems are the somatic nervous system (SNS) and the autonomic (visceral) nervous system (ANS). The somatic nervous system includes the *somatic motor neurons*, which innervate skeletal muscles. The SNS is under *conscious control*. Cell bodies of somatic motor neurons are in the CNS. Axons travel through peripheral nerves, innervating skeletal muscle fibers at neuromuscular junctions.

The ANS is not under conscious control, with **visceral motor neurons** stimulating all peripheral effectors except for skeletal muscles. They innervate cardiac and smooth muscle, adipose tissue, and glands. Visceral motor axons in the CNS innervate other visceral motor neurons in the peripheral *autonomic ganglia*. The cell bodies of neurons innervate and control peripheral effectors. *Preganglionic fibers* are axons that extend from the CNS to autonomic ganglia. *Postganglionic fibers* are axons that connect ganglion cells with peripheral effectors.

Interneurons

Interneurons lie between sensory and motor neurons. There are about 20 billion of them, also known as *association neurons*. Most are in the CNS, but others are in the autonomic ganglia. Interneurons make up more than 99% of all neurons in the body, with their primary function being integration. They carry sensory data and regulate motor activity. More interneurons become activated when a response to stimuli needs to be complex. Interneurons are involved in all higher functions, including learning, memory, cognition, and planning. Most interneurons are multipolar, and have many different sizes, with different fiber branch patterns.

> **Point to remember**
> Immunohistochemical markers of neurons and glial cells are valuable tools for neuroscience. Using antibodies against various cell components, cells expressing a neuronal phenotype can be identified. There are hundreds of different neuronal cell types, with variances in size, soma shape, dendritic branch patterns, axonal projections, and electrophysiological properties. The basic histology of neurons includes Nissl bodies, rough endoplasmic reticulum, ribosomal RNA, neurofilaments, neurotubules, neurofibrils, pigment granules, and structural proteins. The basic histology of glial cells includes cell bodies with extended dendrite-like processes. *Specific tumor markers will be discussed in the individual tumor chapters.*

Membrane potential

Membrane potential and resting membrane potential are unique features of cells. The same concepts also apply to many other types of cells. The cells of the human body have a membrane potential that is always changing, due to cellular activities. This may be explained as a difference in electrical charges across the plasma membrane. It is called a *potential* since it is a type of stored energy known as *potential energy*. When opposing electrical charges, including ions, are separated by a membrane, they have the potential to move towards each other—if the membrane allows them to cross. A membrane with this potential is *polarized*, with a negative pole and a positive pole. The negative pole is where there are more negative ions. The positive pole is where there are more positive ions. The potential difference's magnitude, in between, is measured in millivolts (mV) or volts (V). This voltage is measured with a *voltmeter*. The sign of a membrane's voltage indicates the charge present on the inner surface of a polarized membrane. A negative value, such as -60 mV, indicates the potential difference has a 60 mV magnitude. Therefore, the inner membrane is negative compared to its outer surface. If the voltage is positive, the inner membrane is positive while the outer membrane is negative.

Resting membrane potential

The **resting membrane potential** is the potential of a cell while at rest. All neural activities start with a change in the resting membrane potential, which is temporary and localized. This effect is a **graded potential** and decreases over distance from the stimulus. When it is large enough, an action potential is triggered in the membrane of the axon. An action potential is an electrical signal involving nerve cells. A neuron that is not conducting electrical signals is "resting", usually at about -70 mV, though this varies. The three most important factors about resting membrane potential are as follows:

- *Large differences in the ionic composition of ECF and intracellular fluid (cytoplasm)*. The ECF has high concentrations of sodium (Na^+) and chloride ions (Cl^-). The cytoplasm has significant concentrations of potassium ions (K^+) and negatively charged proteins.
- *Selective cell membrane permeability*. Because of selectively permeable membranes, there is no "even" distribution of ions. Lipid areas of the plasma membrane keep ions from crossing easily. Ions only enter or leave a cell through a membrane channel. At resting membrane potential, ions move through *leak channels*, which are membrane

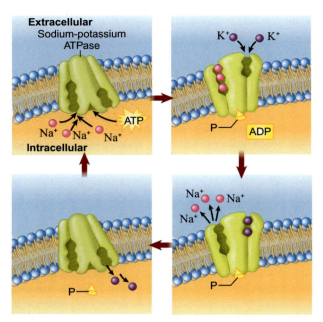

FIGURE 2.4 Sodium-potassium exchange pump.

channels that stay open. Some ions are moved in or out of cells by active transport mechanisms, including the sodium-potassium exchange pump.
- *Ions have different membrane permeabilities.* This means that a cell's passive and active transport mechanisms do not have an equal distribution of charges across the plasma membrane. Negatively charged proteins are unable to cross the membrane due to their large size. The inner membrane surface has excessive negative charges compared to the outer surface.

Passive and active forces determine membrane potential across the plasma membrane.

Sodium-potassium exchange pump

At normal resting membrane potential, sodium is forced outwards and potassium is carry into the cells (see Fig. 2.4). This requires the sodium-potassium exchange pump, which uses ATP in order to operate. Three intracellular sodium ions will be exchanged for every two extracellular potassium ions. At normal resting membrane potential, the sodium ions are ejected as fast as they can enter. The exchange pump then balances the passive forces of diffusion. Since the ionic concentration gradients remain balanced, resting membrane potential remains stable.

Membrane channel effects on resting membrane potential

The resting membrane potential exists because:

- Cytoplasm has different chemical and ionic compositions from ECF;
- Plasma membranes have selectively permeability. Potential rises and falls due to temporary permeability changes. Opening or closing membrane channels causes changes in membrane potential.

Membrane channels control the flow of ions over the plasma membrane. Sodium and potassium ions determine membrane potential of neurons and many other cells. Sodium and potassium ion channels are either passive or active.

Passive ion channels

Leak channels are passive ion channels that are always open. Their permeability changes as proteins that form them change shape, because of local conditions. Leak channels are needed to establish normal resting membrane potential of cells.

Active ion channels

Gated ion channels are active channels in the plasma membranes, opening or closing in response to stimuli. The three types include:

- *Chemically gated ion channels*, or *ligand-gated ion channels*: open or close as they bind certain chemicals, called *ligands*, such as receptors binding acetylcholine (ACh) at the neuromuscular junction. These channels are most common on dendrites and neuron cell bodies, which is where most synaptic communication occurs.
- *Voltage-gated ion channels*: open or close from changes in membrane potential. They are specialized areas of *excitable cell membrane*, able to generate and propagate action potentials. Examples include axons of multipolar and unipolar neurons, and the sarcolemma and T tubules of skeletal and cardiac muscle cells. Sodium, potassium, and calcium ion channels are the most important examples of these channels. Sodium ion channels have two independent gates. The first is an *activation gate*, requiring stimulation to open and allow sodium ions to enter. The second is an *inactivation gate*, closing to stop entry of sodium ions. There are three different functions:
- Closed, but able to open;
- Activated and open;
- Inactivated—closed, and unable to open.
- *Mechanically-gated ion channels*: open or close from changes to the membrane surface, such as pressure. They are needed for sensory receptors that respond to pressure, touch, or vibration. Distribution of membrane channels is different between areas of the plasma membrane, affecting cell responses to stimuli, and the parts of the cell that are involved. In neurons, chemically gated ion channels are found on dendrites and the cell body. Along axons are voltage-gated sodium and potassium ion channels. Voltage-gated calcium ion channels are located at axon terminals.

All gated channels are closed during resting membrane potential. When they open, ion movement across the plasma membrane increases, and the membrane potential is changed.

> **Point to remember**
> Membrane potential is an important biophysical signal in nonexcitable cells, modulating activities such as proliferation and differentiation. It is well known that cancer cells have distinct bioelectrical properties, such as a depolarized membrane potential that favors cell proliferation. Ion channels and transporters control cell volume and migration, and new evidence has revealed that the level of membrane potential has functional roles in neuronal and glial cancer cell migration. Membrane depolarization may be important for the emergence and maintenance of cancer stem cells, resulting in sustained tumor growth. Membrane potential is likely a valuable clinical marker for tumor detection with prognostic value. In the future, with artificial modification, tumor growth and metastasis could be inhibited.

Graded potentials

Graded potentials are produced by stimuli opening a gated channel and are *local potentials*. They cannot spread over long distances away from the stimulation.

1. *Sodium ions enter cells, attracted to negative charges on inner membrane surfaces.* As positive charges move outward, membrane potential moves toward 0 mV. **Depolarization** is from any change between resting membrane potential to a less negative potential. Depolarization reveals changes in potential from -70 mV to lesser negative values (-65, -45, and -10 mV), and to membrane potentials above 0 mV ($+10$, $+30$ mV). For all of these, membrane potential becomes more positive.
2. *As the plasma membrane depolarizes, the outer surface releases sodium ions.* With other extracellular sodium ions, the ions move to open channels, and replace ions already in the cell. *Local current* describes movement of positive charges parallel to the inside and outside of the depolarizing membrane.

In graded potentials, the amount of depolarization is reduced over distance, away from the stimuli. Local current is greatly reduced, because cytoplasm has a large resistance to ion movement. Some sodium ions entering cells move out across the membrane, through sodium leak channels. At a far enough distance away from the point of entry, effects on membrane potential are not detectable. The greatest change in membrane potential is based on stimulus size, which determines how many sodium ion channels are open. With more open channels, larger amounts of sodium ions enter, more of the membrane is affected, and there is more depolarization.

Membrane potential soon returns to resting levels when the chemical stimulus is removed, and normal membrane permeability is restored. *Repolarization* is restoration of normal resting membrane potential after depolarization. Repolarization usually requires combinations of ion movement through membrane channels. Efflux of potassium is mostly responsible for repolarization, requiring ion pumps, mainly the sodium-potassium exchange pump. As a gated potassium ion channel opens from stimuli, effects are opposite. Potassium ion outflow increases. The inner part of the cell loses positive ions, becoming more negative. **Hyperpolarization** is due to the loss of positive ions. It is an increase in negativity of resting membrane potential, such as from −70 to −80 mv, or more. A local current distributes this effect to neighboring areas of the plasma membrane, and the effect decreases over distance from the open channels.

Graded potentials happen in membranes of epithelial cells, fat cells, nerve and muscle cells, gland cells, and sensory receptors. Potentials often begin various cell functions, such as when a graded potential at a gland cell surface initiates exocytosis of secretory vesicles. Another example is when a neuromuscular junction's motor end plate is stimulated by a graded potential, due to ACh. This may trigger an action potential in nearby areas of the sarcolemma. While graded potentials are supported by the motor end plate, the rest of the sarcolemma has of excitable membrane. These areas are different because they have voltage-gated ion channels. When a graded potential causes hyperpolarization in a neuron, an action potential is not likely. If a graded potential causes depolarization, an action potential is more likely.

Passive processes: the electrochemical gradient

Over the plasma membrane, passive processes involve chemical and electrical gradients, and an attraction between positive and negative charges. If not separated, ions with opposite charges move together to stop the potential difference. This movement is a *current*. If a plasma membrane or other barrier separates oppositely charged ions, current strength is based on how easily the ions cross the membrane. *Resistance* of the membrane measures restriction of ion movement. With high resistance, the current is small, since less ions can cross. When resistance is low, the current is large, since more ions can cross. As ion channels open or close, resistance changes. This may cause differences in amounts of ions carried in or out of the cytoplasm. Electrical gradients may oppose or reinforce every chemical gradient of the ions. An *electrochemical* gradient is the sum of chemical and electrical forces acting upon the ion, across the plasma membrane. Resting membrane potential of most cells, including neurons, is mostly affected by electrochemical gradients for K^+ and Na^+.

Potassium ions have higher concentrations within cells, while concentrations are very low outside. Potassium easily moves out of cells due to the chemical gradient. This is opposed by the electrical gradient. Potassium ions in and out of the cell are attracted to negative charges in the plasma membrane. The ions are pushed away by positive charges outside the plasma membrane. Though the chemical gradient can outperform the electrical gradient, the electrical gradient reduces the force driving K^+ out of the cell. **Equilibrium potential** is the membrane potential for an ion when there is no movement of it across the plasma membrane. For potassium, this occurs at a membrane potential of approximately −90 mV. For neurons, resting membrane potential is usually −70 mV, close to the equilibrium potential for potassium ions. The difference is usually from continuous sodium ion leakage into the cell. Equilibrium potential shows the ion contribution to the resting membrane potential.

Sodium ions are highly concentrated outside cells, but lower in concentrations inside. A strong chemical gradient forces sodium ions into cells. Excessive negative charges in the plasma membrane attract extracellular sodium ions. Electrical and chemical forces force sodium ions into cells. Equilibrium potential for Na^+ is about +66 mV. Resting membrane potential is different, since permeability to sodium ions is very low. This is true since ion pumps in the plasma membrane force sodium ions out as fast as they cross. An electrochemical gradient is an example of *potential energy*, or *stored energy*, similar to a fully charged battery. Electrochemical gradients would be ended by diffusion without a plasma membrane. Each stimulus increasing plasma membrane permeability to sodium or potassium ions causes intense, quick movement of ions. The stimulus does not regulate ion movement, but is accomplished by the electrochemical gradient.

Action potentials

Action potentials are *nerve impulses* and are not graded potentials. They are changes in membrane potential. Once they begin, all parts of an excitable membrane will be affected. Action potentials are *propagated* along axon surfaces. They are not reduced while moving further away from their source, and impulses travel along axons to axon terminals. Generation and propagation are related. Action potentials must be generated at one location before they can be propagated away.

Generation of action potentials

Steps involved in generating an action potential from a resting state are shown in Fig. 2.5. Activation gates of voltage-gated sodium ion channels are closed at resting membrane potential. If the membrane reaches threshold, voltage-gated sodium ion channels open in one area, usually the initial axon segment. Then these steps occur:

- Depolarization to threshold;
- Activation of voltage-gated sodium ion channels;
- Rapid depolarization;
- Inactivation of voltage-gated sodium ion channels;
- Activation of voltage-gated potassium ion channels;
- Repolarization;
- Closing of voltage-gated potassium ion channels;
- A short hyperpolarization;
- A return to resting membrane potential.

Several characteristics that differentiate local potentials from action potentials are summarized in Table 2.3.

The refractory period

With more depolarization, the plasma membrane does not respond normally. This happens from when action potentials occur until resting membrane potential is reestablished. This is the *refractory period* of the membrane. When voltage-gated sodium ion channels open at threshold, until inactivation ends, the membrane cannot respond to any more stimulation. This is because all voltage-gated sodium ion channels are either open or are inactivated. This **absolute refractory period** occurs first, and lasts 0.4–1.0 msec.

The **relative refractory period** starts when sodium ion channels return to normal resting condition, continuing until membrane potential has stabilized, at resting level. If the membrane has enough depolarization, another action potential may occur in this period. The depolarization requires a larger-than-normal stimulus because local current must bring in

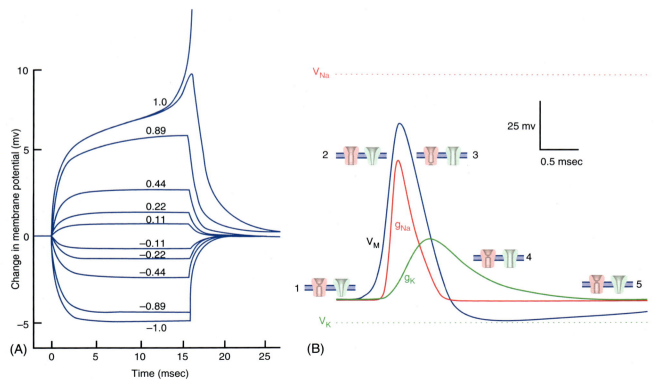

FIGURE 2.5 Generation of action potentials.

TABLE 2.3 A comparison of graded potentials and action potentials.

Graded potentials	Action potentials
Zero threshold level	Depolarization to threshold must occur before an action potential starts
Caused by gated channels on soma and dendrites	Caused by voltage-gated channels on an axon and a trigger zone
Passive outward spread, from the stimulation site	An action potential at one spot depolarizes nearby locations to threshold level
Are depolarizing (positive) or hyperpolarizing (negative)	Always depolarizing
Reversible; return to resting membrane potential when stimulation stops before threshold being reached	Irreversible; continue to completion once starting
Local; effects upon membrane potential decrease over distance from stimulation site	Self-propagated over entire membrane surface; no strength decrease
Depolarization or hyperpolarization levels are based on stimulus intensity; summation occurs	All-or-none event; all stimuli exceeding threshold produce identical action potentials; zero summation occurs
Zero refractory period	No refractory period
Happen in most plasma membranes	Only occur in excitable membranes of specialized cells (neurons, muscle cells)
Decremental; signals are reduced over distance	Nondecremental; signals are equally strong regardless of distance traveled

enough sodium ions to oppose exiting positive potassium ions through open voltage-gated potassium ion channels. It also occurs because the membrane is hyperpolarized to a degree through most of the relative refractory period.

Functions of the sodium-potassium exchange pump

For action potentials, depolarization occurs via inward movement of sodium ions. Repolarization requires a loss of potassium ions. Prestimulation ion levels are later reestablished by actions of this pump. For every action potential, there is only a few ions. Thousands of action potentials have to occur before there is a large change in intracellular ion concentrations. This means that the exchange pump is not important to any individual action potential.

Approximately 1000 action potentials per second can be generated by a completely stimulated neuron. The exchange pump is required to maintain ion concentrations of adequate limits. One ATP molecule is broken down every time the pump exchanges two extracellular potassium ions for three intracellular sodium ions. The membrane protein of the pump is Na^+/K^+ ATPase. Energy is sent to ions of the pump by dividing one phosphate group from a molecule of ATP, forming adenosine diphosphate (ADP). The neuron's function can stop if the cell uses all of its ATP, or if a metabolic poison inactivates Na^+/K^+ ATPase.

Action potential propagation

Events generating action potentials occur in a small part of the plasma membrane surface. Different from graded potentials, action potentials spread along all parts of the excitable membrane. Action potentials are relayed from one site to another in steps, and the message is repeated at every step. Since the same events occur repeatedly, *propagation* is the preferred term—instead of *conduction*, which means a "flow" of charges.

Types of propagation

The opening of other voltage-gated sodium ion channels is triggered when sodium ions move into an axon, depolarizing adjacent areas. This spreads over the membrane surface, and the action potential is propagated along the length of the axon, eventually reaching the axon terminals. Action potentials travel along axons by *continuous propagation* (if unmyelinated) or by *saltatory propagation* (if myelinated).

Continuous propagation

Action potentials move via **continuous propagation** in unmyelinated axons. The axolemma is organized into segments. Continuous propagation occurs as follows:

- Membrane potential briefly turns positive at the peak of the action potential;
- A local current develops, as sodium ions move into the cytoplasm and ECF;
- The local current spreads outward in all directions, depolarizing nearby membrane areas. The axon hillock cannot respond with an action potential, because it has no voltage-gated sodium ion channels;
- The process is continuous and resembles a chain reaction.

Every time there is a local current, the action potential moves *forward*, because the previous axon segment is still in its absolute refractory period. Action potentials move away from the generation site, and never reverse direction. The furthest parts of the plasma membrane will be eventually affected.

Messages are relayed between locations, with distance not affecting the process. An action potential reaching the axon terminal is identical to the one generated at the first axon segment. Though each event takes about a millisecond, they must be repeated at every step. For another action potential to occur in the same location, another stimulus is required.

Saltatory propagation

Action potentials move by **saltatory propagation** through myelinated axons. In the CNS and PNS, saltatory propagation carries action potentials over axons much more quickly than continuous propagation can do. Continuous propagation is impossible occur along a myelinated axon, since myelin increases resistance to the ion flow across the membrane. Ions can cross the axolemma easily, but only at the *nodes*. The nodes with voltage-gated ion channels are the only ones that respond to depolarizing stimuli.

When an action potential arrives at the first portion of a myelinated axon, the local current jumps over the internodes, depolarizing the nearest node to threshold. Nodes can be $1-2$ mm apart in a large myelinated axon. This means that an action potential "skips" from node to node, and does not move along the axon in small steps. Saltatory propagation is faster and uses lower amounts of energy, because there is less surface area involved. Also, less sodium ions need to be pumped out of the cytoplasm.

Threshold and the all-or-none principle

Action potentials are stimulated when a graded potential depolarizes the axolemma to a specific level. A **threshold** is a membrane potential at which an action potential is initiated. An axon threshold is usually -60 to -55 mV. This relates to a depolarization of $10-15$ mV. A stimulus altering the resting membrane potential from -70 to -62 mV produces a graded depolarization, not an action potential. Removal of a stimulus causes membrane potential to return to its resting level. Local currents are initiated by graded depolarization of the axon hillock. They cause depolarization of the initial axon segment.

For excitable membranes such as axons, a graded depolarization is similar to pressure on a gun's trigger. The action potential is similar to the firing of a gun. Every stimulus bringing the membrane to threshold creates identical action potentials. If a stimulus exceeds threshold, the action potential is independent of the intensity of the depolarizing stimulus. This *all-or-none principle* applies to all excitable membranes. The stimulus either triggers a typical action potential or none at all.

Axon diameter and propagation speed

Axon diameter also affects propagation speed, but not as significantly. Axon diameter is important since ions must move through the cytoplasm to depolarize surrounding areas of the plasma membrane. Cytoplasm offers lower resistance to ion movement than the axon membrane. An axon with a larger diameter has less resistance. Axons are classified by diameter, myelination, and propagation speed:

1. *Type A fibers*—the largest myelinated axons, with diameters between 4 and 20 micrometers (μm). Action potentials move up to 120 m per second, equivalent to 268 mph.
2. *Type B fibers*—smaller myelinated axons, with diameters of $2-4$ μm. Propagation speeds average about 18 m per second (about 40 mph).
3. *Type C fibers*—unmyelinated, and less than 2 μm in diameter. They propagate action potentials very slowly, at 1 m/sec (only 2 mph).

TABLE 2.4 Types of synapses in the central nervous system.

Type	Presynaptic Structure	Postsynaptic Structure	Functions
Axodendritic	Axon terminal	Dendrite	Mostly excitatory
Axosomatic	Axon terminal	Cell body	Mostly inhibitory
Axoaxonic	Axon terminal	Axon terminal	Presynaptic inhibition, which regulates transmitter release in the postsynaptic axon
Dendrodendritic	Dendrite	Dendrite	Local interactions that can be excitatory or inhibitory, in neurons that lack axons, such as the retina

Myelin has obvious advantages, illustrated by comparing Type A to Type C fibers. The diameter is increased by 10 times, but propagation speed is 120 times faster.

Type A fibers are motor or sensory, carrying information to the CNS concerning balance, position, plus delicate pressure and touch sensations from the skin surfaces. Motor neurons controlling skeletal muscles send commands by using Type A axons. Type B and Type C fibers carry information back and forth from the CNS, which includes generalized touch, pressure, pain, and temperature sensations. They additionally carry signals to cardiac muscle, glands, other peripheral effectors, and smooth muscles. Approximately 33% of all axons carrying sensory information are myelinated. Most sensory information is via the thin Type C fibers. Sensory information concerning survival and motor commands that avoid injury use Type A fibers. Less critical information is relayed by Type B or Type C fibers.

Synapses

Effective messages move along an axon and are transferred to another cell, occurring at a *synapse*, a specialized area where a neuron communicates with another cell. Synapses are also referred to as *synaptic junctions*. At a synapse, information moves from the *presynaptic neuron* to the *postsynaptic neuron*. Synapses can involve various postsynaptic cells. An example is the neuromuscular junction, a synapse in which the postsynaptic cell is a skeletal muscle fiber. There are more than 100 trillion synapses in the human brain alone. The types of synapses in the CNS are summarized in Table 2.4.

Structure

A synapse consists of three components: the presynaptic membrane formed by the terminal button of an axon, the postsynaptic membrane formed from a segment of dendrite or cell body, and the space in between these two structures, the **synaptic cleft**. Some neural cells have up to two hundred thousand synaptic connections.

Classification

There are two types of synapses: *electrical* and *chemical*, with unique functions. At *electrical synapses*, there is direct physical contact between cells. Presynaptic and postsynaptic membranes of cells are joined at *gap junctions* (see Fig. 2.6A). Lipid areas of nearby membranes are separated by just two nanometers (nm), kept in position by essential membrane proteins called **connexons**. Pores formed by these proteins allow ions to pass between the cells. Changes in membrane potential of one cell cause local currents to affect another. An electrical synapse propagates action potentials between cells efficiently and quickly because of this. In adults, electrical synapses are not common in the CNS or PNS. They occur in brain areas such as the vestibular nuclei, involved in balance, also in the eyes, and in one or several pairs of PNS ganglia (ciliary ganglia). They are also present in embryonic structures.

In a **chemical synapse** (Fig. 2.6B), one neuron signals another. This uses the axon terminal of the **presynaptic neuron**, which sends the message, and the **postsynaptic neuron**, which receives it. The cells are separated by the narrow synaptic cleft. A presynaptic cell is most often a neuron. Specialized receptor cells create synaptic connections with dendrites. Postsynaptic cells are neurons or other forms of cells. Communications between neurons occur at synapses

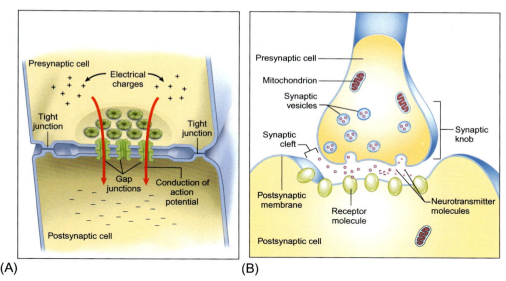

FIGURE 2.6 (A) Electrical synapse; (B) Chemical synapse.

on dendrites, cell bodies, or along axons of receiving cells. *Axoaxonic synapses* are between axons of two neurons. *Axosomatic synapses* have junctions, at an axon terminal of a neuron, and the cell body of another neuron. In an *axodendritic synapse*, synaptic contact is between an axon terminal of one neuron and a dendrite of another. Chemical synapses are very common compared to electrical synapses. Communications across chemical synapses occur from presynaptic membranes to postsynaptic membranes, and not in the reverse direction.

A **neuromuscular junction** is a synapse between a neuron and a skeletal muscle cell. At a **neuroglandular junction**, a neuron regulates activities of a secretory cell. Neurons *innervate* many other types of cells, such as *adipocytes*. Axon terminals of presynaptic cells release neurotransmitters into the synaptic cleft. The neurotransmitters are within *synaptic vesicles* in the axon terminal. Synaptic vesicles contain a small collection, or *quanta*, of neurotransmitter. When the terminal is depolarized from an action potential of its "parent" axon, there is a calcium influx that leads to phosphorylation of proteins called *synapsins*. After this phosphorylation, the vesicle links to the presynaptic membrane that faces the synaptic cleft, and the neurotransmitter is released. Another method of neurotransmitter release is based on *transporter molecules*, which usually act by taking up neurotransmitters from the synaptic cleft.

Each postsynaptic cell has its own type of axon terminal. A round axon terminal is present if a postsynaptic cell is another neuron. At a neuromuscular junction, the axon terminal is of much more complex. Axon terminal structures include mitochondria and thousands of vesicles containing neurotransmitters. Axon terminals reabsorb breakdown molecules of neurotransmitters at the synapse, then reproduce the neurotransmitters. Axon terminals also constantly receive neurotransmitters created by the cell body, along with enzymes and lysosomes, via anterograde flow.

Function of electrical synapses

In electrical synapses, signal transmission occurs as electrical signals, without using molecules. Signals are not modified during transmission. Electrical signals pass over gap junctions. The space between presynaptic and postsynaptic neurons is very small. Signal transmission occurs in both directions. It happens without use of energy and is therefore passive. Signal transmission via electrical synapses is very fast.

Function of chemical synapses

Electrical events trigger release of neurotransmitter release, flooding the synaptic cleft, and binding to receptors on the postsynaptic plasma membrane. This changes membrane permeability, producing graded potentials. The process is similar to operation of neuromuscular junctions. Chemical synapses do not use direct cellular joining, so there is more variation in results. At a chemical synapse, arriving action potentials may or may not release enough neurotransmitter to bring the postsynaptic neuron to threshold.

Cholinergic synapses release acetylcholine (ACh) at all neuromuscular junctions involving skeletal muscle fibers. They release ACh at many CNS synapses, all PNS neuron-to-neuron synapses, and all neuromuscular and neuroglandular junctions in the parasympathetic nervous system. At cholinergic synapses between neurons, a presynaptic cleft lies between presynaptic and postsynaptic membranes. Most ACh in an axon terminal collects in synaptic vesicles, each having thousands of neurotransmitter molecules. One axon terminal may contain a million such vesicles. When an action potential arrives at the presynaptic axon terminal, it depolarizes the membrane, opening its voltage-gated calcium ion channels for a short time. Extracellular calcium ions enter through these calcium channels. The ions attach to the vesicles containing ACh. Attachment of calcium ions to the vesicles causes release of ACh in the synaptic cleft. The ACh is released in groups of nearly 3000 molecules, the average number of molecules in just one vesicle. Release of ACh stops quickly, since active transport removes calcium ions quickly from cytoplasm, in the axon terminal, back to the extracellular space. Ions are pumped out of the cell, or moved to the mitochondria, awaiting another action potential.

ACh binds to receptors on the postsynaptic membrane, depolarizing it. Across the synaptic cleft, ACh diffuses to postsynaptic membrane receptors. The ACh receptors have chemically gated sodium and potassium ion channels. The main response is increased permeability to sodium ions, causing a depolarization in the postsynaptic membrane of about 20 msec. The cation channels move potassium ions outward from the cell. Sodium ions are moved by a stronger electrochemical gradient. A slight depolarization of the postsynaptic membrane occurs, which is a graded potential. The more ACh released at the presynaptic membrane, the more open cation channels in the postsynaptic membrane. Therefore, there is more depolarization. If depolarization brings an adjoining section of excitable membrane such as the initial axon segment to threshold, an action potential occurs in the postsynaptic neuron. The enzyme *Acetylcholinesterase* removes ACh from the synaptic cleft. The effects of ACh on the postsynaptic membrane are temporary because acetylcholinesterase (also called AChE or *cholinesterase*) is contained in the synaptic cleft and postsynaptic membrane. About half of all ACh released at the presynaptic membrane is degraded before reaching the postsynaptic membrane receptors. Only about 20 msec are needed for ACh molecules that bind to receptor sites to be broken down. Through hydrolysis, AChE breaks down ACh into *acetate* and *choline*. The water-soluble, vitamin-like nutrient choline is easily absorbed by axon terminals, then used to synthesize more ACh, via acetate provided by *coenzyme A (CoA)*. Coenzymes from vitamins are needed for many enzymatic reactions. Acetate moving away from the synapse can be absorbed and metabolized by postsynaptic cells, or by various cells and tissues.

Synaptic delay

A **synaptic delay** is the time needed for a signal to cross a synapse between two neurons. There is only $0.2 - 0.5$ msec between arrival of an action potential at the axon terminal and its effect upon the postsynaptic membrane. Most of the delay is due to the time needed for calcium ion influx and release of the neurotransmitters. The delay is not due to neurotransmitter diffusion. The synaptic cleft is thin, and neurotransmitters diffuse across it quickly.

If a delay of 0.5 msec occurs, an action potential can travel over 7 cm (about 3 inches) along a myelinated axon. When information is passed down CNS interneurons, increased synaptic delay may exceed propagation time along the axons. When fewer synapses are involved, cumulative synaptic delay is shorter, and responses are faster. The quickest reflexes have one synapse, with a sensory neuron directly controlling a motor neuron.

Synaptic fatigue

Synaptic fatigue, or *short-term synaptic depression*, is a temporary inability of neurons to fire, and to transmit input signals. Synaptic fatigue is a type of *synaptic plasticity*, a form of negative feedback. It is mostly presynaptic in its function. Since ACh molecules are recycled, axon terminals do not completely depend upon ACh delivered through axonal transport from the cell body. With intense stimulation, resynthesis and transport mechanisms may be unable to handle neurotransmitter demands. Then, synaptic fatigue occurs. The response of the synapse is weakened until ACh is replenished.

> **Point to remember**
> Diphenylhydantoin, antidepressants classified as selective serotonin reuptake inhibitors (SSRIs), and caffeine can affect synaptic transmission. Diphenylhydantoin reduces frequency of action potentials reaching the axon terminal. The SSRIs block serotonin transport into presynaptic cells, increasing stimulation of postsynaptic cells. Caffeine stimulates activity of the nervous system by lowering synaptic thresholds, causing postsynaptic neurons to be excited more easily.

Neuronal integration of stimuli

Each neuron receives information across thousands of different synapses. Effects on cell body membrane potential, mostly near the axon hillock, determine how the neuron eventually responds. With depolarization at the axon hillock, the membrane potential will be affected at its initial segment. Once it reaches threshold, an action potential is generated and propagated. The axon hillock integrates stimuli affecting the cell body and dendrites, determining rates of action potential generation at the initial segment. This is the most basic level of *information processing* within the nervous system. Once the signal reaches the postsynaptic cell, responses depend on activities of stimulated receptors, and which stimuli influence the cell at that time. Excitatory and inhibitory stimuli are regulated via interactions between *postsynaptic potentials*.

Postsynaptic potentials

Postsynaptic potentials are graded potentials developing in the postsynaptic membrane due to neurotransmitter effects. An **excitatory postsynaptic potential (EPSP)** is a graded depolarization, caused by a neurotransmitter arriving at the postsynaptic membrane. An EPSP develops from chemically gated ion channels opening in the plasma membrane, resulting in membrane depolarization. One example is the graded depolarization caused by ACh binding. Since it is a graded potential, an EPSP affects only the portion that closely surrounds the synapse.

An **inhibitory postsynaptic potential (IPSP)** occurs when a postsynaptic membrane has a graded hyperpolarization. Opening of chemically gated potassium ion channels may cause an IPSP. With continued hyperpolarization, the neuron is *inhibited* since a larger-than-usual depolarizing stimulus is required to bring the membrane potential to threshold. An action potential usually occurs from a stimulus changing the membrane potential by 10 mV (from -70 to -60 mV). If the membrane potential is changed to -85 mV by an IPSP, the identical stimulus would depolarize it to only -75 mV, which is below threshold.

Integrating postsynaptic potentials: summation

From one EPSP, membrane potential only receives a small effect. It usually produces depolarization of nearly 0.5 mV at the postsynaptic membrane. Before an action potential develops in the initial segment, local currents requiring at least 10 mV must depolarize the region. One EPSP does not cause an action potential, even when a synapse is on an axon hillock. Single EPSPs combine via *summation*, integrating effects of graded potentials influencing one part of the plasma membrane. Graded potentials can be EPSPs, IPSPs, or both. Summation is either *spatial* or *temporal*.

Spatial summation

Spatial summation occurs when stimuli are applied at the same time, but in different areas, with a cumulative effect upon membrane potential. Spatial summation uses *multiple synapses* acting simultaneously. Each synapse brings sodium ions over the postsynaptic membrane, to cause a graded potential with localized effects. At any active synapse, sodium ions producing the EPSP spread along the inner membrane surface, mixing with those entering at other synapses. Effects on the initial segment are cumulative. The extent of depolarization is based on the number of synapses active at each moment, and the distance from the initial segment. Like temporal summation, an action potential occurs when the membrane potential at the initial segment reaches threshold.

Temporal summation

Temporal summation is the addition of stimuli occurring quickly, at a *single repeatedly-active synapse*. Usually, an EPSP lasts about 20 msec. With maximum stimulation, action potentials can reach axon terminals every millisecond, and the two effects are combined. Each time an action potential arrives, vesicles discharge ACh into the synaptic cleft. Additional chemically gated ion channels open when more ACh molecules arrive at the postsynaptic membrane, and the degree of depolarization increases. Several small steps bring the initial segment to threshold.

Summation of EPSPs and IPSPs

The IPSPs also combine in a similar way. Activation of different chemically gated ion channels occurs with both EPSPs and IPSPs, but with opposite effects upon membrane potential. Antagonism between IPSPs and EPSPs is required in cell information processing. Similar interactions determine membrane potential at the point where the axon hillock and initial segment meet. Neuromodulators, hormones, or both can alter the postsynaptic membrane's sensitivity to excitatory or inhibitory neurotransmitters. By altering balances between EPSPs and IPSPs, the compounds cause facilitation or inhibition of CNS and PNS neurons.

Presynaptic regulation

Inhibitory or excitatory responses occur at postsynaptic neurons and presynaptic neurons. An axoaxonic (axon-to-axon) synapse at the axon terminal can decrease or increase neurotransmitter release at the presynaptic membrane. Sometimes in *presynaptic inhibition*, release of gamma-aminobutyric acid (GABA) restricts voltage-gated calcium ion channels opening in axon terminals (Fig. 2.7A). This reduces the quantity of neurotransmitter released when an action potential arrives, as well as the effects of synaptic activity on the postsynaptic membrane. In *presynaptic facilitation*, axoaxonic synapse activity increases amounts of neurotransmitters released as action potentials that reach the axon terminal (Fig. 2.7B). This strengthens and lengthens neurotransmitter effects on the postsynaptic membrane. The neurotransmitter *serotonin* aids in presynaptic facilitation. At axoaxonic synapses, serotonin is released, and voltage-gated calcium ion channels stay open for a longer time.

Rate of action potential generation

Nervous system messages are usually processed because of frequency of action potentials. Action potentials arrive at neuromuscular junctions as one every second. Once there, isolated, grouped twitches may occur in related skeletal

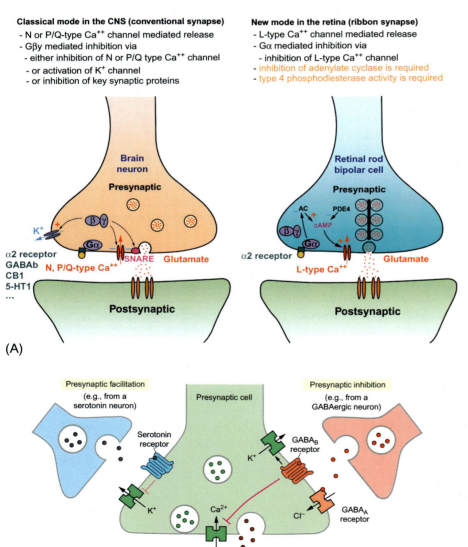

FIGURE 2.7 (A) Presynaptic inhibition; (B) Presynaptic facilitation.

muscle fibers. A sustained tetanic contraction occurs with a rate of 100 action potentials arriving each second. A few action potentials per second, along sensory fibers, may be felt as a light touch, but hundreds of action potentials per second may be felt along the same axon as intense pressure. Frequency of action potentials is generally related to the level of sensory stimulation or the strength of motor responses.

An action potential is generated when a graded potential depolarizes an axon hillock briefly, and the initial segment reaches threshold. More action potentials are produced when the initial segment stays over threshold for a longer time. The *frequency* of action potentials is based on amounts of depolarization above threshold. With more depolarization, there is greater frequency of action potentials. When the absolute refractory period ends, the membrane can respond to a second stimulus. Keeping the membrane potential above threshold causes an identical effect to applying a second, greater-than-normal stimulus. Action potentials may be generated at a maximal rate when the relative refractory period is eliminated. The maximum *theoretical* action potential frequency is achieved by the length of the absolute refractory period. The absolute refractory period is shortest in large-diameter axons, having a theoretical maximum frequency of 2500 action potentials every second. The highest frequencies of axons range between 500 and 1000 per second.

Nerve cell degeneration and regeneration

Nerve cell degeneration involves a loss of functional activity and trophic degeneration of nerve axons plus their terminal branches. This follows destruction of the cells of origin, or interruption of continuity with these cells. It is a pathologic characteristic of neurodegenerative diseases. Nerve degeneration is studied in relation to neuroanatomical correlation of the physiology of neural pathways. *Neurodegenerative disease* describes conditions mostly affecting brain neurons. Since neurons are not normally reproduced or replaced when damaged or when they die, the body cannot replace them. Neurodegenerative diseases are incurable, debilitating conditions resulting in progressive degeneration, with or without the death of nerve cells. There are abnormalities of movement (ataxias) or mental functioning (dementias), with dementias making up the greatest amount of neurodegenerative diseases.

The brain, spinal cord, and peripheral nerves can be injured from trauma or from neurodegenerative diseases that cause progressive deterioration. Due to structural complexity, little spontaneous regeneration, repair, or healing happens. Often, there is permanent and incapacitating brain damage, paralysis from spinal cord injury, and peripheral nerve damage. Patients with serious nervous system injuries or strokes often require lifelong assistance. In the CNS, regeneration is restricted by inhibitory influences of glial and extracellular environments. This is partially due to migration of myelin-related inhibitors, astrocytes, oligodendrocytes and their precursors, and microglia. There will be no expression or re-expression of growth factors. Glial scars form soon, and the glia produce factors inhibitive to *remyelination* and axon repair. Axons lose potential for growth because of aging. When glial scars form, axons are unable to grow across them. In some cases, CNS axons have regrown, in certain environments. Therefore, research into CNS nerve regeneration is focused on crossing or eliminating inhibitory lesion sites. In both the CNS and PNS, *collateral sprouting* may occur when an innervated structure has become partially denervated. Remaining axons form new collaterals that will reinnervate the denervated area, showing that the nervous system's axons are able to control synaptic areas that were formerly controlled by other axons.

In the PNS, neuroregeneration occurs much more often. Axonal sprouts form at proximal stumps, growing until they enter a distal stump. Growth of the sprouts is controlled by chemotactic factors emitted from Schwann cells. Injury to the PNS very quickly results in migration of phagocytes, Schwann cells, and macrophages to lesion sites, to remove debris such as damaged tissue. When a nerve axon is severed, the end attached to the cell body is proximal segment, while the other end is the distal segment. After injury, the proximal end swells, and some retrograde degeneration will occur. Once the debris is removed, it sprouts axons, and growth cones are seen. Proximal axons can regrow when the cell body is intact, and they have contacted Schwann cells in the endoneurial channel or tube. Axon growth rates may be up to 1 mm per day in small nerves, and 5 mm per day in large nerves. The distal segment experiences Wallerian degeneration in just a few hours of the injury. Axons and myelin degenerate, yet the endoneurium remains. In later stages, the remaining endoneurial tube directs growth of axons back to the correct targets.

In Wallerian degeneration, Schwann cells grow as regular columns along the endoneurial tube, creating a **band of Bungner** that protects and preserves the endoneurial channel. Axonal regeneration through this band of Schwann cell tubes surrounded by basal lamina in a distal stump reveals different amounts of regeneration after a *nerve crush*, compared to a *nerve transection*. After a crushing injury, though axons may be severed, the Schwann cells, basal lamina, and perineurium are still continuous through the lesion. This allows regeneration of axons through the injured nerve. If the nerve is transected, the continuity of such pathways is disrupted. Even very precise surgery is usually unable to align the proximal and distal portions of each axonal pathway, making successful regeneration unlikely to occur. Macrophages and Schwann cells release neurotrophic factors, helping to enhance regrowth.

Today, researchers are studying effects of restoring cerebrovascular function, via transplantation of *induced pluripotent stem cell-derived progenitor cells* related to amyloid pathology and cognitive function in *Alzheimer's disease*. Progenitor cells converted from skin fibroblasts by transducing various transcription factors, can generate all tissues in the body, including vascular cells. For *amyotrophic lateral sclerosis (ALS)*, researchers are using adipose-derived mesenchymal stem cells from the patient's body. The cells are modified in the laboratory and put back into the patient's nervous system, promoting neuron regeneration.

For *multiple sclerosis*, some patients experience spontaneous repair of myelin and nerves. Therapies are now being developed that simulate this repair and promote recovery of lost function. Regeneration of the myelin sheath is boosted by folded DNA molecules known as **aptamers**. For *Parkinson's disease*, skin and progenitor cells are being studied regarding their ability to generate cells that die due to the disease. Actual *regrowth* of nerve cells (*axogenesis*), is being studied using *zebrafish*, providing understanding of how nerve cells grow during nervous system development, and how nerve regeneration could be improved after injury. For stroke neuroregeneration, mesenchymal stem cells may save damaged neurons after exposure to oxygen-glucose deprivation stress.

Neurogenesis

Recent studies have revealed that a small amount of neuronal precursor cells are capable of dividing, and then differentiating into neurons. These rare cells may exist in the *subventricular zone* of the forebrain. There is evidence of postnatal neurogenesis within the dentate gyrus of the hippocampus. The speed of generation of new neurons within this critical area may be able to be accelerated. Though the number of new neurons that can be produced in an adult brain is still under study, these precursor cells may influence methods of restoring function after a CNS injury. Intense studies are being conducted today.

Key terms

Absolute refractory period
Action potentials
Anterograde flow
Aptamers
Astrocytes
Axolemma
Axonal transport
Axon hillock
Axons
Axon terminals
Axoplasm
Band of Bungner
Blood – brain barrier (BBB)
Centrioles
Chemical synapse
Cholinergic synapses
Collaterals
Connexons
Continuous propagation
Dendrites
Dendritic zone
Depolarization
Electrochemical gradient
Ependymal cells
Equilibrium potential
Excitatory postsynaptic potential (EPSP)
Ganglion
Gated ion channels
Glial cells
Golgi apparatus
Graded potential
Hyperpolarization
Inhibitory postsynaptic potential (IPSP)
Interneurons
Kinesin
Leak channels
lemnisci
Lipofuscin
Local potentials
Membrane potential
Microfilaments
Microglia
Microtubules
Motor neurons
Myelin
Nervous tissue
Neurilemma
Neurofibrils
Neurofilaments
Neuroglandular junction
Neuroglia
Neuromuscular junction
Neuron
Neurotubules
Nissl substance
Nodes of Ranvier
Nuclear pores
Oligodendrocytes
Perikaryon
Postsynaptic neuron
Postsynaptic potentials
Presynaptic neuron
Propagation
Refractory period
Relative refractory period
Resting membrane potential
Retrograde flow
Saltatory propagation
Satellite cells
Schwann cells
Spatial summation
Summation
Synapse
Synaptic cleft
Synaptic delay
Synaptic fatigue
Synaptic junctions
Telodendria
Temporal summation
Threshold
Tight junctions
Visceral motor neurons
Visceral sensory neurons
Wallerian degeneration

Further reading

Canepari, M., & Zecevic, D. (2011). *Membrane potential imaging in the nervous system: Methods and applications*. Springer.
Cibas, E. S., & Ducatman, B. S. (2014). *Cytology: Diagnostic principles and clinical correlates* (4th Edition). Saunders.
Dubois, M. L. (2010). *Action potential: Biophysical and cellular context, initiation, phases and propagation (Cell biology research progress)*. Nova Science Publishers, Inc.
Fain, G. L., Fain, M. J., & O'Dell, T. (2014). *Molecular and cellular physiology of neurons* (2nd Edition). Harvard University Press.
Fan, F., & Damjanov, I. (2017). *Cytopathology review* (2nd Edition). Jaypee Brothers Medical Publishers Pvt. Ltd.
Felten, D. L., O'Banion, M. K., & Maida, M. E. (2015). *Netter's Atlas of neuroscience (Netter basic science)* (3rd Edition). Elsevier.
Gerstner, W. (2014). *Neuronal dynamics: From single neurons to networks and models of cognition*. Cambridge University Press.
Gonzalez-Perez, O. (2012). *Astrocytes: Structure, functions and role in disease (Neuroscience research progress: Neurology — Laboratory and Clinical Developments)*. Nova Biomedical.
Jones, H. R., Jr., Burns, T., Aminoff, M. J., & Pomeroy, S. (2013). *The netter collection of medical illustrations: Volume 7, nervous system, part I, Brain* (2nd Edition). Saunders.
Kumar, K., & Kumar Sahel, D. (2018). *Role of hippocampal GABAergic interneurons in learning and memory*. Lap Lambert Academic Publishing.
Levitan, I. B., & Kaczmarek, L. K. (2015). *The neuron: Cell and molecular biology* (4th Edition). Oxford University Press.
Lyons, D. A., & Kegel, L. (2019). *Oligodendrocytes: Methods and protocols (Methods in molecular biology)*. Humana Press.
Mai, J. K., & Paxinos, G. (2011). *The human nervous system (3rd Edition)*. Academic Press.
Mody, D. R., Thrall, M. J., & Krishnamurthy, S. (2018). *Diagnostic pathology: Cytopathology* (2nd Edition). Elsevier.
Monje, P. V., & Kim, H. A. (2018). *Schwann cells: Methods and protocols (Methods in molecular biology)*. Humana Press.
Nicholls, J. G., Martin, A. R., Fuchs, P. A., Brown, D. A., Diamond, M. E., & Weisblat, D. (2011). *From neuron to brain* (5th Edition). Sinauer Associates/Oxford University Press.
Pannese, E. (2018). *Biology and pathology of perineuronal satellite cells in sensory ganglia (Advances in anatomy, embryology and cell biology)*. Springer.
Pannese, E. (2015). *Neurocytology: Fine structure of neurons, nerve processes, and neuroglial cells*. Springer.
Paulsen, D. F. (2010). *Histology and cell biology: Examination and board review (LANGE basic science)* (5th Edition). McGraw-Hill Education/Medical.
Ransom, B. R., & Kettenmann, H. (2012). *Neuroglia* (3rd Edition). Oxford University Press.
Sheng, M., Sabatini, B., & Sudhof, T. (2012). *The synapse (Cold spring harbor perspectives in biology)*. Cold Spring Harbor Laboratory Press.
Tremblay, M. E., & Sierra, A. (2014). *Microglia in health and disease*. Springer.
Verkhratsky, A., & Butt, A. (2013). *Glial physiology and pathophysiology*. Wiley-Blackwell.
Verkhratsky, A., Ho, M. S., Zorec, R., & Parpura, V. (2019). *Neuroglia in neurodegenerative diseases (Advances in medicine and biology)*. Springer.
von Bernhari, R. (2016). *Glial cells in health and disease of the CNS (Advances in medicine and biology)*. Springer.
Yao, L. (2018). *Glial cell engineering in neural regeneration*. Springer.

Chapter 3

Embryology

Chapter Outline

Neural tube 65
Neural crest 69
Cranial placodes 70
Medulla spinalis 71
Myelencephalon 71
Metencephalon 75
Mesencephalon 75
Diencephalon 75
Telencephalon 75
Central nervous system malformation 77
Key terms 78
Further reading 79

Neural tube

The **neural tube** has three cell layers: the ectoderm, mesoderm, and endoderm (see Fig. 3.1). Eventually, these layers form various body tissues. The *neuroectoderm* develops from the **ectoderm**, forming skin and neural tissues. The **mesoderm** forms the bones and muscles. The **endoderm** forms the linings of gastrointestinal and respiratory cells. By the 18th day of gestation, chemicals are released by the notochord (which is derived from the **mesoderm**), signaling the ectoderm to form the **neural plate**. This plate folds inward, forming a longitudinal and midline **neural groove**. A parallel **neural fold** on each side surrounds this groove. The neural groove deepens as the neural folds move together in the dorsal midline. By the end of week 3, neural folds grow over the midline and form the neural tube.

The open ends of the tube are called the cranial and caudal **neuropores**. Fusion continues in both directions, with the entire neural tube closing by the end of the fourth week, in a process known as **primary neurulation**. The **rostral** neuropore closes by day 25, while the caudal neuropore closes by day 27. The closing neural tube keeps separating from the neuroectodermal surface, and leaves groups of cells from the crest of the neural folds behind. These **neural crest cells** are derived from the neuroectoderm, and develop into many cell types, including those of the peripheral nervous system (PNS). The neural tube develops into almost all of the central nervous system (CNS). The neural cavity becomes the ventricular system of the brain, and the central canal of the spinal cord. When the neural tube has closed, another cavity extends into the solid cell mass, but at its caudal end. This occurs in weeks 5 and 6, and is called **secondary neurulation**. During the processes of neurulation, the embryo is referred to as a **neurula**.

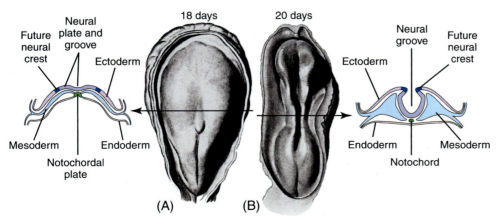

FIGURE 3.1 The neural plate and beginning neural groove at about 18 days of development (A), and the neural groove 2 days later (B), shortly before the neural tube begins to close. The schematic cross sections to the left and right are at the levels indicated by arrows in (A) and (B), respectively. [(A) and (B) from Arey, L.B. (1941). *Developmental anatomy* (4th ed.), Philadelphia: WB Saunders].

Epidemiology of Brain and Spinal Tumors. DOI: https://doi.org/10.1016/B978-0-12-821736-8.00024-8
© 2021 Elsevier Inc. All rights reserved.

Differentiation and migration

Larger neurons differentiate before smaller ones, and most large neurons are motor neurons. Sensory and smaller neurons, plus most glial cells, develop later, continuing to do so up to the time of birth. New neurons may migrate significantly though areas of older neurons. Once the glial cells develop, they can form a framework and guide developing neurons to their target areas. Since a neuron's axonal process may start growing toward its target area during migration, an adult brain's nerve processes often have a curved shape instead of being straightened.

Sensory and motor areas defined by the sulcus limitans

Dorsal-ventral patterns of the brain stem and spinal cord influence development of sensory or motor functions. The ectoderm (near the future dorsal neural tube surface) and the mesodermal notochord (near the ventral surface) produce various signaling molecules early in embryonic development. The nucleus pulposus, which is the inner core of the vertebral disk, also forms from the **notochord**. The signaling molecules have opposite concentration gradients, and cause distinct patterns of further development in these neural tube regions. In week 4, a longitudinal groove called the **sulcus limitans** develops in the lateral wall of the neural tube. It separates the neural tube into a dorsal half or **alar plate**, and a ventral half or **basal plate**. Derivatives of the alar plate are mostly involved in sensory processing, and derivatives of the basal plate are mostly involved in motor processing.

The sulcus limitans is not present in the adult spinal cord, but the central gray matter of the cord is divided into a **posterior (dorsal) horn** and an **anterior (ventral) horn** on either side, forming an H-shaped gray matter structure. Sensory and motor neurons travel by unique pathways in the spinal cord. Central processes of sensory neurons form from neural crest cells. They mostly end in the posterior horn, which has cells with axons that form ascending sensory pathways. These sensory pathways are called *ascending*, because they deliver information from the periphery to the brain. The motor pathways are called *descending* because they deliver information from the brain to the skeletal muscles. The anterior (ventral) horn has cell bodies of autonomic as well as somatic motor neurons. Their axons emerge from the spinal cord, innervating autonomic ganglia and skeletal muscles. The same relationship exists between the sensory alar plate derivatives and the motor basal plate derivatives in the brain stem.

Bulges and flexures of the neural tube

Longitudinal development of the neural tube is also linked to different signaling molecules. Even before completely closing, the tube has bulges in its rostral end, close to where the brain will develop. Additionally, bends in the tube begin to appear. The neural tube is never a simple or straight cylinder.

Primary vesicles

In week 4, there are three bulging *vesicles* that appear in the neural tube. These are the **primary vesicles**. They are described, from rostral to caudal locations, as follows:

- **Prosencephalon**—which develops into the forebrain. Through a series of cleavages, it also develops the optic and olfactory apparatus, and divides transversely into the telencephalon and diencephalon.
- **Mesencephalon**—which develops into the midbrain, and connects the hindbrain and forebrain. A number of nerve tracts run through the midbrain that connect the cerebrum with the cerebellum and other hindbrain structures. A major function of the midbrain is to aid in movement as well as visual and auditory processing.
- **Rhombencephalon**—a rhomboid structure that merges with the caudal (spinal) part of the neural tube. It is also called the *hindbrain*, and forms the pons, medulla, and cerebellum. The hindbrain cavity is the fourth ventricle.

Derivatives of the neural tube vesicles are explained in Table 3.1.

Secondary vesicles

By week 5, the primary vesicles known as the prosencephalon and rhombencephalon subdivide, forming **secondary vesicles** (see Fig. 3.2). These are as follows:

- **Telencephalon**—the *end-brain*, which forms from the prosencephalon. The telencephalon then becomes the cerebral hemispheres in the adult brain. It also includes the cerebral cortex, subcortical structures, and important fiber bundles such as the corpus callosum. The inferior boundaries of the telencephalon are found at the diencephalon and brain stem. Posteriorly, it is bordered by the cerebellum.

TABLE 3.1 Derivatives of the neural tube vesicles.

Primary vesicle	Secondary vesicle	Cavity	Neural derivatives
Prosencephalon (forebrain)	Telencephalon	Lateral ventricles	Cerebral hemispheres
	Diencephalon	Third ventricle	Thalamus, hypothalamus, retina, other structures
Mesencephalon (midbrain)	Mesencephalon	Cerebral aqueduct	Midbrain
Rhombencephalon (hindbrain)	Metencephalon	Part of fourth ventricle	Pons, cerebellum
	Myelencephalon	Part of fourth ventricle, part of central canal	Medulla oblongata

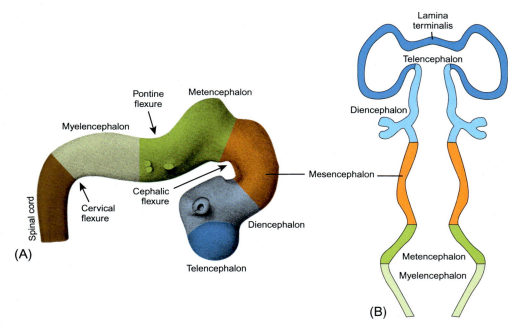

FIGURE 3.2 Secondary vesicles during the sixth week. (A) Lateral view of the neural tube, showing vesicles and flexures. (B) Schematic longitudinal section, as though the flexures are straightened out. (The "A" portion was modified from Hochstetter, F. (1919). Beiträge zur Entwicklungsgeschichte des menschlichen Gehirns, Teil, I. (Ed.). Vienna: Franz Deuticke).

- **Diencephalon**—the *in-between-brain*, also forming from the prosencephalon. It consists of structures that are on either side of the third ventricle, including the thalamus, hypothalamus, epithalamus, and subthalamus.
  - *Thalamus/hypothalamus*—forming from the diencephalon, along with the retina and several small structures. The **thalamus** is a large gray matter mass, between the cerebral cortex and other structures. The **hypothalamus** helps to control autonomic functions. The **epithalamus** is a dorsal segment, and includes the habenula and their interconnecting fibers, the *habenular commissure*, the *stria medullaris*, and the *pineal gland*. The **subthalamus**, also called the *prethalamus*, primarily contains the subthalamic nucleus. The subthalamus connects to the globus pallidus, which is the basal nucleus of the telencephalon.
- **Mesencephalon**—does not divide into secondary vesicles. It develops into the midbrain, as mentioned above. The mesencephalon is associated with vision, hearing, motor control, sleeping and waking, alertness, and temperature regulation.
- **Metencephalon**—forms from the rhombencephalon, to become the *pons* and *cerebellum*. Between the metencephalon and myelencephalon, the **pontine flexure** appears in the dorsal brain stem surface. It is needed for the caudal brain stem structure. The walls of the neural tube spread, to form the diamond-shaped **rhombencephalon**. A thin

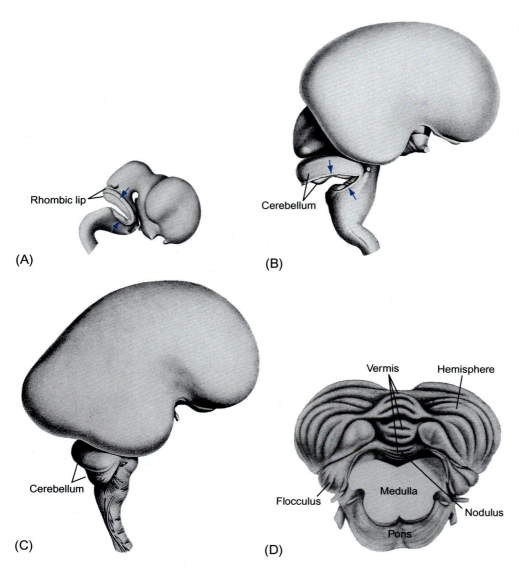

FIGURE 3.3 Development of the cerebellum. (A) During the second month of development lateral parts of the alar plate in the rostral metencephalon thicken to form rhombic lips. These continue to enlarge during the third and fourth month (B and C), forming the cerebellum. (D) By about 5 months, a series of deep fissures develops in the cerebellar surface, and both midline (vermis) and lateral (hemisphere) zones are apparent. (The nodulus is a special part of the vermis, continuous laterally with the flocculus). Arrows in (A) and (B) indicate the cut edge of the thin roof of the fourth ventricle, which becomes overgrown and mostly hidden from view as the cerebellum develops. [(A)-(C) from Hochstetter, F. (1919). Beiträge zur Entwicklungsgeschichte des menschlichen Gehirns. I. Teil, Vienna: Franz Deuticke. (D) from Hochstetter, F. (1929). Beiträge zur Entwicklungsgeschichte des menschlichen Gehirns. I. Teil, Vienna: Franz Deuticke].

"roof" of membranes remains over the future **fourth ventricle**. The metencephalon contains part of the *trigeminal nerve, abducens nerve, facial nerve*, and *vestibulocochlear nerve*.

- **Myelencephalon**—also forms from the rhombencephalon, to become the medulla oblongata, the area of the brain stem that merges with the spinal cord. The myelencephalon is also called the *afterbrain*.

The alar and basal plates are separated by the sulcus limitans. They are eventually located in the floor of the fourth ventricle. The future corresponding adult brain stem, the rostral medulla and caudal pons, will have sensory nuclei that are lateral and not posterior to the motor nuclei. Lateral areas of the alar plate, in the rostral metencephalon, will become much more thick, and form the **rhombic lips**. Parts of these lips develop into the brain stem, to form the cerebellum-related nuclei. Other parts enlarge, fusing at midline to form a transverse ridge. This is the future *cerebellum* (see Fig. 3.3).

> **Point to remember**
>
> The neural tube is the embryonic structure that eventually forms the brain and spinal cord, via the processes known as primary and secondary neurulation. These processes involve folding in the embryo, and include the transformation of the neural plate into the neural tube. At this stage, the embryo is actually called a "neurula". Neurulation begins when the notochord includes formation of the CNS by signaling the ectoderm germ layer above it to form the neural plate, which then folds in upon itself, forming the neural tube.

Neural crest

As the neural tube closes, up to the future diencephalon, the neural crest pinches off as continuous bilateral cell strands. Neural crest cells form most of the PNS. Included are autonomic ganglia, sensory cells, microglia, Schwann cells, adrenal medulla secretory cells, melanocytes, arachnoid cells, and microglial cells. The features involved in development of the neural crest, along with their developmental potentials, are shown in Fig. 3.4. The neural crest cells arise from the embryonic ectoderm cell layer. After gastrulation, they are specified at the border of the neural plate and nonneural ectoderm. During neurulation, the neural folds converge at the dorsal midline to form the neural tube. Therefore, neural crest cells from the roof plate of the neural tube transition from epithelial to mesenchymal cells. They delaminate from the neuroepithelium and migrate,

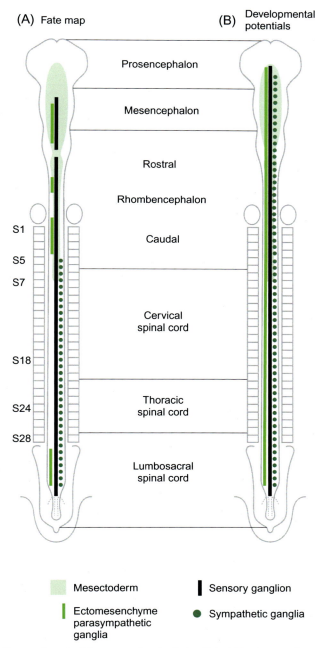

FIGURE 3.4 (A) A fate map along the neural crest of the presumptive territories that yield ectomesenchyme, sensory, parasympathetic and sympathetic ganglia, and neural crest-derived mesenchyme in normal development. (B) Developmental potentials for the same cell types. **S**, somite.

differentiating into various types of cells. Underlying development of the neural crest is a *gene regulatory network* that involves interacting signals, transcription factors, and downstream effector genes conferring cell characteristics such as **multipotency** and migratory abilities. It is important to understand the molecular mechanisms of neural crest formation, in relation to diseases, because of its contributions to many cell lineages. Abnormalities in neural crest development cause **neurocristopathies**, including **frontonasal dysplasia, Waardenburg-Shah syndrome**, and DiGeorge syndrome. Examples of **neurocristopathies** that involve cancer cells include *pheochromocytoma, paragangliomas, Merkel cell carcinoma*, and *melanoma*.

Cranial placodes

A series of **cranial placodes** form via a connection of ectoderm near the brain stem's neural crest. This series continues around the rostral end of the neural plate, thickening in some areas, to form the placodes. The placodes, along with areas of the neural crest, develop into the sensory PNS components of the head (see Fig. 3.5). The single

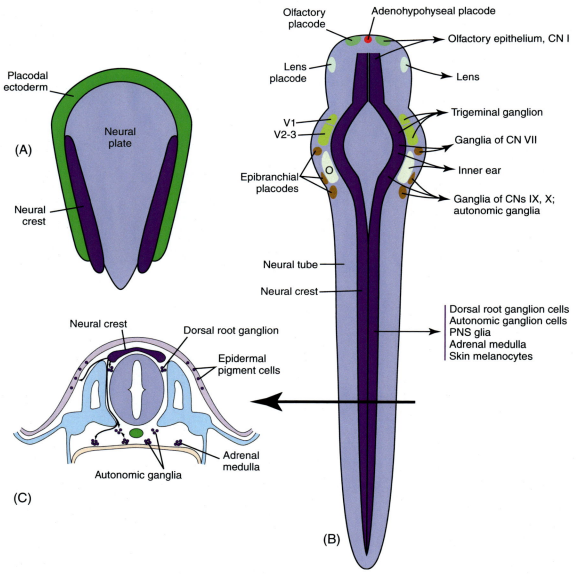

FIGURE 3.5 Formation of the PNS by neural crest and cranial placodes. (A) Schematic view of the neural plate during the third week of development, flanked by strands of neural crest. (B) Migratory paths and some fates of neural crest and cranial placode cells. (C) Cross section of the developing neural tube and neural crest cells, which give rise to the dorsal root ganglia (also known as spinal ganglia), autonomic ganglia, the adrenal medulla, and epidermal pigment cells. (Based in part on Grocott, T., Tambalo, M., & Streit, A. (2012). *Developmental Biology 370* 3.).

adenohypophyseal placode forms the anterior pituitary gland. On each side, the **lens placode** forms the lens of the eye. Migrating neural crest cells forms most or all of the glia in the cranial ganglia and sensory organs. The cells work with the placodes, in various amounts, to create the following:

- The olfactory epithelium and its nerve
- The trigeminal ganglion
- The inner ear
- The sensory ganglia of cranial nerves VII, IX, and X.

The cranial placodes which give rise to neurons are divided into two groups: the **dorsolateral placodes** and the **epibranchial placodes**. The dorsolateral placodes include the *trigeminal placode* and the *otic placode*. The trigeminal placode consists of ophthalmic and maxillomandibular portions, and gives rise to cells of the trigeminal ganglion. The otic placode forms the **otic pit** and *otic vesicle*, eventually giving rise to the organs of hearing and equilibrium.

The epibranchial placodes are also called the *epipharngeal placodes*, and generate the distal portion of the cranial nerves mentioned above. The geniculate placode associated with the first pharyngeal groove generates the geniculate ganglion and distal portions of cranial nerve VII. The **petrosal placode** associated with the second pharyngeal groove generates the inferior ganglion of the glossopharyngeal nerve as well as portions of cranial nerve IX. The **nodosal placode** associated with the third *branchial cleft* generates the nodose ganglion and distal portions of cranial nerve X. The *olfactory placode* gives rise to the olfactory epithelium of the nose.

Medulla spinalis

The remainder of the neural tube forms the spinal cord, which extends from the medulla oblongata to the lumbar vertebrae. The **medulla spinalis** is the clinical name for the *spinal cord*. There is symmetric neural tube progression as it develops, in layers. The spinal cord also has alar and basal plates. Ventral growth of the two basal layers and related marginal layers past the *floor plate* level creates a separation called the **ventral median fissure**. The mantle and marginal layers of the alar plates grow dorsally with the dorsal marginal layers fusing on the median plane. This forms a **dorsal median septum** that is often poorly defined, with its external margin forming the **dorsal median sulcus**. This midline growth displaces the roof plate region ventrally, to reduce the neural canal and form the smaller central canal of the spinal cord. This canal is lined with **ependymal cells**. The mantle layer of the alar plate forms the *dorsal gray column* or *horn*. The mantle layer of the basal plate forms the *ventral gray column* or *horn*. The mantle zone at the plane of the sulcus limitans develops into the *intermediate gray column*. This column is only present in the thoracic, upper lumbar, and sacral spinal cord segments (see Fig. 3.6). Neural crest cells provide the neurons forming the spinal ganglia at each spinal cord segment.

In the mantle layer of each plate, the neurons are arranged in functional columns. General visceral *afferent* and *efferent* neurons close together, in their respective gray columns on each side of the dorsal plane, through the sulcus limitans. General somatic afferent and **proprioceptive** neuronal columns are dorsal, within the alar plate of the mantle layer. The general somatic efferent column is ventral, within the basal plate of the mantle layer. At the limb levels, the spinal cord segments responsible for their innervation are enlarged, forming the cervical and lumbosacral **intumescences**. There are dorsal, lateral, and ventral funiculi (white matter processes) on each side of the spinal cord.

> **Point to remember**
> The neural tube forms regions around the wall along its length, which includes the spinal cord. The floor and roof plate are specialized developmental regions. The floor plate overlies the notochord, while the basal plate lies on either side of the bottom of the floor plate. The alar plate lies on either side of the top of the floor plate. The roof plate underlies the dorsal ectoderm epithelium. Finally, the lumen is a neuroepithelium-lined, fluid-filled space that is continuous with the brain's ventricular system.

Myelencephalon

The medulla oblongata is the lowest section of the brain stem, and is also referred to simply as the *medulla*. The other, superior parts of the brain stem are the midbrain and pons. The structures of the brain stem contain 10 cranial nerve pairs, which do not include the olfactory and optic nerves, and control most of the unconscious activities of the brain.

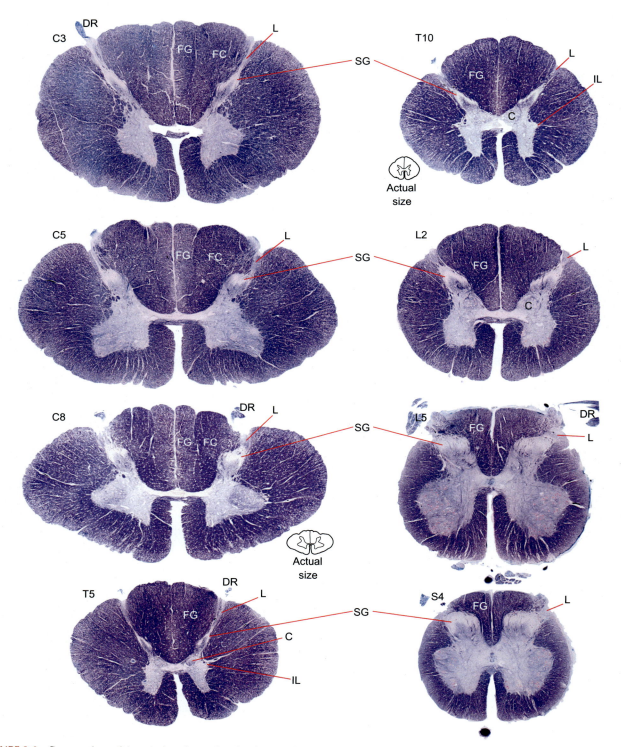

FIGURE 3.6 Cross sections of the spinal cord at various levels; note the large lateral extensions of the anterior horns in C5, C8, and L5. *C*, Clarke's nucleus; *DR*, dorsal root; *FC*, fasciculus cuneatus; *FG*, fasciculus gracilis; *IL*, intermediolateral cell column; *L*, Lissauer's tract; *SG*, substantia gelatinosa.

The neurons of the medulla generate nerve centers responsible for pain relay, along with movements of the cardiovascular, respiratory, and gastrointestinal systems. The medulla also contains the nuclei of these cranial nerves:

- Glossopharyngeal—cranial nerve IX. It is a mixed nerve that carries afferent sensory and efferent motor information, exiting the brain stem out from the sides of the upper medulla, just anterior to the vagus nose. The motor division of the glossopharyngeal nerve derives from the basal plate of the embryonic medulla, and the sensory division originates from the cranial neural crest.
- Vagus nerve—CN X. It was previously referred to as the *pneumogastric nerve*, and interfaces with parasympathetic control of the heart, lungs, and digestive tract. The vagus nerves are paired, but are normally referred to in the singular form, and constitute the longest nerves of the autonomic nervous system. The end portion of the vagus nerve is called the *spinal accessory nucleus.*
- Accessory nerve—CN XI. This nerve supplies the sternocleidomastoid and trapezius muscles. The sternocleidomastoid muscle tilts and rotates the head, while the trapezius muscle connects to the scapulae, and acts to shrug the shoulders. The accessory nerve is traditionally divided into spinal and cranial portions. The cranial portion soon joins the vagus nerve, and it is debated whether or not the cranial portion should even be considered part of the accessory nerve. Therefore, the term "accessory" usually refers just to the nerve that supplies the sternocleidomastoid and trapezius muscles, also called the *spinal accessory nerve*. Injury can cause wasting of the shoulder muscles, winging of the scapula, and weakness of shoulder abduction and external rotation. The accessory nerve derives from the basal plate of the embryonic spinal segments C1 to C6.
- Hypoglossal nerve—CN XII. This nerve innervates all extrinsic and intrinsic muscles of the tongue, except for the *palatoglossus muscle*, which is innervated by the vagus nerve. The hypoglossal nerve only has motor functions, and arises from the hypoglossal nucleus in the medulla as small rootlets. These pass through the hypoglossal canal and downwards, through the neck, but eventually pass upwards again over the tongue muscles that are supplied. There are left and right hypoglossal muscles. The hypoglossal nerve controls tongue movements needed for speech, swallowing, sticking out the tongue, and moving it from side to side. Damage to the nerve or its neural pathways can affect the ability to move the tongue, and even alter its appearance.

Therefore, the medulla controls tongue movements (via CN XII), swallowing (CN IX and XII) and the autonomic nervous system (via CN X). The medulla develops from the secondary brain vesicle called the **myelencephalon** (see Fig. 3.2), which formed from the earlier primary brain vesicle called the rhombencephalon.

The medulla oblongata has two primary parts. Its upper, open, and *superior* area is where the fourth ventricle forms its dorsal surface. Its lower, closed, or *inferior* area is where the fourth ventricle narrows at the **obex** of the caudal medulla, surrounding a portion of the central canal. The *anterior median fissure* has a fold of *pia mater*, and extends along all of the medulla. It ends at the lower border of the *pons*, at the small, triangular **foramen cecum**. Raised **medullary pyramids** lie on either side of this fissure. Inside the pyramids are *pyramidal tracts*, which are the corticospinal and corticobulbar tracts of the nervous system. The corticospinal tract, or descending pathway, is the main motor tract controlling movement of body parts below the neck. Similarly, the corticobulbar pathway is also a descending tract, and controls movements of the head and neck muscles. At the caudal medulla, the tracts cross over and are known as the *decussation of the pyramids*. This obscures the fissure in this location. Other fibers originating from the anterior median fissure above the decussation, which continue laterally across the surface of the pons, are called the *anterior external arcuate fibers*.

The final differentiation of the medulla occurs at week 20 of gestation. Neuroblasts from the alar plate produce sensory nuclei, while basal plate neuroblasts will produce motor nuclei. The alar plate neuroblasts form the following structures:

- **Solitary nucleus**—which contains general visceral afferent fibers for taste; and the special visceral afferent column. The solitary nucleus is also known as the *nucleus of the solitary tract*, the *nucleus solitarius*, and the *nucleus tractus solitarii*. It is purely sensory and forms a vertical column of gray matter. Through its center there is the solitary tract, a white bundle of nerve fibers that include fibers from the facial, glossopharyngeal, and vagus nerves. The solitary nucleus projects to the reticular formation, parasympathetic preganglionic neurons, hypothalamus, and thalamus. It forms circuits contributing to autonomic regulation. Cells along its length are roughly arranged by function. For example, cells involved in taste are in the rostral part, while those receiving cardio-respiratory and gastrointestinal information are in the caudal part.
- **Spinal trigeminal nerve nuclei**—which contains the general somatic afferent column. They receive information about deep or crude touch, pain, and temperature from the ipsilateral face. The trigeminal, facial,

glossopharyngeal, and vagus nerves convey pain information to these nuclei. The spinal trigeminal nerve nuclei are made up of the *subnucleus oralis (pars oralis), subnucleus caudalis (pars caudalis)*, and *subnucleus interpolaris (pars interpolaris)*. The subnucleus oralis is associated with transmission of fine tactile sense from the orofacial region. It is continuous with the principal sensory nucleus of the trigeminal nerve. The subnucleus interpolaris is also associated with transmission of fine tactile sense along with dental pain. The subnucleus caudalis is associated with transmission of nociception and thermal sensations from the head. The spinal trigeminal nerve nuclei project to the **ventral posteriomedial nucleus** in the contralateral thalamus, via the ventral trigeminal tract.

- **Cochlear and vestibular nuclei**—which contains the special somatic afferent column. The *cochlear nuclei* make up two cranial nerve nuclei in the brain stem: the ventral cochlear nucleus and the dorsal cochlear nucleus. The ventral cochlear nucleus is unlayered while the dorsal cochlear nucleus has layers. Auditory nerve fibers travel through the auditory nerve to carry information from the inner ear (cochlea) to the nerve root in the ventral cochlear nucleus. At the nerve root, fibers branch to innervate the ventral cochlear nucleus and deep layer of the dorsal cochlear nucleus. The *vestibular nuclei* are the cranial nuclei for the vestibular nerve and are grouped in the pons and medulla. Their four subnuclei are the *medial vestibular nucleus, lateral vestibular nucleus, inferior vestibular nucleus*, and *superior vestibular nucleus*.
- **Inferior olivary nucleus**—which relays to the cerebellum. It coordinates signals from the spinal cord to the cerebellum, to regulate motor coordination and learning. Degeneration of the cerebellum results in degeneration of this nucleus, and vice versa.
- **Dorsal column nuclei**—which contains the gracile and cuneate nuclei. Both contain second-order neurons of the dorsal column-medial lemniscus pathway, which carries fine touch and proprioceptive information from the body to the brain. Each nucleus has an associated nerve tract.

The basal plate neuroblasts form the following structures:

- **Hypoglossal nucleus**—which contains general somatic efferent fibers. It lies close to the midline and is visible when the medulla is surgically opened as the hypoglossal trigone, a raised area that protrudes slightly into the fourth ventricle. The hypoglossal nucleus lies between the dorsal motor nucleus of the vagus and the midline of the medulla. Its axons pass anteriorly through the medulla, forming the hypoglossal nerve, which exits between the pyramid and olive in the anterolateral sulcus.
- **Nucleus ambiguous**—which forms the special visceral efferent fibers. It is also sometimes spelled "nucleus ambiguus", and consists of large motor neurons situated deep in the medullary reticular formation. It contains cell bodies of neurons innervating the muscles of the soft palate, pharynx, and larynx, which are associated with speech and swallowing. The nucleus ambiguous also contains preganglionic parasympathetic neurons that innervate postganglionic parasympathetic neurons in the heart. The nucleus has histologically different cells lying just dorsal to the inferior olivary nucleus, and it receives upper motor neuron innervation via the corticobulbar tract. The nucleus gives rise to the branchial efferent motor fibers of the vagus nerve.
- **Dorsal nucleus of vagus nerve**—which forms general visceral efferent fibers. It is also called the *posterior motor nucleus of vagus*, and lies ventral to the floor of the fourth ventricle. Mostly, it serves parasympathetic vagal functions in the GI tract, lungs, and other abdominal and thoracic vagal innervations. Cell bodies for the preganglionic parasympathetic vagal neurons that innervate the heart lie in the nucleus ambiguous.
- **Inferior salivatory nucleus**—which also forms general visceral efferent fibers. It is also called the *nucleus salivatorius inferior*, a cluster of neurons in the pontine tegmentum, just above its junction with the medulla. This nucleus is the general visceral efferent component of the glossopharyngeal nerve that supplies the parasympathetic input to the parotid gland, for salivation. The nucleus lies immediately caudal to the superior salivatory nucleus, and just above the upper end of the dorsal nucleus of the vagus nerve in the medulla.

Point to remember

The early development of the medulla oblongata is similar to that of the spinal cord. The sulcus limitans divides each lateral wall into a dorsal or alar lamina, and a ventral or basal lamina. The thin roof plate greatly widens, and the alar laminae develop dorsolateral to the basal laminae. Both laminae are within the floor of the developing fourth ventricle. Cells developing in the lateral part of each alar lamina migrate ventrally, reaching the marginal layer. The remaining cells of the alar lamina develop into sensory nuclei of the cranial nerves related to the medulla. The motor nuclei of these nerves are derived from the basal lamina. The white matter of the medulla is composed of fibers making up the ascending and descending tracts that pass through the medulla.

Metencephalon

The **metencephalon** (see Fig. 3.2) is the embryonic portion of the hindbrain. It differentiates into the pons and cerebellum, and contains part of the fourth ventricle. Nuclei of the trigeminal nerve (CN V), abducens nerve (CN VI), facial nerve (CN VII), and vestibulocochlear nerve (CN VIII) are within the pons. The metencephalon develops from the higher rostral half of the embryonic rhombencephalon. It is differentiated from the myelencephalon by about week 5 of development. By the third month, the metencephalon has become the pons and cerebellum. The pons regulates breathing through particular nuclei that regulate the breathing center of the medulla oblongata. The cerebellum coordinates muscle movements, maintains posture, and integrates sensory information from the inner ear and proprioceptors in muscles and joints.

Mesencephalon

The **mesencephalon**, or *midbrain*, is involved in vision, hearing, motor control, sleeping, waking, alertness, and the regulation of temperature (see Fig. 3.2). The nuclei of cranial nerves III (ocular nerve) and IV (trochlear nerve) lie within the midbrain, which makes up the **tecum**, tegmentum, cerebral peduncles, and several nuclei and fasciculi. The midbrain also adjoins the metencephalon caudally, and adjoins the diencephalon rostrally. The *substantia nigra* is located within the midbrain, with left and right regions. Parts of the substantia nigra are darker than nearby areas due to high levels of *neuromelanin* in dopaminergic neurons. The substantia nigra is important for coordination of movements. The *tectum* is the dorsal area of the metencephalon. It contains the superior and inferior colliculi, which help with processing of audio and visual information. The *isthmus* is the primary control center for the tectum. The mesencephalon arises from the second vesicle of the neural tube, remaining undivided for the rest of the neural development period. Cells in the midbrain continually multiply, compressing the cerebral aqueduct as it is forming. Partial or total obstruction of the aqueduct during development can cause *congenital hydrocephalus*.

Diencephalon

The **diencephalon** is made up of structures lying on either side of the third ventricle (see Fig. 3.2). These include the thalamus, hypothalamus, posterior pituitary, epithalamus, pineal body, and subthalamus. The hypothalamus performs many vital functions, mostly involved in regulation of visceral activities. The diencephalon is a primary brain vesicle. Remember that the neural tube forms the mesencephalon, rhombencephalon, and prosencephalon, from which the telencephalon and diencephalon form.

Telencephalon

Growth of the telencephalon determines future events. This neural tube structure forms from two enlarged areas that are connected at midline by the thin, membranous **lamina terminalis**. The basal wall of the **telencephalon** is close to the diencephalon (see Fig. 3.2). The wall thickens, forming the *primordia* of gray masses. These masses are the basal ganglia or *basal nuclei*. At the same time, the diencephalon walls thicken to form the thalamus and hypothalamus. These are separated by the **hypothalamic sulcus**. Eventually, the telencephalon folds down along the length of the diencephalon, and they fuse together. The area of the telencephalon's surface overlying the location of fusion forms the part of the cerebral cortex known as the insula. Over the next months, the cortex adjoining the insula greatly expand until it is hidden under the other structures. Both cerebral hemispheres eventually have an arched "C" shape, encircling the insular cortex. Areas of each hemisphere that started dorsal to the insula are pushed into the temporal lobe.

The cortical area expansion, which began with the vesicles of the telencephalon, continues as the "C" shape develops. It ends with numerous folds developing on the hemisphere surfaces. Each hemisphere begins with a smooth surface, but becomes more convoluted with continued development. Growth of the cerebral cortex, cerebellum, and other areas of the CNS involves much proliferation and migration of neurons as well as glial cells. This primarily occurs during the third, fourth, and fifth months of development. However, formation of neuronal connections continues long after birth, and the production of myelin sheaths mostly occurs after birth.

76 PART | I Structure, function, and development

> **Point to remember**
> Embryonal tumors of the CNS are malignant. They begin in the embryonic cells of the fetal brain. Embryonal tumors can occur at any age, but usually, in infants and young children. Types of embryonal tumors include: medulloblastomas, tumors with multilayered rosettes, medulloepitheliomas, atypical teratoid or rhabdoid tumors, and others.

Ventricular system

The neural tube cavity develops into the ventricular system of the adult brain, as well as the central spinal cord canal. The third ventricle takes shape from the cavity of the diencephalon. The fourth ventricle forms from the cavity of the pons and rostral medulla. Each cerebral hemisphere has a large, C-shaped *lateral ventricle* communicating with the third ventricle via an **interventricular foramen**. The third ventricle and fourth ventricle communicate via the **cerebral aqueduct** of the midbrain. The roof of the fourth ventricle is thin at the point where the rhombencephalon walls spread apart to form this ventricle. Additionally, an area covering the third ventricle's roof extending onto the telencephalon's surface becomes thinner. At both of these locations, groups of small blood vessels penetrate the roof of the ventricle and form the **choroid plexus**, which produces most of the cerebrospinal fluid that fills the ventricles. When the cerebral hemispheres become C-shaped, the choroid plexus also forms this shape and protrudes into the lateral ventricles (see Fig. 3.7).

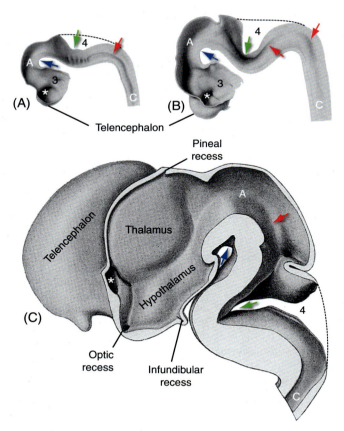

FIGURE 3.7 Development of the ventricular system at 37 days (A), 41 days (B), and 50 days (C). By 50 days the thalamus and hypothalamus form the walls of the slit-shaped midline third ventricle (3); several small recesses protrude from this ventricle. As each cerebral hemisphere grows around in a C shape, so too does its lateral ventricle. The location of the thin, membranous roof of the fourth ventricle (4) is indicated by dotted lines. Blue arrows, cephalic flexure; green arrows, pontine flexure; red arrows, sulcus limitans; *, interventricular foramen; A, cerebral aqueduct; C, central canal of the spinal cord and caudal medulla. (A) and (B) from Hines, M. 1922. *Journal of Comparative Neurology 34* 73. (C) from Hochstetter, F. 1929. Beiträge zur Entwicklungsgeschichte des menschlichen Gehirns. I. Teil, Vienna: Franz Deuticke.

Central nervous system malformation

If the sensitive development processes of the nervous system are interrupted, a congenital malformation may occur. Table 3.2 summarizes the malformations and times in development that are interrelated. However, the actual causes of many malformations are not fully understood. Continued research is aiding in understanding of more information about environmental, genetic, and molecular factors. Normal PNS development is also based on similar genes and molecules to those needed for normal CNS development. In the PNS, some mutations that affect migration or differentiation of neural crest cells may result in abnormalities of the autonomic nervous system, hearing, skin pigment, and cardiac function.

When there is a defective neural tube closure, many congenital malformations of the nervous system can occur. About one of every 1000 live births has one of these malformations. If the neural tube fails to totally close, a fatal deformity called **cranioarchischisis** occurs, in which the CNS is opened on the dorsal surfaces of the head and spinal area. Other CNS malformations cause anencephaly, spina bifida, and other outcomes.

Neural tube defects are detected via elevated levels of **alpha-fetoprotein** (AFP), or through clinical imaging. AFP is an important component of the fetal serum. When the neural tube is open, some AFP leaks into the

TABLE 3.2 CNS malformations and related developmental periods.

Primary developments	Week	Malformations
Neural folds and groove Visibility of three primary vesicles Cephalic and cervical flexures Appearance of motor neurons	3	Neural tube defects
Neural tube begins to close (day 22) Rostral end of neural tube closes (day 24) Caudal end of neural tube closes (day 26) Neural crest cells start to migrate Secondary neurulation begins Emergence of motor nerves	4	Holoprosencephaly and neural tube defects
Optic vesicle and pontine flexure develop Visibility of five secondary vesicles Sulcus limitans and sensory ganglia develop Sensory nerves grow into the CNS Rhombic lips developBasal nuclei begin developing Thalamus and hypothalamus begin developing Autonomic ganglia, lenses of eyes, cochlea begin developing	5	Holoprosencephaly and sacral spinal cord abnormalities
Enlargement of telencephalon Prominence of basal nuclei Completion of secondary neurulation Cerebellum and optic nerve develop Choroid plexus and insula develop	6 and 7	No malformations occur
Neuronal proliferation occurs Cerebral and cerebellar cortices develop Anterior commissure and optic chiasm developInternal capsule develops Appearance of reflexes	8-12	Migration, with or without proliferation problems, such as abnormal cortex or gyri
Neuronal migration and proliferation Glial differentiation Corpus callosum develops	12-16	Migration, with or without proliferation problems, such as abnormal cortex or gyri
Neuronal migration Cortical sulci develop Glial proliferation Mostly postnatal, but some other myelination Formation of synapses	16-40	Hemorrhage or other harmful events

amniotic fluid and eventually reaches the maternal circulation, where it can be detected. Neural tube defects often cause fetal death. Fortunately, most neural tube defects are preventable if the mother has sufficient *folic acid* in her diet at the time the neural tube is closing—which is at the end of the first month of gestation. Since the mother may not know she is pregnant at this time, all women of childbearing age must take routine folic acid supplements. Folic acid is regularly added to fortified cereals and many other grain products in the United States.

Since folic acid supplementation began, the prevalence of conditions such as spina bifida has decreased by 28%. Today, it is recommended that 400 µg of folic acid should be supplemented every day. However, the recommended dosage for a pregnant woman is 1 mg per day. Incidence of congenital malformations is 1.9 out of every 10,000 live Caucasian or Hispanic births, and 1.7 out of every 10,000 live African-American births. Research has shown that neural tube defects are linked to interactions between genetics, teratogens, and folic related metabolic disorders. Folic acid supplements are the safest and best way of preventing these defects. Couples who previously had a child with spina bifida have a recurrence rate of between 3% and 8%. Rates are also higher in children with *trisomy* 13 or 18, as well as in *chromosome 13q deletion syndrome*. Teratogens related to neural tube defects include alcohol, excessive vitamin A, aminopterin, valproic acid, carbamazepine, clomiphene, glycol ether, herbicides, and lead. The two primary groups of neural tube defects are called *occulta*, meaning "hidden", and *aperta*, meaning "visible". About 75% of vertebral defects occur in the lumbosacral region, usually at the L5 to S1 levels. Loss of motor function is varied, with different effects upon the spine and limbs.

> **Point to remember**
> Congenital abnormalities of the CNS are divided into developmental malformations and disruptions. Developmental malformations result from abnormal brain development. These malformations are usually midline or bilateral and symmetric, and do not show gliosis. Disruptions result from destruction of the normally developing or developed brain, and are caused by environmental or intrinsic factors, such as fetal infections, exposure to harmful chemicals, radiation, and fetal hypoxia. Disruptions do not recur unless the exposure recurs or continues.

Key terms

Adenohypophyseal placode
Alar plate
Alpha-fetoprotein
Anterior (ventral) horn
Basal plate
Cerebral aqueduct
Choroid plexus
Cranial placodes
Cranioarchischisis
Diencephalon
DiGeorge syndrome
Dorsal median septum
Dorsal median sulcus
Dorsolateral placodes
Ectoderm
Endoderm
Epibranchial placodes
Epithalamus
Foramen cecum
Fourth ventricle
Frontonasal dysplasia
Habenula
Hypoglossal trigone
Hypothalamic sulcus

Hypothalamus
Interventricular foramen
Intumescences
Lamina terminalis
Lens placode
Medulla oblongata
Medulla spinalis
Medullary pyramids
Mesoderm
Mesencephalon
Metencephalon
Multipotency
Myelencephalon
Neural crest cells
Neural fold
Neural groove
Neural plate
Neural tube
Neurocristopathies
Neuropores
Neurula
Nodosal placode
Notochord
Obex

Otic pit
Otic placode
Petrosal placode
Pontine flexure
Posterior (dorsal) horn
Primary neurulation
Primary vesicles
Proprioceptive
Rhombencephalon
Rhombic lips
Rostral
Secondary neurulation
Secondary vesicles
Subthalamus
Sulcus limitans
Tecum
Telencephalon
Thalamus
Ventral median fissure
Ventral posteromedial nucleus
Waardenburg-Shah syndrome

Further reading

Avilable at https://www.cdc.gov/ncbddd/birthdefects/anencephaly.html
Avilable at https://www.chop.edu/pages/fetal-surgery-spina-bifida-case-study
Avilable at www.delmarlearning.com/companions/content/1401897118/casestudies/01_newborn.pdf
Avilable at https://embryology.med.unsw.edu.au/embryology/index.php/Neural_-_Medulla_Oblongata_Development
Avilable at https://embryology.med.unsw.edu.au/embryology/index.php/neural_system_development
Avilable at www.ijss-sn.com/uploads/2/0/1/5/20153321/ijss_oct_cr17.pdf
Avilable at https://www.meduweb.com/threads/27922-holoprosencephaly-and-strabismus-case-with-photos
Avilable at https://www.sciencedirect.com/topics/veterinary-science-and-veterinary-medicine/spinalis
Avilable at www.upmc.com/services/neurosurgery/brain/conditions/brain-tumors/pages/meningocele.aspx
Avilable at https://library.med.utah.edu/webpath/labs/embrylab/embrylab.html
Ammar, A. (2017). *Hydrocephalus: What do we know? And what do we still not know?* Springer.
Benzel, E. C. (2012). *Spine surgery: Techniques, complication avoidance, & management* (3rd ed.). Saunders.
Bianchi, D. W., Crombleholme, T. M., D'Alton, M. E., & Malone, F. (2010). *Fetology: Diagnosis and management of the fetal Patient* (2nd ed.). McGraw-Hill Education/Medical.
Carlson, B. M. (2013). *Human embryology and developmental biology* (5th ed.). Saunders.
Cochard, L. R. (2012). *Netter's atlas of human Embryology* (Updated ed.). Saunders.
Cramer, G. D., & Darby, S. A. (2013). *Clinical anatomy of the spine, spinal cord, and ANS* (3rd ed.). Mosby.
Daroff, R. B., Jankovic, J., Mazziotta, J. C., & Pomeroy, S. L. (2015). *Bradley's neurology in clinical practice* (7th ed.). Elsevier.
Fritsch, M. J., Meier, U., & Kehler, U. (2014). *NPH – Normal pressure hydrocephalus: Pathophysiology—Diagnosis—Treatment*. TPS.
Gabbe, S. G., Niebyl, J. R., Simpson, J. L., Landon, M. B., & Galan, H. L. (2016). *Obstetrics: Normal and problem pregnancies* (7th ed.). Elsevier.
Gilbert, S. F., & Barresi, M. J. F. (2016). *Developmental biology* (11th ed.). Sinauer Associates/Oxford University Press.
Hogge, W. A., Wilkins, I., Hill, L. M., & Cohlan, B. (2016). *Sanders' structural fetal abnormalities* (3rd ed.). McGraw-Hill Education/Medical.
Kaur, C., & Ling, E. (2013). *Glial cells: Embryonic development, types/functions and role in disease*. Nova Science Publishers, Inc.
Klein, A. (2013). *Neural tube defects: Prevalence, pathogenesis and prevention – neurodevelopmental diseases, laboratory and clinical research*. Nova Science Publishers, Inc.
Moore, K. L., Persaud, T. V. N., & Torchia, M. G. (2015). *Before we are born: Essentials of embryology and birth defects* (9th ed.). Saunders.
Moore, K. L., Persaud, T. V. N., & Torchia, M. G. (2015). *The developing human: Clinically oriented embryology* (10th ed.). Saunders.
Norton, M. E. (2016). *Callen's ultrasonography in obstetrics and gynecology* (6th ed.). Elsevier.
Ozek, M. M., Cinalli, G., & Maixner, W. (2008). *Spina bifida: Management and outcome*. Springer.
Rubenstein, J., & Rakic, P. (2013). *Patterning and cell type specification in the developing CNS and PNS: Comprehensive developmental neuroscience*. Academic Press.
Sadler, T. W. (2014). *Langman's medical embryology* (13th ed.). LWW.
Schoenwolf, G. C., Bleyl, S. B., Brauer, P. R., & Francis-West, P. H. (2014). *Larsen's human embryology* (5th ed.). Churchill Livingstone.
Woodward, P. J., Kennedy, A., & Sohaey, R. (2016). *Diagnostic imaging: Obstetrics* (3rd ed.). Elsevier.

Part II
Epidemiology

Chapter 4

Global epidemiology of brain and spinal tumors (in comparison with other cancers)

Chapter Outline

World population	83
Aging of the population	83
Life expectancy	85
Distribution of brain and spinal tumors by age, gender, and race	87
Attributable deaths	91
Global prevalence	92
Disability adjusted life years	93
Quality adjusted life years	94
Global mortality from brain and spinal tumors	94
Burden of brain and spinal tumors in the United States	95
Mortality from brain and spinal tumors in the United States	96
Comparison of brain and spinal tumors with other cancers	97
Key terms	100
Further reading	100

World population

The population of the world is growing at approximately 1.07% per year. The current annual average population increase is estimated to be 82 million people. The highest annual growth rate occurred in the late 1960s, when it was about 2% per year. It is estimated to keep decreasing annually, reaching only 1% growth by the year 2023. Global population doubled between the years 1959 and 1999. The United Nations projects that world population will have reached 10 billion people by the year 2056. This data comes from the United Nations Population Division, the World Population Prospect, and the United States Census Bureau. As of 2019, census bureaus have estimated the world population exceeding 7.7 billion people. The top 10 largest countries in terms of population are:

- China (1.4 billion);
- India (1.3 billion);
- United States (328 million);
- Indonesia (269 million);
- Brazil (212 million);
- Pakistan (203 million);
- Nigeria (200 million);
- Bangladesh (167 million);
- Russia (143 million);
- Mexico (132 million).

Aging of the population

The world population is also aging rapidly, because humans are simply living longer, due to healthier lifestyles and better medical diagnosis and treatments. A second reason for this is that the fertility rate is decreasing, resulting in less women becoming pregnant. Average life expectancy in the United States is higher now than during any other time in history. According to the United Nations, people 65 years and older increased from 8% of the total population in 1950 to 12% of the total population in 2000. This is expected to increase to 20% by the year 2050 and will probably continue to rise. The phenomenon is primarily because of large improvements in health care, investment into medical research, and better health insurance availability. Less people are dying from diseases such as breast cancer, colon cancer, prostate cancer, heart disease, and HIV.

According to the World Health Organization, there will soon be higher numbers of elderly people than children, and more people at extreme old age than ever in history (see Fig. 4.1). In 1900, the major health threats were infectious and parasitic diseases, which often caused the deaths of infants and children. Infectious diseases such as *typhoid* were associated with poor hygiene and sanitation, while diseases such as *goiter* and *pellagra* were associated with poor nutrition. There was also poor maternal and infant health, plus diseases and injuries related to unsafe workplaces and hazardous jobs. Because of health care being much less advanced at the time, these diseases often caused early mortality—many people died before ever reaching their elderly years. Successes in early public health were achieved by incorporating antibiotics and vaccinations, as well as the development of health education programs, and the impact of these diseases declined. However, chronic diseases such as cancer and cardiovascular disease began to increase in prevalence. Towards the last half of the 20th century, public health identified risk factors for many chronic diseases, and interventions occurred that reduced mortality.

A noncommunicable disease is a noninfectious health condition that cannot be spread from person to person, which lasts for a long period of time, and is also referred to as a *chronic disease*. A combination of environmental, genetic, lifestyle, and physiological factors can cause these diseases. Risk factors include unhealthy diets, lack of physical activity, smoking, and excessive use of alcohol. Noncommunicable diseases kill approximately 40 million people every year, which is about 70% of all deaths worldwide. They affect people of all age groups, countries, and religions. Today, noncommunicable diseases mostly affecting adults and the elderly have the greatest impact. The most common of these diseases include cardiovascular disease, cancer, chronic respiratory disease, diabetes mellitus, arthritis, and obesity. With increased lifespans due to better health care, people live much longer than in previous times, and therefore, such chronic diseases are able to manifest over time. Health and economic burdens of age-related disability are affected by environmental factors that determine individual independence, even though there may be physically limitations. Costs for long-term care are lower because of longer amounts of time that people can remain mobile and take care of themselves.

Decreasing fertility, on a global basis, is also very important. In more developed countries, fertility has fallen below the "replacement" rate of two live births per woman (in the 1970s), while women in the 1950s averaged three live births. In less developed countries, fertility rates have decreased even faster. In 1950, women averaged six live births. By 2006, this rate was at or below two live births. Many countries with large populations are now towards the lower end of the scale. The United States now ranks 135th on the list, with 1.8 children per woman. The lowest fertility rate is in Taiwan, with 1.2 children per woman, followed by Moldova, Portugal, Singapore, Poland, Greece, South Korea, Hong Kong, Cyprus, and Macau. Fig. 4.2 shows fertility rates throughout the world, ranging from 0 to 1 child in the lowest fertility areas, up to a maximum of 6 to 7 children per woman in two African countries. Though the legend on this map shows a maximum of 7 to 8 children per woman, there are no countries with that as their average fertility rates.

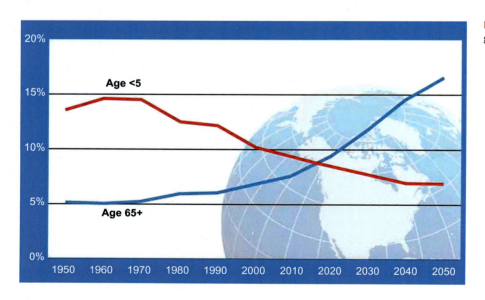

FIGURE 4.1 Changes in the age of the global population, 1950–2050.

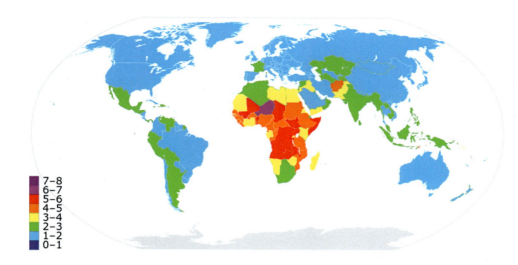

FIGURE 4.2 Global fertility rates as of 2018.

Countries that are losing population quickly include Ukraine, which will decrease 22% by 2050, followed by Poland, the Russian Federation, Italy, and Spain. The population of the European Union is expected to peak by 2050 and then decline slowly. Germany has had a demographic decline for over a generation, and will likely drop in population 7.7% by 2050, not considering recent immigration into the country. Bulgaria is expected to reduce in population 27% by 2050 and Romania will shrink by 22%. Japan will reduce 15%, and by 2030 will actually have more people over the age of 80 than under the age of 15. China's very low fertility rate means that it will have 28 million less people by 2050.

Life expectancy

Life expectancy is also known as *longevity* and is calculated by creating a **life table**, which records deaths and survivors of illness in a specific year, for successive life span intervals. Deaths, survivors, and **age-specific death rates** are established for various age groups, such as 0 − 1 year, 1 − 5 years, and for successive 5-year age groups after that. They are used to create a second life table that represents the total mortality rates from birth to death, for 100,000 hypothetical live births. This involves age-specific death rates in the population being studied for a particular year, and is used to calculate life expectancy as the average life years for all societal members since birth. Life expectancy equals total years of life for all members of the life table, divided by the total number of persons at birth. Longevity at birth describes the *mean years of life*, based totally on age-specific death rates for the population and the year of interest.

Most people born in 1900 did not live past the age of 50 because of the prevalence of infectious diseases and the lack of cures for them. According to the Centers for Disease Control and Prevention, life expectancy at birth for American people born in 2012 was 78.8 years. Today, 10% of females will live past the age of 100 years, and nearly 5% of males will live past 100. Life expectancy for females is 81.2 years. For males, it is 76.4 years. The difference between them is 4.8 years, which has been the same since 2011. A woman reaching age 65 in 2019 can expect to live, on average, until the age of 86.6. A man reaching age 65 can expect to live, on average, until the age of 84.3. In 2015, the average American woman reaching age 65 had more than a 33% chance of reaching the age of 90. This has increased from a 25% chance that existed 50 years ago. The countries having the oldest populations include Monaco, Japan, Germany, Italy, Greece, Sweden, Spain, Austria, Bulgaria, and Estonia (see Fig. 4.3). In Monaco, 22.8% of the population is age 65 or older. In 2019, 13.1% of the United States' population was age 65 or older. The reasons for women generally living longer than men include faster development in utero, a lower likelihood of risky behaviors, later development of heart disease, reduced stress levels because of better social interactions, and better self-management of their own health care.

Health care providers to the elderly and disabled patients increased in the United States from more than 621,000 workers in 2007 to over 911,000 workers in 2012. Revenues from this care increased from $25.3 billion in 2007 to $34.4 billion in 2012. Today, governments must plan decades ahead, with new methods to better manage the quickly changing health care situation. Some cities are building age-friendly housing and other infrastructure. Sweden has implemented low cost approaches to caring for elderly citizens, while offering extremely high quality of care. Most

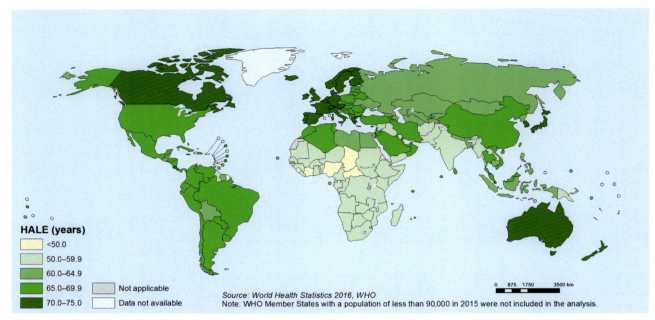

FIGURE 4.3 Healthy life expectancy at birth, both sexes, 2016.

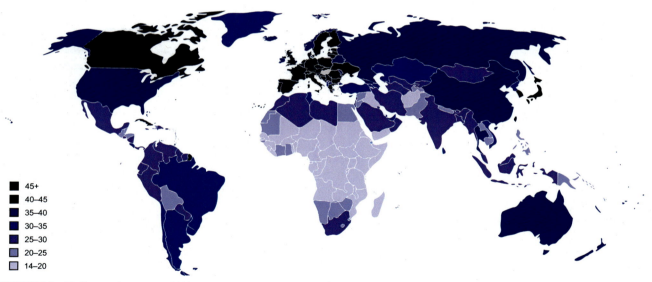

FIGURE 4.4 Median age by country, 2018.

elderly health care in Sweden is funded by municipal taxes and government grants. In 2014, total costs for care were equivalent to $12.7 billion, yet 4% of the cost was financed by patient charges. Privatization of elderly care is growing, allowing private care companies to control operations, which provide over 24% of all elderly in-home care. All patients can choose if they want their care to be provided by public or private operators.

The "oldest old" (people 85 or older) make up 8% of the global elderly population. The term *elderly* means, "age 65 or older." This is 12% of the population in more developed countries and 6% in less developed ones. The median age of countries throughout the world are illustrated in Fig. 4.4, with higher overall ages appearing darker in color. Reasons for less developed countries having overall earlier deaths include unemployment, poor standards of living, malnutrition, unsafe or contaminated food sources, low quality housing, and lack of protection from climate-related factors. In many countries, the 85-and-older group is the fastest growing population segment. Globally, this group will likely increase 351% between 2010 and 2050. This is compared to a 188% increase in the 65-and-older group, and a 22% increase in the population under age 65. People reaching 100 years of age are will probably increase by 10 times in number between 2010 and 2050. Over human history, odds of living to age 100 have risen from one of every 20 million people to *one of every 50 people*—with this figure being for females in low-mortality countries such as Japan and Sweden. Of the percentage change in world population by age shown in Fig 4.5, the largest is the 100-and-older age group, with a *1004% increase*.

> **Point to remember**
>
> Globally, humans are living longer and more productive lives than any time previously in history, with oldest average life expectancy being 83.7 years, in Japan. This figure takes into account males and females to achieve an "average" age. Switzerland and Singapore rank second and third. The United Kingdom is ranked 20th, and the United States is ranked 31st, at 79.3 years. The three lowest average life expectancies are Sierra Leone (50.1 years), Angola (52.4 years), and Central African Republic (52.5 years).

Distribution of brain and spinal tumors by age, gender, and race

Primary brain tumors arise from different cells of the central nervous system (CNS). Gliomas are the most common malignant primary brain tumors at 80% of all tumors, and make up 28% of all brain and spinal tumors. Glioblastoma accounts for 54% of all gliomas, while astrocytomas (including glioblastomas) account for about 75% of all gliomas. Meningiomas are mostly nonmalignant, and make up 35.8% of brain tumors. In children up to age 14, malignant gliomas make up 25.7% of cases, followed by pilocytic astrocytomas (17.5%) and embryonal tumors (15.7%). The American Cancer Society estimated that in 2015, there were about 22,850 new cases of malignant primary CNS tumors. In 2013, there were an estimated 22,620 new cases of malignant primary CNS tumors and 43,110 nonmalignant primary CNS tumors in the United States, with about 4300 of these cases occurring in children. Annual incidence for malignant brain cancer, for all races, between 2006 and 2010 was 7.27 per 100,000 person-years, and 13.77 per 100,000 person-years for primary nonmalignant brain tumors. Incidence of childhood primary malignant and nonmalignant brain and spinal tumors was 5.1 cases per 100,000 person-years in 2013.

Most tumors involving the spinal column are metastases of visceral organ cancers. They mostly occur in older patients. Primary musculoskeletal sarcomas involving the spinal column are rare. Benign tumors, and tumor-like lesions are mostly seen in young patients, often resulting in instability and compromise of the spinal canal. Surgeons have adapted their approach to spinal tumors over the last 20 years. Advances in imaging, surgical techniques, and implant technologies have provided better diagnostic and treatment options—especially for primary spinal tumors. Modern

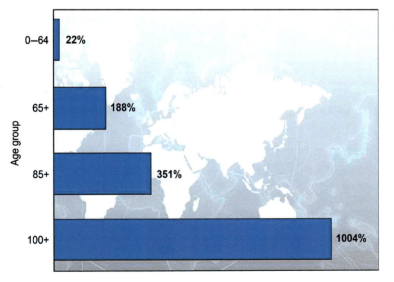

FIGURE 4.5 Percentage change in world population by age.

chemotherapeutic drugs plus newer radiotherapy and radiosurgical options have results in better local and systemic control, even for primary sarcomas of the spinal column. Though primary spinal tumors are rare, spinal column metastases are present in as many as 70% of cancer patients. **Adenocarcinomas** that mostly originate from the lungs, breasts, prostate, kidneys, GI tract, and thyroid often metastasize to the spine. The percentage of cancer patients with bone metastasis before death is 50% − 70%. In the case of breast cancer, this percentage is as high as 85%. Up to 10% of patients with symptomatic spinal metastases can be treated by surgery. About 70% of spinal metastases occur in the thoracic and thoracolumbar spinal areas. The lumbar spine and sacrum receive more than 20% of metastatic lesions, but the cervical spine is less often involved.

> **Point to remember**
>
> A primary malignant brain tumor is a rare type of cancer. The most common brain tumors are called *secondary tumors*, meaning they have metastasized to the brain from other parts of the body. The cause of brain cancer is usually known, and it often does not have obvious symptoms. It can occur at any age, and survival rates vary. Though long-term studies are ongoing and many people believe otherwise, there is no definitive evidence that cell phone usage increases risks.

Primary malignant and nonmalignant brain tumors

Primary malignant brain tumors account for 2% of all cancers. They are 1/5 as common as lung or breast cancer, but result in significant amounts of illnesses and deaths. The 5-year survival rate for primary malignant brain tumors is about 35% of cases. This has increased from a percentage of 22% over the past 30 years. The 5-year survival rates are further broken down as follows:

- 79.1%—oligodendroglioma;
- 26.5%—anaplastic astrocytoma;
- 4.7%—glioblastoma.

There is a slight male predominance in the incidence of malignant CNS tumors. This is 7.4 per 100,000 person-years for men, and 5.5 per 100,000 person-years for women, as of 2010.

However, combining nonmalignant with malignant CNS tumor types, this changes. Men have an incidence rate of 19.1 per 100,000 person-years, while women have a rate of 22.8 per 100,000, person-years. This may be explained by the much higher predominance of meningiomas in women compared to men. Women have meningiomas at the rate of 9.97 per 100,000 person-years, while in men, this rate is only 4.44 per 100,000 person-years. The age-adjusted incidence rates of malignant brain tumors, compared between Caucasians and African-Americans, men, and women is illustrated in Fig. 4.6. Caucasians have a higher incidence of malignant brain tumors for both sexes, with men having an incidence rate of 8.2 per 100,000 person-years compared to African-American males at 6.1 per 100,000 person years. Caucasian women have an annual incidence rate of 4.2 per 100,000 person-years, compared to 3.9 per 100,000 person-years in African-American females. Mortality in Caucasians is at 4.6 per 100,000 population, compared to 2.5 per 100,000 African-Americans. The 5-year survival rate is lower in Caucasians (34%) than in African-Americans (41%). Incidence and mortality rates of malignant brain tumors are lower in Asian-Americans and Native-Americans.

Parenchymal brain metastases

The precise incidence of parenchymal brain metastases is unknown, but they are 10 times more common than primary tumors. It is estimated that parenchymal brain metastases are present in 20% − 40% of terminal cancer cases. This means that out of 1.3 million Americans who have cancer, between 100,000 and 170,000 will develop brain metastases. There are only about 35,000 primary brain tumors diagnosed annually in the United States. Incidence of brain tumor metastases vary with the type of tumor. The likelihood of developing brain metastases with **melanoma** ranges between 18% and 90%. For lung cancer, the range is 18% − 63%, and for breast cancer, 20% − 30%. While 3% of women with ovarian cancer develop brain metastases, just 1% of men with prostate cancer do. Another consideration is the breakdown of various cancers and the actual percentage of brain metastases that they are responsible for, out of all types of cancer, as follows:

- Lung cancer—40% − 50% of all brain metastases;
- Breast cancer—15% − 20%;
- Melanoma—5% − 10%;

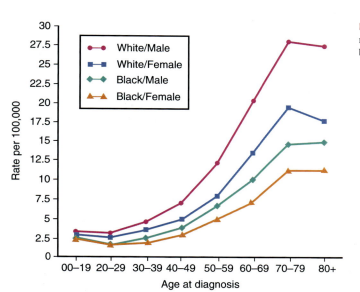

FIGURE 4.6 United States age-adjusted incidence rates for malignant brain cancer and other nervous system cancer for 1992–2010, by race and sex (http://www.seer.cancer.gov).

- Renal cell carcinoma—5% − 10%;
- Gastrointestinal tumors—5% − 10%.

Overall incidence of brain metastases is increasing. This may be explained by improved survival of systemic diseases, which provides more time for the development of these metastases. Also, the **blood-brain barrier** stops systemic medications from entering the CNS, and tumor cells can proliferate without medications affecting them.

Brain metastases can occur anywhere in the organ, but their development is related to the amount of cerebral blood flow. As a result, about 80% develop in the supratentorial area. It is not understood why pelvic and GI primary tumors usually metastasize to the posterior fossa instead of the supratentorial area. Also, brain metastases are the first manifestation of an underlying primary tumor in 10% − 30% of patients. Less than 25% of these patients have clinical features that suggest the tumor location. Regardless, about 80% eventually have the primary tumor site identified. The most common cause of brain metastasis without a known primary tumor is lung cancer (66% of cases). When a lung tumor metastasizes to the brain, 2 of every three are nonsmall-cell lung cancers. Studies reveal that computed tomography (CT) and magnetic resonance imaging (MRI), used as initial methods of diagnosis, are able to identify a site for biopsy in 97% of cases of newly-detected intracranial masses. It is more cost effective to restrict initial radiology studies to the chest, since there is a high chance of primary lung tumors, and because many patients with other primary tumors have lung metastases by the time of developing brain metastases. Most brain metastases are multiple, and the majority of patients have a known cancer. Therefore, only 15% of single intracranial masses, when the patient has not been diagnosed with cancer, are metastatic tumors.

Leptomeningeal metastases

About 5% of cancer patients have **leptomeningeal** metastases. Incidence varies with different types of tumors. These metastases occur in 11% − 70% of leukemia cases, 5% − 29% of **non-Hodgkin lymphoma** cases, and up to 8% of solid tumor cases. Of solid tumor cases, adenocarcinomas are most likely to metastasize to the leptomeninges. Of lung tumors, patients with **small-cell lung cancers** have a greater likelihood of leptomeningeal metastases than those who have other lung cancers. Breast cancer accounts for 11% − 64% of patients with leptomeningeal metastases, followed by lung cancer (14% − 29%), melanoma (6% − 18%), and GI cancers (4% − 14%). Medulloblastoma, high-grade gliomas, and other primary brain tumors often spread to the CNS. Cancers of the head, neck, thyroid, prostate, carcinoid, and bladder rarely metastasize to the leptomeninges.

Adequate CNS prophylaxis has reduced incidence of leptomeningeal metastases from **acute lymphoblastic leukemia** (ALL) in children from 66% to 5%. However, with similar prophylaxis, incidence of this in adults is still high. With CNS prophylaxis, patients with **acute myelogenous leukemia** (AML) have a 5% risk for leptomeningeal metastases, and 10% if CNS prophylaxis is not administered. Cases of **chronic myelogenous leukemia** (CML) of **hairy cell leukemia** rarely metastasize to the leptomeninges. During an autopsy, leptomeningeal involvement is found in up to

50% of patients with **chronic lymphocytic leukemia** (CLL). Usually, these patients are asymptomatic during life. With non-Hodgkin lymphoma, seeding of the leptomeninges occurs in about 6% of cases. Highest risks are in patients with histologies that are diffuse, lymphoblastic, resembling those of **Burkitt lymphoma**, or when there is bone marrow, extranodal site, or testicular involvement. In patients with Hodgkin's disease, **multiple myeloma**, and mycosis fungoides, leptomeningeal disease only rarely occurs.

Primary central nervous system lymphoma

Primary CNS lymphoma is a type of non-Hodgkin lymphoma. There has been a significant increase in this lymphoma over the past few decades. This is similar to the doubled incidence of systemic non-Hodgkin lymphoma over the past 40 years. Incidence peaked early in the 1990s and then declined. The main factor may be the acquired immune deficiency syndrome (AIDS) epidemic. About 2% − 6% of people with AIDS develop this lymphoma. Incidence of AIDS has decreased, as it has for this lymphoma, since the early 1990s. However, AIDS does not fully explain the total increase in incidence of primary CNS lymphoma. Newer studies have suggested a continual increase in this lymphoma among non-AIDS populations—especially in patients over age 65. Primary lymphoma of the brain will be discussed in detail in Chapter 16, Pineal parenchymal tumors.

> **Point to remember**
> Generally data about occupational exposures and brain tumors are inconsistent. Certain occupations have been linked to higher rates of brain tumors, however. These include laboratory researchers, health care workers, and people working with radiation and chemicals. Farmers also have an elevated risk of brain tumors. Agricultural workers are frequently exposed to fungicides, herbicides, and pesticides. Reports of increased occurrences of brain tumors also involve people working in the petrochemical, rubber, and vinyl chloride industries. There is an increased risk for brain tumors among aircraft pilots, firefighters, glass manufacturers, metal cutters, tile makers, and welders, but incidence rates fluctuate over time.

Primary tumors of the spine

Primary spinal tumors are rare and usually asymptomatic. This means that their real incidence is unknown. The incidence of hemangiomas and enostoses, the most common primary spinal tumors, is between 11% and 14%. This is based on lesions detected while performing diagnostic procedures for other reasons. While osteoblastomas and chordomas especially affect the spine, the tumors originating from the skeleton do not usually do this. In all bone and soft tissue sarcomas, just 10% are related to the spine. Hematological malignancies are usually seen after 50 years of age. The most common malignant tumors in younger patients are osteosarcoma and **Ewing's sarcoma**. Plain radiographs help identify almost 80% of benign tumors having a more specific appearance, along with some malignant tumors and metastatic lesions. Plain radiographic findings are present in 40% of patients with spinal metastasis. At least 50% loss of **trabecular bone** must have occurred in order for a destructive spinal lesion to be seen via plain radiography. In 10% of spinal metastases with spinal canal involvement, there is neurological compromise of adjacent or distant levels.

Primary benign tumors of the spine are more common than primary malignant tumors. Benign aggressive tumors, such as osteoblastoma, giant cell tumor of the bone, and aneurysmal bone cyst often relapse. The most common primary benign spinal tumors include the following:

- **Hemangioma**—a vascular tumor derived from blood vessel cell types. The most common benign spinal tumor, they are seen in 10% of the general population and are usually asymptomatic.
- **Osteoid osteoma**—a benign bone tumor that arises from osteoblasts, commonly seen in adolescents and young adults, with painful secondary scoliosis and pain that worsens at night, which is relieved by nonsteroidal anti-inflammatory agents.
- **Osteoblastoma**—an uncommon osteoid tissue-forming primary neoplasms of bones. It mostly originates from posterior portions of the spinal cord, but can grow into the spinal canal, and may cause dural sac compression.
- **Osteochondroma**—the most common benign tumor of bones, in the form of cartilage-capped bony projections or outgrowths of bone surfaces. It mostly originates from posterior areas of the spinal cord, becoming symptomatic by spinal canal compromise or nerve root compression.
- **Giant cell tumor of the bone**—also called *osteoclastoma*, it is a relatively uncommon bone tumor characterized by multinucleated osteoclast-like cells. It is most common in the sacrum, often recurring, and capable of lung metastasis.

- **Aneurysmal bone cyst**—an osteolytic bone neoplasm characterized by several sponge-like blood or serum filled, generally nonendothelialized spaces of various diameters. It is most common in patients under age 20, usually involving more than one spinal segment, and recurring in 25% of patients. The name is not highly accurate because the lesion is not actually an aneurysm or a cyst.
- **Eosinophilic granuloma**—a form of Langerhans cell histiocytosis that is most common in children and adolescents, always healing spontaneously.
- **Neurofibroma**—a benign nerve-sheath tumor of the peripheral nervous system. It is often asymptomatic and extremely similar to **schwannoma**; about 5% − 10% undergo malignant changes (see Chapter 11, Neurofibroma).

Primary musculoskeletal system sarcomas in the spinal column are usually osteosarcoma, Ewing's sarcoma, and chondrosarcoma. They can affect any part of the spinal column. Osteosarcoma and Ewing's sarcoma are most common in children and adolescents, while chondrosarcomas are more common in adults. Ewing's sarcoma is more frequent between the ages of 5 and 20. Spinal column involvement is seen in only 5% of patients with Ewing's sarcoma. About 2% of all osteosarcomas originate in the spine, with the classic form usually seen in the second decade of life. There may be a second peak incidence is the sixth decade of life. Today, 70% − 80% of patients survive with no evidence of disease in later life. For chordomas, 60% begin in the sacrum, with 25% seen at the skull base and the remaining 15% seen in the rest of the axial skeleton. About 50% of all chordomas are palpable via digital rectal examination.

> **Point to remember**
>
> Spinal tumors are rare, and usually are signified by constant back or neck pain that is not provoked by activity, and is not relieved by resting. Besides surgical removal, a biopsy is the only way to tell if a spinal tumor is benign or malignant. Not all of them require immediate surgical removal. Today's improved surgical techniques allow for better visualization of tumors and less risks for nerve damage. Newer treatment options such as stereotactic radiosurgery allow oncologists to deliver higher doses of radiation with extreme preciseness.

Attributable deaths

In 2015, malignant CNS tumors accounted for about 15,320 deaths in the United States, according to the American Cancer Society. As shown in Fig. 4.7, age-specific mortality rates, in all races, show slight increases with each decade up to age 55, and then they sharply increase. Younger age and lower grading are favorable prognostic factors for primary brain tumors. A more favorable prognosis may also be linked to longer symptom duration, lack of mental changes upon diagnosis, cerebellar tumor location, small tumor size before surgery, and the ability of the surgical resection to be completely performed.

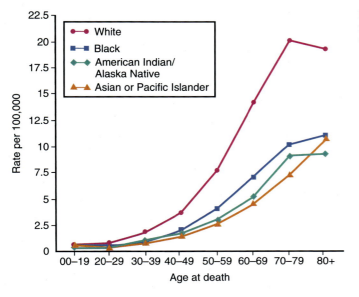

FIGURE 4.7 United States age-adjusted mortality rates for malignant brain and other nervous system cancer for 1990−2010, by race (Source: http://www.seer.cancer.gov).

Global prevalence

Globally, primary brain tumors have a prevalence of about 250,000 people per year. In children under age 15, brain tumors are second only to acute lymphoblastic leukemia as the most common form of cancer. Incidence rates for many types of brain tumors are increasing in most industrialized countries—usually in the elderly population. There are no identified ethnic, gender, or geographical differences. In the United States, overall incidence rates in Caucasians have leveled off, but showed a 1.6% annual increase in past decades. Part of the factors behind the global prevalence increases is due to better diagnostic technology and access to health care, including clinical specialization. Increased brain tumor incidence is related to newer noninvasive diagnostic technologies. This began in the 1970s with the development of CT and in the 1980s with the development of MRI. The highest rates of increased brain tumor incidence are in the United States, Canada, Australia, and the United Kingdom. Developing nations have lower incidence rates of brain tumors. Incidence is linked to levels of economic development, availability of diagnostic technologies, and health care access. In the United States, there are wide differences between various states. Maine has the highest incidence rate of primary brain tumors (8.3 per 100,000 person-years), while Hawaii has the lowest (5.0 per 100,000 person-years). Fig. 4.8 illustrates the states with the highest and lowest age-adjusted incidence rates for malignant brain and other nervous system cancers.

It must be understood that less-developed countries may show lower incidences of brain tumors because many tumor-related deaths are undiagnosed. Patients who are extremely poor often do not get diagnosed because they do not have access to modern diagnostic facilities. They are also more likely to die from other poverty-related causes that can be fatal before a tumor develops, or before a tumor becomes life-threatening. Incidence of low-grade astrocytoma does not vary greatly between different countries. However, there is evidence of variations in incidence of malignant CNS tumors. These variations are significant because some high-grade lesions arise from low-grade tumors. Incidence of CNS is higher in countries such as the United States and Israel, while lower in Japan and other Asian countries. This may reflect biological differences as well as differences in diagnostic methods and reporting.

In the United States, as of 2015, about 166,039 people were living with some type of CNS tumor. In 2018, there were about 23,880 new cases of brain tumors and 16,830 deaths as a result, accounting for 1.4% of all cancers, and 2.8% of all cancer deaths. The median age of diagnosis was 58 years, while the median age of death was 65 years. In the United States, brain cancer is the 10th leading cause of cancer death. The overall lifetime risk for developing brain cancer is 0.6% for both men and women. More than 28,000 people under age 20 are estimated to have a brain tumor, with about 3720 new cases diagnosed in children under age 15 every year. In children, CNS tumors make up 20% − 25% of all cancers. Their average survival rate, for all primary brain cancers, is 74%. Brain cancers are the most common type of cancer in children, causing more deaths than leukemia, with younger patients having a worsened prognosis. In children up to 14 years of age, the most common brain tumors are pilocytic astrocytoma, malignant glioma, medulloblastoma, neuronal and mixed neuronal-glial tumors, and ependymoma. In children under age 2, about 70% are

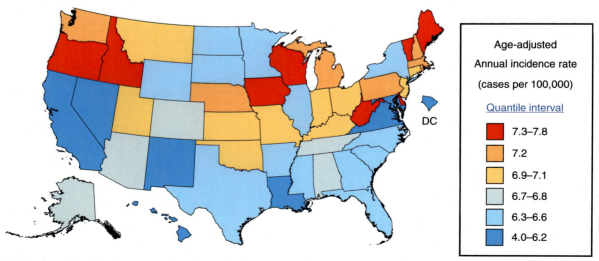

FIGURE 4.8 Age-adjusted incidence rates for malignant brain and other nervous system cancer in the United States for 2006−10, by state (http://statecancerprofiles.cancer.gov).

medulloblastomas, ependymomas, and low-grade gliomas. Less often, and usually in infants, there are teratomas and atypical **teratoid rhabdoid tumors**. Germ cell tumors, which include teratomas, make up only 3% of pediatric primary brain tumors, with varied worldwide incidence. See Chapter 17, Melanocytic tumors, for a more detailed discussion of brain tumors in children. In the United Kingdom, CNS tumors are the ninth most common cancer, with about 10,600 people diagnosed annually. These tumors are also the eighth most common cause of cancer death, with about 5200 dying annually.

Spinal tumors usually occur outside the dura mater lining, which is described as *extradural*. Other spinal tumors are within the dura (intradural), inside the spinal cord (intramedullary), and inside the dura but outside the spinal cord (extramedullary). Spinal tumors are rare in children, but most diagnoses in this group of patients occurs around 11 years of age. Most (55%) spinal tumors are extradural, followed by 40% being intradural/extramedullary, and only 5% being intramedullary. Overall prevalence is estimated at one spinal tumor for every four intracranial tumors.

> **Point to remember**
>
> Since the early 1990s, CNS tumors have increased in incidence by an average of 33% globally. In the CNS, most malignant tumors occur in the brain, while most benign tumors occur in the meninges. Tumors of the CNS are the 10th leading cause of death for both sexes in the United States. About 32.9% of patients survive for five or more years after being diagnosed with a CNS tumor. Fortunately most patients (77.5%) are diagnosed when the cancer is localized, which has the best prognosis. Once the cancer has metastasized, only 2% of patients will survive for five or more years.

Disability adjusted life years

Disability adjusted life years (DALYs) link information about disease occurrence to health outcomes. They are a useful aid to establish country-specific agendas concerning the control of cancer. They are calculated by adding **years of life lost** (YLL) with **years of life lived with a disability** (YLD). One lost DALY equates to one lost year of healthy life, as a result of premature death from the disease, or due to disease-related illnesses or disability. For brain and spinal tumors, YLLs are calculated by multiplying the number of cancer-specific deaths at a given age by the standard life expectancy for that age, or age group. The YLDs are calculated from the product of the number of new cases, average duration of the disability, and the disability weightings for the condition. Disability weights represent a value preference scaling the disease from "full health" (0) to "death" (1). The countries with the highest burden of DALYs for brain and spinal tumors are shown in Fig. 4.9. The United Kingdom ranks at number 45, with 137.6 per 100,000 population. The United States ranks at number 72, with 116.0 per 100,000 population. This is highly attributable to better diagnostic technologies and better overall access to appropriate health care.

Since the *Global Burden of Disease Study* in 2010, age weighting is now not a default value choice for DALYs. Today, DALY values can be calculated with or without using age weighting, and also with or without discounting the rates for future life. Instead, there are three options:

- Using age weights as well as discounting;
- Using either age weights or discounting;
- Using neither age weights nor discounting.

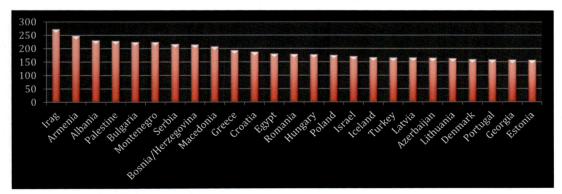

FIGURE 4.9 Countries with the highest disability-adjusted life years for central nervous system tumors (per 100,000 population).

Another consideration is *healthy life years lost per 1000 population per year (HeaLY)*. This differs from DALYs in that the start point for HeaLY is the beginning of a disease. It is the loss of healthy life based on the natural history of the disease, modified by interventions. When disease incidence is changing, such as regarding HIV, the DALY calculation may understate the actual disease situation. Both HeaLY and DALY calculations measure *health gaps*. They are basically equivalent if the disease in question is constant over time, if age weighting is not applied, or if the same measures of disability are used.

Quality adjusted life years

The quality adjusted life year (QALY) provides guidance in selecting alternative tertiary treatment. Basically, one year of life spent in a certain health status may be preferable to one year spent in a different health status. A QALY uses a scale of 0.00 (dead) to 1.00 (perfect health) for each health status. It is the product of duration of life and a measurement of quality of life. For example, 2 years of perfect health = 2 QALYs. Therefore, 2 years in a status measured as 0.5 of perfect health followed by 2 years of perfect health = 3 QALYs. Since being introduced, many QALY measurements have been developed. The largest measure used is the *European Quality of Life with Five Domains (EQ-5D)*, which has three levels of quality in each domain. The most significant use of QALYs has been as a common denominator, measuring utility in cost-utility analyses, and the effectiveness in cost-effectiveness analyses. Originally, the QALY was basically equal to the YLD portion of a DALY. A QALY would be equivalent to a YLD as long as there is no discounting, no age weighting, and the same disability weights are used. In recent years, QALYs have added life expectations of patients into their calculations.

Global mortality from brain and spinal tumors

Mortality has always been the primary indicator of health status of populations, and was first studied in the early 1600s in England. Mortality rates based on age, gender, location, and etiology give us vital information about health statuses and the burden of disease. Most countries require information about deaths that includes age, gender, location, and cause, via the use of *death certificates*. This information is not often as available in lower income countries. However, in higher income countries, this information is regularly collected and usually extremely accurate. *Verbal autopsies* have been used to judge likely causes of death primarily for children under the age of 5 years. This utilizes detailed questions between trained interviewers and family members. The information is later reviewed via computer or by physicians to Asian a cause of death, based on algorithms that use death statistics. There are large variations in cause-specific mortality in many low- and middle-income countries. Mortality from communicable diseases is a primary reason for differences between mortality rates in lower and higher populations. For many years, the cause of death certification system used in many countries has been based on the World Health Organization's *International Classification of Diseases (ICD)*. Basically, mortality is expressed by *mortality rate* and *case fatality ratio*. The mortality rate is expressed as the number of deaths in a specific population over a certain time period. The case fatality ratio is the proportion of people with a specific disease who die from that disease, at any time, unless specified. The mortality is equal to the case fatality ratio multiplied by the incidence rate of disease within the population.

Globally, the mortality rates for brain and spinal tumors are relatively similar between countries, but different between the actual types of tumors. Mortality is also based on the ages of patients. Most mortality rates throughout the world are based on a 5-year disease course. Overall, CNS tumors are the 10th leading cause of death for men and women globally. Once a patient reaches stage 4 cancer, the cancerous cells are reproducing quickly, and are often widespread throughout the brain, spinal cord, and other nearby organs. Symptoms at this point include headaches, seizures, convulsions, memory and speech problems, numbness, and difficulty walking. Personality changes may include moodiness, withdrawal, bizarre behaviors, belligerence, and even violence. Even the treatments for CNS tumors can cause symptoms, including nausea and vomiting.

When treatments fail, the patient begins to have less energy and often sleep more. By this stage, death usually occurs during sleep. When the patient begins to sleep for most of the day, it is a sign that death is imminent. Table 4.1 summarizes the global mortality rates for various CNS tumors for three different age groups. The percentages shown reveal predicted deaths from each type of tumor.

TABLE 4.1 Global mortality rates over 5 years for central nervous system tumors.

Tumor type	Age 20–44 (%)	Age 45–54 (%)	Age 55–64 (%)
Glioblastoma	81	92	95
Anaplastic astrocytoma	46	68	86
Diffuse astrocytoma	32	56	78
Anaplastic oligodendroglioma	29	39	54
Meningioma	13	23	29
Oligodendroglioma	12	9	32
Ependymoma/anaplastic ependymoma	8	11	14

Burden of brain and spinal tumors in the United States

According to the National Brain Tumor Society, the incidence rate of all primary malignant and nonmalignant brain and spinal tumors was 22.64 cases per 100,000 population in the United States. This meant a total of 379,848 cases during the time period studied, which was between 2010 and 2014. An estimated 78,980 new cases of primary malignant and nonmalignant brain and other CNS tumors were diagnosed in 2018. The average annual mortality rate was 4.33 per 100,000 population. There were an estimated 16,616 deaths attributed to primary brain and spinal tumors in the United States in 2018. The 5-year survival rate in the United States after a diagnosis of a primary malignant brain or spinal tumor was only 34.9%. For primary nonmalignant brain tumors, the 5-year survival rate was 90.47%.

Of brain tumors, about 69.1% are benign while 30.1% are malignant. Brain tumors have the highest per-patient initial cost of care for any type of cancer. Annual average costs of care, in the United States in 2010, were more than $100,000 per patient. Brain cancer also had the highest annual mean net costs for last-year-of-life care, relative to other cancers, at $135,000 to $210,000, based on age and gender, per patient. Cancer care can result in significant financial hardships on patients and their families. Brain cancer is exceptionally expensive to treat because of the many medical interventions requires as part of care. Patients with malignant brain tumors often accrue health care costs that are 20 times greater than demographically matched control subjects without cancer.

Years of potential life lost (YPLL) measures the average time a patient would have lived had he or she not died prematurely. This supplements measurements such as overall survival rates and mortality, by measuring how much a patient's life is likely to be shortened by the disease. Compared to other cancers, malignant brain and spinal tumors have the greatest mean YPLL, reflecting the short survival time after diagnosis. Malignant brain tumors have the highest mean YPLL in both genders, averaging about 20 years. The mean YPLL for other common adult cancers range between 14 and 18 years. Benign brain tumors have a mean YPLL of 14.78 in men, which is more than lung cancer (14.47) and prostate cancer (9.64).

Point to remember

The Central Brain Tumor Registry of the United States (CBTRUS) is a not-for-profit corporation committed to gathering and disseminating current epidemiologic data on all primary benign and malignant brain and other CNS tumors, for the purpose of accurately describing incidence and survival patterns. It also strives to evaluate diagnosis and treatment, facilitate etiologic studies, establish awareness of the disease, reduce deaths and costs, and ultimately, achieve prevention of all CNS tumors.

Spinal cord tumors, or *intradural tumors*, form in the spinal cord itself, or in the dura of the cord. A tumor affecting the vertebrae of the spine is called a *vertebral tumor*. Spinal cord tumors may be *intramedullary*, beginning in the cells of the spinal cord itself, such as gliomas, astrocytomas, or ependymomas. They also may be *extramedullary*, growing in the membrane surrounding the cord, or the nerve roots reaching out from it. These types of tumors can affect spinal cord function because they cause spinal cord compression and other abnormalities. Examples of extramedullary tumors include meningiomas, neurofibromas, schwannomas, and nerve sheath tumors. Also, tumors from other parts of the body can metastasize to the vertebrae, the supporting network around the spinal cord, or rarely, the spinal cord itself. Spinal cord tumors can cause pain, neurological problems, paralysis, permanent disability, and even death. Treatments include surgery, radiation therapy, chemotherapy, and other medications.

The cost burden of spinal cord tumors, like brain tumors, includes copayments and coinsurance of 10% − 50%, or more, which usually reaches the yearly out-of-pocket maximum amount. Fortunately, treatment for these tumors is usually covered by health insurance. If a patient is not covered by health insurance, treatment costs can range from approximately $50,000 in the United States, for a small benign tumor that is easy to access during surgery, to over $700,000 if the tumor is malignant and requires surgery, radiation therapy, and chemotherapy. Surgical procedures alone can cost $50,000 or more. Stereotactic radiosurgery ranges between $12,000 and $55,000. Additional treatments can easily reach $100,000. According to the *Brain Tumor Foundation*, costs of treating glioblastomas are estimated to be over $450,000, and some tumors cost up to $700,000 over a patient's lifetime. Additional costs are incurred when a patient is recovering from surgery and requires physical, speech, or occupational therapy.

> **Point to remember**
>
> Costs of brain and spinal tumors across the world generally range between (in United States currency) $50,000 and $700,000. Cost of chemotherapy alone varies widely, based on the drug used, how it is administered, and the number of treatments needed. Costs for surgery also are varied, with countries such as India offering lower-cost surgeries for CNS tumors compared to the costs in countries such as the United States, United Kingdom, and Canada. The average brain tumor surgery cost in India is only $8,000 to $9,000, and success rates are amongst the highest in the world.

Mortality from brain and spinal tumors in the United States

Brain tumors account for over 2% of all adult cancer deaths, and incidence is slightly higher in males. These tumors have increased over time in the United States. Overall, about 6 of every 100,000 adults die from a brain tumor, and about 1 of every 100,000 children do. In general, Caucasians have a higher rate of brain cancer mortality than other ethnic groups. According to the *Central Brain Tumor Registry of the United States (CBTRUS)*, the average annual mortality rate in the United States, between 2011 and 2015, was 4.37 per 100,000 people, with 77,375 deaths attribute to primary malignant brain and spinal tumors. In 2018, there were approximately 16,830 deaths from these tumors, of which 9490 were males and 7340 were females.

From birth, a person in the United States has a 0.62% chance of being diagnosed with a primary malignant CNS tumor, and a 0.47% chance of dying from it. For males, the risk of developing a CNS tumor is 0.70%, and the risk of dying from it is 0.53%. For females, the risk of developing a CNS tumor is 0.54%, and the risk of dying from it is 0.41%. Based on studies between 2000 and 2015, the 5-year relative survival rate in the United States following diagnosis of a primary CNS tumor is 33.8% for males and 36.4% for females. For individual age groups, statistics about CNS tumors vary widely (see Fig 4.10).

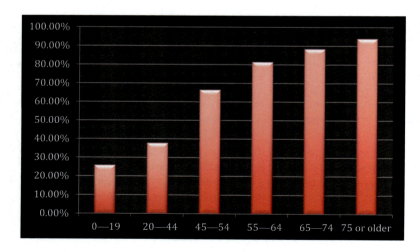

FIGURE 4.10 Percentage of fatalities from central nervous system tumors by age group.

TABLE 4.2 Top 20 most common types of cancer, globally, 2018.

Rank	Type	New cases, 2018	Percentage of all cancers, excluding nonmelanoma skin cancer
1	Lung	2,093,876	12.3
2	Breast	2,088,849	12.3
3	Colorectal	1,800,977	10.6
4	Prostate	1,276,106	7.5
5	Stomach	1,033,701	6.1
6	Liver	841,080	5.0
7	Esophagus	572,034	3.4
8	Cervical	569,847	3.3
9	Thyroid	567,233	3.3
10	Bladder	549,393	3.2
11	Non-Hodgkin lymphoma	509,590	3.0
12	Pancreas	458,918	2.7
13	Leukemia	437,033	2.6
14	Kidney	403,262	2.4
15	Uterus	382,069	2.2
16	Lip/mouth	354,864	2.1
17	Brain/spine	296,851	1.7
18	Ovaries	295,414	1.7
19	Skin (melanoma)	287,723	1.7
20	Gallbladder	219,420	1.3

Comparison of brain and spinal tumors with other cancers

According to the National Brain Tumor Society, there are over 120 different types of brain and spinal tumors that have been documented. The standard classification used today by most medical facilities comes from the World Health Organization (WHO). Globally, there are more than 260,000 new cases of CNS tumors each year, making brain and spinal tumors the 22nd most common type of cancer worldwide. These tumors make up 1.8% of the total number of new cancers, yet they are ranked number 12 in mortality compared to other types of cancer. According to the *World Cancer Research Fund* and the *American Institute for Cancer Research*, in the United States, brain and spinal tumors rank number 17 out of all other types of cancer. The total global number of diagnosed cases of cancer, of all forms, was 17,036,901 in 2018. Table 4.2 summarizes the top 20 most common types of cancer, as of 2018.

However, in males, brain and spinal tumors rank at number 13 of all cancer types, with 162,534 cases in 2018, making up 1.8% of all male cancer cases. In females, brain and spinal tumors rank at number 16 of all cancer types, with 134,317 cases in 2018, making up 1.6% of all female cancer types.

Globally, according to the *International Agency for Research on Cancer* and the *World Health Organization*, brain and spinal tumors rank at number 17 compared to all types of cancer, but ranks at number 12 out of all cancer deaths. Table 4.3 summarizes the incidence of brain and spinal tumors throughout various regions of the world.

To summarize further, the highest cumulative risk for developing brain and spinal tumors is in Southern Europe (0.62%), and the lowest is in Western Africa (0.11%). The highest cumulative risk for death from brain and spinal tumors is in both Southern Europe, and in Australia/New Zealand, with both areas at 0.46%. The lowest cumulative risk for death from brain and spinal tumors is in Eastern Africa, Middle Africa, Western Africa, and Melanesia (all at

TABLE 4.3 Global brain and spinal tumor incidence, 2018.

Location	New cases	Cumulative risk (%)	Deaths	Cumulative risk (%)
Eastern Africa	3302	0.12	2843	0.11
Middle Africa	1240	0.12	1108	0.11
Northern Africa	8177	0.38	6787	0.33
Southern Africa	948	0.16	747	0.14
Western Africa	3289	0.11	2740	0.11
Caribbean	1714	0.34	1293	0.26
Central America	4909	0.26	3797	0.22
South America	22,916	0.48	17,222	0.37
North America	27,062	0.52	19,973	0.36
Eastern Asia	86,407	0.39	69,690	0.31
South-Eastern Asia	15,264	0.22	13,075	0.20
South-Central Asia	42,828	0.23	36,815	0.20
Western Asia	11,718	0.49	9903	0.45
Central and Eastern Europe	21,586	0.52	19,204	0.45
Western Europe	18,066	0.58	14,149	0.43
Southern Europe	15,924	0.62	12,604	0.46
Northern Europe	9063	0.57	7070	0.42
Australia and New Zealand	2339	0.56	1929	0.46
Melanesia	92	0.12	77	0.11
Polynesia	7	0.18	11	0.27

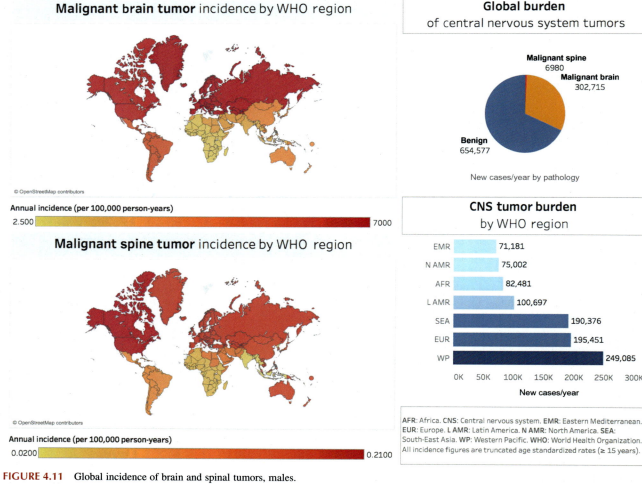

FIGURE 4.11 Global incidence of brain and spinal tumors, males.

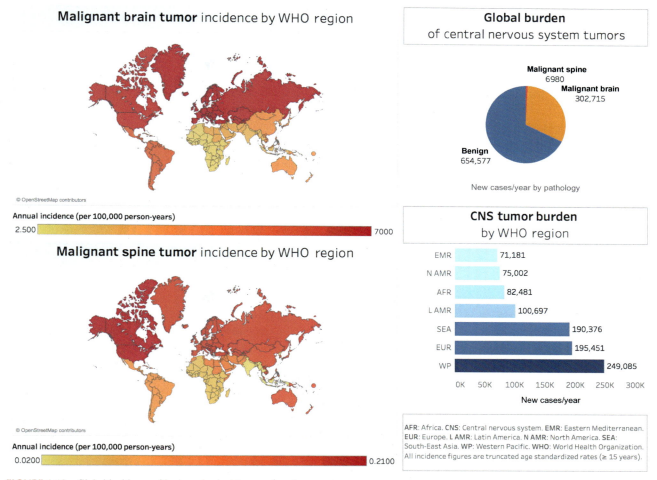

FIGURE 4.12 Global incidence of brain and spinal tumors, females.

0.11%). Overall, Asia is the continent with the highest incidence of development for these tumors (52.6%) as for fatalities from them (53.7%). Asia also has the largest percentage (50.2%) of 5-year prevalence, for both sexes. Fig. 4.11 shows the global incidence of brain and spinal tumors for males, and Fig 4.12 shows the same for females.

CNS tumors are much less common than other types of cancer. For men, the top three most common cancers are those of the prostate, lung/bronchus, and colon/rectum. For women, the top three are cancers of the breast, lung/bronchus, and colon/rectum. For both sexes, lung/bronchus cancer causes the most deaths. Fig. 4.13 illustrates the estimated new cases and estimated deaths from various cancers for both sexes.

> **Point to remember**
>
> The global average for surviving CNS tumors is better than the survival rates for pancreatic, liver, lung, esophageal, and stomach cancers. However, CNS tumors are more likely to cause death than many other cancers, including those of the ovarian cancer, leukemia, and tumors of the larynx, mouth, colon, bones, rectum, cervix, kidneys, bladder, uterus, breast, skin, thyroid, and prostate.

Estimated new cases

	Males				Females		
Prostate	164,690	19%		Breast	266,120	30%
Lung & bronchus	121,680	14%		Lung & bronchus	112,350	13%
Colon & rectum	75,610	9%		Colon & rectum	64,640	7%
Urinary bladder	62,380	7%		Uterine corpus	63,230	7%
Melanoma of the skin	55,150	6%		Thyroid	40,900	5%
Kidney & renal pelvis	42,680	5%		Melanoma of the skin	36,120	4%
Non-Hodgkin lymphoma	41,730	5%		Non-Hodgkin lymphoma	32,950	4%
Oral cavity & pharynx	37,160	4%		Pancreas	26,240	3%
Leukemia	35,030	4%		Leukemia	25,270	3%
Liver & intrahepatic bile duct	30,610	4%		Kidney & renal pelvis	22,660	3%
All sites	856,370	100%		All sites	878,980	100%

Estimated deaths

	Males				Females		
Lung & bronchus	83,550	26%		Lung & bronchus	70,500	25%
Prostate	29,430	9%		Breast	40,920	14%
Colon & rectum	27,390	8%		Colon & rectum	23,240	8%
Pancreas	23,020	7%		Pancreas	21,310	7%
Liver & intrahepatic bile duct	20,540	6%		Ovary	14,070	5%
Leukemia	14,270	4%		Uterine corpus	11,350	4%
Esophagus	12,850	4%		Leukemia	10,100	4%
Urinary bladder	12,520	4%		Liver & intrahepatic bile duct	9,660	3%
Non-Hodgkin lymphoma	11,510	4%		Non-Hodgkin lymphoma	8,400	3%
Kidney & renal pelvis	10,010	3%		Brain & other nervous system	7,340	3%
All sites	323,630	100%		All sites	286,010	100%

FIGURE 4.13 Estimated new cases of cancer and estimated deaths, 2018.

Key terms

Acute lymphoblastic leukemia
Acute myelogenous leukemia
Adenocarcinomas
Age-specific death rates
Aneurysmal bone cyst
Blood-brain barrier
Burkitt lymphoma
Chondrosarcoma
Chronic lymphocytic leukemia
Chronic myelogenous leukemia
Disability adjusted life years
Enostoses
Eosinophilic granuloma
Ewing's sarcoma
Giant cell tumor
Hairy cell leukemia
Hemangioma
Leptomeningeal
Life expectancy
Life table
Melanoma
Multiple myeloma
Neurofibroma
Non-Hodgkin lymphoma
Osteoblastoma
Osteochondroma
Osteoid osteoma
Osteosarcoma
Schwannoma
Small-cell lung cancers
Teratoid rhabdoid tumors
Trabecular bone
Years of life lived with a disability
Years of life lost
Years of potential life lost

Further reading

Abla, O., & Attarbaschi, A. (2019). *Non-Hodgkin's lymphoma in childhood and adolescence*. Springer.
Adami, H. O., Hunter, D. J., Lagiou, P., & Mucci, L. (2018). *Textbook of cancer epidemiology* (3rd Edition). Oxford University Press.

Ahmad, A. (2016). *Introduction to cancer metastasis*. Academic Press.
Ames, C. P., Boriani, S., & Jandial, R. (2013). *Spine and spinal cord tumors: advanced management and operative techniques*. CRC Press.
Arnautovic, K. I., & Gokaslan, Z. L. (2019). *Spinal cord tumors*. Springer.
Aschengrau, A., & Seage, G. R. (2013). *Essentials of epidemiology in public health* (3rd Edition). Jones & Bartlett Learning.
DeMonte, F., Gilbert, M. R., Mahajan, A., McCutcheon, I. E., Sawaya, R., & Yung, W. K. A. (2007). *Tumors of the brain and spine (MD Anderson Cancer Care Series)*. Springer.
Diggle, P. J., & Giorgi, E. (2019). *Model-based geostatistics for global public health: methods and applications*. Chapman and Hall/CRC Press.
Emadi, A., & Karp, J. E. (2017). *Acute leukemia: An illustrated guide to diagnosis and treatment*. Demos Medical.
Ferner, R. E., Huson, S., & Evans, D. G. R. (2011). *Neurofibromatoses in clinical practice*. Springer.
Gajjar, A., Reaman, G. H., Racadio, J. M., & Smith, F. O. (2018). *Brain tumors in children*. Springer.
Gerstman, B. B. (2013). *Epidemiology kept simple: An introduction to traditional and modern epidemiology* (3rd Edition). Wiley-Blackwell.
Gokaslan, Z. L., Boriani, S., Fisher, C. G., & Gomes Vialle, L. R. (2014). *AOSpine masters series volume 1: Metastatic spinal tumors*. Thieme.
Harris, R. E. (2015). *Global epidemiology of cancer*. Jones & Bartlett Learning.
Harris, R. E. (2019). *Epidemiology of chronic disease: global perspectives* (2nd Edition). Jones & Bartlett Learning.
Hayat, M. A. (2014). *Tumors of the central nervous system, volume 13: Types of tumors, diagnosis, ultrasonography, surgery, brain metastasis, and general CNS diseases*. Springer.
Hayat, M. A. (2015). *Brain metastases from primary tumors, volume 2: Epidemiology, biology, and therapy*. Academic Press.
Mahajan, A., & Paulino, A. (2018). *Radiation oncology for pediatric CNS tumors*. Springer.
Medical Ventures Press. (2011). *Brain and spinal cord tumors — neuroectodermal, medulloblastoma, glioma, astrocytoma, craniopharyngioma, CNS tumors, others: pediatric cancer guide*. Progressive Management.
National Comprehensive Cancer Network. (2017). *NCCN guidelines for patients: Diffuse large B-cell lymphoma*. National Comprehensive Cancer Network.
Newton, H. B. (2018). *Handbook of brain tumor chemotherapy, molecular therapeutics, and immunotherapy* (2nd Edition). Academic Press.
Norden, A. D., Reardon, D. A., & Wen, P. Y. C. (2011). *Primary central nervous system tumors: Pathogenesis and therapy (current clinical oncology)*. Humana Press.
Robertson, E. S. (2013). *Burkitt's lymphoma (current cancer research)*. Springer.
Rothman, K. J. (2012). *Epidemiology: An introduction* (2nd Edition). Oxford University Press.
Rothman, K. J., Lash, T. L., & Greenland, S. (2012). *Modern epidemiology* (3rd Edition). LWW.
Sahgal, A., Lo, S. S., Ma, L., & Sheehan, J. P. (2016). *Image-guided hypofractionated stereotactic radiosurgery: A practical approach to guide treatment of brain and spine tumors*. CRC Press.
Schiff, D., & Van den Bent, M. J. (2018). Metastatic disease of the nervous system, volume 149 *(Handbook of clinical neurology)*. Elsevier.
Sciubba, D. M. (2019). *Spinal tumor surgery: A case-based approach*. Springer.
Sik Kang, H., Woo Lee, J., & Lee, E. (2017). *Oncologic imaging: spine and spinal cord tumors*. Springer.
Skolnik, R. (2015). *Global health 101, third edition (Essential public health)*. Jones & Bartlett Learning.
Thun, M., Linet, M. S., Cerhan, J. R., Haiman, C. A., & Schottenfeld, D. (2017). *Cancer epidemiology and prevention* (4th Edition). Oxford University Press.
Tonn, J. C., Reardon, D. A., Rutka, J. T., & Westphal, M. (2019). *Oncology of CNS tumors* (3rd Edition). Springer.
Vimal Kumar, V., Hemdal, G., & Murgod, S. (2019). *Expression of CD34 & CD68 in peripheral & giant cell granuloma: Assessment of CD34 and CD68 in oral central and peripheral giant cell granuloma: An immunohistochemical study*. Lap Lambert Academic Publishing.
Warmuth-Metz, M. (2017). *Imaging and diagnosis in pediatric brain tumor studies*. Springer.
Warwick-Booth, L., & Cross, R. (2018). *Global health studies: A social determinants perspective*. Polity.

Chapter 5

Prevalence and incidence in epidemiology

Chapter Outline

- Prevalence — 103
- Incidence — 104
- Incidence rate — 104
- Cumulative incidence — 105
- Measures of disease frequency — 106
- Relationship between prevalence and incidence — 106
- Prevalence of brain and spinal tumors — 106
- Key terms — 108
- Further reading — 108

Prevalence

The **prevalence** of disease describes the proportion of the total population that is affected at one point or during a specific period of time. This can be mathematically expressed, as follows:

$$\frac{\text{The number of existing cases of the disease}}{\text{The number of the total population}}$$

The actual amounts are divided by the number of people in the specific population in order to calculate the percentages. It is also appropriate to use the *confidence interval formula* for cumulative incidence when measuring prevalence. If, for example, the prevalence of autism was found to be 25 of every 1000 children in a city such as Orlando, Florida, it is possible to calculate the approximate 95% confidence interval for autism prevalence:

$$\text{Confidence interval (upper and lower boundaries)} = 25/1000 \pm 1.96$$
$$\text{Multiplied by the square root of: } \frac{25/1000(1 - 25/1000)}{1000}$$

Therefore, the 95% confidence interval is between 15.3 per 1000 and 34.7 per 1000. This reveals that the best estimate for the actual, true prevalence of autism among children in Orlando is 25 per 1000. However, we are 95% confident that the true prevalence is between 15.3 and 34.7 per 1000.

Point prevalence refers to all cases or deaths from a disease, at a specific point in time that is related to the specific measured population. This is calculated similarly. A proportion is formed by dividing the number of cases in a population by the size of the same population, then multiplied by a specific value. When 100 is the value used in the multiplication, a percentage is created. **Period prevalence** shows the total number of disease cases during a specific period of time. **Lifetime prevalence** refers to disease cases that are diagnosed at any time during a person's life. Overall, prevalence describes the scope and distribution of disease in certain populations. The **burden of disease** is the amount of scope of disease in the population. Prevalence measures variation in disease occurrences. It helps create hypotheses that can be used for analytic studies. However, prevalence obscures causal relationships for disease since it combines incidence with survival. Even so, prevalence helps to estimate needs of medical facilities, and the allocation of resources for treating people who currently have a specific disease. Also, studying birth defects and chronic arthritis must use prevalence, since these conditions' beginnings are hard to pinpoint and interpret.

> **Point to remember**
> In epidemiology, *prevalence* is the proportion of a specific population that is affected by a medical condition. It is calculated by comparing the number of people with the condition against the total number of people studied. It is usually expressed as a fraction, a percentage, or as the number of cases per 1000, per 10,000, or per 100,000 people.

Incidence

Incidence describes the occurrences of new cases of a disease in a specific population over a certain time period. It encompasses three primary ideas:

- Measurement of *new disease events*—regarding diseases that occur more than one time, incidence usually measures the first occurrence.
- New cases in a **candidate population**—this is an "at-risk" population, involving individuals who have the body organ that is affected by the disease, is not immune to the disease, and other factors. For example, a woman with an intact uterus can potentially get uterine cancer. A man with an intact prostate gland can potentially get prostate cancer. A child who has not been fully immunized may contract diseases such as measles. While possible to measure and define incidence in non-at-risk populations, this is not a realistically revealing measurement.
- The *specific length of time* of follow-up—this means how long the population's members are followed until they develop the specific disease. Time must pass for this to occur and be observed, since incidence measures the transition from good health to a diseased state.

Incidence is most useful for the evaluation of effectiveness of programs of disease prevention. Researchers studying the causes of disease usually study new cases instead of existing ones. This is because they are most interested in exposures that lead to development of the disease. Also, many researches prefer the use of incidence since timing of exposures related to disease occurrence can be more accurately determined.

> **Point to remember**
> In epidemiology, *incidence* is the measure of the probability of occurrence of a specific medical condition in a population, within a certain period of time. It may be expressed simply as the number of new cases during the time period, but is best expressed as a proportion or a rate with a denominator.

Incidence rate

The occurrence of new cases of disease that arising during at-risk person-time of observation is called the *incidence rate*. This can also be mathematically expressed:

$$\frac{\text{The number of new cases of disease}}{\text{The person-time of observation}}$$

When the denominator is the sum of the person-time of the "at-risk" population, it is also known as the **incidence density rate** or the **person-time incidence rate**. Using the *Six Cities Study*, conducted by the Harvard School of Public Health, we can illustrate an approximate 95% confidence interval for measuring disease frequency. Incidence rate of death was 291 per 17,914 person-years, which is equivalent to 16.24/1000 person-years. The data comes from the city known as Steubenville, Ohio, and the study focused on deaths linked to air pollution. The exact confidence intervals for incidence rates are calculated by using the **Poisson distribution**. For large sample sizes, a normal distribution approximation formula can be used to quantity random errors that may occur, as follows:

Confidence interval (upper and lower boundaries) = Number of actual cases (A)/Time at risk (R) ± 1.96
Multiplied by the square root of: A/R^2

This allows for calculation of the 95% confidence interval for the incidence rate of mortality in relation to air pollution in Steubenville. Here is the formula:

Confidence interval (upper and lower boundaries) = $291/17,914 \pm 1.96$
Multiple by the square root of: $291/17,914^2$

Therefore, we are 95% confident that the true mortality rate for adults in Steubenville, in relation to air pollution, is in the range of 14.37 − 18.11 per 1000 person-years.

Additionally, it should be understood that the numerator for incidence rate is the same as the **cumulative incidence**, which will be explained later in this chapter. The difference between the two measures, however, is found in the denominator. The incidence rate's denominator focuses on time (t), and therefore, it is a true rate. Therefore, its dimension is $1/t$ or t^{-1}, and its values may range from zero to infinity. An incidence rate of infinity would be possible if every member of a population died at the same time.

The idea of **person-time** is complicated. It is calculated only among the people affected by a disease. A person contributes time to the denominator of an incidence rate only until he or she is diagnosed with the specific disease being measured. Unlike cumulative incidence, the incidence rate is not calculated from the idea that everyone in the "diseased" population has been monitored for a certain time period. Person-time is calculated only while the individual is being followed. The calculation stops when the person dies or is lost to follow-up, such as when he or she moves to another location and can no longer be monitored. The incidence rate can be calculated for a fixed population or a dynamic population. However, since it directly incorporates population changes such as births, deaths, and migrations, it is highly useful in measuring transitions between health and disease, in dynamic populations.

A specific time unit used to measure person-time can be varied, but is based on how long it takes for a disease to develop. Person-years are often used for diseases, such as cancer, that require many years to develop. Person-months and person-days are used for diseases, such as infections, that develop more quickly. The number of person-time units used in the denominator is arbitrary. To understand further, the same incidence rate can be expressed in terms of one person-year, 10 person-years, or 100-person years. Epidemiologists usually use 100,000 person-years for rare diseases, as well as for those that take extremely long times to develop.

> **Point to remember**
>
> An *incidence rate* can be understood as the risk of developing a certain disease during a specific period of time. The numerator of the rate is the number of new cases during the time period, and the denominator is the population at risk during that period.

Cumulative incidence

Cumulative incidence is the proportion of a specific population that develops a disease over a certain period of time. It is also sometimes referred to as *incidence proportion*. This can be mathematically expressed as:

$$\frac{\text{Number of new cases of disease}}{\text{Number in the specific population}}$$

It is calculated over a specific time period. The numerator (new cases) is a subgroup of the denominator (specific population). This means that the possible value of cumulative incidence ranges from 0 to 1. If expressed as a percentage, it ranges from 0% to 100%. Time is not an essential part of this proportion. Instead, it is expressed by words accompanying the numbers of the measurement of cumulative incidence. The differences between cumulative incidence, incidence rate, and prevalence are further summarized in Table 5.1.

TABLE 5.1 Cumulative incidence−incidence rate−prevalence comparison.

Factor	Number type	Units	Range	Numerator	Denominator	Uses
Cumulative incidence	Proportion	None	0 to 1	New cases	At-risk population	Research about disease causes, prevention and treatment
Incidence rate	True rate	1/time, or t^{-1}	0 to infinity	New cases	At-risk person-time	Research about disease causes, prevention and treatment
Prevalence	Proportion	None	0 to 1	Existing cases	Total population	Planning for resources

Cumulative incidence explains the average risk of developing a disease over a certain time period. A "risk" is the probability of developing the disease. A good example that is often referred to is a woman's "lifetime risk of developing breast cancer." This is currently estimated to be one of every eight women in the United States. It means that about 12% of women in this country will develop breast cancer sometime in their lives. Cumulative incidence is affected by various lengths of time. Usually, cumulative incidence over more years is higher than over a few years.

Cumulative incidence is primarily used for fixed populations when there are no losses, or only small losses, to follow-up. An example would be to assess long-term effects upon a population from a nuclear accident. Though the initial effects of such an accident could be severe to a large amount of people, the long-term residual, cumulative effects would be even larger. It is important to select an appropriate time period in order to effectively measure cumulative incidence. Another example of the use of cumulative incidence is when a new drug is developed. When longer periods of time after taking a new drug are followed, and the drug's effectiveness over time is assessed, this can be compared with an existing drug and its effectiveness. It is important that everyone in the study population is followed for the same length of time, regardless of how short or long the time period is.

> **Point to remember**
> In epidemiology, *cumulative incidence* is a measure of disease frequency during a period of time. When the period of time considered is an entire lifespan, it is called *lifetime risk*. Cumulative incidence is defined as the probability that a particular disease has occurred before a certain time. It is equivalent to the incidence, calculated by using a period of time in which all individuals in the population are considered to be at risk for the disease.

Measures of disease frequency

The two primary ways to measure disease frequency are incidence (occurrence of new disease) and prevalence (existence of current disease). Both ways describe parts of natural disease courses. Incidence focuses on the transition from health to disease. Prevalence deals with the period of time during which a person lives with a specific disease.

Relationship between prevalence and incidence

Prevalence is interrelated with incidence. A disease's prevalence is proportional to the incidence of the disease, which is multiplied by the duration of the disease. As a result, when incidence increases, prevalence also increases. The duration of a disease begins at diagnosis and ends when the individual is cured, or dies from the disease. Therefore, the mathematical formula used for this calculation is as follows:

$$\text{Prevalence}/(1 - \text{Prevalence}) = \text{Incidence Rate} \times \text{Average Duration}$$

In the equation, "prevalence" actually means the proportion of the total population with the disease, and "1−Prevalence" means the proportion of the total population without the disease. The "average duration" is the length of time that an individual has the disease. The formula assumes that the population is "steady," with equal inflow and outflow of people, and that the incidence rate and duration do not change over time. If the disease's frequency is rare (less than 10%), the equation can be simplified to:

$$\text{Prevalence} = \text{Incidence Rate} \times \text{Average Duration}$$

The number of people who currently have the disease is influenced by the rate at which new cases develop, and by the rate at which other people are cured, or die from the disease.

Prevalence of brain and spinal tumors

According to various sources, which include the American Brain Tumor Association, the Centers for Disease Control and Prevention, and National Institute of Neurological Disorders and Stroke, there is wide variance in the prevalence of different brain and spinal tumors. It should be noted that about 90% of diagnosed spinal tumors are metastatic and do not originate in the spine. Brain and spinal tumors are ranked from highest to lowest prevalence as follows:

- Meningiomas—about 36% of all primary brain tumors, and 25% of spinal cord tumors;
- Pituitary tumors—about 17% of all primary brain tumors;
- Glioblastomas—about 15% of all primary brain tumors;

- Vertebral hemangiomas—most are found incidentally at autopsy; though benign, they are present in about 10% of the global population; also, *cavernous hemangiomas* of the brain are present in about 0.5% of the global population;
- Acoustic neuromas—8% of all primary brain tumors;
- Schwannomas—8% of all primary brain tumors, but 29% of spinal nerve root tumors;
- Astrocytomas—wide-ranging prevalence of less than 1% up to 15% (average of 8%) based on grade: *grade I* makes up only 2% of recorded astrocytomas; *grade II* makes up 8%; *grade III* makes up 20%; however, the highest graded astrocytoma is called *glioblastoma*;
- Oligodendrogliomas—4% of all primary brain tumors;
- Spinal osteosarcomas—3.6% to 14.5% of primary spinal tumors;
- Craniopharyngiomas—2% − 5% (average of 3.5%) of all primary brain tumors; 5% − 10% of all childhood brain tumors;
- Ependymomas—2% − 3% (average of 2.5%) of all primary brain tumors; however, they are the sixth most common brain tumor in children, with about 30% diagnosed in children younger than 3 years of age;
- Lymphomas—2% − 3% (average of 2.5%) of all primary brain tumors;
- Hemangioblastomas—about 2% of all primary brain tumors, with about 10% of patients also having *Von Hippel-Lindau disease*;
- Medulloblastomas—less than 2% of all primary brain tumors; they make up 18% of all pediatric brain tumors, with over 70% diagnosed in children between 1 and 10 years of age;
- Oligoastrocytomas—1% of all primary brain tumors;
- Pineal parenchymal tumors—less than 1% of all primary brain tumors; however, 3% and 8% of childhood brain tumors occurs in this area;
- Germ cell tumors—classified as rare; 3% − 5% of all childhood brain tumors;
- Atypical teratoid rhabdoid tumors—classified as rare; about 3% of all childhood brain tumors;
- Teratomas—though overall very rare, they make up 26% to 50% of all fetal brain tumors;
- Choroid plexus tumors—classified as rare; 2% − 4% of all tumors in children under age 15 years; but 10%−20% of all brain tumors in infants less than 12 months of age;
- Mixed germ cell tumors—classified as rare; less than 1% of all childhood brain tumors;
- Germinomas—classified as rare; less than 1% of all childhood brain tumors;
- Embryonal tumors—classified as rare; less than 1% of all childhood brain tumors;
- Yolk sac tumors (also called endodermal sinus tumors)—classified as rare; less than 1% of all childhood brain tumors;
- Choriocarcinomas—classified as rare; less than 1% of all childhood brain tumors;
- Peripheral nerve sheath tumors—classified as rare; less than 1% of all childhood brain tumors;
- Multiple myeloma of the spine—classified as rare; less than 1% of all spinal tumors;
- Neurofibromas—classified as rare; less than 1% of all brain or spinal tumors;
- Cranial nerve tumors—classified as rare; less than 1% of all brain tumors;
- Paraspinal nerve tumors—classified as rare; less than 1% of all spinal tumors.

The average prevalence of primary brain tumors is illustrated in Fig. 5.1

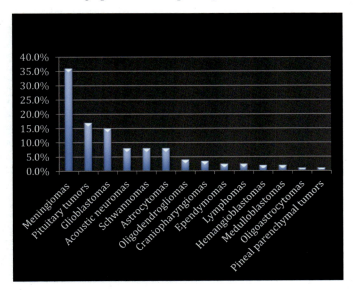

FIGURE 5.1 Average prevalence of primary brain tumors.

Key terms

Burden of disease
Candidate population
Cumulative incidence
Incidence
Incidence density rate
Incidence rate
Lifetime prevalence
Period prevalence
Person-time
Point prevalence
Poisson distribution
Prevalence
Von Hippel-Lindau disease

Further reading

Celentano, D. D., & Szklo, M. (2018). *Gordis epidemiology* (6th Edition). Elsevier.
Cheyne, Sir W. W. (2016). *The defensive arrangements of the body as illustrated by the incidence of disease in children and adults*. Palala Press.
Elmore, J. G., Wild, D., Katz, D. L., & Nelson, H. D. (2019). *Jekel's epidemiology, biostatistics, preventive medicine, and public health* (5th Edition). Elsevier.
Fletcher, R. H., Fletcher, S. W., & Fletcher, G. S. (2012). *Clinical epidemiology: The essentials* (5th Edition). LWW.
Fos, P. J., Fine, D. J., & Zuniga, M. A. (2018). *Managerial epidemiology for health care organizations (Public health/epidemiology and biostatistics)* (3rd Edition). Jossey-Bass.
Friis, R. H. (2017). *Epidemiology 101 (Essential public health)* (2nd Edition). Jones & Bartlett Learning.
Friis, R. H., & Sellers, T. (2013). *Epidemiology for public health practice* (5th Edition). Jones & Bartlett Learning.
Hayat, M. A. (2013). *Tumors of the central nervous system, volume 10: Pineal, pituitary, and spinal tumors*. Springer.
Hayat, M. A. (2015). *Tumors of the central nervous system, volume 14: Glioma, meningioma, neuroblastoma, and spinal tumors*. Springer.
Hayat, M. A. (2016). *Brain metastases from primary tumors, volume 3: Epidemiology, biology, and therapy of melanoma and other cancers*. Academic Press.
Hebel, J. R., & McCarter, R. J. (2011). *Study guide to epidemiology and biostatistics*. Jones & Bartlett Learning.
Kaye, A. H., & Laws, E. R. (2011). *Brain tumors E-book: An encyclopedic approach* (3rd Edition). Saunders.
Krieger, N. (2013). *Epidemiology and the people's health: Theory and context*. Oxford University Press.
Lautenbach, E., Malani, P. N., Woeltje, K. F., Han, J. H., Shuman, E. K., & Marschall, J. (2018). *Practical healthcare epidemiology* (4th Edition). Cambridge University Press.
Merrill, R. M. (2016). *Introduction to epidemiology* (7th Edition). Jones & Bartlett Learning.
Motulsky, H. (2017). *Intuitive biostatistics: A nonmathematical guide to statistical thinking* (4th Edition). Oxford University Press.
Porta, M. (2014). *A dictionary of epidemiology* (6th Edition). Oxford University Press.
Ramirez, W. (2017). *Rare diseases: Prevalence, treatment options and research insights (New developments in medical research)*. Nova Science Publishers Inc.
Rasmussen, S. A., & Goodman, R. A. (2018). *The CDC field epidemiology manual*. Oxford University Press.
Remington, P. L., Brownson, R. C., & Wegner, M. V. (2016). *Chronic disease epidemiology, prevention, and control* (4th Edition). American Public Health Association.
Richa, M. D., Singh, G. P., & Mishra, C. P. (2012). *Burden of non communicable diseases: Prevalence, spectrum and correlates of risk factors of non communicable diseases in rural area of Varanasi district*. Lap Lambert Academic Publishing.
Rothman, K. J. (2012). *Epidemiology: An Introduction* (2nd Edition). Oxford University Press.
Rothman, K. J., Lash, T. L., & Greenland, S. (2012). *Modern epidemiology* (3rd Edition). LWW.
Schiff, D., & Van den Bent, M. J. (2018). *Metastatic disease of the nervous system, volume 149 (Handbook of clinical neurology)*. Elsevier.
Szklo, M., & Nieto, F. J. (2018). *Epidemiology: Beyond the basics* (4th Edition). Jones & Bartlett Learning.
Warmuth-Metz, M. (2017). *Imaging and diagnosis in pediatric brain tumor studies*. Springer.
Weaver, A., & Goldberg, S. (2012). *Clinical biostatistics and epidemiology made ridiculously simple*. Medmaster.
Westreich, D. (2019). *Epidemiology by design: A causal approach to the health sciences*. Oxford University Press.
Woodward, M. (2013). *Texts in Statistical Science Epidemiology: Study design and data analysis* (3rd Edition). Chapman and Hall/CRC.

Part III

Classifications of tumors

Chapter 6

Astrocytoma

Chapter Outline

Overview	111	Anaplastic astrocytoma	121
Astrocytoma classifications	111	Clinical cases	123
Pilocytic astrocytoma	112	Key terms	126
Subependymal giant cell astrocytoma	115	Further reading	126

Overview

Astrocytomas are the most common gliomas. They are primarily graded as I or II by the World Health Organization, though anaplastic astrocytomas are grade III. The differences between the grades are based on the rate of tumor growth—the higher the grade, the faster they grow. Pilocytic astrocytomas (grade I) usually do not spread and are primarily benign. Subependymal giant cell astrocytomas (also grade I) are usually linked to tuberous sclerosis. Low-grade astrocytomas (grade II) were previously known as *diffuse* or *fibrillary* astrocytomas, and while invading nearby tissues, usually grow slowly. Anaplastic astrocytoma (grade III) is rare but more aggressive, and glioblastoma (often evolving into grade IV) is highly aggressive. Generally, symptoms of astrocytomas include headaches due to increased intracranial pressure, seizures, memory loss, behavioral changes, head tilt, incoordination, neck stiffness, nausea, vomiting, irritability, and visual problems. Astrocytomas may be treated by combinations of surgery, radiation therapy, observation, follow-up imaging, chemotherapy, and ventricular shunting. Each tumor grade requires accurate treatment based on individual characteristics.

Astrocytoma classifications

Astrocytomas are tumors arising from astrocytes within the central nervous system. They may be benign, low grade or malignant, and are the most common type of **glioma**. Astrocytomas are graded, from I to IV, based on histologic qualities of their cells. *Low-grade* astrocytomas are most common in children, but *high-grade* astrocytomas are most common in adults. Low-grade astrocytomas are usually localized, with slow patterns of growth. High-grade astrocytomas can grow much more quickly and require radiation and chemotherapy intervention. Grading is based on the tumor's appearance under a microscope, immunohistochemistry, and genomic and metabolomic classifications. The types of astrocytomas are as follows:

- **Pilocytic astrocytoma**—this Grade I tumor is also known as juvenile pilocytic astrocytoma (JPA) and usually does not spread; this is the most benign type. It is important to understand that the term "benign" refers to the degree of cellular abnormality, and does not mean that there will be a complete and effective cure. Pilocyticastrocytoma (in addition to certain grade II astrocytomas) may be associated with **neurofibromatosis type 1 (NF-1)**. This type of tumor is distinguished from the other astrocytomas by their gross and microscopic appearance and mostly benign activities.
- **Subependymal giant cell astrocytoma (SEGA)**—a ventricular tumor associated with tuberous sclerosis that is benign, and also classified by the World Health Organization as Grade I. These almost always occur in younger patients who have **tuberous sclerosis** and can be incidentally found on imaging or become symptomatic because of obstructive hydrocephalus based on tumor location. *Tuberous sclerosis complex* (TSC) is a rare multisystem autosomal dominant genetic disease that causes noncancerous tumors in the brain, kidneys, heart, liver, eyes, lungs, and skin.

- **Low-grade astrocytoma** *(formerly termed "diffuse" or "fibrillary")*—these Grade II tumors usually invade surrounding tissue, but typically exhibit slow patterns of growth. Low-grade astrocytomas usually form in the cerebral hemispheres. Their lobar distribution is similar to the amount of white matter present in either lobe. The highest frequency occurs in the frontal lobes (24.5% of cases) and temporal lobes (19.6% of cases).
- **Anaplastic astrocytoma**—a rare type that requires more aggressive treatments; it is a Grade III tumor. This type, as well as glioblastoma, is sometimes referred to as a malignant or high-grade glioma.
- **Glioblastoma (formerly known as glioblastoma multiforme or GBM)**—if primary, the tumors are very aggressive, while secondary tumors start as lower-grade, and evolve into grade IV. Overall, these are the deadliest of all primary brain and CNS cancers. Grade III and IV astrocytomas are usually located in the frontal or temporal lobes statistically (in adults), or the cerebral hemispheres. Glioblastoma is discussed in detail in Chapter 7, Glioblastoma.

Pilocytic astrocytoma

Pilocytic astrocytomas are usually well circumscribed, though there may be some degree of microscopic infiltration. They are most common in the cerebellum, hypothalamus, third ventricle, optic nerve, spinal cord, and dorsal brain stem. However, the cerebrum is involved in some cases. Diffuse gliomas may also be present in these locations. Most adult cases are extremely similar those in children. Outcome is based on the location of the tumor, and many patients can be cured by surgical resection alone.

Epidemiology

The majority of glial tumors are astrocytomas. *Pilocytic astrocytoma* accounts for 0.6% − 5.1% of all intracranial neoplasms and constitutes 5% of all glioma diagnoses. It is the most common primary brain tumor in children, accounting for 70% − 85% of all cerebellar astrocytomas. About 75% of pilocytic astrocytomas occur in patients under 20 years of age.

Etiology and risk factors

The cause of astrocytomas is unknown, but they originate in the star-shaped cerebral cells called *astrocytes*. They develop either narrow or diffuse zones of infiltration. Suspected etiology of astrocytomas is based on a genome-wide pattern of DNA copy-number alterations. Most pilocytic astrocytomas have a unique *KIAA1549L-BRAF* fusion gene, mutations of which are involved in altered cell growth. The BRAF fusion gene encodes a protein called *B-Raf*. Aging is a general risk factor for astrocytomas. Risk factors for astrocytomas include neurofibromatosis, Li-Fraumeni syndrome, tuberous sclerosis, and *Turcot syndrome*, along with previous radiation therapy. Additional risk factors include smoking, exposure to pesticides, and working in petroleum refining or rubber manufacturing.

Pathology

Pilocytic astrocytomas may appear anywhere in the central nervous system, but are most common in the frontal lobes, cerebellum, and cerebral hemispheres (see Fig. 6.1). However, they rarely occur in the spinal cord, or manifest as drop metastases. They cause regional effects by compression, invasion, and destruction of brain parenchyma. These tumors

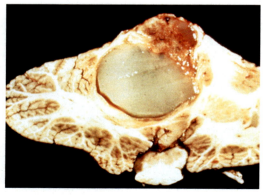

FIGURE 6.1 Pilocytic astrocytoma in the cerebellum with a nodule of tumor in a cyst.

can also cause arterial and venous hypoxia, competition for nutrients, release of metabolic end products such as free radicals, altered electrolytes, and neurotransmitters, plus the release and recruitment of **cytokines** and other cellular mediators that disrupt the normal parenchymal function. Secondary pathological developments of pilocytic and other astrocytomas may be due to increased intracranial pressure linked to direct mass effect/four ventricular obstruction, increased blood volume, or increased CSF volume. The pilocytic astrocytomas only rarely have **tumor protein 53 (TP53)** mutations or molecular signatures of the infiltrating astrocytomas. When NF-1 is concurrent, pilocytic astrocytomas have a functional loss of **neurofibromin**, not seen in sporadic forms. With pilocytic astrocytoma, there are two types of alterations in the BRAF signaling pathway. The first involves translocations that separate the kinase domain from the inhibitory domain. The second involves an activating point mutation (*valine* substituted for *glutamate* at *codon 600*, or *V600E*), which is also found in many other types of tumors. When pilocytic astrocytomas are solid in composition, they have significantly circumscribed appearances. Less often, they are infiltrative. The bipolar cells are GFAP-positive, forming dense fibrillary meshes. The tumors are often biphasic, with loose, microcystic and fibrillary portions. Their increased blood vessels often have thickened walls or vascular cell proliferation, but lack necrosis or quick mitotic activity. There is limited infiltration of surrounding brain tissues.

Cerebellar pilocytic astrocytomas tend to be hemispheric and usually made up of a large fluid-filled cyst, with an enhancing mural nodule. If located in the hypothalamus or optic nerve, the tumor is usually solid. In the optic nerve, they appear as a focal segmental nerve swelling. Unilateral and bilateral optic nerve tumors are highly common. Most tumors are indolent; they are distinct because of having a biphasic pattern and compact pilocytic areas mixed with loose, microcystic, or spongy areas. Histologically, the dense areas have hair-like or bipolar astrocytes and long, spindle-shaped processes, often with **Rosenthal fibers** that are masses of intracellular astrocytic filaments. They have a fusiform or corkscrew shape and a hyaline appearance. Berry-shaped eosinophilic granular bodies are also common (see Fig. 6.2). These fibers and granules are usually indicative of an indolent process and offer clues that differentiate pilocytic astrocytomas from diffuse astrocytomas. NF-1 gene inactivation is much more common in NF1-related pilocytic astrocytomas than in the sporadic type.

The *pilomyxoid* type of these tumors exhibits grade II aggressive behaviors, primarily affecting children under 3 years of age, usually affecting the hypothalamus. It does not have a biphasic appearance or Rosenthal fibers. Instead, it has **piloid** cells that often have an **angiocentric** arrangement within a myxoid matrix. There are usually no eosinophilic granular bodies, or they are only apparent in small numbers. Over years, tumor resection may reveal features of a classic pilocytic astrocytoma, which is believed to show a "maturation" process of the tumor.

Clinical manifestations

Early symptoms of astrocytomas include headaches, seizures, loss of memory, and behavioral changes. Posterior fossa signs include head tilt, incoordination, and neck stiffness. In children with pilocytic astrocytoma, symptoms may include lack of normal weight gain, weight loss, headache, nausea, vomiting, irritability, torticollis, difficulty coordinating movements, and visual problems such as nystagmus. Symptoms vary based on the location and size of the neoplasm. The most common symptoms are related to increased intracranial pressure from the size of the neoplasm.

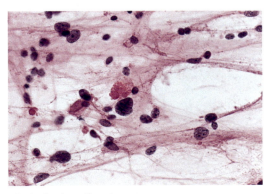

FIGURE 6.2 Berry-shaped eosinophilic granular bodies in a pilocytic astrocytoma.

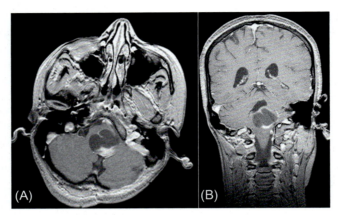

FIGURE 6.3 Magnetic resonance imaging of a pilocytic astrocytoma. Source: *https://commons.wikimedia.org/wiki/File:Pilocytic.jpg*.

Diagnosis

Diagnosis of most astrocytomas usually takes 3 – 6 months after onset of first clinical manifestations. This is because the patient often does not realize the cause of symptoms. On imaging, pilocytic astrocytomas are well circumscribed, usually cystic and gadolinium (contrast) enhancing, appearing close to the ventricle or subarachnoid space. They generally do not have surrounding edema, which is a more common feature of higher-grade astrocytomas. There is rarely any malignant transformation. Pilocytic astrocytomas are usually diagnosed via clinical, neurological, and ophthalmological examinations, plus CT scan or MRI. Contrasted imaging studies are very useful to identify various features of the tumor (see Fig. 6.3). The tumor may be biopsied by a neurosurgeon prior to maximum feasible surgical resection, but a characteristic approach will be to proceed with resection after consideration of whether a preemptive ventricular drain will be required. Microscopically, it appears to be composed of bipolar cells with long, hair-like processes, hence the term "pilocytic," which means "fiber-like." There is often the presence of Rosenthal fibers, eosinophilic granular bodies, and **microcysts**. Sometimes, **myxoid foci** and oligodendroglioma-like cells are present—these are nonspecific. Lesions that have been present for a long time may have hemosiderin-filled macrophages and calcifications.

Treatment

Treatment of astrocytomas includes surgery, radiation therapy, observation, follow-up scans, and chemotherapy. About 75% of patients receive surgical resection. For pilocytic astrocytomas, this is the preferred treatment, leading to a cure or long-term survival for most patients. Chemotherapy can be used before radiation therapy or for tumor progression. After surgery, pediatric patients may experience side effects that require steroid treatments to control tissue swelling. Generally, symptoms of the tumor subside gradually after surgery.

Prognosis

For pilocytic astrocytoma, the overall 15-year survival rate is 80%. For patients with completely resected tumors within the cerebellum, the 15-year survival rate is 95%. The 25-year survival rate is 50% – 94% after surgical resection. Malignant transformation is very rare. However, for the *pilomyxoid* subtype, prognosis is worse due to the more aggressive nature of the tumor.

> **Point to remember**
>
> A pilocytic astrocytoma is usually found in the cerebellum and is typically slow-growing. This benign tumor is fluid-filled (cystic) and not a solid mass. It is often successfully removed surgery and has an excellent prognosis. Most symptoms are related to increased pressure from ventricular compression and obstructive hydrocephalus in the brain, including headaches that are worse in the morning, nausea, vomiting, seizures, and changes in mood or personality. Other symptoms vary based on tumor location and size, such as clumsiness, weakness, vision changes, nystagmus, and in children, affects upon normal growth and stature, behavior, and hormonal levels.

Subependymal giant cell astrocytoma

SEGAs are grade I tumors usually associated with tuberous sclerosis. Patients that have these tumors, but not the features of tuberous sclerosis, have a *forme fruste* of the disorder. Since many types of **hamartomas** characterize tuberous sclerosis, SEGAs may also be considered hamartomas instead of neoplasms. This is linked to the benign activities of these tumors.

Epidemiology

Subependymal giant cell tumors affect 5% − 20% of patients with tuberous sclerosis and are only occasionally found in people over age 20.

Etiology and risk factors

SEGAs are related to alterations in the **hamartin** (TSC1) and **tuberin** (TSC2) genes. This involves abnormal signaling through the downstream growth regulating pathways. The primary risk factor for SEGAs is pediatric tuberous sclerosis. Other risk factors include neurofibromatosis, radiation, chemicals, oil refining exposures, and rubber manufacturing exposures.

Pathology

On imaging, SEGAs usually appear as an intraventricular mass close to the **foramen of Monro**, larger than 1 cm, with calcifications, a heterogenous MRI signal, and significant contrast enhancement. They are usually elongated, similar to a "sausage" in appearance, but may be lobulated. If tuberous sclerosis is linked, there will be smaller masses that look like wax drippings from a candle on the walls of the lateral ventricle. Obstruction of the foramen of Monro may lead to hydrocephalus. Grossly, the significant vascularity of this tumor causes cut surfaces to look like red beef. Calcification is nearly always present, and may cause the mass to appear as if it were made of stone. There is slight cellularity, with astrocytes that are closely packed, and a lot of cytoplasm. Tumors cells often form fascicles, or around blood vessels, similar to the **pseudorosettes** of ependymomas. Some cases show a morphology that is spindled, or similar to **gemistocytes**. Tumor cells are obviously from astrocytic origin. The cytoplasm is filled with **glial fibrillary acidic proteins** (GFAP). Some tumor cells look like neurons, with prominent nucleoli, astrocytoma-like cytoplasm, and neuronal-like nuclei.

Clinical manifestations

For SEGA, there may be no symptoms if CSF flow remains open. Once there is obstruction of CSF flow, symptoms may include nausea, vomiting, headache that is often positional, abnormal drowsiness, blurred or double vision, new or worsened seizures, and personality changes. Hydrocephalus emerges before the patient reaches his or her 30 s as the tumor in the wall of the lateral ventricles or within the interventricular foramen block CSF outflow. The pattern may be changed by long-standing epilepsy.

Diagnosis

SEGAs are diagnosed with contrast-enhanced MRI or CT scan (see Fig. 6.4). Children with tuberous sclerosis should be screened for this type of astrocytoma via neuroimaging everyone to 3 years. Neuronal differentiation of these tumors is suggested by positive immunohistochemical staining for neuronal markers. The cells may stain with neuronal and glial markers—or with neither of them. They may be mixed with large cells appearing similar to **gemistocytic astrocytes**, giant multinucleated pyramidal cells, and elongated tumor cells.

Treatment

For SEGAs, two drugs (rapamycin and everolimus) can shrink or stabilize the tumor. They are both *mechanistic target of rapamycin* (mTOR) immunosuppressants that should be used cautiously to avoid severe infection. When the tumor is growing or causing symptoms, or the patient has a high risk of seizures, it should be surgically removed. For patients with a rapidly growing tumor or symptoms of hydrocephalus, deferring surgery can lead to vision loss, the requirement for a ventricular shunt, and death. Total surgical removal of the tumor is curative. Potential complications of surgery include transient memory impairment, **hemiparesis**, infection, requirement for chronic **ventriculoperitoneal shunt** placement, stroke, and death. Rapamycin may induce tumor regression. Studies have revealed novel proteins that are believed to be mTOR effectors.

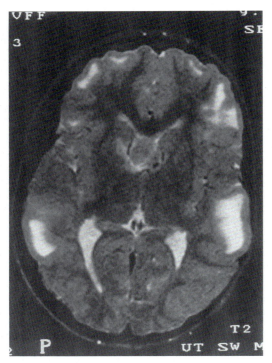

FIGURE 6.4 Noncontrast T2-weighted magnetic resonance imaging scan from a child with tuberous sclerosis demonstrates extensive high-signal cortical lesions typical of tuberous sclerosis.

Prognosis
For SEGA, after complete surgical removal, the tumor will not recur and there is no metastasis to other areas. However, the patient is at risk for new tumors arising from subependymal nodules in other areas of the ventricular system.

> **Point to remember**
> Subependymal giant cell astrocytomas ("SGCAs" or "SEGAs") are benign tumors that are asymptomatic/incidentally found, or become symptomatic because of obstructive hydrocephalus. Surgical removal is often curative. Upon imaging, these tumors classically appear as an intraventricular mass close to the foramen of Monro, and are larger than 1 cm in size. They show calcifications, a heterogenous MRI signal, and marked contrast enhancement.

Infiltrative low-grade astrocytomas
Infiltrative low-grade astrocytomas primarily occur in young adulthood, with symptoms appearing in the first decade of life. However, these astrocytomas also occur in the third and fourth decades. These tumors are slightly more common in males, and overall, make up less than 5% of all brain tumors.

Epidemiology
Low-grade astrocytomas are relatively uncommon in comparison to higher-grade astrocytomas. Only about 1500 cases are believed to occur annually in North America.

Etiology and risk factors
The cause of low-grade astrocytomas is unknown, but current research indicates that the environment does not appear to play any role in their origin. Families with neurofibromatosis are at increased risk of developing low-grade astrocytomas.

Pathology

In most cases, infiltrative low-grade astrocytomas form in the frontal or temporal lobes of the cerebral hemispheres. The cells are cytologically benign yet do infiltrate surrounding brain tissues. Usually, the *fibrillary* subtype is present, followed in prevalence by the *protoplasmic* and *gemistocytic* subtypes. The gemistocytic subtype often undergoes anaplastic progression after 4 – 5 years.

Clinical manifestations

Infiltrative low-grade astrocytomas are indicated by headache, nausea, vomiting, and lethargy. This type of astrocytomas is often signaled by seizures. Fatigue and depression are common in patients with low-grade astrocytomas, whether or not they are receiving radiation therapy. Continued tumor growth often causes cognitive deficits to develop.

Diagnosis

A careful patient history will often disclose symptoms that have been present for years or even decades. In MRI studies, there are poorly demarcated hypointense **T1-weighted lesions** and hyperintense **T2-weighted lesions**. They do not enhance after gadolinium has been administered. Asymptomatic, nonresectable, and diffuse astrocytomas are followed with quarterly MRI scans, which later can be performed every 6 months instead.

Treatment

Surgical resection is indicated for these astrocytomas, which will reveal the subtype of the tumor for correct diagnosis. The goal is to remove the tumor to the largest possible extend without risking losses of neurologic function. Symptoms will subside, including any related to mass effect, hydrocephalus, hemorrhage, cyst formation, or seizure activity. Surgical resection will reduce the amount of cells at risk for malignant degeneration. It will also remove possibly aggressive foci within tumors that appear radiographically benign. Following surgery, early radiation prolongs survival without recurrence to 5.3 years, as compared to no radiation (3.4 years), but does not greatly affect overall survival. Escalation of radiation doses does not impact survival. When procarbazine/CCNU (lomustine)/vincristine chemotherapy is used, there is an improvement in progression-free survival, but not in overall survival. Such treatments are reserved for patients who have progressive symptoms, tumor expansion, uncontrolled seizures, or steroid dependence. The radiation field is made up of the area of radiographically identifiable tumor plus a margin of up to 2 cm. The most widely used chemotherapeutic agent is temozolomide.

Prognosis

Favorable prognoses occur after gross-total resection of tumors in younger patients. Unfavorable prognoses are linked to age above 40 years, tumors larger than 6 cm, tumors that cross the midline, and neurological deficits being present before surgery. The majority of patients suffer anaplastic transformation of these tumors.

Pleomorphic xanthoastrocytoma

Pleomorphic xanthoastrocytoma is a rare, distinctive astrocytoma that has a good prognosis, and was previously and regularly misdiagnosed as a glioblastoma. The tumor usually involves the cerebral cortex and meninges, most commonly in the temporal lobe (see Fig. 6.5). Its superficial location lends a better prognosis than other astrocytomas. The masses are found in the temporal lobes, extending into the **leptomeninges** and **Virchow-Robin spaces**. These tumors usually are unilobular, with only 20% of cases affecting more than one brain lobe.

Epidemiology

About 66% of cases of pleomorphic xanthoastrocytomas are diagnosed before age 25. These tumors comprise less than 1% of all astrocytomas. The median age at time of presentation is about 15 years. There is no predilection for either gender, and no particular geographic distribution.

Etiology and risk factors

Mutations of the V600E type in the BRAF gene have been identified, regardless of grading. This illustrates the importance of the *mitogen-activated protein kinase (MAPK) signaling pathway* in their pathology. This is a chain of proteins in the cells that communicate signals from receptors on cell surfaces to the cellular DNA nuclei. There are no known

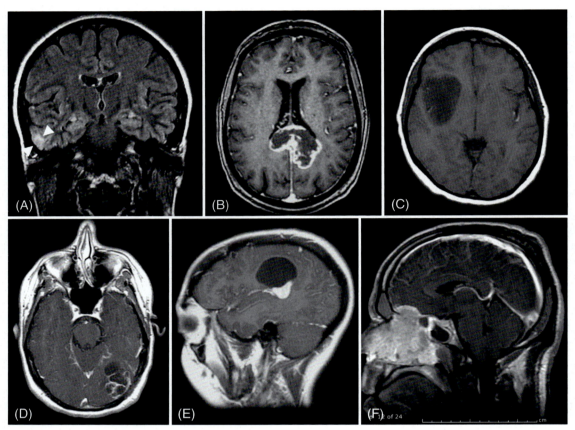

FIGURE 6.5 Pleomorphic xanthoastrocytoma. (A) The tumor is seen on this fluid-attenuated inversion recovery magnetic resonance image as an area of increased signal intensity in the right temporal lobe (arrowheads). (B) Glioblastoma multiforme, (C) Oligoastrocytoma, (D) Anaplastic ependymoma, (E) Ganglioglioma and (F) Esthesioneuroblastoma.

risk factors for pleomorphic xanthoastrocytomas. However, there are associations with NF-1 and with *cortical dysplasias*, which occur when the top layer of the brain does not properly form. The most common type of cortical dysplasia is **focal cortical dysplasia**, of which there are three types:

- Type I—usually involving the temporal lobes. It is difficult to see on a brain scan, and the patient usually does not start having seizures until adulthood.
- Type II—usually involves the temporal and frontal lobes of the brain. It is a more severe type and is seen most often in children.
- Type III—may involve damage in another part of the brain, or be due to some type of brain injury early in life. This type includes either Type I or Type II as a component.

Pathology

The pathology of pleomorphic xanthoastrocytoma includes hypercellularity, with many atypical, pleomorphic tumor astrocytes. These tumors usually develop in the supratentorial region, above the tentorium cerebelli. They are generally located superficially in the cerebral hemispheres, involving the leptomeninges, and only rarely arising from the spinal cord. Though mitoses are unusual, there are oddly-shaped giant cells. Eosinophilic granular bodies make it easy to differentiate these tumors from glioblastomas. Lipidized astrocytes, also called **xanthomatous cells**, containing a "foamy" cytoplasm filled with lipids only occur in about 25% of cases. Pleomorphic xanthoastrocytomas are well demarcated from other tissues, and cysts may be present. The cells vary in size and shape and have single as well as multinucleated giant cells, with the larger cells accumulating lipids. Mitoses and necrosis are seen, and malignant forms have been identified. It is believed that the tumor cells originate from **subpial** astrocytes. Anaplastic transformation occurs in 15% − 20% of cases.

Clinical manifestations

For pleomorphic xanthoastrocytoma, a history of seizures often precedes diagnosis, usually of about 3 years. Headaches are a common complaint. Additional symptoms include **anosmia**, pigmented skin tumors, arthralgia, photosensitivity, fatigue, nausea, auditory hallucinations, polyuria, papilledema, hemiparesis, aphasia or dysphasia, changes in consciousness, and limb weakness.

Diagnosis

The average age at diagnosis is 26. Solid portions of these tumors usually enhance significantly in MRI with gadolinium used as the contrast agent. In un-enhanced T1-weighted images, the tumor will by *hypointense* or *isointense*, with "T1" referring to the **spin-lattice effect** of magnetization in the same direction as the static magnetic field. Patients may have the tumor molecularly profiled, which can help identify opportunities for targeted treatments.

Treatment

For pleomorphic xanthoastrocytoma, complete resection is possible in most patients. Radiation or chemotherapy is given for aggressive or recurrent tumors. Chemotherapeutic agent choices may include targeted therapy with dabrafenib and vemurafenib.

Prognosis

Pleomorphic xanthoastrocytoma has a relatively favorable prognosis. Postoperative survival averages 81% at 5 years and 70% at 10 years. It is estimated, however, that up to 20% of these tumors will undergo malignant transformation into grade III tumors, worsening the clinical course. Well-resected lesions have the best prognosis.

Diffuse astrocytoma

Diffuse astrocytomas are a group of primary, slow-growing, low-grade brain tumors. They usually occur in adults. Arising from neoplastic astrocytes, they most commonly occur in the cerebral hemispheres, but can develop anywhere in the brain or spinal cord. The term "diffuse" describes the fact that they may not be clearly visible in imaging studies, since the borders of the tumor usually send out microscopic fibrillary tentacles into the surrounding brain tissue. Their entanglement throughout healthy brain cells makes complete surgical removal difficult. As of 2016, "fibrillary astrocytomas" were no longer considered by the World Health Organization to be distinct types, and were incorporated into the term "diffuse astrocytomas."

Epidemiology

Diffuse astrocytomas make up about 10% of all gliomas. They usually occur in adults between the ages of 20 and 50 and are the most common type of diffuse low-grade astrocytoma. The mean age at diagnosis is 35. Males develop diffuse astrocytomas 1.5 times as often as females.

Etiology and risk factors

Unlike adult types of gliomas, diffuse astrocytomas usually lack p53 mutations, but may be more likely to have mutations in IDH-1 or IDH-2, which is a factor similar to mutations arising in oligodendrogliomas. There are no specific risk factors for diffuse astrocytomas. General risk factors may include smoking, radiation exposure, family history, neurofibromatosis, tuberous sclerosis, von Hippel-Lindau disease, Li-Fraumeni syndrome, **Gorlin syndrome**, Turcot syndrome, Cowden syndrome, and being immunocompromised.

Pathology

For diffuse (low-grade) astrocytomas, there is increased cellularity and slight nuclear pleomorphism, in comparison to normal brain tissue. There may be cysts and microcalcifications, plus a firm consistency. These diffusely infiltrative tumors are mostly limited to the white matter, with the tumor matrix being rich in neuroglial fibrils. The matrix has neoplastic fibrillary astrocytes, with mild nuclear atypia. They appear enlarged, irregular in contour, and show **hyperchromasia** plus coarse nuclear chromatin patterns (see Fig. 6.6). The nuclei have a low cellular density and are elongated. Microcystic spaces containing mucinous fluid are commonly present.

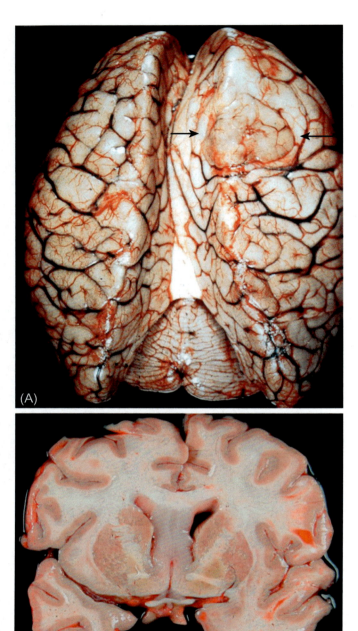

FIGURE 6.6 Diffuse (fibrillary) astrocytoma.

Clinical manifestations

Diffuse astrocytomas usually cause seizures, frequent mood changes, and headaches early in the course of the disease. Hemiparesis is also common. The most common presenting feature is seizures, which are present in 40% of patients—especially in adults. Additional symptoms include hypersomnia, vascular disease, progressive vision loss, neck pain, facial pain, indecisiveness, feelings of worthlessness, and hand tremor.

Diagnosis

Diffuse astrocytomas are diagnosed with a continuous EEG recording of electrical activity in the brain, to help identify and localize seizure activity—primarily in children. CT and MRI scans of the brain may reveal a diffuse mass that does not "light up" when a contrast dye is administered. For some patients, a biopsy is required to confirm the nature of the tumor.

Treatment

For diffuse astrocytomas, most experts prefer complete tumor excision, to at least the greatest possible degree, without any compromise of neurologic function. For patients with well-controlled seizures and smaller asymptomatic lesions, treatment may involve observation and close monitoring, with interventions occurring if the tumor grows, the radiographic appearance changes, or if new or uncontrolled symptoms develop. Delayed treatment will postpone surgical risks and adverse effects of radiation therapy. Though most patients will eventually receive radiation therapy, its timing is under debate. One randomized study revealed improved progression-free survival of 5.3 years when early radiotherapy was given, versus 3.4 years for those who receive delayed radiotherapy. Survival rates were different, however, with 7.4 year survival for the early treated group and 7.2 year survival for the later-treated group. There is no established role for chemotherapy for diffuse (low-grade) astrocytomas. These tumors are rarely cured because of the inability for complete surgical excision. Many tumors in this grade transform to a higher grade over time. The difficulty in treating diffuse astrocytomas is that the microscopic "tentacles" of the tumor make complete surgical removal difficult or impossible without injuring the brain. Surgery can, however, still reduce or control the size of the tumor. Possible side effects of surgery include brain swelling, which is treated with steroids, and seizures. The tumor may recur if resection is incomplete. Standard radiotherapy requires 20 − 30 sessions, based on the tumor subtype. This may be performed alone, or after surgical resection, improving outcome and survival rate. Side effects include local inflammation, and headaches, which can be treated with oral medications. Radiosurgery uses computer modeling to focus a minimal amount of radiation at the exact location of the tumor, minimizing the dose to surrounding brain tissue but this is infrequently applied to this diffuse and infiltrative tumor type. Chemotherapy for diffuse astrocytoma is effective for only about 20% of patients. Techniques under investigation include gene therapy, immunotherapy, and newer types of chemotherapies.

Prognosis

For diffuse astrocytoma, because of their slow growth, the median survival is 5 − 8 years. Only 15% − 20% of patients survive for 10 years, though patients having had compete surgical excision have a better prognosis. Characteristics associated with a worsened prognosis include symptom duration less than 6 months before diagnosis, abducens palsy upon presentation, location within the pons, and engulfment of the basilar membrane.

> **Point to remember**
> Diffuse astrocytoma is a tumor with poorly defined boundaries. Small clusters of tumor cells usually grow into neighboring healthy tissue, making the tumor difficult to completely remove during surgery. The goal with surgery is to achieve "maximal safe resection," removing as much of the tumor as possible while still protecting critical brain function. Diffuse astrocytoma usually arises in the cerebral hemispheres of the brain. Grade II tumors have a better prognosis than Grade III tumors, which progress more rapidly.

Anaplastic astrocytoma

Anaplastic astrocytoma is a rare type of astrocytoma that is considered to be grade III. If untreated, it is often fatal. This tumor usually develops slowly over time, but is able to develop rapidly. It usually occurs sporadically, but can be linked to a genetic disorder. The symptoms caused by this tumor are quite varied based on its actual location.

Epidemiology

Anaplastic astrocytoma has a mean diagnostic age of 40 years. Anaplastic astrocytoma makes up 7.5% of all gliomas and 5% of all primary brain tumors. In the United States, anaplastic astrocytoma has an incidence rate of 0.44 per every 100,000 people. Low-grade astrocytomas are rare in people older than age 50. The overall age-standardized relative survival rates are 23.6% at 5 years and 15.1% at 10 years.

Etiology and risk factors

Most high-grade gliomas, including anaplastic astrocytomas, occur sporadically or without any identifiable cause. However, less than 5% of people with malignant astrocytomas have a definite or suspected hereditary predisposition. The primary hereditary predispositions are NF-1, *Li-Fraumeni syndrome*, hereditary nonpolyposis colorectal cancer, and tuberous sclerosis. Anaplastic astrocytomas have also been linked to previous exposure to vinyl chloride, and to high doses of radiation therapy to the brain. Risk factors for anaplastic astrocytomas include the related genetic disorders as well as environmental factors that may act as triggers for tumor development. Additional risk factors include immune system abnormalities, radiation therapy to the brain, and exposure to ultraviolet rays as well as to certain chemicals. Workers in the oil refining and rubber manufacturing industries have higher incidence of anaplastic astrocytomas.

Pathology

Grade III and IV astrocytomas may also be located in the brain stem. These astrocytomas are usually large, well circumscribed, and have a variegated pattern. Their peripheral rims are pink-gray and solid. They have a yellow, soft necrotic center and hemorrhaged points. When viewed under a microscope, they show an increase in cellularity, pleomorphic astrocytes, vascular proliferation, and necrosis. The main histologic difference between anaplastic grade III tumors and grade IV glioblastoma tumors is necrosis and vascular proliferation. An astrocytoma with hemorrhage and necrosis is grade IV by definition. Their molecular pathology shows high vascularization and extensive heterogenic infiltration. Sometimes, they grow large enough to extend from the meningeal surface, and push through the ventricular wall (see Fig. 6.7). At the time of the patient's death, about 50% of these tumors are bilateral, or occupy more than one lobe microscopically. In rare cases, grade IV astrocytomas have been found outside of the brain and spinal cord. Anaplastic astrocytomas have histologic features that resemble those of low-grade astrocytomas and mitotic activity.

Clinical manifestations

For anaplastic astrocytoma, initial symptoms are usually headache, depression, focal neurological deficits, and sometimes, seizures. Symptoms onset and diagnosis are often 1.5 − 2 years apart, and may also include visual changes, vomiting, personality changes, and problems with walking. The patient often has memory, concentration, and thinking abnormalities. When the tumor is in the frontal lobe, there are usually gradual changes in mood and personality, then paralysis on one side of the body. If the tumor is in the temporal lobe, there are problems with coordination, speech, and memory. A parietal lobe tumor causes problems with sensations, writing, and fine motor skills. A cerebellar tumor causes coordination and balance problems. If in the occipital lobe, the tumor causes problems with vision and visual hallucinations.

Diagnosis

Anaplastic astrocytomas are diagnosed similarly to other types of astrocytomas, including the use of CT and MRI. Biopsy helps confirm diagnosis. An MRI usually shows heterogeneously enhancing lesions. If nonenhancing, lesions are rarely anaplastic. However, this is not as reliable in older patients. **Diffusion-weighted imaging** is able to identify areas of increased cellularity in low-grade neoplasms when therefore may be useful in identifying foci of early anaplastic changes.

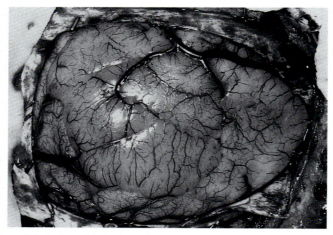

FIGURE 6.7 Expansion and distortion of the brain gyri due to an anaplastic astrocytoma.

Treatment

The standard initial treatment for anaplastic astrocytomas is to remove as much of the tumor as possible, and radiation therapy is administered to prolong survival. While there is no proven benefit to adjuvant chemotherapy, recent studies show that irradiation or chemotherapy may be used upon initial diagnosis, giving similar results to when other modalities are applied for progression of the tumor. The median time period to treatment failure is 42 months. Temozolomide is effective for treating anaplastic astrocytoma, but its role as an adjuvant to radiation therapy has not been fully elucidated. Quality of life following treatment is heavily based on the location of the tumor. The high molecular variability of anaplastic astrocytomas is linked to a lower success rate of treatment. Because of tumor location, it may be difficult to effectively deliver treatment. Many patients experience various types of paralysis, speech impediments, difficulties with cognition, and altered sensory perceptions. However, most of them benefit from speech, vision, physical, and occupational therapies. It is hoped that in the future, more knowledge about the activities of anaplastic astrocytomas will improve treatment outcomes.

There are many new therapies being investigated for anaplastic astrocytomas. These include classes of drugs such as the protein kinase inhibitors, biological response modifiers, and angiogenesis inhibitors. Recurrent/progressive astrocytoma is frequently considered "malignant glioma" and treated similarly to cases of glioblastoma. Because anaplastic astrocytoma is much less common glioblastoma, specific clinical trials mostly focus on treatment of glioblastoma.

Prognosis

For grade III and IV astrocytomas, survival over 5 years is only between 5% and 10%. Anaplastic astrocytoma has a better prognosis that glioblastoma, but patients with AA are 46 times more likely to die than matched members of the overall population. Prognosis across age groups is varied, especially during the first 3 years postdiagnosis. Median survival for anaplastic astrocytoma is 2 – 3 years. Relative survival rate is 42% at 2 years and 26% at 5 years. The elderly population has the worst prognosis, with younger patients surviving longer. Secondary progression to glioblastoma multiforme is common. Prognosis is best if radiation therapy and surgery are used, for younger patients, and for the female gender. Also, prognosis is better when the MGMT promoter region is methylated, and if somatic mutations of the IDH1 gene are noted.

> **Point to remember**
>
> According to the National Organization for Rare Disorders, anaplastic astrocytoma is a WHO grade III aggressive malignancy. It may develop in the frontal, temporal, parietal, and occipital lobes of the cerebrum. Exact causes have not been identified, but it is believed that genetic and immunologic abnormalities, ultraviolet rays, chemicals, radiation, diet, and even stress contribute to anaplastic astrocytoma. It is more common in adults than in children, and primarily treated with surgery, radiation, and chemotherapy.

Clinical cases

Clinical case 1

1. Aside from headaches and vomiting, what are the other signs and symptoms of pediatric pilocytic astrocytomas?
2. Pilocytic astrocytomas are considered to be low-grade astrocytomas, but sometimes, they recur; what are the treatments for this situation in pediatric patients?
3. How common are pilocytic astrocytomas in children, and where do they most often develop?

A mother brought her 2-year-old daughter to her pediatrician because of a forehead-centered headache that had been present for 3 weeks. In the last few days, the girl had vomited in the morning and complained of her "tummy" feeling bad. Physical and neurological examinations revealed no clinical abnormalities. There were no signs of any infections. The girl's pupils were symmetric and normally responding to light, and the eye movements were normal. Babinski's sign was negative. Other reflexes were symmetric and normal. There were no signs of motor balance disturbances. The vascularity of the base of the eye appeared normal related to the girl's age and there were no visible hemorrhages. The pediatrician advised the mother to give her daughter some pediatric analgesics, and closely monitor the headache, as a step to rule out elevated brain pressure. Initially, the analgesics provided some relief. Soon, however, the headaches returned and another examination revealed the presence of papilledema in both eyes.

A brain MRI was performed, showing brain edema and a tumor of the cerebellum that was blocking the spinal fluid spaces. With a few days, successful surgery was undertaken, and the tumor—a pilocytic astrocytoma—was removed. Afterwards, the child was free of any symptoms and the papilledema subsided. Over months of follow-up, the girl had no recurrent symptoms and imaging has revealed no additional tumors.

Answers:

1. **Children with pilocytic astrocytomas, aside from headaches and vomiting, may also have a lack of appropriate weight gain or weight loss, nausea, irritability, torticollis, difficulty coordinating movements, and nystagmus or other visual complaints. The symptoms vary based on the location and size of the tumor. Usually, symptoms are linked to increased intracranial pressure due to the size of the tumor.**
2. **Treatment of recurrent childhood low-grade astrocytomas includes additional surgery, radiation therapy, chemotherapy, targeted therapy using a monoclonal antibody with or without chemotherapy, clinical trials to check tumor samples for certain gene changes, or a clinical trial investigating agents such as lenalidomide, dabrafenib, everolimus, or selumetinib.**
3. **Pediatric pilocytic astrocytomas are relatively rare. They can develop anywhere in the brain and less often, in the spinal cord. The majority of cases arise in the cerebellum, brain stem, hypothalamus, or optic nerve pathways. The condition is usually signified by increased intracranial pressure, either from the tumor itself, or because of blockage of the ventricles and accumulation of CSF.**

Clinical case 2

1. What are the clinical features of this tumor caused by in most pediatric patients?
2. Was the treatment undertaken for this patient the usual one recommended for her age group?
3. How does this tumor appear when it is biopsied in the majority of patients, and how is it classified by the World Health Organization?

A 10-year-old girl was brought to her pediatrician because of recurrent convulsive seizures, yet no signs of increased intracranial pressure. A CT scan with contrast revealed a voluminous mass in her perilateral ventricle, with similar attenuation to the cortical gray matter, and marked enhancement. Surgery was performed for a complete resection of the tumor. Though a ganglioglioma was considered, staining and microscopic examination revealed the tumor was a SEGA. Histologically, it was composed of fibrillated spindle cells and globular large cells. The cells had abundant eosinophilic cytoplasm, large eccentric-looking nuclei, and also large nucleoli. Additionally, there were calcifications and perivascular lymphocytes. The spindle cells were positive for glial fibrillary acidic protein and S-100 protein. Body imaging revealed no signs of tuberous sclerosis.

Answers:

1. **The clinical features of SEGA are due to hydrocephalus, increased intracranial pressure, and seizures. Hydrocephalus occurs because of obstruction of the CSF pathway by the tumor itself. Since this tumor and tuberous sclerosis are commonly related, the characteristic symptoms are often present and must be investigated. The patient's age, plus the location of the mass are indicative of this tumor, since CT and MRI are usually nonspecific.**
2. **Yes, radical and early surgery is the treatment of choice for SEGA because it is associated with a better prognosis without complications. This is partly due to the intracranial hypertension and the link between tuberous sclerosis and the surgical procedure, which affects the functional and vital prognosis. Incomplete surgery is linked to recurrence of this tumor.**
3. **SEGAs are multiobulated, well-circumscribed tumors arising from the wall of the lateral ventricles, near the foramen of Monroe. They often contain cysts and calcifications. They are believed to arise from a subependymal nodule present in the ventricular wall when the patient has tuberous sclerosis, but this is not fully established. The World Health Organization classifies this type of tumor as a "grade I lesion."**

Clinical case 3

1. What other symptoms are common for pleomorphic xanthoastrocytoma?
2. Where do most pleomorphic xanthoastrocytomas occur?
3. How do these tumors appear, macroscopically and microscopically?

A 24-year-old woman presented with seizures and loss of consciousness that lasted for 15 min. A CT scan and MRI scan revealed a mass in her left parietal lobe. The tumor was characterized by pleomorphic cells and prominent vascularity. The angiomatous region varied from a sinusoidal pattern to a venous malformation. In the vessel wall, there were focal fibrinoid necrosis, hyalinization, and moderate infiltration by lymphocytes and plasma cells. The cells were close to adjacent small vessels. Capillaries near to or extending between tumor cells were present. The majority of the tumor cells were positive for glial fibrillary acidic protein as well as oligodendrocyte lineage transcription factor 2. Based on patient history, clinical data, and pathological findings, she was diagnosed with angiomatous pleomorphic xanthoastrocytoma, WHO grade II.

Answers:

1. Other common symptoms of pleomorphic xanthoastrocytoma include headache, vomiting, and visual disturbances. However, the most common symptom, by far, is the development of seizures, which occur in about 75% of all cases.
2. Pleomorphic xanthoastrocytomas are located supratentorially in about 98% of cases, usually peripherally near the leptomeninges, involving the cortex and overlying leptomeninges. Actual dural involvement is rare. About 50% are in the temporal lobe, with the remainder being more common in the frontal lobe.
3. Pleomorphic xanthoastrocytomas, macroscopically, appear well-circumscribed. They often have a cystic component and involve the overlying leptomeninges. Microscopically, their margins are not as well defined. The histological features are varied, with spindle, polygonal, and multinucleated cells, as well as lipid-laden xanthomatous astrocytes. The nuclei are also pleomorphic, with common nuclear includes, and highly varied sizes.

Clinical case 4

1. What is the pathology of diffuse astrocytoma in the majority of patients?
2. Which type of imaging is the procedure of choice for this type of astrocytoma?
3. What factors are related to a worsened survival rate from this type of astrocytoma?

A 50-year-old woman has been diagnosed with diffuse astrocytoma after having experienced several seizures and having a constantly present headache. Her husband reports to the physician that she has been "acting differently" towards him but doesn't realize her behaviors have changed. There is superior extension of the tumor into the periventricular white matter of the left frontal lobe. There is also a mild mass effect with partial effacement of the involved sulci, and mild irregular diffusion restriction located centrally. The size of her brain ventricles is appropriate for her age, and no other focal lesions are identified. A stereotactic biopsy is performed. Examination of the tissue reveals a mild increase in cell density, and some architectural disturbances. There is mild atypia of the fibrillary astrocytes, and scattered mature neurons. There is no vascular endothelial cell hyperplasia and no necrosis.

Answers:

1. Diffuse astrocytomas are mostly composed of a tumor matrix that is rich in neuroglial fibrils that give the tumor its name and firm consistency. In the tumor matrix there are neoplastic fibrillary astrocytes with mild nuclear atypia. They are enlarged, have an irregular contour, hyperchromasia, and coarsened nuclear chromatin patterns. There is also a low cellular density and elongation of the nuclei. Often, microcystic spaces are present, which contain mucinous fluid.
2. MRI is the radiographic modality of choice for diffuse (fibrillary) astrocytomas, since on CT scan; smaller tumors may be difficult to see due to them not enhancing. In fact, the presence of enhancement would suggest a high-grade tumor instead. Calcification is only seen in 10% to 20% of cases.
3. For all diffuse astrocytomas, the factors most associated with a worsened survival rate were: symptoms duration less than 6 months before diagnosis, abducens palsy upon presentation, location within the pons, and engulfment of the basilar membrane. Complete surgical excision is associated with a better prognosis, though these tumors may recur if excision is incomplete.

Clinical case 5

1. Why is anaplastic astrocytoma distinctly unique compared to other types of astrocytomas?

2. What other types of brain tumors must be excluded as part of the diagnostic process for anaplastic astrocytomas?
3. In reality, is there an existing cure for grade III anaplastic astrocytoma?

A 46-year-old woman was diagnosed with a WHO grade III (anaplastic) astrocytoma of the optic chiasm after three months of visual deterioration. Imaging revealed superior extension into her hypothalamus. Before the symptoms had begun, the patient was healthy, with a family history of renal cancer, hypertension, and Parkinson's disease. The patient had quit smoking 10 years before her diagnosis and only occasionally drank wine. However, she and other members of her immediate family experienced period migraine headaches. The vision deterioration began in her left eye, until it reached 20/400, while her right eye was still 20/20.

The patient was treated with corticosteroids to reduce inflammation. In 1 month, the tumor had tripled in size and the patient was referred to a neurooncologist, who determine the tumor to be unresectable. Radiotherapy and eight rounds of chemotherapies were ordered, to which the patient initially responded well. Further imaging revealed that there was distal tumor progression down white matter tracts. The patient's mobility and consciousness began declining quickly, but chemotherapy was continued. She was moved into an emergency hospital's intensive care unit and intubated for respiratory arrest. An MRI revealed severe cortical, subcortical, and cerebellar involvement. The patient never regained consciousness and expired within 10 days.

Answers:

1. **Anaplastic astrocytomas are distinct regarding their histology because they are characterized by many pleomorphic astrocytes, with evidence of mitosis. Their high molecular variability is linked to a lower success rate of treatment. Because of the location of these tumors, it can be difficult to deliver treatment to the tumor site. Hopefully, increased knowledge of anaplastic astrocytoma molecular "behaviors" will result in improved treatments and outcomes in the future.**
2. **The brain tumors that are distinguished from anaplastic astrocytomas include metastatic tumors, lymphomas, hemangioblastomas, craniopharyngiomas, teratomas, ependymomas, and medulloblastomas. Also, conditions that can resemble anaplastic astrocytomas include meningitis and benign intracranial hypertension, also known as pseudotumor cerebri.**
3. **Grade III (anaplastic) astrocytoma is not considered curable, but various treatment possibilities are able to extend survival time. Even after surgery, chemotherapy, and radiation, there is always the probability of a recurrence. In some patients, the cancer may not recur for 10 or 20 years, but this is well outside of the normal range of what can be reasonably expected.**

Key terms

Anaplastic astrocytoma
Angiocentric
Anosmia
Astrocytomas
Cortical dysplasias
Cytokines
Diffuse astrocytoma
Diffusion-weighted imaging
Foramen of Monro
Forme fruste
Gemistocytes
Gemistocytic astrocytes
Glial fibrillary acidic proteins
Glioblastoma multiforme
Glioma

Hamartin
Hamartomas
Hyperchromasia
Hypointense
Isointense
Leptomeninges
Li-Fraumeni syndrome
Microcysts
Myxoid foci
Neurofibromatosis
Neurofibromin
Pilocytic astrocytoma
Piloid
Pleomorphic xanthoastrocytoma
Pseudorosettes

Rosenthal fibers
Spin-lattice effect
Subependymal giant cell astrocytoma
Subpial
T1-weighted lesions
T2-weight lesions
Tuberin
Tuberous sclerosis
Tumor protein p53
Turcot syndrome
Virchow-Robin spaces
Xanthomatous cells
Ventriculoperitoneal shunt

Further reading

Adesina, A. M., Tihan, T., Fuller, C. E., & Young Poussaint, T. (2016). *Atlas of pediatric brain tumors* (2nd Edition). Springer.

Akbar, W. (2015). *Paediatric cerebellar astrocytoma on MRI*. Lap Lambert Academic Publishing.
Baranska, J. (2012). *Glioma signaling (Advances in experimental medicine and biology)*. Springer.
Bigner, D. D., Friedman, A. H., Friedman, H. S., McLendon, R., & Sampson, J. H. (2016). *The Duke Glioma handbook: Pathology, diagnosis, and management*. Cambridge University Press.
Brem, S., & Abdullah, K. G. (2016). *Glioblastoma: Expert consult*. Elsevier.
Burrows, G. (2013). *Brain tumours: Living low grade – the patient guide to life with a slow growing brain tumour*. NGO Media.
Burrows, G. (2017). *Glioblastoma – a guide for patients and loved ones: Your guide to glioblastoma and anaplastic astrocytoma brain tumors*. NGO Media.
Carver, A. R. (2018). *Childhood astrocytomas: Patient care journal*. Carver.
Chernov, M. F., Muragaki, Y., Kesari, S., McCutcheon, I. E., & Lunsford, L. D. (2017). *Intracranial gliomas part I – surgery (Process in neurological surgery, volume 30)*. S. Karger.
Crimi, A., Bakas, S., et al. (2019). *Brain lesion: Glioma, multiple sclerosis, stroke and traumatic brain injuries (Lecture notes in computer science)*. Springer.
Duffau, H. (2017). *Diffuse low-grade gliomas in adults* (2nd Edition). Springer.
Fatterpekar, G. M., Naidich, T. P., & Som, P. M. (2012). *The teaching files: Brain and spine: Expert consult*. Saunders.
Furtado, L. V., & Husain, A. N. (2018). *Precision molecular pathology of neoplastic pediatric diseases (Molecular pathology library)*. Springer.
Gurer, B., Kertmen, H., & Yilmaz, E. R. (2012). *Glioblastomas and anaplastic astrocytomas: The neurosurgical perspective review*. Lap Lambert Academic Publishing.
Hayat, M. A. (2015). *Tumors of the central nervous system, volume 14: Glioma, meningioma, neuroblastoma, and spinal tumors*. Springer.
Kaye, A. H., & Laws, E. R. (2011). *Brain tumors E-book: An encyclopedic approach* (3rd Edition). Saunders/Elsevier.
Kwiatkowski, D. J., Holets-Whittemore, V., & Thiele, E. A. (2010). *Tuberous sclerosis complex: Genes, clinical features, and therapeutics*. Wiley-Blackwell.
McKhann, G. M., II, & Duffau, H. (2019). *Low-grade glioma, an issue of neurosurgery clinics of North America (The clinics: Surgery)*. Elsevier.
Medical Ventures Press. (2011). *2011 pediatric cancer toolkit: Brain and spinal cord tumors – neuroectodermal, medulloblastoma, glioma, astrocytoma, craniopharyngioma*. Progressive Management.
Medifocus.com Inc. (2014). *Medifocus guidebook on glioblastoma – a comprehensive guide to symptoms, treatment, research, and support*. CreateSpace Independent Publishing Platform.
Moliterno-Gunel, J., Piepmeier, J. M., & Baehring, J. M. (2017). *Malignant brain tumors: State-of-the-art treatment*. Springer.
Newton, H. B. (2016). *Handbook of neuro-oncology neuroimaging* (2nd Edition). Academic Press.
Newton, H. B. (2018). *Handbook of brain tumor chemotherapy, molecular therapeutics, and immunotherapy* (2nd Edition). Academic Press.
Placantonakis, D. G. (2018). *Glioblastoma: Methods and protocols (Methods in molecular biology)*. Humana Press.
Prados, M. D. (2012). *Advances in the pathogenesis and treatment of glioblastoma multiforme*. Future Medicine Ltd.
Scarabino, T., & Pollice, S. (2019). *Imaging gliomas after treatment: A case-based Atlas* (2nd Edition). Springer.
Somasundaram, K. (2017). *Advances in biology and treatment of glioblastoma (Current research)*. Springer.
Sughrue, M. E. (2019). *The glioma book*. Thieme.
Sughrue, M. E., & Yang, I. (2019). *New techniques for management of 'inoperable' gliomas*. Academic Press.
Winn, H. R. (2011). *Youmans neurological surgery, 4-volume set: Expert consult* (6th Edition). Saunders.

Chapter 7

Glioblastoma

Chapter Outline

Glioblastoma 129
   Isocitrate dehydrogenase-wild-type 139
   Isocitrate dehydrogenase-mutant 141
   Gliosarcoma 141

Clinical cases 141
Key terms 144
Further reading 144

Glioblastoma

Glioblastoma is a malignant, diffusely invasive, rapidly growing pulpy or cystic tumor of the cerebral hemispheres, cerebellum or spinal cord. It is classified as a Grade IV astrocytic tumor by the World Health Organization. Most glioblastomas arise *de novo* (known as *primary glioblastomas*) and are fully malignant. A smaller amount emerge from grade II or III astrocytic precursors, and are known as *secondary glioblastomas*. Glioblastoma is one of the most lethal forms of brain cancer, with untreated patients usually living only three to six months after diagnosis. Glioblastoma is sometimes abbreviated as "GBM". These tumors occur most often in the temporal lobe, followed by the parietal, frontal, and occipital lobes.

Epidemiology

Glioblastoma, previously known as *glioblastoma multiforme*, mostly affects adults between ages 55 and 85, with the median age being 64. However, it may occur in younger adults and in children, but much less often. Glioblastoma is the most common adult primary brain tumor, accounting for over 80% of diffuse **gliomas**. In adults aged 55 or older, glioblastoma is the second most common type of intracranial neoplasm. Glioblastomas account for about 15% of all intracranial neoplasms, and about 45% − 50% of all primary malignant brain tumors. Astrocytomas of all types are overall more common in Caucasians than in other races, and for all tumor grades, are slightly more common in males than in females. However, Grade III and IV astrocytomas occur two times more often in men than in women. Primary glioblastomas are 1.33 times more common in males than females, but secondary glioblastomas are 0.65 times less common in males than females. Overall, glioblastomas are the third most common cancer between ages 15 and 34, and the fourth most common cancer between ages 35 and 54. The annual incidence rate in children in the United States is 0.14 new cases per 100,000 populations annually.

Grade IV astrocytomas make up about 55% of all types of gliomas. In the United States, approximately 13,000 new diagnoses of glioblastoma were estimated for 2019. There are about 3.19 cases per 100,000 population. The age-adjusted incidence rates of glioblastoma by region in the United States are shown in Fig. 7.1. In Europe, North America, and Australia, the incidence rate is 3 − 4 cases of every 100,000 people. It is less common in eastern Asia, with only 0.59 cases out of every 100,000 people. However, it is believed that lower incidence and prevalence in Asia, as well as Africa, may be due to inaccurate reporting. A study in Switzerland revealed a higher rate of 3.55 out of every 100,000. However, in England, the frequency of glioblastoma doubled between the years 1995 and 2015. Similarly, between 1989 and 2010 in the Netherlands, the incidence of glioblastoma has more than doubled, from 1.5 per 100,000 people to 3.4 per 100,000.

Etiology and risk factors

The cause of most cases of glioblastomas is unclear, but about 5% develop from a lower grade astrocytoma. A small proportion is inherited as part of a **Mendelian syndrome** (such as neurofibromatosis 1, NF1). Glioblastomas develop from astrocytes, which support nerve cells. As part of the Cancer Genome Atlas, glioblastoma was the first brain tumor

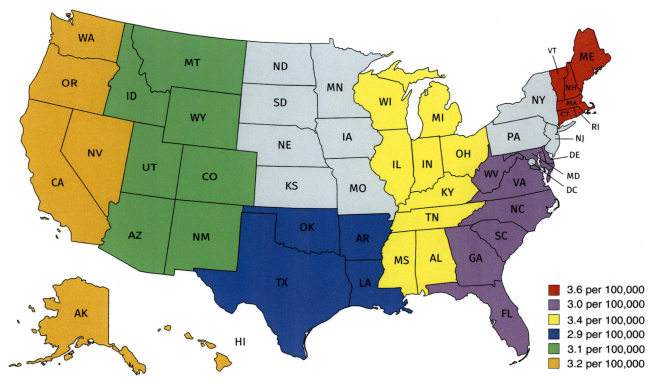

FIGURE 7.1 Age-adjusted incidence rates of glioblastoma by region in the United States.

to be sequenced, which identified its subtypes and their different responses to treatment. According to the American Brain Tumor Association, the vast majority of all brain tumors, including glioblastomas, are not hereditary.

Besides the risk factors for astrocytomas in general, glioblastoma has additional risk factors that include **simian vacuolating virus 40** (SV40), human herpesvirus 6 (HHV-6), and cytomegalovirus. The polyomavirus simian virus 40 is a known oncogenic DNA virus that induces primary brain and bone cancers, malignant mesothelioma, and lymphomas. Latent infections with HHV-6 have been identified in patients with various gliomas, including glioblastomas. A strain of HHV-6A has been isolated from glioma cysts, and HHV-6 DNA and protein in tissue is present in 42.5% of gliomas compared to just 7.7% of normal brain tissue. Elevated levels of several cytokines that were specifically promoted by HHV-6 infections in astrocyte cultures have also been seen in HHV-6-positive cyst fluid samples from glioma tissues. Cytomegalovirus is also implicated as a potential factor in gliomas, with its proteins found in about 50% of glioblastoma samples, and this may be an underestimate based on methodology limitations. Additional risk factors include age over 50 (varies by decade), and Caucasian or Asian heritage. Ionizing radiation to the head and neck has been validated as an independent risk factor for glioblastomas. Intriguingly, people with a history of allergies or atopic diseases such as eczema have a decreased risk. Unproven risk factors for glioblastomas that are still under study include cellular phone usage (varying reports by country and cellular modality), household chemicals, viruses, environmental toxins, hydrocarbon compounds and even nutritional factors.

Pathology

Glioblastoma shares the characteristics of anaplastic astrocytomas, but also shows necrosis, with or without vascular proliferation, involves trafficking monocytes, immature and mature astrocytes, pyriform cells, and neural ectodermal cells having protoplasmic or fibrous processes (see Fig. 7.2). The tumor has a heterogeneous appearance, in gross and microscopic views. A single mass replaces the affected brain area. The mass can appear well circumscribed, but microscopically, there is widespread infiltration. Multifocal tumors can develop, sometimes based on infiltrative growth via nerve fiber pathways or metastatic spread in the cerebrospinal fluid (CSF). Usually, multifocal tumors represent different regions of malignant clonal transmogrification. This is to be discriminated from the term **gliomatosis cerebri**, which is a radiographic connotation of confluent multilobar tumor involvement.

Incidence of multiple, independent gliomas outside the setting of inherited neoplasia are unknown. Brain autopsies do not always show a connection between multifocal gliomas, since cells infiltrating through myelinated pathways are usually small, undifferentiated, and polar. It is believed that about 2.4% of glioblastomas are actually multiple independent tumors. About 7.5% of gliomas, including oligodendrogliomas are multiple independent tumors, and in about 3%

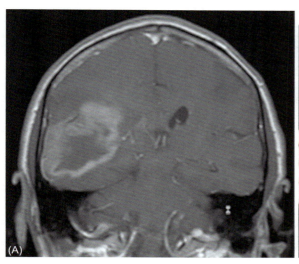 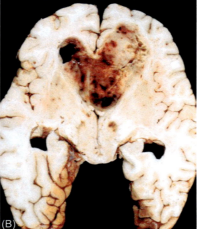

FIGURE 7.2 Glioblastoma.

of these, the tumor foci have different histologic appearances. True multifocal glioblastomas are probably polyclonal if occurring **infratentorially** and supratentorially, such as outside easily accessed routes (the median commissures or cerebrospinal fluid pathways). Therefore, multiple independently arising gliomas must be of polyclonal origin. Only be applying molecular markers can one prove this; molecular characterization is necessary to delineate distinction of tumors from common or independent origins.

Most glioblastomas (31%) develop in the temporal lobe, followed by 24% in the parietal lobe, 23% in the frontal lobe, and 16% in the occipital lobe. Once advanced, the tumor may involve the meninges or a ventricle. Most glioblastomas present in the frontal horn and adjacent to the lateral ventricle, not in the trigone. Malignant gliomas usually recur locally, with more than 90% recurring within 2 cm of the original site of the tumor. They spread more quickly when there is a breach of the ventricle/subventricular zone. They can infiltrate the brain and quickly spread through the neuroaxis. Rarely, they spread to distant locations that are outside the CNS. Infiltrating cells are the most likely source of local recurrences since they escape surgical resection, do not receive the highest dose of radiation therapy, and are in areas with an intact blood−brain barrier, potentially lowering chemotherapeutic bioavailability. The progression of glioblastomas can be fast enough so that size doubles over just 10 days.

Glioblastoma does not usually extend into the subarachnoid space, or spread through the cerebrospinal fluid, though this occurs more often in children than in adults. Penetration of the dura, venous sinus, and bone is rare. Invasion of the vessel lumen is also uncommon. Extracranial metastasis is uncommon without previous surgical intervention, but has occurred in patients having undergone interstitial treatments or with ventricular shunts. Circulating tumor cells have been discovered in the blood of patients with glioblastoma. This suggests that immune reactions or hostile environments of distant organs may suppress metastasis and growth. Immunosuppressed patients who received organ transplants from donors with glioblastoma have developed glioblastoma in the transplanted organs.

Glioblastoma tumors are characterized by small areas of necrotizing tissue surrounded by anaplastic cells in a serpentine pattern, along with the presence of hyperplastic blood vessels. The cut surfaces of these tumors are variegated, with central yellow or white zones of necrosis and hemorrhage. These are surrounded by endothelial hyperplasia (a hyperemic ring), and edematous tissue, with varied amounts of **vasogenic edema**, gliosis, and tumor infiltrations (see Fig. 7.3). Glioblastomas usually form in the subcortical white matter, deeper basal ganglia, or thalamus. In children, they are highly common in the basal ganglia and thalamus. Glioblastomas of the brain stem are uncommon, and usually occur in children. Only rarely to these tumors form in the cerebellum or spinal cord. Glioblastomas grow quickly and can become very large before any symptoms are produced. Less than 10% form more slowly, after degeneration of a low-grade astrocytoma or anaplastic astrocytoma. Rarely, glioblastomas may be found in the cerebellum or spinal cord. Microscopically, the tumor resembles an anaplastic astrocytoma, but also has the characteristic endothelial hyperplasia or necrosis. The hyperplasia involves thickened or **glomeruloid** vessels that are multilayered (see Fig. 7.4).

Point to remember
Similar to a "butterfly glioma," rapid spread of glioblastomas may be seen along white matter tracts of the fornix, internal capsule, anterior commissure, and optic radiation. Other types of infiltration may form secondary **structures of Scherer**, such as **perineuronal satellitosis, perivascular aggregation**, or **subpial spread**.

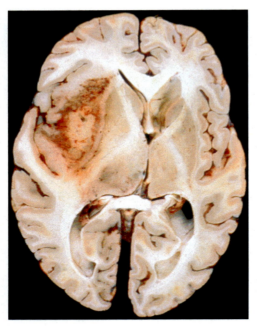

FIGURE 7.3 Glioblastoma (cut in axial imaging plane).

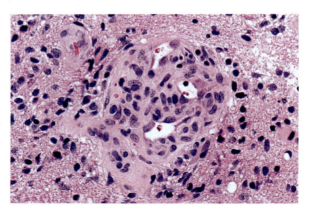

FIGURE 7.4 Glioblastoma. Focus of endothelial hyperplasia with glomeruloid multilayered vessel (hematoxylin-eosin stain, ×400).

The necrosis typically appears **serpiginous** with **nuclear pseudopalisading** along the perimeter (see Fig. 7.5). There are several histological variants, including giant cell glioblastoma, small-cell glioblastoma, and **gliosarcoma**. With gliosarcoma, there is sarcomatous mesenchymal metaplasia in the backdrop of the glioblastoma. There are no significant clinical differences between these histological variants, compared with standard glioblastoma, though gliosarcomas develop more often in the temporal lobes. There are significant biological differences between primary glioblastomas and lower-grade forms, which occur because of a presence or absence of **isocitrate dehydrogenase (IDH)** mutations. Point mutations in IDH1 or its homolog, IDH2, occur in 70% − 90% of astrocytomas, oligoastrocytomas, oligodendrogliomas, and secondary glioblastomas. However, these mutations are rare in primary glioblastomas. Mutations of IDH are believed to represent molecular changes separating rapidly progressive, diffuse gliomas from less indolent forms.

Primary glioblastomas often have many cytogenetic and molecular genetic changes. Mutations in IDH1, IDH2, and **ATP transcriptional regulator**, **x-linked** (ATRX) are nowhere as common in glioblastomas than in grade II or III gliomas and secondary glioblastomas. Between 30% and 40% of primary glioblastomas have **epidermal growth factor receptor (EGFR)** amplifications, often with a related activating mutation (EGFRvIII). The EGFR amplifications are also present in about 70% of cases of small-cell glioblastomas. In recent studies, it has been found that glioblastomas

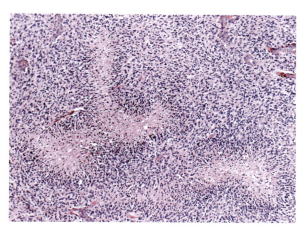

FIGURE 7.5 Glioblastoma. Nuclear pseudopalisading surrounding foci of central necrosis (hematoxylin-eosin stain, ×100).

have somatic mutations of multiple genes. Methylation of the *O-6-methylguanine-DNA methyltransferase* (MGMT) gene promoter is a prognostic, predictive biomarker for glioblastoma. This DNA repair enzyme, when methylated, is believed to reduce the ability of cancer cells to correct DNA damage caused by chemotherapy.

Secondary glioblastomas usually occur in middle age, and may extend into the meninges or ventricular wall to cause high protein content in the CSF, and occasional **pleocytosis**. Malignant cells in the CSF rarely spread to the spinal cord to cause meningeal gliomatosis. About half of the cases of glioblastoma have tumors in more than one lobe of a cerebral hemisphere, or are bilateral. They usually infiltrate across the corpus callosum to produce a **butterfly glioma**. There are four subtypes of glioblastoma, with differing pathologies:

- *Classical*—about 97% carry extra copies of the EGFR gene. Most have higher than normal expression of EGFR, while the TP53 gene, which is often mutated in general glioblastomas, is rarely mutated in this subtype. Loss of **heterozygosity** in chromosome 10 is commonly seen, along with chromosome 7 amplification. Also common is a diminution of cyclin dependent kinase inhibitor 2A (CDKN2A). The gene "signature" of this form of glioblastoma resembles that of astrocytomas.
- *Proneural*—often has high rates of gene alterations related to TP53, platelet-derived growth factor receptor type A (PDGFRA), and *IDH1*. This form has a developmental gene signature and carries a prognostically better outcome.
- *Mesenchymal*—has high rates of mutations or other alterations in the NF1 gene encoding *neurofibromin 1*, and fewer alterations in the EGFR gene. It has less expression of EGFR than other subtypes. There are mutations in phosphatase and tensin (PTEN) genes. The gene signature of this form also resembles that of astrocytomas but carries a poor prognostic outcome.
- *Neural*—typified by the expression of neuron markers, suggesting infiltrating cells converting a normal environment. Often presents as normal cells upon pathological assessment.

Many other genetic alterations are linked to glioblastomas, with most of them in the *retinoblastoma protein* (RB) and *phosphoinositide 3-kinase/protein kinase B* (PI3K/PKB) pathways. Glioblastomas have alterations in 68% − 88% of these pathways. Another significant alteration is the **methylation** of MGMT, which is a "suicide" DNA repair enzyme that impairs DNA transcription and expression of the MGMT gene.

The mechanisms that promote invasion by glioblastomas involve cell motility, cell − matrix interactions, cell − cell interactions, extracellular matrix remodeling, and **microenvironmental** factors. The tumor cells produce extracellular matrix components that enhance migration, and also secrete proteolytic enzymes allowing invasion, including matrix metalloproteinase-2 (MMP2), MMP9, urokinase-type plasminogen activator (uPA) with its receptor (uPAR), and **cathepsins**. All gliomas express integrin receptors, which regulate interactions with molecules of the extracellular space, leading to changes of the cellular cytoskeleton and activation of intracellular signaling networks. These networks include the PKB or AKT, mechanistic target of rapamycin (mTOR), and mitogen-activated protein kinase (MAPK) pathways. Growth factors, including fibroblast growth factor, epidermal growth factor, hepatocyte growth factor and VEGF, stimulate migration by activating related receptor tyrosine kinases, plus downstream mediators that have a direct promotion of migration. These mediators include focal adhesion kinase (FAK) and the *Rho family* hydrolase enzymes that bind to guanosine triphosphate (called GTPases), which include Ras-related C3 botulinium toxin substrate (Rac), Ras homolog family member A (RhoA), and cell division control protein 42 homolog (CDC42).

In EGFR-amplified glioblastomas, cells with amplification locate at the infiltrating edges. This suggests they may play a role in peripheral expansion. Overall mass movement of glioblastoma is radially outward and away from central necrosis and accompanying severe hypoxia. Migration occurs much more quickly than the rates of prenecrotic gliomas. Hypoxia promotes invasion via activation of HIF1 and other transcription factors that induce hypoxia. This is from proangiogenic mechanisms and direct effects that enhance glioma cell migration. Hypoxic tumor cells have elevated expression of extracellular matrix components and intracellular proteins linked with cell motility. Activation of promigrational transcription appears to be linked with decreased proliferation, which can affect therapies.

> **Point to remember**
> The IDH proteins are metabolic enzymes. Normally, they convert isocitrate to alpha-ketoglutarate, as core components of the citric acid cycle. In relation to gliomas, IDH mutations change enzymatic activity, resulting in generation of large amounts of the metabolite *2-hydrocyglutarate (2HG)*. This greatly alters cellular **epigenomes** and DNA methylation. Patients with IDH-mutant tumors are usually younger, and have longer overall survival than those with IDH-wild-type tumors.

Clinical manifestations

Glioblastomas usually cause seizures, and headaches that can become severe and pulsating, but specific clinical characteristics are lacking. Additionally, nausea, vomiting, loss of appetite, lethargy, memory loss, localized neurological problems such as one-sided body weakness, ataxia, and changes in personality, mood, or concentration are commonly noted symptoms. Up to 50% of patients have an initial seizure that results in the diagnosis. An initial seizure in an adult always suggests a brain tumor or structural lesion to be ruled out, and is the most common initial manifestation of primary and metastatic brain tumors. These seizures usually have a focal onset, and then become generalized. They can also only occur one time. However, most patients have multiple seizures, which may precede or follow other clinical manifestations. Headaches range widely in severity, but generally occur during the night or upon awakening in the morning. If vomiting occurs at the peak of a headache, a tumor is more likely to be present. It is usually a sign that the tumor has become large, and is more frequent if the tumor is located in the posterior fossa. The symptoms are mostly dependent on the tumor location. Symptoms may develop quickly, but sometimes, glioblastoma is asymptomatic until the tumor becomes extremely large. Additional symptoms include dysphasia, hemiparesis, cranial nerve palsies, dizziness, dyspraxia, and visual field defects such as double vision or blurred vision, as well as generalized signs of intracranial pressure and hydrocephaly because of tumor-associated edema. Syncope is often frequent. Because of the lack of specific symptoms, glioblastomas may be misdiagnosed as infections, inflammatory processes, or either circulatory or immunological diseases. Cerebral edema and increased intracranial are the most common causes of death.

Diagnosis

Glioblastomas, seen during MRI or CT, often appear as irregular, ring-enhancing lesions, but may resemble an abscess, metastasis, or **tumefactive** multiple sclerosis in the right clinical setting. They tend to have a darker, central area of necrosis that is surrounded by the ring. Definitive diagnosis requires a **stereotactic biopsy** or craniotomy with tumor resection and pathologic confirmation. This is often based on tissue patterns instead of specific cell types. The target is usually the contrast-enhancing ring that is seen during imaging. Biopsy or subtotal tumor resection can result in undergrading of the lesion because grading is based upon the most malignant portion of the tumor. Imaging of tumor blood flow with perfusion MRI, and measuring tumor metabolite concentration with magnetic resonance spectroscopy (MRS), positron emission tomography (PET) scans, ad cerebral blood volume measurements may improve results (see Fig. 7.6). In general, PET scans are less often performed than the other methods, and pathology is still the gold standard. It is important to distinguish primary and secondary glioblastomas. This can be done by searching for IDH1 gene mutations, since these are much more common in secondary glioblastomas. Brain tumors that are distinguished from anaplastic astrocytomas include metastatic tumors, lymphomas, **hemangioblastomas, craniopharyngiomas, teratomas**, ependymomas, and **medulloblastomas**.

Glioblastomas are often very large at the time of presentation, depending on their location, and able to occupy much of a brain lobe. Though usually unilateral, those occurring in the brain stem and corpus callosum may be bilaterally symmetrical. Supratentorial bilateral extension occurs from growth along myelinated structures, especially over the corpus callosum and commissures. Most cerebral hemisphere glioblastomas are intraparenchymal, with epicenters in the

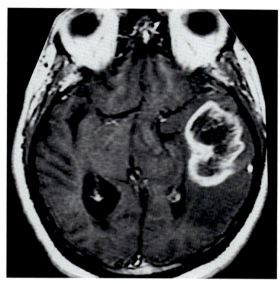

FIGURE 7.6 MRI showing glioblastoma in the temporal lobe. *Source: https://upload.wikimedia.org/wikipedia/commons/a/a3/AFIP-00405558-Glioblastoma-Radiology.jpg (PUBLIC DOMAIN, FREE TO USE).*

white matter. Less often, they are superficial, contacting the leptomeninges and dura, and may be interpreted as metastatic carcinoma or as meningioma. Cortical infiltration may result in a **gyriform** rim of thick gray cortex that overlies a necrotic zone in the white matter. Necrotic tissue may border nearby brain structures with no intermediate area of macroscopically detectable tumor tissue. Central necrosis may involve upwards of 80% or more of the total volumetric mass of the tumor. Extensive hemorrhages may cause symptoms that resemble those of stoke, which may be the first clinical sign, although smaller scale hemorrhagic conversion is more frequently seen. Macroscopic cysts may be present, containing a turbid fluid, and resemble liquefied necrotic tumor tissue.

Glioblastomas are usually extremely cellular gliomas, and prominent microvascular proliferation with or without necrosis is essential for diagnosis. Histopathology is highly variable. Some lesions have much cellular and nuclear polymorphism with many multinucleated giant cells. Others are highly cellular but comparatively monomorphic. In some tumors, the astrocytic nature is easily identified focally. This is difficult to see in tumors with poor differentiation. Regional heterogeneity is significant, making **histopathological** diagnosis difficult in **stereotaxis** needle biopsy specimens.

In glioblastomas, differentiated neoplastic astrocytes are usually seen focally, though poorly differentiated, fusiform, **pleomorphic**, or round cells can prevail. This is especially true of glioblastomas due to progression of grade II diffuse astrocytomas. Transitions between areas with recognizable astrocytic differentiation and extremely anaplastic cells can be abrupt or continuous. In gemistocytic lesions, anaplastic cells may be mixed with differentiated gemistocytes. An abrupt change in morphology may reveal emergency of a new tumor clone, from acquiring a genetic alteration or alterations. Pleomorphic cells involve small, lipidized, granular, giant, and undifferentiated cells. Bipolar, fusiform cells may form intersected bundles, with fascicles prevailing. If polymorphic tumor cells with delineated plasma membranes accumulate, and there is a lack of cell processes, the tumor may resemble melanoma or metastatic carcinoma. Some gliomas have clear patterns with specific cell types, which include the following:

- Small cells—extremely monomorphic small, round to slightly elongated, hyperchromatic nuclei. There is only slight visible cytoplasm or nuclear atypia. Often, mitotic activity is brisk. This appearance of glioblastoma may overlap with anaplastic oligodendroglioma (AO) because of their nuclear regularity, although microcalcifications, clear halos, and microvasculature that resembles chicken wire tend to be see in AO. Small cell glioblastomas often (70%) have EGFR amplification, and more than 95% have chromosome 10 losses.
- Primitive neuronal cells—a diffuse glioma with one or several solid-appearing primitive nodules that show neuronal differentiation. The glioma is usually astrocytic, but rare primitive neuronal foci have been reported in oligodendrogliomas. The primitive foci are greatly demarcated from adjacent glioma. They have significant increased cellularity, a high nuclear-to-cytoplasmic ratio, and high mitosis-karyorrhexis index. Variable features include cell wrapping, Homer Wright rosettes, and anaplastic cytology that resembles medulloblastoma. There may be

immunoreactivity for neuronal markers, including **synaptophysin**, loss or reduction of GFAP expression, and a greatly elevated monoclonal antibody Ki-67 proliferation index, in comparison with adjacent foci of glioma. This subtype presents de novo or in progression from a known diffuse glioma. Survival time and genetic background are similar to general glioblastoma, but there is a 30%−40% rate of CSF dissemination, and about 40% frequency of amplification of the MYCN or MYC genes. New chromosome 10q losses, or larger regions of 10q loss are also found. Alterations of all these genes or chromosomes may be linked to MYNC-controlled murine forebrain tumors.

- Oligodendroglioma components—these sometimes appear in glioblastomas, of varied sizes and frequencies. Necrosis is linked to a worsened prognosis in an anaplastic glioma that has **oligodendroglial** as well as astrocytic components. If the tumor has necrosis, survival time is shorter. If classified as glioblastomas with an oligodendroglial component, there may be a better prognosis time than for standard glioblastoma. The World Health Organization does not consider glioblastoma with an oligodendroglioma component to be distinct, diagnostically. 1p/19q co-deletion (1p19q codel) is diagnostic of oligodendroglioma.
- Gemistocytes—cells with prominent glassy, nonfibrillary cytoplasm that displaces nuclei to the periphery. Processes radiate out from the cytoplasm. These are short and "stubby" in appearance. The central hyaline organelle-rich area stays mostly unstained, while GFAP staining largely affects the cell peripheries. Perivascular lymphocytes often fill the gemistocytic regions. Often, with a preexisting glioma, the cells may signal a lower-grade precursor lesion in a secondary IDH-mutant glioblastoma. Areas of higher differentiation may be identified radiologically as noncontrast-enhancing peripheral regions. In whole-brain sections, they may be identified as grade II to III astrocytomas that are distinct from glioblastoma foci. Immunohistochemical studies reveal low proliferation rates of neoplastic gemistocytes, even though grade II or III gemistocytic astrocytomas progress more quickly to glioblastomas. The proliferating component presents as a cell population with minimal cytoplasm and larger hyperchromatic nuclei.
- Multinucleated giant cells—large multinucleated cells are common with glioblastomas. They occur with increasing size as well as pleomorphism. Their presence is not necessarily related to a more aggressive disease course. They are considered a regressive type of change, and if they dominate histopathologically, the descriptive term *giant cell glioblastoma* is used.
- Granular cells—large cells with a granular, periodic acid-Schiff-positive cytoplasm may be scattered in the glioblastoma. They rarely dominate to resemble a similar yet unrelated granular cell tumor of the pituitary stalk. In the cerebral hemispheres, transitioning forms between neoplastic astrocytes and granular cells may be identified. The cells also resemble macrophages, but are larger and exhibit more coarse granularity. They can be misinterpreted as a demyelinating disease rich in macrophages. The cells may be immunoreactive for macrophage markers, and some cells are peripherally **immunopositive** for GFAP, yet most are negative. Diffuse astrocytic tumors have extensive granular cell change. They have been referred to as *granular cell astrocytoma* or *granular cell glioblastoma*. The lesions have a unique appearance, characterized by aggressive glioblastoma-like development.
- Lipidized cells—cells with foamy cytoplasm; they are rare, but heavily lipidized, and can be extremely enlarged. If the lesion is superficial and the patient is young, a diagnosis of *pleomorphic xanthoastrocytoma* is considered—especially if xanthomatous cells are surrounded by basement membranes that stain positively for **reticulin**, and are companied by eosinophilic granular bodies. There may be epithelioid cytology of other lesions, and the lobules of fully lipidized but not foamy cells can resemble adipose tissue. If the epithelioid appearance dominates histopathologically, the descriptive term *epithelioid cell glioblastoma* is used.
- Metaplasia—there can be aberrant differentiation in neoplasms, such as when glioblastoma foci display features similar to squamous epithelial cells, such as epithelial whorls, with keratin pearls and expression of cytokeratin. Sometimes, glioblastomas have foci with glandular and ribbon-like epithelial structures. There are large, oval nuclei, prominent nucleoli, and rounded, defined cytoplasm. The term *adenoid glioblastoma* may be used. There may be reduced expression of GFAP, replaced by expression of epithelial markers. A diagnosis of *gliosarcoma* should be considered if the mesenchymal component is extensive—especially a spindle cell sarcoma-like component. In metaplastic glioblastomas, a mucinous background and a mesenchymal component are common. Adenoid and squamous epithelial metaplasia are seen in gliosarcomas more than in glioblastomas.
- Secondary structures—the ability of glioblastoma cells to migrate in the CNS is more obvious when they reach any barrier. The cells form a line and accumulate in the subpial cortex, within the subependymal region, and around neurons and blood vessels. These patterns, or *secondary structures*, form from glioma cell interactions with brain structures. They are very diagnostic of an infiltrating glioma. Secondary structures may also exist in oligodendrogliomas and other infiltrative gliomas. Tumor cells adapt to myelinated pathways, often creating a polar, fusiform shape. The identification of neoplastic astrocytes in the **perifocal** zone of edema and at distant sites can be difficult—especially in stereotaxic biopsies. Secondary structures and most multifocal glioblastomas basically arise as a result of how

glioma cells migrate in the CNS. In terminal disease stages, the subependymal region may have additional diffuse infiltration.
- Proliferation—this is usually prominent, and atypical mitoses are often detected. There is a wide **inter-** and **intratumoral** variation in mitoses. Growth fraction also has great regional variation. Though values are usually 10% − 15%, tumors can have a proliferation index of over 50% focally. If there are small, undifferentiated, and fusiform cells histologically, the tumor often has significant proliferative activity.
- Microvascular proliferation—this is a hallmark of glioblastomas, and in light microscopy, it usually appears as glomeruloid groupings of multilayered, mitotically active endothelial and smooth muscle cells or pericytes. A less common form is a hypertrophic proliferating endothelial cell in medium-sized vessels. The glomeruloid type of microvascular proliferation usually occurs near the area of necrosis and reflects responses to **vasostimulatory** factors released from ischemic tumor cells. Vascular thrombosis is common, and may be apparent, appearing as grossly visible "black veins." Basal lamina and an incomplete pericyte layer surround hyperplastic endothelial cells. Inconspicuous vessels have a proliferation index of 2% − 4%, while proliferating tumor vessels are at 10%. Vascularization occurs through vessel co-opting by migrating tumor cells, creating angiogenesis/vasculogenesis. Intussusception and cancer stem cell-derived vasculogenesis also can occur. Hypoxia frequently drives glioblastoma angiogenesis, leading to intracellular stabilization of the regulator HIF1A, leading to transcriptional activation of over 100 hypoxia-linked genes that encode proteins controlling angiogenesis, cellular metabolism, survival or apoptosis, and migration. The most important mediator of glioma-associated vascular functions appears to be vascular endothelial growth factor A (VEGFA), which is mostly produced by palisading cells, driven by hypoxic and hypoglycemic stressors. In turn, tumor angiogenesis is induced by VEGFA, increasing edema. It also regulates the homing of bone marrow-derived cells. Use of monoclonal antibodies to block VEGFA, in the treatment of recurrent glioblastoma, appears to mostly target small and immature vessels, leading to vascular normalization with improved perfusion and oxygenation which almost counterintuitively typically helps the clinical and radiographic picture. Other signaling pathways needed for glioblastoma angiogenesis include **angiopoietin/Tie2** receptor, interleukin 8 (IL8) and its receptor, platelet-derived growth factor (PDGF) and its receptor, wingless-interleukin (WNT)-beta-catenin, ephrin (Eph), and transforming growth factor beta signaling. Vascular remodeling also involves pericytes/smooth muscle cells and perivascular bone marrow-derived cells, along with endothelial cells.
- Necrosis—tumor necrosis is a basic feature of many tumors and specifically for glioblastomas, greatly predicts aggressive tumor behavior. There is almost always a hypodense core in a contrast-enhancing rim, with necrosis making up areas of nonviable tumor tissue that may be very small or imperceptible on MRI, to more than 80% of total tumor mass. On MRI, more necrosis is capriciously linked to shorter survival. Necrosis grossly appears as white or yellow granular coagulum. Glioma cells as seen under the microscope reveal **necrobiotic** debris and faded contours of large, dilated necrotic vessels. Another hallmark is the palisading form, with many small, irregularly shaped foci that are surrounded by radially oriented and densely packed glioma cells. Palisading cells are hypoxic, strongly expressing HIF1A and other hypoxia-inducing transcription factors. Microvascular proliferation occurs very close to the palisading cells and provides a biological link between necrosis and microvascular hyperplasia.
- Apoptosis—this programmed cell death is characterized by nuclear fragmentation and condensation. The apoptotic cell bodies are packaged in an intact membrane, with the process started by release of mitochondrial factors, or by death receptor ligation by members of the tumor necrosis factor (TNF) family. Higher levels of apoptosis in palisading cells around necrotic areas may be from increased expression or ligation of death receptors. TNFs induce apoptosis by ultimately activating caspase-8. Most TNF expression is in the palisading cells, and physical interactions between tumor cells that express TNFs may promote apoptosis. For malignant gliomas, overall amounts of cell death from apoptosis are low, compared with coagulative necrosis. Apoptotic rates are not necessarily related to prognosis.
- Inflammation—there are varied numbers of inflammatory cells in glioblastomas. Sometimes, significant perivascular lymphocyte "cuffing" occurs, usually in areas that have a homogeneous gemistocytic component. In small cell glioblastomas, inflammatory cells are rare or totally absent. Inflammatory cells are mostly CD8 + T-lymphocytes, but CD4 + lymphocytes are also present. Also, B-lymphocytes are detected in under 10% of cases. Long-term glioblastoma survivors may have more extensive CD8 + T-cell infiltrates. Microglia and histiocytes are also present, but histiocytes laden with lipids are not common.
- Immunophenotype—there is often expression of GFAP, though degrees of reactivity vary widely. Gemistocytic areas are often highly positive, while primitive cellular components are usually negative. There may be expression of GFAP in the gliomatous component of gliosarcoma, but not in the sarcomatous component. Expression of **cytokeratins** is based on the class of intermediate filaments present and antibodies used. Keratin positivity is usually

detected with the keratin antibody mixture called AE1/AE3. **Nestin** is often expressed and can aid in diagnosis between glioblastomas from other high-grade gliomas. Glioblastomas that have missense TP53 mutations show strong, diffuse overexpression of p53 in up to 53% of cases. The expression of EGFR occurs in up to 98% of glioblastomas and is in some way related with the presence of gene amplification. Tumors with histone 3 (H3), K27M gene mutations can be detected via an antibody that is specific for K27M-mutant H3, a useful tool to distinguish the tumors from other astrocytic tumors. Notably, H3K27M gliomas tend to be midline gliomas, have a different biological behavior and outcomes than glioblastoma.

Treatment

For glioblastoma, treatment usually includes maximum feasible surgical resection, radiation therapy, biochemotherapy, and alternating electric fields. Glioblastoma cannot be totally removed with surgery, but patients undergoing extensive resection may have a slight survival advantage. In most cases, only a part of the glioblastoma can be removed because of its location and infiltrative nature. Debulking, or partial resection, prolongs survival in many cases, and is often combined with radiation and chemotherapy. Subtotal excision means that substantive portions of the tumor can be removed, for cytoreduction and possible improvement of neurological function. Gross total excision means that all of the tumor can be removed, even though diffuse and infiltrative microscopic cells remain. There are many cortical electrophysiologic imaging and mapping techniques that allow for maximal resection without harming nearby brain tissues. Accurate biopsy of tumor tissue is extremely critical. Open (or stereotactic) biopsy involves only partial tumor removal.

Radiation is a mainstay of treatment, and is given 5 days per week. Regional cranial irradiation increases survival for an average of five months, even in elderly patients who have had only a biopsy and no resection. Use of chemotherapy for anaplastic astrocytoma is contingent on the molecular characterization of the tumor. Generally, maximum surgical resection with radiation therapy is used, and use of temozolomide (TMZ) or combination procarbazine, lomustine (CCNU), ± vincristine (PCV) chemotherapy is reserved for recurrences. For glioblastoma, there is evidence that newly diagnosed tumors can be treated with temozolomide plus radiation therapy, followed by adjuvant temozolomide. This increases median survival and greatly improves 2-year survival rates. Another form of treatment in the post chemoradiation phase of glioblastoma patients (adjuvant treatment) involves **alternating electric field therapy** in combination with temozolomide. This is an FDA approved NCCN guideline category I recommendation for patients with glioblastoma. Studies have shown that this treatment, along with temozolomide and potentially continuation of the device through second line therapy, appears to improve outcomes with respect to time to radiographic survival and overall survival.

Radiographic worsening may appear soon after radiotherapy is started, especially in patients who are MGMT-methylated. This may reflect pseudoprogression of the disease, and not true tumor progression. If pseudoprogression is identified, adjuvant temozolomide should be continued, since there will be resolution of pseudoprogression in 1 − 3 months. Because there is no reproducibly validated biomarker for glioblastoma and advanced imaging with MR spectroscopy and MR perfusion may not be completely indicative of the complete tumor biology. Another consideration is the temozolomide causes thrombocytopenia or leukopenia in up to 10% of patients, along with *Pnemocystis jirovecii (carinii)* pneumonia, a rare outcome. The difficulties in treating glioblastomas are increased because the tumor cells are very resistant to many therapies, which can also damage the brain. Nitrosoureas such as carmustine or lomustine may help extend survival slightly, and carboplatin or irinotecan may offer similar small benefits.

Tyrosine kinase inhibitors such as erlotinib and gefitinib have been developed in relation to upregulation of EGFR, with no reproducible or predictable tested success to date. Though not proven successful, high-dose focused radiation, electron or proton based, are under continuing study in subgroups of patients. High-dose focused radiation, whether single dose or fractionated, is also called *stereotactic radiosurgery*. Palliative treatments include brief use of dexamethasone or other corticosteroids, when there are symptoms of mass effect such as headache or drowsiness. Surrounding edema and local signs usually also improve. If there have been seizures, antiepileptic medications are used. Serious skin reactions such as erythema multiforme and **Stevens − Johnson syndrome** can develop if the patient is receiving phenytoin along with cranial radiation.

For recurrence, there is no standard of care for treatment. It is guided by molecular characterization of the tumor, tumor location and pattern of growth, plus the age of the patient and his or her state of health. Most glioblastomas recur within 2 cm of the original site. About 10% develop additional lesions in distant locations. Additional surgery may be undertaken, with the most aggressive approach being a second surgery combined with chemotherapy. This can be effective and is generally used for patient under age 40 whose initial operation was performed many months previously. Often, the medical oncologist recommends procarbazine-lomustine (CCNU)-vincristine as a combination treatment.

However, this combination is usually not as well tolerated as temozolomide and the trade-off for gains against the tumor are difficult to weigh. The combination regimen may reduce symptoms but does not have a significant effect upon survival for patients with glioblastoma. Drugs targeting tumor vasculature are promising. Antiangiogenic drugs include the VEGF inhibitor bevacizumab. They may be given along with chemotherapy, delaying progression and reducing cerebral edema, which attenuate if not mitigate the need for corticosteroids. However, they also do not extend survival in the "up front" setting but do typically help with quality and quantity of life at recurrence, particularly when there is prominent vasogenic edema. A new study showed that valganciclovir—usually used for concurrent cytomegalovirus infections—has prolonged survival in glioblastoma patients.

Prognosis

Unfortunately, glioblastoma is usually fatal—the deadliest of all primary brain and CNS cancers. However, there is improved overall survival if the tumors demonstrate the DNA repair enzyme known as MGMT. If the tumor's promoter region is methylated, the mass is less able to repair damage caused by chemotherapy, giving a 2-year survival of 27.2% instead of only 10.9% in tumors lacking MGMT when implementing a standard chemotherapy plan without addition of alternating electric field therapy. Likewise, the 5-year survival is about 10% for those with MGMT compared to only 1.9% in those without the enzyme. There are no other predictive biomarkers that aid in determining if primary or secondary therapies should be undertaken. Most patients with glioblastomas survive only 14 − 18 months following diagnosis, if radiation therapy and chemotherapy are used. The most common adjunct biochemotherapeutic agents include bevacizumab, nitrosourea compounds, platinum drugs, or other interventions in the context of a clinical trial. Recent vaccine and immunotherapy treatments are showing promise. Combining different vaccines with *checkpoint inhibitors* may aid in better outcomes. Survival is usually shorter, about 12.1 years, when only surgery and radiotherapy are used. This increases to about 24 months when alternating electric field therapy is used appropriately. About 3 of every 10 people will survive for 2 years, but less than 3% − 5% of patients survive longer than 5 years. Without treatment, survival is usually 3 months. Increased age, over 60 years, has a worsened prognosis, with death usually being from widespread tumor infiltration, cerebral edema, and increased intracranial pressure. Primary glioblastomas have a poorer prognosis than secondary glioblastomas. A cure is only considered to have occurred if there has been no recurrence within 10 years following treatment. In extremely rare cases, patients have survived 20 years and beyond; this subset of patients are continually being studied and surveilled.

Isocitrate dehydrogenase-wild-type

The IDH-wild-type glioblastomas are the most common and most malignant glioblastomas, making up about 90% of these tumors. They are also called *IDH-wild-type primary glioblastomas*. Unlike the IDH-mutant glioblastomas, they do not arise from lower-grade IDH-mutant astrocytomas. In fact, they have no lower-grade precursor tumors. Many of their genetic alterations are also present in most grade II and III wild-type astrocytomas. These alterations may involve loss of chromosomes *10p* and *10q*, *telomerase reverse transcriptase (TERT) promoter* mutations, and mutations of *EGFR* and *phosphatase/tensin homolog (PTEN)*. A higher likelihood for IDH-wild-type glioblastomas is linked to inherited cancer syndromes such as NF1, **Li-Fraumeni**, and **Turcot syndrome**.

IDH-wild-type glioblastomas are distributed throughout the cerebral hemispheres, involving the subcortical and deep periventricular white matter. They spread across compact tracts, including the corpus callosum and corticospinal tracts, and the butterfly glioma pattern is common (see Fig. 7.7). This form occurs with fewer predilections for any certain brain area, compared to general glioblastomas. Sizes of these glioblastomas vary. Since they spread quickly and significantly, along the compact white matter tracts, 20% may be multifocal lesions upon diagnosis. Between 2% and 5% of multifocal glioblastomas are synchronous tumors that develop independently. The time period from first symptoms to diagnosis is less than 3 months in 68% of cases, and less than 6 months in 84% of cases.

In most cases they have a reddish-gray outer layer that surrounds a central necrotic core. There is significant mass effect and **peritumoral edema**, and often, **intratumoral hemorrhage** and increased vascularity. The necrosis and microvascular proliferation distinguishes these tumors from anaplastic astrocytomas. On the cellular level, there are pleomorphic fibrillary astrocytes, gemistocytes, bipolar cells that have a bland appearance but are mitotically active, and larger bizarrely multinucleated giant cells. There is a high **proliferation index** of usually more than 10%.

IDH-wild-type glioblastomas usually occur in older adults between the ages of 60 and 75 years, with a mean age of 62 years. The male to female ratio is 1.60:1 in the United States, and lower in Switzerland (1.28:1). The most common symptoms are seizures, focal neurologic deficits, and mental status changes. Headache, because of elevated intracranial

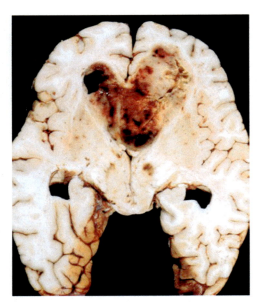

FIGURE 7.7 Glioblastoma appearing as a necrotic, hemorrhagic, and infiltrating mass.

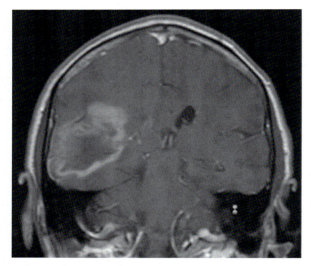

FIGURE 7.8 Contrast T-1 weighted coronal magnetic resonance image shows a large mass in the right parietal lobe with "ring" enhancement.

pressure, is common. Mean survival is less than 12 months. Usually, there is a **hypodense** central mass surrounded by an isodense to slightly hyperdense rim. In rare cases, no dominant mass is present, with the tumor extending diffusely through the cerebral white matter. Confluent, patchy white matter hyperintensities resemble small vessel vascular disease. Hemorrhage is common while calcification is rare. There is strong but heterogeneous irregular rim enhancement. Fig. 7.8 shows a contrast T-1 weighted coronal MRI of a large IDH-wild-type-glioblastoma in the right parietal lobe with "ring" enhancement. An extremely rare variant known as *primary diffuse leptomeningeal gliomatosis* extends diffusely around brain surfaces, but mostly between the pia and glia limitans of the cortex.

Via magnetic resonance imaging (MRI), a poorly marginated mass with mixed signal intensity is seen, often with subacute hemorrhage. Heterogenous hyperintensity is present with indistinct tumor margins and vasogenic edema. Sometimes, cysts, necrosis, hemorrhage at different stages of development, fluid/debris levels, and **flow voids** because of extensive **neovascularity** are seen.

Isocitrate dehydrogenase-mutant

The IDH-mutant glioblastomas are much less common than the IDH-wild-type glioblastomas, making up only 5% − 10% of cases. These *secondary glioblastomas* usually develop from lower-grade IDH-mutant astrocytomas. An IDH mutation occurs early in **gliomagenesis**, continuing throughout tumor progression. There is no EGFR amplification, but *tumor protein 53 (TP53)* mutations and *alpha-thalassemia/mental retardation syndrome, X-linked (ATRX)* loss are common. The IDH-mutant glioblastomas not only appear different, but for unknown reasons usually occur in the frontal lobe, usually in the area that surrounds the rostral lateral ventricles. Such glioblastomas may project into the lateral ventricle. They infiltrate into the nearby cortex and through the corpus callosum, into the contralateral hemisphere. They diffusely infiltrate the brain, and hemorrhage. Large areas of central necrosis are usually not present. Focal areas of oligodendroglioma-like components are often seen. These grade IV tumors arise at a mean age of 45 years. They have a better prognosis than the IDH-wild-type glioblastomas, with a mean overall survival of two to three years. On imaging, significant nonenhancing areas are usually present, and there is usually no thick outer layer surrounding a necrotic core. However, molecular profiling is required for a definitive diagnosis, since exceptions to this are common. When symptom development takes a long time, this form of glioblastoma should be considered.

Gliosarcoma

Gliosarcomas are characterized by a biphasic tissue pattern. There are areas of **gliomatous** and sarcomatous differentiation. In these areas, there are similar cytogenetic alterations, and they are monoclonal in origin. Sarcomatous areas are probably from phenotypic changes in glioblastoma cells instead of concurrent development of two separate neoplasms that are genetically distinct. Gliosarcomas can be primary or secondary, but most are primary, arising de novo. Secondary gliosarcomas occur after previous resection or irradiation of glioblastomas, or as radiation-induced tumors when there is no prior history of glioblastomas. Genetically, gliosarcomas are mostly similar to primary glioblastomas. There are mutations or deletions of PTEN, mutations of TP53, and amplification of *cyclin dependent kinase 4 (CDK4)* and *mouse double minute 2 (MDM2)*. Amplification of EGFR is rare, as is MGMT methylation and IDH1 mutations.

Nearly half of all cases of gliosarcomas occur in the temporal lobe, followed by 20% in the frontal lobe and 15% in the parietal lobes. There are two gross phenotypic subtypes. One resembles a meningioma, having a superficial location and a relatively well circumscribed, solid, and mostly sarcomatous tumor mass (see). The other subtype is located deeper in the brain, resembling a glioblastoma with a necrotic center and hypervascular outer layer. The mesenchymal features may be fibroblastic, cartilaginous, osseous, muscular, or adipose. These grade IV neoplasms only make up about 2% − 4% of all glioblastomas. They usually occur between the fifth and seventh decades with a mean age of 60 years. They occur 2.5 times more often in males than females. Overall survival is only 13 months, but median survival if there is a meningioma-like phenotype is slightly longer. After surgical resection, local recurrence is common (about 90% of cases). Extracranial metastases occur in 10% − 15% of cases.

Gliosarcomas usually appear as a heterogeneously enhancing solid mass with various amounts of surrounding edema. These tumors often reach the meninges, but may not have dural attachment or any obvious tissue invasion. Many develop deep in the cerebral hemispheres, away from the dura-arachnoid. They have a central necrotic core surrounded by a thick outer layer of enhancement that may be nearly identical to the IDH-wild-type glioblastomas. Major differential diagnoses include anaplastic meningioma and glioblastoma with dural invasion. Other dura-based lesions, with variable amounts of brain invasion, include other sarcomas, dural metastases, lymphoma, **neurosarcoid**, and **plasmocytoma**.

Clinical cases

Clinical case 1

1. Where do most glioblastoma tumors occur in the brains of adults?
2. How are glioblastoma tumors diagnosed in the majority of patients?
3. What are the facts about survival rates for glioblastoma multiform of the cerebrum?

A 44-year-old man was examined after having been diagnosed with a cerebral glioblastoma four months previously. This was initially seen on brain MRI and confirmed by needle biopsy. Since diagnosis, he had developed numbness in his left hand and on the left side of his face. The epicenter of the tumor was within the right thalamus, and the lesion was expanding toward the brainstem and hippocampus. Unfortunately, this tumor was considered inoperable due to its deep location. The patient was given steroids, radiotherapy, and chemotherapy, which provided slight initial improvement. Eventually, the tumor progressed and the patient declined additional chemotherapy. His past history included smoking for 10 years, but he had quit 5 years before his diagnosis. The patient's medications included oxycodone for headache, pregabalin for tumor pain, lactulose for constipation, metoclopramide for nausea, and escitalopram for depression. Significant symptoms included intermittent vomiting and constipation, dizziness, headaches, stiffness in both arms, and a feeling of pressure in the posterior neck. A year after diagnosis, the patient was still living a relatively normal life, with modest lifestyle limitations.

Answer:

1. **In most cases, glioblastoma tumors occur in the cerebral hemispheres—mostly in the temporal lobes, followed by the frontal, parietal, and occipital lobes—in adults. They can arise from lower-grade astrocytomas.**
2. **The diagnosis of glioblastoma is always a pathology based one. Preoperatively, lesion(s) may be seen via CT, MRI ± enhanced software such as MRS and cerebral blood volume measurements, or less commonly PET scans.**
3. **For adults, glioblastoma has a 5-year survival rate of about 10%. When all cases of this tumor are considered, the survival rate averages 14–18 months with standard of care therapy. Median survival rate decreases to 12.1 months when only surgery and radiotherapy are used and increases to roughly 24 months when alternating electric field therapy is appropriately applied. Overall, glioblastoma is the deadliest of all primary brain and CNS cancers.**

Clinical case 2

1. How do glioblastomas develop in the brain over time?
2. What is the primary noninvasive diagnostic tool for detecting glioblastoma?
3. What are the common diagnostic factors of glioblastoma?

A 60-year-old man from England was admitted to a hospital after experiencing convulsions. Physical examination revealed that his pupillary light reflex was slow; there was high muscular tension of his limbs, and positive bilateral Babinski's signs. An MRI revealed abnormal signals from his bilateral temporal-occipital-parietal lobes, resulting in suspicion of a brain tumor. The patient had experienced head trauma and a basal skull fracture 10 years earlier. Additional imaging revealed a solid cystic lesion about 7 mm in diameter, with clear edges. The inferior horn of the lateral ventricles was compressed. The patient was given mannitol to lower intracranial pressure, diazepam, carbamazepine to manage the seizures, and ceftriaxone for infection control. Though symptoms were managed well for a time, 3 weeks later another MRI revealed multiple lesions that had greatly enlarged. Glioblastoma was suspected, and resection of the tumor in the left temporal occipital lobe was performed. Seven months later, the patient had a sudden headache and then could not control his arms or legs. A CT scan revealed recurrence of the glioblastoma, which ultimately proved to be fatal.

Answers:

1. **Glioblastomas develop as a result of infiltrative growth, via nerve fiber pathways, or metastatic spread in the cerebrospinal fluid. Therefore, satellite lesions form in nearby areas of the brain. The infiltrative growth of tumor increases intracranial pressure and may then lead to hydrocephaly. Aggravation of clinical symptoms may be due to the rapid tumor growth, which damages local tissue structures and causes impaired brain function.**
2. **MRI is the primary diagnostic tool for glioblastoma. The rapid morphological progression of cystic lesions is indicative of glioblastoma. Progression of tumor growth can be fast enough so that the size doubles over just 10 days.**
3. **Glioblastoma has no specific clinical characteristics. Based on localization and progression, clinical symptoms include headache, ataxia, dizziness, visual disturbances, and frequent syncope. Occurrence of seizures in patients not previously diagnosed with epilepsy may be an indication for neuroimaging because of suspected glioblastoma. This tumor, because of the lack of specific symptoms, is often misdiagnosed as an infection, inflammatory process, or a circulatory or immunological disease. Imaging can be quite useful to guide diagnostic intervention and histological and molecular profiling of tissue is diagnostic for glioblastoma.**

Clinical case 3

1. How does this patient's age impact the treatment for glioblastoma?
2. What are the links between pesticides and glioblastoma?
3. What are the hallmark features of glioblastoma seen in imaging studies?

A 65-year-old woman with glioblastoma entered a neuro-oncology clinic. She had previously been diagnosed with chronic obstructive pulmonary disease (COPD) and a thyroid tumor. She had smoked cigarettes for 23 years of her life, but had quit 8 years previously. Her work history included 20 years at a pesticide manufacturing facility. Over time, she had become fatigued, easily angered, and then developed weakness in her left arm and leg, as well as slurred speech. An MRI revealed a right fronto-parietal lesion that enhanced with contrast, and edema. The next day, the patient underwent a stereotactic-guided right fronto-parietal craniotomy with radical subtotal resection of the lesion. She was later given phenytoin, followed by levetiracetam, both of which caused her to experience adverse reactions, and eventually valproic acid was used successfully. Her left-sided weakness improved and she began to be able to walk without help. Over time, she still was fatigued and her left arm and leg were somewhat weak, but otherwise there were no significant symptoms.

Answers:

1. **Older age is one of the most significant factors associated with a poor prognosis of glioblastoma. This tumor mainly occurs in people between ages 55 and 85. With increased age, the survival prognosis worsens with no clear clinical cutoff age. Survival is short even when the elderly patient receives multiple therapies. Surgical resection prolongs survival regardless of age.**
2. **Glioblastoma is linked to environmental exposures to pesticides, cigarette smoke, vinyl chloride, petroleum refining, and synthetic rubber manufacturing. Occupational exposures to pesticides are also linked to other types of brain tumors, such as meningiomas.**
3. **The hallmark features of glioblastoma seen during imaging studies include enhancement with gadolinium contrast, and an irregularly shaped mass that has a dense ring of enhancement with a hypointense center of necrosis. The presence of necrosis is required for the tumor to be grade IV or to be classified as a glioblastoma using the World Health Organization's classification system. Surrounding vasogenic edema, hemorrhage, and ventricular distortion or displacement may also be seen in diagnostic imaging.**

Clinical case 4

1. How common is a glioblastoma that is located within the lateral ventricle?
2. How does glioblastoma rank compared to other cancers for people of this patient's age?
3. What are the three surgical modalities for glioblastoma?

A 32-year-old woman with no significant previous medical history presented for evaluation of headaches and intermittent short-term memory loss, plus mild nausea. The headaches had started three months before presentation. No neurological deficits were present. A head CT scan found a large, poorly enhancing right occipital-parietal mass within the lateral ventricle. Subsequent MRI confirmed that the tumor was located in the trigone with local expansion of the ventricle. The patient underwent surgical resection of the tumor, which was gray in color, firm, and not highly vascular. After surgery, the patient experienced no neurological problems. The tumor was diagnosed as a glioblastoma based on its necrosis and endothelial cell proliferation. Two years after surgery, the patient still had occasional headaches, but there were no other symptoms or any tumor progression.

Answers:

1. **Glioblastomas have often been reported to project into the lateral ventricle, along with many other types of brain tumors. Overall, though, intraventricular neoplasms are relatively rare in the global population. Most glioblastomas present in the frontal horn and adjacent to the lateral ventricle, not in the trigone.**
2. **Overall, glioblastomas are the third most common cancer between ages 15 and 34, ranked behind breast and skin cancers. However, in younger adults, brain cancers such as glioblastoma are the most common cause of cancer death. After age 34, thyroid cancer becomes more common than brain tumors.**
3. **The three surgical modalities for glioblastoma are biopsy, subtotal, and gross total. Biopsy (open or stereotactic) means that only a part of the tumor can be removed. Subtotal means that substantive portions can be removed for cytoreduction and potential improvement of neurological function. Gross total means that the entire tumor can be removed grossly, though diffuse and infiltrative microscopic cells remain.**

Key terms

Alternating electric field therapy
Angiopoietin/Tie2 receptor
ATP transcriptional regulator, x-linked
Brachytherapy
Butterfly glioma
Cathepsins
Craniopharyngiomas
Cytokeratins
de novo
Epidermal growth factor receptor (EGFR)
Epigenomes
Flow voids
Glioblastoma
Gliomagenesis
Gliomas
Gliomatosis cerebri
Gliomatous
Gliosarcoma
Glomeruloid
Gyriform
Hemangioblastomas
Heterozygosity
Histopathological
Hypodense
Immunopositive
Infratentorially
Intertumoral
Intratumoral hemorrhage
Isocitrate dehydrogenase (IDH)
Li-Fraumeni
Medulloblastomas
Mendelian syndrome
Methylation
Microenvironmental
Necrobiotic
Neovascularity
Nestin
Neurosarcoid
Nuclear pseudopalisading
Oligodendroglial
Perifocal
Perineuronal satellitosis
Peritumoral edema
Perivascular aggregation
Plasmocytoma
Pleocytosis
Pleomorphic
Pseudotumor cerebri
Reticulin
Serpiginous
Simian vacuolating virus 40
Stereotactic biopsy
Stereotaxis
Stevens-Johnson syndrome
Structures of Scherer
Subpial spread
Synaptophysin
Teratomas
Tumefactive
Turcot syndrome
Vasogenic edema
Vasostimulatory

Further reading

Baranska, J. (2012). *Glioma signaling (Advances in experimental medicine and biology, Book 986)*. Springer.
Berger, M. S., & Wilson, C. B. (1999). *The gliomas*. Saunders.
Bigner, D. D., Friedman, A. H., Friedman, H. S., McLendon, R., & Sampson, J. H. (2016). *The duke glioma handbook: Pathology, diagnosis, and management*. Cambridge University Press.
Brem, S., & Abdullah, K. G. (2016). *Glioblastoma*. Elsevier.
Burrows, G. (2017). *Glioblastoma – a guide for patients and loved ones: Your guide to glioblastoma and anaplastic astrocytoma brain tumors*. NGO Media.
Chernov, M. F., Muragaki, Y., Kesari, S., McCutcheon, I. E., & Lunsford, L. D. (2017). *Intracranial gliomas part I – Surgery (Progress in neurological surgery, Volume 30)*. S. Karger.
De Vleeschouwer, S. (2017). *Glioblastoma*. Codon Publications.
Duffau, H. (2017). *Diffuse low-grade gliomas in adults*. Springer.
Fleming, K. (2019). *Jack's story: A journey through the nightmare of a glioblastoma multiforme brain tumor*. K&D Publishing.
Germano, I. (2015). *Glioblastoma multiforme: Symptoms, diagnosis, therapeutic management and outcome*. Nova Science Publishers, Inc.
Hanicke, J. (2019). *GBM calling: A life journey with terminal cancer*. Amazon Digital Services LLC.
Hatrak, S. (2012). *A statistic of one: My walk with glioblastoma multiforme*. iUniverse.
Hayat, M. A. (2011). *Tumors of the central nervous system, volume 1: Gliomas: Glioblastoma (Part 1)*. Springer.
Hayat, M. A. (2015). *Tumors of the central nervous system, volume 14: Glioma, meningioma, neuroblastoma, and spinal tumors*. Springer.
McKhann, G. M., II, & Duffau, H. (2019). *Low-grade glioma, an issue of neurosurgery clinics of North America*. Elsevier.
Medifocus.com Inc. (2014). *Medifocus guidebook on glioblastoma – a comprehensive guide to symptoms, treatment, research, and support*. CreateSpace Independent Publishing Platform.
National Comprehensive Cancer Network. (2017). *NCCN guidelines for patients: Brain cancer – gliomas*. NCCN.
Nickerson, T., & Nickerson, J. (2014). *John: Funny and strong – our glioblastoma story*. CreateSpace Independent Publishing Platform.
Oberg, I. (2019). *Management of adult glioma in nursing practice*. Springer.
Parker, P. M. (2019). *The 2020–2025 world outlook for glioblastoma multiforme (GBM) treatments*. ICON Group International, Inc.
Pirtoli, L., Gravina, G. L., & Giordano, A. (2016). *Radiobiology of glioblastoma: Recent advances and related pathobiology*. Humana Press.
Placantonakis, D. G. (2018). *Glioblastoma: Methods and protocols (Methods in molecular biology)*. Humana Press.
Pope, W. B. (2019). *Glioma imaging: Physiologic, metabolic, and molecular approaches*. Springer.
Prados, M. D. (2012). *Advances in the pathogenesis and treatment of glioblastoma multiforme*. Future Medicine Ltd.
Ray, S. K. (2010). *Glioblastoma: Molecular mechanisms of pathogenesis and current therapeutic strategies*. Springer.

Simmons, A. M. (2018). *Embracing life from death: A caregivers journey through glioblastoma, grief and healing.* CreateSpace Independent Publishing Platform.
Somasundaram, K. (2017). *Advances in biology and treatment of glioblastoma (Current cancer research).* Springer.
Sughrue, M. E. (2019). *The glioma book.* Thieme.
Sughrue, M. E., & Yang, I. (2019). *New techniques for management of 'inoperable' gliomas.* Academic Press.
Vainboim, T. B., Nagahashi-Marie, S. K., & Pereira-Franco, M. H. (2017). *Quality of life among caregivers of patients with glioblastoma multiforme: Psychoeducational program.* Lap Lambert Academic Publishing.

Chapter 8

Oligodendroglioma

Chapter Outline

Oligodendroglioma	147	NOS anaplastic oligodendroglioma	156
IDH-mutant and 1p/19q-codeleted oligodendroglioma	147	Clinical cases	157
NOS oligodendroglioma	151	Key terms	158
Anaplastic oligodendroglioma (IDH-mutant and 1p/19q-codeleted)	153	Further reading	159

Oligodendroglioma

Oligodendroglioma is usually a grade II type of glioma, and is one of the slowest growing of all gliomas. These tumors are believed to originate from oligodendrocytes or glial precursor cells. At the cellular level, they appear similar to normal oligodendrocytes, being mostly composed of cells with small to slightly enlarged, round nuclei, with dark and compact nucleoli, and a small amount of eosinophilic cytoplasm. Oligodendrogliomas are most common in the forebrain, and especially common within the frontal lobes (see Fig. 8.1). They are characterized by deletion of the *p-arm of chromosome 1* (known as a *1p deletion*) and deletion of the *q arm of chromosome 19* (a *19q deletion*). CODEL1p19q and IDH mutation are both required to render the diagnosis of oligodendroglioma. These deletions aid in diagnosis, responses to treatment, and prognosis. Allelic losses on 1 P and 19q are commonly seen in this classic form of oligodendroglioma, and loss in one or the other ascribes the diagnosis as another variant form or astrocytoma. Oligodendrogliomas can also evolve into more aggressive forms, such as *anaplastic oligodendrogliomas*, which are grade III tumors. There is controversy about the grading of these tumors between various organizations. The differentiation between grade II and grade III tumors is based on a variety of factors, including hypercellularity, mitotic activity, pleomorphism, necrosis, and vascular proliferation. Seizures are a common feature of oligodendrogliomas, occurring in 70%-90% of cases, with nonspecific seizure patterns. There may be generalized, partial, and mixed seizure types. Younger patients are diagnosed earlier than adults, since the seizures are the dominating symptom, while adults usually have other neurologic symptoms. Surgery is aimed at gross-total resection while brain function is protected, but this requires special techniques since the tumors are often present near brain areas controlling movement, language, and sensation. *Awake surgery*, with brain mapping, is often performed if the tumor is near the language or movement related regions. When procarbazine-1-(2-chloroethyl)-3-cyclohexyl-1-nitrosourea (CCNU) (lomustine)-vincristine (PCV) chemotherapy is given before or after standard radiation therapy, patients usually have better odds of improved overall survival. For those lacking chromosome 1p and 19q deletions, survival benefit is not statistically significant. A survival advantage with combination therapy has been seen in patients with *O-6-methylguanine-DNA methyltransferase (MGMT)* methylated tumors and IDH mutation; all CODEL1p19q and IDH mutated oligodendrogliomas can be assumed to be MGMT methylated.

> **Point to remember**
> Oligodendrogliomas arise from oligodendrocytes, and can be grade II or grade III. They are most common within the frontal and temporal lobes of the brain, are soft, and often contain calcifications, hemorrhagic areas, with or without cysts. They usually growly slowly and may be present for years before being diagnosed. Most patients survive for a long time with these tumors.

IDH-mutant and 1p/19q-codeleted oligodendroglioma

This form of oligodendroglioma is diffusely infiltrative. It has the features of **isocitrate dehydrogenase** (IDH) type 1 or 2 mutations, and codeletion of chromosomal arms *1p* and *19q*. Chromosome 1 is the designation for the largest

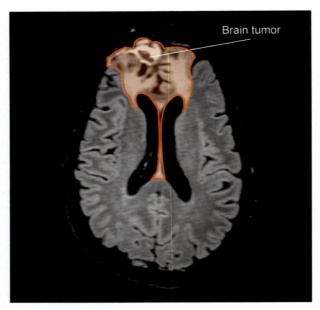

FIGURE 8.1 MRI of an oligodendroglioma in the brain. *(Courtesy: National Cancer Institute-CONNECT Staff.)*

human chromosome, and it has about 249 million nucleotide base pairs, which are the basic units of information for DNA. Chromosome 1 makes up about 8% of the total DNA in human cells. The chromosomal "p-arm" includes genes for breast carcinoma and Parkinson's disease. Chromosome 19 has more than 58.6 million base pairs, with genes related to programmed cell death and prostate-specific antigen.

The IDH-mutant and 1p/19q-codeleted oligodendroglioma is made up of tumor cells that morphologically resemble oligodendrocytes. They have isomorphic, rounded nuclei, with swollen and clear cytoplasm seen in processed paraffin sections. There are often microcalcifications and a thin branching capillary network. The histologic appearance of an astrocytic tumor component is diagnostically compatible with, but superseded by a diagnosis of oligodendroglioma, when an IDH mutation is combined with 1p/19q codeletion. Most of these oligodendrogliomas occur in adults, usually in the frontal lobe of a cerebral hemisphere.

Epidemiology

The IDH-mutant and 1p/19q-codeleted oligodendroglioma is a newer classification of these types of tumors. According to the Central Brain Tumor Registry of the United States (CBTRUS), oligodendroglioma incidence is 0.26 cases per 100,000 population. Oligodendroglial tumors make up 1.7% of all primary brain tumors, of which oligodendrogliomas are 1.2%, and anaplastic oligodendrogliomas are 0.5%. Oligodendroglial tumors also make up 5.9% of all gliomas, and they are still considered to be classified as "rare tumors". Most develop in adults, with peak incidence between ages 35 and 44.

Oligodendrogliomas are rare in children (only 0.8% of all brain tumors in children under age 15, and 1.8% in teenagers between ages 15 and 19). These tumors often lack IDH mutation or 1p/19q codeletion, and by strict definition cannot be considered as oligodendrogliomas. Rarely, children who have 1p/19q-codeleted oligodendrogliomas are usually older than 15 years. Generally, males are affected 1.3 times as often as females. In the United States, oligodendrogliomas are 2.5 times more common in Caucasians than in African-Americans.

Etiology and risk factors

Rarely, oligodendroglial tumors have been diagnosed following brain irradiation for other conditions. There is no convincing evidence of medications such as **ethyl-** or **methylnitrosourea** inducing oligodendrogliomas in humans. In some oligodendrogliomas, polyomavirus genome sequences and proteins have been found. In one case, a pediatric patient had concurrent *human herpesvirus 6 A* (HHV6A) infection with diffuse leptomeningeal **oligodendrogliomatosis**. There have been isolated oligodendrogliomas in immunodeficient patients, including those with HIV infection, and following organ transplantation. Rarely, oligodendroglioma has been associated with demyelinating diseases. There may be genetic component to these tumors, though most develop sporadically. Only isolated cases of familial oligodendrogliomas with 1p/19q codeletion have been reported. There are no proven risk factors.

Pathology

IDH mutational status is related to outcomes and prognosis. IDH-mutant and 1p/19q-codeleted oligodendrogliomas mostly arise in the white matter and cortex of the cerebral hemispheres. The frontal lobe is involved in more than 50% of all cases, followed by the temporal, parietal, and occipital lobes. Commonly, more than one cerebral lobe is affected, or bilateral tumor spread occurs. Rare locations for oligodendrogliomas are the posterior fossa, basal ganglia, and brain stem. Only a few patients experience **leptomeningeal** spread. Primary leptomeningeal oligodendrogliomas or oligodendroglial gliomatosis cerebri are also very rare. Primary oligodendrogliomas of the spinal cord only make up 1.5% of all oligodendrogliomas, and just 2% of all spinal cord tumors. There have been rare cases of primary spinal intramedullary oligodendroglioma with secondary meningeal dissemination, with one case involving a 1p/19q-codeleted tumor.

The IDH-mutant and 1p/19q-codeleted oligodendrogliomas usually extend diffusely into adjacent brain tissues. They rarely manifest upon first presentation with a gliomatosis cerebri pattern of significant CNS involvement. The affected area ranges from three or more lobes of one cerebral hemisphere to both cerebral hemispheres, with further involvement of the deep gray matter structures, brain stem, cerebellum, and spinal cord. Over 90% of IDH mutations in grade II oligodendrogliomas are the isocitrate dehydrogenase 1 (IDH1) R132H gene mutation. In less than 10% of cases, other IDH1 codon 132 or IDH2 codon 172 mutations are present. Most IDH-mutant and 1-codeleted oligodendrogliomas have chloride channel **capicua** (CIC) mutations.

Clinical manifestations

About 66% of patients with oligodendrogliomas experience seizures. Common symptoms include headache, focal neurological deficits, cognitive changes, and mental changes. Modern neuroimaging has substantially reduced the time between initial clinical manifestations and diagnosis.

Diagnosis

Correct diagnosis of these oligodendrogliomas requires demonstration of an IDH mutation. This is usually via **immunohistochemistry**, using the mutation-specific antibody against *R132H-mutant IDH1*, followed by DNA genotyping if this immunostaining is negative. Diagnosis is also via demonstrating a 1p/19q codeletion by **fluorescence in situ hybridization**, chromogenic in situ hybridization, or by molecular genetic testing. The World Health Organization recommends that 1p/19q assays should be able to detect whole-arm chromosomal losses. Pathologists must be experienced, and also be aware of possible errors in methodology and interpretation. If the IDH and codeletion factors cannot be detected, but there are classic features of an oligodendroglioma, a classification of the tumor as an *NOS oligodendroglioma* is recommended as a diagnosis of exclusion. This means that the tumor may be a *classic oligodendroglioma*, and will potentially have similar clinical behaviors to those of the IDH-mutant and 1p/19q-codeleted oligodendrogliomas.

In pediatric cases, the tumor may appear histologically as a classic oligodendroglioma, but will likely need to be evaluated further to exclude differential considerations such as **dysembryoplastic** neuroepithelial tumor, clear-cell ependymoma, neurocytoma, and pilocytic astrocytoma. If these can be excluded, the tumor can be classified as an oligodendroglioma without IDH mutation or 1p/19q codeletion, which is also known as a *pediatric-type oligodendroglioma*. In children, most of these tumors are grouped as low-grade diffuse gliomas with alterations of the fibroblast growth factor receptor 1, **myeloblastosis virus** (MYB), or MYB proto-oncogene-like-1 genes. A tumor with evidence of 1p/19q codeletion, but without a detectable IDH mutation, requires further evaluation to exclude incomplete 1p and/or 19q deletions. These changes may exist within subsets of IDH-wild-type, usually high-grade gliomas, which include glioblastomas that are linked a more aggressive course.

On CT scans, the IDH-mutant and 1p/19q-codeleted oligodendrogliomas usually appear as well-demarcated hypodense or isodense mass lesions. They are usually seen in the cortex and subcortical white matter. Though calcification is common, it is not a diagnostic factor. An MRI will reveal a lesion that is T1-hypointense and T2-hyperintense. It will be often well-demarcated, with only slight perifocal edema. Heterogeneous features may be present, due to intratumoral hemorrhages, with or without areas of cystic degeneration. **Gadolinium** enhancement has been found in less than 20% of grade II oligodendrogliomas, in comparison to more than 70% of grade III oligodendrogliomas. As such, contrast enhancement seen on brain MRI in the tumor bed is not necessarily predictive of grade of tumor but it is a useful data point. In low-grade tumors, contrast enhancement is generally linked with a worsened prognosis and may be subject to sampling error in terms of tissue obtained for tumor grading. Elevated 2-hydroxyglutarate levels, seen in **magnetic resonance spectroscopy** are believed to be a good way to detect

IDH-mutant tumors. This procedure has revealed differences in various features between 1p/19q-codeleted and intact low-grade gliomas. However, reliable determination of either form via neuroimaging is still being developed.

Macroscopically, the tumor is usually well-defined, soft, and gray-to-pink in color. It is usually seen in the cortex and white matter, resulting in blurring of the boundary between the gray and white matter. There may be local invasion into the leptomeninges. Calcification is common and may make the tumor appear to be rough in texture. Sometimes, dense calcifications appear as intratumoral stones. Areas of cystic degeneration and intratumoral hemorrhages are common. Rarely, extensive mucoid degeneration makes the tumor appear gelatinous.

Microscopically, there is moderate cellularity. The classic form of the tumor has monomorphic cells, with regular, rounded nuclei, and variable perinuclear halos. These have an appearance resembling a fried-egg or a honeycomb (see Fig. 8.2). Other features may include microcalcifications, a densely branched capillary network, and mucoid or cystic degeneration. The grade II form has nuclear atypia and sometimes an occasional mitosis. The grade III form has brisk mitotic activity, significant microvascular proliferation, and spontaneous necrosis. Wide sampling of resected specimens is needed for an accurate diagnosis.

Scattered cells are identifiable by a characteristic nuclei appearance, and when the most common IDH mutation is present, by immunostaining for R132H-mutant IDH1. Sometimes, there are small gemistocytes with rounded, eccentric cytoplasm that is positive for glial fibrillary acidic protein (GFAP). These cells are called **minigemistocytes** or **microgemistocytes**. Various cellular phenotypes do not prevent a diagnosis for oligodendroglioma if the tumor has an IDH mutation and 1p/19q codeletion. Fibrillary or gemistocytic astrocytic morphology is compatible with this diagnosis if testing confirms these mutations and codeletion. This means that diffuse gliomas with oligoastrocytomas histology, or with vague histological features, should be diagnosed as IDH-mutant and 1p/19q-codeleted oligodendrogliomas—if molecular testing reveals this particular genotype. There may be reactive astrocytes scattered throughout, that are prominent at tumor borders.

Most oligodendrogliomas have strong, regular immunoreactivity with an antibody that is specific for R132H-mutant IDH1, and positive R132H-mutant IDH1 staining aids greatly in differential diagnosis of these tumors versus other CNS clear-cell tumors, as well as for non-neoplastic and reactive lesions. Lack of R132H-mutant IDH1 immunopositive, however, does not completely exclude oligodendroglioma, since less common IDH mutations may not be detectable and require DNA sequence analysis. The IDH-mutant and 1p/19q-codeleted oligodendrogliomas usually have nuclear expression of *transcriptional regulator alpha thalassemia/mental retardation syndrome X-linked (ATRX)*. Also, these tumors usually lack large amounts of nuclear p53 staining. GFAP and vimentin immunostaining is usually positive in the astrocytic components of IDH-mutant and 1p/19q-codeleted oligodendrogliomas that are histologically similar to oligoastrocytomas. There are no cytokeratins. Some antibody mixtures such as AE1/AE3 may cause false-positive staining, because of cross-reactivity. The IDH-mutant and 1p/19q-codeleted oligodendrogliomas may have neoplastic cells expressing synaptophysin with or without other neuronal markers such as **hexaribonucleotide** binding protein-3 (NeuN) and neurofilaments. Immunostaining for proneural alpha-**internexin** protein is common. This cannot, however, be considered to be a reliable marker for 1p/19q codeletion. Also, **neurite outgrowth inhibitor** A (NOGO-A) positivity is non-exclusively typical of these tumors. The Ki-67 proliferation index is usually less than 5%.

These tumors are morphologically similar, especially on imaging to many reactive and neoplastic lesions, which is clinically relevant. Demyelinating diseases and cerebral infarcts should be easily distinguished via immunostaining for

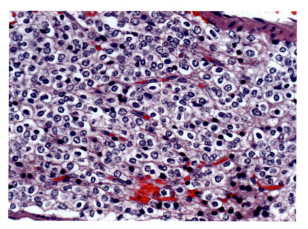

FIGURE 8.2 Oligodendroglioma. Cells have uniform round nuclei with bland chromatin and a clear perinuclear halo, producing a "fried-egg" appearance. The rich, branching capillary network has been compared to chicken wire (hematoxylin−eosin stain, ×200).

macrophage markers, and lack of IDH mutations. Lack of IDH mutations can also distinguish reactive changes, including increased numbers of oligodendrocytes that can be seen in lobectomy specimens performed for intractable seizures. Differential diagnosis with diffuse astrocytoma is based on histological, molecular, and immunohistochemical features. Significantly, IDH-mutant diffuse astrocytomas do not have 1p/19q codeletion. Also, **telomerase reverse transcriptase** (TERT) promoter mutations are common in IDH-mutant and 1p/19q-codeleted oligodendrogliomas, but not in IDH-mutant diffuse astrocytomas. The TERT promoter mutation may be a useful marker for 1p/19q codeletion, but its general use over 1p/19q codeletion testing is not recommended. Evidence of IDH mutations, usually shown by positive R132H-mutant IDH1 immunostaining, excludes other differential diagnoses.

In children, molecular distinction between oligodendrogliomas and pilocytic astrocytomas is difficult since pediatric oligodendrogliomas usually do not have combined IDH mutations with 1p/19q codeletion, yet sometimes have BRAF fusion genes. Therefore, differential diagnosis relies mostly on histological and ancillary radiological features, unless there are advanced molecular testing procedures that are based on significant methylation with or without mutation profiling, that provide accurate diagnostic profiles.

Treatment

The treatment of IDH-mutant with 1p/19q-codeleted oligodendrogliomas may vary according to location of the tumor, associated symptomatology, for example, intractable seizures, and age of the patient. After maximum feasible tumor resection, radiotherapy and chemotherapy may be withheld for a course of watchful waiting with imaging and clinical evaluation until there is tumor progression, since neurotoxicity is a major concern with respect to potential adverse event profiles of therapy. If the patient has symptomatic residual and progressive tumors following surgery, there is usually radiotherapy and/or chemotherapy. Studies, overall, show that these combined treatments usually afford a better outcome.

Prognosis

Studies have shown that grade II oligodendrogliomas have a better prognosis than grade III oligodendrogliomas, especially if the tumors have concurrent IDH mutation and TERT promoter mutations. The grade II tumors have a median survival of about 17 years, while grade III tumors have a median survival of about 10.5 years. The CBTRUS has documented 5-year survival rates of 79.5% and 10-year survival rates of 62.8% for these oligodendrogliomas. There is an absence of molecular information on IDH mutations and 1p/19q codeletions. Along with differing treatment approaches, these influence the variability of disease studies and their reported outcomes. Oligodendrogliomas often recur locally. Rarely, **oligosarcomas** may arise from IDH-mutant and 1p/19q-codeleted oligodendrogliomas.

Features associated with better prognoses include younger patient age during surgery, frontal lobe location, presentation with seizures, a high postoperative **Karnofsky score**, no contrast enhancement during neuroimaging, and complete macroscopic surgical removal. A higher degree of resection is associated with longer, progression-free survival. However, this does not prolong the time to malignant progression. A worsened prognosis is linked to high mitotic activity, necrosis, increased cellularity, cellular pleomorphism, nuclear atypia, and microvascular proliferation. For patients with identified IDH-mutant and 1p/19q-codeleted tumors, prognosis, based on histology, should be reevaluated. A higher Ki-67 proliferation index is linked to a worse prognosis.

> **Point to remember**
> The IDH-mutant and 1p/19q-codeleted oligodendrogliomas resemble classic oligodendrogliomas, with round nuclei and clear cytoplasm. They are most common in the frontal lobe of the brain, and usually affected adults. The tumors often invade nearby brain tissues. The most common symptom is seizures, which occur in about 66% of all cases.

NOS oligodendroglioma

Also referred to as *oligodendroglioma, not otherwise specified (NOS)*, this form of oligodendroglioma is diffusely infiltrative, and has a classic oligodendroglial histology; with molecular testing for combined IDH mutation and 1p/19q codeletion is incomplete or inconclusive. The abbreviation "NOS" stands for *not-otherwise-specified*. It should be noted that the NOS type of oligodendroglioma was first reported in the 2016 World Health Organization's classification of CNS tumors. It is equivalent to the diagnosis of a general oligodendroglioma made before 2016, based on the lack of (or the inability to detect) an IDH mutation or 1p/19q codeletion.

Epidemiology

Like other types of oligodendrogliomas, this type is exceedingly rare. With only about 670 new oligodendrogliomas of all types diagnosed in the United States in 2017, this type's difficult diagnosis makes it one of the least common. These tumors mostly occur in middle-aged adults, and are rare in children. They are slightly more common in males than in females, like all other oligodendrogliomas.

Etiology and risk factors

The cause of NOS oligodendrogliomas is unknown, though some studies have linked some oligodendrogliomas, which may or may not have IDH mutations or 1p/19q codeletions, to a viral cause. There are also no specific risk factors for NOS oligodendrogliomas.

Pathology

With NOS oligodendroglioma, there is limited tissue availability for diagnosis, and low tumor-cell content. These tumors mostly arise in the cerebral hemispheres, and only rarely develop in the brain stem, cerebellum, or spinal cord. The NOS oligodendrogliomas are well circumscribed, gelatinous, and gray in color. They are often calcified, and have focal hemorrhage or cystic changes. Slow growing tumors such as these can expand a gyrus, causing remodeling of the skull.

Clinical manifestations

Symptoms of NOS oligodendrogliomas are generally the same as for all other types, including seizures, headaches, body weakness, and speech and language changes. There may also be changes in personality, changes in behavior, and difficulty with body movements.

Diagnosis

The diagnosis of NOS oligodendroglioma is reserved for a diffusely infiltrative grade II glioma that has classic oligodendroglial histology, but no confirmed IDH mutation and 1p/19q codeletion. Diagnosis is further impaired by the possibility of inconclusive test results or any circumstances that impede molecular testing. Generally, molecular testing for IDH mutation and 1p/19q codeletion is vital for the World Health Organization classification of oligodendroglial tumors. This may mean that diagnosis of NOS oligodendroglioma is limited to only a small amount of cases, and is a diagnosis of exclusion.

Diagnosis is supported by immunohistochemical demonstration of an IDH mutation—especially an IDH1 R132H mutation—and nuclear positivity for ATRX. It should be understood that unless there is successful testing for a 1p/19q codeletion, gliomas that have an oligodendroglial histology, IDH mutation, and nuclear ATRX positivity are still likely to be NOS oligodendrogliomas. The 1p/19q codeletion testing cannot be replaced by immunohistochemical positivity for markers such as alpha-internexin and NOGO-A. It also cannot be replaced by an immunohistochemical demonstration of loss of nuclear CIC or *far upstream binding protein 1 (FUBP1)* expression. The presence of an obvious astrocytic component is not compatible with a diagnosis of NOS oligodendroglioma. However, NOS oligodendrogliomas show more homogeneous signal on T1 and T2 images and have sharper borders than other types of oligodendrogliomas.

Treatment

The uncertainty of the abbreviation "NOS" presents treating physicians with challenges in deciding adequate treatment. Therefore, the World Health Organization recommends to avoid its use and seek as precise a diagnosis as possible. As with other oligodendrogliomas, treatments include surgery, radiation, and chemotherapy.

Prognosis

As for other oligodendrogliomas, complete surgical removal gives the best prognosis. With all grade II oligodendrogliomas, the patient is often able to remain tumor-free for many years. Regular monitoring is essential to find any progression to a higher grade tumor. Like other oligodendrogliomas, about 50% of patients survive over 10 years.

> **Point to remember**
>
> An NOS (not-otherwise-specified) oligodendroglioma has the histologic features of a classic oligodendroglioma, and lacks IDH mutations or 1p/19q codeletions. This may be because tests for these mutations or deletions have not been performed, or because the markers for them are lacking.

Anaplastic oligodendroglioma (IDH-mutant and 1p/19q-codeleted)

This type of oligodendroglioma shows focal or diffuse histological features of anaplasia. Especially, there is pathological microvascular proliferation with or without brisk mitotic activity. Necrosis, which may be accompanied by palisading, does not necessarily indicate a progression to a glioblastoma. An astrocytic tumor component is compatible with diagnosis if molecular testing shows combined IDH mutation and 1p/19q codeletion. This anaplastic tumor is a grade III neoplasm according to the World Health Organization.

Epidemiology

Most IDH-mutant and 1p/19q-codeleted anaplastic oligodendrogliomas occur in adults. This may be because the tumors can exist undiagnosed for a long time. According to CBTRUS, anaplastic oligodendroglioma has an annual incidence of 0.11 cases out of every 100,000 people. It accounts for 0.5% of all primary brain tumors and 2.5% of all gliomas. About 33% of all oligodendroglial tumors are anaplastic oligodendrogliomas. The median age at diagnosis is 49 years. According to the World Health Organization, the reported annual incidence rates of anaplastic oligodendroglioma ranges from 0.07 to 0.18 per 100,000 person-years, and comprise only 0.5%-1.2% of all primary brain tumors. Only about 30% of oligodendroglial tumors have anaplastic features.

Patients diagnosed with these tumors are about an average of 6 years older than patients diagnosed with grade II oligodendrogliomas. In CBTRUS studies, these tumors have never been seen in children from 0 to 14 years of age. Only 32 patients between ages 15 and 19 have ever been documented with anaplastic oligodendrogliomas with IDH mutation and 1p/19q codeletion (30%-65% of all anaplastic oligodendrogliomas). Mutation of the IDH1 or IDH2 genes is seen in all of these tumors that also show the 1p/19q codeletion. Overall, there is a slight male predominance (1.2 males to 1 female) for anaplastic oligodendrogliomas. In the United States, this tumor is 2.4 times more common in Caucasians than in African-Americans. Global studies are not available.

Etiology and risk factors

The etiology of these tumors is identical with the IDH-mutant and 1p/19q-codeleted oligodendrogliomas. The hallmark alteration is combined whole-arm deletion of 1p and 19q, usually because of unbalanced translocation between chromosomes 1 and 19. Concurrent **polysomy** of 1p and 19q appears to be more common in these tumors, and is linked with a higher Ki-67 proliferation index, plus a worsened outcome. In most of these tumors, promoter mutations in TERT are also present, and often, so are CIC mutations. The average number of chromosomes involved with copy number abnormalities is higher in these tumors than in low-grade oligodendrogliomas. Epigenetically, there is *glioma (Cytosine-phosphate-guanine) island methylator phenotype* (G-CIMP) type A, and often, a proneural glioblastoma-like profile of gene expression. Usually, the MGMT promoter is **hypermethylated**. There are no proven risk factors.

Pathology

Anaplastic oligodendroglioma with IDH mutation and 1p/19q codeletion is usually located in the cerebral hemispheres, most often in the frontal lobe, followed by the temporal lobe. There have been rare cases of this tumor developing in the intramedullary parts of the spine. The histological features linked to high-grade malignancy include high cellularity, high mitotic activity, significant cytological atypia, necrosis with or without palisading, and pathological microvascular proliferation. Several of these features are seen in anaplastic oligodendrogliomas with IDH mutation and 1p/19q codeletion. Microvascular endothelial proliferation and mitotic activity at 6 or more mitoses per 10 high-power fields are important indicators of anaplasia. The anaplastic oligodendrogliomas develop de novo, usually with a short preoperative history, or from progression following a grade II oligodendroglioma. The mean time for such progression is about 6 to 7 years. Extracranial metastases are extremely rare, only occurring in 116 report cases. In these cases, metastases to the bones were 40% by glioblastomas, 27% from medulloblastomas, 16% from ependymomas, 10% from astrocytomas, and 5% from oligodendrogliomas.

Clinical manifestations

Patients with anaplastic oligodendroglioma usually present with headache, focal neurologic deficits, mental status changes, and cognitive deficits or signs of increased intracranial pressure. Seizures are less common compared to their occurrence in patients with low-grade oligodendrogliomas.

Diagnosis

Correct diagnosis requires, at very least, conspicuous microvascular proliferation, with or without brisk mitotic activity. If the case is borderline diagnostically, immunostaining for MIB1 and paying attention to rapid symptomatic growth and contrast enhancement, may assist in diagnosis. In imaging studies, there may be heterogeneous patterns variable necrosis, cystic degeneration, calcification, and intratumoral hemorrhages. Upon CT and MRI, contrast enhancement is usually seen. It can be homogeneous or patchy in appearance (see Fig. 8.3). Lack of contrast enhancement, however, does not exclude anaplastic oligodendroglioma. Ring enhancement is less common in these tumors than in malignant gliomas—especially glioblastomas without such molecular markers.

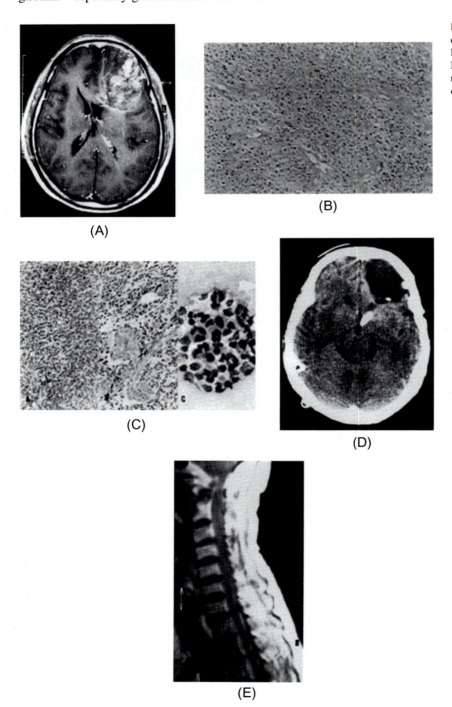

FIGURE 8.3 Anaplastic oligodendroglioma with drop metastasis to the spinal cord. (A) Frontal lobe anaplastic oligodendroglioma. (B and C) Microscopic views of tumor tissue. (D) Post-resection image. (E) Drop metastasis to the spinal cord with compression.

Macroscopically, these tumors resemble grade II oligodendrogliomas except for areas of necrosis. Microscopically, they are cellular, diffusely-infiltrating gliomas with various morphologies. Most cells demonstrate features similar to oligodendroglial cells. These include round hyperchromatic nuclei, only a few cellular processes, and perinuclear halos. Focal microcalcifications are common. Mitotic activity is usually significant. Sometimes, there is marked cellular pleomorphism and multinucleated giant cells. This is called the **polymorphic variant of Zulch**. Rare cases involve sarcoma-like areas, referred to as *oligosarcomas*. In some cases, **gliofibrillary** oligodendrocytes and minigemistocytes proliferate. While branched capillaries are often seen, pathological microvascular proliferation is, in most cases, prominent. In areas of cortical tumor infiltration, secondary structures such as perineuronal satellitosis are common. As long as the tumor has the IDH-mutant and 1p/19q-codeleted genotype, the diagnosis is not difficult. Presence of an obvious astrocytic component is also compatible with this diagnosis.

Though these anaplastic oligodendrogliomas have the same immunoprofile as grade II oligodendrogliomas, their proliferative activity is usually higher, and is usually determined by MIB1 immunostaining. The MIB1-positive cells usually make up more than 5% of the tumor cells. Differential diagnosis includes many other clear-cell tumor entities distinguishable by unique immunohistochemical profiles and lack of the IDH mutation and 1p/19q codeletion. It is important to distinguish between these tumors and small-cell astrocytic tumors, which have much more aggressive courses. Small-cell astrocytomas and glioblastomas lack combined IDH mutation and 1p/19q codeletion.

Treatment

Treatments are often different between low-grade versus high-grade oligodendrogliomas. Potential bias cannot be excluded in analyses that do not control various treatments and prognostic parameters. Anaplastic oligodendrogliomas with IDH-mutation and 1p/19q codeletion are strongly associated with improved responses to adjuvant radiation therapy and chemotherapy, and longer survival. After treatment, patients rarely develop CSF spread or systemic metastases.

Prognosis

According to many studies, these tumors offer a much better prognosis than the IDH-mutant but 1p/19q-intact or IDH-wild-type anaplastic astrocytic gliomas. However, the prognostic value of the World Health Organization's grading of patients with IDH-mutant, 1p/19q-codeleted oligodendroglial tumors are not as clear. Survival times for this form of anaplastic oligodendroglioma average 8.5 years in comparison to 3.7 years for patients with other types of oligodendrogliomas. Historical studies found a 5-year survival rate of 52% and a 10-year survival rate of 39%, but these studies did not account for IDH mutations or 1p/19q codeletions. More recent data reveals that the 5-year survival rate for anaplastic oligodendrogliomas is 74.1% (see Fig. 8.4). The Radiation Therapy Oncology Group and European Organization for Research and Treatment of Cancer trials indicate 10 or more years as the median survival times, when treatment includes combined radiotherapy and PCV chemotherapy. Local tumor progression is the most common cause of death. Factors associated with longer survival times include higher Karnofsky performance score, younger age at diagnosis, and greater amount of surgical resection. A more favorable prognosis also comes from previous resection for a lower-grade tumor. Anaplastic oligodendrogliomas also have a better prognosis than for all glioblastomas, with a longer progression time and greater response to therapy.

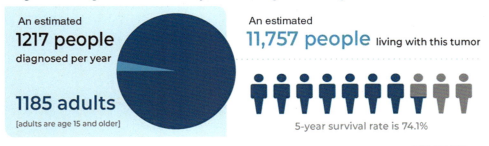

FIGURE 8.4 Diagnoses and survival rates of oligodendroglioma and anaplastic oligodendroglioma.

> **Point to remember**
> Anaplastic oligodendrogliomas are important to recognize because they have unique clinical, histologic, and molecular features. Any patient with new-onset seizures or focal neurologic deficits must be referred for a brain MRI with contrast. If it suggests a malignant tumor, the maximal tumor resection that is possible should be undertaken, followed by radiation and chemotherapy.

NOS anaplastic oligodendroglioma

There is also a *not-otherwise-specified* form of anaplastic oligodendroglioma. It is a diffusely infiltrative glioma that has the classic oligodendroglial histology. This type is defined by molecular testing in which combined IDH mutation and 1p/19q codeletion cannot be completed, or is inconclusive. It is also a grade III tumor, and the World Health Organization also first identified this form in 2016.

Epidemiology

This type of anaplastic oligodendroglioma also occurs mostly in adults during middle age, with the same prevalence for the male gender and Caucasians. No significant epidemiological data is yet available, however.

Etiology and risk factors

There are also no proven causes or risk factors for this type of anaplastic oligodendroglioma.

Pathology

The histological features are similar to those of the IDH-mutant and 1p/19q-codeleted anaplastic oligodendroglioma. However, a conspicuous astrocytic component, if present, is not compatible with diagnosis of NOS anaplastic oligodendroglioma. Palisading necrosis and other areas of necrosis are compatible with diagnosis as long as the tumor has the usual cytological features and other hallmarks. These include a branched capillary network and microcalcifications.

Clinical manifestations

The clinical manifestations of this type of anaplastic oligodendroglioma are identical to those of the anaplastic oligodendroglioma (IDH-mutant and 1p/19q-codeleted) form.

Diagnosis

Diagnosis is supported by immunohistochemical demonstration of an IDH mutation—especially IDH1 R132H and nuclear positivity for ATRX. However, this does not substitute for 1p/19q codeletion testing since nuclear ATRX expression is in place in a small amount of IDH-mutant but 1p/19q-intact anaplastic astrocytomas. Also, **immunopositivity** for the intermediate filament called *alpha-internexin*, and loss of nuclear CIC or FUBP1 expression supports an NOS anaplastic oligodendroglioma diagnosis, but not the other form. Differential diagnosis of glioblastoma, especially the small-cell type, must be carefully considered if the hallmark signs are present. This is especially true when there are no IDH mutations, and there are detected genetic features of glioblastoma, which include EGFR amplification, gain of chromosome 7, and loss of chromosome 10.

Treatment

Treatments are identical to those of the anaplastic oligodendroglioma (IDH-mutant and 1p/19q-codeleted) form.

Prognosis

The prognosis is also identical.

> **Point to remember**
> The NOS anaplastic oligodendrogliomas are grade III tumors that usually occur in adults. These tumors are especially signified by immunopositive for alpha-internexin, plus loss of nuclear CIC or FUBP1 expression. Alpha-internexin is seen transiently in neuronal development, and may be a component of the CNS as well as the PNS.

Clinical cases

Clinical case 1

1. From where do oligodendrogliomas likely originate?
2. Why is the grading of oligodendrogliomas controversial?
3. What is the unique genetic profile of an oligodendroglioma?

A 53-year-old man underwent a gross-total resection of a grade II oligodendroglioma. Eleven years later, he experienced a local tumor recurrence. A second gross-total resection was performed, and histopathological examination revealed residual features of classical oligodendroglioma, plus new sarcomatous characteristics. Both the primary and recurrent tumors showed a 1p/19q codeletion plus an IDH1 mutation.

Answers:

1. Oligodendrogliomas are believed to originate from brain oligodendrocytes or from glial precursor cells. On the cellular level, they appear similar to normal oligodendrocytes, mostly composed of cells with small to slightly enlarged round nuclei that have dark, compact nucleoli and a small amount of eosinophilic cytoplasm.
2. The histopathologic grading of oligodendrogliomas is controversial because authoritative bodies such as the World Health Organization do not all agree on the subjective criteria. Differentiating between grade II and III tumors is a matter of debate, and many different factors must be considered.
3. The most common structural deformity is codeletion of chromosomal arms 1p and 19q, and the high frequency of this codeletion is the genetic profile of an oligodendroglioma. Allelic losses on 1p and 19q, either separately or together, are most common in classic oligodendrogliomas.

Clinical case 2

1. How common are seizures in relation to oligodendrogliomas?
2. What are the difficulties concerning surgery for oligodendrogliomas?
3. Does PCV combination chemotherapy improve outcomes for oligodendroglioma cases?

A 29-year-old man was hospitalized after experiencing a generalized tonic-clonic seizure. Initial MRI revealed a poorly defined mass with subtle enhancement in the left frontal lobe. Gross-total resection was carried out, and the tumor was revealed to be a grade II oligodendroglioma. Seven years later, a definitive single tumor was found in the same left frontal area via MRI. Another surgery was performed, but this time, the lesion was diagnosed as an anaplastic oligodendroglioma (WHO grade III). The patient received three cycles of PCV chemotherapy, followed by radiation treatment. Almost 8 years later, there was yet another recurrence, and surgical resection was performed, but this time, the lesion was interpreted as anaplastic astrocytoma. The patient was re-challenged with PCV chemotherapy.

Answers:

1. Oligodendrogliomas present with seizures in 70%-90% of cases. Seizure patterns are nonspecific. The types are evenly divided into generalized, partial, and mixed seizure types although it stands to reason that a focal tumor will yield focal epileptogenic foci that can secondarily generalize. Younger patients are diagnosed earlier than adults because seizures are predominant, which brings a tumor to clinical attention while the tumor is smaller; whereas in adults, there are often other neurologic symptoms.
2. Maximal safe gross-total resection of an oligodendroglioma is undertaken to remove as much of the tumor as possible while still protecting critical brain functions. Special measures may be required, since oligodendrogliomas can occur near areas of the brain that control body movement, language, or sensation. *Awake surgery* with brain mapping is often used when tumors are located in brain regions that control language or movement.
3. People with newly diagnosed oligodendroglial tumors often live much longer if PCV chemotherapy is added before or after standard radiation therapy. However, the survival benefit is not statistically significant in patients lacking chromosome 1p *and/or* 19q deletions. Of course, this would render the diagnosis of oligodendroglioma in some doubt vs. oligodendroglioma NOS, oligoastrocytoma, or astrocytoma. A survival advantage with combination therapy was also seen in patients with MGMT methylated tumors and IDH mutations, though further study is still needed.

Clinical case 3

1. How rare are extracranial metastases of anaplastic oligodendrogliomas?
2. How common is the 1p/19q codeletion in relation to anaplastic oligodendrogliomas?
3. What is the usual prognosis for this type of oligodendroglioma?

A 36-year-old man had been previously diagnosed with a grade III 1p/19q-codeleted anaplastic oligodendroglioma. He underwent an incomplete resection, because complete resection was prevented by massive cerebral edema. A second surgical resection was performed one month later, which was also incomplete. The patient was treated with cranial radiation therapy and temozolomide chemotherapy. This was continued for approximately 11 months. After another 9 months, a local tumor recurred and another subtotal resection was performed, followed by PCV combination chemotherapy. Eventually, in a relatively rare occurrence, the tumors metastasized, affecting the patient's bones, and proved to be fatal.

Answers:

1. **Extracranial metastases of anaplastic oligodendrogliomas are exceedingly rare. In a compendium of 116 cases of primary brain tumor with metastases to the bones, 40% were by glioblastomas, 27% from medulloblastomas, 16% from ependymomas, 10% from astrocytomas, and only 5% from oligodendrogliomas.**
2. **The 1p/19q codeletion occurs in 30%-65% of anaplastic oligodendrogliomas. Also, the mutation of IDH1 or IDH2 genes is seen in all anaplastic oligodendrogliomas with the 1p/19q codeletion.**
3. **The prognosis for this type of oligodendroglioma is better than for glioblastoma, with a longer progression time and a greater response to therapy. The genetic loss of 1p/19q chromosomes has a high sensitivity to both radiotherapy and chemotherapy in most cases.**

Clinical case 4

1. How common are anaplastic oligodendrogliomas in children compared to adults?
2. What are the most common presenting symptoms?
3. What are the historical survival rates regarding anaplastic oligodendrogliomas over 5 and 10 years?

A 9-year-old boy presented with severe headache, vomiting, flexion of his head in a backwards direction, nystagmus, and ataxia. An MRI revealed brain stem enlargement, and he was initially diagnosed with an inflammatory process, that was believed to be caused by *Mycoplasma pneumoniae*. Corticosteroids quickly improved his symptoms, but within two months, fulminant disease progression led to brain death. A final neuroradiological examination indicated meningoencephalitis. Autopsy revealed brain swelling and brain stem softening with a superficial gelatinous mass along the spinal cord. A disseminating anaplastic oligodendroglioma with allelic loss of the D19S246 tumor suppressor candidate locus of chromosome 19 was diagnosed. This was the first such case ever seen in a child.

Answers:

1. **Anaplastic oligodendrogliomas usually present in adults rather than in children (in which they are rare), possibly because they may have existed, but were undiagnosed for a long time. They mostly arise in the cerebral hemispheres. These tumors make up 0.5% of primary brain tumors and 2.5% of all gliomas, with a median age at diagnosis of 49 years.**
2. **The most common presenting symptoms include headache, focal neurologic deficits, or mental status changes. Presentation with seizures is less common than in grade II oligodendrogliomas.**
3. **The historical studies found a 5-year survival rate of 52% and a 10-year survival rate of 39% for anaplastic oligodendrogliomas. However, these studies did not account for IDH mutations or 1p/19q codeletions. More recent data reveals survival times of greater than 10 years for patients with IDH-mutant, 1p/19q-codeleted anaplastic oligodendroglioma that is treated with combined radiotherapy and procarbazine-carmustine-vincristine (PCV) chemotherapy.**

Key terms

Anaplastic oligodendroglioma
Capicua
Dysembryoplastic
Ethylnitrosourea
Fluorescence in situ hybridization
Gadolinium
Gliofibrillary
Hexaribonucleotide
Hypermethylated
Immunohistochemistry
Immunopositivity
Internexin
Isocitrate dehydrogenase
Karnofsky score
Leptomeningeal
Magnetic resonance spectroscopy
Methylnitrosourea
Microgemistocytes
Minigemistocytes
Myeloblastosis virus
Neurite outgrowth inhibitor A
Oligodendroglioma
Oligodendrogliomatosis
Oligosarcomas
Polymorphic variant of Zulch
Polysomy
Telomerase reverse transcriptase

Further reading

Baranska, J. (2012). *Glioma signaling (Advances in experimental medicine and biology, Book 986)*. Springer.
Berger, M. S., & Weller, M. (2016). *Gliomas, volume 134 (Handbook of clinical neurology)*. Elsevier.
Bigner, D.D., Friedman, A.H., Friedman, H.S., McLendon, R., and Sampson, J.H., *The Duke Glioma Handbook: Pathology, Diagnosis, and Management*. (2016) Cambridge University Press.
Chen, T.C., and Chamberlain, M. Controversies in neuro-oncology: Avastin and malignant gliomas. (2018) Bentham Science Publishers.
Chernov, M. F., Muragaki, Y., Kesari, S., McCutcheon, I. E., & Lunsford, L. D. (2017). *Intracranial gliomas part i—Surgery (progress in neurological surgery, volume 30)*. S. Karger.
Duffau, H. (2017). *Diffuse low-grade gliomas in adults, 2nd edition*. Springer.
Hayat, M. A. (2012). *Tumors of the central nervous system, volume 8: Astrocytoma, medulloblastoma, retinoblastoma, chordoma, craniopharyngioma, oligodendroglioma, and ependymoma*. Springer.
Hekmatpanah, J. (2012). *Gliomas: Current concepts in biology, diagnosis and therapy (Recent results in cancer research, Book 51)*. Springer.
Icon Group International. (2010). *Oligodendroglioma: Webster's timeline history, 1929-2007*. Icon Group International, Inc.
Karajannis, M. A., & Zagzag, D. (2015). *Molecular pathology of nervous system tumors: Biological stratification and targeted therapies*. Springer.
McKhann, G. M., II, & Duffau, H. (2019). *Low-grade glioma, an issue of neurosurgery clinics of North America (The clinics: Surgery)*. Elsevier.
Medical Ventures Press. (2011a). *Childhood astrocytomas, oligodendrogliomas, oligoastrocytomas, glioblastoma multiforme: Pediatric cancer guide to symptoms, diagnosis, treatment, prognosis, clinical trials*. Progressive Management.
Medical Ventures Press. (2011b). *2011 pediatric cancer toolkit: Childhood astrocytomas, oligodendrogliomas, oligoastrocytomas, glioblastoma multiforme*. Progressive Management.
Moliterno Gunel, J., Piepmeier, J. M., & Baehring, J. M. (2017). *Malignant brain tumors: State-of-the-art treatment*. Springer.
National Comprehensive Cancer Network. (2017). *NCCN guidelines for patients: Brain cancer—Gliomas*. National Comprehensive Cancer Network (NCCN).
Newton, H. B. (2016). *Handbook of neuro-oncology neuroimaging (2nd edition)*. Academic Press.
Oberg, I. (2019). *Management of adult glioma in nursing practice*. Springer.
Paleologos, N. A., & Newton, H. B. (2019). *Oligodendroglioma: Clinical presentation, pathology, molecular biology, imaging, and treatment*. Academic Press.
Pope, W. B. (2019). *Glioma imaging: Physiologic, metabolic, and molecular approaches*. Springer.
Reeves, C. (2016). Oligodendrogliomas: Diagnosis, outcomes and prognosis. Nova Science Publishers, Inc.
Sedo, A., & Mentlein. (2014). *Glioma cell biology*. Springer.
Sughrue, M. E. (2019). *The glioma book*. Thieme.
Sughrue, M. E., & Yang, I. (2019). *New techniques for management of 'inoperable' gliomas*. Academic Press.
Taillandier, L., Capelle, L., & Duffau, H. (2013). *New therapeutic strategies in low-grade gliomas*. Nova Science Publishers, Inc.
Wiranowska, M., & Vrionis, F. D. (2013). *Gliomas: Symptoms, diagnosis and treatment options (cancer etiology, diagnosis and treatments)*. Nova Science Publishers, Inc.

Chapter 9

Ependymal tumors

Chapter Outline

Classic ependymoma	161	RELA fusion-positive ependymoma	171
Subependymoma	165	Anaplastic ependymoma	172
Myxopapillary ependymoma	168	Ependymoblastoma	173
Papillary ependymoma	171	Clinical cases	174
Clear cell ependymoma	171	Key terms	176
Tanycytic ependymoma	171	Further reading	176

Classic ependymoma

Classic ependymoma is a circumscribed glioma made up of uniform small cells. There are round nuclei within a fibrillary matrix, and perivascular anucleate zones called *pseudorosettes*. In about 25% of cases, ependymal rosettes are also found. There is a low cell density and low mitotic count. Rarely, the tumor invades nearby central nervous system (CNS) parenchyma to a substantive degree. Ultrastructural examination reveals cilia and microvilli. The tumor is usually intracranial, but sometimes occurs in the spinal cord, predominantly as the *myxopapillary* variant. Classic ependymomas develop in adults as well as children. In childhood, the tumor is usually within the posterior fossa. Ependymomas have varied clinical outcomes, based mostly on the amount of surgical resection that can be achieved, use of adjuvant radiation therapy, and the molecular group involved. The three distinct variants include papillary, clear cell, and tanycytic ependymomas, which are discussed later in this chapter. Classic ependymomas are classified as grade II by the World Health Organization. There is no established association between described grade II ependymal tumor histologies, biological behavior, and survival.

Epidemiology

In the United States, ependymomas make up 6.8% of all neuroepithelial neoplasms. Incidence decreases with increased patient age upon diagnosis as follows:

- 0-14 years—5.6%
- 15-19 years—4.5%
- 20-34 years—4.0%

In children under 3 years of age, up to 30% of all CNS tumors are ependymomas. The incidence according to CBTRUS data is 0.43/100,000 population constituting 1.8% of all primary CNS tumors. Ependymomas are most common in Caucasians than in African-Americans by a 1.67 to 1 ratio. There is no such difference between Hispanic and non-Hispanic populations. Within the spinal cord, ependymomas are the most common type of neuroepithelial neoplasms. They make up 50%-60% of all spinal gliomas in adults. They are, however, rare in children. Ependymomas have developed in people from birth to 81 years, according to the Central Brain Tumor Registry of the United States. Incidence varies between the location of the tumor, the histological variant, and the molecular group. For adults, infratentorial and spinal ependymomas occur at nearly the same frequency. However, in the United States, there are only about 200 new cases of ependymoma annually, including children and adults.

For childhood posterior fossa ependymomas, there is a mean patient age of 6.4 years. Spinal tumors dominate a secondary age peak between 30 and 40 years of age, being the most common spinal cord glioma. Supratentorial ependymomas affect children as well as adults. Males develop ependymomas 1.77 times as often as females. The ratio is varied widely over different molecular groups and anatomical locations. Children develop infratentorial ependymomas more often than spinal ependymomas. In children under age 3, about 80% of ependymomas are in the posterior fossa, within the fourth ventricle. They sometimes involve the cerebellopontine angle. Within the fourth ventricle, 60% originate in the floor, 30% in the lateral aspect, and 10% in the roof. In 60% of cases, supratentorial ependymomas arise from the lateral or third ventricles, as well as from the cerebral hemispheres. They have no clear connection to a ventricle in 40% of cases. Within the spinal cord, cervical or cervicothoracic localization often occurs. Also, in rare cases, ependymomas have developed in the broad ligaments, ovaries, mediastinum, pelvic and abdominal cavities, and lungs.

In Canada, there is a mean annual incidence of only 4.6 ependymoma cases per every 100,000 infants. About 2 of every 100 primary brain tumors in England are ependymomas. This is similar to the rates of ependymomas in most countries throughout the world, though statistics are not easily obtained for individual countries.

> **Point to remember**
> For all types of CNS tumors, patient considerations must include access to care, socioeconomic factors, cultural approaches to potentially stigmatizing conditions, and the level of available care. For example, non-Hispanic black patients are eight times more likely to die from a CNS tumor than non-Hispanic white patients. This is related to factors of available care. Risk of mortality is also higher in children from the non-Hispanic black ethnic group.

Etiology and risk factors

Like most brain tumors, there are no proven causes for ependymomas, but there is familial clustering, suggesting an inherited risk. Spinal ependymomas commonly occur along with neurofibromatosis type 2. This indicates a role of the NF2 gene. Other hereditary types of ependymoma are rare. *Turcot syndrome* and ependymomas have been rarely linked, and parental colon cancer is linked to increased risk of ependymomas in offspring, with a standardized incidence ratio of 3.7. In affected siblings, there has been a common allelic loss at 22q11.2-qter. Existence of a tumor suppressor gene on chromosome 22 may be related to familial ependymomas. There are no proven risk factors for classic ependymoma.

Pathology

Ependymomas may develop through the ventricular system, spinal canal, cerebral hemispheres, or extra-CNS locations (see Fig. 9.1). About 60% develop within the posterior fossa, 30% in the supratentorial compartment, and 10% within the spinal canal, according to the *Surveillance, Epidemiology, and End Results* study by the U.S. National Cancer Institute. Ependymomas usually arise within, or close to the ventricular system. They appear tan in color, and have a soft, spongy texture, sometimes containing rough calcium deposits. If arising in the caudal fourth ventricle, the tumors often grow through the foramina of Luschka and Magendie. They wrap around the cranial nerves and vessels of the brain stem, often being called *plastic ependymomas*. In rare cases, they develop in the cerebral hemisphere. In rare cases, they infiltrate the cerebral parenchyma, demonstrating histopathological overlap with small cell glioblastoma.

Microscopically, classic ependymoma has monomorphic cells of varied densities, and round-to-oval nuclei that have speckled nuclear chromatin (see Fig. 9.2). Key histological features include pseudorosettes and true ependymal rosettes. Pseudorosettes are made up of tumor cells that are radially arranged around blood vessels. This creates perivascular

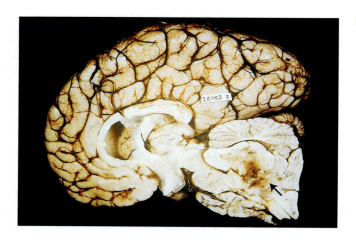

FIGURE 9.1   Classic ependymoma.

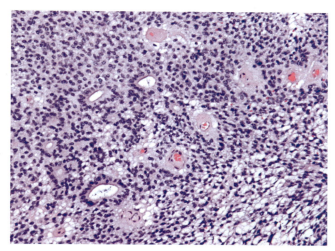

FIGURE 9.2 Cellular appearance of classic ependymoma.

anucleate zones, with fine fibrillary processes. The ependymal rosettes and tube-like canals are made up of bland cuboidal or columnar cells around a centralized lumen. Cell density varies widely. A high nuclear-to-cytoplasmic ratio may not be related with anaplastic features such as brisk mitotic activity, especially in supratentorial vascular ependymoma. This form has distinctive branches of thin capillary vessels and focal clear cell changes. In the posterior fossa, some ependymomas have nodules with high tumor cell density, often with higher mitotic counts. This biphasic pattern may occur with a distinct *cerebriform* folding of the surfaces of the tumor.

Additional histological features include intratumoral hemorrhage, areas of myxoid degeneration, dystrophic calcification, and sometimes, metaplastic bone or cartilage. There may be prominent hyalinization of the tumor vessels—especially in spinal and posterior fossa ependymomas. Areas of geographic necrosis may be seen. Palisading necrosis and microvascular proliferation are focal features only. Usually, the connection between the tumor and CNS parenchyma is well-demarcated. However, brain tissue infiltration is sometimes seen. Rarely, there has been *lipomatous* metaplasia, extensive *vacuolation* of tumor cells, widespread pleomorphic giant cells, signet ring cells, *neuropil*-like islands, and melanotic differentiation.

Ependymomas have the ultrastructural properties of ependymal cells, including cilia in a 9 + 2 microtubular pattern, microvilli and *blepharoblasts* on the luminal surface, junctional complexes on the lateral surface, and no basement membrane on the internal surface. *Microrosettes* may be formed by the cells, with cilia and microvilli projecting into them. The *zonulae adherentes* are junctional complexes with irregular links by gap junctions or *zonulae occludentes*, and cell processes filled with intermediate filaments. At the connection between tumor cells and vascularized stroma, a basal lamina may be present. In the pseudorosettes, immunoreactivity for glial fibrillary acidic protein (GFAP) is usually seen. This is more variable in other tumor areas, including the papillae and rosettes. Ependymomas usually express vimentin and the S100 protein. Most of these tumors have epithelial membrane antigen (EMA) immunoreactivity, with expression over the luminal surface of some ependymal rosettes. There may also be dotted perinuclear or ringed cytoplasmic structures. The expression of oligodendrocyte transcription factor 2 is usually sparse in comparison to other types of gliomas. In some cases, focal cytokeratin immunoreactivity is seen. Rarely, there is expression of neuronal antigens. In supratentorial ependymomas that have a C11orf95 rearrangement, L1CAM express is seen.

Stem cells that have been isolated from ependymomas show a radial glia phenotype. This suggests that radial glia cells are the histogenetic source. Distinct groups of stem cells specific to the anatomical site have been identified. Cerebral neural and adult spinal neural stem cells are possible origination cells for the cerebral and spinal ependymomas, respectively. These origination areas explain the main locations of the tumors in various age groups. Histogenesis is reflected by the molecular groups of the disease, as defined by methylome profiling.

Molecular alterations are common, with cytogenetic, *epigenetic*, genetic, and *trascriptomic* alterations. There are many different cytogenetic aberrations, predominantly in chromosomes 1q, 5, 7, 9, 11, 18, and 20; with losses of chromosomes 1p, 3, 6q, 6, 9p, 13q, 17, and 22. Supratentorial tumors have a loss of chromosome 9, and especially, homozygous deletion of cyclin dependent kinase inhibitor 2A (CDKN2A) is seen. In several studies, a gain of chromosome 1q has been seen, and is linked to a poor outcome of a posterior fossa tumor. In spinal cord tumors and those related to neurofibromatosis type 2, monosomy 22 and deletions or translocations of chromosome 22q are highly common.

Mutations of the NF2 gene are common in spinal ependymomas, and this gene is involved in ependymoma tumorigenesis. DNA methylation profiling has revealed distinct molecular groups, with strong relationships to specific anatomical sites. In a large study, three groups have been identified in the supratentorial, posterior fossa, and spinal compartments. Tumors having a subependymomatous morphology are classified into spinal (SP-SE), posterior fossa (PF-SE), and supratentorial (ST-SE) groups. Fusion genes involving reticuloendotheliosis homolog A (RELA) or yes-associated protein 1 (YAP1) characterize the ST-EPN-RELA and ST-EPN-YAP1 supratentorial groups. The two posterior fossa groups called PF-EPN-A and PF-EPN-B match the groups that were previously known as group A and group B. The last two molecular groups for the spinal cord, which includes the cauda equina, contain myxopapillary ependymomas (SP-MPE) and classic ependymomas (SP-EPN).

In infants and young children, most posterior fossa ependymomas are in the PF-EPN-A group, while PF-EPN-B tumors occur mostly in teenagers and adults. The PF-EPN-A ependymomas have few copy number alterations, while the PF-EPN-B ependymomas have these alterations—mostly gains and losses of entire chromosomes and chromosome arms. The ST-EPN-RELA and PF-EPN-A groups have a much worse prognosis. Posterior fossa ependymomas have an extremely low mutation rate, lacking recurrent somatic mutations. Supratentorial ependymomas have a recurrent structural variant, the C11orf95-RELA fusion gene. This results from chromothripsis, occurring in 70% of pediatric supratentorial ependymomas. As such, RELA ependymomas can be considered either grade II or III contingent on these considerations.

Clinical manifestations

The clinical manifestations of classic ependymomas are based on localization. In the posterior fossa, ependymomas often cause signs and symptoms of hydrocephalus and increased intracranial pressure. These include headache, dizziness, nausea, and vomiting. If the cerebellum and brain stem are involved, there may be ataxia, paresis, cranial nerve deficits, and visual disturbances. Supratentorial ependymomas may cause epilepsy, focal neurological deficits, and the features of increased intracranial pressure. In young infants, head enlargement or separation of the cranial sutures can be obvious. Spinal ependymomas can cause back pain, focal motor deficits, focal sensory deficits, and paraparesis.

Diagnosis

With gadolinium-enhanced magnetic resonance imaging (MRI), there are well-circumscribed masses that have varying degrees of contrast enhancement (see Fig. 9.3). Common accompanying features include ventricular obstruction, brain

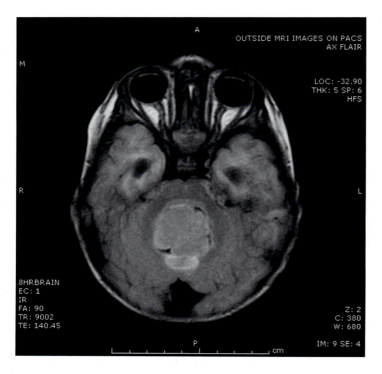

FIGURE 9.3 Classic ependymoma MRI.

stem displacement, and hydrocephalus. Cystic components are often seen in supratentorial tumors. Sometimes, there is intratumoral hemorrhage and calcification. Rarely, gross infiltration of nearby brain structures and edema occur. The use of MRI is very good in determining relationships with surrounding structures, syrinx formation, and invasion along the cerebrospinal fluid pathway. Spread of CSF is important to assess staging, treatment planning, and prognosis.

Treatment

The first treatment option for classic ependymoma is neurosurgery, with the goal being complete tumor resection. This may be impossible due to the location of the tumor. If hydrocephalus is present, a shunt may be implanted in the brain to help drain excessive CSF, usually to the abdomen, where it is harmlessly reabsorbed. The shunt is a tube-like device. Radiation therapy may be used to help shrink the tumor if complete surgical resection is impossible. Highly targeted radiation beams (electron or proton based) can reach just the tumor and not the surrounding tissues. Chemotherapy may also be used to help shrink the tumor, particularly in pediatric age groups. If the tumor has spread, radiation (over 3–5 years of age) or chemotherapy may be the most successful options.

Prognosis

Children with ependymomas have a worse prognosis than adults. This may be due to their higher incidence of posterior fossa tumors in comparison to the spinal location of adult tumors. Children under 1 year of age have an overall 5-year survival rate of 42.4%. With increased rate, this rate improves to 55.3% in 1 to 4 year olds, 74.7% in 5 to 9 year olds, and 76.2% in 10 to 14 year olds. A reliable indicator of outcome is the extent of surgical resection possible. Gross total resection is linked to a much-improved survival rate. In children under 3 years of age, complete resection has a 43% five-year survival rate, while incomplete resection only has a 36% five-year survival rate. The tumor location is also important. Supratentorial ependymomas have better survival rates than posterior fossa tumors, especially for children. Spinal ependymomas have a much better outcome than intracranial tumors, though late recurrences, 5 or more years after surgery, can occur. Metastasis is linked to a poor prognosis. An accurate definition of anaplasia for these tumors is needed since there is an inconsistent relationship between pathology and outcome in grades II and III ependymomas.

Subependymoma

A *subependymoma* is a slow-growing, *exophytic*, intraventricular tumor that is usually detected via MRI, without clinical manifestations. It has discrete, lobulated ependymal nodules in the walls of the posterior fourth ventricle (65%-70% of cases), or in the anterior third ventricles. In extremely rare cases, it can develop in the spine. When developing in the spine, it has the potential to be life threatening, which is unlike its development in the brain. Subependymoma is typified by bland to slightly pleomorphic and mitotically inactive cells within a dense fibrillary matrix showing numerous microcystic changes. It is considered to be a grade I tumor by the World Health Organization. Subependymoma in years gone past was often discovered after death as an incidental finding during an autopsy. This type of ependymal tumor was inaccurately described as a *subependymal astrocytoma* prior to the current grading system.

Epidemiology

About 90% of subependymomas occur in adults, usually after age 40. They appear 2.3 times as often in males than in females. Subependymomas, like all ependymal tumors, are more prevalent in Caucasians than in other racial or ethnic groups. For example, in one study between 2004 and 2009, conducted by the Central Brain Tumor Registry of the United States, Caucasians developed subependymomas in 87% of all cases, followed by African-Americans (7.6%), Asian-Pacific Islanders (2.4%) and American-Indians / Alaska-Natives (0.7%)—see Fig. 9.4. Since these tumors are usually asymptomatic, true incidence is hard to determine. Overall, subependymomas are extremely rare. Peak incidence of subependymoma is between 40 and 60 years of age. Based on various studies of prevalence, they are believed to make up 8% of ependymal tumors, and 0.51% of all central nervous system tumors. There are no accurate figures on the actual amounts of global cases, which is also true for cases within the United States.

Subependymomas occur rarely in families. However, there have been reports of simultaneous occurrence, within the fourth ventricle in some adult identical twins. In one family with several brain tumors, three siblings developed this tumor within the fourth ventricle. In another family, a sibling who was only 13 years of age at the time, developed a tumor in the pons, while a much older brother developed a different type of ependymoma in the fourth ventricle. Another occurrence within a family in which the father and a son had brain tumors, suggests a possible genetic susceptibility.

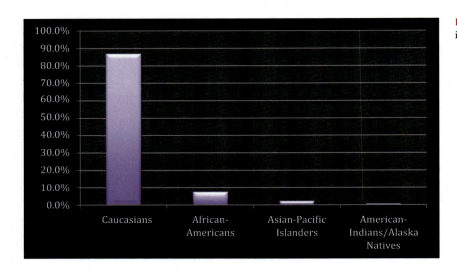

FIGURE 9.4 Percentages of subependymomas in various ethnic groups (USA).

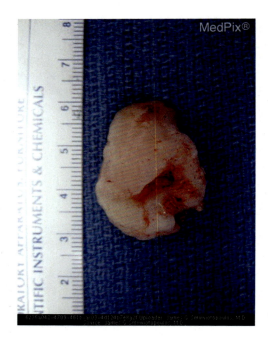

FIGURE 9.5 Subependymoma.

Etiology and risk factors

The etiology of subependymoma is not fully understood. Like other forms of ependymal tumors, causes include *glioma susceptibility 1*, *mismatch repair cancer syndrome*, *neurofibromatosis II*, and both types 1 and 2 of tuberous sclerosis. Less proven causes may include other subtypes of ependymal tumors, *fetal hydantoin syndrome*, oligodendroglioma, and *Turcot syndrome*. The only proven risk factor for subependymoma is older age.

Pathology

Subependymomas are small in size and incidental. They appear as shiny, pearled-white tumors with lobules and intraventricular protuberances (see Fig. 9.5). Subependymomas usually develop in the fourth ventricle (in up to 60% of cases), and sometimes in the lateral ventricles (in up to 40% of cases). They have also occurred, much less frequently, in the septum pellucidum, third ventricle, and cervical spinal cord. Within the spinal cord, they develop as cervical or cervicothoracic intramedullary masses, or uncommonly, as extramedullary masses. The least common occurrence of

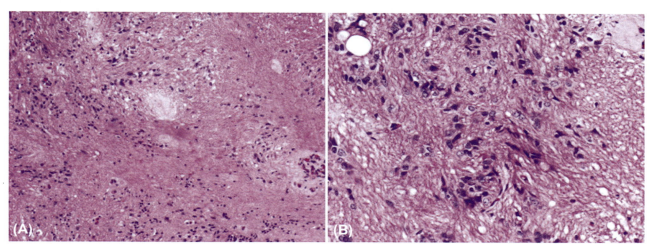

FIGURE 9.6 Cellular appearance of subependymoma. (A) Gross view. (B) Fine focus view.

subependymomas is intraparenchymally in the cerebrum. Subependymomas may affect the *cerebellar tonsils*, which are rounded lobules on the underside of the cerebellar hemisphere.

Subependymomas have sharp demarcation, grow slowly, and are usually noninvasive. Larger tumors may obstruct the ventricle and cause noncommunicating hydrocephalus. Their proliferative index is usually low, and histogenesis is still uncertain, but includes astrocytes, ependymal cells, subependymal glia, or a mixture of these. Genetic changes are not widely understood. In rare cases, there are foci of classic ependymoma, named as *mixed ependymoma/subependymoma*.

Subependymomas are firm nodules of varied sizes that bulge into the ventricular lumen. They usually do not exceed 1-2 cm in size. The intraventricular and spinal forms are predominantly well-demarcated. In the fourth ventricle, large tumors may compress the brain stem. Rarely, they form in the cerebellopointine angle, in adults as well as children. On the microscopic level, these tumors have clustered, small, and uniform nuclei within the dense fibrillary matrix of glial cell processes (see Fig. 9.6). Often, there are small cysts - usually in lesions beginning within the lateral ventricles. Nuclei appear isomorphic, like those of the subependymal glia. When the tumors are solid, there may be pleomorphic nuclei, though variances in nuclei are common in multicystic tumors. Low levels of mitotic activity are not common, but calcifications and hemorrhage are often seen. Significant tumor vasculature is only rarely present along with microvascular proliferation. Cell processes sometimes surround vessels to form ependymal pseudorosettes.

Subependymomas may be a superficial aspect of a classic ependymoma or, less often, a tanycytic ependymoma. They are graded based on the ependymoma component. In one case, the subependymomatous factor was more dominant, giving a better overall prognosis. There have also been cases involving melanin formation, *rhabdomyosarcomatous* differentiation, and even sarcomatous changes of the vascular stromal elements. Subependymomas may have cells with common ependymal characteristics, which include cilia and microvilli, or even many intermediate filaments. In most cases, there is immunoreactive for GFAP present in various amounts. Immunoreactivity may exist for poorly specific neural markers such as neural cell adhesion molecule 1 (NCAM1) and *neuron-specific enolase*. There is only rarely any expression of EMA. In one study of possible therapeutic targets, it was found that subependymomas have frequent expression of mouse double minute 2 homolog, DNA topoisomerase 2-beta, hypoxia-inducible factor 1-alpha, *nucleolin*, and phosphorylated signal transducer and activator of transcription 3.

Subependymomas may originate from subependymal glial cells, ependymal cells, astrocytes in the subependymal plate, and mixtures of astrocytes with ependymal cells. In recent studies, nine molecular groups of ependymomas have been identified in three sites—the supratentorial compartment, posterior fossa, and spinal compartment. Posterior fossa and spinal subependymomas have chromosome 6 copy number changes. Supratentorial tumors have nearly no such changes. Excellent overall survival occurs with the supratentorial and posterior fossa subependymomas.

Clinical manifestations

Though often asymptomatic, subependymomas within the brain cause headaches because of increased intracranial pressure (89% of patients), and about 44% of patients experience weakness, numbness, nausea, vomiting, dizziness, and visual problems. These clinical manifestations depend on the location of tumor in the brain. In some patients, ataxia and hydrocephalus

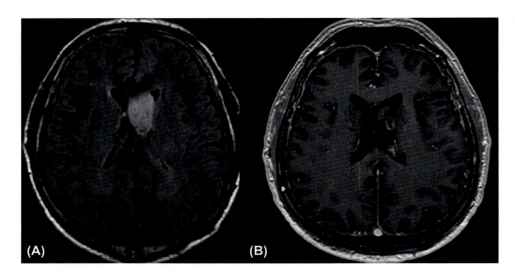

FIGURE 9.7 Subependymoma MRI. (A) With T1-enhancement. (B) With T2-enhancement.

may be seen. Common presenting symptoms for spinal subependymomas are limited, though severe itching has been reported. Patients have reported severe fatigue, back pain, focal sensory and motor problems, paraparesis, and sexual dysfunction. Following surgery, the most common symptom reported, for either brain or spinal subependymomas, is weakness.

Diagnosis

Upon imaging studies, subependymomas appear as nodular masses, and are usually non-enhancing. There may be obvious calcification and foci of hemorrhage. Intramedullary subependymomas are usually in oddly differing locations, and not centrally positioned, which is typical of intraspinal ependymomas. Subependymomas are hypointense to hyperintense, with minimal to moderate enhancement, on T1- as well as T2-weighted MRI—see Fig. 9.7. These processes utilize a *gradient echo sequence*, which is generated by using two bipolar gradient pulses during MRI.

Treatment

Subependymomas are graded and treated based on the possibly more aggressive ependymoma component. Complete excision is not always possible if the tumor arises from the floor of the fourth ventricle. Recurrences are usually not expected with subtotally resected tumors, but this has happened in rare cases, and CSF spread has also been seen. There is usually low mitotic activity. Recurrences may be related to the extremely low proliferation index.

Prognosis

Subependymomas have an excellent prognosis compared to other types of ependymal tumors. After gross total resection, no recurrences have ever been reported. Debulking alone usually means an excellent prognosis, since the tumors grow slowly and residual tumors may require decades to manifest any symptoms.

Myxopapillary ependymoma

Myxopapillary ependymoma is a glial tumor that nearly always occurs in the conus medullaris, cauda equina, or filum terminale. It is characterized by long fibrillary processes arranged radially, around cores that are fibrovascular, mucoid, and vascularized. It grows slowly, and usually occurs in young adults. This type of ependymal tumor makes up 9%-13% of all ependymomas. It is a grade I tumor according to the World Health Organization.

Epidemiology

Though very rare overall, myxopapillary ependymoma is more common in young adults, and is highly vascular, with a red color. In younger patients, it almost always originates from the lumbosacral nervous tissue. This type of ependymal tumor is the most common tumor of the lumbosacral canal, making up about 90% of all lesions in this area. In the conus medullaris or cauda equina areas, these tumors are the most common intramedullary neoplasm. Average age upon presentation is 36 years, though myxopapillary ependymomas have been seen in a large range of ages, from one

3-week-old infant to adults as old as 82 years. In one particular study, 83% of ependymomas occurring in the filum terminale were of the myxopapillary type. The male-to-female ratio was 2.2:1. More Caucasian patients have been diagnosed with myxopapillary ependymomas than other races, but data is very scarce about ethnicities and these tumors. Annual incidence rates are approximately 0.08 per 100,000 males, and about 0.05 per 100,000 females. There is no accurate data about prevalence rates. According to the Surveillance, Epidemiology, and End Results Program, as of 2016, there have been only 773 total reported cases of myxopapillary ependymoma in the United States. There are no statistics on global cases of this type of tumor.

Etiology and risk factors

The etiology of myxopapillary ependymomas is unknown. There are also no known risk factors.

Pathology

Myxopapillary ependymomas may originate from the ependymal glia of the filum terminale and then involve the cauda equina. Only in rare cases do they invade nerve roots or cause erosion of sacral bone. The tumor may be multifocal. Myxopapillary ependymomas may have a gelatinous, cystic, or hemorrhagic appearance. A thin, collagenous capsule usually surrounds these tumors. They appear lobulated, soft, and tan, gray, or pinkish red in color (see Fig. 9.8). There are different papillary structures, with many *hyalinized* vessels surrounded by mucin, plus an outside tumor cell layer. Microscopically, these tumors have cuboidal to elongated cells that are radially arranged in a papillary fashion around their cores. However, sometimes there is only a slight papillary structure, and joined cell sheets consisting of three or more straight sides. There may also be fascicles of spindled cells (see Fig. 9.9). A myxoid material is accumulated

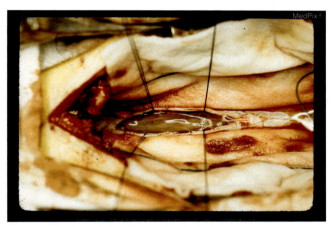

FIGURE 9.8   Myxopapillary ependymoma.

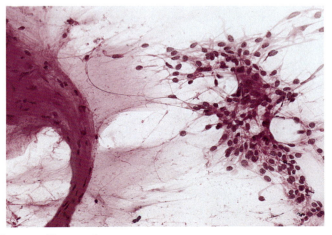

FIGURE 9.9   Cellular appearance of myxopapillary ependymoma.

between tumor cells and blood vessels, and also within microcysts. This material, if staining positive with *Alcian blue*, aids in identification. Balloon-like eosinophilic structures may be seen, which are periodic acid-Schiff-positive. They also are speculated reticulin staining. There is low mitotic activity and a *Ki-67 protein* proliferation index. The histological features of *anaplasia* are significant.

The tumor cells do not exhibit polarity. They have adherens junctions with thickened cytoplasm, and wide spaces that contain loose filaments or amorphous material. Cells that have basal membranes delineate the extracellular spaces. The spaces contain projected villi. There are not significant amounts of cilia, basement membrane structures, or complex *interdigitations*. A distinct feature is aggregation of microtubules inside endoplasmic reticulum complexes. Recent studies have confirmed presence of intracytoplasmic lumina with microvilli, and adherens junctions.

Myxopapillary ependymomas are distinguished from chordomas, metastatic carcinomas, myxoid chondrosarcomas, paragangliomas, and schwannomas by their diffuse immunoreactivity for GFAP. They have typical labeling for the S100 cellular antigen protein or the structural protein called *vimentin*, plus reactivity for the cluster of differentiation 99 antigen and NCAM1. A common feature is immunoreactivity for the mixture of anion exchangers (AE1 and AE3). Labeling for CAM5.2 (also called cytokeratin-8), cytokeratin 5/6 (CK5/6), CK7, CK20, or the keratin called *34betaE12* may be absent or exceptional. Myxopapillary ependymomas are genetically characterized by *polyploidy*, especially across multiple chromosomes.

Myxopapillary ependymoma is usually smaller than 2 cm across, and its size is the best way to distinguish it from a subependymal giant cell astrocytoma. Soft tissue myxopapillary ependymomas may metastasize to the lungs and other sites, while appearing to be benign. Myxopapillary ependymomas are sometimes seen in the cervicothoracic spinal cord, fourth ventricle, lateral ventricles, or brain parenchyma. A distinct subgroup consists of subcutaneous sacrococcygeal or presacral myxopapillary ependymomas. These are believed to form from ectopic ependymal remnants. When occurring in the *intrasacral* region, they may mimic chordomas. Distant spinal metastases occur in about 9.3% of patients, and brain metastases occur in about 6%. Children are especially likely to have spinal metastatic dissemination upon initial presentation.

Clinical manifestations

In most cases, myxopapillary ependymomas cause chronic back pain that may radiate to the legs. Other symptoms include *radiculopathy*, focal neurological deficits, cranial nerve palsies, leg paralysis, dysuria, skin ulcerations, subcutaneous nodules, and foot drop.

Diagnosis

Diagnosis of myxopapillary ependymomas is primarily via MRI, and findings usually include an intradural mass that is hypointense or isointense with the spinal cord on T1-weighted images. On T2-weighted images, it is hyperintense with intense homogenous enhancement after contrast material is administered. The lesions often affect two to four vertebral body levels. In imaging studies, these tumors are usually contrast enhancing and sharply circumscribed. There may be extensive hemorrhage and cystic changes. Alcian blue and periodic acid-Schiff staining is used to aid in diagnosis. Confirmation is based on biopsy. They may mimic ependymoma or metastatic carcinoma histologically, but have an immunohistochemical profile that is distinct. The nuclei usually form clusters. In imaging studies, these tumors are usually contrast enhancing and sharply circumscribed. There may be extensive hemorrhage and cystic changes. Alcian blue and periodic-acid-Schiff staining is used to aid in diagnosis.

Treatment

Complete surgical excision, often followed by radiation therapy, usually achieves a cure. However, it is difficult to resect completely. If the tumor is entangled with the spinal nerves, recurrence is more likely. Failures of treatment are more likely in younger patients, when initial adjuvant radiotherapy is not administered, and when there is incomplete excision. Use of chemotherapy is controversial. It has been widely used in pediatric patients with more aggressive tumors.

Prognosis

Prognosis of myxopapillary ependymoma is usually favorable, especially when the collagenous tumor capsule is intact when operated upon. If the tumor is removed bit by bit and mucin is spilled, there is a higher chance for recurrence. These tumors are more aggressive when they occur in children. There is a 5-year survival rate of 98.4% after total or partial resection. In one study of 183 patients, treatment failed in about 33%. Recurrence was mostly local, and more common in younger patients or those not given early adjuvant radiotherapy. Gross total resection improves outcomes. Adjuvant radiotherapy improves survival without tumor progression. Aside from age being the strongest predictor of recurrence, expression of EGFR is also a possible biomarker of recurrence.

> **Point to remember**
>
> The various subtypes of ependymomas discussed in this chapter can also evolve into a *cellular ependymoma*. This sub-subtype shows conspicuous hypercellularity without other properties of anaplasia, and therefore, is a grade II tumor according to the World Health Organization. The pseudorosettes may be inconspicuous, and true ependymal rosettes may be absent. There is a lack of elevated mitotic activity or vascular proliferation.

Papillary ependymoma

Papillary ependymoma is a rare histological variant that is characterized by papillae that are well formed. The papillae result if growths arise that have projections that resemble "fingers", which are lined by one layer of cuboidal tumor cells that have smooth contiguous surfaces, and GFAP-immunopositive tumor cell processes. Different from these, *choroid plexus papillomas* and metastatic carcinomas form *bumpy cell* surfaces without extensive GFAP reactivity. Unlike choroid plexus tumor papillae, the papillae of these ependymomas do not have a basement membrane under the neuroepithelial cells. Instead, fibrillary processes connect downward to a vascular core, with the same arrangement as in pseudorosettes. There are no sources of accurate epidemiological data for papillary ependymoma.

Clear cell ependymoma

Clear cell ependymoma is characterized by its appearance, which resembles that of an oligodendrocyte. Because of clearance of cytoplasm, there are perinuclear "halos". Clear cell ependymomas usually occur in young patients, within the supratentorial compartment. Occasionally, these tumors are found in the posterior fossa or spine. It is important to distinguish clear cell ependymomas from *central neurocytomas*, oligodendrogliomas, clear renal cell carcinomas, and hemangioblastomas. Correct diagnosis can help to differentiate ependymal and perivascular rosettes, immunoreactivity for EMA and GFAP, and *ultrastructural* studies. Clear cell ependymoma is believed to be more aggressive than other ependymomas. Previously, a clear cell tumor of the lateral ventricles was classified as *ependymoma of the foramen of Monro*. Today, this is usually recognized as a *central neurocytoma*. There are no sources of accurate epidemiological data for clear cell ependymoma.

Tanycytic ependymoma

Tanycytic ependymoma is characterized by tumor cells that are arranged in fascicles of different widths and cell densities, as well as by elongated cells that have spindle-shaped nuclei. This tumor usually occurs in the spinal cord. *Rosettes* are usually not present, and pseudorosettes may be thin and delicate. The nuclei, like other types of ependymomas, have speckled chromatin, and usually there is no anaplasia. This tumor is named because its elements resemble tanycytes, which are the paraventricular cells that have long cytoplasmic processes extending to ependymal surfaces. Because of appearance, tanycytic ependymomas have been mistaken for pilocytic astrocytomas and other types of astrocytomas. Their ultrastructural characteristics are ependymal, aiding in correct diagnosis. Tanycytic ependymoma is perhaps the rarest of all subtypes of ependymomas, and there are no sources of accurate epidemiological data. According to the International Journal of Clinical & Experimental Pathology, there have only been 14 reported cases in history, with nine cases being ventricular and five cases being subcortical in location.

RELA fusion-positive ependymoma

RELA fusion-positive ependymoma is a type that is supratentorial, and is characterized by a *RELA fusion gene*. It accounts for about 70% of all childhood supratentorial tumors, but a lower percentage of supratentorial tumors in adults. There is about a 2:1 male-to-female ratio. The term "RELA" describes the *REL-associated* protein involved in nuclear factor kappa-light-chain-enhancer of activated B-lymphocytes, a protein complex that controls transcription of DNA, cytokine production, and cell survival. The RELA fusion gene is not a component of ependymomas in the posterior fossa and spinal compartments. This gene encodes *transcription factor p65*. RELA fusion-positive ependymomas have a range of different histopathologies, with or without anaplasia. The World Health Organization classifies them as grade

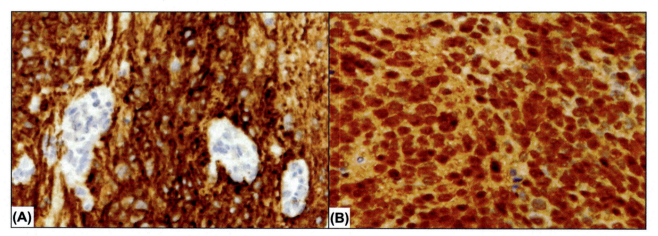

FIGURE 9.10 Cellular appearance of RELA fusion-positive ependymoma. (A) Clear cell changes. (B) Branched capillaries.

II or III. There has never been a grade I ependymoma with this particular genetic alteration. There is also no specific morphology. RELA fusion-positive ependymomas have the same basic structure and cytology as supratentorial ependymomas, but often with a vascular pattern of branched capillaries or clear cell changes (see Fig. 9.10). Tanycytic ependymomas and the other uncommon variants previously discussed are usually not RELA fusion-positive.

The RELA fusion-positive ependymomas have the same *immunoreactivities* for EMA and GFAP as other ependymomas. Expression of the transmembrane protein *L1CAM* is well linked with presence of RELA fusions in supratentorial ependymomas, though different types of brain tumors also express L1CAM. In ependymomas, the *C11orf95-RELA fusion* is the most common structural variant. It is formed via *chromothripsis*, in which the genome that rearranges genes and produces oncogenic gene products is shattered and reassembled. RELA fusion-positive ependymomas have constitutive activation of the *NF-kappaB pathway*, in which the RELA-encoded transcription factor p65 is a key effector. As a result of chromothripsis, there may be rare cases of C11orf95 or RELA being fused with other genes. The presence of the C11orf95-RELA fusion gene is easily detected using *interphase fluorescence in situ hybridization* with break-apart probes around both genes, working with formalin-fixed and paraffin-embedded tissue samples. Via chromothripsis, rearrangement splits dual-color signals in probe sets for C11orf95 and RELA. There is a single study that suggests that RELA fusion-positive ependymomas have the worst prognosis of the three-supratentorial molecular types. There are no additional sources of accurate epidemiological data for this subtype.

Anaplastic ependymoma

Anaplastic ependymoma is a circumscribed glioma made up of nearly identical small cells. They have a greater tendency to disseminate in the CNS via the cerebrospinal fluid versus grade II ependymoma. Anaplastic ependymoma often grow back locally if they recur following treatment. Perivascular pseudorosettes and ependymal rosettes are key histological features. They often express GFAP, the S100 protein, and vimentin. Vital brain structures are often involved, making total resection difficult, and possible in just 30%-40% of cases. They have round nuclei in a fibrillary matrix. There are perivascular anucleate zones (pseudorosettes), a high nuclear-to-cytoplasmic ratio, and a high mitotic count. In about 25% of cases, there are ependymal rosettes present. Anaplastic ependymomas are usually not linked to genetic mutations or pathogenic variants that are inherited. Instead, they are linked to somatic mutations, in which body cells change over time. In rare cases, anaplastic ependymomas are related to genetic diseases such as neurofibromatosis type 2, which increase risks for CNS tumors.

Signs and symptoms of anaplastic ependymomas include headaches, nausea, vomiting, seizures, lethargy, changes in thinking or concentration, nystagmus, hydrocephalus. If an anaplastic ependymoma develops in the spinal cord, symptoms may include pain, weakness, paralysis, sensory changes, neck stiffness, and loss of reflexes. Diagnosis is usually confirmed when there is a high cell density, elevated mitotic count, widespread microvascular proliferation, and necrosis. Anaplastic ependymoma rarely invades nearby CNS parenchyma to any large degree. Ultrastructural examination reveals cilia and microvilli. These ependymomas are usually intracranial, and rare in the spinal cord. They occur in people of all ages, though most posterior fossa tumors occur in children. There can be clear cell, papillary, or tanycytic morphologies. Clinical outcome is variable, mostly based on the molecular group and the extent of surgical resection.

To define the molecular group, *transcriptome* or *methylome* profiling has identified the importance of the anatomical site. Anaplastic ependymoma is a grade III tumor according to the World Health Organization. These tumors, when the cell density is extremely high, may be mistaken for embryonal tumors. They always have brisk mitotic activity. This is associated with a poor prognosis for posterior fossa tumors. The mitotic activity is often accompanied by microvascular proliferation and palisading necrosis. Though pseudorosettes are definitive, in a poorly differentiated supratentorial anaplastic ependymoma, they may be difficult to visualize. Anaplastic ependymomas have the same *immunoprofile* as classic ependymomas, but tumor growth fraction indices, including the Ki-67 proliferation index, are higher. There are no accurate sources of epidemiological data for anaplastic ependymomas.

Ependymoblastoma

Ependymoblastoma is a highly malignant brain tumor of childhood. It is usually seen in very young children or infants. Other terms used for this tumor include *childhood ependymoma*, *ependymal tumor*, and *primitive neuroectodermal tumor*. Ependymoblastoma occurs supratentorially in both hemispheres in 70% of cases. It is infratentorial, in the cerebellum and brain stem in the remaining 30%.

It is rare among brain tumors generally, but is the second most common malignancy in very young patients, second only to leukemia. Symptoms include abnormal speech, loss of balance, double vision, and general weakness or weakness on one side of the face. *Infratentorial ependymoblastomas* of the lower back portion of the brain present with increased intracranial pressure and coordination deficits. Ependymoblastomas involve cysts in the tumor periphery in 55% of cases, and 77% have signs of intratumoral hemorrhage. Hypointense to hyperintense signal intensities are present in 86% of cases, and the predominant T2 signal is isointense in 55% of cases. About 86% of cases show sharp tumor margins (see Fig. 9.11). *Supratentorial ependymoblastomas* of the upper brain usually cause focal headaches and motor signs. History and physical examination may suggest the area of the brain that is involved. The child will have worsened symptoms in the morning hours, then improve as the day progresses. Though a CT scan of the head may reveal calcification that is not visible on an MRI, it should include the entire spine as well as the brain in order to identify any tumor spread. The imaging appearance is of a large, well-demarcated by heterogeneous mass with variable contrast enhancement. Diagnosis is based on positive tests of the tumor. There is a high morbidity associated with whole brain or neuroaxis radiation in young children. Therefore, therapy is divided into different methods, for children older than 3 years, or for those 3 years and younger.

For children older than 3 years, standard treatment is usually surgery, followed by radiation therapy to the brain and spinal cord. Sometimes, chemotherapy is administered along with radiation therapy, or after it. For children 3 years or younger, standard treatment is usually surgery, followed by chemotherapy. Other treatments may include surgery

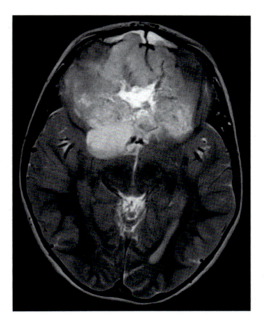

FIGURE 9.11 Ependymoblastoma MRI.

followed by high doses of chemotherapy, with bone marrow or stem cell transplant, and surgery followed by chemotherapy plus low doses of localized radiation therapy. Treatment for these younger patients often occurs as part of a clinical trial. Unfortunately, prognosis is poor. Five-year survival rates range from 0% to 30% of patients.

Clinical cases

Clinical case 1

1. What is the general appearance of a subependymoma?
2. What is the basic epidemiology of these tumors?
3. What is the outcome for this type of tumor?

A 47-year-old man was referred to a neurologist because of headaches, numbness, nausea, vomiting, and dizziness. He had experienced the headaches for three years, but they suddenly became much worse and occurred more often over the last 2 weeks. His entire head was affected, with only slight relief using over-the-counter analgesics. An MRI revealed a mass in the inferior fourth ventricle, extending through the foramen magnum, and pressing upon the pontomedullary junction and medulla oblongata, with posterior compression of the cerebellar tonsils. The mass had similar intensity to the brain parenchyma on T1-weighted images, and showed scattered heterogeneous enhancement. The T2-weighted images revealed increased signal intensity compared to the brain parenchyma, and no adjacent edema. A gradient echo sequence showed scattered foci of hypointensity, which indicated T2 susceptibility, and favored the presence of calcific deposits and possibly blood. Via craniotomy, the mass was completely resected. Histologic evaluation confirmed that it was a subependymoma. The patient recovered fully.

Answers:

1. **Subependymomas appear as shiny, pearled-white tumors with lobules and intraventricular protuberances. They have sharp demarcation. Microscopically, these tumors have clusters of bland, round nuclei within a fibrillary matrix, and microcysts as well as calcification.**
2. **About 90% of subependymomas occur in adults, usually after age 40. They appear 2.3 times as often in males than in females. Like all ependymal tumors, they are more prevalent in Caucasians than in other racial or ethnic groups. True incidence is hard to determine as they may never cause clinical symptomatology and discovered incidentally or at postmortem examination. Overall, they are extremely rare. Peak incidence is between 40 and 60 years of age. Based on various studies of prevalence, they are believed to make up 8% of ependymal tumors, and 0.51% of all central nervous system tumors.**
3. **Subependymomas are WHO grade I tumors that have an excellent prognosis compared against other ependymal tumors. After gross total resection, no recurrences have been reported. Debulking alone usually provides an excellent prognosis, since the tumors grow slowly, and residual tumors may require decades to cause symptoms.**

Clinical case 2

1. What is the general epidemiology of myxopapillary ependymoma?
2. What are the common clinical manifestations of this tumor?
3. How are these tumors usually diagnosed?

A 19-year-old man with a long history of back pain had previously been diagnosed with soft tissue injuries. Two years later, he began to experience erectile and bowel dysfunction, and was referred to an orthopedic surgeon. An MRI revealed a large myxopapillary ependymoma that extended from the T12 to the L4 parts of his spine. A pathologist confirmed this. The tumor was surgically resected, followed by adjuvant radiotherapy. After one year, the patient still needed catheterization. However, his back pain was almost totally gone. Follow-up MRIs revealed no disease progression or any new spinal lesions, over four years after initial diagnosis.

Answers:

1. **Spinal ependymal tumors are glial tumors derived from ependymal cells in the spinal cord. They represent 40%-60% of primary spinal cord tumors. The myxopapillary ependymomas mostly occur in the**

thoracolumbar region, and are the most common form of ependymoma in the lumbar spine. They make up 13% of all spinal ependymomas and 90% of tumors in the conus medullaris.
2. Clinically, the most common finding is lumbosacral radicular pain that is often worse when lying down—hence many patients complain of nighttime pain. To a lesser degree, there are sensory changes, motor deficits, bladder abnormalities, and sexual dysfunction. Average duration of symptoms before diagnoses ranges from 13 months to about 8 years, due to the slow growth of the tumor.
3. Diagnosis of myxopapillary ependymomas is primarily via MRI, and findings usually include an intradural mass that is hypointense or isointense with the spinal cord on T1-weighted images. On T2-weighted images, it is hyperintense with intense homogenous enhancement after contrast material is administered. The lesions often affect two to four vertebral body levels. Confirmation is only confirmed with pathologic examination of excised tumor tissues and biopsy. Electromyography and nerve conduction velocity studies may be helpful to plan surgical approach and extent of deficit.

Clinical case 3

1. How can anaplastic ependymomas develop?
2. What are the possible signs and symptoms of this type of tumor?
3. How do inheritance factors play a role in the diagnostic evaluation of anaplastic ependymomas?

A 2-year-old boy was brought to the emergency department. His mother described the following symptoms: a chronic headache, focal seizures on the left side of his body, weakness of his left arm and leg, and vomiting. Examination revealed him to be alert, but funduscopy revealed papilledema. A brain CT scan showed an heterogeneously enhancing intra-axial mass in the right cerebral hemisphere that arose from the superolateral aspect of the right lateral ventricle, and extended into the nearby brain parenchyma. The mass had multiple cystic areas. An MRI of his brain revealed that the margins of the mass were slightly better appreciated, and it was lobulated. Biopsy and histopathological study of the mass was performed. There were tumor cells with extensive necrosis, and nests of cells were arranged around blood vessels, forming rosettes. The Ki-67 proliferation index was 10%, and a grade III anaplastic ependymoma was diagnosed. Complete surgical resection of the tumor was undertaken and the child recovered well from surgery. Radiotherapy followed, and the boy had no recurrence.

Answers:

1. Anaplastic ependymomas have a tendency to disseminate in the CNS via the cerebrospinal fluid. They often grow back locally if they recur following treatment. Perivascular pseudorosettes and ependymal rosettes are key histological features. They often express GFAP, the S100 protein, and vimentin. Vital brain structures are often involved, making gross total resection difficult, and possible in just 30%-40% of cases.
2. Signs and symptoms of anaplastic ependymomas include headaches, nausea, vomiting, seizures, lethargy, changes in thinking or concentration, nystagmus, hydrocephalus. If an anaplastic ependymoma develops in the spinal cord, symptoms may include pain, weakness, paralysis, sensory changes, neck stiffness, and loss of reflexes.
3. Anaplastic ependymomas are usually not linked to genetic mutations or pathogenic variants that are inherited. Instead, they are linked to somatic mutations, in which body cells change over time. In rare cases, anaplastic ependymomas are related to genetic diseases such as neurofibromatosis type 2, which increase risks for CNS tumors.

Clinical case 4

1. Which anatomic locations are usually involved in ependymoblastoma?
2. What is the clinical presentation of ependymoblastoma?
3. Describe the pathology of the majority of these tumors.

A 4-year-old girl was hospitalized with nausea, vomiting (especially in the morning), spasticity, ataxia, and bilateral papilledema. A brain MRI revealed a firm mass in the fourth brain ventricle. A craniotomy was performed, and a grayish red mass was removed. The tumor had extended from the aqueduct of Sylvius through the fourth ventricle to the level of the lamina of the second cervical vertebrae. The tumor was diagnosed as an ependymoblastoma, and surprisingly for such as young patient, it showed calcification. The tumor cells were uniform, with regular oval nuclei and long, coarse processes. Follow-up treatment included radiation therapy, and the girl luckily survived without recurrence.

Answers:

1. Ependymoblastoma occurs supratentorially in both hemispheres in 70% of cases. It is infratentorial, in the cerebellum and brain stem in the remaining 30%.
2. Clinically, ependymoblastomas are described as highly aggressive tumors that occur mainly in young children, with rapid growth and craniospinal dissemination. The imaging appearance is of a large, well-demarcated by heterogeneous mass with variable contrast enhancement.
3. Pathologically, ependymoblastomas involve cysts in the tumor periphery in 55% of cases, and 77% have signs of intratumoral hemorrhage. Hypointense to hyperintense signal intensities are present in 86% of cases, and the predominant T2 signal is isointense in 55% of cases. About 86% of cases show sharp tumor margins.

Key terms

Alcian blue
Anaplasia
Anaplastic ependymoma
Blepharoblasts
Bumpy cell
Central neurocytomas
Cerebellar tonsils
Cerebiform
Childhood ependymoma
Choroid plexus papillomas
Chromothripsis
Classic ependymoma
Clear cell ependymoma
Ependymoblastoma
Exophytic
Fetal hydantoin syndrome
Foramen of Luschka
Foramen of Magendie
Funduscopy
Glioma susceptibility 1
Gradient echo sequence
Hyalinized
Immunoprofile
Immunoreactivities
Interdigitations
Interphase fluorescence in situ hybridization
Intrasacral
Lipomatous
Methylome
Microrosettes
Mismatch repair cancer syndrome
Myxopapillary ependymoma
Myxopapillary variant
Neurofibromatosis II
Neuron-specific enolase
Neuropril
Nucleolin
Papillary ependymoma
Polyploidy
Radiculopathy
RELA fusion-positive ependymoma
Rhabdomyosarcomatous
Rosettes
Subependymoma
Tanycytic ependymoma
Transcriptome
Transcriptomic
Turcot syndrome
Ultrastructural
Vacuolation
Vimentin
Zonulae adherentes
Zonulae occludentes

Further reading

Adesina, A. M., Tihan, T., Fuller, C. E., & Young Poussaint, T. (2016). *Atlas of pediatric brain tumors* (2nd edition). Springer.
Chaichana, K., & Quinones-Hinojosa, A. (2019). *Comprehensive overview of modern surgical approaches to intrinsic brain tumors.* Academic Press.
Fatterpekar, G. M., Naidich, T. P., & Som, P. M. (2012). *The teaching files: Brain and spine: Expert consult.* Saunders.
Fountas, K., & Kapsalaki, E. Z. (2019). *Epilepsy surgery and intrinsic brain tumor surgery: A Practical atlas.* Springer.
Fowler, R. (2016). *Ependymomas: Prognostic factors, treatment strategies and clinical outcomes.* Nova Science Publishers, Inc.
Gajjar, A., Reaman, G. H., Racadio, J. M., & Smith, F. O. (2018). *Brain tumors in children.* Springer.
Gunderson, L. L., & Tepper, J. E. (2011). *Clinical radiation oncology: Expert consult* (3rd edition). Saunders.
Gursoy Ozdemir, Y., Bozdag Pehlivan, S., & Sekerdag, E. (2017). *Nanotechnology methods for neurological diseases and brain tumors—Drug delivery across the blood-brain barrier.* Academic Press.
Hayat, M. A. (2012). *Tumors of the central nervous system, volume 8: Astrocytoma, medulloblastoma, retinoblastoma, chordoma, craniopharyngioma, oligodendroglioma, and ependymoma.* Springer.
Hoppe, R., Phillips, T. L., & Roach, M. (2010). *Leibel and Phillips textbook of radiation oncology—E-Book: Expert consult* (3rd edition). Saunders.
Icon Group International. (2010). *Ependymomas: Webster's timeline history, 1949-2007.* Icon Group International, Inc.
Jain, R., & Essig, M. (2015). *Brain tumor imaging.* Thieme.
Karajannis, M. A., & Zagzag, D. (2015). *Molecular pathology of nervous system tumors: Biological stratification and targeted therapies.* Springer.
Kaye, A. H., & Laws, E. R. (2011). *Brain tumors e-book: An encyclopedic approach* (3rd edition). Saunders.
Kesharwani, P., & Gupta, U. (2018). *Nanotechnology-based targeted drug delivery systems for brain tumors.* Academic Press.
Kim, D. H., Chang, U. K., Kim, S. H., & Bilsky, M. H. (2008). *Tumors of the spine.* Saunders.
Lopez Savon, E. (2018). *Intraspinal ependymomas: Their association with metastasis/disseminations in patients over a period of 22 years.* Scholars' Press.

Medical Ventures Press. (2011a). *Childhood ependymoma and subependymoma: Pediatric cancer guide to symptoms, diagnosis, treatment, prognosis, clinical trials*. Progressive Management.

Medical Ventures Press. (2011b). *2011 Pediatric cancer toolkit: Childhood ependymoma and subependymoma*. Progressive Management.

Mucci, G. A., & Torno, L. R. (2015). *Handbook of long term care of the childhood cancer survivor (specialty topics in pediatric neuropsychology)*. Springer.

National Comprehensive Cancer Network. (2017). *NCCN guidelines for patients: Brain cancer—Gliomas*. National Comprehensive Cancer Network.

Newton, H. B. (2018). *Handbook of brain tumor chemotherapy, molecular therapeutics, and immunotherapy* (2nd edition). Academic Press.

Norden, A. D., Reardon, D. A., & Wen, P. C. Y. (2011). *Primary central nervous system tumors: Pathogenesis and therapy (current clinical oncology)*. Humana Press.

Ozek, M. M., Cinalli, G., Maixner, W., & Sainte-Rose, C. (2015). *Posterior fossa tumors in children*. Springer.

Perry, A., & Brat, D. J. (2010). *Practical surgical neuropathology: A diagnostic approach: the pattern recognition series, expert consult*. Churchill Livingstone.

Quinones-Hinojosa, A., Raza, S. M., & Laws, E. R. (2013). *Controversies in neuro-oncology: Best evidence medicine for brain tumor surgery*. Thieme.

Sampson, J. H. (2017). *Translational immunotherapy of brain tumors*. Academic Press.

Shiminski-Maher, T., Woodman, C., & Keene, N. (2014). *Childhood brain & spinal cord tumors: A guide for families, friends & caregivers* (2nd edition). Childhood Cancer Guides.

Warmuth-Metz, M. (2016). *Imaging and diagnosis in pediatric brain tumor studies*. Springer.

Chapter 10

Schwannoma

Chapter Outline

Schwannoma 179
    Cellular schwannoma 184
    Plexiform schwannoma 185
    Melanotic schwannoma 188
Schwannomatosis 190
Clinical cases 192
Key terms 195
Further reading 195

Schwannoma

A *schwannoma* is a benign, slow-growing tumor composed entirely of well-differentiated Schwann cells. It is usually encapsulated, and affects the myelin nerve sheaths that cover the peripheral nerves. Despite that, they can emerge from the optic and olfactory nerves, which do not have myelin sheaths. Schwannomas are attached to the cranial or spinal nerves from which they originate, with nearly 45% occurring in the head and neck area. There is a loss of *merlin* expression in the conventional forms. Merlin is the gene product of *neurofibromatosis 2* (NF2). The tumor is homogeneous, always remaining on the outside of the nerve, though the tumor can push the affected or surrounding nerve(s) aside or against a bone, causing damage. Schwannomas are relatively slow-growing. Less than 1% becomes malignant, degenerating into *neurofibrosarcoma*. Surgical removal is often successful, due to the encapsulated nature of schwannomas, allowing a surgical cleavage plane.

Epidemiology

Schwannomas make up 85% of cerebellopontine angle tumors, 29% of spinal nerve root tumors, and 8% of all intracranial tumors. Intracerebral schwannoma accounts for less than 1% of all intracranial schwannomas. The intracerebral type, for some reason, mostly occurs in children and younger people, with a slight male preponderance. Most intracerebral schwannomas are supratentorial, usually within the superficial sections of the brain parenchyma or near the ventricle. They have occurred in the frontal and temporal lobes of the cerebral hemisphere, the cerebellar hemisphere, the cerebellar vermis, and fourth ventricle.

About 90% of cases of all schwannomas are solitary and sporadic. Approximately, 4% arise in the setting of NF2. Out of the 5% of schwannomas that are multiple in development, but not linked to NF2, some are associated with *schwannomatosis*. Patients may be of any age, though cases during childhood are rare. The highest incidence is within the third and fourth sixth decades of life. Though most studies show no significant difference in occurrence between males and females, there have been few studies revealing female predominance within the setting of intracranial tumors. Cerebral intraparenchymal schwannomas are related to a male predominance, and a younger patient age. Schwannomas of the spinal cord parenchyma are so rare, that they cannot be adequately assessed regarding their epidemiology.

Etiology and risk factors

Apart from NF2 and schwannomatosis, the causes of schwannomas are unknown. Individuals with family history of spinal cancer are more likely to develop a spinal schwannoma, which suggests a genetic link. Exposure to radiation is considered to be another cause. However, there are no actual identified risk factors for schwannomas.

Pathology

Most schwannomas occur outside the central nervous system, usually within the peripheral nerves of the skin and subcutaneous tissue. Intracranial schwannomas most often affect the eighth (vestibulocochlear) cranial nerve, in the cerebellopointine angle—especially accompanying NF2. The tumors arise at the transition area between the central and peripheral myelination, affecting the vestibular division, and from the cervical or brachial plexuses. Overall, brachial plexus tumors are rare, and less than 5% of upper extremity schwannomas come from this location. The nearby cochlear division is hardly ever the site of origination. Because of the characteristic location of these tumors, and resultant enlargement of the internal auditory meatus, neuroimaging is easily able to detect and diagnose them. *Intralabyrinthine* schwannomas are rare.

The intraspinal schwannomas often affect the sensory nerve roots, and much less often, the motor and autonomic nerves. Sometimes, central nervous system schwannomas are not linked to a specific nerve. As such, there have been 70 cases of spinal intramedullary schwannomas, along with 40 cases of intraventricular or cerebral parenchymal schwannomas. These tumors rarely affect the dura. Unlike neurofibromas, peripheral nerve schwannomas usually are attached to nerve trunks, often involving the head and neck region or extremity flexor surfaces. Visceral and osseous schwannomas are also rare.

Clinical manifestations

There are several clinical manifestations of peripherally located schwannomas, contingent on their location, geometry, and rate of growth. They may be incidental, asymptomatic paraspinal tumors, or spinal nerve tumors causing *radicular pain*, plus signs of nerve root or spinal cord compression. Most schwannomas present with painless swelling, though eventual symptoms are usually based on the tumor site. In some patients, significant lower back or knee pain, muscles, bones, and dyspnea upon exertion, and other breathing difficulties, are reported. There may be a tingling sensation and numbness along the course of the involved nerve. Lower back pain or partial lower limb paralysis may be linked to lumbosacral schwannomas.

They may also be cranial nerve tumors with symptoms of hearing loss, tinnitus, and sometimes, horizontal nystagmus, slow corneal reflexes, vertigo, nausea, vomiting, and difficulty in walking. Often, clinical presentation is related to trigeminal nerve dysfunction including neuralgia, neurasthenia, or numbness. Mass effect symptoms may be present if the tumor is large. If the nose is involved, there can be breathing difficulty, painful swallowing, nose bleeds, and "crackling" of the voice. Intracerebral schwannoma has no specific clinical manifestations or age classifications. It may cause epilepsy, increased intracranial pressure, and local neurological dysfunction. In patients under age 25, headaches over epilepsy are most common, but in elderly patients, there is usually acute local neurological dysfunction.

Since schwannomas favor the sensory nerve roots, motor symptoms are uncommon. The hallmark feature of NF2 is development of bilateral vestibular tumors. For patients with schwannomatosis, pain is the most common symptom of a schwannoma. Significantly, there is lower back and knee pain in their muscles and bones, dyspnea upon exertion, and chronic fatigue. For the rare lumbosacral schwannomas, there is usually persistent back pain that can last for years, which may become extremely severe, requiring treatment.

Diagnosis

Upon magnetic resonance imaging, a schwannoma appears as a well circumscribed, often heterogeneously enhancing, and sometimes-cystic mass. An MRI, the preferred imaging method, often indicates an abnormal cystic signal, with nodular shadows above the cystic wall, within the parenchyma. A vestibular schwannoma (also called an *acoustic neuroma*) often appears like an "ice cream cone", with a tapering intraosseous "cone" that exits the internal auditory canal, and expands out into a round cerebellopontine angled mass. A tumor that is paraspinal or in a head or neck site may be linked to bone erosion, which can sometimes be seen on a plain X-ray. An MRI of an intracranial schwannoma provides high contrast resolution along with detailed view of brain structures, allowing for precise tumor localization. Differential diagnoses include acoustic schwannoma, ependymoma, meningioma, chondrosarcoma, or metastasis. If the tumor is confined to the Meckel's cave, and is small, the differential diagnoses also include aneurysm, pituitary adenoma, and vascular malformation.

Paraspinal tumors may also have a "dumbbell" shape, and a point of constriction located at the neural exit foramen. An MRI showing a schwannoma of the right trochlear nerve, near the brain stem, is shown in Fig. 10.1. Lumbosacral schwannomas are solid, appearing well-enhanced in comparison to other lesions. They cannot be diagnosed based on clinical examination alone, so an MRI scan with contrast is helpful, though a histopathological examination would be definitive.

Macroscopically, most schwannomas are globoid and less than 10 mm in size. They are usually encapsulated, except for tumors of the intraparenchymal central nervous system sites, skin, bone, and viscera. For peripheral tumors, a nerve of origin is identified in less than 50% of cases and is often covering the tumor capsule. The external surface of a schwannoma reveals the parent nerve of origin, and a variably translucent capsule (see Fig. 10.2). A cut-tumor surface usually shows glistening tissue that is light tan in color, interrupted with bright yellow areas (see Fig. 10.3). There may

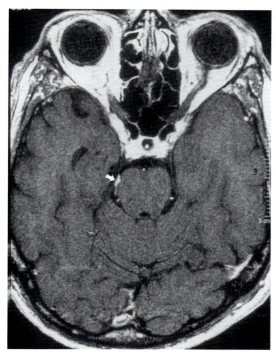

FIGURE 10.1 T1-weighted axial magnetic resonance imaging with gadolinium at the level of inferior colliculus shows enhancing lesion consistent with a schwannoma (arrow) of right trochlear nerve as it courses ventrally around brain stem.

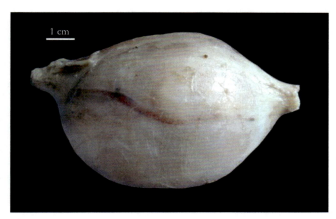

FIGURE 10.2 External surface of a schwannoma.

or may not be cysts or hemorrhage. Degenerative vascular changes may be related to infarct-like necrosis, such as in larger tumors. Examples include, the giant lumbosacral, pelvic, and retroperitoneal schwannomas that often cause erosion of nearby vertebral bodies. Lumbosacral schwannomas are rare, and often resemble a meningocele or lumbar degenerative disk disease.

Microscopically, a conventional schwannoma is totally made up of neoplastic Schwann cells that may be focally numerous, and not inflammatory cells. There are two primary structural patterns. These include areas of compacted, lengthened cells, and occasional nuclear palisading. This is known as the *Antoni A pattern*. Antoni A tissue has normochromic round or spindle-shaped nuclei, about the same size as those of smooth muscle cells (see Fig. 10.4). However, they are tapered instead of having blunted endings. The Antoni A pattern has densely packed, spindled cells that are arranged in fascicles going in different directions. The nuclei may align in alternating parallel rows to form nuclear palisades. When they are marked, the nuclear palisades are called *Verocay bodies*. All schwannoma cells have a

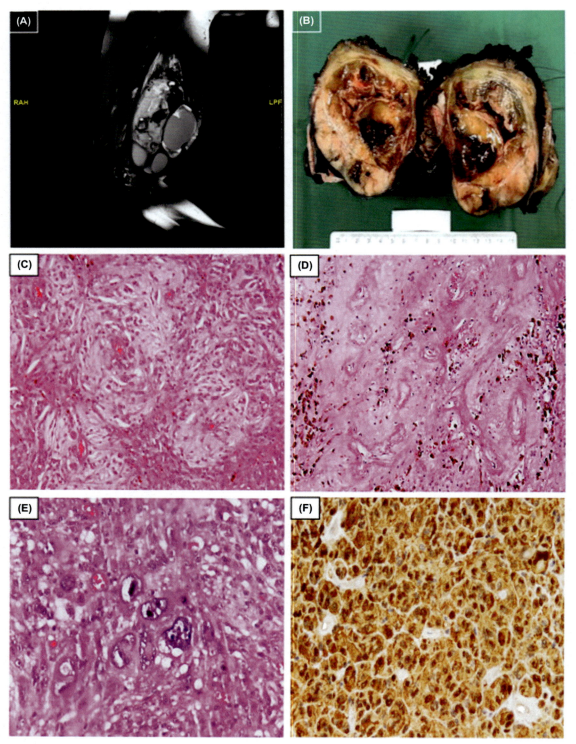

FIGURE 10.3 Schwannoma, showing the cut-tumor surface and its tissues. (A) MRI of the tumor. (B) Resected tumor, showing a complete surrounding capsule and focal hemorrhages. (C and D) Two views of hypocelluar and hypercelluar areas of mostly epithelioid cells. (E) Focal moderate atypia and a few mitotic figures. (F) Strong, diffuse S100 and SOX10 immunoreactivity plus immunostains for melanocytic markers.

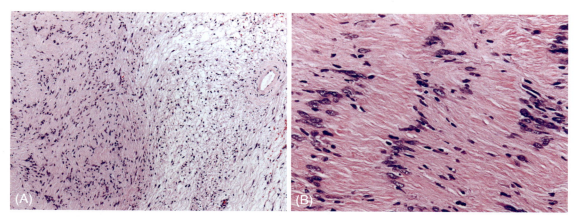

FIGURE 10.4 (A) Schwannomas often contain dense pink Antoni A areas (left), and loose, pale Antoni B areas (right), as well as hyalinized blood vessels (right). (B) Antoni A area with the nuclei of tumor cells aligned in palisading rows.

pericellular reticulin pattern that corresponds with the surface basement membranes. Related to NF2, vestibular schwannomas may have a predominance of Antoni A tissue, in a whorl formation, and a lobular, grape-like growth pattern seen in low-power microscopic studies. Molecularly, these are polyclonal, and probably make up the majority of multiple small schwannomas. Malignant transformation, usually extensive and *transcapsular* instead of microscopic, is rare in conventional schwannomas.

There may also be a less cellular, loose texturing of cells that have indistinct processes and variable amounts of *lipidization*. This is known as the *Antoni B pattern*. *Retiform* patterns are only rarely seen. The tumor cells have moderate amounts of eosinophilic cytoplasm and no discernible cell borders. In Antoni B tissue, the cells have smaller nuclei that may be round or ovoid. The cells are loosely arranged. Groups of lipid-laden cells can be present, as they can in the Antoni A tissue. Vasculature is usually thick-walled and hyalinized. Dilated blood vessels are common, and surrounded by hemorrhage.

In a subtype called *ancient schwannoma*, nuclear pleomorphism occurs. This includes strangely-appearing forms with cytoplasmic-nuclear inclusions. A mitotic shape may be seen occasionally, but this should not be misunderstood as an indicator of malignancy. Like other schwannomas, there is usually diffuse S100 protein and collagen IV positivity. Collagen IV is the type of collagen found mostly in the basal lamina. There is also usually extensive transcription factor *SOX10* expression, and a low Ki-67 protein proliferation index in enlarged and atypical cells. Monitoring of schwannomas usually involves regular MRI scans with thin cuts through the relevant anatomy, volumetric measurements, and clinical correlation (e.g., auditory testing).

Treatment

Schwannomas are usually removed surgically, and can often be scraped-off of nerves without permanent nerve injury. Recovery time and remaining symptoms are widely varied based on tumor size and location. The patient may be given a corticosteroid injection into the facet joints as applicable, along with local anesthetics in spinal involvement to decrease pain. If the tumor is small and not causing any significant problems, it may simply be monitored for signs of growth or change. If the tumor is cancerous or there are coexisting conditions that make surgery dangerous, the physician may recommend *stereotactic radiosurgery*. A strong dose of radiation is provided directly to the tumor, to shrink it over one or multiple fractionated treatments, contingent on the specifics of the tumor and patient. This approach has fewer adverse effects than traditional radiation, which involves smaller doses of radiation over a longer period of time. Malignant schwannomas may also be treated with chemotherapy and with immunotherapy medications. For lumbosacral schwannomas, there is controversy about treatment, which can include complete excision that involves sacrificing the nerve root. Oncology management results are usually good, but surgical trauma causes postoperative neurological deficits in up to 15% of cases. Deficits may remain for several months to a few years, with minimal but residual possibility of tumor recurrence.

Prognosis

The prognosis for schwannomas is based largely on tumor size, location, and whether it is malignant. Since most schwannomas are benign and may never produce symptoms, overall prognosis is usually excellent. It is important to monitor any changes regarding symptoms in order to achieve the best treatment goals and improve outcomes.

> **Point to remember**
> Schwannomas are nerve sheath tumors, that involve the coating around nerve fibers transmitting message to and from the brain and spinal cord, and the rest of the body. They are usually benign, and most often found in the head and neck. Though of unknown cause, schwannomas are more common in people who have the inherited disorder called neurofibromatosis. Schwannomas are signified by painless or painful growths, numbness, weakness, or paralysis.

Cellular schwannoma

A *cellular schwannoma* is a benign form of schwannoma that is hypercellular, and made up partially or totally of Antoni A tissue, while lacking well developed Verocay bodies. Cellular schwannomas are usually located in paravertebral sites of the pelvis, retroperitoneum, and mediastinum. Cranial nerves may be affected, which are usually the fifth (trigeminal) and eighth cranial nerves.

Epidemiology

Cellular schwannoma usually develops between ages 30 and 60 with even distribution between the genders. However, with NF2, these tumors usually occur before age 30, and affect females more than males. These tumors have no geographical, racial, or ethnic preference, and occur worldwide.

Etiology and risk factors

There are no causative factors identified for the formation of cellular schwannomas. In the case of multiple tumors, which are rare, risk factors include the presence of NF2, a positive family history, schwannomatosis, and *Gorlin-Koutlas syndrome*.

Pathology

Clinically, cellular schwannomas are similar to classic schwannomas, but histology reveals hypercellularity, fascicular cell growth, and sometimes, nuclear hyperchromasia and atypia. There is low-to-medium mitotic activity that is usually less than 4 mitoses per 10 high-power fields. However, sometimes there are 10 or more mitoses per 10 high-power fields. This can result in a misdiagnosis of malignancy, as a peripheral nerve sheath tumor. Cellular schwannomas are believed to be the result of sporadic mutations, with abnormalities in chromosome 22 usually found. The tumor may be caused by overproduction of Schwann cells that wrap around a nerve. In one study, cellular schwannomas had Schwannian whorls, subcapsular lymphocytes, a peritumoral capsule, and macrophage-rich infiltrates. There was a lack of fascicles, and strong, wide expression of S100, SOX10, CDKN21, and neurofibromin (see Fig. 10.5). In most cases, the Ki-67 proliferation index is below 20%. The tumor ranges in size from 2 to 10 centimeters. In very rare cases, malignancy can develop.

Clinical manifestations

Cellular schwannomas are usually asymptomatic. However, if symptoms develop, they are usually based on the location of the tumor, and can be wide-ranging, depending on the nerve or region affected. The tumor can compress the nerve on which it lies, causing nerve dysfunction. It is slow-growing, and may be detected incidentally during examination for another condition. Possible complications include nerve compression, loss of nerve function, nerve damage, infrequent recurrence, and rare malignant transformation.

Diagnosis

Upon imaging, the tumor appears as a firm mass that may or may not be attached to a nerve. If close to a bone, the tumor can invade surrounding bone and cause significant damage. Diagnosis is based on complete physical examination, through evaluation of patient and family history, a neurological examination, CT or MRI of the affected region, nerve conduction studies, and tissue biopsy.

Treatment

In most cases, complete surgical excision is curative, if the tumor can be removed without damaging the underlying nerve. Generally, they are difficult to completely remove due to their nerve involvement. Asymptomatic tumors may

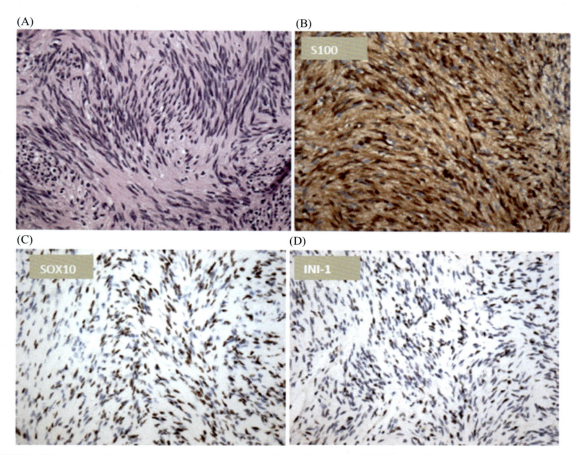

FIGURE 10.5 Histologic and immunochemical characteristics in cellular schwannoma. (A) Schwann cells arranged in fascicles. (B) Appearance in S100 staining. (C) Appearance in SOX10 staining. (D) Appearance of a mosaic pattern of INI-1 immunoreactivity.

require no treatment unless they cause discomfort, affect quality of life, or affect nerve function. Close monitoring is usually done to determine the need for treatment. Surgical options include microsurgery (subtotal, near total, or total tumor removal), craniotomy, stereotactic radiosurgery, and for spinal schwannomas, decompressive laminectomy is applicable. For malignant tumors, additional treatments may include chemotherapy, radiation therapy, treatment of underlying NF2, good postoperative care, and follow-up care with regular screening and check-ups.

Prognosis

The prognosis of cellular schwannoma is generally excellent, but it can be influenced by tumor location, NF2, and malignant transformation. Recurrences of cellular schwannomas have been seen occasionally, mostly in intracranial, spinal, and sacral locations. However, no cellular schwannoma is known to have metastasized, or to have a malignant course that caused patient death. There have been two examples of malignant transformation, with one of these cases linked to NF2. If a cellular schwannoma becomes malignant, prognosis is based on tumor stage, location, overall patient health, and response to treatments.

Plexiform schwannoma

A *plexiform schwannoma* has a plexiform or multinodular growth pattern, and can either be conventional or cellular in structure (see Fig. 10.6). Most of these tumors develop in the skin or subcutaneous tissues of extremities, the head, neck, or trunk. They are believed to involve multiple nerve fascicles or a nerve plexus, and are sometimes located more deeply. These tumors are also referred to as *nonmelanotic plexiform schwannomas* and *plexiform neurilemomas* (sometimes spelled *neurolemomas*).

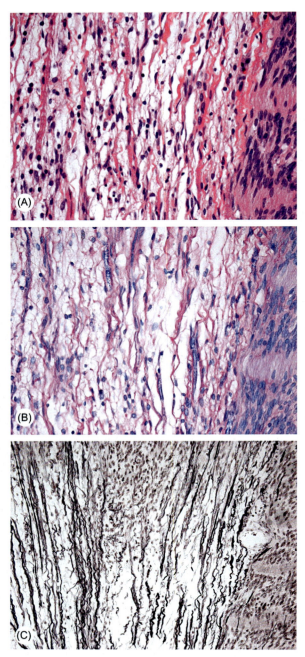

FIGURE 10.6 Cellular appearance of plexiform schwannoma. (A) Plexiform pattern of intraneural growth with multinodularity. (B) Hyperchromatic nuclei. (C) Thin, convoluted cytoplasmic processes.

Epidemiology

Plexiform schwannomas may occur at birth or during childhood, and only make up about 5% of all types of schwannomas. Both males and females are equally affected. There are no known geographical, racial, or ethnic preferences. The tumor occurs throughout the world. In 93% of tumors in patients with familial schwannomatosis, a mosaic pattern of SMARCB1 (INI1) expression is seen. This is also seen in 83% of NF2-associated tumors, in 55% of tumors in patients with sporadic schwannomatosis, and only 5% of single, sporadic schwannomas.

Etiology and risk factors

No causative factors have been identified. There are also no identifiable risk factors, though trauma, NF2, family history, schwannomatosis, and Gorlin-Koutlas syndrome may be related. These tumors are rarely linked with NF2, and also with schwannomatosis. They are not linked to *neurofibromatosis 1* (NF1). The NF2 gene is a tumor suppressor that is integral to formation of sporadic schwannomas. Inactivating mutations of the NF2 gene have been found in about 60% of all schwannomas. There are mostly small *frameshift mutations*, resulting in truncation of proteins. Though not described for exons 16 or 17, mutations have been seen throughout the gene-coding sequence, and at *intronic* sites. Usually, these mutations accompany a loss of the remaining wild-type allele on chromosome 22q. In other patients, there is loss of chromosome 22q with an absence of detectable NF2 gene mutations. Regardless, most schwannomas have a loss of merlin expression, detected via western blotting of immunohistochemistry. This is regardless of the type of mutation or the allelic status. Therefore, it is believed that abrogation of merlin function is a vital step in tumorigenesis. In cellular schwannomas, there is also a loss of chromosome 22. Other genetic changes, however, are rare, but a small amount of patients have had a loss of chromosome 1p, and gains of 9q34 or 17q.

While most schwannomas are sporadic, multiple tumors may occur along with two tumor syndromes. The bilateral vestibular schwannomas are pathognomonic of NF2. Multiple, predominantly non-vestibular schwannomas without other NF2 features are typical with schwannomatosis. In some cases, there is a segmental tumor distribution. Germline SMARC1 mutations at 22q11.23 have been seen in 50% of all familial schwannomatosis cases, and in less than 10% of all sporadic schwannomatosis cases. While somatic NF2 inactivation has been detected, germline NF2 mutations are absent. Germline loss-of-function mutations with LZTR1, predispose patients to an autosomal dominant inherited disorder, with multiple schwannomas. This is identified in about 80% of 22q related cases of schwannomatosis, with no mutations in SMARCB1. The LZTR1 mutations also give an increased risk for vestibular schwannoma, and additional overlapping with NF2.

Pathology

Plexiform schwannomas often grow quickly, are hypercellular, and have increased mitotic activity. However, they tend to behave like benign tumors and may recur locally and have no metastatic potential. The cranial and spinal nerves are usually unaffected. The cells have thin, convoluted cytoplasmic processes with only a few pinocytotic vesicles. They are lined by a continuous basal lamina. In a conventional plexiform schwannoma, *Luse bodies*, made of stromal long-spaced collagen are commonly found. These are less common in the cellular plexiform schwannoma. There is nuclear palisading, and sometimes a non-prominent biphasic pattern. The nuclei may be hyperchromatic. There is no necrosis and myxoid changes.

The tumor cells diffusely and largely express the S100 protein. They often express SOX10, LEU7, and *calretinin*. Sometimes, they focally express GFAP. Because the cells have the surface basal lamina, membrane staining for collage IV and laminin is significant, and usually pericellular. Especially in cellular plexiform schwannomas, there is low-level p53 protein immunoreactivity. Though NFP-positive axons are usually absent, there may be small amounts of them — mostly in tumors linked to NF2 or schwannomatosis.

Clinical manifestations

In most cases, the tumor is asymptomatic. If signs and symptoms are present, they are usually based on tumor location, and can be wide-ranging. If the tumor compresses the nerve upon which it lies, this can cause nerve dysfunction. With schwannomatosis, the patient often presents with more than one schwannoma, often with pain. Complications may include nerve compression, reduced nerve function, additional brain tumors, nerve damage, infrequent recurrence, and rare malignant transformation.

Diagnosis

Diagnosis is based on the ultrastructural features of these tumors, which have a characteristic appearance. Other diagnostic methods include complete physical examination, evaluation of patient and family medical history, neurological exam, CT or MRI, nerve conduction studies, other specific tests based on tumor location, and tissue biopsy. The tumor may have a congenital presentation. Like other schwannomas, diagnosis often occurs when the tumor is found during an examination for another medical condition. Most plexiform schwannomas are less than 2 cm in size, along the longest dimension, and appear as a firm mass. The tumor may or may not be attached to a nerve. These tumors are often found concurrently with a glioma or meningioma. Important differential diagnoses include *malignant peripheral nerve sheath tumor (MPNST)* and *plexiform neurofibroma*.

Treatment

In most cases, a complete excision by surgery is curative. However, it is very important to preserve nerve function that is affected by the tumor. Surgical options include microsurgery, craniotomy, stereotactic radiosurgery, and laminectomy. These tumors are often difficult to completely remove, like other schwannomas. For malignancies, other treatments include chemotherapy, radiation therapy, treatment of underlying NF2, adequate postoperative care, and follow-up care with regular screening. No treatment may be necessary if the tumor is very small and the patient is asymptomatic. Period observations are maintained through regular check-ups.

Prognosis

Usually, the prognosis of plexiform schwannoma is excellent. However, it can be influenced by tumor extent and location, NF2, and malignant transformation. If malignant, prognosis is based on tumor stage, location, overall patient health, and response to treatments.

Melanotic schwannoma

A *melanotic schwannoma* is a rare, circumscribed, unencapsulated tumor with gross pigmentation. It is made up of cells that have the ultrastructure and immunophenotype of Schwann cells, yet contain melanosomes. The tumor is reactive for melanocytic markers. There are *psammomatous* and *non-psammomatous* forms of these tumors (see Fig. 10.7). Melanotic schwannoma is also known as *melanocytic neurilemmoma, melanotic neurinoma*, and *pigmented schwannoma*. Most of these tumors are benign. However, malignant tumors may metastasize to organs such as the lungs and brain.

Epidemiology

The peak age of incidence of melanotic schwannoma is about 10 years younger than that of conventional schwannoma, making it more common in young adults (aged 30 to 40 years). There is a slight female predominance. The tumors occur worldwide, with no known geographical, racial, or ethnic preferences. About 50% occur in the context of *Carney complex*, a congenital condition.

Etiology and risk factors

An allelic loss of the PRKAR1A region on 17q has been seen in tumors from patients who have Carney complex. This has not been found in non-psammomatous melanotic schwannomas. However, in the psammomatous melanotic schwannomas with Carney complex, the patient has loss-of-function germline mutations of the PRKAR1A gene on chromosome 17q that encodes the cAMP-dependent protein kinase type I-alpha regulatory subunit. Abnormalities of chromosome 2 have been observed, with additions or deletions in the 2p16 region. This factor is present in four of

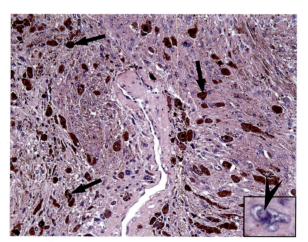

FIGURE 10.7 Cellular appearance of melanotic schwannoma.

every five tumors. The CNC2 gene is also involved. Mutations are seen in the familial as well as sporadic types. Risk factors for melanotic schwannomas include family history, and the presence of Carney complex.

Pathology

In melanotic schwannomas, there are often cytological changes such as hyperchromasia and *macronucleoli*. As opposed to conventional schwannomas, collagen IV and laminin usually surround cell nests—instead of having extensive pericellular deposition. There are true melanosomes, but less envelopment of individual cells by the basal lamina. Most non-psammomatous melanotic schwannomas affect the paraspinal ganglia and spinal nerves. Most psammomatous tumors also affect the autonomic nerves of the viscera, including the heart and intestinal tract. The cranial nerves may also be affected. Just over 10% of all melanotic schwannomas become malignant.

Clinical manifestations

Signs and symptoms of melanotic schwannoma are linked to tumor location. There may be pain and mass effect if the tumor is large and compresses nearby structures. Since the nerves are affected, there may be alterations of sensations. About 50% of these tumors affect the spinal nerves and paraspinal ganglia, with the cervical and thoracic spinal nerves more often involved than the lumbar and sacral spinal nerves. The second most common site involves autonomic nerves of the GI tract. The tumor is usually a firm mass. They are mostly solitary, though multiple tumors occur with Carney complex. Complications may include nerve compression, poor nerve function, nerve damage, bone erosion, infrequent recurrence, and metastasis. This tumor's metastasis sites include the lungs, brain, stomach, liver, and adrenal glands. If the focus of metastasis is significant, it may cause liver and adrenal gland functional abnormalities.

Diagnosis

It is important to distinguish between the psammomatous and nonpsammomatous melanotic schwannomas, since approximately 50% of patients with psammomatous tumors also have Carney complex. This is an autosomal dominant disorder characterized by *cardiac myxoma*, *lentiginous* facial pigmentation, and endocrine hyperactivity. This hyperactivity includes Cushing syndrome, linked to adrenal hyperplasia and acromegaly due to a pituitary adenoma. Complete diagnosis involves physical examination, evaluation of patient and family history, neurological exam, CT or MRI, nerve conduction studies, other specific tests based on tumor location, and tissue biopsy. It is important to understand that from a biopsy specimen, it is difficult to differentiate between a benign or malignant melanotic schwannoma. Therefore, work-up of the patient for metastasis is highly important.

Treatment

Treatments for melanotic schwannoma include observation with regular check-ups, monitoring and treatment of underlying Carney complex, complete surgical excision of the tumor, chemotherapy, radiation therapy, good postoperative care, and follow-up screenings. Though not preventable, if the tumor occurs with Carney complex, genetic testing of expecting parents and related family members, and prenatal diagnosis via molecular testing during pregnancy may be undertaken. If there is a family history of Carney complex, genetic counseling helps assess risks for family planning and monitoring. If a patient is diagnosed with Carney complex, close blood relatives must be screened as well.

Prognosis

The prognosis of melanotic schwannoma is excellent when the tumor is benign, and Carney complex is not present. In these cases, surgical removal is curative. Prognosis for malignant tumors depends on staging and metastasis. Syndromic tumors with Carney complex have a higher chance of metastasis. Both the nonpsammomatous and psammomatous forms can metastasize. Outcome of Carney complex depends on tumor type and location and combined signs and symptoms.

> **Point to Remember**
> Carney complex is an autosomal dominant condition that can manifest as connective tissue myxoid tumors, hyperpigmentation, and endocrine overactivity. Cardiac myxoid tumors (or *myxomas*) may lead to embolic strokes and heart failure. Hyperpigmentation usually affects the face as well as the conjunctiva and oral mucosa. Endocrine tumors may manifest as Cushing syndrome. Other cancers linked to Carney complex include testicular, thyroid, and pancreatic cancers.

Schwannomatosis

Schwannomatosis is usually a sporadic, but sometimes autosomal dominant disorder. It is characterized by multiple benign schwannomas, usually in the spinal, cutaneous, and cranial locations, as seen in MRI. It is also characterized by multiple cranial and spinal meningiomas, linked to inactivation of the NF2 gene in tumors, but not in the germline. In the past, schwannomatosis was referred to as *neurilemmomatosis, multiple schwannomas*, and *multiple neurilemmomas*. Generally, schwannomatosis is considered by many experts to be a form of neurofibromatosis, with features similar to those of NF2, usually *without* inner ear (vestibular) schwannomas that are hallmarks of NF2.

Epidemiology

Schwannomatosis has been estimated to be roughly as common as NF2. There is an estimated annual incidence of 1 case per 40,000–80,000 people. Familial schwannomatosis makes up only 10%–15% of all cases. Globally, the exact frequency of schwannomatosis is unknown. Some populations have noted frequencies as few as 1 case per 1.7 million people.

Etiology and risk factors

Schwannomatosis is caused by mutations in SMARCB1 on 22q or LZTR1 on 22q. Various studies reveal that SMARCB1 germline mutations predispose patients to development of multiple meningiomas. There is a preferred location of cranial meningiomas at the falx cerebri. The proportion of schwannomatosis with meningioma is only 5%. Sometimes, patients present with more than one meningioma. Most cases of schwannomatosis are sporadic but substantively, 15% of patients have a positive family history of the disease. In this form, there is an autosomal dominant pattern of inheritance, and incomplete penetrance. As of 2007, the SMARCB1 gene on chromosome arm 22q was identified as a gene predisposing for familial schwannomatosis. In additional studies, it was found to be involved in approximately 50% of familial cases, but only in 10% or less of sporadic cases. A second causative gene, LZTR1 on 22q, was identified in 2014. In cases of schwannomatosis without a SMARCB1 germline mutation, LZTR1 mutations were found in approximately 40% of familial cases and 25% of sporadic cases. Additional causative genes are believed to be implicated, since most schwannomatosis cases are not linked to SMARCB1 or LZTR1.

A germline missense mutation has been recently identified in the COQ6 gene on chromosome arm 14q, in familial schwannomatosis. Germline mutations of the NF2 gene are excluded from involvement in schwannomatosis. However, other somatically acquired mutations in the NF2 gene are characteristic in schwannomatosis-related schwannomas. About 33% of all patients with sporadic schwannomatosis have segmental tumor distribution. This suggests somatic mosaicism for the causative gene, but this has not been shown for SMARCB1 or LZTR1.

Based on detailed studies of the SMARCB1 gene, tumorigenesis in schwannomatosis is believed to have three steps, as follows:

- Inherited SMARCB1 germline mutation occurs
- The other chromosome 22 is lost with the wild-type copy of SMARCB1 and one copy of NF2
- There is a somatic mutation of the remaining copy of the NF2 gene

This may explain why different NF2 mutations have been seen in multiple schwannomas, in a single patient with schwannomatosis. These steps also occur in meningiomas that have a SMARCB1 germline mutation. SMARCB1 germline mutations may also predispose patients to development of highly aggressive rhabdoid tumors during childhood. This is known as *rhabdoid tumor predisposition syndrome 1*. Most cases are sporadic, with the child dying before age 3. The same steps that occur with the SMARCB1 gene in schwannomatosis are believed to occur with the LZTR1 gene, though there are differences concerning mitotic recombination, which only occurs in 30% of cases.

Pathology

There have been reports of multiple non-vestibular schwannomas, but this has been debated as to whether it is a type of attenuated NF2, or a separate disease. The tumors may develop in spinal roots, cranial nerves, or skin. Sometimes, they develop unilaterally in the vestibular nerve. Cutaneous schwannomas can be plexiform. An Antoni A tissue pattern is fibrillary, highly polar, and elongated with accompanying ordered cellular component palisading as Verocay bodies. An Antoni B tissue pattern may be adjacent to Antoni A tissue regions, but is distinct, loose, and microcystic, with a myxoid hypocellular component.

In schwannomatosis, the tumors have segmental distribution, in approximately 30% of cases. They may show prominent myxoid stroma and an *intraneural* growth pattern. With familial schwannomatosis, there may be mosaic immunohistochemical staining for SMARCB1 protein, such as a mixture of positive and negative tumor cell nuclei, with significant intertumoral and intratumoral heterogeneity, and less than 10% to more than 50% *immunonegative* nuclei. Mosaic SMARCB1 staining is often present when there is sporadic schwannomatosis and NF2. This signifies an interaction between NF2 and SMARCB1 in tumor pathogenesis. Conversely, sporadic schwannomas rarely have mosaic SMARCB1 staining.

Extraneural tumors are rare, which are not like those linked to NF1 and NF2. In one case, a uterine leiomyoma was developed along with schwannomatosis. It had a molecular profile that was similar to a schwannoma, and related mosaic staining for SMARCB1 protein. This indicates that the SMARCB1 defect in schwannomatosis may sometimes contribute to oncogenesis of extraneural neoplasms.

Clinical manifestations

The patient with schwannomatosis usually has more than one schwannoma. Severe pain from the tumors is characteristic of this disease. This is in distinct juxtaposition to schwannomatosis from NF2 as a syndrome, which only rarely causes pain, but usually causes neurological deficits and polyneuropathy. Signs and symptoms usually appear in early adulthood, involving chronic pain in any part of the body.

Diagnosis

For diagnosis, it is important to exclude other forms of neurofibromatosis, via confirmation of a lack of vestibular schwannomas on MRI, and an absence of other manifestations of NF1 or NF2. Since 2012, diagnosis has included molecular studies. Exclusion of NF2 by clinical criteria and imaging of the vestibular nerves is diagnostically essential. This distinction may be very difficult for pediatric patients, since vestibular schwannomas may only develop later in the disease course. There can be some overlapping between features of early NF2, its mosaics, and schwannomatosis. The current *molecular criteria* for definite schwannomatosis are as follows:

- Two or more schwannomas or meningiomas that are pathologically proven, plus genetic studies of two or more tumors with LOH for chromosome 22, and two different NF2 mutations, or
- One schwannoma or meningioma, pathologically proven, plus SMARCB1 germline mutation

The current *clinical criteria* for definite schwannomatosis include the following:

- Two or more schwannomas that are non-dermal, with one pathologically proven, plus no bilateral vestibular schwannomas by using thin-slice MRI, or
- One schwannoma or meningioma, pathologically proven, plus a first-degree relative who has schwannomatosis

The current *clinical criteria* for *possible* schwannomatosis includes two or more schwannomas without pathological evidence, or severe chronic pain that is associated with a schwannoma. Schwannomatosis is sometimes misdiagnosed as a neurofibroma or a *MPNST*. Multiple spinal tumors in schwannomatosis are illustrated in Fig. 10.8.

Treatment

Because schwannomatosis patients have multiple tumors, risks of repeated surgeries tend to outweigh their benefits in many cases. Genetic counseling is important in relation to schwannomatosis. The risk of transmission to children is about 50% for patients with a germline mutation of SMARCB1 or LZTR1, and in familial cases in which the germline is unknown. Risk of transmission to children of patients with sporadic cases, with no SMARCB1 or LZTR1 mutation, is unknown. Schwannomatosis is found in 2.4% to 5% of patients undergoing surgical resection for a schwannoma. Intraoperative sonography helps to localize small schwannomas. After surgical excision, tumor-related pain usually subsides. However, pain from damaged nerves and scar tissue from surgery can be chronic. Tumors sometimes recur in the same location after surgery. Stereotactic radiosurgery can be provided, with greater than 90% of patients needing no further oncology treatment intervention beyond that. When surgery is not indicated, pain management will be required, since schwannomatosis can cause severe, chronic and debilitating pain. There are no other medical biological treatments specifically recommended for schwannomatosis.

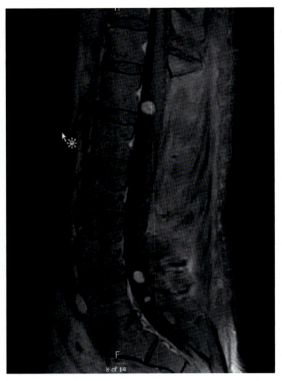

FIGURE 10.8 Multiple spinal tumors as seen in schwannomatosis.

Prognosis

There is not a sufficient threshold of understanding about schwannomatosis, making it difficult to determine the prognosis. Usually, prognosis is based on each patient's specific symptoms. Prognosis is improved when the patient is managed by a medical provider or team experienced with this disease.

> **Point to Remember**
>
> Schwannomatosis is characterized by multiple schwannomas on nerves throughout the body, often associated with intense pain. Additional symptoms include numbness, tingling, weakness, and less often, difficulty with urination, bowel dysfunction, headaches, lumps, swollen areas, and vision changes. Genetic testing for the causative IN1 gene is available and can be done for families in which testing is appropriate.

Clinical cases

Clinical case 1

1. How common is a lumbosacral schwannoma?
2. Why is correct diagnosis of this condition difficult to achieve short of pathology confirmation?
3. Are there clear treatment options for spinal schwannomas?

A 51-year-old man presented after suffering from frequent, intense back pain and neuralgia in his legs. He complained of significant lower back and knee pain in his muscles and bones, dyspnea upon exertion, and chronic fatigue. Initial assessment resulted in a diagnosis of lumbar degenerative disk disease. However, after a CT scan, homogeneous and distinct masses were found on his spinal column, raising the possibility of unilateral meningoceles with canal stenosis. Heavy weight lifting was restricted, and he was told to get plenty of bed rest and take OTC non-steroidal anti-inflammatory drugs. His symptoms did not improve over 6 weeks. An MRI was ordered, using intravenous gadolinium contrast, and a solid heterogeneous enhancing mass lesion was found, arising from the L1 to L2 area, likely representing a schwannoma. There was marked mass effect upon the psoas muscle and some diffuse disk protrusions. The patient was given a corticosteroid injection into the facet joints along with local anesthetics. His pain reduced from 10 on the pain scale to 3 over 12 days and no complications occurred. He was followed up monthly, for 6 months with no further problems, and is now being checked twice per year.

Answers:

1. A lumbosacral schwannoma is one of the rare types of schwannoma, often resembling lumbar degenerative disk disease or a meningocele. It commonly causes persistent back pain that can last for years, eventually becoming severe enough to seek treatment. These solid benign tumors appear well-enhanced compared to other lesions that usually lack such a structure as this.
2. Usually, schwannomas may manifest only as back pain or partial lower limb paralysis when they are in the lumbosacral location. They cannot often be diagnosed based on clinical examination alone. An MRI scan with contrast aids in accurate diagnosis, but a histopathological examination would provide a definitive identification against differential diagnostic conditions such as meningioma, metastases or ependymoma.
3. Controversy exists regarding approaches to take for treatment of spinal schwannomas. Some people recommend complete excision, which also involves sacrificing the nerve root. Although results are usually good in the lumbosacral area, chances for residual chronic postoperative neurological deficits because of surgical trauma are upwards of 15%. A clinician must always balance oncologic versus neurologic management strategies in treatment decision making algorithms such as these. In this situation, there is a rare possibility of tumor recurrence.

Clinical case 2

1. From where do most schwannomas arise and develop anatomically?
2. What are the most typical clinical presentations of schwannoma?
3. Is a woman of this patient's age among the most common age group for developing schwannomas?

A 74-year-old woman presented with chronic swelling of her left collar bone area, the supraclavicular region. She had experienced the swelling for over seven years, and did not complain of weakness, abnormal sensation, or numbness in her left hand. Clinical examination revealed that the swelling was firm, not tender, and not pulsatile. It was located at the level 5 region, and measured 4×4 cm. Neck CT, MRI with contrast, and fine-needle aspiration cytology were performed. There was a large, well-defined mass in the left supraclavicular region. It has a low T1 signal with a high T2 signal, and heterogeneous enhancement. The tumor was in continuity with peripheral nerves at the C5 and C6 levels, close to the upper part of the left brachial plexus. The cytology study revealed clusters of cohesive epithelioid histiocytes in the background of lymphocytes, with no malignant cells, suggesting a benign spindle cell neoplasm. Conservative management was initiated since the patient was basically asymptomatic. Regular follow-ups were ordered.

Answers:

1. Schwannomas are benign, slow-growing tumors usually encapsulated and attached to the nerves from which they originate. They typically emerge from a cranial or spinal nerve, with a sheath, except for the optic and olfactory nerves, which lack a myelin sheath. Nearly 45% of all schwannomas occur in the head and neck area.
2. Most schwannoma patients present with painless swelling. However, if present, symptoms usually are based on the site of the tumor. Other than pain, a tingling sensation and numbness along the course of the involved nerve has been reported. Other symptoms, based on location, can include breathing difficulty, especially if the tumor involves the nose, painful swallowing, nose bleeding, and "crackling" of the voice.
3. No, the most common age of occurrence of schwannomas involves the third and fourth decades of life. However, schwannomas have affected people of any age. This patient's tumor probably arose from the cervical or brachial plexus. The MRI findings indicated it had emerged from the peripheral nerves in close proximity to the upper brachial plexus. Overall, brachial plexus tumors are rare, with less than 5% of upper extremity schwannomas coming from this location.

Clinical case 3

1. What are the common manifestations of cranial nerve schwannomas?
2. For intracranial schwannomas such as this, what is the imaging study of choice?
3. What are differential diagnoses for the schwannoma discussed in the case study?

A 28-year-old woman had developed a tumor arising from the fifth cranial nerve. The patient had severe pain over the left half of her face, which mimicked trigeminal neuralgia. Preoperative images showed it to be large, and its location made surgical removal difficult. The tumor abutted on and compressed the brain stem. However, using microsurgery, it was successfully removed via a retro mastoid craniotomy approach. The surgery took 6 hours. However, within 4 hours, the patient was conscious, could talk, and even walked to the restroom. Over time, she made a complete recovery. Postoperative contrast-enhanced MRI showed no tumor and no areas of abnormal enhancement. The diagnosis was a trigeminal schwannoma, the second most common type of intracranial schwannoma, behind *acoustic schwannomas*.

Answers:

1. **Typically, clinical presentation is related to trigeminal nerve dysfunction, such as neuralgia, neurasthenia, or numbness. If the tumor is large, mass effect symptoms may be present. There may be a slight female predilection for these tumors.**
2. **An MRI is the imaging study of choice for the assessment of intracranial schwannomas. This is not only due to greater contrast resolution, but also highly detailed views of brain structures, allowing for precise localization of the tumor.**
3. **Differential diagnoses include acoustic schwannoma, meningioma, metastasis, or chondrosarcoma. When an intracranial schwannoma is confined to the Meckel's cave, and is small in size, the differential diagnoses include aneurysms, vascular malformation, and pituitary adenoma. A brain MRI is quite sensitive but less specific, and clinical correlation is always required.**

Clinical case 4

1. How common is intracerebral schwannoma?
2. Anatomically, where do most intracerebral schwannomas develop?
3. What are the clinical manifestations of these schwannomas?

A 12-year-old girl was hospitalized for chronic vertigo, headache, nausea, vomiting, and the need for help in order to walk. Though her facial sensation was normal, there was binocular horizontal nystagmus and slow reflexes of the corneas of both eyes. Preoperative MRI of the head showed an abnormal cystic signal in the cerebellar vermis. The parenchyma presented nodular shadows above the cystic wall. Before surgery, the patient was diagnosed with a glioma in the cerebellar vermis. Therefore, preparations were made to excise the cerebellar hemisphere tumor via median suboccipital approach. A brain stem parenchymal tumor was not considered, since there were no obvious symptoms of hydrocephalus. No neurophysiological monitoring or ventricular drainage was performed. However, during surgery, it was found that the tumor had originated from brain stem parenchyma. The fourth ventricle was malformed because of tumor compression. The surface of the tumor consisted of a layer of parenchymal tissue. About 10 mL of light yellow liquid was extracted by puncturing the cyst. The tumor tissue was rich in blood supply, hard in texture, and grayish-red in color. There was calcification inside and a clear border. The resected solid tumor was $1.5 \times 1.5 \times 1.5$ cm in size. The patient did not develop any transient post-surgical hydrocephalus. An MRI 10 days later showed complete tumor resection, and that the brain stem was now in its original position. Final diagnosis was a schwannoma within the brain stem.

Answers:

1. **While intracranial schwannoma is relatively common, intracerebral schwannoma accounts for less than 1% of all intracranial schwannomas. Brain schwannomas are predominantly identified in children and younger people, with a slight male preponderance.**
2. **The majority of intracerebral schwannomas are supratentorial, which typically occur in the superficial sections of the brain parenchyma, or near the ventricle. However, previous studies have shown they are also likely to occur in the frontal and temporal lobes of the cerebral hemisphere, and in the cerebellar hemisphere, cerebellar vermis, and fourth ventricle.**

3. **Intracerebral schwannoma has no specific clinical manifestations. It is not classified by age. The main clinical manifestations may include epilepsy, increased intracranial pressure, and local neurological dysfunction. In patients under 25 years of age, headaches and epilepsy are most often seen, while in elderly patients, there is usually precipitous local neurological dysfunction.**

Clinical case 5

1. What is the typical profile of schwannomatosis?
2. Describe the relationship between schwannomatosis and neurofibromatosis?
3. What are the features of Antoni A and Antoni B tissue patterns?

A 21-year-old woman was hospitalized because of an incidentally found mediastinal mass. There were palpable soft tissue masses on the right posterior neck and right ankle, which had been there for about 4 years. Tumor aspiration had revealed benign soft tissue tumors. There were no clinical stigmata of NF1 or NF2, or any other tumors. Family history was negative for neurofibromatosis. An MRI of the brain revealed no intracranial tumors. There was a well-marginated mass in the right carotid space and a smaller mass between the right quadratus lumborum and the psoas muscle, at the level of the second lumbar vertebra. Contrast-enhanced chest CT showed a well-defined soft tissue mass in the subcutaneous tissue of the right posterior neck, and an enhancing mass in the right mediastinum, lateral to the heart. An MRI of the ankle revealed an enhancing mass below the right medial malleolus. Excisional biopsies were done, and a resection of the mediastinal mass was undertaken. All surgically removed tumor specimens were schwannomas with areas of Antoni A and B. Immunohistochemical staining revealed that nearly all the tumor cells reacted strongly for the S100 protein. The patient was diagnosed with schwannomatosis.

Answers:

1. **Schwannomatosis is characterized by multiple benign tumors called schwannomas, which grow on nerves. Schwannomas develop when Schwann cells grow uncontrollably and form a tumor. Signs and symptoms usually appear in early adulthood, usually involving chronic pain that can affect any part of the body.**
2. **Schwannomatosis is considered by many experts to be a form of neurofibromatosis, which is a group of disorder characterized by nervous system tumors. The features of schwannomatosis can be very similar to NF2. However, schwannomatosis almost never includes inner ear tumors called vestibular schwannomas, which are a hallmark of NF2.**
3. **The Antoni A tissue pattern is fibrillary, extremely polar, and appears elongated. It has a highly ordered cellular component that palisades, as Verocay bodies. The Antoni B tissue pattern lies adjacent to Antoni A regions, but is described as distinct, loose, microcystic tissue. It has a myxoid hypocellular component.**

Key terms

Ancient schwannoma
Antoni A pattern
Antoni B pattern
Calretinin
Cardiac myxoma
Carney complex
Cellular schwannoma
Extraneural
Frameshift mutations
Gorlin-Koutlas syndrome
Immunonegative

Intralabyrinthine
Intraneural
Intronic
Lentiginous
Lipidization
Luse bodies
Macronucleoli
Melanotic schwannoma
Merlin
Neurofibrosarcoma
Non-psammomatous

Pericellular
Plexiform schwannoma
Psammomatous
Radicular pain
Retiform
Schwannoma
Schwannomatosis
Stereotactic radiosurgery
Transcapsular
Verocay bodies

Further reading

Arnan, M. (2011). *Cancer biology, a study of cancer pathogenesis: How to prevent cancer and diseases.* Xlibris.
Barrett, Q., & Lum, L. (2016). *Wnt signaling: Methods and protocols (methods in molecular biology).* Humana Press.
Boardman, L. A. (2018). *Intestinal polyposis syndromes: Diagnosis and management.* Springer.

Bonavida, B. (2015). *Nitric oxide and cancer: Pathogenesis and therapy*. Springer.

Chang, E. L., Brown, P. D., Lo, S. S., Sahgal, A., & Suh, J. H. (2018). *Adult CNS radiation oncology: Principles and practice*. Springer.

Ellis, C. N. (2010). *Inherited cancer syndromes: Current clinical management (2nd Edition)*. Springer.

Frank, D. A. (2012). *Signaling pathways in cancer pathogenesis and therapy*. Springer.

Goss, K. H., & Kahn, M. (2011). *Targeting the Wnt pathway in cancer*. Springer.

Gregory, C. D. (2016). *Apoptosis in cancer pathogenesis and anti-cancer therapy: Perspectives and opportunities (advances in experimental medicine and biology)*. Springer.

Gregory, J. E. (2012). *Pathogenesis of cancer*. Literary Licensing, LLC.

Gupta, N., Banerjee, A., & Hass-Kogan, D. A. (2017). *Pediatric CNS tumors (pediatric oncology)* (3rd ed.). Springer.

Hayat, M. A. (2013). *Tumors of the central nervous system, volume 13: Types of tumors, diagnosis, ultrasonography, surgery, brain metastasis, and general CNS diseases*. Springer.

Hoppler, S. P., & Moon, R. T. (2014). *Wnt signaling in development and disease: Molecular mechanisms and biological functions*. Wiley-Blackwell.

Kelley, M. R., & Fishel, M. L. (2016). *DNA repair in cancer therapy: Molecular targets and clinical applications* (2nd ed.). Academic Press.

Lindon, J. C., Nicholson, J. K., & Holmes, E. (2018). *The handbook of metabolic phenotyping*. Elsevier.

Litchman, C. (2012). *Desmoid tumors*. Springer.

Low, V. H. S. (2012). *Gastrointestinal imaging: Case review series* ((3rd ed.)). Saunders.

Mahajan, A., & Paulino, A. (2018). *Radiation oncology for pediatric CNS tumors*. Springer.

Mercier, I., Jasmin, J. F., & Lisanti, M. P. (2012). *Caveolins in cancer pathogenesis, prevention and therapy (current cancer research)*. Springer.

Musella, A. (2014). *Brain tumor guide for the newly diagnosed* ((9th ed.)). Musella Foundation for Brain Tumor Research & Information, Inc.

National Comprehensive Cancer Network. (2018). *NCCN guidelines for patients: colon cancer*. National Comprehensive Cancer Network (NCCN).

National Comprehensive Cancer Network. (2019). *NCCN guidelines for patients: Rectal Cancer*. National Comprehensive Cancer Network (NCCN).

Nose, V., Greenson, J. K., & Paner, G. P. (2013). *Diagnostic pathology: Familial cancer syndromes*. Lippincott Williams & Wilkins.

Nose, V. (2019). *Diagnostic pathology: Familial cancer syndromes* ((2nd ed.)). Elsevier.

Pennisi, C. P., Prasad, M. S., & Rameshwar, P. (2017). *The stem cell microenvironment and its role in regenerative medicine and cancer pathogenesis (series in research and business chronicles: Biotechnology and medicine)*. River Publishers.

Rajendran, J., & Manchanda, V. (2010). *Nuclear medicine cases (McGraw-Hill radiology series)*. McGraw-Hill Education/Medical.

Strickland, J., & Green, E. (2011). *Lynch syndrome: Tests, causes and treatments*. CreateSpace Independent Publishing Platform.

Tonn, J. C., Reardon, D. A., Rutka, J. T., & Westphal, M. (2019). *Oncology of CNS tumors* ((3rd ed.)). Springer.

Vincan, E. (2008). *Wnt signaling: Volume 1: Pathway methods and mammalian models (methods in molecular biology)*. Humana Press.

Vogelsang, M. (2013). *DNA alterations in lynch syndrome: Advances in molecular diagnosis and genetic counseling*. Springer.

Wahlsten, D. (2019). *Genes, brain function, and behavior: What genes do, how they malfunction, and ways to repair damage*. Academic Press.

Williams, C. K. O. (2019). *Cancer and AIDS: Part II: Cancer pathogenesis and epidemiology*. Springer.

Chapter 11

Neurofibroma

Chapter Outline

Neurofibroma	197	Clinical cases	211
Atypical neurofibroma	200	Key terms	213
Neurofibromatosis type 1	202	Further reading	213
Neurofibromatosis type 2	207		

Neurofibroma

A **neurofibroma** is a fibrous tumor of nerve tissue resulting from abnormal proliferation of Schwann cells. It is a benign, well-demarcated tumor. A neurofibroma may be an intraneural or diffusely infiltrative extraneural nerve sheath tumor. Neurofibroma is made up of neoplastic, well-differentiated Schwann cells that are mixed with nonneoplastic fibroblasts, perineurial-like cells, mast cells, a matrix that can range from myxoid to collagenous, and residual ganglion cells or axons. Multiple and plexiform neurofibromas are usually related to neurofibromatosis type 1 (NF1). However, sporadic neurofibromas are commonly seen. They are usually cutaneous tumors that can affect people of all ages, and develop on any body area. A congenital condition transmitted as an autosomal dominant trait, characterized by neurofibromas of the nerves and skin.

Epidemiology

Neurofibromas are common tumors, and may be sporadic, single nodules that are not related to any obvious syndrome, or much less often, as single, multiple, or numerous lesions accompanying NF1. Patients of all ages, races, and genders can be affected. Neurofibromas are the most prevalent benign peripheral nerve sheath tumor. They affect men and women equally, with no racial or ethnic predilection. The age of onset is extremely variable, but localized lesions are most common in adults between 20 and 40. The diffuse and plexiform types are more common in children, and the plexiform type rarely occurs after age 5. Overall, the incidence of diffuse neurofibroma with neurofibromatosis occurs in about 10% of all cases.

Etiology and risk factors

The tumorigenesis of neurofibroma encompasses genetics, cell signaling, histology, and the cell cycle. The NF1 gene is made up of 60 exons that span 350 kilobytes of genomic data, mapping to chromosomal region 17aII.2. This gene codes for **neurofibromin**. The functional part of neurofibromin is GTPase-activating protein (GAP), which influences cell growth and survival. Nonmyelinating Schwann cells are the neoplastic factor in neurofibromas. **Conglomeration** of nonmyelinating Schwann cells and axons is called a **Remak bundle**. When nonmyelinating Schwann cells are the origin of neurofibromas, mutations making them susceptible to this transformation occur in Schwann cell precursors, early in nerve development. Mutated nonmyelinating Schwann cells do not create normal Remak bundles. It is not understood why only the nonmyelinating type give rise to neurofibromas.

These tumors arise from nonmyelinating Schwann cells that only express the inactive version of the NF1 gene. This leads to total loss of expression of functional neurofibromin. Loss of **heterozygosity** must occur before a neurofibroma can form. This is known as the *two-hit hypothesis*. This loss of heterozygosity occurs via the same mechanisms, such as oxidative DNA damage, that causes mutations in other cells. When a nonmyelinating Schwann cell has had inactivation of its NF1 genes, it starts to proliferate quickly. Even with this hyperplasia, there is no neurofibroma yet. In order for one to develop, cells that are heterozygous for the NF1 gene must be recreated to the site. It is believed that the cells secrete **chemoattractants** such as **stem cell factor**, and angiogenic factors such as the heparin-binding growth factor referred to as **midkine**. These promote migration of various cells that are heterozygous for the NF1 gene, into the hyperplastic lesions created by the nonmyelinating Schwann cells. These include fibroblasts, endothelial cells, mast cells, and perineurial cells. The mast cells secrete **mitogens** or survival factors that change the developing tumor's microenvironment, and result in formation of a neurofibroma. Dermal and plexiform neurofibromas differ in later stages of development, but this is not fully understood.

Though most neurofibromas occur sporadically, with a very low risk of malignant transformation, the plexiform type is pathognomonic for NF1 and has an increased risk of malignant transformation. There are no definitive risk factors for sporadic neurofibromas that develop without the presence of NF1. Spinal tumors are less likely with NF1, while NF2 is likely to cause multiple neurofibromas, meningiomas of the CNS, and ependymomas of the spinal cord.

Pathology

Neurofibroma is graded by the World Health Organization as a grade I tumor. Macroscopically, cutaneous neurofibromas are nodular to polyploid. They may be circumscribed or diffuse (a less common form), involving the skin and subcutaneous fat tissues. Fat and other normal structures may become entrapped. Diffuse neurofibromas are distinctive in appearance, usually large, but can be of varied sizes. There is significant dermal and subcutaneous thickening. When a neurofibroma is confined to nerves, it is fusiform, but may also be ovoid in shape (see Fig. 11.1). In all but the proximal and distal margins, they are well circumscribed. Plexiform neurofibromas have multinodular tangles that resemble a bag of worms when they involve trunks of a neural plexus. However, they resemble rope-like lesions when they affect multiple fascicles of larger, nonbranched nerves such as the sciatic nerve. Upon cut surface, they appear greyish-tan, firm, and glistening. The cut surface is usually pale, and wavy to gelatinous.

Microscopically, neurofibromas are made up of neoplastic Schwann cells with thin, curved or elongated nuclei, and little cytoplasm. They are not capsulated. There are also fibroblasts with a collagen fiber matrix, and Alcian blue-positive myxoid substance. The nuclei are much smaller than those of schwannomas. Cell processes are delicate, and often cannot be seen in routine light microscopy. There are often residual axons inside neurofibromas. These can be highlighted with **Bodian silver** or neurofilament immunohistochemistry. Large and diffuse tumors often have characteristic tactile-like components—especially pseudo-Meissner corpuscles. They may also have melanotic cells. Formation of stromal collagen is highly varied, sometimes forming dense and refractile bundles that look like shredded carrots. Intraneural neurofibromas are often confined to a nerve and surrounded by its thick epineurium. Oppositely, tumors beginning in smaller cutaneous nerves often experience diffuse spreading into the nearby dermis and soft tissues. Blood vessels generally do not have hyalinization. Though neurofibromas may appear like the Antoni B regions of a schwannoma, they usually do not have Antoni A-like regions or Verocay bodies. A form known as *ancient neurofibroma* has degenerative nuclear atypia that is similar to *ancient schwannoma*. It must be distinguished from atypical neurofibroma since there is no clinical relevance. This form lacks any other features of malignancy.

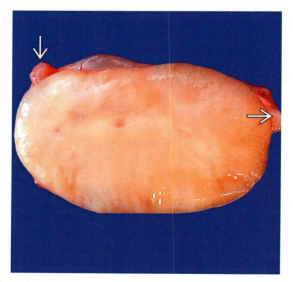

FIGURE 11.1 Grossly, neurofibroma often appears as an ovoid or fusiform mass and usually shows a pale, wavy to gelatinous cut surface. The gross degenerative changes often seen in schwannomas are seldom present in neurofibroma. Note the transected nerve fibers (arrows) at both ends of the mass.

Clinical manifestations

Neurofibroma usually presents as a cutaneous nodule, known as a *cutaneous neurofibroma*. Less often, it presents as a circumscribed mass within a peripheral nerve, called *localized intraneural neurofibroma*, or as a plexiform mass inside a major nerve trunk or plexus. The least common form is diffuse but localized effects upon skin and subcutaneous tissue, called *diffuse cutaneous neurofibroma*. There may also be extensive or massive effects upon soft tissues, known as *localized gigantism* or **elephantiasis neuromatosa**. Diffuse neurofibromas are characterized by infiltrative growth in the subcutaneous fat, and entrapment of the fat and other normal structures. They are uncommon but distinctive forms of neurofibromas, of variable size, but usually large. There is marked dermal and subcutaneous thickening.

Neurofibromas only rarely involve spinal roots on a sporadic basis, but often involve these roots in patients with NF1. When this occurs, there are multiple, bilateral tumors often related with scoliosis and risks of malignant transformation. For some reason, these nearly ever affect the cranial nerves. Neurofibromas are rarely painful, and present simply as masses. When deeper, such as the paraspinal form, they cause motor and sensory deficits related to the originating nerve. Multiple neurofibromas are hallmarks of NF1, along with many other common manifestations.

Diagnosis

The presence of a neurofibroma may be detected via a blood test for protein melanoma inhibitory activity. *Melanoma-derived growth regulatory protein* is encoded by the *MIA gene*, and is also a marker for malignant melanoma. The diagnosis of neurofibroma is usually suspected clinically, though it is confirmed by skin biopsy. The biopsy will reveal a nonencapsulated tumor that is made up of fascicles of slender, spindle-shaped cells. The surrounding matrix is pale staining, and has thin, "wavy" collagen. Neurofibromas have increased numbers of mast cells, and stain positive for the S100 protein. Both subcutaneous and plexiform neurofibromas are usually encapsulated and surrounded by perineurium or epineurium. They have many large nerve fascicles within a cellular matrix of mucin, collagen, fibroblasts, and Schwann cells. To determine extend of involvement of plexiform neurofibromas, magnetic resonance imaging (MRI) may be required. If clinical suspicion exists, but there are no clinical criteria, genetic testing for NF1 is available. Differential diagnoses for cutaneous neurofibromas include dermal melanocytic nevus, neuroma, lipoma, fibroma, and **acrochordon**. For example, for a neurofibroma of the scalp, axial T2 weighted imaging often reveals the tumor to be hyperintense using the *fluid-attenuated inversion recovery (FLAIR)* technique of MRI. However, axial precontrast T1 weighted imaging will reveal tumor enhancement in a more targeted fashion. Differential diagnoses for plexiform neurofibromas include congenital melanocytic nevus, *Klippel-Trenaunay syndrome*, and *Proteus syndrome*.

Treatment

The treatment of cutaneous neurofibromas may be required in order to relieve symptoms such as pain, bleeding, or itching, along with functional impairments such as being irritated by clothing, and to improve cosmetic appearance. For solitary neurofibromas, simple excision via shave biopsy is performed. Multiple lesions may also be removed by simple excision. For large numbers of lesions that accompany NF1, carbon dioxide laser vaporization and electrosurgical excision have been successful. There is a risk, with all of these methods, for hypertrophic scarring as well as recurrence. For diffuse neurofibromas, treatment is partial or complete surgical excision, though these tumors can recur due to their infiltrative growth patterns. For some patients, complete excision is not successful with extensive surgery that in some cases results in permanent tissue mutilation and scarring. Diffuse neurofibromas require yearly follow-up.

Plexiform neurofibromas, since they are infiltrative, are harder to complete remove, and often recur. It is important that an MRI be conducted before surgical excision in order to determine the extent of each lesion and the chances for successful removal. Major risks include nerve damage and hemorrhaging. The medical treatment of plexiform neurofibromas is empiric, including interferon-alpha, methotrexate, vincristine, and thalidomide. Currently, clinical trials are evaluating the *mammalian target of rapamycin (mTOR) inhibitor* called sirolimus, the *proto-oncogene c-kit inhibitor* called imatinib, the *vascular endothelial growth factor (VEGF) inhibitor* called ranibizumab, the *antifibrotic agent* called pirfenidone, the *tyrosine kinase receptor inhibitor* called sorafenib, and photodynamic therapy. All of these studies are focused on treating progressive or problematic plexiform neurofibromas.

Prognosis

The prognosis of neurofibroma is based on the specific type, and the presence of NF1 or NF2. Regardless of the patient's life expectancy, neurofibromas below or on the skin occurs more often with aging, causing psychological and cosmetic issues. Prompt diagnosis and appropriate treatment are able to greatly affect overall prognosis. In relation to neurofibromatosis, patients with NF1 may have a reduced life expectancy by up to 15 years, usually because of malignant tumors. Prognosis for those with NF2 depends on age of onset and the numbers and locations of tumors.

Atypical neurofibroma

Atypical neurofibroma is a form that has high cellularity, monomorphic cytology, scattered mitotic figures, and sometimes, fascicular growth that accompanies atypical cytology. This tumor is also called **bizarre neurofibroma**. It may have premalignant features and is hard to differentiate from low-grade malignant peripheral nerve sheath tumors. Nuclear atypia on its own is generally insignificant, but loss of architecture, high cellularity, and high mitotic activity are problematic, because these factors are linked to future malignancy.

Epidemiology

The lack of available studies on atypical neurofibroma has resulted in little epidemiological information about how often this form of neurofibroma actually occurs throughout the world. However, there appears to be no general male-to-female prevalence ratio, and the median age of tumor development is about 27 years.

Etiology and risk factors

There are no specific causes or risk factors for atypical neurofibromas, though the limited amount of studies have revealed a deletion at chromosome 9p21.3, which includes genes CDKN2A and 2B. This has been seen in 94% of study cases, as well as in 70% of high-grade malignant peripheral nerve sheath tumors, but not in plexiform neurofibromas.

Pathology

Atypical neurofibroma cells have large, hyperchromatic, and irregular nuclei. They are arranged in a distinctive lamellar or fibrillary pattern. Some tumors are hypercellular, though the significant density that is characteristic of malignant peripheral nerve sheath tumors is not present. Also, there is higher cellularity, high mitotic activity, and necrosis in malignant peripheral nerve sheath tumors. All atypical neurofibromas are nonplexiform, and mitoses are mostly absent. While increased cellularity does not necessary suggest malignant transformation, if accompanied by generalized nuclear atypia and very rare mitoses, it is suggestive of transformation into a malignant peripheral nerve sheath tumor (Fig. 11.2). The pleomorphic cells express the S-100 protein but are negative for p53. In 70% of cases, MIB-1 is negative. Epithelial membrane antigen (EMA) and p16 expression is variable. Superficial atypical neurofibromas are

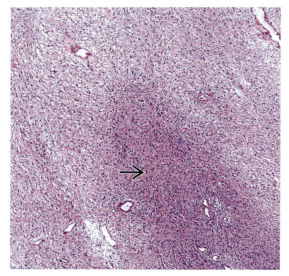 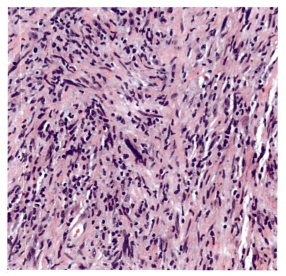

FIGURE 11.2 (Left) Large examples of localized soft tissue neurofibroma may show centralized increases in cellularity (arrow). These areas should be evaluated at higher power for other features that might indicate possible malignant transformation. (Right) Increased cellularity in atypical (bizarre) neurofibroma does not necessarily imply malignant transformation. However, marked cellularity with generalized nuclear atypia and more than rare mitoses suggests transformation into a malignant peripheral nerve sheath tumor.

morphologically unusual, yet have no apparent link to NF1. Atypical neurofibromas can metastasize to the osteoblasts, CSF, and **leptomeninges**, which are the pia mater and arachnoid mater considered together.

Clinical manifestations

Atypical neurofibroma usually appears as distinct, nodular lesions that are avid for F-fluorodeoxyglucose (FDG). They may cause pain, are often palpable of visible, and have been linked to motor weakness. Other signs and symptoms include headache, blurred vision, and general lymphadenopathy. In some patients, they are totally asymptomatic.

Diagnosis

Since atypical neurofibromas have been described as precursor lesions for malignant peripheral nerve sheath tumors, early diagnosis and treatment could potentially prevent these nerve sheath tumors from developing. One component of atypical neurofibromas is a high blood level of alkaline phosphatase. Diagnostic methods also include biopsy, CT scan, MRI, and cerebrospinal fluid cytologic examination. Atypical neurofibromas, unlike plexiform neurofibromas, are not as commonly diagnosed in early childhood.

Treatment

Total surgical resection of atypical neurofibromas is usually successful, with no recurrence. However, these tumors have transformed into malignant peripheral nerve sheath tumors, sometimes in a different location than the original atypical neurofibromas. Early surgical removal of atypical neurofibromas is an important method to prevent the development of malignant peripheral nerve sheath tumors (MPNST).

Prognosis

Though only small groups of patients with atypical neurofibroma have been available for studies, there has been no tumor recurrence and none of them developed malignancy, over study periods ranging between 6 and 63 months. Prognosis is worsened if atypical neurofibromas develop into malignant peripheral nerve sheath tumors.

Plexiform neurofibroma

Plexiform neurofibroma is defined by multiple fascicles that are expanded by collagen and tumor cells. They often have bundled and residual nerve fibers in their centers. Rare cases may show limited perineurial differentiation or develop into a hybrid neurofibroma-**perineurioma**. The term "plexiform" describes a histologic pattern that resembles a network or plexus. Plexiform neurofibromas are considered to be the prototype of the plexiform tumor pattern. They are considered by the World Health Organization to be grade I tumors.

Epidemiology

Plexiform neurofibromas are found in about 30% of patients with neurofibromatosis 1. There have been only limited studies, but one study of 104 patients between 1 and 17 years of age revealed a nearly identical male to female ratio. The tumors are of highest prevalence in children between 13 and 17 years.

Etiology and risk factors

The occurrence of plexiform neurofibromas is a hallmark of NF1. These tumors are present at birth, but may not be identified until later. They are believed to be congenital defects of uncertain origination. There are no documented risk factors for these tumors except for the link to NF1.

Pathology

Microscopically, there is a mixture of cell types—primarily Schwann cells and perineurial-like cells, which have long, thin cell processes, an interrupted basement membrane, and pinocytotic vesicles. Fibroblasts, mast cells, and hematopoietic cells in a collagenous or myxoid fluid are also seen. The growth pattern of plexiform neurofibromas looks like a plexus or network, instead of only showing atypical growth. Staining for the S-100 protein is usually positive. The amount of reactive cells is less, however, than with schwannomas. There is a similar pattern between plexiform neurofibromas and schwannomas regarding SOX10 positivity, but expression of basement membrane markers is more varied than with schwannomas. Plexiform neurofibromas contain limited amounts of EMA-positive cells. Reactivity is most visible in the residual perineurium. Scattered cells—probably the perineurial-like cells—show **claudin** or glucose transporter 1 (GLUT1) positivity. Neurofilament staining will show entrapped axons, but *KIT receptor staining* shows recruited mast cells. Certain stromal cells may stain for cluster of differentiation 34 (CD34). Also, GFAP expression may be seen. Plexiform neurofibromas may cause extensive deformation and enlargement of body parts or regions, which is known as *elephantiasis neuromatosa*, such as the "cow-eye" buphthalmos when the eye orbit is affected. Often, congenital lesions are made up of sheets of neurofibromatous tissue that can infiltrate and encase major nerves, blood vessels, and vital structures. Plexiform neurofibromas often have many enlarged and worm-like tumor fascicles, and differ from schwannomas because they infiltrate between fascicles, while schwannomas displace nerve fascicles. Plexiform neurofibroma may involve the cervical and thoracolumbar nerve roots.

Clinical manifestations

Initially, plexiform neurofibromas present as subcutaneous masses that feel like "bags of worms." There is usually a superficial cutaneous or subcutaneous lesion that can occur nearly anywhere on the body, but mostly on the face, neck, back, and inguinal areas. Symptoms may be related to localized mass effect. Large tumors are attached to major nerve trunks in the neck or extremities. Symptoms may include gradual hearing loss, tinnitus, poor balance, headache, numbness and weakness in the extremities, pain, facial droop, vision problems, development of cataracts, radial dysplasia, tibial dysplasia, freckling in skin folds, **Lisch nodules** on the irises of the eyes, learning disabilities, and muscle wasting.

Diagnosis

Plexiform neurofibromas are usually diagnosed in early childhood. Via CT scan, the tumors are nonspecific, infiltrative lesions. On MRI, reported signal characteristics include T1: hypointense, T2: hyperintense plus or minus a hypointense central focus, which is a target sign. The form of MRI known as *short-T1 inversion recovery (STIR)* is often used. On T1 C+, there is mild enhancement. Differential diagnoses include sarcomas, plexiform schwannomas, lymphatic malformations, and venous malformations.

Treatment

When complete resection is possible, it should be performed, and is usually curative. However, due to the infiltrative nature of plexiform neurofibromas, complete resection is usually not possible. **Stereotactic radiosurgery** is sometimes used. Once a plexiform neurofibroma has become malignant, radiation and chemotherapy can be used, though radiation is of debated use since it may actually promote malignant transformation. Medications for nerve pain include gabapentin and pregabalin. Other medications may include tricyclic antidepressants, serotonin and norepinephrine reuptake inhibitors, and epilepsy medications.

Prognosis

Plexiform neurofibromas are predictive of malignant peripheral nerve sheath tumors, with malignant transformation occurring in 5%−10% of large tumors. A large plexiform neurofibroma raises the change for the patient to have NF1, and must be investigated.

> **Point to Remember**
>
> Plexiform neurofibromas are uncommon, benign tumors of the peripheral nerves that arise from a proliferation of all neural elements. However, they have a significant risk of eventual malignant transformation. These tumors are usually diagnosed early in childhood, and they are found in about 30% of all patients with neurofibromatosis type 1.

Neurofibromatosis type 1

NF1 is an autosomal dominant disorder that is characterized by neurofibromas, café-au-lait spots, cutaneous lesions, optic gliomas, axillary and inguinal freckling, Lisch nodules (hamartomas of the irises), and osseous lesions (see Fig. 11.3). Patients with NF1 have a higher risk for malignant peripheral nerve sheath tumors, juvenile chronic myeloid leukemia, gastrointestinal stromal tumors, duodenal carcinoids, rhabdomyosarcomas, C-cell hyperplasia with medullary thyroid carcinomas, and **pheochromocytomas**. This disease is also known as **von Recklinghausen's disease**, *multiple neuroma*, and *neuromatosis*.

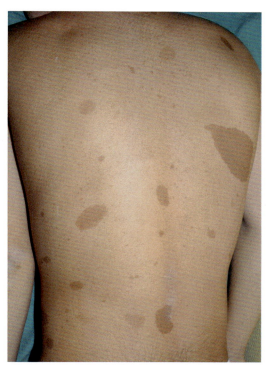

FIGURE 11.3 Photograph showing multiple café-au-lait spots in NF1.

Epidemiology

NF1 occurs in about one of every 3000 births, globally. As of 2015, more than 100,000 people in the United States were documented as having NF1. It affects males and females in equal numbers, and affects all races and ethnic groups equally. Most signs are visible during infancy, though many symptoms develop with aging as hormonal changes occur. In a recent study, Lisch nodules of the irises occur in 100% of patients with NF1 who are beyond the second decade of life.

Etiology and risk factors

NF1 is caused by mutations of the NF1 gene on chromosome 17q11.2. This gene is large and contains 59 exons. The intron 27b includes coding sequences for three embedded genes transcribed in a reverse manner. These are called EVI2A, EVI2B, and OMG. Twelve nonprocessed NF1 pseudogenes are localized on 8 chromosomes. None extend beyond exon 29. The product of the NF1 gene is neurofibromin, a cytoplasmic protein that is mostly expressed in the CNS, PNS, and adrenal glands. Neurofibromin is linked to the *RAS proteins* and various related pathways involved in the proliferation, differentiation, migration, apoptosis, and angiogenesis or tumors cells. Neurofibromin also controls activities of adenylyl cyclase and levels of intracellular cAMP. There is a very high degree of variable expressivity in families with NF1. There are only two genotype-phenotype correlations that have been discovered. Approximately 5% to 10% of cases have *NF1 microdeletion syndrome*, which is caused by unequal, homologous recombinations of NF1 repeats. This causes a loss of DNA on 17q, including all of the NF1 gene and 13 genes that surround it. In this form, the patient has a more severe phenotype that includes mental retardation, facial dysmorphisms, more neurofibromas, developmental delays, and a higher risk of developing malignant peripheral nerve sheath tumors. This indicates that other genes are likely involved. A deletion in exon 17 of the NF1 gene is linked to an absence of cutaneous neurofibromas as well as obvious plexiform tumors.

A segmental or regional type of NF1 is caused by somatic mosaicism at the NF1 gene locus. In this form, the patient often has pigment changes on a limb or in a body region, though sometimes there are classic plexiform neurofibromas or optic pathway gliomas. Spinal neurofibromatosis involves many bilateral spinal neurofibromas, with cutaneous manifestations being only slight or totally absent. No clear genotype-phenotype link has been found. Neurofibromatosis 1 is part of a heterogeneous group of developmental syndromes called **RASopathies**. These are caused by germline mutations of the genes that encode for regulators or members of the **RAS/MAPK pathway**. The RASopathies may have overlapping phenotypes, such as when NF1 occurs without neurofibromas, which may be nearly identical to **Legius syndrome**.

Pathology

The dermal and plexiform variants of neurofibromas are characteristic of NF1. Deeply located intraneural neurofibromas are less common, but can cause neurological manifestations. Plexiform neurofibromas may develop by age 2 as single swellings in the subcutaneous layer, having poorly defined margins. They can result in severe disfigurement later in life, over large areas of the body. When developing in the head or neck, they may cause impairment of vital functions. If a malignant peripheral nerve sheath tumor develops in a young patient, they may be multiple in number, and may have **rhabdomyoblastic** or glandular elements. These nerve sheath tumors play an important role in reduction of life expectancy. If gliomas are present, they are usually pilocytic astrocytomas in the optic nerve, often growing bilaterally (see Fig. 11.4). An optic nerve glioma with NF1 may remain stable for years, and can even regress. Other gliomas occurring more often in NF1 include diffuse astrocytomas and glioblastomas.

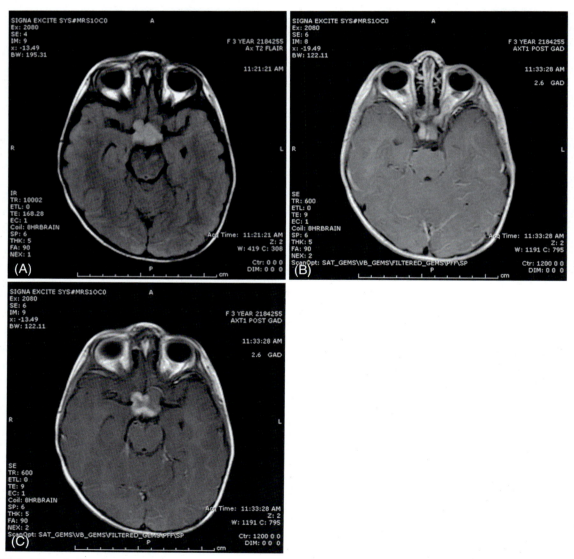

FIGURE 11.4 Bilateral pilocytic astrocytoma of the optic nerve.

Clinical manifestations

The most common sites involved in NF1 include the CNS, PNS, skin, eyes, and bones. Manifestations that are more frequent include macrocephaly, attention deficit hyperactivity disorder, learning disability, aqueductal stenosis, epilepsy, hydrocephalus, and symmetrical axonal neuropathy. Alterations of melanocytes result in café-au-lait spots, Lisch nodules, and freckling (see Fig. 11.5). The pigmented café-au-lait spots often are the first manifestation seen in newborns, increasing in number and size during infancy, but often stabilizing or decreasing during adulthood. They vary in color from light to dark brown. In patients with NF1, the ratio of melanocytes to keratinocytes is higher in the unaffected skin, and is more significant in the café-au-lait spots. Melanocytes within these spots often contain somatic NF1 mutations along with the germline mutation. Freckling occurs in most NF1 patients, with the freckles being identical in appearance to the café-au-lait spots. Freckles may appear diffusely over the trunk, extremities, upper eyelids, and base of the neck. The presence of small, elevated, and pigmented Lisch nodules on the iris surfaces is very useful for diagnosis, since they are present in nearly all adults with NF1 (see Fig. 11.6). They are hamartomas, also called *melanocytic nevi*, of varied size and with a smooth, dome-shaped structure. They may sometimes be seen in the trabecular meshwork. When multiple, Lisch nodules may be indicative of NF1.

The eye orbits often show sphenoid wing dysplasia. Spinal deformities often cause severe scoliosis that may require surgery. These deformities include thoracolumbar scoliosis, kyphoscoliosis, and sometimes, cervical spine kyphosis. Cervical kyphosis can develop with or without laminectomy, though in children, laminectomy may contribute to this deformity. These changes have the ability to cause significant chronic pain that may become disabling. The tibia and other long bones may be thin, bent, and **pseudarthrotic**. The patient may be short in stature and have osteopenia or osteoporosis. In some cases, there has been cerebral aneurysm, fibromuscular dysplasia of the renal (and other) arteries, and stenosis of the internal carotid or cerebral arteries. The NF1 patient has increased risks for juvenile chronic myeloid leukemia, rhabdomyosarcomas, GI stromal tumors, juvenile **xanthogranulomas**, C-cell hyperplasia with medullary thyroid carcinomas and other carcinomas, duodenal carcinoids, and pheochromocytomas.

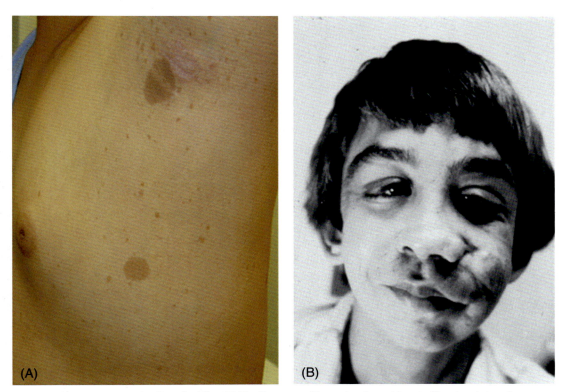

FIGURE 11.5 (A) Café-au-lait spots and axillary freckling (B) Plexiform neurofibroma in a boy with neurofibromatosis type 1.

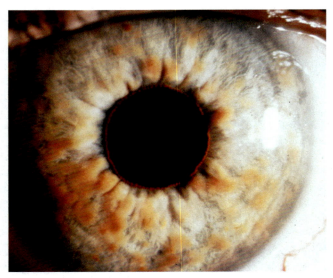
FIGURE 11.6 Lisch nodules of the iris in a patient with neurofibromatosis type 1.

Diagnosis

Diagnosis of NF1 requires the presence more than two of the following features, according to the National Institutes of Health (NIH):

- Café-au-lait macules—six or more, of over 5 mm in largest diameter (in prepubertal patients) and over 15 mm in postpubertal patients;
- Neurofibromas—two or more, of any type, or 1 plexiform neurofibroma;
- Lisch nodules (iris hamartomas)—two or more;
- Optic glioma;
- Freckling in the axillary or inguinal regions;
- One distinctive osseous lesion—such as sphenoid dysplasia, or tibial pseudarthrosis;
- One first-degree relative with NF1—a parent, sibling, or offspring, with the disease features as defined above.

Treatment

There is no specific treatment for NF1. Patients are followed by a team of specialists to manage any symptoms or complications. Genetic counseling is suggested for patients with the disorder. For every offspring of a diagnosed individual, the likelihood of inheriting the pathogenic NF1 gene is 50%. Disease penetrance is 100% to varying degrees and clinical manifestations. Both prenatal and preimplantation mutation testing are available. Genetic testing may help to distinguish the NF1 phenotype from Legius syndrome and other conditions. Mutation screening of the NF1 gene is complicated. This is because of the large size of the gene, diversity of mutations, and the presence of pseudogenes. More than 1000 different mutations have been documented. Today's improved screening techniques allow up to 95% of gene mutations to be detected in patients that fulfill the National Institutes of Health criteria for NF1. Over 80% of mutations either encode a shortened protein or cause a lack of protein production. Selumetinib is an oral selective inhibitor of methyl ethyl ketone (MEK) 1 and 2 that has shown activity in children with NF1 and inoperable plexiform neurofibromas. These patients benefited from long-term dose-adjusted treatment with selumetinib without having excess toxic effects. Because selemetinib may have mechanistic advantages, treatment of low-grade gliomas in NF1 patients has also shown success, and investigation of this agent as a "disease modifying agent" for persons with NF1 is being investigated. Treatment of NF1 associated conditions such as optic nerve sheath glioma, low grade glioma, or neurofibroma is many times different than outside of the context of nonNF1 tumors because of mutagenicity factors that need consideration and that may secondarily impact prognosis. Because of the relatively high prevalence of NF1, the Department of Defense is a major funding source for research to help this population of patients.

Prognosis

The prognosis of NF1 is difficult to predict since the condition is progressive and diverse. Most patients are concerned about the disfigurement that the disease causes, but the most severe outcomes involve chronic, disabling pain. A substantial percentage of patients have a diminished cognitive profile in comparison to normal healthy control subjects. This can affect socioeconomic access to healthcare in a negative way, which in turn may drive down prognostic outcomes with respect to both quality and quantity of life.

> **Point to Remember**
> NF1 is a multisystem neurocutaneous disorder, and an RASopathy. It is also one of the most common inherited CNS disorders, autosomal dominant disorders, and inherited tumor syndromes. About half of all cases are of autosomal dominant inheritance, with the remainder being due to a *de novo* mutation—one that is present for the first time in one family member, as a result of a variant or mutation in a germ cell of one of the person's parents. A de novo mutation may also arise in a fertilized egg during early embryogenesis. There is variable expression of NF1, but 100% penetrance by age 5.

Neurofibromatosis type 2

NF2 is an autosomal dominant disease, involving dysplastic and neoplastic lesions, which mostly affect the central nervous system. This condition is also known as *MISME syndrome*, which stands for "multiple inherited schwannomas, meningiomas, and ependymomas."

Epidemiology

NF2 affects one of every 25,000–40,000 people. Nearly 50% of cases are sporadic. They occur with no family history of the condition and are caused by newly acquired germline mutations. In previous times, the many different clinical manifestations of this condition caused it to be underdiagnosed.

Etiology and risk factors

NF2 is caused by mutations of the NF2 gene located on chromosome 22q12. This gene is believed to function in tumor suppression. Germline NF2 mutations are different from somatic mutations seen in sporadic schwannomas and meningiomas. Usually, germline mutations are point mutations, altering splice junctions or creating new stop codons. Germline mutations occur in all gene areas except alternatively sliced exons. They usually occur, however, in exons 1 to 8. The NF2 gene is expressed in the brain and most other body tissues. The similarity of the NF2-encoded protein to the moesin, ezrin, and radixin proteins resulted in the first letters of these proteins being used to create the term *merlin*. Though the actual mechanism of tumor suppression by merlin is not known, it is believed that merlin provides linkage between membrane-associated proteins and the actin cytoskeleton. Tumor suppression likely occurs via regulation of signal transmission from the extracellular environment to the cell, and activation of many downstream pathways. New data suggest that merlin suppresses signaling in cell nuclei. There are no known risk factors for NF2.

Pathology

With NF2, vestibular schwannomas (acoustic neuromas) may entrap the seventh cranial nerve fibers, and proliferate quickly. Tumor growth ranges from mild to aggressive, occurring at different stages of life. The vestibular division of the eighth cranial nerve, the fifth cranial nerve, and the spinal dorsal roots may be affected. Even the 12th cranial nerve and other motor nerves can be affected. Intracranial tumor locations are also common, including benign meningiomas and ependymomas. Peripheral manifestations can occur, which are usually larger and less numerous than in NF1. Schwannomas may appear like clusters of grapes, both macroscopically and microscopically. Multiple **tumorlets** may develop on individual nerves—especially on the spinal roots. The pathology of NF2 varies widely between and within families. There may be many tumors and a high tumor load, or later-developing vestibular schwannomas only. All families with NF2 show linkage of the disease to chromosome 22. **Nonsense mutations**, and **frameshift mutations** are linked to a more severe phenotype. Milder disease is seen with **missense mutation***s*, **large deletions**, and **somatic mosaicism**. **Splice-site mutations** result in phenotypic variability.

Clinical manifestations

The clinical manifestations of NF2 mostly involve bilateral, vestibular schwannomas (see Fig. 11.7). Initial complaints often include tinnitus, vertigo, imbalance, and hearing loss. Additional features are schwannomas of various cranial nerves, spinal and peripheral nerves, and the skin. There may be intracranial and spinal meningiomas, gliomas—primarily spinal ependymomas, and various nontumor and dysplastic or developmental lesions. These include glial hamartomas, **meningioangiomatosis**, and eye abnormalities such as posterior subcapsular cataracts (often developing during childhood), **epiretinal** membranes, and retinal hamartomas. Neuropathies may also be present. Patients with NF2 often have a family history of brain tumors. Schwannomas linked to NF2 are grade I tumors made up of neoplastic Schwann cells. These are different from sporadic schwannomas because the tumors present in younger patients, in the third decade of life. Sporadic tumors usually present in the sixth decade. Many patients with NF2 develop bilateral vestibular schwannomas by the fourth decade of life. Cutaneous schwannomas are common, and may be plexiform.

Multiple meningiomas are the second hallmark, occurring in 50% of patients with NF2. They occur earlier in life than sporadic meningiomas, and are often the presenting feature—especially in children with NF2. Most of these meningiomas are grade I tumors, but they may have a higher mitotic index than sporadic meningiomas. The most common subtype is fibroblastic, but all major subtypes of meningiomas can occur. They develop through the cranial and spinal meninges, and can affect the cerebral ventricles.

About 80% of gliomas in NF2 cases are spinal intramedullary or cauda equina tumors. Another 10% occur in the medulla oblongata. Ependymomas are the most often diagnosed types of gliomas in NF2. Usually, NF2 spinal ependymomas are multiple, grow slowly, with an intramedullary location. They usually cause no symptoms. While diffuse and pilocytic astrocytomas have been seen with NF2, many of these are believed to actually be misdiagnosed tanycytic ependymomas. Cutaneous neurofibromas have also been seen with NF2, but histological studies often prove them to be plexiform neurofibromas instead.

There are also other nervous system lesions that may be present with NF2. These include the following:

- *Schwannosis*—proliferation of Schwann cells without actual tumor formation, but sometimes with entangled axons. This is often found in the spinal dorsal root entry zones. It may be related with a schwannoma of the dorsal root. Schwannosis may also be in the perivascular spaces of the central spinal cord, with the nodules appearing like small, traumatic neuromas.
- *Meningioangiomatosis*—a cortical lesion with plaque-like proliferation of **meningothelial** and fibroblast-like cells around small vessels. This may be sporadic or accompanying NF2. There is usually one **intracortical** lesion, but the lesion can be multifocal or noncortical. Meningioangiomatosis may be mostly vascular or meningothelial (with an associated meningioma). A sporadic form of the condition involves just one lesion, usually in young adults or children presenting with persistent headaches or seizures. Oppositely, NF2-related meningioangiomatosis may be multifocal. It is often asymptomatic, and only found at autopsy.

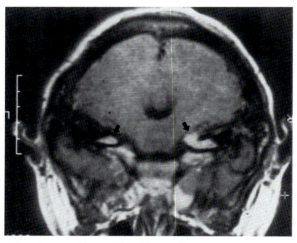

FIGURE 11.7 Cranial magnetic resonance imaging scan from a child with neurofibromatosis type 2.

- *Glial hamartias (microhamartomas)*—circumscribed cell clusters, with medium to large atypical nuclei, in the cerebral cortex. The lesions are scattered through the cortex and basal ganglia, with strong immunoreactivity for S100, yet only focal positivity for GFAP. These hamartias are not related to astrocytomas or mental retardation. They are usually intracortical, in the molecular and deeper cortical layers. However, they also have developed in the basal ganglia, cerebellum, spinal cord, and thalamus. Merlin expression is retained, suggesting that glial hamartias may be due to **haploinsufficiency**.
- *Intracranial calcifications*—often seen in neuroimaging studies, usually in the cerebral and cerebellar cortices, choroid plexus, and periventricular areas.
- *Peripheral neuropathies*—these are not related to tumor masses. Mononeuropathies may be the presenting symptoms in pediatric patients. Progressive polyneuropathies are more often seen in adult patients. Sural nerve biopsies show that NF2 neuropathies are predominantly axonal, and can be secondary to focal nerve compression by tumorlets, or by Schwann cell (or perineurial cell) proliferations with an onion-bulb appearance but no associated axons.
- *Extraneural manifestations*—including posterior lens opacities. Retinal hamartomas, dysplasias, and tufts may be seen. Ocular abnormalities may be diagnostic for pediatric patients. Café-au-lait spots have also been seen.

The risk of transmission of NF2 to offspring is 50%.

Diagnosis

The hallmark of diagnosis involves bilateral vestibular schwannomas or the classic NF2 symptoms, as well as if the patient has a first-degree relative with NF2. Prenatal diagnosis by mutation analysis and testing of children of NF2 patients can be done when the mutation is known. According to the National Institutes of Health, the diagnostic criteria for NF2 include one or more of the following:

- Bilateral vestibular schwannomas;
- A first-degree relative with the condition *and* unilateral vestibular schwannoma, *or* any two of the following:
  - Meningioma;
  - Glioma;
  - Schwannoma;
  - Neurofibroma;
  - Posterior subcapsular lenticular opacities of both eyes.
- Unilateral vestibular schwannoma *and* any two of the following:
  - Meningioma;
  - Glioma;
  - Schwannoma;
  - Neurofibroma;
  - Posterior subcapsular lenticular opacities of both eyes.
- Multiple meningiomas *and* unilateral vestibular schwannoma *or* and two of the following:
  - Glioma;
  - Schwannoma;
  - Neurofibroma;
  - Cataracts of both eyes.

Diagnosis is based on clinical features and may be difficult due to the variability of symptoms and time of disease onset. Genetic mosaics make up 30% of sporadic cases, and are very difficult to diagnose. Pediatric cases are also hard to diagnose, since full disease manifestation has not occurred. It is also difficult to differentiate NF2 from NF1 and schwannomatosis in some patients. This is because of a clinical phenotypic overlap between these diseases. If a patient does not meet the clinical criteria for a definite NF2 diagnosis, but the phenotype is suggestive of the disease, testing for NF2 mutations may be confirmative. A mosaic pattern of immunostaining for *SWI/SNF-related matrix-associated actin-dependent regulator of chromatin subfamily B member 1 (SMARCB1)* expression, which indicates patchy loss, has been seen in most NF2 schwannomas. This also occurs with schwannomatosis. Contrast-enhanced MRI, T1-weighted imaging may be used for diagnosis of NF2 (see Fig. 11.8). Distinct radiographic features include meningiomas, schwannomas usually affecting an inferior vestibular division of CN VIII but also may be from the facial nerve, and ependymomas. These are usually spinal intramedullary, and not intracranial or intraventricular.

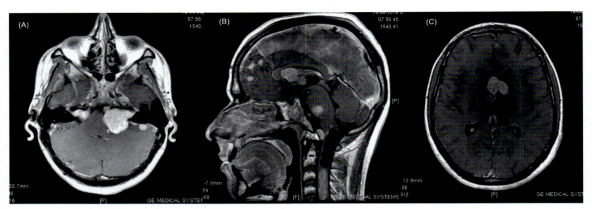

FIGURE 11.8 T1-weighted MRI images of neurofibromatosis type 2. *Source: Copyright © 2011, Director General, Armed Forces Medical Services. Published by Reed Elsevier India Pvt. Ltd. on behalf of Director General, Armed Forces Medical Services. All rights reserved.*

Treatment

Surgical resection of symptomatic tumors is the most common approach for NF2. For small vestibular schwannomas, surgical resection or stereotactic radiosurgery may be used, seeking to preserve hearing and facial nerve function for as long as possible. An emerging literature supporting the administration of low dose pulsed bevacizumab (anti-VEGF antibody) prior to surgery or radiation in addition to after radiation supports improvement of quality of life by allowing for better and longer duration of hearing preservation. On the other hand, large tumors may require surgical resection even if hearing loss will be irreversible—primarily when there is brain stem compression, facial nerve palsy, or hydrocephalus. A debulking procedure can result in preservation of hearing, or at least, prolongation of auditory decompensation. There have been a few cases in which a vestibular schwannoma spontaneously regressed after another, contralateral vestibular schwannoma was resected. Nonvestibular cranial nerve schwannomas have been successfully treated with microsurgery combined with radiosurgery. For intracranial meningiomas, surgical resection is only considered if there are serious, disabling symptoms. Resection of spinal cord tumors is often more difficult, and should be done when neurologic symptoms appear. Complete resection may not be possible. Single fraction radiosurgery may be used for spinal cord schwannomas. Surgical resection of cutaneous or subcutaneous growths is possible, but plastic surgical consultation should occur.

For some patients, auditory brain stem implants have been used successfully when there was hearing loss from a vestibular schwannoma. Often, these implants do not restore hearing but improve the ability to appreciate environmental sounds, and improve communication. The implants do not allow for high levels of speech recognition, probably because of cochlear nerve damage from NF2. Before surgery, the patient should be educated about realistic expectations, and the support of family members should be evaluated.

In rare cases, radiation and/or chemotherapy are used, usually for disabling ependymomas. There are concerns about additional risks of radiation therapy, however. For unresectable and progressive vestibular schwannomas, the drug called erlotinib has been used, resulting in tumor shrinkage and improved auditory function. However, studies of this chemotherapeutic drug are ongoing. A trial of bevacizumab, which is an endothelial growth factor monoclonal antibody, showed shrinkage of vestibular schwannomas and auditory improvement. Early in-vitro studies suggested that imatinib, a tyrosine kinase inhibitor, may be successful in treating vestibular as well as spinal cord schwannomas.

Prognosis

The prognosis of NF2 is based on the patient's age of onset as well as the number and location of tumors, some of which can be life threatening. Patients with NF2 often have a substantially shorter life span than the general population. For those with acoustic nerve tumors, the vast majority becomes completely deaf over time. Because of balance, muscle, and visual problems, many people become wheelchair-bound. In previous decades, many patients lived only into their mid-30s, but with better screening and health care, average survival is improving slightly. While the disease is not curable, there are ongoing clinical trials and improved technologies that can be of help in promoting longevity of patients.

Point to Remember

NF2 is characterized by development of multiple schwannomas and meningiomas. The schwannomas often affect both vestibular nerves, leading to hearing loss and deafness. Most patients present with hearing loss that is usually unilateral at onset, and can be accompanied or preceded by tinnitus. These schwannomas may also cause dizziness or imbalance as a first symptom. Most patients face substantial morbidity and reduced life expectancy.

Point to Remember

There is also a rare form of neurofibromatosis known as *type 3*, or *mixed type neurofibromatosis*. It resembles Von Recklinghausen's disease, but also present with cutaneous neurofibromas. Symptoms may include café-au-lait spots, freckles, acoustic neuromas, meningiomas in the upper neck, spinal or paraspinal neurofibromas, and CNS tumors. Some of these tumors have the ability to become malignant. Skin tumors usually develop on the palms of the hands.

Clinical cases

Clinical case 1

1. What are the characteristics of diffuse neurofibromas?
2. What is the incidence of diffuse neurofibroma associated with NF1?
3. Discuss the likely treatment plan for this neurofibroma?

A 10-year-old boy presented with scalp swelling that had persisted for 4 years. The boy had fallen headfirst to the ground while playing, prior to the onset of the swelling. At the time of the injury, he was evaluated by a neurosurgeon for head trauma. There was no bony erosion, and a hematoma was removed through aspiration. The edematous swelling of the scalp had remained since that time. Physical examination revealed a swollen mass on the scalp that was 8 cm in diameter. Via CT scan, there was diffuse thickening and increased density of the mass, and a provisional diagnosis of lipoma was made. However, an MRI showed soft-tissue infiltration with superficial mass-like lesions, and deeper, poorly defined infiltration. The lesion had intermediate signal intensity on axial T1-weighted imaging. A biopsy revealed fusiform cells with elongated nuclei, surrounded by a myxoid matrix with collagen fibers, and S-100 protein antibody staining was positive, resulting in a diagnosis of diffuse neurofibroma.

Answers:

1. **Diffuse neurofibromas are characterized by infiltrative growth in the subcutaneous fat, and entrapment of the fat and other normal structures. They are uncommon but distinctive forms of neurofibromas, of variable size, but usually large. There is marked dermal and subcutaneous thickening.**
2. **The incidence of diffuse neurofibroma in the context of neurofibromatosis type I has been reported to be approximately 10% of cases.**
3. **The treatment of this diffuse neurofibroma is partial or complete surgical excision. Even after complete excision, the tumor can recur because of its infiltrative growth pattern. Yearly follow-up is therefore recommended. However, for the patient in this case, complete excision is unlikely to achieve success without extensive surgery that would yield undesirable cosmetic results.**

Clinical case 2

1. What is the principal cellular difference between atypical neurofibroma and malignant peripheral nerve sheath tumors?
2. How do atypical neurofibromas differ from plexiform neurofibromas?
3. Describe cellular features of atypical neurofibromas?

A 35-year-old man presented with discolored tumors on his left leg, plus a headache, limb weakness that had persisted over the past 4 months, and a 2-week history of blurred vision. Upon examination, he was pale and had general lymphadenopathy plus lower motor neuron type weakness of his legs and left arm. His right arm was not affected. A blood sample revealed high alkaline phosphatase. A chest radiograph revealed an osteosclerotic metastatic lesion of the humerus. Biopsy of a leg lesion revealed atypical neurofibroma, and a CT scan of his thorax revealed osteoblastic metastasis. An MRI of his brain and spinal cord revealed metastatic leptomeningeal deposits, and CSF cytology was positive for malignant cells.

Answers:

1. **Atypical neurofibromas are noted to have atypical nuclei and higher cellularity, whereas malignant peripheral nerve sheath tumors are highly cellular but also with high mitotic activity and areas of necrosis.**
2. **Plexiform neurofibromas are much more likely to become malignant peripheral nerve sheath tumors (MPNSTs) than are atypical neurofibromas. Also, plexiform neurofibromas are usually diagnosed in early childhood.**
3. **Atypical neurofibromas are composed of cells with hyperchromatic nuclei and a lack of mitoses.**

Clinical case 3

1. Describe the cellular composition of plexiform neurofibromas.
2. Are plexiform neurofibromas common in people of this patient's age?
3. What are the differential diagnoses of plexiform neurofibromas?

A 56-year-old woman presented with slowly progressive painless swelling of the right side of her face, which had persisted for a few years. She had no symptoms except for the change in her facial appearance. Examination revealed extensive subcutaneous soft tissue swelling extending inferiorly from the right supraorbital region to the right periauricular region. The texture of the swelling resembled a "bag of worms." There was a localized depression over the right temple. The mass was not tender and had a poorly defined border. The patient had no family history of neurofibromatosis. A CT scan was performed, revealing an extensive soft tissue mass, and some degree of periosteal reaction of the underlying right zygomatic bone. A biopsy confirmed diagnosis of a plexiform neurofibroma.

Answers:

1. **Plexiform neurofibromas are benign peripheral nerve sheath tumors composed of a variable mixture of Schwann cells, perineurial elements, fibroblasts, and hematopoietic cells, in a collagenous or myxoid fluid. They are usually poorly defined subcutaneous or dermal masses.**
2. **No, plexiform neurofibromas are most common in children between the ages of 13 and 17 years.**
3. **Differential diagnoses include sarcomas, plexiform schwannomas, lymphatic malformations, and venous malformations.**

Clinical case 4

1. Are spinal deformities a typical phenotypic expression of NF1?
2. Are pigmented lesions common manifestations of NF1?
3. Describe Lisch nodules, which are often seen in NF1 but only rarely seen in NF2.

A 15-year-old boy with NF1 had previously undergone a C4 to C6 laminectomy for tumor excision. He fell while skateboarding and suffered 10 min of quadriplegia, but no neck pain. Examination reveals mild myelopathy and multiple pigmented skin lesions. The boy also had a severe kyphotic deformity and myelomalacia. The patient did not want another spinal surgery.

Answers:

1. **Yes, patients with NF1 often develop spinal deformities such as thoracolumbar scoliosis, kyphoscoliosis, and sometimes, cervical spine kyphosis, as occurred in this patient. Cervical kyphosis may develop with or without laminectomy as in this case, though a laminectomy in a child certainly may contribute to the deformity.**
2. **Yes, pigmented lesions are common manifestations of NF1, usually appearing during the first years of life, or being present at birth, either as café-au-lait spots or freckles. Café-au-lait spots are hyperpigmented maculae varying in color from light to dark brown. Freckling may occur diffusely over the trunk, extremities, upper eyelids, and base of the neck.**
3. **Lisch nodules are hamartomas of the iris, also known as melanocytic nevi, that are variable in size, with a smooth, dome-shaped structure. They may also be seen in a trabecular meshwork. In a recent study, incidence in patients with neurofibromatosis beyond the second decade of life approaches 100%. They can be indicative of NF1 when multiple.**

Clinical case 5

1. How do the tumors of NF2 differ from those of NF1?
2. How does NF2 usually manifest clinically?
3. What are the distinct radiographic features of NF2?

A 68-year-old man was hospitalized because of recent weight loss, difficulty swallowing, blurry vision, and generalized weakness. Past medical history included allergies, gastritis, left deep vein thrombosis, and NF2. He had a previous surgery to remove an acoustic neuroma and multiple surgeries to remove neurofibromas from his left leg and arm. He also had several laminectomies to remove neurofibromas in this thoracic and cervical spine. The patient complained of left-sided facial weakness and numbness around his mouth.

Answers:

1. **NF-2 is a hereditable or spontaneous disease that induces nervous system tumors with its hallmark presentation of bilateral acoustic neuromas and other intracranial tumors — especially benign meningiomas and ependymomas. Peripheral manifestations may also occur, but these are usually larger and less numerous than with NF1.** Tumor growth with NF2 ranges from mild to aggressive, and occurs at different stages throughout life.
2. **Initial complaints of NF2 often include tinnitus, vertigo, imbalance, and hearing loss associated with the growth of acoustic neuromas. Many NF2 sufferers have a childhood history of cataracts or family members with a history of brain tumors. A clinical diagnosis can be made if the patient has bilateral acoustic neuromas or classic symptoms of NF2, and has a first-degree relative diagnosed with the condition.**
3. **Distinct radiographic features of NF2 include meningiomas, schwannomas that usually affect an inferior vestibular division of cranial nerve VIII (can also involve the facial nerve), and ependymomas that are usually intramedullary along the spinal cord (typically *not* intracranial or intraventricular).**

Key terms

Acrochordon
Atypical neurofibroma
Bizarre neurofibroma
Bodian silver
Chemoattractants
Claudin
Conglomeration
Elephantiasis neuromatosa
Epiretinal
Frameshift mutations
Haploinsufficiency
Heterozygosity
Intracortical
Klippel-Trenaunay syndrome
Large deletions
Legius syndrome
Leptomeninges
Lisch nodules
Meningioangiomatosis
Meningothelial
Midkine
Missense mutations
Mitogens
Neurofibroma
Neurofibromatosis type 1
Neurofibromatosis type 2
Neurofibromin
Nonsense mutations
Perineurioma
Pheochromocytomas
Plexiform neurofibroma
Proteus syndrome
Pseudarthrotic
RAS/MAPK pathway
RASopathies
Remak bundle
Rhabdomyoblastic
Somatic mosaicism
Splice-site mutations
Stem cell factor
Stereotactic radiosurgery
Tumorlets
von Recklinghausen's disease
Xanthogranulomas

Further reading

Abouelmagd, A., & Ageely, H. M. (2013). *Basic genetics: A primer covering molecular composition of genetic material, gene expression and genetic engineering, mutations and human genetic disorders* (2nd Edition). Universal Publishers.

Argenyi, Z., & Jokinen, C. H. (2011). *Cutaneous neural neoplasms: a practical guide (current clinical pathology).* Humana Press.

Baykal, C., & Yazganoglu, K. D. (2014). *Clinical atlas of skin tumors.* Springer.

Billings, S. D., Patel, R. M., & Buehler, D. (2019). *Soft tissue tumors of the skin.* Springer.

Burt, E. (2018). Type 1 neurofibromatosis: A definitive guide.

Chang, E. L., Brown, P. D., Lo, S. S., Sahgal, A., & Suh, J. H. (2018). *Adult CNS radiation oncology: Principles and practice.* Springer.

Dicato, M. A., & Van Cutsem, E. (2018). *Side effects of medical cancer therapy: Prevention and treatment* (2nd Edition). Springer.
Elder, D. E., & Murphy, G. F. (2010). *Melanocytic tumors of the skin (AFIP Atlas of tumor pathology, Fourth Series Fascicle)*. American Registry of Pathology.
Ferner, R. E., Huson, S., & Evans, D. G. R. (2011). *Neurofibromatoses in clinical practice*. Springer.
Gupta, N., Banerjee, A., & Hass-Kogan, D. A. (2017). *Pediatric CNS tumors (pediatric oncology)* (3rd Edition). Springer.
Hayat, M. A. (2015). *Brain metastases from primary tumors. Volume 2: epidemiology, biology, and therapy*. Academic Press.
Hayat, M. A. (2014). *Tumors of the central nervous system, volume 13: types of tumors, diagnosis, ultrasonography, surgery, brain metastasis, and general CNS diseases*. Springer.
Hopkins, K. (2010). *Thriving with neurofibromatosis*. CreateSpace Independent Publishing Platform.
Icon Group International. (2010). *Neurofibroma: Webster's timeline history, 1913-2007*. Icon Group International, Inc.
Kazakov, D. V., McKee, P., Michal, M., & Kacerovska, D. (2012). *Cutaneous adnexal tumors*. LWW.
Kesharwani, P., & Gupta, U. (2018). *Nanotechnology-based targeted drug delivery systems for brain tumors*. Academic Press.
Liu, D. (2017). *Tumors and cancers: skin — soft tissue — bone — urogenitals (pocket guides to biomedical sciences)*. CRC Press.
Longo, C. (2019). *Diagnosing the less common skin tumors: clinical appearance and dermoscopy correlation*. CRC Press.
Lyck, R., & Enzmann, G. (2017). *The blood brain barrier and inflammation (progress in inflammation research)*. Springer.
Mahajan, A., & Paulino, A. (2017). *Radiation oncology for pediatric CNS tumors*. Springer.
Michaels, A. S. (2014). *Neurofibromatosis: Causes, tests and treatment options* (2nd Edition). CreateSpace Independent Publishing Platform.
Norden, A. D., Reardon, D. A., & Wen, P. C. Y. (2011). *Primary central nervous system tumors: Pathogenesis and therapy (current clinical oncology)*. Humana Press.
Olch, A. J. (2013). *Pediatric radiotherapy planning and treatment*. CRC Press.
Ozdemir, Y. G., Pehlivan, S. B., & Sekerdag, E. (2017). *Nanotechnology methods for neurological diseases and brain tumors — drug delivery across the blood-brain barrier*. Academic Press.
Plaza, J. A., & Prieto, V. G. (2016). *Applied immunohistochemistry in the evaluation of skin neoplasms*. Springer.
Requena, L., & Kutzner, H. (2014). *Cutaneous soft tissue tumors*. LWW.
Rongioletti, F., Margaritescu, I., & Smoller, B. R. (2015). *Rare malignant skin tumors*. Springer.
Scheineman, K., & Bouffet, E. (2015). *Pediatric neuro-oncology*. Springer.
Tonn, J. C., Reardon, D. A., Rutka, J. T., & Westphal, M. (2019). *Oncology of CNS tumors* (3rd Edition). Springer.
Upadhyaya, M. (2014). *The molecular biology of neurofibromatosis type 1 (Colloquium series on genomic and molecular medicine)*. Morgan & Claypool Life Sciences.
Watts, C. (2013). *Emerging concepts in neuro-oncology*. Springer.
Wodajo, F. M., Gannon, F., & Murphey, M. (2010). *Visual guide to musculoskeletal tumors: a clinical-radiologic-histologic approach*. Saunders.
Zalaudek, I., Argenziano, G., & Giacomel, J. (2015). *Dermatoscopy of non-pigmented skin tumors: pink — think — blink*. CRC Press.

Chapter 12

Meningioma

Chapter Outline

Meningioma	**215**	Chordoid meningioma	222
Meningothelial meningioma	219	Atypical meningioma	223
Fibrous meningioma	220	Clear cell meningioma	225
Psammomatous meningioma	220	Anaplastic meningioma	225
Transitional meningioma	220	Rhabdoid meningioma	227
Metaplastic meningioma	221	Papillary meningioma	227
Angiomatous meningioma	221	**Clinical cases**	**227**
Microcystic meningioma	222	**Key terms**	**230**
Secretory meningioma	222	**Further reading**	**231**
Lymphoplasmacyte-rich meningioma	222		

Meningioma

A **meningioma** is a slow-growing, benign tumor of the meninges usually derived from meningothelial cells within the arachnoid layer and villi. The tumors may compress nearby brain tissue. Meningiomas are of three primary grades, each with specific biological pathologies. The majority of meningiomas are grade I according to the World Health Organization. Some subtypes have worsened outcomes and are graded as II or III tumors. These tumors can occur in any part of the skull, since they originate from the arachnoid villi. Meningiomas were among the first solid tumors identified with cytogenetic alterations — usually in monosomy 22.

Epidemiology

Meningiomas are amongst the most common intracranial tumors, making up almost 11% of all brain tumors. They are different from many other brain tumors, in that they occur more often in females than males. Females have an annual incidence of 10.5 cases per 100,000 population, while males have 4.8 cases per 100,000 population. The largest difference occurs prior to menopause. Females between 35 and 44 years are 3.15 times more likely to develop a meningioma, compared to males in the same age group. According to the American Society of Clinical Oncology and the Brain Science Foundation, meningiomas in all ages make up approximately 36% of all *primary* brain tumors. Malignant meningiomas account for 2% to 3% of all meningiomas.

Annual incidence is six cases per 100,000, with a peak incidence in the sixth and seventh decades (median age 65 years), and risks increase with age. In children, meningiomas only make up 2.8% of all *primary* brain tumors. Meningiomas affect about 1% of the overall population, but they are still the most common brain tumors in the United States. Over 90% of all meningiomas are single tumors. Between 20% and 25% of meningiomas are grade II tumors, and 1% to 6% are grade III. For unknown reasons, these higher tumor grades are more common in males. There is also a significant racial component for meningiomas, as follows (see also Fig. 12.1):

- 9.1 cases per 100,000—African-Americans
- 7.4 cases per 100,000—Caucasians
- 4.8 cases per 100,000—Asian-Americans/Pacific Islanders

Etiology and risk factors

While the actual cause of meningiomas is uncertain, the only identified environmental risk factor is ionizing radiation. Individuals who were exposed to this radiation have a higher risk, if the exposure occurred during childhood. Also,

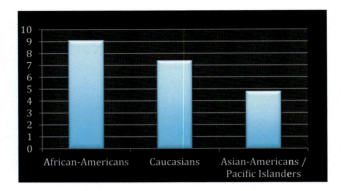

FIGURE 12.1 Meningiomas in various ethnic groups (cases per 100,000 population).

high doses of therapeutic radiation to the head have been implicated, and lower doses of radiation are similarly related to meningiomas. In only two studies, CT scan and other high-dose radiation methods have shown links to eventual meningiomas and gliomas. There is a significant genetic susceptibility to meningioma following ionizing radiation exposure. Hormones are also linked to development of meningiomas, supported by higher incidence in women, the fact that most of these tumors contain progesterone receptors, and slightly increased risks linked to use of various hormones. There are also risks for meningiomas related to smoking and higher body mass index. However, there is a decreased risk when breastfeeding occurs for only up to 6 months. Women who have had uterine fibroids, endometriosis, and breast cancer also have a higher risk of later development of meningiomas. For unknown reasons, allergic conditions such as asthma and eczema are related to a reduced risk of developing a meningioma. For both genders, the risk for meningioma increases with age.

Aside from monosomy 22, other cytogenetic changes include deletion of chromosome 1p, and losses of chromosomes 6q, 9p, 10, 14q, and 18q. In the higher-grade meningiomas, chromosomal gains include those of 1q, 9q, 12q, 15q, 17q, and 20q. Mutations of the neurofibromatosis type 2 gene are found in most meningiomas linked to NF2, and up to 60% of sporadic meningiomas. Usually, there are small insertions or deletions, or nonsense mutations, affecting splice sites, creating top codons, or causing **frameshifts**. A nonfunctional merlin protein usually results. Fibroblastic and transitional meningiomas usually have NF2 mutations. Most non-NF2 meningioma types develop meningothelial tumors. In the atypical and anaplastic meningiomas, NF2 mutations occur in about 70% of cases. Other possibly implicated genes include *adapter protein complex 1 subunit beta-1, like-acetylglucosaminyltransferse, meningioma (disrupted in balanced translocation) 1*, and *SWI/SNF-related matrix-associated actin-dependent regulator of chromatin subfamily B member 1* (SMARCB1). The vast majority of meningiomas with NF2 mutations have just one gene mutation. Multiple meningiomas usually have a clonal origin. Though most meningiomas are sporadic, they rarely occur as part of tumor predisposition syndromes, with NF2 being the leading example.

Pathology

Meningiomas may develop in any of the dural layers within the skull vault, and are believed to be derived from arachnoid (meningothelial) cells. The parasagittal area is the most common location, followed by the falx, cavernous sinus, tuberculum sellae, lamina cribosa, foramen magnum, and at the zones in which the great cranial venous sinuses meet. Meningiomas often form over the cerebral convexities near the venous sinuses, along the base of the skull, in the posterior fossa. Intracranial, intraspinal, and orbital development is most common. The tumors can develop in the olfactory grooves, parasellar or suprasellar regions, sphenoid ridges, optic nerve sheath, petrous ridges, or tentorium. The tumors are attached to the dura mater, and arise within the intracranial cavity, the spinal cavity, or rarely, one of the orbits. They are rare within the ventricles, and multiple meningiomas can also develop.

These tumors compress, but do not invade the brain parenchyma. They may invade and distort surrounding bones. Though histologies vary, they all develop similarly, and sometimes become malignant. When occurring in the spine, most meningiomas develop in the thoracic region. Atypical and anaplastic meningiomas usually affect nonskull base areas and the convexities. Malignant meningiomas usually involve the lungs, bones, pleura, or liver. Many patients with meningiomas have little or no edema, even after the tumors have become quite large. This is different from high-grade gliomas and other brain tumors, with which edema is very common. Benign meningiomas usually invade nearby structures—primarily the dura—though this is usually slight in comparison to the more aggressive forms.

Therefore, certain meningiomas cause significant morbidity and mortality. It is very rare for an extracranial metastasis to occur—only 1 of every 1000 meningiomas are usually related to a grade III tumor. Rare metastases of benign meningiomas usually occur following surgery, but have occurred with no identified cause. Cellular proliferation generally increases proportionally, based on tumor grade. Meningiomas with a 4% or higher proliferation index have a higher risk of recurrence that is similar to the atypical meningiomas. A 20% of higher proliferation index is linked to death rates similar to those occurring form anaplastic meningiomas.

Clinical manifestations

Signs and symptoms are based on the brain area that is compressed by the tumor. In elderly patients, midline tumors can cause dementia without abundant focal neurologic problems. Neurological manifestations develop due to compression of tissues, with deficits based on tumor location. Though nonspecific, headaches and seizures are commonly seen. Additional symptoms include hemiparesis, cranial neuropathies that include vision loss, chronic low back pain, loss of smell and taste, and bilateral absence of pathological reflexes.

Diagnosis

Diagnosis of meningioma is via CT and MRI, with a paramagnetic contrast agent. They usually are isodense and uniformly contrast-enhancing masses, though most meningiomas do not have greatly restricted diffusion when diffusion-weighted images are assessed. In CT scan or plain x-rays, bone abnormalities may be revealed, along with intratumoral calcifications. These abnormalities include brain atrophy, changes to the tuberculum sellae, and hyperostosis around the cerebral convexities. On CT scan, meningiomas appear as well-circumscribed (defined) extra-axial masses, with brain displacement present. Meningioma is often accompanied by a tail-like structure called **dural tail**, which indicates the anchor point of the tumor and dura. However, this structure does not predict the foci of dural involvement. There may be prominent cerebral edema—usually with some of the variant tumor forms and the higher-grade tumors. Cysts may form in the tumor, or at its margins. Neuroimaging is not entirely diagnostic, and also does not definitively predict tumor pathology or exclude differential diagnoses. Currently, it is difficult or even impossible to distinguish between grade I, II, and III meningiomas only by imaging prior to surgery. Definite diagnosis of all brain lesions requires biopsy.

Macroscopically, meningiomas usually appear firm or rubbery, with clear demarcation and a "mottled" structure. This reveals the high vascularization that is present. Sometimes, the tumors are lobulated and rounded, with wide dural attachments (see Fig. 12.2). While dura or dural sinus invasion is common, less often, they invade the adjacent skull. If hyperostosis is caused, this greatly indicates bone invasion. Meningiomas can encase or attach to the cerebral arteries. Rarely, they infiltrate the walls of the arteries. However, they can infiltrate the skin, orbits, and other extracranial compartments. Though the nearby brain tissues are usually compressed, they are only rarely invaded. Along the sphenoid wing, the tumors may resemble flattened masses referred to as *en plaque meningiomas*. Some of the tumors have a gritty appearance, related to many psammoma bodies being present. Bone formation is much less common. The atypical and anaplastic meningiomas are usually larger, often with necrosis. Microscopically, meningiomas usually display an extremely sharp border between the brain and the tumors themselves. The microscopic appearance of the tumors determine grading. The monomorphic cells are present in clusters and sheets, usually having bland oval-shaped nuclei and delicate chromatin. Sometimes, there are intranuclear pseudoinclusions. The cytoplasm is abundant, but "feathered" in

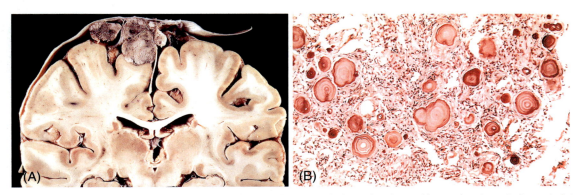

FIGURE 12.2 Meningioma. (A) Parasagittal multi-lobular meningioma attached to the dura with compression of the underlying brain. (B) Meningioma with a whorled pattern of cell growth and psammoma bodies.

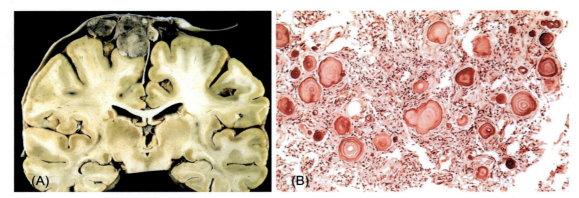

FIGURE 12.3 (Left figure): Meningothelial whorls (large arrow) are a characteristic feature of meningiomas. Intranuclear pseudoinclusions (small arrow) are also a common finding, but are not specific to meningiomas. Both findings are well visualized by hematoxylin & eosin (H&E) stain. (Right figure): Meningothelial cells have bland oval nuclei with delicate chromatin and feathery or wispy cytoplasm, as seen by H&E stain. In low-grade meningiomas, nucleoli or mitoses are uncommon.

appearance, and the cell borders are indistinct. Classic findings include psammoma bodies and meningothelial whorls (see Fig. 12.3).

The most common subtypes of meningiomas are meningothelial, fibrous, and transitional. They are usually benign, but there are four distinct variants (grades II and III) that recur more often, and are more aggressive. Most meningiomas stain for epithelial membrane antigen (EMA), but this is not as prevalent in the atypical and malignant tumors. All meningiomas have vimentin positivity. Usually, somatostatin receptor 2 A is strongly and diffusely expressed. The S-100 protein is usually positive in the fibrous meningiomas. Also, Ki-67 and progesterone receptor may be useful as immunohistochemical markers. There is usually abundant intermediate filaments, **desmosomal** intercellular junctions, and complicated **interdigitating** cellular processes—usually in the meningothelial subtypes. However, in the fibrous meningiomas, these features are not as common, and the cells are separated by collagen. The secretory meningiomas have single or multiple lumina, resembling epithelial cells, in individual cells. The cell surfaces have short, apical microvilli, surrounding secretions that are electron-dense. The microcystic meningiomas have long cytoplasmic processes surrounding an intercellular electron-lucent matrix, and cells that are connected by desmosomes.

Treatment

For symptomatic or enlarging meningiomas, excision or radiation therapy is performed. Serial neuroimaging usually allows for monitoring of asymptomatic small meningiomas, especially in older patients. Excision is preferred when possible. If the meningioma is large, encroaches on surrounding veins or other blood vessels, or is close to the brain stem or other critical brain areas, surgery may cause more damage than that caused by the tumor. Therefore, surgery is deferred. **Stereotactic radiosurgery** is undertaken for surgically inaccessible meningiomas, and performed electively for others. It is sometimes used when there is remaining tumor tissue after surgical excision, or when the patient is elderly. Radiation therapy may be helpful if stereotactic radiosurgery is impossible, or if the meningioma recurs.

Prognosis

The large majority of meningiomas are asymptomatic and do not require surgical removal. With total removal being possible in about 80% of patients, approximately three of every four individuals survive at least 10 years with no recurrence. Only about 20% of gross totally resected benign meningiomas recur within 20 years. Recurrence likelihood is mostly based on extent of resection, tumor site, extent of invasion, attachment to intracranial structures, and surgical skill. Overall, benign meningiomas have recurrence rates of about 7%, but atypical meningiomas recur in 29%−52% of patients, and anaplastic meningiomas recur in 50%−94%. Over 10 years, recurrence is between 9% and 20% for completely resected tumors, and between 18.4% and 50% for subtotal resections. Mean time to recurrence is 2.5−5 years. In another study, grade II meningiomas recurred within an average of 5 years, while grade III tumors recurred within an average of 2 years. However, this information is based on averages and does not predict outlook for individual patients.

> **Point to remember**
> Nearly all meningiomas are benign, while about less than 10% are atypical or malignant. However, even benign meningiomas can grow and constrict the brain, causing disability and even threatening life. Benign meningiomas are less likely to recur than the atypical and malignant types. Atypical meningiomas have a faster growth rate than benign meningiomas, and malignant meningiomas grow the fastest. It is also possible, but rare, to have more than one meningioma at a time.

Meningothelial meningioma

Meningothelial meningioma is a common type of meningioma in which there are medium-sized epithelioid cells that form lobules. Some lobules are demarcated, in places, by thin, collagenous septa. Meningothelial meningioma is a grade I tumor that is sometimes referred to as an **endothelial meningioma**.

Epidemiology

Meningothelial meningioma is the most common subtype of meningioma. It is found in about 60% of all meningiomas, usually combined with fibrous meningioma in about 40% of cases, or isolated in about 17% of cases. Otherwise, epidemiology is highly similar to those of classic meningiomas, discussed above. Often, meningothelial and other types of meningiomas are not discovered until autopsy.

Etiology and risk factors

The causes and risk factors of meningothelial meningiomas are generally indistinct from classic meningiomas. However, increased body fat has been linked to a slightly higher occurrence of these tumors, and excessive exposure to radiation for other medical procedures has also been implicated.

Pathology

In meningothelial meningioma, the tumor cells are mostly uniform. They have oval-shaped nuclei, delicate chromatin, and various types of nuclear holes, such as clear spaces that appear empty (see Fig. 12.4). Nuclear **pseudoinclusions** include cytoplasmic invaginations. There is significant eosinophilic cytoplasm. Whorls and psammoma bodies are uncommon, but do occur. If present, they are usually not as well-formed as in the fibrous, psammomatous, and transitional subtypes. Larger lobules can be confused with structural changes of atypical meningiomas. Meningothelial meningiomas usually develop in the anterior skull base. They are less likely influenced by NF2 mutations. Since tumor cells closely resemble cells in the normal arachnoid cap, sometimes, reactive meningothelial hyperplasia looks similar to meningothelial meningioma. Meningothelial hyperplasia usually occurs near optic nerve gliomas, hemorrhage, and other tumor types. This hyperplasia is also found along with extreme age, **arachnoiditis ossificans**, chronic kidney disease, spontaneous intracranial hemorrhage, and sometimes, with diffuse dural thickening and contrast enhancement during neuroimaging.

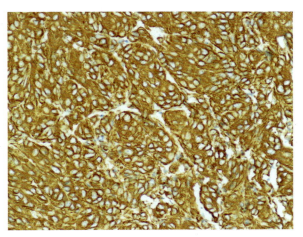

FIGURE 12.4 Whorls of meningothelial cells are present in a meningothelial meningioma. In this micrograph, the tumor cells are highlighted by immunohistochemical staining for vimentin. Primary objective magnification is 20×.

Clinical manifestations

Signs and symptoms of meningothelial meningiomas include chronic headaches, focal seizures, progressive spastic weakness in the legs, incontinence, increased intracranial pressure, double vision, and uneven pupil size.

Diagnosis

Under light microscopy, delicately woven tumor cell processes cannot be seen, though they are ultrastructurally visible. In previous decades, therefore, meningothelial meningiomas were called *syncytial meningiomas*. In most cases, contrast CT, MRI with gadolinium, and arteriography are able to easily visualize these tumors. The tumors are also easily seen in conventional skull x-rays. If lumbar puncture is used to obtain cerebrospinal fluid (CSF), protein levels are usually elevated.

Treatment

Treatment for meningothelial meningiomas is similar to those for classic meningiomas, including observation (if asymptomatic), surgical resection, and radiation therapy. Current chemotherapies are not effective.

Prognosis

The prognosis for meningothelial meningiomas is generally good as long as the tumor remains asymptomatic. Basically, these meningiomas have the same outlook as for classic meningiomas. However, meningothelial meningiomas have recurred as atypical meningiomas and anaplastic meningiomas.

Fibrous meningioma

Fibrous meningiomas consist of spindle-shaped cells. These create bundles within a collagen-rich matrix. The bundles are interlaced, and parallel, with a storiform (irregularly whorled) pattern. The nuclei are elongated but bland, and have delicate chromatin. The cytoplasmic processes sometimes give these tumors a glial appearance. Psammoma bodies are rare. There are often similar nuclear features to those of meningothelial meningiomas. Tumor cells create fascicles with varied amounts of intercellular collagen. The fascicles may increase differential diagnoses of a solitary fibrous tumor or **hemangiopericytoma**, though this type of tumor expresses nuclear *signal transducer and activator of transcription 6*. Fibrous meningiomas are the second most common histological subtype of meningiomas, found in about 50% of all meningiomas. They usually accompany meningothelial histology, in 40% of cases, or in isolation (7%). For some reason, they are the most common intraventricular meningioma histological subtype. In rare cases, fibrous meningiomas have nuclear palisades that look like Verocay bodies. This complicates diagnoses because of increased S-100 expression, though these tumors do not stain as diffusely as the majority of schwannomas. Most fibrous meningiomas express EMA, and convexities as well as NF2 mutations are common. Fibrous meningiomas are grade I tumors.

> **Point to remember**
> There are five key imaging features that help differentiate actual meningiomas from other types of tumors. These include: significant T2 hypointensity, significant T2 hyperintensity, osseous destruction, leptomeningeal extension, and the lack of a dural tail. Differentiating between meningiomas and other tumors is essential, since treatments may be very different.

Psammomatous meningioma

Psammomatous meningiomas are also usually transitional. They contain more psammoma bodies compared to tumor cells. These bodies are often confluent, and form irregular calcified masses. Sometimes, they form bone tissue. Tumor cells may be totally replaced by psammoma bodies in some tumors. Meningothelial cells are difficult to visualize. Most psammomatous meningiomas occur in middle-aged to elderly females, in the thoracic spinal region. In fact, the psammomatous type of meningioma is the most common subtype occurring in the spinal cord. Most tumors have an indolent pattern of development. They are also grade I tumors.

Transitional meningioma

Transitional meningiomas make up a common subtype, and are sometimes referred to as **mixed meningiomas**. These tumors are 2.1 times more common in females compared to males. The mean age for transitional meningiomas is

50.5 years. Along with their transitional features, they have meningothelial and fibrous patterns. Lobular and fascicular foci occur in a side-by-side pattern. There are tight, obvious whorls and psammoma bodies. Also, origin from the convexity and NF2 mutations are common. These tumors are also grade I.

Metaplastic meningioma

Metaplastic meningiomas have significant focal or widespread mesenchymal factors. These include cartilaginous, osseous, lipomatous, myxoid, and xanthomatous tissues that can occur individually or in various combinations. There is no known clinical significance to these changes. Many of them likely do not signify actual **metaplasia**, such as accumulation of lipids instead of true lipomatous metaplasia. There may be a need for clinical determination of ossified meningiomas from meningiomas that show bone invasion. These are also grade I tumors. Metaplastic meningiomas have a mean age of 50.6 years, and usually cause headache, dizziness, seizures, vision impairment, and weakness of both legs.

Angiomatous meningioma

Angiomatous meningiomas are also known as *vascular meningiomas*, and make up only 2.1% of all meningiomas. They have numerous blood vessels, often on the surface, which usually make up more of the tumor mass than the meningioma cells do (see Fig. 12.5). In many of these tumors, 50% of the tumor mass is made up by blood vessels. Blood flow within the vessels is usually visible. However, it may be hard to visualize the meningothelial cells, and the cytology may resemble that of microcystic meningioma. The extreme vascularization of these tumors make surgery potentially difficult and dangerous, since cutting a large artery or vein could lead to severe hemorrhage. Vascular channels are usually hyalinized, small to medium in size, and have walls that are usually thin, but can be thick. Angiomatous meningiomas are soft, and usually in fleshy red color. The tumor boundaries are clear, and it is often delineated from brain tissues.

Degenerative nuclear atypia is common, and moderate to significant. However, most angiomatous meningiomas are benign, both clinically and histologically. The median age for these tumors is 51.8 years. Most patients present with headache and dizziness (55.6%), followed by temporary loss of consciousness (33.3%) and epilepsy (25.9%). Additional manifestations have included nausea, vomiting, stool incontinence, visual impairment, facial hypoesthesia, tinnitus, deafness, hoarseness, swallowing problems, papilledema, and unilateral limb muscle strength declines.

Differential diagnoses include hemangioblastomas and vascular malformations. The difference between angiomatous meningiomas and hemangioblastomas is that there are meningothelial markers, including EMA and somatostatin receptor 2 A. In previous decades, these tumors were called *angioblastic meningiomas*, but this term is not correct. The tumors are not aggressive, but nearby cerebral edema may be excessive in comparison to the size of the tumors. These tumors are usually aneuploid, and often have polysomies — especially for chromosomes 5, 13, and 20. Angiomatous meningiomas are also grade I tumors, and have a 5-year recurrence rate of 5.3%.

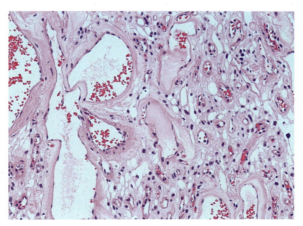

FIGURE 12.5 Cellular appearance of an angiomatous meningioma.

> **Point to remember**
> Angiomatous meningiomas demonstrate abundant blood vessels—more than 50% of each tumor—with areas of classic meningothelial differentiation. They are grade I tumors, but cellular atypia or anaplasia are not present. They appear similar to the meningothelial and fibrous subtypes, and are slightly hyperdense to nearby brain parenchyma, with vivid contrast enhancement. They may be difficult to distinguish during MRI from microcystic meningiomas and chordoid meningiomas.

Microcystic meningioma

Microcystic meningiomas have thin and long processes, with microcysts, and make up only 1.3% of intracranial meningiomas. They form a structure resembling a spider's cobweb, and have significant nuclear pleomorphism, and often, vacuolation. Sometimes, **macrocysts** may be visualized. Though microcystic meningiomas are usually benign, there may be degenerative nuclear atypia. As in the angiomatous subtype, microcystic meningiomas often accompanies cerebral edema. These tumors are also grade I.

Secretory meningioma

Secretory meningiomas have focal epithelial differentiation and make up less than 3% of meningiomas. This differentiation involves the intracellular lumina having **pseudopsammoma** bodies, with periodic acid-Schiff-positive eosinophilic secretions. The bodies have immunoreactivity for carcinoembryonic antigen, and many epithelial and secretory markers. The tumor cells that surround the bodies have carcinoembryonic antigen and cytokeratin. Secretory meningiomas may be linked to raised blood levels of carcinoembryonic antigen that reduce with surgical resection, but increased with tumor recurrence. There may be many mast cells. Peritumoral edema is also common. These tumors are genetically characterized by **Kruppel-like factor 4** and *TNF receptor associated factor 7* mutations. These tumors are also grade I.

Lymphoplasmacyte-rich meningioma

Lymphoplasmacyte-rich meningiomas are rare (less than 1% of all meningiomas), with many chronic inflammatory infiltrates. The meningothelial component is often hard to visualize. There is often peritumoral edema, and sometimes, multifocality. There may be diffuse, rug-like meningeal involvement that resembles pachymeningitis. Less often, systemic hematological problems such as **hyperglobulinemia** and iron-refractory anemia have been seen. Another term for this type of meningioma, *inflammation-rich meningioma*, comes from the predominance of macrophages and plasma cells that are not always obvious. These are also grade I tumors.

Chordoid meningioma

Chordoid meningiomas are also rare subtypes. They resemble chordomas, with cords or trabecular consisting of eosinophilic and often vacuolated cells within a large mucoid matrix (see Fig. 12.6). The chordoid areas are usually mixed with typical meningioma tissues. When present, chronic inflammatory infiltrates are usually patchy, but can be prominent. These tumors are usually large and supratentorial. There is an extremely high recurrence rate after subtotal surgical resection. Less often, there are related hematological conditions such as anemia or **Castleman's disease**. These tumors are often grade I, but also may be grade II, with a more aggressive course. The aggressiveness is usually higher when the chordoid pattern is predominant and strongly developed. Chordoid meningiomas may develop as primary orbital meningiomas, which make up about 2% of all orbital lesions, and about 5% are bilateral. They develop from meningothelial cells in the eye orbits. Manifestations include subjective sensory abnormalities. Though treatments vary between individual patients, surgery, radiation, and close observation are options. Surgery is likely if the patient is symptomatic and the tumor is growing, often involving decompression of the optic canal. There may be difficulty in surgery because of the tumor being close to the optic nerve and internal carotid artery. There is a mean presentation age for all primary orbital meningiomas, including the chordoid type, of 45 years. Females are affected in 59% of cases. After surgery, some patients experience visual "spots" of various colors and sometimes, eyelid ptosis.

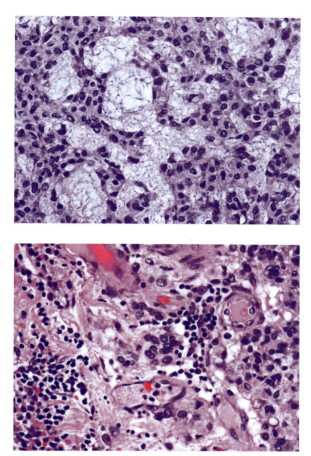

FIGURE 12.6 Two different cellular views of a chordoid meningioma.

Atypical meningioma

Atypical meningiomas have increased mitotic activity. Histology reveals brain invasion, or one of these three features: increased cellularity, spontaneous necrosis, and small cells that have a high nuclear-to-cytoplasmic ratio, prominent nucleoli, and sheeting. In this example, "sheeting" refers to sheet-like growth or uninterrupted patternless grown. When necrosis is spontaneous, it means that it is not iatrogenic. Increased mitotic activity has been 4 or more mitoses per 10 high-power fields. There may be hypercellularity with a mitotic count of 5 or more mitoses per 10 high-power fields. However, nuclear atypia is not used for grading since it usually occurs because of degenerative changes. Risk factors for atypical meningioma include the male gender, previous surgeries, and a location that is not in the skull base. There is a high recurrence rate for atypical meningiomas, even following gross total resection. Increased recurrence is linked to bone involvement. Brain invasion usually involves irregular protrusions of tumor cells into the parenchyma. Reactive astrocytosis then occurs, and there are entrapped sections of glial fibrillary acidic protein-positive parenchyma at the edges of the tumor. Though perivascular spreading and hyalinization can resemble meningioangiomatosis, this is not actually true brain invasion. These occurrences are most common in children (see Fig. 12.7). Brain invasion may happen when the meningioma appears benign, atypical, or anaplastic. It is linked to higher changes for recurrence. Since benign and atypical meningiomas that invade the brain have similar recurrence and mortality rates to those of atypical meningiomas, brain invasion is an actual criterion for atypical meningioma. These are grade II tumors, and make up 20% to 30% of all meningiomas.

> **Point to remember**
>
> The use of the word "atypical" to signify a higher grade of tumor, refers to its histological appearances. It is important not to confuse atypical meningioma with histological variants, which often have unusual imaging and histological features, but are still

(Continued)

(Continued)
grade I tumors. Atypical meningiomas are characterized by histologically by 4 to 19 mitoses per 10 high-power fields, direct invasion of brain parenchyma, and three or more of these features: necrosis, sheet-like growth, small cell changes, increased cellularity, or prominent nucleoli.

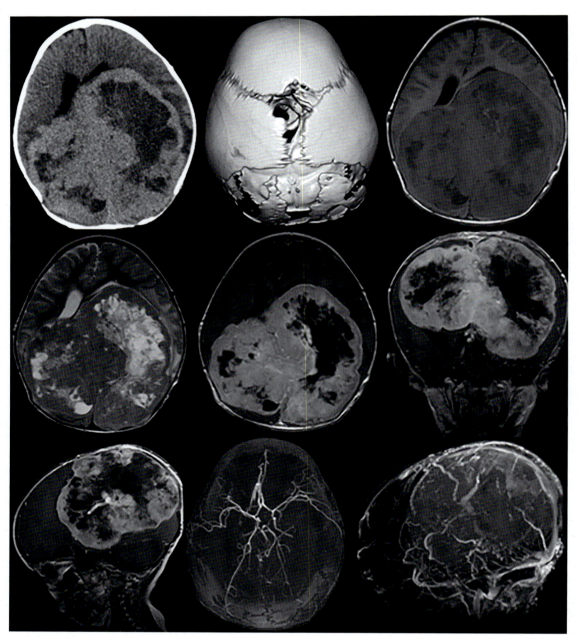

FIGURE 12.7 Multiple views of an atypical meningioma in a 2-year-old patient.

Clear cell meningioma

Clear cell meningiomas are rare meningiomas that usually have no patterns, though they may have sheet-like designs (see Fig. 12.8). The cells are round to polygonal, with clear cytoplasm that is rich in glycogen. Perivascular and interstitial collagen is prominent, appearing in blocks. This collage may form large acellular zones, or brightly eosinophilic collagen that is **amianthoid**-like. It is prominently period acid-Schiff-positive and diastase-sensitive. Whorls and psammoma bodies are hard to be visualized. These tumors often occur in the cerebellopointine angle and spine—primarily in the cauda equina. They usually affect children and young adults. Clear cell meningiomas have more aggressive pathologies, with frequent recurrence and sometimes, CSF seeding. In families, the tumors have been linked to *SWI/SNF-related, matrix associated, actin-dependent regulator of chromatin, subfamily E, member 1* mutations. These aggressive tumors are grade II, with higher aggressiveness attributed to a predominant, strongly developed clear cell pattern. This subtype of meningiomas were first described in 1980 and officially recognized in 1988.

Anaplastic meningioma

Anaplastic meningiomas resemble the cytologies of carcinomas, melanomas, or high-grade sarcomas (see Fig. 12.9). They may have greatly elevated mitotic activity, up to over 20 mitoses per 10 high-power fields. The tumors make up 1%-3% of overall meningiomas (see Fig. 12.10). They usually have extensive necrosis and a Ki-67 proliferation index above 20%. These tumors are usually in the intraventricular region. Rarely, there is true epithelial or mesenchymal metaplasia. To confirm a meningothelial origin, with diffuse anaplasia, there must be a history of meningioma at the identical site, or genetic, immunohistochemical, or ultrastructural evidence. Anaplastic meningiomas are often fatal. Average survival ranges between 2 and 5 years, based on the extent of surgical resection. Clinical risk factors include the male gender, previous surgery, and a nonskull base origin. It may be difficult to differentiate anaplastic from atypical meningiomas. These tumors are grade III.

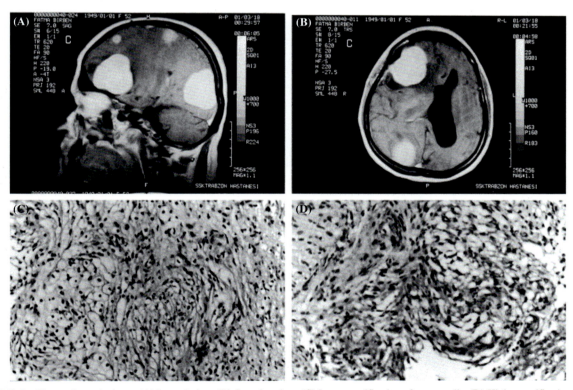

FIGURE 12.8 Clear cell meningioma. (A) Lateral view. (B) Superior view. (C) Low magnification of tumor cells. (D) High magnification of tumor cells.

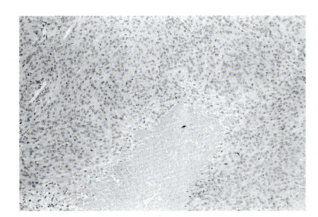

FIGURE 12.9 Anaplastic meningioma.

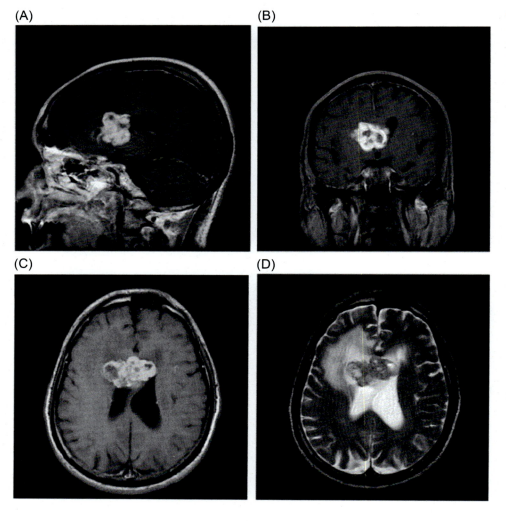

FIGURE 12.10 Imaging studies of a patient with an anaplastic meningioma. (A) Lateral view. (B) Anterior view. (C) Superior view, T1-weighted imaging. (D) Superior view, T2-weighted imaging.

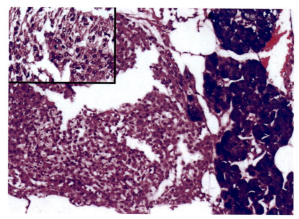

FIGURE 12.11 Rhabdoid meningioma.

> **Point to remember**
>
> Anaplastic meningiomas have aggressive local growth and a high recurrence rate. They also have more heterogeneous imaging appearances. They have obvious malignant features that are similar to those of melanomas, carcinomas, and high-grade sarcomas, plus 20 or more mitoses per 10 high-power fields. The most reliable feature suggesting a non-grade-I tumor is the presence of lower apparent diffusion coefficient, a measure of the magnitude of diffusion of water molecules within tissue, which reflects higher cellularity.

Rhabdoid meningioma

Rhabdoid meningiomas are mostly made up of **rhabdoid** cells, and are uncommon. The cells appear "fat", with odd nuclei, a prominent nucleolus, open chromatin, and obvious eosinophilic **paranuclear inclusions** (see Fig. 12.11). These appear either compact and waxy, or as whorled fibrils. The cells resemble those of atypical teratoid or rhabdoid brain tumors. However, rhabdoid meningiomas retain SMARCB1 expression. The cells may become more obvious if the tumors recur. Most tumors have a high proliferation index, with other malignancy histologies. Some tumors have a papillary structure with rhabdoid cytology. If the rhabdoid features are completely developed and appear with other signs of malignancy, the course is usually aggressive, with a grade III tumor classification. These tumors occur in the middle-age group about 10 years earlier than typical meningioma. There is a slight predominance in females and in African-Americans. Some tumors only have focal rhabdoid feature or lack additional features of malignancy—these are usually less aggressive. Rarely, the tumors appear rhabdoid in histology, yet have interdigitating processes instead of the common paranuclear groupings of intermediate filaments.

Papillary meningioma

Papillary meningiomas are rare subtypes, making up less than 1% of all meningiomas, featuring a perivascular **pseudopapillary** pattern over most of the tumor tissues. There is reduced cohesion, as tumor cells cling to blood vessels. There is also a perivascular zone that lacks a nucleus, which looks like pseudorosettes of ependymomas. With recurrence, this feature is often increased. Other higher-grade histologies are usually found. Papillary meningiomas are more common in children and younger adults. The tumors are invasive including into the brain, in 75% of cases. They may also be primary intraspinal tumors (see Fig. 12.12). Papillary meningiomas recur 55% of the time, and metastasize 20% of the time, usually to the lungs. About 50% of patients with papillary meningiomas will die from the tumors. Some of them have a papillary architecture along with rhabdoid cytology. The tumors are highly aggressive, grade III, which usually occurs when there is a predominant, well developed papillary pattern.

Clinical cases

Clinical case 1

1. How common are meningiomas in general?
2. How do meningiomas appear on CT scans?

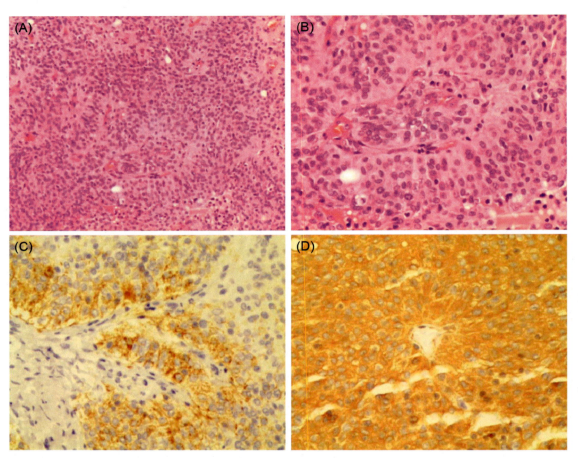

FIGURE 12.12 Primary intraspinal papillary meningioma. (A) Microscopy, low magnification, with hematoxylin/eosin (H&E) staining. (B) Microscopy, high magnification, with H&E staining. (C) Microscopy with vimentin staining. (D) Microscopy, with positive S100 immunostaining.

3. What is the likelihood of recurrence?

A 49-year-old man complained of chronic low back pain as well as a loss of smell and taste that had progressed over 2 years. Physical examination revealed that pathological reflexes were absent bilaterally. The patient's chronic low back pain was attributed to degenerative disk disease. A neurologist ordered a CT scan, revealing a meningioma that was compressing his brain. The tumor was 6 cm in diameter. It was well-circumscribed, with no osseous abnormalities. The tumor was completely resected via surgery, and the pathological analysis confirmed the diagnosis. After surgery, the patient's sense of taste returned, but his sense of smell remained diminished.

Answers:

1. **Meningiomas are amongst the most common intracranial tumors, making up almost 11% of all brain tumors.**
2. **Meningiomas appear on CT scans as well-defined extra-axial masses, associated with brain displacement, and the images allow for identification of characteristics such as hyperostosis and intratumoral calcification.**
3. **The recurrence rate at 10 years ranges between 9% and 20% for totally resected meningiomas, and from 18.4% to 50% for meningiomas following subtotal resection. The mean time to recurrence is 2.5–5 years.**

Clinical case 2

1. Why can meningiomas be located in any part of the skull?
2. Where are meningiomas most common, in regards to tumor location?
3. What are the characteristic features of meningiomas in CT and MRI?

A 51-year-old woman presented with complaints of a chronic headache that had been troubling her for several years, on and off. A new symptom had also developed — impaired vision in her right eye. Cranial MRI revealed a 5 cm hard mass in her right orbitofrontal area. A portion of the tumor invaded into the posterior segment of the right eye. Diagnosis was confirmed via skull x-rays and a CT scan. Because of the local bone, dura, and right eye infiltration, a subtotal resection was carried out by right frontal craniotomy. The tumor was confirmed to be a grade I meningothelial meningioma. Three years later, the patient returned, with a small local recurrent lesion in the right orbitofrontal convexity. A second subtotal resection was performed but this time the diagnosis was a grade II atypical meningioma. The patient then had 6 months of gamma knife radiotherapy, but a large recurrent tumor mass, with frontal and temporal dural metastasis, eventually developed. Another surgical resection was conducted, but the patient soon fell into a coma, and died 8 days later. Autopsy revealed that the final tumor was a grade III anaplastic meningioma.

Answers:

1. **Because meningiomas originate from the arachnoid villi, the location of the tumor mass can be in any part of the skull. Symptoms, therefore, can be extremely varied, including headaches, seizures, hemiparesis, and cranial neuropathies, which include loss of vision.**
2. **Meningiomas are usually located in the skull vault and skull base, with the parasagittal area most common, followed by the falx, cavernous sinus, tuberculum sellae, lamina cribrosa, foramen magnum, and zones where the great cranial venous sinuses meet.**
3. **Meningiomas have highly characteristic features in CT and MRI images. A 'dural tail' indicates the anchor point of the tumor mass and dura, and a 'mottled' structure indicates the high vascularization of the tumor mass.**

Clinical case 3

1. Are grade II meningiomas more common in women or men?
2. Is edema a common feature of most meningiomas?
3. Is it possible to distinguish between grade I, grade II, and grade III meningiomas with preoperative imaging?

A 58-year-old man presented to his primary care physician with right-sided subjective sensory abnormalities that had only recently developed. He had no history of cancer, and no history of any significant radiation exposure or any tumor predisposition syndrome. His general physical and neurological examinations were normal, and both CT and MRI images of his brain were ordered. Imaging revealed a large mass that arose from the posterior falx with surrounding edema. The tumor extended to the posterior third of the sagittal sinus. The patient underwent extensive subtotal resection of a grade II meningioma. A small amount of the tumor had invaded the posterior portion of the sagittal sinus, and was unresectable. The patient's sensory symptoms resolved after surgery. Radiation therapy was recommended because of the remaining tumor tissue. Over follow-ups, the patient had no recurrence or tumor tissue regrowth.

Answers:

1. **Grade II meningiomas, as well as grade III meningiomas, are more common in men than in women, but only by a small percentage. For some reason, grade I meningiomas are more common in women. Meningioma risk increases with age for both genders.**
2. **Many patients with meningiomas have little or no edema, even when the tumor is quite large. This is contrast to patients with other brain tumors such as high-grade gliomas, with which edema is a highly common manifestation.**
3. **Currently, it is difficult or impossible to distinguish between grade I, II, and III meningiomas simply via preoperative imaging. Most meningiomas do not have greatly restricted diffusion in diffusion-weighted images.**

Clinical case 4

1. How common are primary orbital meningiomas such as this chordoid tumor?
2. What are the treatment options?
3. What is the epidemiology of these types of meningiomas?

A 36-year-old woman complained to her physician that her left eye appeared swollen and bulging outward in comparison to her right eye. She also described occasional headaches, but no vision changes, eye pain, or double vision. The patient had no history of cancer, but two of her grandparents had died from cancer. Imaging was performed first with CT, followed by MRI, and a meningioma was found in the left sphenoid wing. Surgical subtotal resection of the tumor was performed, and a grade II chordoid meningioma was diagnosed. After surgery, the patient's headaches resolved, but she experienced "gray spots" in her left eye and left upper lid ptosis.

Answers:

1. **Primary orbital meningiomas account for about 2% of all orbital lesions. Approximately 5% of them are bilateral. They are derived from meningothelial cells within the eye orbits.**
2. **Treatment options vary for individual patients, but include surgery, radiation, and close observation. Surgery is considered when the patient has symptoms and the tumor is growing. It often includes decompression of the optic canal. Surgery can be difficult since the tumor is often close to vital structures such as the optic nerve and internal carotid artery.**
3. **Primary orbital meningiomas, including this chordoid type, have a mean age at presentation of 45 years, and females are affected in 59% of cases. Though this patient had a grade II tumor, the majority of these tumors are grade I.**

Clinical case 5

1. Why is surgery for an angiomatous meningioma potentially dangerous?
2. How do angiomatous meningiomas usually appear?
3. What is the 5-year recurrence rate for these tumors in comparison to general meningiomas?

A 37-year-old male soldier was evaluated for recurrent frontal headaches, acute nausea, and vertigo. The headaches had recently worsened quite drastically. Otherwise, the patient was in extremely good health, and physical examination revealed no other anomalies. He was asked to walk a short distance, and when he did, his path curved to the left without him being able to control it. The physician ordered a CT scan of his brain, which revealed a 6 cm mass in the right frontoparietal area. Then, an MRI revealed that the mass was delineated from the brain tissue. It was successfully resected, though there was extensive bleeding during surgery. A pathological study revealed the tumor to be an angiomatous meningioma.

Answers:

1. **Often with angiomatous meningiomas, large arteries or veins have the potential to be cut, leading to catastrophic hemorrhage. These tumors have an abundant blood supply, with a large amount of blood vessels, often on the surface.**
2. **Angiomatous meningiomas have a soft texture, and are usually a fleshy red color, with a thin capsule and clear boundaries. Blood flow within the main thinly-walled vessels is usually visible. The vessels exceed 50% of the area of the tumor.**
3. **The 5-year recurrence rate of angiomatous meningiomas is only 5.3%, while for general meningiomas, it is 7%.**

Key terms

Amianthoid
Anaplastic meningiomas
Angiomatous meningiomas
Arachnoiditis ossificans
Atypical meningiomas
Castleman's disease
Chordoid meningiomas
Clear cell meningiomas
Desmosomal
Dural tail
Endothelial meningioma
Fibrous meningiomas
Frameshifts

Hemangiopericytoma
Hyperglobulinemia
Interdigitating
Kruppel-like factor 4
Lymphoplasmacyte-rich meningiomas
Macrocysts
Meningioma
Meningothelial meningioma
Metaplasia
Metaplastic meningiomas
Microcystic meningiomas
Mixed meningiomas

Papillary meningiomas
Paranuclear inclusions
Psammomatous meningiomas
Pseudoinclusions
Pseudopapillary
Pseudopsammoma
Rhabdoid cells
Rhabdoid meningiomas
Secretory meningiomas
Stereotactic radiosurgery
Transitional meningiomas

Further reading

Adesina, A. M., Tihan, T., Fuller, C. E., & Young Poussaint, T. (2016). *Atlas of pediatric brain Tumors* ((2nd ed.)). Springer.
Arnautovic, K. I., & Gokaslan, Z. L. (2019). *Spinal cord tumors*. Springer.
Cappabianca, P., & Solari, D. (2018). *Meningiomas of the skull base: Treatment nuances in contemporary neurosurgery*. Thieme.
Carter, L. K., & Ewens, S. (2010). *My brain tumour: One woman's uplifting story*. CreateSpace Independent Publishing Platform.
Chang, S. D., & Veeravagu, A. (2015). *Current management of central nervous system meningiomas (neuroscience research progress)*. Nova Science Publishers, Inc.
DeMonte, F., McDermott, M. W., & Al-Mefty, O. (2011). *Al-Mefty's meningiomas* ((2nd ed.)). Thieme.
Figueroa, D. (2016). *Meningiomas: Risk factors, treatment options and outcomes*. Nova Science Publishers, Inc.
Fountas, K., & Kapsalaki, E. Z. (2019). *Epilepsy surgery and intrinsic brain tumor surgery: A practical atlas*. Springer.
Hayat, M. A. (2012). *Tumors of the central nervous system, . Meningiomas and schwannomas* (Volume 7). Springer.
Hayat, M. A. (2015). *Tumors of the central nervous system, . Glioma, meningioma, neuroblastoma, and spinal tumors* (Volume 14). Springer.
Jain, R., & Essig, M. (2015). *Brain tumor imaging*. Thieme.
Kaye, A. H., & Laws, E. R. (2011). *Brain tumors e-book: An encyclopedic approach, expert consult* ((3rd ed.)). Saunders.
Lacruz, C. R., de Santamaria, J. S., & Bardales, R. H. (2018). (Essential in Cytopathology) *Central nervous system intraoperative cytopathology* ((2nd ed.)). Springer.
Lucas, D. (2017). *The memoirs of a brain tumour survivor: Going into surgical & psychological warfare against a meningioma brain tumor*. Lucas.
Markovic, M. (2014). *Peritumoral edema and angiogenesis in intracranial meningioma surgery*. Lap Lambert Academic Publishing.
Martin, M. (2015). *Meningiomas: Management outlook and surgery*. Hayle Medical.
Minyahil, W. (2014). *Nanoparticulates for imaging and treatment of malignant brain tumor*. Lap Lambert Academic Publishing.
Moliterno Gunel, J., Piepmeier, J. M., & Baehring, J. M. (2017). *Malignant brain tumors: State-of-the-art treatment*. Springer.
Moore, D. (2010). *Meningioma: Two years later*. Null.
Musella, A. (2014). *Brain tumor guide for the newly diagnosed* ((9th ed.)). Musella Foundation for Brain Tumor Research & Information, Inc.
Nema, I. S. (2018). *Parasagittal meningiomas: Surgical treatment and outcome*. Lap Lambert Academic Publishing.
Norden, A. D., Reardon, D. A., & Wen, P. C. Y. (2011). *Primary central nervous system tumors: Pathogenesis and therapy. (current clinical oncology)*. Humana Press.
Pamir, M. C., Black, P. M., & Fahlbusch, R. (2010). *Meningiomas: Expert consult—a comprehensive text*. Saunders.
Ramina, R., Pires de Aguiar, P. H., & Tatagiba, M. (2014). *Samii's essentials in neurosurgery* ((2nd ed.)). Springer.
Saad, A. G. (2018). *Meningiomas in children and adults: A reference textbook*. Nova Science Publishers, Inc.
Sampson, J. H. (2017). *Translational immunotherapy of brain tumors*. Academic Press.
Speck, K. (2014). *Diary of a meningioma: A patient's perspective*. CreateSpace Independent Publishing Platform.
Van Rhee, F., & Munshi, N. C. (2017). *Castleman disease, an issue of hematology/oncology clinics (the clinics: internal medicine)*. Elsevier.
Winn, H. R. (2011). *Youmans neurological surgery, 4-volume set: Expert consult* (6th Edition). Saunders.
Zada, G., & Jensen, R. L. (2016). *Meningiomas, an issue of neurosurgery clinics of north america (the clinics: internal medicine)*. Elsevier.

Chapter 13

Choroid plexus tumors

Chapter Outline

Choroid plexus papilloma	233	Clinical cases	239
Atypical choroid plexus papilloma	236	Key terms	241
Choroid plexus carcinoma	238	Further reading	241

Choroid plexus papilloma

Choroid plexus papilloma is a benign, intraventricular papillary neoplasm. The tumor forms from **choroid plexus** epithelium and has extremely low mitotic activity, which in some cases is totally absent. It is classified by the World Health Organization as a grade I tumor. The choroid plexus as a normal structure consists of tufts of villi within the ventricular system, which produces cerebrospinal fluid (CSF) analogous to how kidneys produce urine. Choroid plexus papillomas are usually present in the infratentorial region in adults, and in the supratentorial space in children, which are the opposite of what might normally be expected when compared to the locations of other brain tumors.

Epidemiology

Choroid plexus tumors make up 0.3%–0.8% of all brain tumors in patients of all ages. However, they make up 2%–4% of brain tumors in children under the age of 15 years, and 10%–20% of brain tumors within the first year after birth. In adults, they make up on 0.5% of all brain tumors. According to the *Surveillance, Epidemiology, and End Results (SEER) program*, choroid plexus tumors make up 0.77% of all brain tumors, but make up 14% of those that occur in the first year of life. Choroid plexus papillomas make up 58.2% of all choroid plexus tumors, according to SEER, making them 5–10 times more common than choroid plexus carcinomas. Males develop choroid plexus tumors 1.2 times as often as females. However, the male-to-female ration for lateral ventricle tumors is equal, and about 80% of these tumors occur in patients under 20 years of age. For reasons heretofore undiscovered, fourth ventricle tumors develop in a 3:2 male-to-female ratio, and are distributed across all age groups in an even manner. Median patient age is 1.5 years for tumors within the lateral and third ventricles. Median age is 22.5 years within the fourth ventricle, and 35.5 years for tumors within the cerebellopontine angle.

Etiology and risk factors

Choroid plexus papillomas are caused by genes involved in development and biology of the plexus epithelium, such as the *orthodenticle homeobox 2 (OTX2)* and *transient receptor potential cation channel subfamily M member 3 (TRPM3)* genes. Alteration of these genes may initiate choroid plexus oncogenesis. There are no identified risk factors for choroid plexus papilloma, though there have been associations with von Hippel–Lindau syndrome and Li–Fraumeni syndrome.

Pathology

The histologic appearance of choroid plexus papilloma is very similar to the nonneoplastic choroid plexus, with no or only slight **atypia**. The majority of tumor cells contain the potassium channel *KIR7.1*, and its gene is described as *potassium inwardly rectifying channel, subfamily J, member 13 (KCNJ13)*. Choroid plexus papillomas occur in the ventricular system, but primarily in the lateral ventricles, mostly in children. They occur much less often in the fourth ventricle, third ventricle, and cerebellopontine angle. However, in adults, choroid plexus papillomas usually occur in the fourth ventricle, and are usually solid and vascular. Another rare form is the *triventricular choroid plexus papilloma*, in which the neoplasm forms in the third ventricle but extends through the foramen of Munro, into both lateral ventricles. These tumors are rare in the spinal cord or in locations outside the central nervous system (CNS). They are nearly always single lesions, of a **frond**-like structure and usually block CSF pathways. It is uncertain whether overproduction of CSF contributes greatly to any developing hydrocephalus. Tumor cells can be seeded in the CSF, which in rare cases may cause drop metastases within the cauda equina.

Macroscopically, choroid plexus papillomas are circumscribed, with a **polypoid** or cauliflower-like appearance, and tan, yellow, or pink-red in color (see Fig. 13.1). They may be adhered to ventricular walls, and are usually well delineated from the brain parenchyma, sometimes having hemorrhages or cysts. Some of the tumors are highly vascular, increasing the likelihood for bleeding. Microscopically, there are delicate and fibrovascular connective tissue structures. These are covered by one layer of regular cuboidal to columnar epithelial cells with oval or round monomorphic nuclei in a basal location. Mitotic activity is less than 2 mitoses per 10 high-power fields. Single cells or cell clusters may invade the brain, but this is unusual. Also uncommon are high cellularity, necrosis, focal blurring of papillary patterns, and nuclear pleomorphism. Choroid plexus papilloma cells are usually crowded together, lengthened, or stratified when compared to the normal nonneoplastic choroid plexus, which looks like cobblestones. In rare cases, the papillomas may experience oncocytic changes, **melanization**, tubular glandular cell architectures, mucinous degeneration, neuropil-like islands, and connective tissue degeneration. This degeneration may be manifestations of angioma-like blood vessel increase, xanthomatous changes, and formation of adipose tissue, bone, or cartilage.

Most tumors express cytokeratins and vimentin, stain positive for *cytokeratin 7 (CK7)*, and sometimes, are positive for the B-lymphocyte antigen *CD20*. The *epithelial membrane antigen* is usually negative, or has weak, focal positivity. Transthyretin is usually positive, though staining may be negative or varied. Most tumors are variably positive for S-100, which may be related to an older age of the patient, affording a better outlook. Membranous expression of KIR7.1 is usually present. Also, the glutamate transporter *EAAT1* is usually present, which helps differentiate these tumors from the nonneoplastic choroid plexus, which seldom expresses EAAT1. Gene mutations such as *TP53* are present in only 10% of choroid plexus papillomas. The tumors do have **hyperdiploidy**, and *MGMT promoter methylation* is always present. Choroid plexus papilloma is a major diagnostic component of **Aicardi syndrome**, which is rare and severe. This sporadic genetic condition is believed to be related to the X chromosome. It involves a triad of partial or total dysgenesis of the corpus callosum, infantile spasms, and chorioretinal lacunae. Characteristically, it affects the brain, spine, and eyes. Choroid plexus papilloma may be implicated in **hypomelanosis of Ito**.

Clinical manifestations

The signs and symptoms of choroid plexus tumors include hydrocephalus, papilledema, and increased intracranial pressure. In infants, there is increased head circumference due to hydrocephalus. About 91% of patients have increased

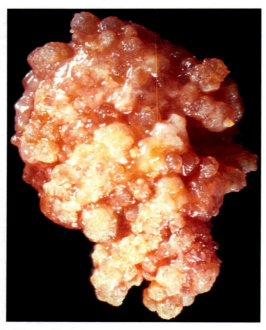

FIGURE 13.1 Choroid plexus papilloma. Cauliflower-like lesion with a nodular and partially calcified surface, taken from the fourth ventricle.

intracranial pressure. Other common symptoms include vomiting, homonymous visual field defects, and headaches. Additional symptoms include tinnitus and dizziness.

Diagnosis

Ultrasonography has had sufficient sensitivity to revealed congenital and fetal choroid plexus tumors. On CT and MRI, choroid plexus papillomas are usually isodense or hyperdense, T1-isointense, and T2-hyperintense (see Fig. 13.2). They have irregular contrast enhancement, and are well delineated in the ventricles. However, in some cases, the tumor has irregular margins and the disease becomes disseminated. Commonly, these tumors show linear and branching internal *flow voids* that reveal their increased vascularity. There may be rare subtypes diagnosed that are totally cystic or involve cystic extra-axial metastases from an interventricular primary tumor. Additionally, diagnosis may be made of a *choroid plexus xanthogranuloma*, usually in the lateral ventricular choroid plexus. These tumors have desquamated epithelial cells and accumulated lipids, macrophages, and multinucleated foreign body giant cells. This diagnosis is usually only seen in middle-aged and older adults.

Treatment

The majority of patients with choroid plexus papilloma are cured via complete surgical resection. Regular follow-up is usually conducted to assess for any possible recurrence, in which case, chemotherapy or radiation therapy may be used. Repeated imaging is advised to assess recurrence. Possible complications of treatments include hydrocephalus and facial paralysis due to tension on the facial nerve.

Prognosis

After surgical resection, the 5-year survival rate is upward of 100%. In one study, 5-year rates of localized control were lower at 84%. Control of relapse occurred within the brain, but at a distant site over 5 years was 92% with overall survival at 97%. The 5-year localized control rate is better following gross total resection (100%) than subtotal resection (68%). Tumors that relapse to distant sites can also recur locally. A metaanalysis revealed differences in survival rates over different time periods when comparing choroid plexus papillomas and carcinomas:

- 1-year survival from choroid plexus papillomas: 90%
- 1-year survival from choroid plexus carcinomas: 71%
- 5-year survival from choroid plexus papillomas: 81%
- 5-year survival from choroid plexus carcinomas: 41%
- 10-year survival from choroid plexus papillomas: 77%
- 10-year survival from choroid plexus carcinomas: 35%

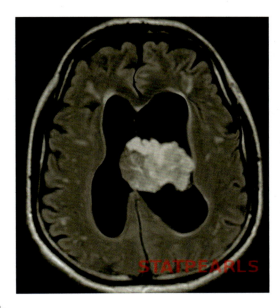

FIGURE 13.2 Choroid plexus papilloma.

In children under 3 years of age, prognosis is excellent after surgery. Malignant progression of choroid plexus papillomas is rare, but has occurred.

> **Point to remember**
> According to the National Cancer Institute, in the United States choroid plexus tumors are considered *primary* because they begin in the brain or spinal cord. They are more common in children in the first year of life, and are usually of unknown cause. These tumors almost always form within the ventricles of the brain, but rarely form in other areas of the CNS. Most of them have irregular borders and a "cauliflower-like" appearance that is seen in imaging studies.

Atypical choroid plexus papilloma

An **atypical choroid plexus papilloma** has higher mitotic activity, yet does not qualify as a choroid plexus carcinoma. This form of papilloma in children over 3 years of age, and in adults, is more likely to recur than the classic form. These tumors are considered grade II by the World Health Organization. Generally, these tumors are more aggressive than the classic choroid plexus papillomas, but not as aggressive as choroid plexus carcinomas.

Epidemiology

According to the SEER database, atypical choroid plexus papillomas make up 7.4% of all choroid plexus tumors. In one study, patients with this form were younger, having a median age of 0.7 years, than patients with classic choroid plexus papilloma or choroid plexus carcinoma. These forms both had a median patient age of 2.3 years in the study.

Etiology and risk factors

There are no established differences in the causes or risk factors between atypical choroid plexus papilloma and classic choroid plexus papilloma. There is a proposed theory that simian vacuolating virus 40 may be linked, but this is not proven.

Pathology

Atypical choroid plexus papillomas form in areas of the normal choroid plexus. They differ from the classic form of papillomas in that they are overall more common in the lateral ventricles. In one large study, 83% of atypical choroid plexus papillomas were in the lateral ventricles, with only 13% in the third ventricle and just 3% in the fourth ventricle. In children, they are always more common in the lateral ventricles and less common in the posterior fossa. However, in adults, they usually occur in the fourth ventricle and its lateral recesses. These tumors also block the CSF pathways. Macroscopically, rare cases reveal high vascularization with increased likelihood of hemorrhage, a feature also seen in classic choroid plexus papillomas. These tumors are often gray in color, soft, and extremely vascular.

Microscopically, there are mitotic counts of 2 or more mitoses per 10 randomly selected high-power fields. One or two features may also be present, out of a group that includes increased cellularity, blurred papillary patterns signifying solid growth, nuclear pleomorphism, and necrosis (see Fig. 13.3). These features are not needed for diagnosis, however. Atypical choroid plexus papillomas are more likely to recur than the classic form. Immunohistochemical results are related to recurrence. Negative S-100 protein staining in more than 50% of tumor cells is linked to higher tumor aggressiveness. One specific study revealed that positive CD44 staining was linked to these tumors as well as choroid plexus carcinomas. This relates to both forms' higher level of infiltration. The atypical and classic forms have similar immunohistochemical profiles. Along with higher mitotic activity, RNA expression profiling and analyses of gene sets have shown increased expression of cell cycle–related genes.

Clinical manifestations

Because of CSF pathway blockage, signs and symptoms include hydrocephalus, increased intracranial pressure, and papilledema. Ataxia is also seen in some patients.

Diagnosis

In imaging studies, atypical and classic choroid plexus papillomas appear identical. At diagnosis, atypical choroid plexus papilloma presents with metastasis in 17% of cases. Diagnosis is based on many immunohistochemical studies, with transthyretin and/or S-100 expression being somewhat supportive, yet not totally reliable. The major diagnostic difference between classic and atypical choroid plexus papillomas is based on the increased mitotic activity in the

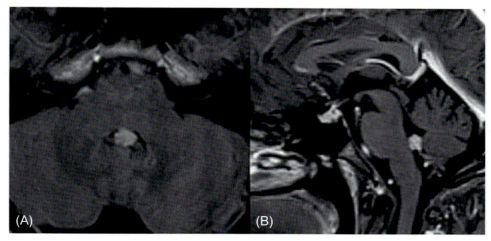

FIGURE 13.3 Choroid plexus papilloma with (A) well-differentiated papillary pattern and (B) transthyretin expression. *Courtesy: Neuropathology by Michael Bienkowski, Johannes A. Hainfellner, et. al., in Handbook of clinical neurology, 2018, Figure 32.36 (A and B).*

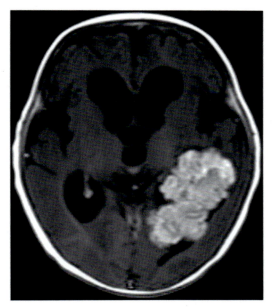

FIGURE 13.4 Choroid plexus papilloma, axial MRI image with gadolinium shows a large intraventricular lobulated mass in the left lateral ventricle with intense enhancement. *Courtesy: Pediatric Brain Tumors by Tina Young Poussaint, in Handbook of neuro-oncology neuroimaging, 2008, Figure 46.5.*

atypical papillomas, of 2 or more mitoses per 10 high-power fields. These tumors enhance intensely on T1 imaging with contrast (C+), fat saturated (see Fig. 13.4).

Treatment

Treatment of atypical choroid plexus papilloma is usually via craniotomy with gross total excision. Chemotherapy and radiation therapy may be used, as needed. The most common agent used in chemotherapy is bevacizumab, though other medications include nitrosourea, ifosfamide, carboplatin, and etoposide—sometimes used in combination. Bevacizumab is a recombinant humanized monoclonal antibody that blocks angiogenesis by inhibiting vascular endothelial growth factor, a protein that stimulates angiogenesis in tumors.

Prognosis

For prognosis, increased mitotic activity is the single histological factor associated with tumor recurrence. Tumors with more than 2 mitoses per 10 high-power fields are 4.9 times as likely to recur after 5 years than tumors with less than

2 mitoses per 10 high-power fields. For these tumors, 5-year overall survival is 89%, and event-free survival is 83%. Diagnosis is relevant to prognosis in children of 3 years or older and in adults, but not in children under 3 years of age. These young patients may have a good prognosis even if the tumors have a higher proliferation rate. There is always the risk of residual and potentially incapacitating neurological complications that may occur from the tumor, its direct and indirect impact, as well as potential complications of treatments (surgery, radiation, biochemotherapy).

> **Point to remember**
> The microscopic appearance of the classic and atypical choroid plexus carcinomas is a delicate fibrovascular core, lined with columnar or cuboidal epithelial cells that have vesicular basal nuclei. Their appearance is extremely similar to that of the normal choroid plexus. On imaging, they are characterized as vividly enhancing masses that are usually intraventricular.

Choroid plexus carcinoma

Choroid plexus carcinoma is a highly malignant and aggressive epithelial neoplasm. It is most common in the lateral ventricles in pediatric patients. It is defined by at least four of these five features: increased cellular density, frequent mitoses, blurred papillary patterns with coarse tumor cell sheets, nuclear pleomorphism, and necrosis. This tumor is classified as grade III by the World Health Organization.

Epidemiology

According to SEER, choroid plexus carcinomas make up 34.4% of all choroid plexus tumors. Approximately 80% of these tumors occur in children. They have an incidence of about 0.3 per million people annually, usually in children under age 5. There are no accurate global epidemiological statistics.

Etiology and risk factors

Most choroid plexus carcinomas are sporadic, but can occur with hereditary syndromes such as Li–Fraumeni syndrome, followed by Aicardi syndrome. Malignant rhabdoid tumors may also be implicated. A mutation of tumor suppressor gene is most likely related to tumor development.

Pathology

Choroid plexus carcinoma often invades nearby brain structures such as the parenchyma, metastasizing through the CSF. Most of these tumors reside and infiltrate within and around the lateral ventricles. They block CSF pathways, and can cause hydrocephalus because of this, as well as because of overproduction of CSF, spontaneous hemorrhage, and ventricular expansion. Risks of local recurrence or metastasis are 20 times higher than the risks related to choroid plexus papillomas. Macroscopically, these tumors are highly vascular and often bleed. They are invasive, appearing solid, hemorrhagic, and necrotic. Choroid plexus carcinomas resemble the histology of adenocarcinoma, but do not secrete mucin, which is an important designation between the two types of tumors. Microscopically, they have strong signs of malignancy, and diffuse brain invasion often occurs. They have blurred papillary features, atypical cells, and increased atypical mitotic activity. The cell atypia includes solid cellular sheets and variations in chromatin.

These tumors express cytokeratins but are not as often positive for transthyretin and S-100. TP53 mutation is present in about 50% of cases but p53 dysfunction is present in nearly all cases of these tumors. They have a higher proliferative index than classic and atypical choroid plexus papillomas. The index may range from 4.1% to 60%. About 50% of all cases show distinct membranous staining for KIR7.1, and nearly all cases are nuclear positive for SMARCB1 and SMARCA4. Complex chromosomal alterations, linked to the age of the patient, characterize these tumors. Hyperdiploidy or **hypodiploidy** may be present. Approximately 40% of cases occur with germline TP53 mutations/Li–Fraumeni syndrome.

Clinical manifestations

Choroid plexus carcinomas cause symptoms of hydrocephalus, including enlarged head size, increased intracranial pressure, nausea, and vomiting. Additional symptoms include persistent or new onset headaches, infantile macrocephaly or bulging fontanels, loss of appetite, papilledema, ataxia, strabismus, developmental delays, and altered mental status.

Diagnosis

Correct diagnosis of choroid plexus carcinoma is based on nuclear expression of SMARCB1 and SMARCA4 in early all tumors. This separates this tumor from the differential diagnosis of an atypical teratoid or rhabdoid tumor. There is usually marked enhancement, with a heterogenous nature, with the tumor being usually isodense to hyperdense to the gray matter. Calcifications are present in 20%—45% of all cases. Contrast-enhanced CT reveals the tumors prominently, with necrosis and cysts. In an MRI, these carcinomas usually appear as large intraventricular lesions. They have irregularly enhanced margins as seen in a **Diff-Quik stain**, edema in nearby brain tissues, a heterogeneous signal (T1- and T2-weighted images), hydrocephalus, and dissemination of the tumors. At diagnosis, metastases have occurred in 21% of cases. Any patient with a choroid plexus carcinoma should be tested for a TP53 germline mutation, even with no family history of Li—Fraumeni syndrome.

Treatment

The treatment of choroid plexus carcinoma is based on tumor location and aggressiveness. Shunts to drain CSF may be necessary, and surgical resection (the mainstay of treatment), radiation therapy, and chemotherapy have all been used. Both radiation and chemotherapy may help lengthen survival. It is important to note that radiation cannot be safely used in many younger patients. Best outcomes occur when total tumor resection is combined with adjuvant chemotherapy and radiation therapy. However, gross total resection is difficult because of the tumor's hypervascularity and poor demarcation from the adjacent brain parenchyma. If subtotal resection is all that can be performed, or there is widespread leptomeningeal disease, craniospinal irradiation is often indicated when possible.

Prognosis

The 3-year progression-free survival rate is 58%, while the 5-year rate is 38%. Overall survival rates are 83% over 3 years and 62% over 5 years. This means that their survival rate is about half that of the choroid plexus papillomas. The extent of surgery is a significant factor related to survival as is radiation therapy. Prognosis of TP53-mutant tumors may be improved by using intensive chemotherapy.

> **Point to remember**
> Choroid plexus carcinomas usually occur in young children, and are much less common than choroid plexus papillomas. They often invade adjacent brain tissues, resulting in focal neurological dysfunction. These carcinomas grow rapidly, and survival is based on the ability to achieve gross complete macroscopic resection. When resection is incomplete, survival rates are much lower.

Clinical cases

Clinical case 1

1. From where are choroid plexus papillomas derived, and what is their hallmark manifestation?
2. What do choroid plexus papillomas resemble, in regards to normal tissues?
3. List less common locations for choroid plexus papillomas to occur in children.

A 5-year-old boy was taken to his pediatrician with headache and vomiting that had persisted for a few months, off and on. Neurological examination showed papilledema. Noncontrast CT scan revealed an isodense lesion in the posterior third ventricle, and hydrocephalus was present. An MRI showed a lobulated, intensely enhancing lesion that was isointense on T1 and slightly hypointense on T2. The lesion measured 4.1 cm in greatest diameter. A ventriculoperitoneal shunt was performed, followed by surgical excision. The tumor was soft, red, vascular, and attached to choroidal vessels via a single pedicle. Complete surgical removal was achieved, and histopathologic examination confirmed a diagnosis of choroid plexus papilloma. Afterward, the child had no residual tumor growth and returned to normal health.

Answers:

1. **Choroid plexus papillomas are derived from the choroid plexus lining the brain ventricles, with overproduction of cerebrospinal fluid (CSF) being the hallmark manifestation. Hydrocephalus is seen in most cases. The reasons for this include overproduction of CSF or blockage of CSF pathways.**

2. Microscopically, choroid plexus papillomas resemble the normal choroid plexus with only slight or more commonly, no atypia.
3. Posterior third ventricle, fourth ventricle, and cerebellopontine angle locations are uncommon locations for choroid plexus papillomas to be discovered in affected children.

Clinical case 2

1. Where are choroid plexus papillomas usually identified in children?
2. Are these tumors more common in children or adults?
3. What is the basic description of Aicardi syndrome, as seen in this case study?

A 15-month-old girl was examined after experiencing strabismus, abnormal eye movements, and seizures. A CT scan revealed diffuse ventricular dilation. Her head circumference was larger than normal, due to the presence of hydrocephalus. She had developmental delays, macrocephaly, and bulging fontanels. An MRI demonstrated contrast-enhancing lesions in the lateral and third ventricles. A craniotomy was performed with radical removal of the lesion in the right lateral ventricle. One month later, a left parietal craniotomy was performed with radical excision of the lesion in the left lateral ventricle. Two months later, a right frontal craniotomy was performed, with radical removal of the third ventricle lesion. Pathological examination revealed a choroid plexus papilloma. A ventriculoperitoneal shunt was used to drain the ventricles. Six years later, follow-up revealed cognitive deficits, raising consideration of Aicardi syndrome. However, the patient was otherwise in overall normal physical health.

Answers:

1. **In children, choroid plexus papillomas are usually identified in the lateral ventricles. This differs from their development in adults, in which they usually occur in the fourth ventricle. The tumors are usually solid and vascular.**
2. **Choroid plexus papillomas are more common in children (2%–4%) than in adults (0.5%).**
3. **Aicardi syndrome is a rare and severe developmental disorder that results from an X-linked genetic defect. In infancy, it usually presents with a triad of infantile spasms, corpus callosal dysgenesis, and distinctive chorioretinal lacunae. Its characteristic malformations affect the brain, spine, and eyes**.

Clinical case 3

1. How are atypical choroid plexus papillomas distinguished from classic choroid plexus papillomas?
2. Anatomically speaking, where are atypical choroid plexus papillomas most commonly found?
3. Bevacizumab is the most common agent used for treatment of atypical choroid plexus papillomas, but how does it work?

A 13-year-old girl presented with 2 years of occasional frontal headaches that would awaken her from sleep. The headaches had been worsening over time. In the week before presentation for clinical evaluation, paresthesias began on the left side of her head, followed by central facial palsy and nystagmus. An MRI revealed a fourth ventricular mass that was isointense on T1-weighted imaging, and hyperintense on T2-weighted imaging. She had ventricular enlargement that was consistent with hydrocephalus. After surgical resection, the patient was diagnosed with an atypical choroid plexus papilloma.

Answers:

1. **These tumors are distinguished from the classic form of choroid plexus papillomas mostly based on increased mitotic activity (2 or more mitoses per 10 high-power fields). Other features that may be present include increased cellularity, nuclear pleomorphism, more solid growth patterns with blurring of the papillary patterns, and necrosis.**
2. **Atypical choroid plexus papillomas usually occur in the lateral ventricles. In children, all choroid plexus papillomas are more common in the lateral ventricles and less common in the posterior fossa. In adults, most of these tumors are in the fourth ventricle and its lateral recesses.**
3. **Bevacizumab is a recombinant humanized monoclonal antibody that blocks angiogenesis by inhibiting vascular endothelial growth factor, a protein that stimulates angiogenesis in tumors.**

Clinical case 4

1. What are principal imaging features of choroid plexus carcinomas?
2. What are the characteristic cellular features of choroid plexus carcinomas?
3. What is the mainstay of treatment for these tumors?

A 1-year-old girl had been vomiting nearly every day for just under a month. Her head circumference was enlarged, with bulging fontanels. She had papilledema of both eyes. An MRI revealed a T1-weighted isointense/hypointense mass in the fourth ventricle, and a T2-weighted image depicting brain stem displacement and infiltration. There was also dilatation of the ventricles. A diagnosis of a right cerebellar medulloblastoma was considered. A craniotomy was performed to excise the tumor, and pathological examination revealed it to be a choroid plexus carcinoma.

Answers:

1. Choroid plexus carcinomas exhibit marked heterogeneous enhancement and are usually isodense to hyperdense to the gray matter. Between 20% and 45% have calcifications. Contrast-enhanced CT shows the tumors prominently, with heterogenous qualities and areas of necrosis and cyst formation.
2. The characteristic features of choroid plexus carcinomas are blurring of the papillary features, cellular atypia, invasion to nearby brain parenchyma, necrosis, and increased mitotic activity. Cellular atypia includes variation in chromatin of tumor cells, solid cellular sheets, and atypical mitotic figures.
3. The mainstay of treatment is surgical resection. The extent of tumor resection is important in determining long-term survival. Gross total removal is difficult due to hypervascularity and poor demarcation from nearby brain parenchyma. Radiation therapy and chemotherapy may help increase survival but radiation cannot be used in many younger patients.

Clinical case 5

1. List typical symptoms caused by choroid plexus carcinomas.
2. How can these tumors induce hydrocephalus?
3. What is the incidence and survival associated with these tumors?

A 5-year-old boy presented with headaches that had lasted for about 2 months, but exhibited no seizures or focal neurological deficits. On CT scan, there was an isodense to hyperdense lesion in the white matter of the left occipito-temporo-parietal region, with heterogeneous enhancement pattern noted after contrast administration. The tumor was firm and highly vascular. The overall size was 4 cm in greatest dimension. Histological examination revealed tumor cells in sheets and clusters, a perivascular pseudopapillary pattern, pleomorphic nuclei, high mitotic activity, and necrosis.

Answers:

1. Symptoms include persistent or new onset headaches, macrocephaly, bulging fontanels, loss of appetite, papilledema, nausea, vomiting, ataxia, strabismus, developmental delays, and altered mental status.
2. Choroid plexus carcinomas can induce hydrocephalus by blocking normal CSF flow, overproduction of CSF, spontaneous hemorrhage, and ventricular expansion.
3. Choroid plexus carcinomas have an incidence of about 0.3 per million people per year, mostly in children under age 5. They are much more aggressive and have half the survival rate of choroid plexus papillomas.

Key terms

Aicardi syndrome
Atypia
Atypical choroid plexus papilloma
Choroid plexus
Choroid plexus carcinoma
Choroid plexus papilloma
Diff-Quik stain
Frond
Hyperdiploidy
Hypodiploidy
Hypomelanosis of Ito
Melanization
Polypoid

Further reading

Ahmad, A. (2016). *Introduction to cancer metastasis*. Academic Press.
Alfano, R. R., & Shi, L. (2018). *Neurophotonics and biomedical spectroscopy (Nanophotonics series)*. Elsevier.
Ammar, A. (2017). *Hydrocephalus: What do we know? And what do we still not know?* Springer.
Baehring, J. M., & Piepmeier, J. M. (2006). *Brain tumors: Practical guide to diagnosis and treatment (Neurological disease therapy)*. CRC Press.

Bodey, B., Siegel, S. E., & Kaiser, H. E. (2004). *Molecular markers of brain tumor cells: Implications for diagnosis, prognosis and anti-neoplastic biological therapy*. Springer.

Bouldin Darmofal, K. (2015). *101 Tips for recovering from traumatic brain injury: Practical advice for TBI survivors, caregivers, and teachers*. Loving Healing Press.

Brandao, L. A. (2016). *Adult brain tumors, an issue of neuroimaging clinics of North America (The clinics: Radiology)*. Elsevier.

Chang, E. L., Brown, P. D., Lo, S. S., Sahgal, A., & Suh, J. H. (2018). *Adult CNS radiation oncology: Principles and practice*. Springer.

Cinalli, G., Maixner, W. J., & Sainte-Rose, C. (2004). *Pediatric hydrocephalus*. Springer.

Cuneo, H. M., & Rand, C. W. (2012). *Brain tumors of childhood: American Lectures in Surgery, No. 104*. Literary Licensing, LLC.

Czernicki, Z., Baethmann, A., Ito, U., Katayama, Y., Kuroiwa, T., & Mendelow, A. D. (2010). *Brain edema XIV*. Springer.

Ellison, D., Love, S., Cardao Chimelli, L. M., Harding, J. S., Vinters, H. V., Bradner, S., & Yong, W. H. (2012). *Neuropathology: A reference text of CNS pathology* (3rd ed.). Mosby Ltd.

Fatterpekar, G. M., Naidich, T. P., & Som, P. M. (2012). *The teaching files: Brain and spine: Expert consult*. Saunders.

Guidetti, V., Arruda, M. A., & Ozge, A. (2017). *Headache and comorbidities in childhood and adolescence*. Springer.

Gupta, N., Banerjee, A., & Haas-Kogan, D. A. (2017). *Pediatric CNS tumor (Pediatric oncology)* (3rd ed.). Springer.

Hayat, M. A. (2012). *Tumors of the central nervous system, Volume 9: Lymphoma, supratentorial tumors, glioneuronal tumors, gangliogliomas, neuroblastoma in adults, astrocytomas, ependymomas, hemangiomas, and craniopharyngiomas*. Springer.

Hayat, M. A. (2014). *Tumors of the central nervous system, Volume 13: Types of tumors, diagnosis, ultrasonography, surgery, brain metastasis, and general CNS diseases*. Springer.

Hayat, M. A. (2015). *Brain metastases from primary tumors, Volume 2: Epidemiology, biology, and therapy*. Academic Press.

Hornick, J. L. (2013). *Practical soft tissue pathology: A diagnostic approach: A volume in the pattern recognition series (Expert consult)*. Saunders.

Icon Health Publications. (2004). *Aicardi syndrome — A medical dictionary, bibliography, and annotated research guide to internet references, a 3-in-1 medical reference*. ICON Health Publications.

Igbaseimokumo, U. (2019). *Brain CT scans in clinical practice* (2nd ed.). Springer.

Kocjan, G., Gray, W., Levine, T., Kardum-Skelin, I., & Vielh, P. (2013). *Diagnostic cytopathology essentials: Expert consult*. Churchill Livingstone.

Lacruz, C. R., Saenz de Santamaria, J., & Bardales, R. H. (2018). *Central nervous system intraoperative cytopathology, 2nd edition (Essentials of cytopathology)*. Springer.

Mahajan, A., & Paulino, A. (2017). *Radiation oncology for pediatric CNS tumors*. Springer.

Mucci, G. A., & Torno, L. R. (2015). *Handbook of long term care of the childhood cancer survivor (Specialty topics in pediatric neuropsychology)*. Springer.

Neman, J., & Chen, T. C. (2015). *The choroid plexus and cerebrospinal fluid: Emerging roles in CNS development, maintenance, and disease progression*. Academic Press.

Norden, A. D., Reardon, D. A., & Wen, P. C. Y. (2011). *Primary central nervous system tumors: Pathogenesis and therapy (Current clinical oncology)*. Humana Press.

Reeves, M. (2016). *Hydrocephalus: Prevalence, risk factors and treatment*. Nova Science Publishers, Inc.

Shiminski-Maher, T., Woodman, C., & Keene, N. (2014). *Childhood brain & spinal cord tumors: A guide for families, friends & caregivers* (2nd ed.). Childhood Cancer Guides.

Som, P. M., & Curtin, H. D. (2011). *Head and neck imaging (Expert consult)* (5th ed.). Mosby.

Teicher, B. A., & Ellis, L. M. (2007). *Antiangiogenic agents in cancer therapy (Cancer drug discovery and development)* (2nd ed.). Humana Press.

Tonn, J. C., Reardon, D. A., Rutka, J. T., & Westphal, M. (2019). *Oncology of CNS tumors* (3rd ed.). Springer.

Walker, D., Perilongo, G., Punt, J., & Taylor, R. (2004). *Brain and spinal tumors of childhood (Hodder Arnold Publication)*. CRC Press.

Wolff, J. (2010). *Choroid plexus tumor protocol Cpt-Siop-2000* (2nd ed.). Lulu.com.

Chapter 14

Mesenchymal tumors

Chapter Outline

Overview	243	Angiolipoma	268
Hemangioblastoma	243	Desmoid-type fibromatosis	269
Hemangiopericytoma	246	Inflammatory myofibroblastic tumor	270
Hemangioma	249	Myofibroblastoma	271
Epithelioid hemangioendothelioma	252	Benign fibrous histiocytoma	272
Angiosarcoma	253	Malignant fibrous histiocytoma	273
Chondrosarcoma	255	Leiomyoma	275
Ewing sarcoma	256	Rhabdomyoma	275
Fibrosarcoma	257	Chondroma	276
Kaposi sarcoma	258	Osteochondroma	278
Leiomyosarcoma	260	Osteoma	279
Liposarcoma	261	Hibernoma	280
Osteosarcoma	262	Clinical cases	280
Rhabdomyosarcoma	264	Key terms	283
Lipoma	266	Further reading	283

Overview

Mesenchymal tumors are nonmeningothelial tumors that may be benign or malignant. They originate in the central nervous system (CNS), and have histological features related to similar soft tissue and bone tumors. These similarities are also linked to their clinical names. Mesenchymal tumors arise more often in the meninges, but sometimes in the parenchyma or choroid plexus. However, primary mesenchymal CNS tumors are extremely rare, arising most often in the supratentorial region. Mesenchymal tumors range from grade I to grade IV according to the World Health Organization. The most common forms of mesenchymal tumors include fibrosarcoma and malignant fibrous histiocytoma. Clinical signs and symptoms depend mostly upon tumor location, with treatments and prognoses also highly varied based on individual tumor type.

Hemangioblastoma

A **hemangioblastoma** is a vascular central nervous system (CNS) tumor that originates from the vascular system, most often in middle age. They can also occur in the spinal cord and retinas of the eyes, and are linked to other diseases. Hemangioblastomas are usually made up of **stromal cells** in smaller blood vessels. Within the brain, they are most common in the cerebellum and brain stem. The World Health Organization classifies them as grade I tumors.

Epidemiology

While rare, the actual incidence rates of hemangioblastomas are uncertain—they are among the rarest of CNS tumors, making up less than 2%. Average patient age at diagnosis, for the VHL (von Hippel–Lindau)-associated tumors, is usually 20 years younger than for the sporadic tumors. Though the tumors have been seen in various ages, they are most common in adults around age 40. The male-to-female ratio is nearly even.

Etiology and risk factors

Though the cause of hemangiopericytoma is uncertain, people with VHL disease are more likely to develop hemangioblastomas. About 10% of people with VHL disease also develop hemangioblastomas. Risk factors for hemangioblastomas include diseases such as polycythemia and pancreatic cysts.

Pathology

Hemangioblastomas may occur anywhere within the nervous system. The sporadic forms usually develop in the cerebellar parenchyma, in the hemispheres 80% of the time. Tumors associated with VHL are multiple in 65% of cases, affecting the brain stem, spinal cord, and nerve roots along with the cerebellum. In rare cases, peripheral nervous system and supratentorial lesions are seen. Most hemangioblastomas are highly perfuse compared to normal parenchymal tissues.

Macroscopically, about 60% of hemangioblastomas are well circumscribed, slightly cystic, and highly vascularized, while about 40% are totally solid. Sometimes, the mass is yellow because of significant amounts of lipids. Classically, there is a cyst with a solid vascular nodule that touches a pial surface, and necrosis may be present. There is sometimes a defined cystic cavity with clear fluid that appears hyperintense in T2-weighted imaging. Microscopically, the two primary factors are large stromal cells that are vacuolated, which may have cytological variations, and an abundance of vascular cells (see Fig. 14.1). The cellular and reticular variants are determined by stromal cell abundance and morphology. Many thinly walled vessels are seen that are easily outlined via reticulin stain. Intratumoral hemorrhage may occur because of high vascularization, and often involves many abnormal blood vessels. Astrocytic gliosis and Rosenthal fibers are common in nearby reactive tissues—especially in cystic and **syrinx** walls. Tumor edges are often well demarcated. Infiltration into neural tissues is rare, as are mitotic figures. Stromal cells make up 10%−20% of all cells, and are the neoplastic tumor component. Nuclei vary in size, sometimes being atypical and hyperchromatic. The most significant morphological factors are the lipid-containing vacuoles, which are numerous, causing a typical clear-cell morphology. The tumor may look similar to a metastatic renal cell carcinoma. Also, patients with VHL are more likely to develop renal cell carcinoma. There have been reports of renal cell carcinoma metastasizing into hemangioblastoma.

Stromal cells do not have endothelial cell markers such as **von Willebrand factor** or the transmembrane phosphoglycoprotein CD34. They do not express CD31 and other endothelium-associated adhesion molecules. However, stromal cells may express neuron-specific enolase, *neural cell adhesion molecule 1 (NCAM1)*, ezrin, and S-100. The major intermediate filament expressed is vimentin. Stromal cells also express proteins such as aquaporin-1, *C-X-C chemokine receptor type 4 (CXCR4)*, **brachyury**, carbonic anhydrase isozymes, and *epidermal growth factor receptor*. Both *hypoxia-inducible factor 1-alpha (HIF1A)* and HIF2A have been detected, which may encourage expression of vascular endothelial growth factor. The endothelial cells express receptors for platelet-derived and other angiogenic growth factors.

Many nonneoplastic cells that include endothelial cells, lymphocytes, and pericytes surround stromal cells. The exact cellular origin of hemangioblastomas, however, is uncertain. Stromal cells express proteins that are common to

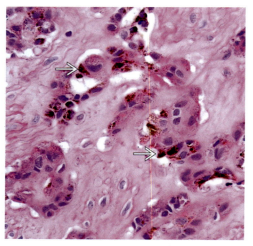

FIGURE 14.1 A mixture of spindled and polygonal cells comprise most cases of hemangioblastoma. A prominent blood-filled capillary vascular network (*arrows*) is often easily visible.

embryonal **hemangioblast** progenitor cells. The distribution of the **SCL protein** within an embryo's nervous system is similar to the distribution of hemangioblastomas in patients. This may mean that stromal cells are hemangioblasts or hemangioblast progenitor cells. In familial hemangioblastomas, **biallelic inactivation** of the VHL gene is common, but not in sporadic tumors. Studies of sporadic tumors have shown loss or inactivation of the VHL gene in up to 78% of patients. A loss of VHL function may be the central event in formation of hemangioblastomas. The VHL tumor suppressor protein forms the **VCB complex** with *transcription elongation factor B1 (TECB1)* and *TECB2*. The VHL tumor-suppress protein regulates cell cycles and degradation of HIF1A and HIF2A. Therefore loss of VHL function results in upregulation of hypoxia-responsive genes even without any tumor hypoxia. This is known as **pseudohypoxia**. Hemangioblastoma is a common manifestation of VHL, in 60% of patients.

Clinical manifestations

The signs and symptoms of intracranial hemangioblastomas are usually due to impaired cerebrospinal fluid flow because of a cyst or solid tumor mass. This results in an increase in intracranial pressure and hydrocephalus. There may be cerebellar deficits that include **dysmetria** and ataxia. Spinal tumors cause symptoms because of localized compression, including loss of touch or sensation, incontinence, and headache. Nausea and vomiting are often seen. The **erythropoietin** produced by hemangioblastomas causes secondary **polycythemia** in about 5% of cases, with hemorrhage being only a rare complication.

Diagnosis

On both CT and MRI, hemangioblastomas have a specific appearance (see Fig. 14.2). With cerebral angiography, they produce a distinctive, prolonged vascular stain. While MRI is preferred, angiography sometimes shows occult vascular

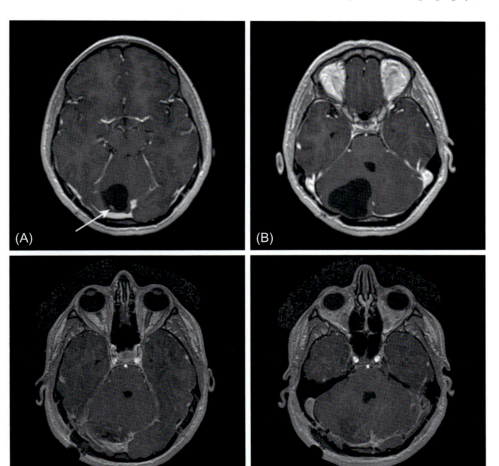

FIGURE 14.2 Four views of a hemangioblastoma. (A and B) Two views of the hemangioblastoma before surgery. (C and D) Two views after surgery.

nodules more clearly. An MRI study usually shows a mass that is gadolinium enhancing, often intensely. About 75% of cases have an associated cyst. The solid mass is usually located peripherally, in the cerebellar hemisphere. There may be flow voids in the nodule because of enlarged blood vessels. A spinal cord hemangioblastoma is usually related to the presence of a syrinx. Immunohistochemistry helps to distinguish hemangioblastoma from renal cell carcinoma, which is positive for epithelial markers such as epithelial membrane antigen (EMA).

Treatment

Genetic counseling and molecular genetic screening for VHL germline mutations are essential for patients. Treatments basically involve surgical resection and stereotactic radiosurgery. Resection surgery may not be possible if the tumor is associated with VHL disease. Stereotactic radiosurgery may require several months of follow-up to determine that the tumor has been completely destroyed. It may be the best option if the tumor is located in an area that would be difficult to treat with traditional surgical resection.

Prognosis

Prognosis of sporadic CNS hemangioblastoma is excellent when surgical resection can be performed successfully, which is common. Permanent neurological deficits are rare. They can be avoided when the tumors are diagnosed and treated early. Sporadic tumors have better outlooks than those associated with VHL since patients with this form usually develop multiple lesions.

> **Point to remember**
> A hemangioblastoma is a benign and highly vascular tumor that can occur in the brain, spinal cord, and retina. As it enlarges, it presses upon the brain and can cause neurological symptoms. Most hemangioblastomas occur sporadically, but some people develop these tumors as part of the genetic syndrome called von Hippel–Lindau syndrome. These people usually develop multiple tumors in the brain and spinal cord over their lifetimes.

Hemangiopericytoma

Hemangiopericytoma is a solitary fibrous, solid tumor that is the most common type of **mesenchymal neoplasm**, though overall in comparison to other CNS tumors, mesenchymal tumors are rare. Hemangiopericytoma is of the fibroblastic type, often having a significant branched vascular pattern. The tumor is believed to originate in the **pericytes** within the walls of capillaries, but this is uncertain.

Epidemiology

Hemangiopericytoma is rare, and makes up less than 1% of all primary CNS tumors throughout the world. It usually affects adults in the fourth to sixth decade of life. According to the 2014 Central Brain Tumor Registry of the United States (CBTRUS) report, the rarity of hemangiopericytomas resulted in these tumors being grouped with other meningeal mesenchymal tumors. Together, these tumors have an average annual age-adjusted incidence of just under 1 case per 100,000 people, globally. Males are affected slightly more than females. These tumors are much rarer in children, but do occur.

Etiology and risk factors

Most hemangiopericytomas have a genomic inversion at the 12q13 locus that fuses the *NGFI-A-binding protein 2 (NAB2)* and *signal transducer and activator of transcription 6 (STAT6)* genes. This results in STAT6 nuclear expression that is detectable via immunohistochemistry. There is no evidence of familial clustering of hemangiopericytomas or any known risk factors.

Pathology

There are two primary morphologies of hemangiopericytomas. The first is a solitary fibrous tumor phenotype with a patternless structure or a short fascicular pattern, along with hypocellular and hypercellular areas and thick collagen bands. The second is a phenotype characterized by high cellularity plus a thin but rich network of **reticulin fibers** that usually invests individual cells. Both phenotypes share thinly branched **staghorn vessels** (see Fig. 14.3).

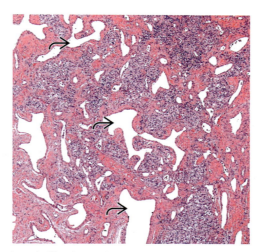

FIGURE 14.3 Scattered, irregularly shaped, and dilated vascular spaces (*arrow*), known as "staghorn vessels," are often prominent in hemangiopericytomas.

Hemangiopericytomas are usually based in the dura, and are often supratentorial, with only about 10% being spinal. The solitary fibrous phenotype is usually benign as long as gross total resection can be performed. The hemangiopericytoma phenotype tumors recur in over 75% of cases at 10 years. There may be resultant extracranial metastasis, mostly in the bones, lungs, and liver, in about 20% of cases.

Grading of hemangiopericytomas is based on hypercellularity, necrosis, and elevated mitotic count of more than 4 mitoses per 10 high-power fields. If the mitotic count is this high, this is important to designate the tumor as malignant regardless of the phenotype. Within the CNS, there are three basic grading criteria, as follows:

- Grade I—benign; usually treated only by surgical resection; a hypocellular, collagenized tumor with a classic solitary fibrous phenotype.
- Grades II and III—malignant; treated also with adjuvant therapies—mostly radiation therapy; more densely cellular tumors of the hemangiopericytoma phenotype; the grade III tumors are subclassified as anaplastic, based on mitotic counts of 5 or more mitoses per 10 high-power fields.

There is a newer grading system proposed that will have four different grades, but its use has not been proven to be more effective than the current system. Most hemangiomas are dural-based and usually supratentorial, with approximately 10% being spinal. Common locations include the base of the skull and the parasagittal and falcine locations. Less common locations are the cerebellopontine angle, pineal gland, and sellar region. Hemangiopericytomas have also occurred in the transverse and sigmoid sinuses.

Macroscopically, hemangiopericytomas are usually firm or hard, white to pink to reddish-brown, and well demarcated. This depends on the amount of collagenous stroma and cellularity. Sometimes, there is infiltrative growth and no dural attachment. There may be hemorrhagic or myxoid changes, and the tumor can bleed extensively. Microscopically, along with the two primary phenotypes, there may be an intermediate-hybrid morphology. Tumor cells may have a round, ovoid, or spindle shape, with a small amount of eosinophilic cytoplasm. There may be round or oval nuclei, with slightly dense chromatin and nucleoli without pseudoinclusions. Mitoses are usually less than 3 per 10 high-power fields. However, the hemangiopericytoma phenotype has closely apposed ovoid cells, erratically arranged, with small amounts of intervening stroma. Though mitotic activity and necrosis may be present, they are varied. There may be many reticulin fibers investing individual cells or small cell groups. Invasion of brain parenchyma or engulfment of nerves or vessels may be present, which makes surgical removal difficult. In rare cases, a varied adipocyte component may be seen.

Hemangiopericytomas are usually diffusely positive for vimentin and CD34, but a loss of CD34 expression is common if the tumor is malignant. Other rare findings include markers such as **desmin**, **cytokeratin**, *smooth muscle actin (SMA)*, *EMA*, and progesterone receptor. ALDH1A1 gene overexpression can be revealed by immunohistochemistry, with ALDH1 positive in 84% of patients. Median Ki-67 proliferation index ranges between 0.6% and 36%. The cell of origin of hemangiopericytomas is still under debate.

Clinical manifestations

Usually, signs and symptoms of hemangiopericytomas are related to their location along with mass effect and increased intracranial pressure. A chronic headache is usually the primary symptom. Rarely, complications include massive intracranial hemorrhage and hypoglycemia caused by release of insulin-like growth factor.

Diagnosis

Detection of STAT6 nuclear expression or NAB2-STAT6 fusion is recommended in order to confirm diagnosis of a hemangiopericytoma. If this is negative, another diagnosis must be considered. If STAT6 immunohistochemistry or NAB2-STAT6 fusion tests cannot be performed, or are not performed, this must be documented. A plain CT scan will reveal solitary, irregularly contoured tumors without calcifications of **hyperostosis** of the skull. In an MRI, the tumor is isointense on T1-weighted imaging, and has high or mixed intensity on T2-weighted imaging, with variable contrast enhancement. Fig. 14.4 shows CT and MRI views of a hemangiopericytoma. Sometimes, flow voids and dural contrast

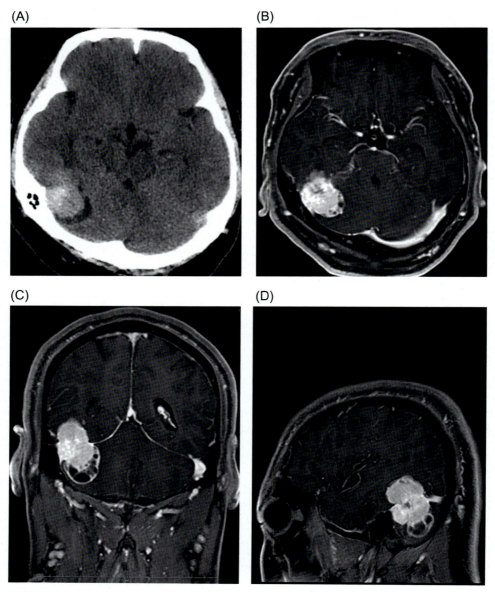

FIGURE 14.4 Hemangiopericytoma: (A) CT scan showing the high-density mass lesion and (B—D) MRI scans showing the lesion has clear margins and is isointense on T1-weighted imaging, plus well enhanced on gadolinium-contrast T1-weighted imaging.

enhancement at the dural tail are seen. Arteriography will show a hypervascular mass with large draining veins that may be branched, with a staghorn appearance.

Differential diagnoses include soft tissue and meningothelial neoplasms. Similar tumors include fibrous meningioma, **Ewing sarcoma**, monophasic synovial sarcoma, mesenchymal chondrosarcoma, and malignant peripheral nerve sheath tumor. However, electron microscopy will show that a hemangiopericytoma has closely apposed long cells with short processes containing small groupings of intracytoplasmic intermediate filaments. An extracellular basement membrane-like substance that is electron-dense surrounds individual cells. This is the equivalent of the reticulin network seen in light microscopy.

Treatment

Based upon the grade of the tumor, hemangiopericytomas are treated with surgical resection, radiation therapy, and for some patients, chemotherapy. More recently, the endoscopic endonasal approach and *Gamma Knife* radiosurgery have become popular techniques since they are noninvasive and comparatively less painful than traditional surgeries. They also allow for a faster recovery time.

Prognosis

With gross total resection and a lack of **atypia**, prognosis for hemangiopericytomas is generally good. Most patients benefit from adjuvant radiation therapy. However, if pulmonary metastasis occurs, it is a major cause of death. Maximum survival rate is 10 years after detection of a hemangiopericytoma, but most patients survive 2–5 years because of multiple malignancies.

> **Point to remember**
> Mesenchymal tissue neoplasms are soft tissue tumors, also referred to as connective tissue tumors, which may be located in the CNS and other areas of the body. They may originate from mesodermal-derived precursor cells that develop into bone, cartilage, blood vessels, adipose tissue, smooth muscles, or fibroblasts. In the CNS, they usually arise from the meninges instead of from the parenchyma.

Hemangioma

A **hemangioma** is a benign, vascular tumor that may be of many different sizes. The majority are primary bone lesions that affect the CNS secondarily. In the CNS they more frequently occur in the brain stem, cerebellum, and spinal cord. Less commonly, dural and parenchymal hemangiomas are seen. Another term for these tumors is *congenital hemangiomas*, in which the tumors are present at birth and do not develop later. Within the brain, additional names for hemangiomas include **cavernous hemangiomas**, *cavernous angiomas, cavernomas,* or *cerebral cavernomas*. Very often, cavernous hemangiomas are referred to as **cerebral cavernous malformations**.

Epidemiology

Hemangiomas usually occur in children, and they are the most common benign tumors in the pediatric population. They are more common in females than males, but actual incidence is difficult to estimate since these tumors are frequently misdiagnosed. Congenital hemangiomas are less common than **infantile hemangiomas**. For cavernous hemangiomas, for unknown reasons, about 50% of Hispanic patients have a familial link. This occurs in 10%–20% of cases in Caucasians. Cavernous hemangiomas, unlike other forms, usually occur in the third to fourth decade of life, with no gender preference. Cavernous hemangiomas that cause vision problems are more prevalent in females, between 20 and 40 years of age, than in males of the same age group. Cerebral cavernous malformations are present in 0.5% of the overall global population, with about 40% of those affected having symptoms. Asymptomatic people usually developed the malformation sporadically, while symptomatic people usually have an inherited genetic mutation. Of cerebral cavernous malformations, 75% of cases are in adults and 25% are in children.

Etiology and risk factors

Many hemangiomas have a congenital link, but may also develop over a person's lifetime. However, there is no definitive known cause. Cerebral cavernous malformations may be inherited as an autosomal dominant disorder, in 20% of cases, yet most cases are sporadic. Inherited cerebral cavernous malformation pathogenesis involves biallelic somatic and germline mutations, within one of the cerebral cavernous malformation genes. The implicated gene mutations concern the genes known as *K-Rev interaction trapped 1 (CCM1), malcavernin (CCM2)*, and *programmed cell death protein 10 (CCM3)*. These genes are located at the chromosomes 7q21.2, 7p13, and 3q25.2-q27.

Radiation treatment for other medical conditions has been suggested as a cause of cavernous malformations. Hemangiomas show rapid proliferation of endothelial cells and pericytic hyperplasia, or tissue enlargement because of abnormal pericyte cell division. Also, a loss of heterozygosity is common in tissues where hemangiomas develop. There are no known risk factors for hemangiomas.

Pathology

Hemangiomas usually have capillary-type growth. Infantile hemangiomas are almost always positive for *glucose transporter 1 (GLUT1)*, which helps distinguish these tumors from arteriovenous malformations in extracranial locations. However, a newer study showed that cerebral cavernous malformations as well as cerebral arteriovenous malformations had endothelial immunoexpression of GLUT1. Hemangiomas consist of blood vessels, and usually grow quickly for up to a few months, often shrinking or involuting thereafter.

Cavernous hemangiomas may develop within the brain, spinal cord, or even the eyes, usually in the muscle cone lateral to the optic nerve, and have significant blood supplies. This type is usually made up of capillary-like vessels and also large, saccular vessels with fibrotic walls. Cavernous hemangiomas are reddish-brown and sponge-like appearance of intraosseous capillary hemangioma. A pathological form is called *intravascular papillary endothelial hyperplasia*, in which there is papillary proliferation of the endothelium, with thrombosis. Cerebral cavernous malformations differ from other cavernous hemangiomas in that there is no tissue within the malformation, and the borders are not encapsulated. This means that they can change in size and quantity. The histologic appearance of cavernous hemangioma is shown in Fig. 14.5.

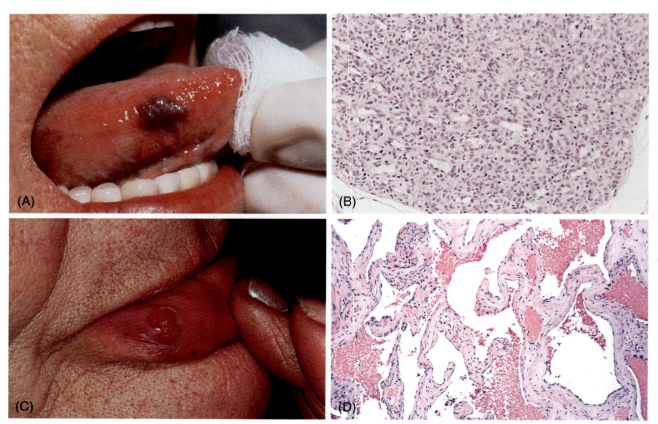

FIGURE 14.5 Histologic appearance of cavernous hemangioma.

Clinical manifestations

Cavernous hemangiomas may cause seizures because of compression of cortical brain tissue, or tumor hemorrhaging resulting in scarring of surrounding tissue. There can be intraparenchymal hemorrhage, double vision, other types of vision problems, language difficulties, loss of memory, and incidental hydrocephalus. Less severe symptoms include headaches, limb numbness or weakness, and ataxia. If the tumor extends from the brain toward the eyes, there may be progressive proptosis.

Diagnosis

For cavernous hemangiomas, gradient-echo T2-weighted magnetic resonance imaging is the most sensitive procedure, and has led to an increase in diagnosed cases since it became available in the 1980s. Radiographic images reveal a "popcorn" shaped tumor (see Fig. 14.6). CT scans are not as sensitive or specific, and angiography is only usually used to rule out other diagnoses. Biopsies may be obtained from the tumor tissue for examination, and correct diagnosis is important because these lesions are less aggressive than tumors such as angiosarcomas. Biopsies are only rarely needed because of the accuracy of the MRI imaging for these tumors.

Treatment

If a cavernous hemangioma is destroying surrounding tissues and causing major symptoms, total surgical excision is usually performed. A common complication of this surgery is significant hemorrhage. There is also a risk of stroke or death. Hemangiomas may recur. Treatments include radiosurgery and microsurgery, based on the tumor site, size, and symptoms as well as any history of tumor hemorrhage. Microsurgery is usually preferred if the tumor is superficial within the CNS, or risks of damage to surrounding tissues from radiation therapy are too high. Further indications for microsurgery are large hemorrhages, seizures, or a coma. The preferred microsurgical technique is gamma-knife radiation, directing a precise radiation dose to the hemangioma while mostly sparing surrounding tissues.

Prognosis

Cavernous hemangiomas have a serious outlook since they may cause bleeding and seizures. Patients who have experienced previous brain bleeding from these tumors have a higher risk of subsequent bleeding. Fortunately, once these tumors are completely excised there is very little risk of growth or rebleeding. Not enough data has been collected to provide an adequate estimate of life expectancy.

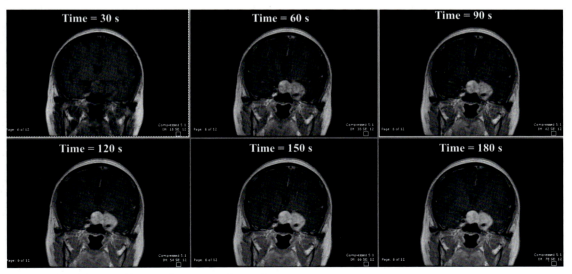

FIGURE 14.6 Six T2-weighted MRI images of a cavernous hemangioma.

Epithelioid hemangioendothelioma

An **epithelioid hemangioendothelioma** is a rare, low-grade malignant vascular tumor with epithelioid endothelial cells that appear in cords as well as single cells within a distinct stroma that may be **chondromyxoid** or hyalinized. In rare cases, these tumors develop in the brain parenchyma, dura, or the skull base. Epithelioid hemangioendotheliomas are histologic intermediates, between angiosarcomas and hemangiomas, yet are distinct from these tumors.

Epidemiology

Epithelioid hemangioendotheliomas usually occur in adults between 20 and 40 years of age, with a slight preference for females over males. However, these tumors are so rare that only 0.01% of cancer patients have them, and their incidence is only 1 in every 1,000,000 people globally. In the United States, only about 20 cases are diagnosed every year.

Etiology and risk factors

The only apparent cause of epithelioid hemangioendothelioma is a gene mutation of some form. About 90% of these tumors have the recurrent t(1;3)-(p36;q25) translocation. This causes a *WW domain-containing transcription regulator protein 1 (WWTR1)—calmodulin-binding transcription activator 1 (CAMTA1)* fusion. Less often, there is a t(x;11)-(p11;q22) translocation, resulting in a *yes-associated protein 1 (YAP1)—transcription factor E3 (TFE3)* fusion. There are no identified risk factors for these tumors.

Pathology

The cells of an epithelioid hemangioendothelioma contain a fairly high level of eosinophilic cytoplasm that may be vacuolated (see Fig. 14.7). Generally, the nuclei are round, or sometimes, vesicular, with indentations, and only slight atypia. There may be mitoses and limited necrosis. While small intracytoplasmic lumina, known as **blister cells** may be present, there are usually no vascular channels of any significance.

Clinical manifestations

The signs and symptoms of epithelioid hemangioendotheliomas include discomfort, pain, nausea, difficulty breathing, fever, inflammation, weight loss, and difficulties with movements or walking. All manifestations are directly based upon the location of the tumor.

Diagnosis

The endothelial components of these tumors are confirmed via immunohistochemical studies, such as those for *cluster of differentiation 31 (CD31)* and *erythroblast transformation-specific related gene (ERG)*. Diagnostic imaging methods include MRI, CT, standard X-rays, and PET scans. If possible due to location, a biopsy may be taken for confirmation.

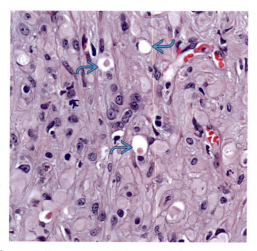

FIGURE 14.7 Vacuolated tumor cells may feature various numbers of intracytoplasmic vacuoles (*arrows*) in an epithelioid hemangioendothelioma.

Treatment
There is no standard treatment for these extremely rare tumors. Surgery is most often used when the tumors have not metastasized, and it can be accompanied by radiation therapy and chemotherapy. In some cases, radiation therapy is used without surgery. A variety of medications can be used to treat epithelioid hemangioendotheliomas. These include rapamycin that slows the growth of abnormal vessels, helping to shrink tumors and improve pain and other symptoms. Tyrosine kinase inhibitors have shown short-term success, and include pazopanib, sorafenib, and sunitinib. Vincristine and interferon are usually successful for treatment as well. In the rare case of a faster growing epithelioid hemangioendothelioma, multiagent chemotherapy may be indicated, but this is usually only for slightly older adults. Newer *antiangiogenic* medications that target the tumor blood vessels are under continued study.

Prognosis
Epithelioid hemangioendotheliomas usually develop slowly over time, and therefore, many patients survive for decades. The extent and number of organs that these tumors may affect do not appear to greatly affect lifespan.

> **Point to remember**
> Epithelioid hemangioendothelioma is a rare vascular tumor that affects the epithelial cells lining the insides of blood vessels. They commonly affect the soft tissues, organs, and bones. These malignant tumors are slow growing and usually do not metastasize as quickly as other cancers. Since they are so rare, they can be undetected for a long time prior to diagnosis, allowing time for tumor spread.

Angiosarcoma
An **angiosarcoma** is a high-grade malignant tumor with endothelial differentiation. In rare cases, it originates in the brain or meninges. Basically, angiosarcoma is a cancer of the cells lining the walls of blood vessels. Since the cells are carried through these vessels, metastasis to sites such as the liver and lungs is more likely. These tumors can range from grade I to grade IV.

Epidemiology
Angiosarcomas are relatively uncommon. Globally, incidence is low, with angiosarcomas making up 3.3% of all soft tissue sarcomas, with an incidence rate of 1.5 per 1,000,000 population. In the United States, about 50% of diagnosed angiosarcomas occur in the head and neck—yet they account for less than 0.1% of head and neck malignancies. Most angiosarcomas occur in the elderly, but they can occur at any age.

Etiology and risk factors
No genetic abnormalities have been documented for CNS angiosarcomas, and the etiology is usually unknown for most cases. Risk factors for angiosarcomas may include exposure to toxins or radiation therapy, exposure to carcinogens, and previous history of lymphedema. Development of angiosarcomas may be linked to radical mastectomy, preexisting bone infarctions, chronic osteomyelitis, **pagetoid bone**, and acquired immunodeficiency syndrome (AIDS).

Pathology
Angiosarcomas that originate in the brain or meninges have varied differentiations, and can appear in a variety of different locations. They may be extremely vascular and have anastomosing vascular channels that are lined with atypical endothelial cells with mitotic activity. In other examples, they may be poorly differentiated, and often are epithelioid, solid lesions. For this pathology, immunoreactivity for various vascular markers must be determined. Angiosarcomas may show signs of extensive hemorrhage and necrosis. The cells have an increased nuclear-to-cytoplasm ratio, nuclear hyperchromasia and pleomorphism, and high mitotic activity. The tumors are aggressive and often recur locally, spread widely, and have high rates of lymph node and systemic metastasis. They usually appear red in color, and are hemorrhagic.

Clinical manifestations

Angiosarcomas are insidious and may not produce symptoms until they are well advanced. Manifestations may include headache, blurred vision or loss of vision, proptosis, exophthalmos, edema, numbness, hemorrhage, memory loss, speech difficulties, confusion, impaired movement and ataxia, seizures, lethargy, dysarthria, hyperactivity of reflexes, and paralysis of various body areas.

Diagnosis

For the angiosarcomas that are poorly differentiated and often solid, definitive diagnosis requires immunoreactivity for vascular markers such as CD31, CD34, ERG, and *friend leukemia integration transcription factor 1 (FLI1)*. These are heterogeneously enhancing during imaging (see Fig. 14.8). It may be difficult to diagnose these angiosarcomas from metastatic carcinoma because of occasional cytokeratin reactivity. Unfortunately, angiosarcomas are often misdiagnosed, resulting in a worsened prognosis and high mortality rate. Differential diagnostic considerations include meningiomas, granulomas, hemangioblastomas, hemangiomas, Kaposi sarcoma, and cancer metastases from other body sites. Laboratory studies of blood samples are usually not significantly diagnostic, though sudden thrombocytopenia may suggest rapid tumor growth or development of metastasis. All angiosarcomas have similar microscopic findings, with vascular spaces lined with atypical tumor cells. Lower grade lesions have vascular spaces lined with large, fat endothelial cells penetrating the stroma and papillary fronds of cells projecting into the lumens. Higher grade lesions are more cellular, with atypia and abnormal mitoses.

Immunohistochemical staining helps to confirm diagnosis by revealing tumor cell structures, with intercellular or intracellular lumina that may or may not show red cells. The cytoplasm contains intermediate filaments of vimentin and keratin, plus pinocytotic vesicles. Sometimes, **Weibel–Palade bodies**, a marker of endothelial differentiation, are seen. Most lesions express vimentin and **factor VIII-related antigen**. In 74% of cases, CD34 is expressed, followed by the endothelial cell marker *BNH9* (72%) and cytokeratins (35%). Some tumors show actin expression.

Treatment

Surgical excision is performed whenever possible and to the greatest extent permitted by the tumor location. Chemotherapeutic agents include dacarbazine, docetaxel, doxorubicin, epirubicin, eribulin, gemcitabine, ifosfamide, mesna, olaratumab, pazopanib, temozolomide, trabectedin, and vinorelbine. Responses to preoperative chemotherapies only range between 40% and 50%, and there is significant toxicity. Therefore preoperative chemotherapy is reserved for only high-grade tumors. The regimen is continued when the patient responds with tumor shrinkage after two to three courses of multiagent chemotherapy following tumor resection. Use of radiation therapy along with surgery results in 80% of localized control, plus excellent outcomes. However, since 50% of angiosarcomas have distant metastases, radiation therapy does not improve survival. Intraoperative radiation, **brachytherapy**, or additional external beam therapy may complement preoperative external beam radiotherapy.

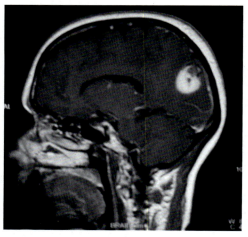

FIGURE 14.8 Angiosarcoma of the right occipital region. *Courtesy: https://doi.org/10.1016/j.jocn.2007.03.028 (https://www.sciencedirect.com/science/article/abs/pii/S0967586807003578) from* Journal of Clinical Neuroscience 15(8), 2008, 927–929.

Prognosis

The rate of death related to angiosarcomas is high. The reported 5-year survival rate is between 20% and 35%. More than 50% of patients develop metastatic disease that most often reaches the lungs. Older patient age and larger tumor size are unfavorable prognostic factors.

Chondrosarcoma

A **chondrosarcoma** is a rare, malignant mesenchymal tumor that has cartilaginous differentiation. The majority of these tumors are primary, but they also can develop in a preexisting, benign cartilaginous lesion. When these tumors affect the CNS, they usually develop at the base of the skull.

Epidemiology

Chondrosarcomas may develop at any age, but are most common during the fourth and fifth decades of life. There is a slight male predominance of 1.5:1 or 2:1.

Etiology and risk factors

The cause of chondrosarcomas is unknown, but patients may have a history of enchondromas or osteochondroma. A small amount of secondary chondrosarcomas occur along with **Maffucci syndrome** and **Ollier disease**. Chondrosarcomas are linked with abnormal isocitrate dehydrogenase 1 and 2 enzymes, which are also associated with gliomas and leukemias. Risk factors for chondrosarcomas include aging, Paget's disease, and Wilms' tumor.

Pathology

There are two microscopically defined major types of chondrosarcomas: *conventional* and *mesenchymal*. Intracranial, **extraosseous** chondrosarcomas of the conventional form are rare, as well as **extraskeletal** myxoid chondrosarcomas of the brain and its leptomeninges. Mesenchymal chondrosarcomas arise more frequently in the bones of the skull or spine, compared to within the dura or brain parenchyma. The histological pattern looks like that of a Ewing sarcoma-like small blue cell neoplasm or a solitary fibrous tumor/hemangiopericytoma. This is due to a distinct staghorn vascular network. Transitions between chondroid and mesenchymal components usually occur abruptly, but can also be more gradual. If gradual, it makes it hard to distinguish from a small cell osteosarcoma.

Clinical manifestations

In most cases, patients with chondrosarcoma present in overall good health, and the diagnosis often occurs accidentally during examination and testing for another condition. If symptoms present, they usually include pain, a palpable lump, local mass effect, or hyperglycemia because of a paraneoplastic syndrome.

Diagnosis

Imaging studies include X-rays, CT scan, and MRI. However, definitive diagnosis is based on biopsy. The most characteristic imaging findings are usually obtained with CT. For highly anaplastic tumors, immunohistochemistry is required. There are no current blood tests that allow for diagnosis. Chondrosarcomas that arise in the skull base—primarily in the midline—must be distinguished from chondroid chordomas. Unlike chordomas, chondrosarcomas do not react to EMA, brachyury, or keratin. For mesenchymal chondrosarcomas, some tumors mostly have a small cell component with occasional islands of atypical hyaline cartilage. In other cases, the chondroid element is present in higher amounts. If cartilage is not easily seen, *transcription factor SOX-9 immunostaining* may be diagnostically helpful, since it is positive in the undifferentiated mesenchymal component. If diagnosis is still difficult, it may be confirmed by *nuclear receptor coactivator 2 (NCOA2)* gene rearrangements.

Treatment

Chondrosarcomas are resistant to chemotherapy as well as radiation therapy. Treatment is based on the tumor location and its aggressiveness, and usually occurs at specialty hospitals. Surgery is the primary treatment—usually complete surgical ablation, but this may be difficult. Proton therapy radiation has begun to show strong effects against chondrosarcomas, with localized tumor control higher than 80%. It can be used in certain tumor locations to increase the benefits of surgery. In recent studies, induction of apoptosis for high-grade tumors has shown to be significantly helpful.

Prognosis

Prognosis for chondrosarcomas is based on early diagnosis and treatment. For the least aggressive tumors, about 90% of patients have 5-year survival. However, for the most aggressive grade, only 10% of patients survive for 1 year. Chondrosarcomas may recur, so follow-up scans are highly important to detect this, or any metastasis, which is most often to the lungs.

> **Point to remember**
> Chondrosarcoma is the most common form of primary bone cancer diagnosed in adulthood. When occurring in mesenchymal tissues, they are very aggressive tumors. Tumor location within the skull or spine may result in difficulty in surgery and other treatments.

Ewing sarcoma

Ewing sarcoma is also known as a *peripheral primitive neuroectodermal tumor*. It is a small, round, blue cell tumor of neuroectodermal origin that may involve the CNS as either a primary dural neoplasm, or via direct extension from contiguous soft tissue or various bones, including the paraspinal soft tissue, vertebra, or skull.

Epidemiology

Patients of many ages are affected by Ewing sarcoma, though it is most common during the second decade of life. The peak incidence is between 10 and 20 years of age. It represents 2% of all childhood cancers. In the United States, Ewing sarcoma affects about 1 million people per year, but this figure includes tumors that are not strictly confined to the CNS. Caucasians are affected 10–20 times more often than any other racial or ethnic group, and males are affected more often than females in a 1.6:1 ratio.

Etiology and risk factors

The cause of Ewing sarcoma is unknown, and most cases appear to occur insidiously. The underlying mechanism often involves a reciprocal genetic translocation. There are no reported risk factors.

Pathology

The histology, biology, and immunophenotype of Ewing sarcoma are very similar to bone and soft tissue tumors. There may be mass effect and destruction of cranial bones. Ewing sarcoma consists of sheets of small and round to oval, primitive looking cells that have limited clear cytoplasm, vesicular nuclei, and indistinct nucleoli (see Fig. 14.9). The nuclei contain fine chromatin and have smooth contours. The tumor cells often form lobules. Homer Wright rosettes may be present, but are mostly not prominent. Most cases of Ewing sarcoma (85%) result from a translocation between chromosomes 11 and 22. Cytokeratin is only seen focally in up to 20% of cases. For most patients, CD99 has strong, diffuse

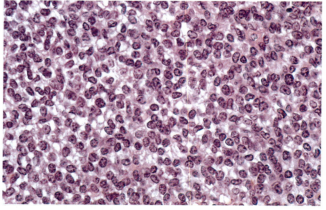

FIGURE 14.9 Ewing sarcoma composed of sheets of small round cells with minimal amounts of clear cytoplasm.

membranous immunoreactivity, and can also be expressed, but usually in a patchy, cytoplasmic fashion. There have been various gene fusions, including *protein capicua homolog (CIC)-double homeobox 4 (DUX4)* and the *BCL-6 corepressor protein (BCOR)-cyclin B3 (CCNB3)* **intrachromosomal inversion**.

Clinical manifestations

Symptoms of Ewing sarcoma in the CNS may include chronic headache, inflammation, anemia, leukocytosis, increased sedimentation rate, nausea, vomiting, loss of consciousness, papilledema, and fever. Ewing sarcoma can cause swelling of the scalp with increasing tumor size, which may be tender yet firm, is often immobile, lacks any pulsatile qualities, and of variable definition. Paraplegia may be a complication of Ewing sarcoma.

Diagnosis

On radiologic imaging (including X-rays, MRI, and CT), Ewing sarcoma may resemble a well-defined meningioma. Postcontrast T1-weighted MRI easily reveals Ewing sarcoma, with the tumors being hypointense. They are heterointense in T2-weighted imaging. Most Ewing sarcomas are heterogeneously hyperdense enhancing, and stain focally, or more significantly, with synaptophysin and neuron-specific enolase. Most Ewing sarcomas must be confirmed by *reverse transcription polymerase chain reaction (RT-PCR)* for a *Ewing sarcoma breakpoint region EWS-1 (EWSR1)*-FLI1 or EWSR1-ERG fusion transcript, or via *fluorescence in situ hybridization (FISH)* for EWSR1 gene rearrangement. In about 25% of cases, Ewing sarcoma has already spread to other parts of the body upon diagnosis. Confirmed diagnosis is based on tumor biopsy.

Treatment

Treatment of Ewing sarcoma often includes surgery, chemotherapy, radiation therapy, and stem cell transplantation. Targeted therapy and immunotherapy are being studied. Multidrug regimens include ifosfamide, etoposide, vincristine, doxorubicin, and cyclophosphamide. In many cases, after about 3 months of chemotherapy, any remaining tumor is surgically resected, irradiated, or both. In many cases, Ewing sarcoma is eradicated by radiation therapy quite easily, but the tumor often recurs dramatically.

Prognosis

The 5-year survival rate for Ewing sarcoma is about 70%, but this is affected by many different factors, such as effective use of chemotherapy. Before the availability of multidrug chemotherapy, long-term survival was less than 10%.

Fibrosarcoma

A **fibrosarcoma** is a rare, malignant sarcoma, with monomorphic spindle cells arranged in a **herringbone** pattern of intersecting fascicles. These tumors have high cellularity, brisk mitotic activity, and often, necrosis. Another name for fibrosarcoma is *fibroblastic sarcoma*.

Epidemiology

Fibrosarcomas are most often seen in males between the ages of 30 and 40, which are slightly younger than for other sarcomas. However, they are also seen in younger adults and adolescents. *Congenital infantile fibrosarcoma* may occur in infants, usually before the age of 2 years. Overall, CNS fibrosarcomas are rare compared to other types of sarcomas. They make up about 10% of all types of sarcomas. However, primary intracranial sarcomas make up only 0.5%—2.7% of all intracranial neoplasms.

Etiology and risk factors

The cause of fibrosarcomas is believed to be linked, in most cases, to a translocation between chromosomes 12 and 15, resulting in formation of the fusion gene *ETV6-NTRK3*. Also, individual cases show trisomy for chromosomes 8, 11, 17, or 20. There are no proven risk factors for fibrosarcomas.

Pathology

Fibrosarcomas are derived from fibrous connective tissues. There are immature proliferating fibroblasts or undifferentiated anaplastic spindle cells in a storiform pattern. The histological features of fibrosarcomas are similar, regardless of patient age. The features are also similar to those of hemangiopericytomas or dermatofibrosarcoma protuberans, which both can transform in a fibrosarcomatous fashion. Fibrosarcomas may present with different degrees of differentiation (low-grade, intermediate, or high-grade). They are often well demarcated, with high mitotic activity. The presence of immature blood vessels raises the likelihood of bloodstream metastases. Hemorrhage is common. Within the CNS, fibrosarcomas are commonly seen in the meninges and cerebral hemispheres (often frontal in location), but also at the skull base, external cranium, and infrequently, within the orbital or mastoid bones. Sclerosing epithelioid fibrosarcoma of the CNS has also been documented.

Clinical manifestations
Fibrosarcomas may cause headaches, vision problems, dizziness, edema, hemorrhage, local mass effect, seizures, papilledema, paralysis, and speech and language difficulties.

Diagnosis
Diagnosis of fibrosarcomas primarily utilizes CT and MRI. There are five diagnostic subtypes of fibrosarcomas. These include **intraparenchymatous** *inflammatory, meningeal inflammatory, mixed intraparenchymatous and meningeal, intraventricular inflammatory*, and tumors that extend into the cranial cavity or sphenoid sinus (also inflammatory in nature). The tumors often strongly enhance are intense on T2-imaging and have strong vimentin immunoreactivity. Differential diagnoses include spindle cell-squamous cell carcinoma, synovial sarcoma, leiomyosarcoma, malignant peripheral nerve sheath tumor, and **biphenotypic sinonasal sarcoma**. Ancillary testing for fibrosarcoma includes immunohistochemistry where vimentin is positive, cytokeratin and S-100 are negative, and actin is variable.

Treatment
The mainstay of treatment for fibrosarcomas is surgical excision or craniotomy, which can be curative. There is the potential for significant bleeding during surgery. If the tumor is inaccessible to surgery, radiation therapy and chemotherapy may be indicated. Gamma-knife radiosurgery is commonly performed for any residual tumor tissues.

Prognosis
Prognosis of fibrosarcomas may be better than for other sarcomas since they are often less aggressive than sarcomas of comparable size. Overall, prognosis is good for complete tumor resection, but the tumor is known to recur. However, surgical complications sometimes include hemiparesis.

> **Point to remember**
> Fibrosarcoma is a rare tumor that affects fibroblast cells that create the body's fibrous tissues. These tumors often take a long time to cause symptoms and are often difficult to diagnose correctly because of the presence of similar features to those of osteosarcomas. Key features, however, signal fibrosarcomas. These include the lack of calcium buildups, revealed on X-rays.

Kaposi sarcoma

A **Kaposi sarcoma** is a malignant tumor with spindle-shaped cells that form very thin blood vessels described as "slit-like." It only rarely occurs as a meningeal or parenchymal tumor, usually accompanying HIV type 1 infection or AIDS. When this occurs, it is difficult to determine if the tumor is primary or metastatic.

Epidemiology
After peaking at about 47 cases per 1,000,000 population (in the United States alone) in the early 1990s, Kaposi sarcoma cases greatly declined because of better treatment for HIV/AIDS. Today, while over 35% of patients with AIDS develop Kaposi sarcoma, there are now only about 6 cases per 1,000,000 per year in this country. However, this number takes into account all forms of Kaposi sarcoma, not just its CNS manifestations. The CNS is a very rare site of Kaposi sarcoma development, and there are no accurate figures as to the amount of cases today throughout the world.

Etiology and risk factors

The cause of Kaposi sarcoma is not fully proven but its viral link is understood because of encoding of oncogenes, microRNAs, and circular RNAs—all of which promote cancer cell proliferation and escape of these cells from the immune system. The tumors may arise as a cancer of the lymphatic endothelium. Risk factors for Kaposi sarcoma include poor immune system function and chronic lymphedema.

Pathology

Kaposi sarcoma tumors are nearly always immunopositive for *human gammaherpesvirus 8 (HHV8)*. The tumor forms masses that are usually purple in color, occurring singularly or widespread. The tumors may become worse very slowly or very quickly. The actual histogenesis of Kaposi sarcoma is controversial. Tumor cells have a characteristic, abnormally elongated shape, known as spindle cells (see Fig. 14.10). Mitotic activity is moderate and there is usually no pleomorphism. The tumors are extremely vascular, with highly dense and irregular blood vessels that leak red blood cells. Inflammation surrounding the tumors may cause pain. Hyaline bodies that are PAS-positive may be present in the cytoplasm or extracellularly.

Clinical manifestations

Like other CNS sarcomas, symptoms of Kaposi sarcoma include headaches, vision and hearing disturbances, mental abnormalities, emotional problems, nausea, vomiting, dizziness, impaired consciousness, seizures, and speech problems.

Diagnosis

There are four diagnostic subtypes of Kaposi sarcoma: *classic, endemic, immunosuppression therapy-related*, and *epidemic*. Diagnosis is confirmed by tumor biopsy, though imaging helps determine the extent of the tumor(s). Discovery of the *latency-associated nuclear antigen (LANA)* protein in tumor cells is confirmative. Differential diagnoses include arteriovenous malformations and pyogenic granulomas.

Treatment

The treatment of Kaposi sarcoma is based on the individual subtype, if the tumors are local or widespread, and the patient's immune function. Kaposi sarcoma is not curable. When the patient has immune system dysfunction, treatments can slow or stop

FIGURE 14.10 Histologic view of the nodular stage of Kaposi sarcoma, demonstrating sheets of plump, proliferating spindle cells, and slit-like vascular spaces. *Courtesy Christopher D.M. Fletcher, MD, Brigham and Women's Hospital, Boston, Massachusetts.*

its progression. For AIDS patients, the cornerstone of therapy is **highly active antiretroviral therapy (HAART)**. Medications used for systemic therapy include interferon alpha, liposomal doxorubicin or daunorubicin, paclitaxel, and thalidomide.

Prognosis

The prognosis of Kaposi sarcoma varies. With widespread disease, death is likely. According to the American Cancer Society, the 5-year relative survival for Kaposi sarcoma varies based on its stage, as follows:

- localized—81%;
- regional—59%;
- distant—45%; and
- combined stages—74%.

Leiomyosarcoma

A **leiomyosarcoma** is a rare malignant tumor that mostly features smooth muscle differentiation. There are a few different morphologies. These include *epithelioid leiomyosarcoma, granular cell leiomyosarcoma, inflammatory leiomyosarcoma*, and *myxoid leiomyosarcoma*. Leiomyosarcomas may appear anywhere in the body. Any benign tumor originating from the same tissue is called a *leiomyoma*.

Epidemiology

Approximately 1 of every 100,000 people, globally, is diagnosed with a leiomyosarcoma making this tumor slightly more common than other soft tissue sarcomas. Leiomyosarcomas make up 10%—20% of new cases.

Etiology and risk factors

The exact cause of leiomyosarcomas is unknown, but there is a link between the Epstein—Barr virus (EBV) and immunosuppression. Risk factors for leiomyosarcomas include neurofibromatosis, **Gardner's syndrome**, **Li—Fraumeni syndrome**, prior radiation therapy, chronic lymphedema, exposure to herbicides, arsenic, vinyl chloride, and dioxin, as well as damage or removal of lymph nodes as part of a previous cancer treatment.

Pathology

Intracranial leiomyosarcomas are similar to other soft tissue tumors in that their intersecting fascicles are situated at 90-degree angles and their cytoplasm is eosinophilic. However, they also have significant nuclear pleomorphism, with increased mitotic activity and necrosis. Most leiomyosarcomas develop within or near the dura, such as in the epidural space or paraspinal region. They only rarely occur in the parenchyma. When a leiomyosarcoma develops in relation to EBV or immunosuppression, the tumors are less obviously malignant, and may respond to reconstitution of the immune system. The majority of leiomyosarcomas express desmin and SMA. The course of these tumors may be extremely unpredictable. They can stay dormant for long periods of time and then recur. Most leiomyosarcomas are hemorrhagic and soft, with 15—30 mitotic figures per 10 high-power fields.

Clinical manifestations

The signs and symptoms of leiomyosarcomas vary, based on tumor location, size, and metastasis. In the early states, there may be no obvious symptoms. Leiomyosarcomas may cause fatigue, fever, weight loss, malaise, nausea, vomiting, pain, and inflammation. The majority of leiomyosarcomas are aggressive, likely to spread to the lungs or liver, and then become life threatening.

Diagnosis

The diagnosis of a leiomyosarcoma is completed via a soft tissue biopsy and histopathological examination. If not diagnosed early, leiomyosarcoma has a high risk of recurring after treatment. In imaging studies, leiomyosarcomas often have a similar appearance to fibroid tumors. Imaging techniques include CT, MRI, ultrasound, and X-rays. Differential diagnoses include spindle cell carcinoma, fibrosarcoma, and malignant peripheral nerve sheath tumor.

Treatment

Leiomyosarcomas are not extremely responsive to chemotherapy or radiation therapy. Therefore the best treatment option is surgical resection, with as wide margins as possible, while the tumor is small and still in situ. If the possible surgery must leave some tumor cells behind, chemotherapy or radiation therapy has been proven to provide a clear survival benefit. If the disease is metastatic, chemotherapy and targeted therapies are the first treatment choices. Chemotherapeutic agents include combinations of docetaxel, doxorubicin, gemcitabine, ifosfamide, pazopanib, and trabectedin. According to the website called *Sarcoma: Information for Leiomyosarcoma Families*, over 50% of patients treated for this type of tumor require additional treatments with 8–16 months after diagnosis.

Prognosis

The prognosis for leiomyosarcoma is based on tumor location, size, type, and metastasis. Some patients with low-grade tumors or with tumors that have not spread beyond stage I have an excellent prognosis. Generally, high-grade tumors that have widely spread have less favorable survival rates. The 5-year survival rates widely range between 20% and 90%, compared to an overall relative 5-year survival rate of other soft tissue sarcomas that is approximately 50%.

> **Point to remember**
>
> Leiomyosarcomas are tumors of smooth muscle cells. They form in one location, but can spread widely. Sometimes, leiomyosarcomas cause no problems for years, and then suddenly "erupt." The tumors are extremely resistant to chemotherapy and radiation, so early, radical surgical resection is indicated in order to achieve the best outcomes for patients.

Liposarcoma

A **liposarcoma** is a rare malignant tumor partly or entirely made up of neoplastic adipocytes. It only occurs rarely within the cranium. There has been one documented case of a liposarcoma occurring along with a subdural hematoma. There is also one documented case of a gliosarcoma that had a liposarcomatous element.

Epidemiology

Liposarcomas are most common in adults age 40 and above. They are the second most common of all soft tissue sarcomas, following *malignant fibrous histiocytomas*. Annually in the United States, there are about 2.5 cases per 1,000,000 people. There are no accurate global figures for liposarcomas.

Etiology and risk factors

The actual cause of liposarcomas is not known. There are also no identified risk factors for these tumors.

Pathology

Liposarcoma microscopically resembles normal adipocytes. The tumor is usually large and bulky, with many smaller satellites extending beyond its margins. There are several pathological subtypes of liposarcomas. Well-differentiated liposarcomas are also called *atypical lipomatous tumors* (see Fig. 14.11A). Myxoid/round cell liposarcomas have an abundant ground substance and rich capillary network (see Fig. 14.11B). Dedifferentiated liposarcomas are well differentiated, high-grade tumors located near more poorly differentiated tumors. Pleomorphic liposarcoma is a high-grade tumor that has lipoblasts admixed with mostly high-grade, pleomorphic-appearing spindle cells.

Clinical manifestations

Liposarcomas may cause a deep-seated mass, and only cause pain or functional disturbances once they become large. In the rare case of an intracranial liposarcoma, manifestations may include nausea, vomiting, headache, photophobia, diplopia, strange sensations, and impaired movements of the upper body areas.

Diagnosis

Diagnosis of liposarcoma is established by biopsy or surgical excision. **Lipoblasts** are often present, with cells having abundant, clear, and **multivacuolated** cytoplasm, plus an eccentric, darkly staining nucleus that is indented by the vacuoles. Ultrasound may be insufficient to distinguish liposarcomas from benign lipomas, so MRI is usually the initial imaging method.

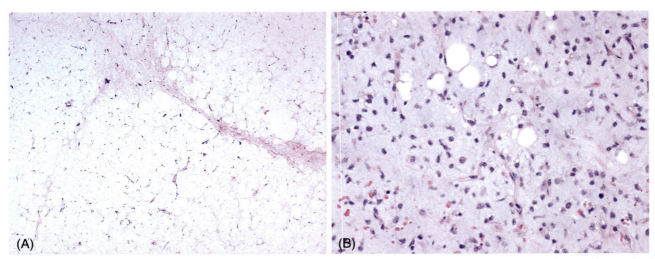

FIGURE 14.11 Liposarcoma. (A) The well-differentiated subtype consists of mature adipocytes and scattered spindle cells with hyperchromatic nuclei. (B) Myxoid liposarcoma with abundant ground substance and a rich capillary network in which there are scattered immature adipocytes and more primitive round-to-stellate cells.

Treatment

The treatment of liposarcomas usually consists of surgical resection of the tumor and its margins. Sometimes, radiation therapy and chemotherapy are indicated. Well-differentiated liposarcomas that are treated with surgery, intraoperative distilled water lavage, and radiation therapy have a low recurrence rate of about 10%, and rarely metastasize.

Prognosis

The prognosis of liposarcomas is varied, based on site of origin, subtype, tumor size, and depth. The 5-year survival rates vary between 56% and 100%, based on the tumor subtype.

Osteosarcoma

An **osteosarcoma** is an aggressively malignant, bone-forming tumor. It forms in the skull in only 8% of all cases, followed less often by formation in the spine, meninges, and brain. Another term used for this tumor is *osteogenic sarcoma*. Also, osteosarcoma is the most common histological form of primary bone cancer.

Epidemiology

Osteosarcoma mostly affects adolescents and younger adults (ages 10–30), with a second peak in incidence in the elderly. Only 800–900 cases are diagnosed in the United States every year, and global rates are not well documented. When an osteosarcoma develops in the skull, it is often situated within the jaw area. Osteosarcoma is the eighth most common form of childhood cancer, making up 2.4% of all pediatric malignancies, and about 20% of all primary bone cancers. In the United States, incidence rates for patients under age 20 are about 5 per 1,000,000 population annually. The tumors are slightly more common in African-Americans (6.8 per 1,000,000) and Hispanic-Americans (6.5 per 1,000,000) than in Caucasians (4.6 per 1,000,000). Also, there is a general male predominance (5.4 per 1,000,000) compared to females (4 per 1,000,000)—see Fig. 14.12.

Etiology and risk factors

The causes of osteosarcoma are varied. They include familial cases involving deletion of chromosome 13q14, which inactivates the retinoblastoma (RB) gene that is linked to a high risk of osteosarcoma development. A predisposing factor is Li–Fraumeni syndrome. Risk factors for osteosarcoma include **Paget's disease** of the bone, enchondromatosis, fibrous dysplasia, hereditary multiple **exostoses**, and **Rothmund–Thomson syndrome**. There may also be a link to radiation therapy for unrelated conditions.

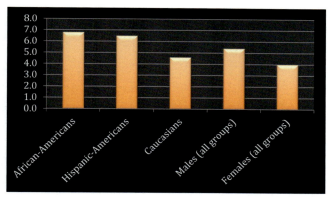

FIGURE 14.12 Cases of osteosarcoma in various populations (per 100,000).

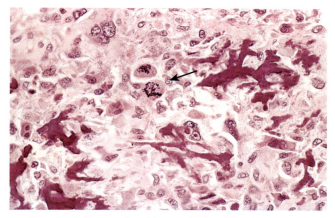

FIGURE 14.13 Fine, lace-like pattern of neoplastic bone produced by anaplastic malignant tumor cells in an osteosarcoma. Note the abnormal mitotic figure (*arrow*).

Pathology

Osteosarcomas tend to occur at sites of bone growth. In rare cases, there may be osteosarcomatous elements that are components of gliosarcomas or germ cell tumors. Osteosarcomas are solid, hard, and have a "moth-eaten" or lace-like appearance of irregular formation (see Fig. 14.13). This is due to tumor spicules of calcified bone that radiate in 90-degree angles, which form a **Codman's triangle**. There is infiltration of surrounding tissues. The characteristic feature is the presence of bone formation within the tumor. The tumor cells are highly pleomorphic, and some are giant, with numerous atypical mitoses. They produce irregular trabeculae with or without central calcification, and tumor cells are included in the osteoid matrix. Osteosarcomas may metastasize to the lungs.

Clinical manifestations

Osteosarcomas may cause pain that worsens at night, which is intermittent and of varying intensity. A large tumor can cause significant localized inflammation, though this is not often externally apparent.

Diagnosis

For diagnosis of osteosarcoma, there must be bone matrix or osteoid deposition by proliferating tumor cells. Diagnosis is based on X-rays, CT, PET, bone scans, MRI, and biopsy. Various MRI images of an osteosarcoma at the base of the skull are shown in Fig. 14.14. The Codman's triangle is highly suggestive of an osteosarcoma. The tumors are sometimes found incidentally when imaging is performed for another reason. The diagnostic variants of osteosarcomas may be described as *conventional, extraskeletal, high-grade surface, low-grade central, parosteal, periosteal, secondary, small cell*, and *telangiectatic*.

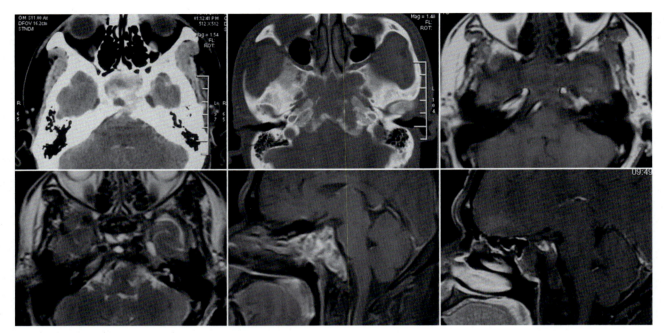

FIGURE 14.14   MRI images of an osteosarcoma at the base of the skull.

Treatment

For osteosarcoma, a complete radical surgical **en bloc resection** is the treatment of choice. Complications may include infection and recurrence. The drug called *mifamurtide* may be used after surgery, along with chemotherapy, to kill any remaining tumor cells. Often, chemotherapy is started before surgery and then continued as needed. Chemotherapeutic agents include methotrexate, leucovorin, adriamycin, bleomycin-cyclophosphamide-dactinomycin, cisplatin, etoposide, ifosfamide, mesna, and muramyl tripeptide.

Prognosis

Prognosis for osteosarcoma is based on the percentage of tumor cell necrosis seen after surgery. In pediatric patients, osteosarcoma has a low survival rate. The highest reported 10-year survival rate is 92% of patients, but overall, 65%–70% of patients will survive for 5 years, based on tumor location, size, and metastasis. To explain this further, staging must be considered as follows:

- Stage I osteosarcoma—5-year survival of more than 90%;
- Stage II osteosarcoma—5-year survival of approximately 40%; and
- Stage III osteosarcoma—5-year survival of approximately 30%.

Fortunately, mortality rates from osteosarcoma are declining by about 1.3% per year.

> **Point to remember**
>
> According to the American Cancer Society, osteosarcomas are most common between the ages of 10 and 30, with teenagers affected more often than any age group. Overall, however, only 800–900 cases are diagnosed in the United States every year. Factors that may be related to the development of osteosarcomas may include growth cycles, height, male gender, African-American race, previous radiation therapy, noncancerous bone diseases, inherited cancer syndromes, and rare genetic conditions.

Rhabdomyosarcoma

A **rhabdomyosarcoma** is an aggressive and highly malignant tumor that mostly has skeletal muscle differentiation. Almost all primary CNS rhabdomyosarcomas are of the embryonal type, regardless of whether they occur in the meninges or parenchyma. Much rarer examples include alveolar rhabdomyosarcoma and a rhabdomyosarcomatous element of a gliosarcoma. Rhabdomyosarcomas can occur anywhere on the body, but are most common in the head and neck, followed by the eye orbits and other non-CNS sites.

Epidemiology

A large majority of rhabdomyosarcomas occur in patients under age 18. Though relatively rare, these tumors make up about 40% of all documented soft tissue sarcomas. They are the most common soft tissue sarcomas in children, and the third most common solid tumors in children as well. About 4.5 cases per 1,000,000 children or adolescents occur annually—which means about 250 new cases in the United States every year. About 66% of cases occur in children under age 10. Males are affected 1.3–1.5 times as often as females. Also, rhabdomyosarcomas occur just slightly more often in Caucasian children compared to other ethnic or racial groups.

Etiology and risk factors

Rhabdomyosarcoma has been associated with some congenital abnormalities such as neurofibromatosis type 1, **Beckwith–Wiedemann syndrome**, Li–Fraumeni syndrome, cardio-facio-cutaneous syndrome, and **Costello syndrome**. Rhabdomyosarcomas are also linked to parental use of marijuana and cocaine. Up to 90% of rhabdomyosarcomas have translocations of t(2;13)-(q35;q14) or less often, t(1;13)-(p36;q15). Both involve translocations of a DNA-binding domain of PAX3 or PAX7. There are no clear risk factors for these tumors, and they usually develop sporadically with no obvious cause.

Pathology

The majority of rhabdomyosarcomas consist of mostly undifferentiated small cells. Less often, **strap cells** with cross striations are seen. The most common subtype is called *embryonal rhabdomyosarcoma*, which makes up about 60% of cases (see Fig. 14.15). Embryonal rhabdomyosarcomas may be further subclassified as either **botryoid** or *spindle cell* forms. The botryoid form may develop in the mucosal linings of the skull, while the spindle cell form usually occurs outside the CNS. The two other subtypes are called *alveolar* and *anaplastic* rhabdomyosarcomas, but these also occur outside the CNS. Rhabdomyosarcomas may metastasize to the lungs, bone marrow, and bones.

Clinical manifestations

Signs and symptoms of rhabdomyosarcomas vary according to tumor location. When developing in the head, face, or neck, earlier signs of disease may appear.

Diagnosis

The diagnosis of rhabdomyosarcoma, usually after an open biopsy, is usually confirmed by immunostaining for desmin and **myogenin**. A hematoxylin & eosin (H & E) stain shows the tumor to appear, small, round, and of blue cell origin. Imaging methods include MRI, ultrasound, and bone scans. Sometimes, a lumbar puncture is performed to rule out

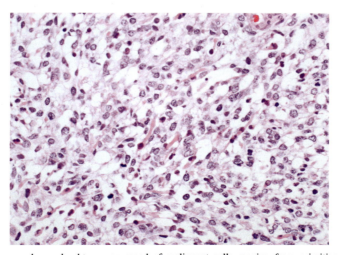

FIGURE 14.15 Rhabdomyosarcoma: embryonal subtype composed of malignant cells ranging from primitive and round to densely eosinophilic with skeletal muscle differentiation.

metastasis to the meninges. It is important to differentiate rhabdomyosarcomas from brain tumors that sometimes show skeletal muscle differentiation. These include germ cell tumors, gliosarcomas, malignant peripheral nerve sheath tumors, **medullomyoblastomas**, and meningiomas. There is also a variant called malignant **ectomesenchymoma**, which is a mixed brain tumor of ganglion cells or neuroblasts, plus one or more mesenchymal elements. A clear diagnosis is critical for effective treatment. However, this can be difficult because of the heterogeneity of the tumor and the lack of strong genetic markers.

Treatment

Treatment for rhabdomyosarcoma usually combines surgery with chemotherapy and radiation, which gives a 60%—70% cure rate if there is no metastasis. Immunotherapy is also sometimes used. Surgery is usually performed first, but less than 20% of the tumors are fully resected with negative margins. However, rhabdomyosarcomas are highly chemosensitive, with about 80% of cases responding well. Multiagent chemotherapy is indicated for all cases. There are two primary methods, known as the **VAC regimen** and the **IVA regimen**. The VAC regimen combines vincristine, actinomycin D, and cyclophosphamide. The IVA regimen combines ifosfamide, vincristine, and actinomycin D, with the drugs in either regimen administered in 9—15 cycles, as needed. Other drugs may include doxorubicin and cisplatin. Radiation therapy is used when resecting an entire tumor would involve disfigurement or loss of an important organ such as the eye. Usually, radiation therapy follows 6—12 weeks of chemotherapy, except when a parameningial tumor has invaded the brain, spinal cord, or skull—meaning that radiation must be started immediately. Immunotherapy for rhabdomyosarcoma is improving and still being actively studied.

Prognosis

The prognosis of rhabdomyosarcomas is closely linked to tumor location and size, patient age, possibility of complete resection, metastasis, and the biological and histopathological tumor cell characteristics. The 5-year survival rates range between 35% and 95% based on the tumor subtype. However, if the tumors have metastasized, less than 20% of patients will be able to be cured. Better prognoses exist for rhabdomyosarcomas of the head, neck, and eye orbits than for tumors in the cranial **parameningial** area.

Lipoma

A **lipoma** is a benign lesion with cells that resemble normal adipocytes. Intracranial lipomas are believed to be congenital malformations, and not actually neoplasms. They result from abnormal differentiation of the undifferentiated mesenchyme known as the *meninx primitive*. Most lipomas are less than 5 cm in size. *Corpus callosum lipoma* is a rare congenital tumor that may or may not cause symptoms. *Neural fibrolipomas* occur along nerve trunks and often lead to nerve compression.

Epidemiology

About 2% of the global population are affected by lipomas of various body sites. They most often occur in adults between 40 and 60 years of age, with males slightly more affected than females. Lipomas are the most common benign soft tissue tumor, and have only been seen in the CNS in a few cases.

Etiology and risk factors

The cause of lipomas is generally not known. Risk factors include family history, lack of exercise, and obesity. Familial multiple lipomatosis may be implicated. There is a link between the *HMG I-C gene* and lipomas.

Pathology

The majority of lipomas are lobulated, with a thin, delicate, and transparent capsule (see Fig. 14.16). Patchy fibrosis is common, and sometimes there is calcification and myxoid changes. A form called *osteolipoma* is extremely rare, sometimes showing zonation with central adipose tissue and peripheral bone. *Complex lipomatous lesions* have different histologies. *Lumbosacral lipomas*, also called **leptomyelolipomas**, occur as part of **tethered spinal cord syndrome** and are usually malformative. They may contain subcutaneous and intradural components that are tethered by a **fibrolipomatous stalk**, sometimes attached to the filum terminale or the dorsum of the spinal cord. In the cerebellopontine angle, which is an uncommon location, lipomas may involve intradural areas of the cranial nerve roots and ganglia.

FIGURE 14.16 A lipoma, surrounded by a thin, delicate, and transparent capsule (*arrow*) that is highly lobulated.

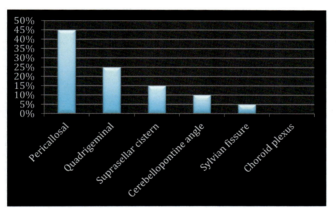

FIGURE 14.17 The most common locations of intracranial lipomas.

Another rare form is called *epidural lipomatosis*, with diffuse hypertrophy of the spinal epidural adipose tissue. This is not actually a neoplasm but a metabolic response, often to chronic use of steroids. Overall, the most common locations of intracranial lipomas are as follows (also see Fig. 14.17):

- *Pericallosal* (45% of cases)—associated with agenesis of the corpus callosum in about 50% of cases, and divided morphologically into **tubonodular** and **curvilinear** types;
- *Quadrigeminal cistern* (25% of cases)—associated with underdevelopment of the inferior colliculus;
- *Suprasellar cistern* (15% of cases);
- *Cerebellopontine angle* (10% of cases)—with the facial nerve and vestibulocochlear nerve often involved;
- *Sylvian fissure* (5% of cases); and
- *Choroid plexus* (less than 1% of cases).

Clinical manifestations

A lipoma in the brain often causes no initial symptoms, and may not be noticed until a scan is performed for another medical reason. However, an intracranial lipoma can cause symptoms such as seizures and hydrocephalus along with its accompanying features, including headache.

Diagnosis

Diagnosis of lipomas is based on physical examination, medical imaging, and tissue biopsy. The imaging method of choice is MRI since it offers better sensitivity to distinguish lipomas from liposarcomas. The signal characteristics are extremely close to those of fat tissues, with high T1 and T2 signal intensities, low-signal fat-saturated sequences, and sometimes, **blooming** that is produced from susceptibility artifacts. The lipomas are often traversed by the cranial nerves and nearby blood vessels, which are easily seen in high-resolution imaging. The presence of multiple lipomas points to a diagnosis of lipomatosis, which is more often seen in males. Liposarcomas are diagnosed in only 1% of all lipomas. On CT scans, lipomas usually appear as nonenhancing masses with uniform fat density, a lobulated soft appearance, and some amount of peripheral calcification. Differential diagnoses include intracranial *dermoids* or *teratomas*; lipomatous transformation of an *ependymoma, glioma,* or *primitive neuroectodermal tumor*; a thrombosed berry aneurysm; or a **white epidermoid**.

Treatment

Treatment of lipomas may be with surgical removal, though sometimes only observation is done when there are no symptoms. After surgical removal, recurrence is rare, and most lipomas are not generally linked to a future risk of malignancy. However, surgical removal of intracranial lipomas has been of little benefit in some cases, and carries with it a fairly high level of possible morbidity. Instead, treatment of seizures or symptoms of hydrocephalus is suggested.

Prognosis

Lipomas are rarely life threatening, and malignant transformation is very rare. Deeper lipomas are more likely to recur than superficial lipomas since complete surgical removal is not always possible.

> **Point to remember**
> Intracranial lipomas are congenital lesions that may be found incidentally during imaging for other conditions. They are usually asymptomatic, but treatment may be indicated if the patient experiences seizures or hydrocephalus. The characteristic finding during imaging is a mass that appears very similar to normal body fat.

Angiolipoma

An **angiolipoma** is a type of lipoma that has prominent vascularization. Its blood vessels are capillaries that are most easily seen in the peripheral regions and often contain dispersed thrombi made of fibrin. Though benign, the primary difference between angiolipomas and lipomas is that angiolipomas are often painful—but this feature is predominantly in superficial angiolipomas. Though most common in the skin, they can occur in the CNS, but this is extremely rare. Spinal angiolipomas occur in the thoracic epidural space of the spine.

Epidemiology

Between 5% and 17% of all lipomas are angiolipomas. For spinal angiolipomas, only about 200 cases have ever been documented. These tumors are most common in young adults, between 20 and 30 years of age, and rarely occur in older adults or children. They are slightly more common in males than in females.

Etiology and risk factors

The exact cause of angiolipomas is not known, but they have occurred in families. There is a link between multiple angiolipomas and the inherited condition called familial multiple angiolipomatosis, but these tumors usually occur on the arms and trunk, not in the CNS. Risk factors for angiolipomas include use of indinavir sulfate and chronic use of corticosteroids.

Pathology

There are varying amounts of adipose cells and vasculature that make up angiolipomas. Over time, interstitial fibrosis may develop. Angiolipomas are usually very small, between 1 and 4 cm in size, and usually occur as multiple tumors.

Clinical manifestations

Intracranial angiolipomas can cause seizures, severe headache, proptosis, reduced eye movements, dilation and nonreactivity of the pupils, vision impairment, visual hallucinations, balance problems, and memory impairment. Spinal angiolipomas can cause weakness or tingling sensations in various parts of the body, and also problems with balance.

Diagnosis

Angiolipomas are believed to be overdiagnosed because hemangiomas are often accompanied by fat cells, but they have a more heterogeneous variety of vascular channels that include cavernous spaces with thick-walled channels that resemble veins. Diagnosis of CNS angiolipomas involves MRI, CT, and biopsy.

Treatment

The only recognized treatment for CNS angiolipomas is surgical removal if possible. Infiltrative angiolipomas are more difficult to remove, and multiple tumors make surgery more complicated. Fortunately, surgical complications are rare.

Prognosis

Angiolipomas do not spread and are not life threatening. There is very little risk of recurrence, though new and unrelated tumors can appear.

Desmoid-type fibromatosis

Desmoid-type fibromatosis is a rare condition that involves locally infiltrative benign lesions that are made up of uniform, myofibroblastic-type cells. There is a significant amount of collagenous stroma that is arranged in intersected fascicles. Other terms for desmoid-type fibromatosis include *desmoid tumor* and *aggressive fibromatosis*.

Epidemiology

Desmoid-type fibromatosis usually affects young adults, with a higher prevalence for females. However, in children and older adults, the male-to-female ratio is basically equal. The condition makes up less than 3% of all soft tissue tumors. Between 10% and 30% of familial adenomatous polyposis (FAP) patients have desmoid-type fibromatosis, and between this type of fibromatosis makes up 7.5%−16% of all cases of fibromatosis. The mean age at diagnosis is between 36 and 42 years. For unknown reasons, males develop FAP, without desmoid-type fibromatosis, three times more often than females. Desmoid-type fibromatosis is rare, affecting 1−2 of every 500,000 people globally. In the United States, between 900 and 1500 new cases are diagnosed annually.

Etiology and risk factors

Though the causes of desmoid-type fibromatosis are not clearly identified, some patients also have a condition called FAP and Gardner's syndrome. Risk factors include the female gender, a *3′ adenomatous polyposis coli (APC)* mutation, and positive family history.

Pathology

Desmoid-type fibromatosis develops from fibroblasts, and can occur anywhere in the body. The condition can be slow growing, and also can resolve on its own without treatment. The histology of this condition involves long fascicles, with thinly walled vessels and microhemorrhages. The cells are bland, with mild-to-moderate cellularity and minimal atypia. The immunohistochemistry is positive for SMA and nuclear beta-catenin. Molecularly, *catenin beta-1 (CTNNB1)* and APC gene mutations may be seen.

Clinical manifestations

When symptoms of desmoid-type fibromatosis appear, there is a risk of functional impairment. The most common symptom is pain caused by tumor growth. Other symptoms include facial nerve paralysis and dysfunction of breathing or swallowing based on tumor location.

Diagnosis

It is important to diagnostically differentiate desmoid-type fibromatosis from *cranial fasciitis of childhood*, which is related to **nodular fasciitis**—a rapidly growing condition in the deep scalp that lacks an intradural component and is not malignant. Another differential diagnosis is *cranial infantile myofibromatosis*, often with a biphasic pattern and alternating areas of hemangiopericytoma-like and myoid components. For a patient with FAP, diagnosis may occur after tests and scans to investigate progression of desmoid-type fibromatosis. Diagnostic imaging techniques include X-rays, ultrasound, CT scan, MRI, and biopsy.

Treatment

Though the mainstay of treatment is surgical resection, there is significant morbidity associated. Radiation therapy alone and surgery plus adjuvant radiation therapy have shown improved local control over surgery alone. However, the long-term effects of radiation therapy prevent standardized use—especially in younger patients. For advanced disease, systemic therapy includes tamoxifen, nonsteroidal antiinflammatory drugs, doxorubicin, dacarbazine, pegylated liposomal doxorubicin, and the tyrosine kinase inhibitors called sorafenib and sunitinib.

Prognosis

Desmoid-type fibromatosis recurs locally in 20%–30% of patients. Prognosis is varied widely based on tumor location, and the lack of sufficient studies does not provide a large amount of prognostic statistics.

Inflammatory myofibroblastic tumor

An **inflammatory myofibroblastic tumor** is extremely rare, and has a distinct neoplastic proliferation. It is generally made up of bland myofibroblastic-type cells that are closely linked with a variable lymphoplasmacytic infiltrate. The cells are arranged in loose fascicles in an edematous stroma. In the CNS, inflammatory myofibroblastic tumor is rare and benign, regardless of its development in the brain or spinal cord. However, it sometimes has aggressive behavior, while remaining a low-grade tumor. Other names for this tumor include *inflammatory fibrosarcoma of the CNS, inflammatory pseudotumor of the CNS*, and *plasma cell granuloma of the CNS*.

Epidemiology

A CNS inflammatory myofibroblastic tumor can occur in patients of all ages. Generally, the predominant age range is 10–60 years, with average age at diagnosis being middle age. However, children as young as 4 years have been affected. Only about 100 cases of these tumors have ever been recorded, with no gender, racial, or ethnic preferences.

Etiology and risk factors

The cause of inflammatory myofibroblastic tumors is generally unknown, and there are no established risk factors.

Pathology

About 50% of inflammatory myofibroblastic tumors have *anaplastic lymphoma kinase (ALK)* gene rearrangement, and also overexpress ALK. Their pathogenesis may involve gene fusions that involve other kinases as well, including *platelet-derived growth factor receptor beta (PDGFRB)*, the proto-oncogene called *RET*, and the proto-oncogene called *ROS1*. In most cases, these tumors develop in the meninges and cerebral hemispheres, but may also occur at the skull base, external cranium, and within the orbital or mastoid bones.

Clinical manifestations

Signs and symptoms of inflammatory myofibroblastic tumors depend on tumor location, which can occur anywhere in the body. Manifestations may include headache, vision problems, dizziness, and speech and language problems.

Diagnosis

In imaging studies, inflammatory myofibroblastic tumors often resemble meningiomas. In the CNS, these tumors may have fasciitis-like, fibromatosis-like, myxoid-nodular, and scar-like features. Differential diagnoses include *hypertrophic intracranial pachymeningitis*, which is a **pseudotumoral** lesion with progressive thickening of the dura because of **pachymeningeal fibrosis** and chronic inflammation. The lesions are often related to autoimmune disorders and some have been reclassified as *IgG4-related disease*. Diagnosis is confirmed by biopsy since medical imaging findings are generally nonspecific.

Treatment

The primary treatment modality is surgical resection, which may be curative. However, when the tumor is inaccessible to surgery, radiation therapy and chemotherapy may be required.

Prognosis

The prognosis, following gross total resection of inflammatory myofibroblastic tumors, is usually favorable. However, these tumors can recur, even 10 years after surgery.

> **Point to remember**
>
> Inflammatory myofibroblastic tumors are made up of myofibroblastic spindle cells. Though usually benign, the tumors can recur, become locally invasive, or metastasize. Fortunately, malignant tumors of this type are rare. Because they contain a lot of immune cells, inflammatory myofibroblastic tumors appear to be inflamed, as if from an infection.

Myofibroblastoma

A **myofibroblastoma** is a rare benign mesenchymal neoplasm made up of spindle-shaped cells. The cells resemble myofibroblasts and are embedded within a stroma containing coarse bands of hyalinized collage, plus conspicuous mast cells, along with varying amounts of adipose tissue. These tumors are sometimes referred to as *benign stromal spindle cell tumors with predominant myofibroblastic differentiation*.

Epidemiology

Myofibroblastoma occurs with equal frequency in males and females, mostly in elderly individuals. However, less than 10 cases of intracranial myofibroblastomas have ever been documented, and some have occurred in middle-age adults. These tumors were mostly in the meninges of the frontal lobe.

Etiology and risk factors

Though of uncertain cause, myofibroblastomas are believed to derive from CD34-vimentin-containing fibroblasts of stroma capable of multidirectional differentiation. They are believed to be related to spindle cell lipomas and solitary fibrous tumors. Risk factors for myofibroblastomas are unknown, but previous radiation therapy may be a contributing factor.

Pathology

In the CNS, myofibroblastomas are similar to those occurring in the breasts. They are part of a spectrum of CD34-positive lesions, and grow slowly, occurring as solitary nodules or multilobar masses. They resemble cellular angiofibromas, fibroadenomas, and spindle cell lipomas. There are common 13q14 losses or losses of *RB* immunoexpression. The tumors are strongly positive for desmin, and usually range between 1 and 4 cm in size, though in one case, the tumor reached 15 cm. They are usually grayish pink to tan in color with a resilient structure and abundant blood supply. There is often significant adhesion to surrounding brain tissues, with edema. The tumor cells are often strongly positive for SMA, slightly positive for epithelial membrane antigen, and usually of a low Ki-67 proliferation index. They have abundant eosinophilic cytoplasm, and a round to oval nucleus with 1 or 2 small nucleoli. There may be mild nuclear pleomorphism and prominent mast cells. Variants include cellular, collagenized, epithelioid, and infiltrative tumors. The infiltrative form has no atypia or mitotic activity. There may be **histiocytoid** cells, prominent vessels, focal cartilaginous differentiation, and smooth muscle or multinucleated giant cells with a flower-like appearance. There is usually no entrapment of ducts or lobules, and no necrosis. The first case of myofibroblastoma was seen in 1987.

Clinical manifestations

Intracranial myofibroblastomas may cause headaches and seizures, but often are asymptomatic at the time of diagnosis.

Diagnosis

Diagnosis of myofibroblastomas is with CT scan and MRI. The tumors usually appear as well-circumscribed masses with clear boundaries. They are hypointense on T1-weighted images and often of mixed intensity on T2-weighted images. *Gadolinium-diethylene triamine pentaacetic acid (DTPA)*–enhanced T1-weighted images reveal heterogeneous enhancement with ring-like boundary enhancement. Sometimes, the cerebrospinal fluid is examined. Myofibroblastomas stain positive for estrogen receptors, progesterone receptors, vimentin, B-cell lymphoma 2 (Bcl2), and often, CD34. They have variably positive staining for desmin, **caldesmon**, and androgen receptors. Myofibroblastomas stain negatively for S-100 and cytokeratin, and may stain negative or low negative for Ki-67. Myofibroblastomas are often found during imaging studies for other conditions. Differential diagnoses include fibromatosis, low-grade myofibroblastic sarcoma, **myoepithelioma**, and nodular fasciitis.

Treatment

The goal of treatment is gross surgical resection, which has been achieved, and was curative without tumor recurrence. If required, medications include bevacizumab and crizotinib.

Prognosis

The prognosis of myofibroblastoma is usually good as long as gross surgical resection is achieved.

Benign fibrous histiocytoma

A **benign fibrous histiocytoma** is a solitary tumor made up of spindled, fibroblast-like cells and plump, **histiocyte**-like cells that have a storiform arrangement. This tumor is also referred to as *fibrous xanthoma* or **fibroxanthoma** that involves the dura or cranial bones. Benign fibrous histiocytomas resemble **dermatofibromas** of the skin.

Epidemiology

Benign fibrous histiocytomas affect all races and ages, but are most common between ages 20 and 49 years. There is a slight female preponderance. About 33% of cases show **metachronous** multiple tumors. In immunosuppressed patients, and those with systemic lupus erythematosus, **synchronous tumors** have been documented.

Etiology and risk factors

The actual cause of benign fibrous histiocytomas is unknown, and the tumors may involve either reactive or neoplastic processes. Risk factors are also unknown.

Pathology

In benign fibrous **histiocytoma**, scattered giant cells with or without inflammatory cells are often seen. In the past, many tumors with an intraparenchymal component were first described as fibrous xanthomas, but later found to be GFAP-positive. They were then reclassified as *pleomorphic xanthoastrocytomas*. Benign fibrous histiocytomas usually are slow-growing nodules, with poorly demarcated cellular fibrous tissue that encloses collapsed capillaries. There are scattered hemosiderin-pigmented and lipid macrophages. There may be clonal proliferation. Fibrin stabilizing factor XIIIa-positive dendritic cells are found more often than for macrophage marker antibody (MAC387)-positive cells. The color of these tumors ranges from pink to tan to bluish black. Variants include aneurysmal, atrophic, deep-penetrating, palisading, clear cell, **cholesterotic**, epithelioid, **hemosiderotic**, and myxoid. These tumors only rarely metastasize. Local recurrence is rare.

Clinical manifestations

In most cases, patients with benign fibrous histiocytomas are asymptomatic, but they can cause headaches and seizures if they become large enough. Benign fibrous histiocytomas may appear with pedunculated or dome-shaped structures. The atrophic variants are flat and depressed. The tumors are firm and may range from a few millimeters to a few centimeters in diameter. Rarely, these tumors spontaneously regress, and may leave behind a postinflammatory hypopigmentation of tissues.

Diagnosis

CNS benign fibrous histiocytomas are diagnosed via CT and MRI, and if possible, biopsy. Differential diagnoses include atypical fibroxanthoma, Kaposi sarcoma, low-grade myofibroblastic sarcoma, neurofibroma, nodular fasciitis, and **Rosai−Dorfman disease**.

Treatment

Wide surgical excision usually prevents tumor recurrence. Radiation therapy may be administered to prevent recurrence.

Prognosis

In most cases, the prognosis for benign fibrous histiocytomas is excellent.

Malignant fibrous histiocytoma

A malignant fibrous histiocytoma is made up of the cells seen in the benign fibrous histiocytomas, arranged in a storiform or fascicular pattern. In most cases, the malignancy is severe, with many mitoses along with significant necrosis. There is an inflammatory variant of this tumor. Another term for this tumor is *undifferentiated pleomorphic sarcoma*. In previous decades, it was described as a *fibrosarcoma*.

Epidemiology

Malignant fibrous histiocytoma is considered to be the most common type of soft tissue sarcoma. Only a few cases of the inflammatory variant of malignant fibrous histiocytoma have actually involved brain tissues. Most of these tumors occur in adults between ages 32 and 80, with a mean age of 59 years. They are slightly more common in males than females, with a 1.2:1 ratio. These tumors make up 25%−40% of all adult soft tissue sarcomas.

Etiology and risk factors

There are no known causes of malignant fibrous histiocytomas, but genetic abnormalities and chemical exposures may be implicated. Risk factors include previous radiation therapy and Paget's disease.

Pathology

Malignant fibrous histiocytoma has an aggressive biological behavior. The tumor is usually large, between 5 and 20 cm, well circumscribed, and unencapsulated. It has a gray, firm, heterogeneous cut surface. Cells include poorly differentiated fibroblasts, histiocyte-like cells with significant cellular pleomorphism, myofibroblasts, and bizarre multinucleated giant cells (see Fig. 14.18).

Histological subtypes of malignant fibrous histiocytomas occur as follows (see also Fig. 14.19):

- *storiform-pleomorphic*—50%−60% of cases;
- *myxoid*—25%;
- *inflammatory*—5%−10%;
- *giant cell*—5%−10%; and
- *angiomatoid*—less than 1% (this form is unlike the others in that it is comparatively nonaggressive, usually does not metastasize, and occurs most often in young adults or adolescents).

Clinical manifestations

Malignant fibrous histiocytoma usually presents as a painless, enlarging mass. There may be headaches, seizures, and neurologic deficits if the tumor becomes large enough. It may be difficult to differentiate these tumors from other high-grade sarcomas.

Diagnosis

In plain radiographs, the tumor appears soft, and may have curvilinear or punctate regions of calcification. On CT scan, there are heterogeneous lower density areas if significant hemorrhage, necrosis, or myxoid material is present. The soft tissue component enhances, and in as many as 20% of cases, there is some amount of mineralization. On

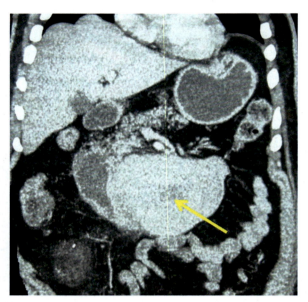

FIGURE 14.18 Histologic appearance of a malignant fibrous histiocytoma. (Top image) 40x magnification, (Bottom image) 200x magnification.

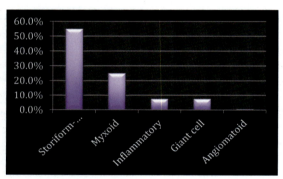

FIGURE 14.19 Histological subtypes of malignant fibrous histiocytomas.

MRI, which is the best way to determine tumor staging, the tumor is usually well circumscribed, with a positive mass effect because of the large size. In T1-weighted imaging, there is intermediate-to-low signal intensity, heterogeneity (based on hemorrhage, calcification, necrosis, and myxoid material), and prominent enhancement of solid components. In T2-weighted imaging, there is intermediate-to-high signal intensity and heterogeneity related to the same components as in T1 imaging. Differential diagnoses include other sarcomas and soft tissue tumors, as well as soft tissue metastases.

Treatment

Most malignant fibrous histiocytomas are grade III or IV tumors that have metastasized by the time of diagnosis in 30%–50% of cases. They locally recur despite aggressive treatment. Treatment usually consists of aggressive en bloc resection with a wide margin. Supplementary adjuvant chemotherapy and radiation therapy is very useful to reduce recurrence.

Prognosis

The prognosis for malignant fibrous histiocytoma is poor. Overall 5-year survival is between 25% and 70%. Prognostic factors include tumor size, location, and grade.

Leiomyoma

A **leiomyoma** is a benign smooth muscle tumor that is usually easily recognized because of its intersecting fascicles. It is composed of eosinophilic spindle cells that have nuclei with blunt ends. Leiomyomas are often referred to as *fibroids*, and rarely become malignant.

Epidemiology
Primary intracranial leiomyomas are extremely rare, with only a few documented cases. They mostly occur in middle-aged men.

Etiology and risk factors
The causes of intracranial leiomyomas are unknown. Risk factors may include the EBV and AIDS.

Pathology
Generally, leiomyomas lack mitotic activity and atypical cells. There are occasional features of nuclear palisading that can be mistake for the same factor in schwannomas. Variants include *diffuse leptomeningeal leiomyoma* and *angioleiomyomatous tumors*. The tumors are usually well encapsulated, with spindle cells arranged in interlaced fascicles and bundles. They usually lack necrosis, mitotic figures, and cell pleomorphism. Immunohistochemistry reveals immunoreactivity to SMA and negativity for S-100, CD34, and epithelial membrane antigen.

Clinical manifestations
Intracranial leiomyomas may cause headaches, nausea, vomiting, hemiparesis, and seizures if they become sufficiently large.

Diagnosis
Usually, intracranial leiomyomas are diagnosed via MRI, with homogeneous low signals on T1-, T2-, diffusion-weighted, and FLAIR sequences. These tumors may mimic meningiomas in imaging.

Treatment
Intracranial leiomyomas can usually be surgically resected with no recurrence. These tumors are radio-resistant, so radiation therapy is not recommended and may even cause damage.

Prognosis
The prognosis of intracranial leiomyomas is usually good with total surgical resection.

Rhabdomyoma

A **rhabdomyoma** is an extremely rare, benign tumor that consists of mature, striated muscle. The two major forms are called *neoplastic* and *hamartomas*. The neoplastic form is subdivided into adult, fetal, and genital subtypes. The **hamartoma** form does not occur in the CNS.

Epidemiology
Rhabdomyomas are most often diagnosed in males between 25 and 40 years of age. The fetal subtype usually affects boys between birth and the age of 3 years. The majority of rhabdomyomas involve the head and neck regions. Worldwide, rhabdomyomas are rare, with no exact data concerning their incidence. Less than 20 total cases of multifocal CNS rhabdomyomas have been documented. There is no difference in occurrence between racial or ethnic groups.

Etiology and risk factors
The causes of rhabdomyomas are not understood, but they are believed to represent genetic variants of striated muscle development. Medications or environmental factors have not been identified as causative. There are no known risk factors.

Pathology

The majority of reported intraneural rhabdomyomas are actually *neuromuscular choristomas*, also known as *benign triton tumors*. There has been one documented case of a rhabdomyoma affecting a cranial nerve that also had a slight adipose tissue component. This is important because of the high risk of postoperative fibromatosis that is related to neuromuscular choristomas. Rhabdomyomas of the CNS must be distinguished from *skeletal muscle heterotopias*, which usually occur in the prepontine leptomeninges. Cross-striation is demonstrated by using hematoxylin, muscle-specific actin, desmin, and myoglobin. **Dystrophin** is expressed by the cell membranes. The cells are eosinophilic and polygonal-shaped, with small, peripherally located nuclei and occasional intracellular vacuoles. There may be expression of ALK and insulin-like growth factor-1 receptor (IGF-1R).

Clinical manifestations

Signs and symptoms of rhabdomyomas vary based on tumor location. Documented manifestations have included difficulty breathing or swallowing, and hoarseness of the voice.

Diagnosis

Diagnosis of rhabdomyomas is with imaging studies such as CT and MRI, and also if possible, biopsy. Differential diagnoses include tuberous sclerosis and granular cell tumors. Staging of rhabdomyomas is by grade and site. The tumor ranges in grading from benign to intracapsular, to extracapsular/intracompartmental. There are three stages of the benign tumors, as follows:

- Benign stage 1—latent—remains static or heals spontaneously;
- Benign stage 2—active—progressive growth, but limited by natural barriers; and
- Benign stage 3—aggressive—progressive growth that is not limited by natural barriers.

Treatment

Treatment of rhabdomyomas is generally with gross surgical resection. Complete or partial tumor removal is usually successful, and reports of malignancy or recurrence are rare.

Prognosis

The prognosis for rhabdomyomas is considered fair to good after surgical removal. Morbidity depends on the subtype and the tumor location. Metastases have not been documented.

Chondroma

A **chondroma** is a benign, well-circumscribed tumor made up of hyaline cartilage with low cellularity. The rare cases of intracranial chondromas most often occur in the dura. The tumors may also develop in the skull, which can displace the brain and dura. They grow very slowly, and sometimes affect the base of the skull.

Epidemiology

Incidence of intracranial chondromas is estimated between 0.2% and 1% of all primary cerebral tumors. About 20% of all chondromas are convex or falcine. Chondromas are most common in young adults, with a peak incidence in the third decade. Overall reported ages of chondroma patients range from 15 months to 60 years, however. The male-to-female ratio is basically equivalent.

Etiology and risk factors

The causes of intracranial chondromas are not understood, and there are no identified risk factors.

Pathology

There has been one documented case of a large CNS chondroma becoming malignant as a chondrosarcoma, over a long period of time. These tumors are encapsulated, and have a lobular growth pattern. The tumor cells, when biopsied, are chondrocytes and cartilaginous cells, which resemble normal cells, within the lacunar spaces. They produce an amorphous and basophilic cartilaginous matrix. Vascular axes inside them, distinguishing them from normal hyaline cartilage, characterize chondromas. They may originate from the falx, convexity dura, tentorium, choroid plexus, or parenchyma. Chondromas are sometimes referred to as *parafalcine*, meaning above or behind a falx. The tumors appear white in color and resemble hard candle wax in consistence. They may be avascular, but adhered to the surrounding arachnoid. The mature hyaline cartilage may have increased cellularity, no atypia, no mitoses, and no necrosis. Chondromas may be solitary, or part of multiple polysystemic enchondromatosis, as well as *Maffucci syndrome*. There may be ectopic chondrocytes or chondroid metaplasia of the meningeal connective tissue. There are also two primary types—one in which the tumor appears isodense and homogeneous during CT scan, and one characterized by hypodense central areas, signifying degenerative cysts often filled with a tan to brown fluid or jelly.

Clinical manifestations

Intracranial chondromas may cause headaches, nausea, vomiting, drowsiness, hemiparesis, papilledema, visual disturbances, and seizures.

Diagnosis

Diagnosis of intracranial chondromas is with MRI, with contrast. T1-weighted imaging reveals hypointensity, as does T2-weighted imaging, which also shows heterogeneous postcontrast enhancement (see Fig. 14.20). The imaging

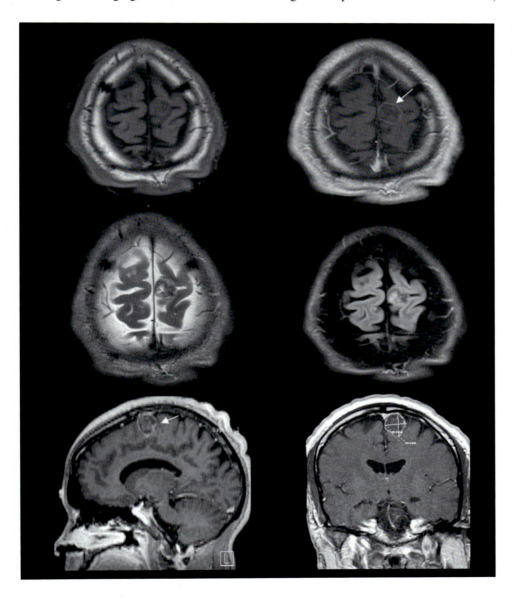

FIGURE 14.20 MRI images of an intracranial dural parafalcine chondroma.

features of chondromas are often similar to meningiomas, though chondromas do not have homogeneous enhancement or a dural tail sign. Diagnosis is confirmed by biopsy.

Treatment
The goal of treatment is total surgical resection, often with the endoscopic endonasal approach. Radiation therapy is not effective for these tumors. In one case, malignant transformation to chondrosarcoma was reported after a partial surgical resection.

Prognosis
The prognosis of intracranial chondromas is generally good, especially when total surgical resection can be performed.

Osteochondroma

An **osteochondroma** is a benign cartilaginous tumor that may be pedunculated or **sessile**. It develops on a bone surface, and is the most common benign type of bone tumor. These tumors commonly develop during childhood or adolescence, which are the periods of the most rapid skeletal growth. Once formed, they remain until they are removed.

Epidemiology
Osteochondromas occur in 3% of the general global population, and make up 35% of all benign tumors as well as 8% of all bone tumors. Malignant transformation is most likely to occur in patients between 28 and 35 years of age.

Etiology and risk factors
The exact causes of osteochondromas are not understood. However, most osteochondromas are nonhereditary, and about 15% occur as hereditary multiple exostoses, which are also called *hereditary multiple osteochondromas*. Multiple osteochondromas may be an autosomal dominant inherited disease. Germline mutations in the EXT1 and EXT2 genes on chromosomes 8 and 11 are believed to be implicated. Though usually sporadic, osteochondromas can be part of hereditary multiple exostoses or *Trevor's disease*, which is also known as **dysplasia epiphysealis hemimelica**. No known risk factors have been discovered.

Pathology
A cartilaginous cap plus a fibrous perichondrium extending to the periosteum of the bone distinguish osteochondromas. These tumors are long and thin and the cartilage cap appears as disorganized growth plate-like cartilage (see Fig. 14.21). The cartilaginous cap is covered by fibrous perichondrium that continues with the periosteum of the related bone. A cap more than 2 cm thick indicates malignant transformation of the tumor. Malignant transformation occurs in 1%–5% of cases in general, but in the setting of hereditary multiple exostoses, this can be up to 25% of cases.

Clinical manifestations
Most osteochondromas are asymptomatic upon discovery. When symptoms develop, they may include headaches, paresthesias, paraplegia, and neuropathies.

Diagnosis
Most intracranial osteochondromas are discovered accidentally during evaluation for other conditions. Diagnostic methods include X-rays, CT, MRI, ultrasound, angiography, genetic testing, and biopsy. An MRI is the preferred method, especially for tumors in the spinal column.

Treatment
The treatment of choice for osteochondromas is surgical removal of a solitary tumor or partial excision of outgrowths from it. For hereditary multiple exostoses, surgery is decided based on patient age, tumor location and number, symptoms, family history, and underlying gene mutations. Local recurrence occurs in 2%–15% of cases.

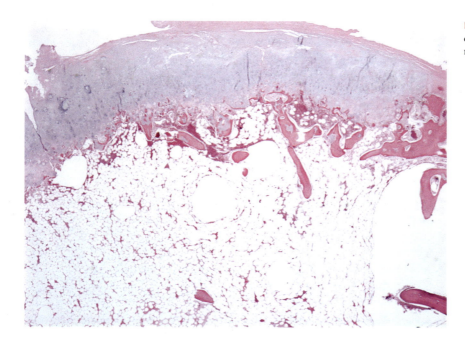

FIGURE 14.21 Osteochondroma: the cartilage cap has the histologic appearance of disorganized growth plate-like cartilage.

Prognosis

The prognosis and survival for osteochondromas are widely varied between individual patients. Since many of these tumors stop growing once the skeleton has matured, outlook can be better than for other CNS tumors.

Osteoma

An **osteoma** is a benign, bone-forming tumor that only has a limited growth potential. These tumors may develop inside or outside the CNS, often in the skull, which secondarily displace the brain tissues and dura. When arising from soft tissue, these tumors are called *heteroplastic osteomas*, and when arising from bone, they are called *homeoplastic osteomas*.

Epidemiology

Only isolated cases of intracranial osteomas have been documented, usually occurring in the dura. The prevalence of all types of osteomas is 3% in the global population, with adults and children being affected. The mean age at diagnosis is 37 years, and males are affected more than females, in a 3:2 ratio. The incidence of osteomas is unknown. There is no racial or ethnic predilection for these tumors.

Etiology and risk factors

The exact causes of osteomas are not understood, and there are no known risk factors.

Pathology

Osteomas histologically resemble similar tumors that arise in bone. It is important to distinguish them from *asymptomatic dural calcification*, ossification that is caused by trauma or metabolic disease, and rarely, astrocytomas or gliosarcomas that have osseous differentiation. Osteomas grow slowly. There are three histologic patterns: *ivory, mature*, and *mixed*. The ivory form appears as extremely radiodense lesions, while mature osteomas may have central bone marrow.

Clinical manifestations

Osteomas are usually asymptomatic, but may cause headaches and inflammation of brain tissues. Hearing and vision problems have been documented, along with sinus infections when the sinuses are affected.

Diagnosis

Osteomas may be incidentally identified as a mass in the skull or mandible, or as the cause of sinusitis or paranasal sinus **mucocele** formation. If multiple, Gardner's syndrome should be considered. Imaging methods include CT and MRI, with CT being preferred because it more clearly shows the tumor and its extensions.

Treatment

Osteomas only require surgical resection if they cause complications or mass effect. Prognosis for osteomas is good with total surgical resection.

Hibernoma

A **hibernoma** is an extremely rare variant of a lipoma within the CNS. The tumor is slow-growing, painless, and solitary. It may take years for any symptoms to develop. These lesions have also been called **fetal lipomas** *or lipomas of embryonic fat.*

Epidemiology

Hibernomas are rare, mostly affecting adults in the fourth decade of life. There is a slight male predominance.

Etiology and risk factors

Though the actual cause of hibernomas is unknown, there are structural rearrangements of 11q13-21. This is detectable by FISH. The MEN1 gene is usually deleted and the GARP gene may also be involved. There are no known risk factors for hibernomas.

Pathology

Hibernomas are made up of uniform granular or multivacuolated cells that are small. They have centrally located nuclei, and look similar to brown fat. The tumor appears well defined, encapsulated, or circumscribed. The color is yellow-tan to deep brown. The texture is soft, and the size ranges from 1 to 27 cm, with a mean size of 10 cm. The *lobular* type has various degrees of differentiation, of uniform and round-to-oval cells that have granular eosinophilic cells and prominent borders. These alternate with coarsely multivacuolated pale fat cells. There are usually small, centrally placed nuclei without pleomorphism and large cytoplasmic lipid droplets throughout. The *myxoid* type has a loose and basophilic matrix with thick fibrous septa and foamy histiocytes. The *lipoma-like* type has univacuolated lipocytes with only isolated hibernoma cells. The *spindle cell* type is a spindle cell lipoma combined with a hibernoma. The cells are S-100 protein positive in about 80% of cases. They show membrane and vacuole CD31 immunoreactivity. Uncoupling protein 1 is also positive.

Clinical manifestations

When a hibernoma develops within the head, neck, or face, symptoms include progressive inflammation, discomfort, and pain when the tumor presses on surrounding tissues.

Diagnosis

Generally, imaging studies reveal a well-defined, heterogeneous mass that is hypointense to subcutaneous fat on MRI T1-weighted images. Serpentine and thin low-signal bands are often seen in the tumor. The differential diagnoses include rhabdomyomas, granular cell tumors, and liposarcomas.

Treatment

Total surgical excision is the treatment of choice, which usually provides an excellent long-term result. The prognosis for hibernomas is good, since these tumors are benign and do not metastasize or convert to a malignancy.

Clinical cases

Clinical case 1

1. What is the general description of a hemangiopericytoma?
2. What are the two primary morphologies of these tumors?
3. What is the epidemiology of hemangiopericytomas?

A 59-year-old man presented with a headache that had persisted for about 4 months. A brain MRI revealed a well-demarcated, solid mass within the left cerebellum, of about 4.5 cm in diameter. It was located around the transverse and sigmoid sinus area. The mass was compressing the fourth ventricle. Surgical resection was performed to remove the mass, and the tumor was of a pink color and hard in consistency. There was massive bleeding from inside of the tumor during the resection. The invasive portion of the tumor could not be removed due to the potential for even greater bleeding. The tumor cells were round to spindle shaped. The branched vessels inside the tumor had a staghorn appearance, and the tumor cells were positive for CD34. The tumor was a grade II hemangiopericytoma. Adjuvant radiation therapy was administered following the surgery, with success.

Answers:

1. **Hemangiopericytomas are solitary fibrous solid tumors and are the most common type of mesenchymal neoplasm. They are of the fibroblastic type, often having a significant branched vascular pattern.**
2. **The first is a solitary fibrous tumor phenotype with a patternless structure or a short fascicular pattern, with hypocellular and hypercellular areas, plus thick collage bands. The second is characterized by high cellularity and a thin but rich reticulin fiber network that usually invests the individual cells. Staghorn vessels are present in both phenotypes.**
3. **Hemangiopericytomas usually affect adults in the fourth to sixth decade of life. There is just under 1 case per 100,000 people, globally with males affected slightly more often than females. The tumors rarely occur in children.**

Clinical case 2

1. From where do hemangioblastomas originate and what is their cellular makeup?
2. In which age group are hemangioblastomas most common?
3. What are the general macroscopic descriptions of hemangioblastomas?

A 26-year-old woman presented with complaints of headache and vomiting over the past 6 months. An MRI revealed a midline cystic mass in the infratentorial region, centered in the cerebellar parenchyma. There were multiple intensely enhancing nodules within the lesion, and some had irregular cystic and necrotic regions. Also, there was a well-defined cystic cavity, with clear fluid that appeared hyperintense on T2-weighted imaging. The lesion showed much more perfusion in comparison to the normal parenchyma. There were foci of bleeding within and around the lesion, with multiple abnormal blood vessels. The patient underwent surgical resection of the cerebellar lesion, which was diagnosed as a hemangioblastoma.

Answers:

1. **Hemangioblastomas originate from the vascular system and are usually made up of stromal cells in smaller blood vessels.**
2. **Hemangioblastomas have been seen in various ages, but are most common in adults around age 40.**
3. **About 60% of hemangioblastomas are well circumscribed, slightly cystic, and highly vascularized, while about 40% are totally solid. Sometimes, the mass is yellow because of significant amounts of lipids.**

Clinical case 3

1. What is the basic description of an angiosarcoma?
2. What are the risk factors for development of angiosarcomas?
3. How aggressive are angiosarcomas, and how do they appear?

A 47-year-old man presented with acute onset of visual loss in his left eye over the past 10 days. He had slight memory problems and trouble finding the correct words while speaking. Imaging revealed a large, heterogeneously enhancing left sphenoid wing mass that extended into the sphenoid sinus. It at first appeared to resemble a meningioma. During surgical resection, there was extensive bleeding, and the margin of the tumor near the dura was removed from the tumor with no evidence of intradural extension. The tumor was diagnosed as a high-grade angiosarcoma. Additional treatments involved radiation therapy and chemotherapy, and the patient survived.

Answers:

1. **An angiosarcoma is a high-grade malignant tumor with endothelial differentiation. Basically, it is a cancer of the cells lining the walls of blood vessels. Since the cells are carried through these vessels, metastasis to sites such as the liver and lungs is more likely.**

2. Risk factors for angiosarcomas may include exposure to toxins or radiation therapy, exposure to carcinogens, and previous history of lymphedema. Their development may be linked to radical mastectomy, preexisting bone infarctions, chronic osteomyelitis, pagetoid bone, and AIDS.
3. Angiosarcomas are aggressive, and often recur locally, spread widely, and have high rates of lymph node and systemic metastasis. They usually appear red in color, and are hemorrhagic.

Clinical case 4

1. What is the other name for Ewing sarcoma, and what is its basic description?
2. When does Ewing sarcoma most commonly occur?
3. When occurring in the CNS, what are the clinical manifestations?

An 11-year-old girl presented with a history of chronic headache that had become much worse in the past month. She also had chronic vomiting, and a left temporal scalp swelling that was increasing in size. Examination revealed bilateral papilledema. The left temporal scalp swelling was slightly tender, firm, immobile, and nonpulsatile, with poorly defined margins. A CT scan showed a round, well-defined, heterogeneously hyperdense enhancing lesion, with a mass effect and destruction of the left temporal bone, which extended into the scalp. A meningioma was at first suspected. There was no calcification within the lesion. An MRI revealed the lesion to be hypointense in T1-weighted imaging, and heretointense in T2-weighted imaging. The lesion was about 7.4 cm in diameter. Surgical resection was successful, and the tumor had round-to-oval cells arranged in lobules, vesicular nuclei, and indistinct nucleoli. The tumor was diagnosed as a Ewing sarcoma, and the child also received chemotherapy as well as radiation therapy, and survived.

Answers:

1. A Ewing sarcoma is also known as a peripheral primitive neuroectodermal tumor. It is a small, round, blue cell tumor of neuroectodermal origin that may involve the CNS as a primary dural neoplasm, or via direct extension from contiguous soft tissue or various bones.
2. Ewing sarcoma is most common during the second decade of life. The peak incidence is between 10 and 20 years of age. It represents 2% of all childhood cancers.
3. The clinical manifestations of Ewing sarcoma include chronic headache, inflammation, anemia, leukocytosis, increased sedimentation rate, nausea, vomiting, loss of consciousness, papilledema, scalp swelling, and fever. Paraplegia may be a complication of Ewing sarcoma.

Clinical case 5

1. What is the age range of patients affected by fibrosarcoma?
2. Which chromosomes are usually involved in the development of fibrosarcoma?
3. Where in the CNS do fibrosarcomas usually form?

A 17-year-old boy presented with seizures and headache. A neurological examination revealed papilledema and decreased visual acuity. A CT scan revealed a large mass with hemorrhage in both frontal areas, and MRI revealed a large mass with strong enhancement and T2-imaging intensity. Both distal anterior cerebral arteries were encapsulated by the tumor. Surgical resection was performed, and profuse bleeding occurred. The tumor was well demarcated, and additional surgeries were required to remove all of it. After surgery, the patient had a left side hemiparesis. The tumor cells showed a storiform pattern. There was high mitotic activity. The cells showed strong vimentin immunoreactivity. The tumor was diagnosed as a fibrosarcoma.

Answers:

1. Fibrosarcomas are most often seen in males between ages 30 and 40, but also in younger adults and adolescents. Congenital infantile fibrosarcoma may occur before the age of 2 years.
2. In most cases, fibrosarcomas are believed to be linked to a translocation between chromosomes 12 and 15. Also, individual cases show trisomy for chromosomes 8, 11, 17, or 20.
3. Within the CNS, fibrosarcomas are commonly seen in the meninges and cerebral hemispheres (often frontal in location), but also at the skull base, external cranium, and infrequently, within the orbital or mastoid bones.

Key terms

Angiolipoma
Angiosarcoma
Atypia
Beckwith—Wiedemann syndrome
Benign fibrous histiocytoma
Biallelic inactivation
Biphenotypic sinonasal sarcoma
Blister cells
Blooming
Botryoid
Brachytherapy
Brachyury
Caldesmon
Cavernous hemangiomas
Cerebral cavernous malformations
Cholesterotic
Chondroma
Chondromyxoid
Chondrosarcoma
Codman's triangle
Costello syndrome
Curvilinear
Cytokeratin
Dermatofibromas
Desmin
Desmoid-type fibromatosis
Dysmetria
Dysplasia epiphysealis hemimelica
Dystrophin
Ectomesenchymoma
En bloc resection
Epithelioid hemangioendothelioma
Erythropoietin
Ewing sarcoma
Exostoses
Extraosseous
Extraskeletal
Factor VIII-related antigen
Fetal lipomas
Fibrolipomatous stalk
Fibrosarcoma
Fibroxanthoma
Gardner's syndrome
Hamartoma
Hemangioblast
Hemangioblastoma
Hemangioma
Hemangiopericytoma
Hemosiderotic
Herringbone
Hibernoma
Highly active antiretroviral therapy (HAART)
Histiocyte
Histiocytoid
Histiocytoma
Hyperostosis
Infantile hemangiomas
Inflammatory myofibroblastic tumor
Intrachromosomal inversion
Intraparenchymatous
IVA regimen
Kaposi sarcoma
Leiomyoma
Leiomyosarcoma
Leptomyelolipomas
Li—Fraumeni syndrome
Lipoblasts
Lipoma
Liposarcoma
Maffucci syndrome
Medullomyoblastomas
Mesenchymal neoplasm
Metachronous
Mucocele
Multivacuolated
Myoepithelioma
Myofibroblastoma
Myogenin
Nodular fasciitis
Ollier disease
Osteochondroma
Osteoma
Osteosarcoma
Pachymeningeal fibrosis
Paget's disease
Parameningial
Pericytes
Polycythemia
Pseudohypoxia
Pseudotumoral
Reticulin fibers
Rhabdomyoma
Rhabdomyosarcoma
Rosai—Dorfman disease
Rothmund—Thomson syndrome
SCL protein
Sessile
Staghorn vessels
Strap cells
Stromal cells
Synchronous tumors
Syrinx
Tethered spinal cord syndrome
Tubonodular
VAC regimen
VCB complex
von Willebrand factor
Weibel—Palade bodies
White epidermoid

Further reading

American Registry of Pathology. (2015). *Tumors of the soft tissue (Atlas of tumor pathology, Book 4)*. American Registry of Pathology.
Bolontrade, M., & Garcia, M. (2016). *Mesenchymal stromal cells as tumor stromal modulators*. Academic Press.
Brennan, M. F., Antonescu, C. R., & Marki, R. G. (2013). *Management of soft tissue sarcoma*. Springer.
Chang, E. L., Brown, P. D., Lo, S. S., Sahgal, A., & Suh, J. H. (2018). *Adult CNS radiation oncology: Principles and practice*. Springer.
Cheung, R. (2015). *Best of SRS and radiotherapy of CNS tumors and public health*. Scholars' Press.
Christian Tonn, J., Reardon, D. A., Rutka, J. T., & Westphal, M. (2019). *Oncology of CNS tumors* (3rd ed.). Springer.
Dicato, M. A., & Van Cutsem, E. (2018). *Side effects of medical cancer therapy: Prevention and treatment*. Springer.
Fisher, C. (2010). *Diagnostic pathology: Soft tissue tumors*. LWW.
Fisher, C., Montgomery, E., & Thway, K. (2015). *Biopsy interpretation of soft tissue tumors (Biopsy interpretation series)* (2nd ed.). LWW.
Fletcher, C. D. M. (2013). *Diagnostic histopathology of tumors: 2 Volume set: Expert consult (Diagnostic histopathology of tumors)* (4th ed.). Churchill Livingstone.
Gabrieli, L., & Sardaro, A. (2018). *The intracranial hemangiopericytoma*. Lap Lambert Academic Publishing.

Goldblum, J. R., Weiss, S. W., & Folpe, A. L. (2019). *Enzinger & Weiss's soft tissue tumors* (7th ed.). Elsevier Science Health Science.
Gupta, N., Banerjee, A., & Haas-Kogan, D. A. (2017). *Pediatric CNS tumors (Pediatric oncology)* (3rd ed.). Springer.
Gursoy Ozdemir, Y., Bozdag Pehlivan, S., & Sekerdag, E. (2017). *Nanotechnology methods for neurological diseases and brain tumors — Drug delivery across the blood-brain barrier*. Academic Press.
Harsh, I. V., Griffith, R., & Vaz-Guimaraes, F. (2017). *Chordomas and chondrosarcomas of the skull base and spine* (2nd ed.). Academic Press.
Hayat, M. A. (2012a). *Tumors of the central nervous system, Volume 5: Astrocytomas, hemangioblastomas, and gangliogliomas*. Springer.
Hayat, M. A. (2012b). *Tumors of the central nervous system, Volume 9: Lymphoma, supratentorial tumors, glioneuronal tumors, gangliogliomas, neuroblastoma in adults, astrocytomas, ependymomas, hemangiomas, and craniopharyngiomas*. Springer.
Hayat, M. A. (2013). *Tumors of the central nervous system, Volume 10: Pineal, pituitary, and spinal tumors*. Springer.
Hayat, M. A. (2014). *Tumors of the central nervous system, Volume 13: Types of tumors, diagnosis, ultrasonography, surgery, brain metastasis, and general CNS diseases*. Springer.
Hayat, M. A. (2015a). *Brain metastases from primary tumors, Volume 2: Epidemiology, biology, and therapy*. Academic Press.
Hayat, M. A. (2015b). *Tumors of the central nervous system, Volume 14: Glioma, meningioma, neuroblastoma, and spinal tumors*. Springer.
Hornick, J. L. (2018). *Practical soft tissue pathology: A diagnostic approach: A volume pattern recognition series* (2nd ed.). Elsevier.
Horvai, A., & Link, T. (2012). *Bone and soft tissue pathology: A volume in the high-yield pathology series*. Saunders.
Icon Group International. (2009). Angiosarcoma: Webster's timeline history, *1906–2007*. Icon Group International, Inc.
Icon Group International. (2010a). Hemangioendothelioma: Webster's timeline history, *1938–2007*. Icon Group International, Inc.
Icon Group International. (2010b). Hemangiopericytoma: Webster's timeline history, *1952–2007*. Icon Group International, Inc.
Karajannis, M. A., & Zagzag, D. (2015). *Molecular pathology of nervous system tumors: Biological stratification and targeted therapies (Molecular pathology library)*. Springer.
Kesharwani, P., & Gupta, U. (2018). *Nanotechnology-based targeted drug delivery systems for brain tumors*. Academic Press.
Kransdorf, M., & Murphey, M. D. (2013). *Imaging of soft tissue tumors* (3rd ed.). LWW.
Lacruz, C. R., Saenz de Santamaria, J., & Bardales, R. H. (2018). *Central nervous system intraoperative cytopathology (Essentials in cytopathology)* (2nd ed.). Springer.
Lahoti, S., Rao, K., & Ashish, K. (2019). *Epithelial mesenchymal interactions*. Lap Lambert Academic Publishing.
Leach, R. H. (2014). *Leiomyosarcoma: Risk factors, diagnosis and treatment options (Etiology, diagnosis and treatments)*. Nova Science Publishers Inc.
Lindberg, M. R. (2019). *Diagnostic pathology: Soft tissue tumors* (3rd ed.). Elsevier Science / Health.
Mahajan, A., & Paulino, A. (2018). *Radiation oncology for pediatric CNS tumors*. Springer.
Matevosyan, N. (2015). *Rhabdomyoma & rhabdomyosarcoma*. CreateSpace Independent Publishing Platform.
Mattassi, R., Loose, D. A., & Vaghi, M. (2015). *Hemangiomas and vascular malformations: An atlas of diagnosis and treatment* (2nd ed.). Springer.
Miettinen, M. (2017). *Modern soft tissue pathology: Tumors and non-neoplastic conditions* (2nd ed.). Cambridge University Press.
Norden, A. D., Reardon, D. A., & Wen, P. C. Y. (2011). *Primary central nervous system tumors: Pathogenesis and therapy (Current clinical oncology)*. Humana Press.
Olch, A. J. (2013). *Pediatric radiotherapy planning and treatment*. CRC Press.
Rakshit (Kambale), K. (2017). *Osteoma — A benign bone forming tumor: Clinical approach and management*. Lap Lambert Academic Publishing.
Sahgal, A., Lo, S. S., Ma, L., & Sheehan, J. P. (2016). *Image-guided hypofractionated stereotactic radiosurgery: A Practical approach to guide treatment of brain and spine tumors*. CRC Press.
Scheinemann, K., & Bouffet, E. (2015). *Pediatric neuro-oncology*. Springer.
Thompson, L. D. R. (2012). *Head and neck pathology: A volume in the series: Foundations in diagnostic pathology* (2nd ed.). Saunders.
Ueda, T., & Kawai, A. (2016). *Osteosarcoma*. Springer.
Warmuth-Metz, M. (2017). *Imaging and diagnosis in pediatric brain tumor studies*. Springer.

Chapter 15

Pituitary tumors

Chapter Outline

Overview	285	Pituitary carcinoma	307
Pituitary adenoma	285	Pituicytoma	312
Somatotroph adenoma	291	Spindle cell oncocytoma	314
Prolactinoma	295	Secondary tumors	317
Thyrotropin-secreting tumors	297	Various parasellar masses	318
Adrenocorticotropic hormone−secreting tumors	299	Clinical cases	319
Craniopharyngioma	302	Key terms	321
Nonfunctioning pituitary tumors	305	Further reading	322

Overview

Pituitary tumors make up as many as one in four intracranial tumors. Their biologic courses classify them into three groups: benign, invasive adenomas, and carcinomas. Adenomas make up the largest subgroup of pituitary tumors, yet only a small amount of these tumors have associated symptoms. Invasive adenomas may invade the dura mater, cranial bones, or sphenoid sinus. Fortunately, pituitary carcinomas make up less than 1% of all pituitary tumors. Most pituitary tumors develop in the anterior pituitary gland.

Pituitary adenoma

A **pituitary adenoma** is a generally benign type of pituitary tumor. In some cases, they can become invasive or develop as carcinomas. Tumors more than 0.39 in. in size are called **macroadenomas**, while those smaller than this size are called **microadenomas**. The majority of pituitary adenomas are microadenomas. They are also classified as *intrasellar* or *extrasellar* based on an ability to expand outside the sella turcica.

Epidemiology

Pituitary adenomas represent 10%−25% of all intracranial neoplasms. Also, in a study of 1120 patients for sellar masses, 91% had a pituitary adenoma. Invasive pituitary adenomas occur in 35% of all pituitary adenoma cases, and only 0.1%−0.2% are carcinomas. Pituitary adenomas represent 10%−25% of all intracranial neoplasms. Also, in a study of 1120 patients for sellar masses, 91% had a pituitary adenoma. The estimated prevalence rate of pituitary adenomas in the global population is approximately 17%. Most of these adenomas are microadenomas, with an estimated prevalence of 16.7%. They are found 22.5% of the time in radiologic studies, and 14.4% of the time in autopsy studies. Overall, pituitary adenomas affect about one of every six people in the general population. However, clinically active pituitary adenomas, requiring surgical treatment, are more rare, and affect about 1 of every 1000 in the general population. In one study of 2598 patients who underwent an MRI, pituitary adenomas made up 82% of all visible lesions.

In Japan, 15.8% of 28,424 cases were confirmed to be pituitary adenomas. In the United Kingdom, there were 63 pituitary tumors out of 89,334 individuals. These were subdivided as **prolactinomas** (57%), nonfunctioning adenomas (28%), growth hormone (GH)-secreting adenomas (11%), and **Cushing's adenomas** (2%). In the United States, overall pituitary tumor incidence has increased from 2.52 per 100,000 population to 3.13 per 100,000, between the years 2004 and 2009. Prolactinomas are the most common secretory pituitary tumors, with an annual incidence of about 30 per 100,000 people. For microprolactinomas, females are diagnosed 20 times more than males, but for macroprolactinomas, the ratio is nearly equivalent.

Etiology and risk factors

The exact cause of most pituitary tumors, including adenomas, is unknown. Most develop in the anterior pituitary gland, which produces GH (also called somatotropin), prolactin (PRL), luteinizing hormone (LH), follicle-stimulating hormone (FSH), adrenocorticotropic hormone (ACTH; corticotropin), and thyroid-stimulating hormone (TSH; thyrotropin). Adenomas of the anterior pituitary gland are an important clinical feature of **multiple endocrine neoplasia type 1 (MEN1)**, which are a rare inherited endocrine syndrome affecting 1 of every 30,000 people. It causes various benign or malignant tumors in various endocrine glands, or can cause these glands to enlarge with no tumor formation. The parathyroid glands, pancreatic islet cells, and anterior pituitary are often affected. Also, MEN1 can cause facial **angiofibromas**, **collagenomas**, lipomas, ependymomas, **leiomyomas**, and meningiomas. About 25% of patients with MEN1 develop a pituitary adenoma. **Carney complex**, also known as *LAMB syndrome* or *NAME syndrome*, is an autosomal dominant condition made up of **myxomas** of the heart and skin, **lentiginosis**, and endocrine overactivity. About 7% of all cardiac myxomas are linked to Carney complex, which causes GH-producing pituitary tumors that sometimes also secrete prolactin. The pituitary gland may have hyperplastic areas, with hyperplasia usually preceding formation of a GH-producing adenoma.

Familial isolated pituitary adenoma (FIPA) is of autosomal dominant inheritance, and signified by two or more related patients who have only pituitary gland adenomas, without the symptoms of MEN1 or Carney complex. Families with FIPA may be homogenous or heterogeneous. In homogeneous families, all affected family members have the same type of adenoma, but in heterogeneous families, each person can have a different type of adenoma. Familial isolated pituitary adenomas are linked to mutations of the *aryl hydrocarbon receptor-interacting protein (AIP)* gene or duplications in chromosome Xq26.3, including the *GPR101* gene. Approximately 20% of AIP mutation carriers will develop a pituitary adenoma. Also, AIP mutations are the most common genetic cause of pituitary gigantism (29% of patients). Another condition called X-LAG occurs rarely, in very early childhood, leading to GH excess, severe overgrowth of the body, and pituitary gigantism. Risk factors for pituitary adenomas include MEN1, Carney complex, and familial isolated pituitary adenoma.

Pathology

Though in most cases pituitary adenomas are benign and noninvasive, an invasive adenoma may enter the dura mater, cranial bone, or sphenoid bone. When the tumor reaches 0.39 in. in diameter, visual manifestations often appear. Pituitary adenomas are classified based on their anatomical, histological, and functional criteria. Cell differentiation, driven by transcription factors, leads to the acidophilic, gonadotrophic, and corticotrophic cell lineages. Transcription factors include *pituitary-specific POU class homeodomain 1 (PIT1)*, *steroidogenic factor 1 (SF-1)*, and *T-box family member TBX19 (T-PIT)*.

Clinical manifestations

Pituitary macroadenomas are the most common cause of **hypopituitarism**. Hormone-secreting pituitary adenomas can also cause one of several forms of **hyperpituitarism**, with specific manifestations based on the type of hormone. Some of these tumors secrete more than one hormone (usually GH plus PRL), resulting in unexpected bone growth and lactation—in men as well as women. There may be visual field defects, most often **bitemporal hemianopsia**, due to the tumor compressing the optic nerve. Specifically, the tumor compresses the **optic chiasma**, producing a defect in the temporal visual field on both sides. If this originates superior to the optic chiasm, usually in the craniopharyngioma of the pituitary stalk, the defect first appears as **bitemporal superior quadrantanopia**. If the adenoma laterally expands, and the abducens nerve is also compressed, there will be a **lateral rectus palsy**. Pituitary adenomas can also cause increased intracranial pressure and its related symptoms. In a study of microadenomas, headache resolved or disappeared after surgery in 90% of patients with nonfunctioning tumors, and in 56% with functioning tumors. In another study of 49 patients, prior to surgery, mean intrasellar pressure was two to three times higher than normal. Prolactinomas often initiate symptoms during pregnancy, as progesterone increases the growth rate of the tumor.

Various types of headaches are also common with pituitary adenomas, which can be the main cause of them, or worsen a headache that was started by other factors. Forms of headaches include chronic and episodic migraine, and less often, unilateral headaches, a primary stabbing headache, short-lasting unilateral **neuralgiform headache** with conjunctival injection and tearing, and a different type of stabbing headache that involves short stabs of pain, cluster headache, and **hemicrania continua**. Nonsecreting adenomas may remain undetected for a long time, since there are no obvious symptoms. Lack of ACTH results in the adrenal glands not producing enough cortisol. This means there will be slow recoveries from illnesses, inflammation, and chronic fatigue. If there is not enough GH, children and teenagers may have a short stature. Psychiatric symptoms include anxiety, depression, apathy, emotional instability, and easy irritability or hostility.

About 90%−95% of cases of **acromegaly** come from a pituitary adenoma that causes the anterior pituitary gland to produce excessive GH. This is most common in middle-aged adults. Acromegaly may result in severe disfigurement, sleep apnea, heart problems, type 2 diabetes mellitus, colon polyps, and premature death. It is associated with **gigantism**, and is difficult to diagnose in its early stages. This means that acromegaly is often missed for many years, until changes in the facial features and other areas become noticeable (see Fig. 15.1). The median time from initial symptom development to diagnosis is 12 years. **Cushing's syndrome**, caused by **Cushing's disease** (from a pituitary adenoma), is resulting in excessive ACTH secretion in 60%−80% of cases. This condition will be discussed later in this chapter. Hyperpituitarism results in hypersecretion of GH, PRL, thyrotropin, LH, FSH, and ACTH. Additional complications include **pituitary apoplexy** and central diabetes insipidus.

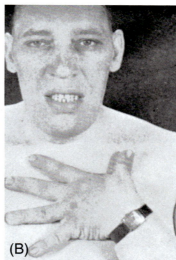
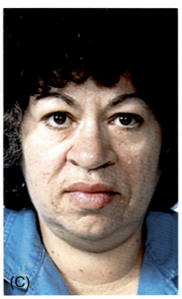

FIGURE 15.1 Effects of growth hormone. (A) Primary gigantism in twins, showing changes in the twin on the left. (B) Facial and hand manifestations of acromegaly. (C) Facial changes with acromegaly. *Courtesy: Lewis S, et al: Medical-Surgical Nursing: Assessment and Management of Clinical Problems, ed 10, St. Louis, 2017, Elsevier.*

Diagnosis

Most pituitary microadenomas remain undiagnosed. Those that are diagnosed are often found as incidental findings, and referred to as *incidentalomas*. Diagnosis is confirmed by testing hormone levels, and via radiographic imaging of the pituitary—usually CT scan or MRI (see Fig. 15.2). Classifications include the following grades:

- Grade I—microadenomas less than 1 cm in diameter, without **sella** expansion, occupying the suprasellar cistern;
- Grade II—macroadenomas 1 or more centimeters in diameter that may extend above the sella, with elevation of the third ventricle;
- Grade III—macroadenomas with enlargement and invasion to the floor, or suprasellar extension, occupying the anterior of the third ventricle; and
- Grade IV—destruction of the sella itself, beyond the foramen of Monro, or grade C with lateral extensions.

The radiographic classifications of pituitary adenomas include five different categories:

- 0—Normal appearance of the pituitary;
- I—Enclosed in the sella turcica, microadenoma; smaller than 10 mm;
- II—Enclosed in the sella turcica, macroadenoma; 10 mm or more;
- III—Invasive, locally, into the sella; and
- IV—Invasive, diffusely, into the sella.

Classification is based on which type of hormone is secreted by the tumor. The screening tests for secreting (functioning) pituitary adenomas are listed in Table 15.1. It is important to understand that about 20%−25% of adenomas do not secrete any easily identifiable active hormone, hence the name *nonfunctioning tumors*. Functional classification is based on endocrine activity, determined by serum hormone levels, and immunohistochemical staining to detect pituitary tissue cellular hormone secretion. Nonsecreting adenomas may be *null cell adenomas*, or a specific form that still is nonsecreting.

Overall, the diagnostic classifications of pituitary adenomas include the following:

- **Lactotrophic adenomas** (prolactinomas)—secrete prolactin, are **acidophilic**, cause **amenorrhea**, galactorrhea, hypogonadism, impotence, and infertility; 30% of pituitary adenomas are in this class.
- **Null cell adenomas**—do not secrete hormones, may stain positive for synaptophysin; make up 25% of pituitary adenomas.
- **Corticotrophic adenomas**—secrete ACTH, are basophilic, and cause Cushing's disease; make up 19.5% of pituitary adenomas.
- **Somatotrophic adenomas**—secrete GH, are acidophilic, causing gigantism in children and acromegaly in adults; make up 15% of pituitary adenomas.
- **Gonadotrophic adenomas**—secrete LH, FSH, and subunits of these hormones, are basophilic, and usually do not cause symptoms; make up 10% of pituitary adenomas.
- **Thyrotrophic adenomas**—a rare form that secretes TSH, are basophilic to **chromophobic**, and usually do not cause symptoms, but may cause hyperthyroidism; less than 1% of pituitary adenomas are of this form.

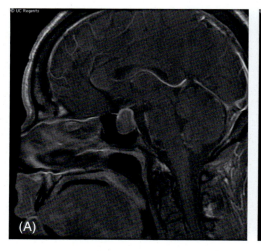
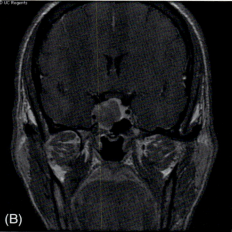

FIGURE 15.2 T1-weighted MRI of pituitary adenoma with gadolinium enhancement. (A) Sagittal view. (B) Coronal view. A heterogeneously enhancing mass arises out of the sella turcica and extends up to involve the region of the optic chiasm. Surgery confirmed a growth hormone-secreting pituitary adenoma.

TABLE 15.1 Screening tests for functioning pituitary adenomas.

Related disorder	Screening test	Additional information
Acromegaly	Insulin-like growth factor type 1 (IGF-1). Oral glucose tolerance test, with growth hormone (GH) obtained at 0, 30, and 60 min.	Interpret IGF-1 relative to control subjects who are age and gender matched. Normal subjects should suppress GH to less than 1 µg/L.
Cushing's Disease	24-h urinary-free cortisol. Nighttime salivary cortisol. Dexamethasone (1 mg) at 11:00 p.m. and fasting plasma cortisol measured at 8:00 a.m. Adrenocorticotropic hormone (ACTH) assay.	Ensure that urine collection is total and accurate; measure urinary creatinine. Free salivary cortisol reflects circadian rhythm, with elevated levels possibility indicating Cushing's disease. Normal subjects suppress to less than 1.8 µg/dL. To assess ACTH levels.
Prolactinoma	Serum prolactin level.	A level greater than 500 µg/L is pathognomonic for macroprolactinoma. If greater than 200 mg/L, prolactinoma is likely, though use of risperidol may cause these levels.
TSH-secreting tumor	Thyroid-stimulating hormone measurement. Free thyroxine by dialysis. Total triiodothyronine.	If thyroxine or triiodothyronine is measurable or elevated, a TSH-secreting tumor may be present.

Pituitary incidentalomas are often discovered by CT or MRI during evaluation for other conditions. They are also often discovered during an autopsy. The rare *ectopic pituitary adenoma* occurs outside of the sella turcica, and usually in the sphenoid sinus, suprasellar region, nasopharynx, or cavernous sinuses. They secrete GH. There are also *peripheral GH-secreting tumors*, and are capable of causing acromegaly. Pituitary carcinomas resemble pituitary adenomas under a microscope, making diagnosis complicated.

Carcinomas that metastasize into the pituitary gland are rare, and usually seen in elderly patients. Most often, the original tumor site was in the lungs or breasts. For breast cancer patients, metastasis to the pituitary gland occurs in 6%—8% of cases. Symptomatic pituitary metastases only make up 7% of reported cases. In symptomatic patients, **diabetes insipidus** occurs in up to 71% of cases. Other symptoms include anterior pituitary dysfunction, headache or head pain, ophthalmoplegia, and visual field defects. The differential diagnosis of any pituitary mass should be focused on excluding pituitary adenoma before other rare sellar lesions are considered. Differential diagnoses include **pituitary tuberculoma**, especially in developing countries, and in those who are immunocompromised.

Treatment

Treatments are based on the type of pituitary adenoma and its size. Prolactinomas are usually treated with the dopamine agonists called cabergoline or quinagolide. They decrease the size of the tumor. Serial imaging is used to detect any increases in size. If the tumor is large, radiation therapy, **proton therapy**, or surgery can be used—usually with good results. Progesterone antagonists have been tried in the treatment of prolactinomas without success. Somatotrophic adenomas often respond to the long-acting somatostatin analog called octreotide. Thyrotrophic adenomas usually do not respond well to treatment with dopamine agonists. For surgery, the most common approach is **transsphenoidal** adenectomy, which usually removes the tumor and leaves the brain and optic nerves unaffected. The indications, adverse effects, and mortality considerations of transsphenoidal pituitary surgery are detailed in Table 15.2. Transsphenoidal surgery is performed through the sphenoid sinus. Many neurosurgeons use a direct transnasal approach. An incision is made in the back wall of the nose, and the sphenoid sinus is entered directly. Another method is to make an incision along the front of the nasal septum, and create a tunnel back to the sphenoid sinus. The final option is to make an incision under the lip and approach through the upper gum, entering the nasal cavity, and then the sphenoid sinus (see Fig. 15.3). Other surgical approaches include endoscopic surgery, which is usually **endonasal**, and craniotomy for rare, invasive, suprasellar masses that extend into the frontal or middle cranial fossa, optic nerves, or for significant posterior **clival** invasion.

Pituitary radiation can also be used, in which high-energy ionizing radiation is sent into the deep tissues using megavoltages. Today, this is much more effective and safe than in previous decades. Radiation is administered in multiple visits over 5—6 weeks. Highly precise techniques include stereotactic confocal radiotherapy, gamma knife with focused

TABLE 15.2 Considerations of transsphenoidal pituitary surgery.

Main indications	Adverse effects	Related mortality (up to 1%)
Compression of visual tract or CNS from within the sella	*Transient:* Diabetes insipidus	Brain or hypothalamic injury Vascular damage
Relief of compressive hypopituitarism due to the tumor	CSF leak with rhinorrhea Abnormal antidiuretic hormone secretion	Meningitis after surgery CSF leak
Tumor recurrence after previous treatments	Arachnoiditis	Pneumocephalus
Pituitary hemorrhage	Meningitis	Acute cardiopulmonary disease
Leak of cerebrospinal fluid	Psychosis after surgery.	Effects of anesthesia
Resistance or intolerance to therapies	Local hematoma.	Seizures
Personal patient choice	Arterial wall drainage	
Desire for immediate pregnancy with macroadenoma	Epistaxis Local abscess	
Need for diagnostic tissue histology	Pulmonary embolism	
Acromegaly	Narcolepsy	
Cushing's disease	Permanent (up to 10%)	
Nonfunctioning macroadenoma	Diabetes insipidus	
Prolactinoma	Hypopituitarism	
Nelson's syndrome (caused by removal of both adrenal glands)	Visual loss Abnormal antidiuretic hormone secretion	
TSH-secreting adenoma	Vascular occlusion	
	Oculomotor palsy, hemiparesis, or encephalopathy	
	Nasal septum perforation	

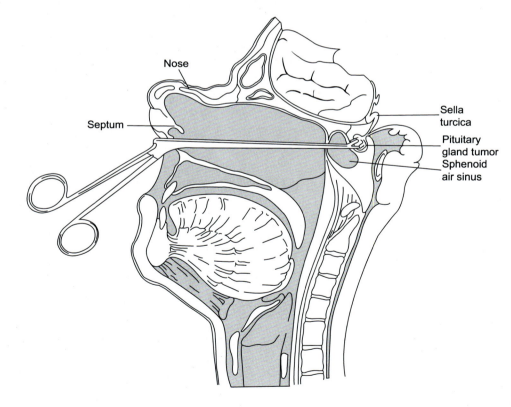

FIGURE 15.3 Transsphenoidal surgical resection of a pituitary tumor. *Courtesy:* Cancer nursing principles and practice, *8th ed. (by Yarbro) (Jones & Bartlett Learning).*

cobalt-60, and proton beams. **Gamma knife radiosurgery** is the best procedure for intrasellar and cavernous lesions that are not close to the optic nerves. Complications of **stereotactic radiosurgery** include worsened hypopituitarism, thyroid problems, cortisol abnormalities, cranial nerve dysfunction, GH abnormalities, gonadotropin abnormalities,

TABLE 15.3 Factors concerning pituitary irradiation.

Indications	Adverse effects	Risks of secondary brain tumors
Pituitary adenoma (acromegaly, Cushing's disease, nonfunctioning adenoma, prolactinoma) Craniopharyngioma Hormone hypersecretion recurrence Nelson's syndrome Nonadenomatous invasive sellar mass Tumor recurrence	Hypopituitarism (inadequate GH, gonadotropin, TSH, and ACTH reserves) Brain: necrosis, cognitive dysfunction, temporal lobe deficits Eyes: visual loss, optic neuritis	Astrocytoma Gliomas Meningioma Meningeal sarcoma

additional surgery or radiation requirements, additional tumor growth, and diabetes insipidus. Use of radiation is extremely individualized per patient. Generally, radiation is used for persistent hormone hypersecretion or residual mass effects following surgery, or when a surgical procedure is contraindicated. Mostly, radiation is adjuvant to surgery. Adverse effects of radiation include hypopituitarism, secondary brain tumors, cerebrovascular disease, visual damage, and brain necrosis. Table 15.3 summarizes the indications, adverse effects, and risks of secondary brain tumors as a result of pituitary irradiation.

Prognosis

The prognosis for pituitary adenomas is based on any growth or spreading of the tumor. Since most tumors are benign, the outlook is generally excellent, unless the tumor develops as a carcinoma. If diagnosis is delayed, even a nonfunctioning tumor can cause problems if it increases in size to press on the brain, carotid arteries, or optic nerves. Survival rates are also based on the patient's overall health and age.

> **Point to remember**
> Tumors of the pituitary gland, including adenomas, are best diagnosed with MRI since it offers the best resolution for identification of soft tissue changes. The procedure should be specifically focused on the pituitary, and not be widely spaced. This will allow for detailed visualization of the tumor mass and its effects upon nearby soft tissue structures such as the cavernous sinus or optic chiasm. The MRI will also show the hypothalamus, pituitary stalk, and the sphenoid sinus. Most pituitary masses are distinguished by high-resolution T1-weighted sections in the coronal and sagittal plane, both before and after **gadolinium/pentetic acid** contrast administration.

> **Point to remember**
> The aggressive cell types of pituitary adenomas include acidophil stem cell, Crooke cell, densely granulated lactotrophic, gonadotrophic, null cell, plurihormonal, silent, sparsely granulated somatotrophic, and thyrotrophic adenomas. Of these the acidophil stem cell adenomas are rapidly growing and undifferentiated. They are immunopositive for prolactin and growth hormone.

Somatotroph adenoma

Somatotroph adenoma is a benign tumor of the anterior pituitary gland that secretes primarily GH. It often occurs from excessive GH, leading to gigantism, with or without acromegaly. There are two subtypes that include *densely granulated somatotroph adenomas (DGSAs)* and *sparsely granulated somatotroph adenomas (SGSAs)*. Somatotroph adenomas are also known as *GH adenomas, GH-cell adenomas, GH-producing adenomas, GH-secreting adenomas, somatotrope adenomas*, and *somatotrophinomas*. Other forms of adenomas may lead to gigantism and acromegaly, and secrete both GH and prolactin. These include the plurihormonal adenomas, **mammosomatotroph** adenomas, and the mixed lactotroph and somatotroph adenomas.

Epidemiology

More than 95% of cases of acromegaly and gigantism are linked to pituitary somatotroph adenomas. Acromegaly has a reported prevalence of 125−137 cases out of every 1 million people. Gigantism caused by excessive GH is rare. It usually occurs in childhood or puberty, before epiphyseal closure.

The prevalence of somatotroph adenomas has not been accurately evaluated. There is a nearly 1:1 male-to-female ratio. There appears to be no significant racial predilection for these tumors. Somatotroph adenomas occur in patients of all ages, but the average patient age at diagnosis is 47 years. These tumors make up 10%−15% of all surgically resected pituitary adenomas.

Etiology and risk factors

The tumorigenesis of somatotroph adenomas involves hormonal stimulation, genetic and epigenetic factors, and the effects of growth factors plus their receptors. In younger people, about 20% of cases of pediatric gigantism and 8%−11% of acromegaly cases are linked to an AIP gene mutation. About 50% of these cases have family history of familial isolated pituitary adenoma. Over 80% of cases of extremely early onset pediatric gigantism, with patients being under 5 years at diagnosis, are linked to X-linked *acrogigantism syndrome*. This syndrome involves a **microduplication** of the *G-protein-coupled receptor 101 (GPR101)* gene. It is important to note that GH excess also occurs along with multiple endocrine neoplasia types 1 and 4, Carney complex, and **McCune−Albright syndrome**. The SGSAs are the most common cause of acromegaly, and usually occur in younger female patients. The DGSAs occur more often in older people. The **mammosomatotroph** adenomas are very common in young acromegalic adults. However, the plurihormonal adenomas are exceedingly rare.

GNAS somatic-activating mutations have been documented in 40%−60% of somatotroph adenomas—usually adults, and often the DGSA subtype. Pituitary disease related to McCune−Albright syndrome has been linked to postzygotic somatic-activating GNAS mutations that cause activation of the cAMP/protein kinase A pathway. This results in increased cAMP levels in the cytoplasm, GH hypersecretion, and secretion of alpha-subunit in the somatotrophs. Not all DGSAs have G protein-mediated high cAMP levels. There are genomic imbalances at chromosomes 10q, 11q, and 13q. There are also documented losses in chromosomes 1, 6, 13−16, 18, and 22, plus gains at chromosomes 12 and X. There are also epigenetic alterations involving *caspase-8 (CASP8), cyclin-dependent kinase inhibitor 2A (CDKN2A, also abbreviated P16), death-associated protein kinase 1 (DAPK1), FGFR2, growth arrest and DNA-damage-inducible protein 45g (GAD-D45G), MGMT, p14, Ras association domain-containing protein 1 (RASSF1A), retinoblastoma 1 (RB1), rhomboid domain containing 3 (RHBDD3, or PTAG), thrombospondin 1 (THBS1, or TSP1)*, and *tumor protein 73 (TP73)*. Most of these alterations also occur in other types of adenomas. There has been documentation of **downregulation** of microRNAs that target transcription factors *E2F1, high-mobility group AT-hook 1 (HMGA1)*, and HMGA2. In somatotroph adenomas, the *interacting protein PTTG1 gene* has higher mRNA levels than in nonfunctioning adenomas. Further study has shown that PTTG1 gene expression was higher in hormone-secreting invasive adenomas.

Dysregulation of *fibroblast growth factor receptors (FGFRs)* may play a part in pituitary adenoma tumorigenesis. The chromatin-remodeling factor *ikaros family zinc finger 1 (IKZF1)*, which is important for hypothalamic GHRH neuron development, targets FGFR4 promoter activity within somatotrophs. Tumor size and excessive hormone in somatotroph adenomas are linked to the *FGFR4-R388* polymorphism. Pituitary adenomas are not always sporadic and may form as part of an isolated pituitary disease or as part of a syndrome.

Germline AIP mutations are seen in 20% of familial isolated pituitary adenomas. In most cases, patients are in their second or third decades of life. Only a few patients are younger than 10 years of age. Most patients have gigantism that is seen more often in males, and the majority of AIP mutations result in somatotroph adenomas. There are also mixed somatotroph−lactotroph adenomas, and pure lactotroph or pure plurihormonal adenomas. When nonfunctioning, pituitary adenomas usually stain for GH, PRL, or both. However, corticotroph, gonadotroph, and thyrotroph adenomas are rarely seen. When patients have germline or somatic *G protein-coupled receptor 101 (GPR101)* microduplication, they usually present at the age of 5 years of less. Pituitary adenoma or hyperplasia is present in 25% of cases. Most patients have hyperprolactinemia. Usually, there are mixed somatotroph−lactotroph adenomas instead of mammosomatotroph adenomas. There are no identified risk factors for somatotroph adenomas.

Pathology

Somatotroph adenoma arises from PIT1-lineage cells. Most of these tumors are intrasellar. Some are macroadenomas that may or may not invade the cavernous sinus. There have also been documented cases of ectopic localization.

Somatotroph adenomas appear soft, and are white to gray-red in color. They often invade nearby structures. When pituitary apoplexy is present, the tumor will contain hemorrhagic necrosis. After treatment, adenomas can become fibrotic.

Both a somatotroph adenoma and mammosomatotroph hyperplasia can cause gigantism or acromegaly. Adenohypophyseal proliferation may be the first pathologic change. Adenomas can be distinguished from hyperplasia because of a disrupted reticulin framework. There has been a documented link between SGSAs with a ganglionic component and AIP gene mutations. Adenomas that only cause GH hypersecretion originate from somatotrophs. The two subtypes of DGSAs and SGSAs are based on density of GH-containing secretory granules and low-molecular-weight secretion of cytokeratin.

The DGSAs are made up of tumor cells containing eosinophilic cytoplasm. This is related to high density of GH-containing secretory granules. The DGSAs have GH, alpha-subunit, and PIT1 positivity. Perinuclear staining occurs for low-molecular-weight cytokeratin, which is CAM5.2 or CK18. The tumor cells have many large secretory granules between 400 and 600 nm in size. The tumors have high levels of GH and IGF-1, and are slightly smaller, with a more benign course than the SGSAs. There is a proposed *intermediate-type somatotroph adenoma* for DGSAs that have occasional fibrous bodies (see Fig. 15.4).

The SGSAs consist of chromophobic to pale eosinophilic cells that are positive for PIT1. There may be nuclear pleomorphism with multinucleated bizarre cells. Reactivity ranges from weak to patchy or focal. There is sparse granularity and variable GH positivity. These tumors do not have alpha-subunit secretion. Via low-molecular-weight cytokeratin staining, there are usually fibrous bodies in 70% or more of the cells. These bodies are also known as **juxtanuclear keratin aggresomes**. Since their nuclei are indented, the fibrous bodies may be recognized via H&E staining. There are sparse secretory granules of 100−250 nm in size. The fibrous bodies are round, with keratin filaments and trapped secretory granules. There is a varying level of endoplasmic reticulum. The SGSAs are fairly large, with lower levels of GH and IGF-1. Sometimes, gangliocytomas are admixed with SGSAs.

Mixed somatotroph−lactotroph adenomas contain densely granulated eosinophilic cells that secrete PIT, and sparsely granulated chromophobic cells. Lactotrophs are usually positive for PRL and estrogen receptor. The densely granulated lactotrophs have diffuse cytoplasmic PRL reactivity, while the sparsely granulated lactotrophs have Golgi-type prolactin staining. Lactotrophs are negative for alpha-subunit. Both GH and alpha-subunit secretion varies between types of somatotrophs. Ultrastructural examination may be required to distinguish the mixed adenomas from the mammosomatotroph adenomas. Visualizing pleomorphic and heterogeneous secretory granules does this.

Plurihormonal adenomas with GH hypersecretion are rare. The adenomas that produce GH are classified in this group, yet acidophil stem cell adenomas plus plurihormonal PIT1-positive pituitary adenomas can also be involved with acromegaly. GH-producing plurihormonal adenomas are mostly DGSAs that have varying mammosomatotroph and thyrotroph differentiation. There is positivity for alpha-subunit, GH, and PIT1. There is variable positivity for estrogen receptor, low-molecular-weight cytokeratin (perinuclear), PRL, and TSH-beta. The ultrastructural factors are the same as the PIT1-lineage adenohypophyseal cells.

Mammosomatotroph adenomas mostly contain eosinophilic cells with secretory granules that contain GH and PRL. These tumors are positive for alpha-subunit, estrogen receptor, GH, low-molecular-weight cytokeratin, PIT1, and PRL.

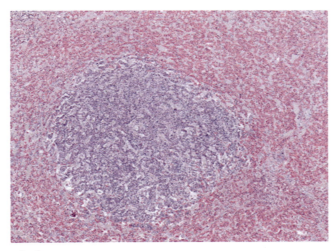

FIGURE 15.4 Histology of somatotroph adenoma.

They have pleomorphic, heterogeneous secretory granules of 150–1000 nm in size. There is extrusion of the granules along the lateral cell surface, an abnormal form of exocytosis seen with PRL secretion.

Plurihormonal adenomas with GH hypersecretion are rare. The adenomas that produce GH are classified in this group, yet acidophil stem cell adenomas plus plurihormonal PIT1-positive pituitary adenomas can also be involved with acromegaly. GH-producing plurihormonal adenomas are mostly DGSAs that have varying mammosomatotroph and thyrotroph differentiation. There is positivity for alpha-subunit, GH, and PIT1. There is variable positivity for estrogen receptor, low-molecular-weight cytokeratin (perinuclear), PRL, and TSH-beta. The ultrastructural factors are the same as the PIT1-lineage adenohypophyseal cells.

While the majority of somatotroph adenomas have biochemical activation, there are also rare silent tumors. Gigantism and/or acromegaly, with amenorrhea and galactorrhea, develop as the result of mammosomatotroph and mixed somatotroph–lactotroph adenomas. This is because of cosecretion of PRL and GH. With acromegaly, multinodular goiter is common, which may or may not be accompanied by hyperthyroidism. In rare cases, central hyperthyroidism occurs because of TSH cosecretion—primarily with plurihormonal adenomas. Signs and symptoms include headache, hypopituitarism, mild hyperprolactinemia, visual field defects, and pituitary apoplexy. These manifestations are linked to invasive somatotroph macroadenomas. In AIP-mutated pediatric cases, symptomatic apoplexy occurs more often than in adults.

Diagnosis

Often, the diagnosis of acromegaly linked to these tumors is relatively easy, yet can be delayed. It has taken between 4 and 10 years from onset of symptoms to accurate diagnosis. In most patients there are increased levels of serum GH and IGF-1, with the GH levels fluctuating, as the IGF-1 levels are stable. The patient with a somatotroph adenoma, when taking oral glucose, will not have full suppression of GH secretion. Diagnosis of GH hypersecretion results from an IGF-1 level that is above the age-specific normal range. It is confirmed via an oral glucose tolerance test, with GH levels not being suppressed to less than 1 ng/mL. If an ectopic source of growth hormone–regulating hormone (GHRH) is suspected in relation to acromegaly, plasma GHRH is good for diagnosis of a tumor producing this hormone, using a level of 250–300 ng/L.

Using an MRI scan, a somatotroph adenoma is usually easily seen. Most of these tumors are macroadenomas. However, DGSAs may be smaller and not invade the cavernous sinus as often. They also are less likely to compress the optic chiasm in comparison with the SGSAs. Responsiveness to somatostatin analogs can be predicted by the T2-weighted intensity of the tumor. The majorities of DGSAs are T2-hypointense and respond to somatostatin analogs. Most SGSAs are invasive macroadenomas with T2-hyperintensity. When the pituitary is of normal or enlarged hyperplastic quality, the correct diagnosis may be done by measuring GHRH and then locating a tumor that secretes this hormone. X-linked acrogigantism syndrome should be evaluated for a young patient with acromegaly but a normal MRI scan.

When a multifocal somatotroph adenoma or one that is linked to hyperplasia exists, there is a chance of a link between Carney complex, GHRH-secreting tumors, or McCune–Albright syndrome. Most cases of MEN1-related pituitary hyperplasia are linked to GHRH-producing neuroendocrine tumors of the pancreas. Somatotroph adenomas have also been documented as double adenomas, with lactotroph, gonadotroph, or corticotroph adenomas.

Treatment

For some somatotroph adenomas treated with somatostatin analogs, there have been various effects. These include enlarged secretory granules, increased amounts of lysosomes, stromal fibrosis, and increased tumor cell apoptosis. There is insufficient data about the effects of the drugs called *cabergoline*, *pegvisomant*, and *temozolomide*. Another treatment option is total or partial surgical resection, without any administration of postsurgical medications.

Prognosis

Prognosis is poor if there is uncontrolled GH or IGF-1 secretion. A cure is usually not possible when a somatotroph adenoma has invaded the cavernous sinus. Outcomes are calculated based upon histological subtype, tumor invasiveness, the biomarker profile, and the Ki-67 proliferation index. The SGSAs are less responsive to somatostatin antagonists, so use of pegvisomant, a GH receptor antagonist may result in a better outcome. Treatment response can be predicted by immunohistochemistry detection of somatostatin receptors.

Prolactinoma

Prolactinomas are the most common type of pituitary adenomas. They produce prolactin, resulting in hyperprolactinemia. Prolactin stimulates the breasts to produce milk, helps regulate mood, and is seen in increased levels during pregnancy and after childbirth. In males, prolactin regulates the sexual refractory period after orgasm, but high levels can cause erectile dysfunction. Prolactinomas are classified as *microprolactinomas* when they are less than 10 mm in diameter, and as *macroprolactinomas* if they are larger than that size.

Epidemiology

Prolactinomas have an annual incidence of about 30 cases per 100,000 people. However, this calculation would be much higher if it included the microadenomas that are discovered in about 11% of autopsies. Of these, 46% **immunostain** positively for prolactin (PRL). Microprolactinomas are 20 times more common in females compared to males, while macroprolactinomas are nearly evenly distributed between the genders. Prolactinomas make up over 75% of all female pituitary adenomas, yet tumor size is larger in men.

Etiology and risk factors

The cause of prolactinomas is unknown. Though stress can greatly increase prolactin levels, it is not considered to be a cause or risk factor for prolactinomas. Most moderately raised prolactin levels are not due to microprolactinomas, and are linked to some prescription medications that include phenothiazines such as chlorpromazine, butyrophenones such as haloperidol, atypical antipsychotics such as risperidone and paliperidone, gastroprokinetic drugs such as metoclopramide and domperidone, and less commonly, alpha-methyldopa, reserpine, estrogens, thyroid-releasing hormone, ramelteon, and etizolam. Other causes include different pituitary tumors, normal pregnancy, and breastfeeding. The **xenoestrogenic** chemical called *bisphenol-A* leads to hyperprolactinemia and the growth of prolactin-producing pituitary cells. Increased, prolonged exposure from childhood may contribute to prolactinoma growth.

Pathology

Prolactin levels and tumor size are usually stable for prolactinomas, though some patients experience decreased prolactin levels over time. Macroprolactinomas are more likely to grow, with tumor size being related to serum prolactin levels. This means that a prolactin level over 200 ng/mL strongly indicates a PRL-secreting tumor. In males, prolactinomas are not only larger, but are also more invasive, with faster growth. Giant prolactinomas are those larger than 4 cm in diameter, with a serum PRL of over 1000 ng/mL, and are more common males.

Over 99% of prolactinomas are benign and sharply demarcated, showing no evidence of invasion. However, about 50% invade local structures. These invasive tumors may have more mitotic activity, and are also more cellular and pleomorphic. Any invasion into adjacent bone, dura, or venous structures may signify an intermediate type of tumor, between the sharply demarcated benign form and the very rare malignant form. An invasive tumor that does not metastasize is considered to be benign. Prolactin immunostaining confirms diagnosis, which is often distinct from nearby normal pituitary tissues, but not actually encapsulated. The tumors have a pseudocapsule made up of compressed adenohypophyseal cells. There is a reticulin fiber network. Approximately 20% of macroprolactinomas have hemorrhagic areas not usually linked with apoplexy. These areas may return to normal.

Prolactinomas usually grow slowly. They arise sporadically, usually occurring as one tumor. They are the most common pituitary tumors linked to MEN1, and occur in about 20% of cases, along with many other types of pituitary tumors. Familial prolactinomas have occurred with no other MEN1 features. In rare cases, these tumors occur in patients who have germline *AIP* gene mutations.

Clinical manifestations

Prolactinomas usually manifest with signs and symptoms related to hyperprolactinemia, or are found because of their size or invasion of other tissues. The signs and symptoms associated with tumor size include blurred vision or decreased visual acuity, headaches, pituitary apoplexy, temporal lobe seizures, symptoms of hypopituitarism, visual field abnormalities, and rarely, cranial nerve palsies, hydrocephalus, or unilateral **exophthalmos**. The signs and symptoms related to hyperprolactinemia include amenorrhea, oligomenorrhea, infertility, decreased libido, erectile dysfunction, impotence, oligospermia, premature ejaculation, galactorrhea, and osteoporosis. As many as 50% of women and 35% of men may have galactorrhea, with the difference in percentages possible due to male mammary tissue being less susceptible to the lactogenic effects of hyperprolactinemia. An increase in vertebral fractures has been found in imaging studies of women with prolactinomas.

Microprolactinomas range from as small as 2—3 mm in diameter to approximately 10 mm in diameter. They can be invasive, regardless of their size. Macroprolactinomas range in size from noninvasive or diffuse tumors of about 1 cm in diameter to huge sizes that can impinge on parasellar structures. The manifestations from large or invasive tumors are often related to compression upon visual structures. The most common complaint in one study of 1000 patients was vision loss. There may be superior bitemporal defects or hemianopsia (also called *hemianopia*), and decreased visual acuity. Headaches are common, though seizures as a result of extension into the temporal lobe are rare. Though the tumors often invade the cavernous sinuses, cranial nerve palsies are not common in occurrence.

Diagnosis

Prolactinomas are often signified by PRL levels above 200 ng/mL, though these levels can also occur due to use of a drug such as risperidol. However, levels over 500 ng/mL only occur with a prolactinoma. Oppositely, a PRL concentration below 200 ng/mL that accompanies a macroprolactinoma means that the tumor is probably not producing PRL. Hyperprolactinemia may occur due to mass pressure upon the pituitary stalk or portal circulation, probably interrupting inhibitor control by dopamine. Microprolactinomas are linked to PRL levels between only slight elevations to hundreds of nanograms per milliliter. Sometimes, prolactinomas are found during an MRI or CT scan performed for another purpose.

All pituitary tumor patients should have serum PRL measurements. Oppositely, patients with elevated serum PRL that is not explained by pregnancy or use of neuroleptic medications should be evaluated for a pituitary tumor. Prolactinomas may coexist with another cause of hyperprolactinemia. Any PRL elevations should be investigated since they can indicate a large pituitary tumor that does not secrete PRL. However, PRL levels are linked to tumor size, and are usually higher in males. An extremely high serum PRL level that appears normal when serum dilutions are not assayed explains the high-dose hook effect. Also, serum PRL can be elevated from high-molecular-weight PRL, which is weaker than the monomeric PRL molecule. Macroprolactinemia can also occur with other pituitary tumors. In 20% of patients with macroprolactinemia, pituitary adenomas are diagnosed. Some of these patients have **galactorrhea**, oligomenorrhea, amenorrhea, erectile dysfunction, or decreased libido. Evaluation of macroprolactinemia using **polyethylene glycol precipitation** should be done for any patient with high levels of PRL but few or no clinical features of hyperprolactinemia.

Careful patient history often reveals signs or symptoms of a space-related tumor, including impaired visual acuity, visual field abnormalities, blurred or double vision, cerebrospinal fluid (CSF) rhinorrhea, diabetes insipidus, headaches, and hypopituitarism. The patient must be carefully asked about sexual history, including beginning of menarche, menses regularity, fertility, libido, potency, and ability to maintain erections. Galactorrhea history should be determined. Galactorrhea with amenorrhea is suggestive of a pituitary adenoma unless this is disproven. In up to 50% of patients with acromegaly, prolactin is elevated. The patient in the early stage, or with mild disease or acidophilic stem cell adenomas may only have a few signs of excessive GH. Prolactinoma can be mimicked by a tumor that only secretes GH, so serum IGF-1 should be assessed. Also, elevated PRL is sometimes seen in patients with TSH-secreting tumors. Various pituitary hormone functions must be determined to assess for hypopituitarism. Definitive diagnosis of a prolactinoma requires an MRI.

Treatment

Treatments for prolactinoma include normalizing prolactin levels, and complete tumor removal or shrinkage to reverse tumor-mass effects. Abnormal sexual function and fertility issues should be restored, galactorrhea should stop, and bone density should improve. Pituitary or hypothalamic function should not be impaired, and vision should return to normal. Dopamine agonists are the drugs of choice, and include *bromocriptine* and *cabergoline*. Bromocriptine is a semisynthetic ergot alkaloid that lowers PRL levels, restores menstrual function in 80%—90% of patients, shrinks prolactinomas, restores impaired sexual function, and resolves galactorrhea. Also, about 90% of patients experience improved vision. Bromocriptine acts upon the cytoplasmic, nuclear, and nucleolar areas of tumor cells. If withdrawn, there may be rapid expansion of the tumor. Some patients are partially or completely resistant to the effects of bromocriptine as well as cabergoline. The main difference between the two medications is that cabergoline has a longer duration of action. It is usually administered once or twice per week, and has become the current first-line medication, except for women desiring to become pregnant.

It is important to medically treat any patient with a small macroprolactinoma and a PRL level of approximately 200 ng/mL. Dopamine agonists should lower PRL levels, and shrink prolactinomas. If it does not shrink, the tumor is probably not secretory, and the hyperprolactinemia is being caused by compression of the pituitary stalk.

Microprolactinomas may disappear when dopamine agonist therapy is discontinued, yet 7%–14% continue to increase in size. Also, smaller prolactinomas sometimes regress following pregnancy and lactation. Adverse effects of bromocriptine and cabergoline include nausea, nasal blockage, depression, and digital vasospasm. Postural hypotension sometimes occurs, influencing dosage size. A small amount of patients have psychotic symptoms or worsening of preexisting psychoses. Rare adverse effects include CSF rhinorrhea, liver dysfunction, and cardiac arrhythmias. High doses of bromocriptine are linked to pleural effusions and thickening, retroperitoneal fibrosis, and restrictive mitral regurgitation. Other medications that may be used include olanzapine.

Additional treatments include radiation therapy, surgery, and chemotherapy. Linear accelerator radiotherapy helps control or reduce tumor size but requires years of therapy, with a mean of 7.3 years. An adverse effect of radiation is hypopituitarism. Gamma knife stereotactic radiosurgery is often successful when the patient is resistant to dopamine agonists or cannot tolerate them. Successful rates of pituitary surgery are linked inversely with tumor size and concentrations of serum PRL. Overall, serum PRL is normalized in 71% of patients with microprolactinomas after surgery. The rate of hyperprolactinemia recurrence is estimated at about 17% of patients who are initially considered to be cured. Complete resection of macroprolactinomas, especially when they are large and invasive, is difficult. Only 32% of patients experience normalization of serum PRL levels after surgery, and there is a 19% recurrence rate. When a prolactinoma can only be partially removed, adjunctive radiation therapy should be evaluated. For a woman with a large prolactinoma that could threaten vision, prophylactic transsphenoidal surgery is considered. Also, endoscopic endonasal transsphenoidal surgery has been used for resection of prolactinomas. The majority of patients with microprolactinomas experience return of PRL levels to normal, with about 50% of macroprolactinoma patients achieving remission following surgery. Chemotherapy is used for aggressive PRL-secreting tumors that do not respond to other treatments. Temozolomide is an alkylating compound that has controlled tumor growth in certain patients. During pregnancy, the pituitary gland enlarges, as does any prolactinoma that may be present. Generally, medications are avoided in pregnant patients with prolactinomas, though bromocriptine has been used without significant fetal harm. Women of childbearing age who have a prolactinoma should be tested for sensitivity to dopamine agonists before becoming pregnant.

Prognosis

People with microprolactinomas usually have an excellent prognosis. In 95% of cases, the tumor shows no signs of growth after a 4- to 6-year period. Macroprolactinomas usually require more aggressive treatment to stop them from growing. Regular monitoring is recommended.

Thyrotropin-secreting tumors

These rare TSH-producing pituitary tumors make up only 0.85%–2.8% of cases. They can cosecrete GH, PRL, and in rare cases, ACTH. They also cause elevated serum IGF-1 or PRL levels. They are sometimes referred to as *thyrotropinomas* or *TSHomas*.

Epidemiology

Thyrotropin-secreting tumors cause less than 1% of all cases of hyperthyroidism. Incidence is about 2.8 of every million people. Epidemiology has not been widely studied throughout the world.

Etiology and risk factors

The causes and risk factors of TSH-secreting tumors are unknown. However, they can be associated with long-standing primary hypothyroidism.

Pathology

Though invasive, these tumors are usually benign. There are rare cases of distant metastases. Immunoreactive antibodies to TSH-beta, alpha-subunit, GH, PRL, and ACTH determine secretion. Over 24-hour sampling, the pulse frequency of TSH increases, and diurnal rhythms are at a higher mean hormone level. The tumors show positive immunostaining for alpha-subunit and TSH-beta in as many as 75% of the cells, and for pituitary-specific transcription factor (Pit-1). The tumors secrete somatostatin receptor type 2 (SSTR2) mRNA. In some cases, they secrete somatostatin receptor type 3 and 5 (SSTR3 and SSTR5) mRNA.

Clinical manifestations

Patient may present with symptoms of tumor growth. These include visual field abnormalities, headache, and cranial nerve palsies. Signs and symptoms of hyperthyroidism include arrhythmias, palpitations, weight loss, tremor, nervousness, or goiter. Rarely, periodic paralysis and postoperative thyroid storm have been seen. Serum TSH is usually elevated. When it is, abnormally high TH levels plus a normal TSH value indicates a TSH-producing pituitary tumor. There may be a long period of hyperthyroidism, often misdiagnosed and treated as Graves' disease, before the tumor is identified. Thyroid hormone insensitivity can present with a similar laboratory profile. The tumors are usually large, with 70%−88% being macroadenomas. More than 60% are locally invasive. The TSH level is highly elevated in about 58% of cases. One case of an ectopic TSH-producing tumor has been seen. Most patients have higher serum thyroxine as well as the glycoprotein hormone alpha-subunit. About 66% have goiters with elevated radioactive iodine uptake. In rare cases, differentiated thyroid cancer is also present. About 30% of patients also have cosecretion of GH or PRL. Therefore features of acromegaly or hyperprolactinemia may be present.

Diagnosis

Serum thyroxine, triiodothyronine, TSH via a high-sensitivity assay, and alpha-subunit are measured. Strong confirmation of a TSH-producing tumor is given by high thyroxine, triiodothyronine, and alpha-subunit with high or inappropriately normal TSH. Stimulation of thyroid-regulating hormone (TRH) can differentiate between overproduction of TSH by a tumor and TH insensitivity. If there is a tumor, the TSH response brought about by TRH will be reduced. TSH usually increases as a response to TRH in normal individuals, and with TH insensitivity. It is good to measure alpha-subunit at every point during the TRH test, since the molar ratio of this to TRH is high, over 1, in nearly 85% of patients with these tumors. Triiodothyronine suppression tests help, since total inhibition of TSH does not happen in patients with TSH-secreting tumors. The tests can differentiate from a tumor and subclinical hypothyroidism following previous radioactive iodine treatments for hyperthyroidism. Elevated TSH may also occur from insufficient TH replacement. A pituitary MRI is done, along with determination of IGF-1 and PRL levels, to exclude acromegaly or hyperprolactinemia. The amount of hyperthyroidism should be evaluated, since it is often severe with the tumor. Perioperative death has occurred, which could be linked to poorly controlled hyperthyroidism.

Treatment

Treatment for TSH-secreting tumors includes preoperative management, surgery, radiation therapy, and somatostatin analogs. Surgery is the first-line therapy. Unless the patient's vision is threatened, the patient is evaluated to determine if hyperthyroidism requires immediate treatment. This may require propranolol, radioactive iodine thyroid ablation, thyroidectomy, tapazol or other antithyroid medications, and somatostatin analogs. Radioactive iodine and antithyroid medications inhibit negative feedback of triiodothyronine on TSH, leading to increased tumor TSH production. Surgery and somatostatin analogs treat hyperthyroidism along with tumor TSH hypersecretion. Somatostatin analogs lower alpha-subunit, thyroxine, and TSH. They are the first-line drugs in initial control of hyperthyroidism from a TSH-secreting tumor, since their onset is faster than other therapies. Plus, tumor shrinkage occurs in as many as 40% of patients. If there is invasive tumor tissue, the patient has abnormal TSH responses to TRH, requiring somatostatin analogs.

For most patients with microadenomas, surgery is curative. However, remission is achieved in less than 60% with macroadenomas. There have been no large controlled studies because of how rare these tumors are. There is usually cavernous or sphenoid sinus invasion. The tumors are often very hard and fibrous. About 33% of patients need radiotherapy for biochemical normalization. About 9% develop pituitary hormone deficiencies, and 3% have tumor recurrence or hyperthyroidism in the first 2 years.

Treating of TSH-secreting tumors using only radiotherapy is not well documented. Radiation is usually used as adjunctive treatment with surgery, especially when the surgery is not curative. The somatostatin analogs include octreotide and lanreotide. Octreotide is used either as primary or adjunctive treatment. It normalizes thyroxine and triiodothyronine, and reduces TSH levels by 50%. Tumors shrink in about 33% of patients. Lanreotide can greatly decrease TSH and thyroxine levels, but usually does not shrink the tumors. The long-acting form of octreotide, given monthly, appears to work as well as the shorter acting form. In one report, the medication suppressed TSH in 90% of patients and reduced tumor size by 50%. Somatostatin analogs are effective in over 90% of cases, though tachyphylaxis sometimes occurs.

Prognosis

Prognosis of TSH-secreting tumors is generally excellent when surgery is undertaken, but this is for microadenomas and not macroadenomas. After a mean follow-up period of 64.4 months, biochemical control is achieved in about 75% of patients. About 58% have normalized pituitary imaging and thyroid function, but a surgical cure occurred in fewer than 40%.

> **Point to remember**
> There is also a subset called *silent TSH-secreting tumors*. They immunostain positively for TSH, but do not hypersecrete it. They also do not cause thyrotoxicosis. In a study of 29 of these tumors, 9 of them were not linked to hyperthyroidism.

Adrenocorticotropic hormone–secreting tumors

ACTH-secreting tumors are also described as *corticotrophic adenomas, ACTH-producing adenomas, corticotropinomas, corticotropic adenomas*, and *corticotrope adenomas*. Along with ACTH, other proopiomelanocortin peptides are secreted.

Epidemiology

ACTH-secreting tumors make up about 15% of all pituitary adenomas, with a peak incidence in patients between 30 and 50 years of age. In prepubertal children, males are more often affected, but in adults, there are more female cases. In children, ACTH-secreting adenomas make up about 55% of all pituitary adenomas under the age of 11 years. They make up 75% of all cases with Cushing's syndrome in children under 5 years. Sometimes an ACTH-secreting tumor, or more rarely *corticotroph hyperplasia*, results in development of Cushing's disease, which has an annual incidence of 3–10 cases per 1 million people. Since many cases of mild, subclinical Cushing's disease are not diagnosed, this figure may be extremely underestimated.

Etiology and risk factors

ACTH-secreting tumors are derived from *T-box factor, pituitary (TPIT)*-lineage adenohypophyseal cells. Formation of ACTH-secreting tumors is due to genetic and epigenetic factors, growth factors and their receptors, and hormonal stimulation. The tumors have normal secretion of stimulatory vasopressin (ADH) and corticotropin-releasing hormone (CRH) receptors and normal glucocorticoid receptor secretion. They have slightly reduced ACTH feedback receptor secretion. No mutations are usually found. There is an alteration of the ratio of 11-beta-dehydrogenase isoenzymes 1 and 2. This deranges cortisol feedback, which also occurs in other types of tumors. There are alterations in secretion of cyclin-dependent kinases, cyclins, and their regulators. Specific alterations include extremely reduced p27 protein secretion and increased cyclin E secretion. There are changes in feedback transcription factors such as *histone deacetylase 2 (HDAC2)* and *SWI/SNF-related, matrix-associated, actin-dependent regulator of chromatin, subfamily A, member 4 (SMARCA4, also known as BRG1)*. Also, there is increased activity of pro-proliferative signal pathways linked to *mitogen-activated protein kinase (MAPK)* and *phosphatidylinositol-4,5-bisphosphate 3-kinase (PI3K)/mechanistic target of rapamycin kinase (mTOR)*.

There are a few specific somatic abnormalities related to ACTH-secreting tumors. The *epidermal growth factor receptor (EGFR)* signal pathway is implicated, since activating somatic mutations in the *ubiquitin-specific peptidase 8 (USP8)* gene have been found in 40%–60% of all of these tumors. The mutations increase mRNA levels of proopiomelanocortin and EGFR secretion. Genotype–phenotype studies reveal that USP8 mutations are not present in silent corticotroph adenomas. The majority of USP8-mutant lesions occur as small, functional tumors at earlier ages than are involved in the wildtype tumors. USP8-related tumors have higher secretion of *O-6-methylguanine-DNA methyltransferase (MGMT)* and *somatostatin receptor 5 (SSTR5)* than tumors without the mutations. Therefore a USP8 mutation may predict tumor responsiveness to *pasireotide*. Increased EGFR secretion results from *protein kinase C delta (PRKCD)* silencing. In a recent study, PRKCD was found to slow the growth of ACTH-secreting tumors, and along with *microRNA-26a (miR-26a)*, plays an important role in controlling tumor cell cycles.

Cushing's disease may or may not be linked to corticotroph hyperplasia, which is very rare and usually features diffuse corticotroph proliferation. This is caused by ectopic CRH-secreting tumors or untreated **Addison's disease**, which is also known as *primary adrenal insufficiency and hypocortisolism*. Examples of the ectopic CRH-secreting tumors include neuroendocrine carcinomas, gangliocytomas, and paragangliomas. There are signs of a corticotroph hyperplasia to neoplasia progression, with chronic untreated Addison's disease and with chronic stimulation of ectopic CRH by a gangliocytoma.

Only a few cases of Cushing's disease have been reported along with familial isolated pituitary adenoma, MEN1, or MEN4 (also called CDKN1B). There were two cases occurring with tuberous sclerosis (TSC1/2). One case occurred with MEN2. Along with Cushing's disease, there have been no documented germline mutations of *GNAS complex locus (GNAS)*, *protein kinase CAMP-dependent type I regulatory subunit alpha (PRKAR1A)*, or USP8. With pituitary blastoma, which causes Cushing's disease in infants (in rare cases), there have been germline *Dicer 1, ribonuclease III (DICER1)* mutations.

Pathology

Most ACTH-secreting tumors are within the pituitary gland, but have developed ectopically in the nasal cavity, cavernous or sphenoidal sinuses, clivus, eye orbits, and suprasellar region. Histologically, they are classified as one of three types: *densely granulated corticotroph adenomas* (most common), *Crooke cell adenoma*, and *sparsely granulated corticotroph adenomas*. These tumors are usually small, soft, and may be less than 2 mm in diameter. They are often yellow-white to grayish-red in color and have petechiae. Most are solitary microadenomas, though 5%—10% are macroadenomas. These can be components of double adenomas or intermingled with gangliocytomas or adrenal cortical choristomas. They have a basophilic appearance and have a positive periodic acid-Schiff (PAS) reaction. The tumors are always diffusely positive for low-molecular-weight cytokeratin (CAM5.2), *neuronal differentiation 1 (NeuroD1)*, and TPIT.

The densely granulated form is made up of basophilic PAS-positive cells. These are strongly and diffusely positive for ACTH. They are also consistent with the amount of secretory granules when the ultrastructure is evaluated. The sparsely granulated form is made up of slightly basophilic or chromophobic cells that are PAS-positive. They have patchy or weak ACTH positivity, which is consistent with the type of granules that are ultrastructurally present. The Crooke cell adenomas consist of Crooke hyaline changes and ring-like cytokeratin secretion. The secretion of ACTH is close to the cell edges and juxtanuclear region. Intermediate filaments appear within a ring-like pattern.

The majority of ACTH-secreting tumors have biochemical function, though 20% do not have an evidence of excessive ACTH or cortisol. These are called *silent corticotroph adenomas*, commonly discovered incidentally, or if there are ophthalmic or neurologic symptoms such as acutely hemorrhagic necrosis and apoplexy. The silent form often presents as macroadenomas that invade the cavernous sinuses. There are two types of silent ACTH-secreting tumors:

- Type 1—densely granulated and
- Type 2—sparsely granulated, with higher expression of biomarkers of cell proliferation, migration, and invasion. These biomarkers include CD29 (integrin beta-1), *FGFR4*, and *matrix metallopeptidase 1 (MMP1)*. In rare cases ACTH secretion is lacking, but TPIT is secreted.

Proliferative activity is varied, but reduced p27 secretion is common. Galectin-3 secretion is weak or absent in all silent corticotroph adenomas. Galectin-3 may also be secreted in aggressive tumors.

Clinical manifestations

Signs and symptoms of Cushing's disease include central upper body obesity with supraclavicular fat accumulation, thin arms and legs, a moon-shaped face, and in children, a slowed growth rate. Other signs and symptoms include acne or skin infections, purple striae on the skin and thinning of the skin, easy bruising, backache, bone pain or tenderness, a buffalo hump of fat on the back between the shoulders, plus bone and muscle weakening. Women may have excessive hair growth, and changes in their menstrual cycles (see Fig. 15.5). Men may have low libido and erection problems. Other symptoms include mental changes, fatigue, frequent infections, headache, increased thirst and urination, hypertension, glucose intolerance, and type 2 diabetes. Hyperpigmentation from overproduction of proopiomelanocortin-derived peptides may develop because of severe Cushing's disease and **Nelson's syndrome**. This syndrome involves fast enlargement of corticotroph adenomas with hyperpigmentation and weakening of the muscles—often after a bilateral adrenalectomy procedure.

Diagnosis

Diagnosis is suggested by hypercortisolism, elevated 24-hour urinary-free cortisol, elevates late night salivary cortisol, and unsuppressed serum ACTH levels. Testing is based on a failure of normal feedback and loss of circadian rhythms. Diagnosis is supported by failure to suppress morning cortisol levels to below 1.8 ng/dL following administration of 1 mg *dexamethasone* at 11:00 p.m. the previous night. If the patient is healthy, glucocorticoid feedback suppresses

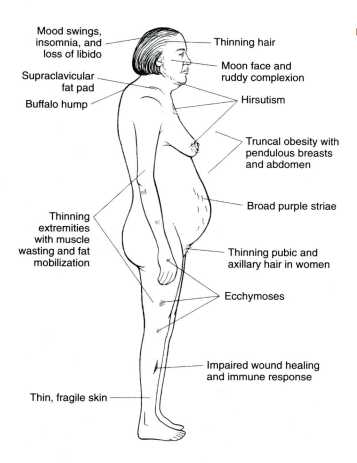

FIGURE 15.5 Cushing syndrome—typical appearance.

ACTH and CRH, which attenuates secretion of cortisol. They can be hard to visualize, or may be localized incorrectly by MRI and venous sampling for ACTH. It is important to realize that a comprehensive endocrine work-up is essential to diagnose Cushing's disease. This is because many of its signs and symptoms commonly occur in the general public even though the condition itself is relatively rare. Accurate diagnosis of Cushing's disease may require *bilateral inferior petrosal sinus sampling*. Often, a causative tumor is hard to find. About 10% of Cushing's disease patients are diagnosed based on endocrine tests, with no visible pituitary tumor seen on standard MRI or even high-definition 3 Tesla MRI. Pseudo-Cushing states that complicate diagnosis include depression, morbid obesity, alcoholism, anxiety, pregnancy, and the effects of certain medication—all associated with varying amounts of hypercortisolism, which simulates Cushing's syndrome. It is also important to assess surrounding anterior pituitary tissue to identify Crooke hyaline corticotroph changes, which reflects hypercortisolism on the hypothalamic-pituitary-adrenal axis. This does not occur with pseudo-Cushing states or the silent form of these tumors. TPIT positive is used to confirm a corticotroph origin. Extremely rarely, a carcinoma may be diagnosed, also via TPIT positivity, but along with metastasis or cerebrospinal spread.

Treatment

The treatment of choice is surgical resection of the ACTH-secreting adenoma. They present an extreme surgical challenge. Before surgery, bilateral petrosal venous ACTH sampling and cavernous sinus venography should be performed. If sellar venous sinus drainage is mostly unilateral, left−right ACTH gradients may not be able to lateralize the lesion. If the gradient is detected, **hemihypophysectomy** may be curative, if there are well-defined biochemical features of Cushing's disease. Intensive surgical exploration of the anterior and posterior lobes of the pituitary is required. They can be accidentally suctioned during surgery. The "normal" side of the gland must also be explored carefully since preoperative lateralization can miss the tumors.

The preferred surgery is transsphenoidal adenoma resection. Remission is achieved in about 75% of patients. Partial hypophysectomy, in patients with an unidentifiable adenoma, can result in biochemical remission. Cortisol levels are monitored after surgery, and a level below 3 µg/dL gives a 95% 5-year remission rate. Pasireotide is a somatostatin analog approved for treatment. About 20% of patients have normalized urinary-free cortisol levels after administration, which are linked to improved symptoms. As many as 50% of patients develop hyperglycemia, so blood glucose must be monitored closely. *Mifepristone* is a glucocorticoid receptor antagonist used for this hyperglycemia. Adverse effects include adrenal failure, excessive vaginal bleeding, and hypokalemia.

Prognosis

Surgery is the treatment of choice for ACTH-secreting tumors, and is often curative. For macroadenomas, the cure rate is usually lower after surgery. Silent and Crooke cell tumors are more aggressive. In an aggressive tumor that has low or absent MGMT secretion, responses to *temozolomide* occur, improving the prognosis. Though not commonly performed, prophylactic radiation therapy during a bilateral adrenalectomy can reduce incidence of Nelson's syndrome, improving prognosis. However, prognosis is very poor for chronic Cushing's disease.

Craniopharyngioma

Craniopharyngioma is a low-grade parasellar tumor that is rare, and derived from pituitary gland embryonic tissue. These tumors are distinct from tumors of *Rathke's pouch* and from intrasellar arachnoid cysts. There are two forms: *adamantinomatous* and *papillary*. The adamantinomatous form is made up of distinct epithelium, forming basal palisading and stellate reticulum. There is also *wet keratin*, also known as eosinophilic keratinous material. The papillary form resembles squamous papilloma and is made up of nonkeratinizing epithelium, plus fibrovascular cores. Usually, craniopharyngioma is benign, but can become malignant in some cases.

Epidemiology

There is an estimated worldwide annual incidence of 1.34 cases per 1 million people for the craniopharyngioma. In the United States, craniopharyngioma occurs in about 0.18 cases out of every 100,000 people. In Denmark, annual incidence was recently estimated at 1.86 cases per 1 million overall population, but 2.14 cases per 1 million children below the age of 15 years. For the adamantinomatous craniopharyngiomas have no gender predilection. The incidence of papillary craniopharyngiomas is unknown, but these tumors only occur in adults, with the median age being 40–50 years, and there is no gender predilection.

Etiology and risk factors

The causes of craniopharyngiomas are linked to CTNNB1 mutations on chromosome 3. These are especially prevalent in the adamantinomatous form. The mutations affect exon 3, and result in nuclear accumulation of beta-catenin as well as activation of the WNT pathway. Rare cases of adamantinomatous craniopharyngiomas have combined BRAF V600E and CTNNB1 mutations. For the papillary form, the BRAF V600E mutations are the primary etiology. Therefore this form may be responsive to BRAF inhibitors.

Pathology

Craniopharyngioma arises from embryonic squamous remnants of Rathke's pouch, which extend dorsally, toward the diencephalon. The tumors may be more than 10 cm in diameter, invading the third ventricle and other structures. More than 60% develop from inside the sella, while others come from parasellar cell groups. When they are intrasellar, these tumors are often distinguished from pituitary adenomas by a separate visible rim of normal pituitary tissue. In rare cases, the tumors are within the posterior fossa or sphenoidal sinus.

Existing cysts may contain immunoreactive human chorionic gonadotropin and calcifications. Histology reveals that the cysts are linked with a squamous epithelium that contains islands with columnar cells, along with a mixed inflammatory reaction accompanying calcification. Adamantinomatous craniopharyngiomas relapse more often than the papillary variant, which is less aggressive. When large, craniopharyngiomas often obstruct CSF flow, but rarely become malignant.

Clinical manifestations

Signs and symptoms of craniopharyngiomas are often linked to sellar or suprasellar mass effects. Increased intracranial pressure causes headache, papilledema, projectile vomiting, obesity, and somnolence—these are most common in children. About 50% of all patients have personality changes or cognitive impairment. Patients over age 40 are often present with asymmetric visual disturbances that include field deficits, optic atrophy, and papilledema. Other cranial nerves can be involved if cavernous sinus invasion is present. Diabetes insipidus is often the earliest feature of craniopharyngioma, and the patient may develop pituitary deficiency. A GH deficiency with shortened stature, gonadal failure, and diabetes inspires is common. Hyperprolactinemia is caused by pituitary stalk compression, or damage to the hypothalamic dopaminergic neurons. Therefore craniopharyngioma may mimic prolactinoma because of a favorable biochemical response to dopamine agonists, intrapituitary imaging, and the presence of hyperprolactinemia.

Diagnosis

An MRI is able to distinguish craniopharyngioma, structurally, from pituitary adenoma (see Fig. 15.6). The lobulated, solid/cystic mass is often filled with viscous dark green-to-brown fluid that is rich in cholesterol. The tumor is usually spongy and well defined. The viscous fluid, which resembles oil, may leak into the CSF, resulting in aseptic meningitis. A computed tomography scan will reveal about 50% of adults have distinctive convex or flocculent calcifications. In rare cases, pituitary adenomas, various parasellar tumors, and vascular lesions in the sella also show calcification.

The adamantinomatous form more often has calcifications than the papillary form, which also usually lacks cysts. The adamantinomatous form has an epithelium with a palisaded appearance and a loose stellate reticulum. The epithelium may have cords, basophilic cell nests along with squamous differentiation, trabeculae, and whorls, plus wet keratin and ghost cells. Staining methods include hematoxylin/eosin (H&E) and beta-catenin (see Fig. 15.7). The tumors may

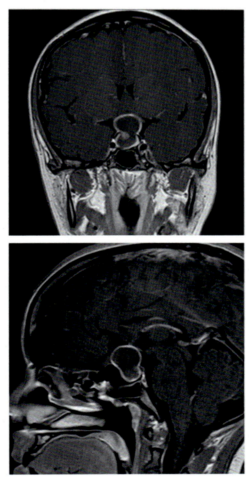

FIGURE 15.6 Preoperative MRI scan of a craniopharyngioma in T1 coronal and sagittal images, with contrast. They show a suprasellar, mostly cystic mass with rim enhancement, with a 2.6-cm craniocaudal size, and suprasellar extension measuring 15 mm.

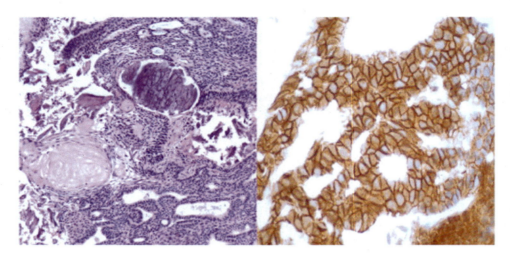

FIGURE 15.7 Surgical specimen of an adamantinomatous craniopharyngioma, showing small basophilic cells, stellate reticulum, areas of focal calcification, and extensive deposition of wet keratin. The left image shows H&E staining at 10× magnification. The right image shows strong beta-catenin staining at 60× magnification.

have cholesterol clefts, gliosis, and granulomatous inflammation. When they infiltrate the brain, Rosenthal fibers are usually present. There is a strong similarity to ameloblastoma. The tumors secrete enamel proteins. Tumors with significant squamous anaplasia, necrosis, and high mitotic activity are considered to be malignant. A malignant variant is called *primary suprasellar malignant odontogenic tumor*. There is also rare adamantinomatous form that somewhat overlaps the papillary form. A reliable marker for the adamantinomatous form is nuclear beta-catenin positivity in cell clusters. There will be variable positivity of *cytokeratin 7 (CK7)*, CK8, and CK14.

The papillary form has visible ciliated and goblet cells in small amounts. Though different molecularly, there has been a hypothesized spectrum that involves Rathke cleft cysts with significant squamous metaplasia and papillary craniopharyngioma. There are also papillary projections of epithelial cords. Wet keratin is not present, and mitoses are rare. The papillary form is positive for claudin 1, p63, and V600E-mutated BRAF. Patterns of staining of claudin 1 and p63 are likely different between the papillary and adamantinomatous forms. With the use of *fluid-attenuated inversion recover magnetic resonance imaging (FLAIR MRI)*, papillary craniopharyngiomas can easily be visualized (see Fig. 15.8).

Treatment

Treatment of primary or recurrent craniopharyngiomas includes radical surgery, radiotherapy, or a combination. After surgery, a primary side effect is obesity, which can be avoided by a procedure that spares the hypothalamus. More visual problems occur when surgery is highly complex. If the patient has diabetes insipidus, there is a higher rate of anterior pituitary hormone deficits, with related obesity. Obesity after resection may complicate the patient's life, and is linked to increased appetite and abnormal levels of leptin and ghrelin, which affect food intake regulation. Preoperative treatment with glucagon-like peptide-1 (GLP-1) analogs has led to weight loss in about 62% of patients. Outcome of treatment is likely related to the tumor's hypothalamic involvement. Hypothalamus-sparing surgery followed by local irradiation is recommended. Transsphenoidal surgery has been used successfully for intrasellar craniopharyngiomas. For suprasellar tumors, however, the expanded endoscopic transnasal approach is preferred. For retrochiasmatic craniopharyngiomas, a subtemporal approach may be used (see Fig. 15.9). Stereotactic irradiation has also been successful.

Prognosis

Survival rates from craniopharyngioma are high (92%), but recurrences and progressions often occur. Postoperative recurrence happens in approximately 20% of cases of radical surgical excision. There are no significant differences in prognosis if adjuvant radiotherapy follows subtotal surgical excision. In rare cases, the adamantinomatous craniopharyngiomas may become large enough to be life threatening. For this form, the prognosis is primarily based on the extent of surgical resection. Larger tumors have a worse prognosis, partly because total resection is more difficult to achieve. If malignancy develops, the prognosis is very poor. For the papillary form, extent of resection also predicts outcome. However, patients with subtotal resection may be helped by radiation therapy, and a favorable prognosis exists.

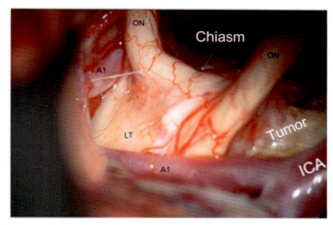

FIGURE 15.8 Surgical resection of a craniopharyngioma. *Courtesy:* Craniopharyngiomas-Comprehensive diagnosis, treatment and outcome, *2015, pp. 183–192.*

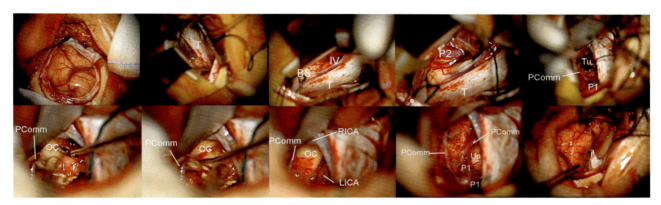

FIGURE 15.9 Subtemporal approach for gross total resection of a retrochiasmatic craniopharyngioma.

Nonfunctioning pituitary tumors

Nonfunctioning pituitary tumors do not actively secrete hormones; yet form from pituitary cells that are able to secrete hormones such as LH, FSH, TSH, and ACTH. **Gonadotrophic cell tumors** are nonfunctional, and arise from gonadotroph cells. They are not related to elevated serum gonadotropins, but usually secrete gonadotropin subunits that can be revealed during immunohistochemistry.

Epidemiology

Between 25% and 35% of pituitary tumors are considered to be nonfunctioning. However, there are not a lot of studies concerning nonfunctioning pituitary tumors. It is estimated that they occur in between 1.8 and 42.3 of every 100,000 people.

Etiology and risk factors

The etiology and risk factors for nonfunctioning pituitary tumors are unknown. They are believed to arise for one mutation, or multiple mutations, in a single pituitary gland cell, but it is unknown why or how this happens—except for a few rare, inherited syndromes.

Pathology

Nonfunctioning adenomas have been immunostained with the following results (also see Fig. 15.10):

- 92% stained for the alpha-subunit,
- 83% stained for the LH beta-subunit,

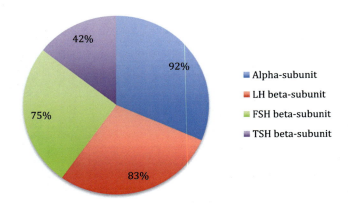

FIGURE 15.10 Percentages of immunostaining results in nonfunctioning adenomas.

- 75% stained for the FSH beta-subunit, and
- 42% stained for the TSH beta-subunit.

Though alpha-subunit, LH, and FSH are released when these tumors are maintained after culture, hormone production is usually not high enough to elevate blood levels. These tumors are classified as *null cell adenomas* when they do not secrete subunits of glycoprotein. A small amount of the tumors secrete enough hormones to elevate serum gonadotropin or alpha-subunit levels, sometimes causing clinical syndromes. A small amount of gonadotroph adenomas causes elevated alpha-subunit, FSH, and LH. These are evaluated as functioning adenomas, and may be linked to certain endocrine syndromes. The majority are macroadenomas.

Clinical manifestations

Often, a gradual visual deficit due to optic chiasmal compression occurs, with the patient unaware of the development. Recognition of deficits may take time since the formal visual fields are not often evaluated unless there is a clinically suspected defect. Most endocrine symptoms are related to deficiency of gonadotropin. Quality of life may be decreased, and daytime somnolence can occur. Female patients with FSH-secreting tumors may have pelvic pain because of ovarian hyperstimulation. In rare cases, an LH-producing tumor causes precocious puberty in children, who may also have hypogonadism because of gonadal downregulation. In pregnant patients, macroadenomas lead to impaired vision more often than microadenomas. Risks of visual impairment during pregnancy are assessed carefully.

Diagnosis

Nonfunctioning pituitary tumors often are detected because of their size, or are found incidentally. Approximately 64% of incidentally discovered pituitary tumors are clinically nonfunctioning. The rest are cystic or parasellar. Large tumors can go unrecognized for years, to be accidentally detected during imaging for other conditions. They may be found during evaluation of sinusitis, pituitary apoplexy, or head trauma. The patient is often deficient of one or several pituitary hormones, seen in 66% of patients with nonfunctioning macroadenomas. For patients with macroadenomas, imaging may reveal increased testicular volume in males and ovarian cysts in females. The only sign that a pituitary tumor is secreting FSH may be a high serum FSH level that usually occurs with a low LH level.

High levels of gonadotropin, associated with menopause or testicular failure, can make the interpretation of these levels difficult. However, LH as well as FSH is high in primary gonadal failure. The rare tumors that produce LH, in males, may be identified because of elevated serum testosterone, plus acne and oily skin. Diagnostic methods to assess tumors include MRI, visual field examination, and measurement of pituitary hormones. Alpha-subunit, FSH, IGF-1, cortisol, LH, PRL, TSH, and thyroid hormones are all evaluated. To exclude secondary adrenal insufficiency, a serum cortisol measurement is taken at 8:00 a.m., and then cortisol response to **cosyntropism** or an insulin tolerance test can be performed. Dehydroepiandrosterone sulfate (DHEAS) and DHEA itself can be measured to assess adrenal function for these patients. A low SHEAS may occur before cortisol levels are lowered. If LH or FSH levels are high, they are interpreted in relation to the physiologic state of the patient, considered menstruation, menopause, and such factors. Gonadotropin elevations with primary gonadal failure usually involve more than one hormone. Circulating alpha-subunit elevation is linked to pituitary tumors, but not to gonadal failure. Stimulation of TRH may differentiate elevations of gonadotropin linked to end-organ failure, or to tumor production that is independent. With gonadotrophic adenomas, increase in FSH, LH, LH beta-subunit, or alpha-subunit occurs in response to TRH. Calculating the molar ratio of LH or FSH to alpha-subunit assists diagnosis.

Treatment

The treatments for gonadotrophic cell tumors include surgery, postoperative radiotherapy, observation, and medications. Transsphenoidal microscopic or endoscopic endonasal transsphenoidal surgery is used when tumors threaten vision, or are macroadenomas that threaten vital structures. For nonfunctioning pituitary adenomas, gross total mass removal is reported in about 79% of patients. In one study, gross total resection was achieved in 65.3%, with improved vision or return to normal vision in 80.2%. Newer flap surgical techniques have reduced incidence of CSF leaks, though permanent diabetes insipidus develops in as many as 1.4% of patients.

After surgery, about 32% of patient experience tumor regrowth over 76 months if not receiving radiotherapy. Other studies have had recurrence rates of between 6% and 46%, and between 0% and 36%. Avoiding radiotherapy after surgery requires careful follow-ups, annual MRIs and visual examinations, and regular medical assessment for overall health. Radiation is used if the tumor reexpands. In one study, 60% of patients with nonfunctioning pituitary tumors, after gamma knife radiosurgery, had decrease in tumor size, while 37% remained unchanged. However, risks of developing new anterior pituitary hormone defects were 32% at 5 years. Tumor mass is stabilized or decreased in 90% of gamma knife radiosurgeries, with median time to tumor progression being 14.5 years. Delayed hypopituitarism is seen in 30% of patients. Some tumors do not grow over years or decades. Since 20% of tumor regrowth happen more than 10 years following surgery, serial MRIs are usually performed over the patient's lifetime. Occasional endocrine evaluation is suggested to detect hypopituitarism. Medications are only used sometimes. The GnRH antagonists and somatostatin analogs slightly shrink tumors in a small amount of patients, but are not used widely. In rare cases, dopamine agonists have caused shrinkage of some nonfunctioning tumors, or have prevented their regrowth.

Prognosis

Outcomes of transsphenoidal surgery for nonfunctioning pituitary tumors are usually good. Most patients experience complete tumor removal, only transient diabetes insipidus, some new anterior pituitary deficits, pituitary function improvement, visual field defect improvement, and improvement in ACTH and TSH levels. Progression-free tumor survival is linked to smaller tumor size and lack of suprasellar extension. New or worsened pituitary failure is seen in 21% of patients.

Pituitary carcinoma

Pituitary carcinoma is a very rare tumor of the adenohypophyseal cells, which metastasizes craniospinally from a primary sellar tumor, or may not be continuous with such tumor. There may be systemic metastasis. Prolactin-secreting pituitary carcinomas are most common, followed by corticotrophic pituitary carcinomas, usually secreting ACTH.

Epidemiology

Pituitary carcinoma only accounts for about 0.12% of all adenohypophyseal tumors according to the *German Pituitary Tumor Registry*. In the United States, up to 6% of invasive adenomas are pituitary carcinomas, but this percentage is believed to be overestimated. The European RARECARE project estimates annual incidence to be less than 1 case per 1 million people. Rates could be underestimated because silent tumors discovered only at autopsy. Pituitary carcinomas in children are extremely rare. According to the *Surveillance, Epidemiology, and End Results (SEER) program*, there is a 1.33:1 female-to-male ratio. Median age at diagnosis is in the sixth decade, though the age range is about 12–70 years. Over 70% of cases are of the lactotroph and corticotroph forms. Clinically nonfunctional tumors occur at younger ages than the benign tumors.

Etiology and risk factors

The causes of pituitary carcinomas are unknown. There is no evidence that environmental factors, including radiation, play a role in the development of pituitary carcinoma from a benign adenoma. There are three risk factors for these tumors, which include MEN1, Carney complex, and familial acromegaly. No predisposition for the tumors has been identified. However, it has been hypothesized that a slow-growing adenoma may transform into an invasive adenoma, and then into a carcinoma. Less often, development of a carcinoma from a normal gland or a common adenoma may occur. In one case, a loss-to-retention pattern of metastatic deposits was compared with a premetastatic adenoma. This indicated that there were two or more clones from the primary tumor, and one clone resulted in metastases.

The genetic profile of pituitary carcinoma is still not fully understood. Point mutations in *Harvey rat sarcoma* exist in the metastatic deposits but not in nonmetastasizing lesions, meaning they may be related to formation, growth, or both of metastases. There has been one case of pituitary carcinoma that was ACTH-positive in which the patient also had a benign corticotroph adenoma, suggesting that the *RB1* gene may have been involved. It was secreted in the adenoma, but not the carcinoma. In one lactotroph carcinoma, there was increased *telomerase reverse transcriptase* expression and increased telomerase activity. In one case of gonadotroph carcinoma, there was oversecretion of *erb-b2 receptor tyrosine kinase 2 (ERBB2,* also known as HER2/neu), with low-level gene amplification in the localized recurrence and metastasis. This indicated a link between ERBB2 and aggressive tumor activity. Genomic hybridization has been used to identify about 8.3 chromosomal imbalances in each pituitary carcinoma, with 7 being gains, and 1.3 being losses. The most common chromosomal gains involve chromosomes 5, 7p, and 14q. The next most common gains are in chromosomes 1q, 3p, 7, 8, 9p, 13q22, 14q, and 21q. In the metastasis of a corticotroph carcinoma compared with its sellar recurrence, there was loss of heterozygosity in chromosomes 1p, 3p, 10q26, 11q13, and 22q12. In three lactotroph carcinomas, there was an allelic loss in chromosome 11q, plus a loss of a region that spanned 139 coding genes.

Aggressive carcinoma behavior may be induced by the genes *cluster of differentiation 44 (CD44), diacylglycerol kinase zeta (DGKZ), general transcription factor 2H subunit 1 (GTF2H1), HIV-1 Tat interactive protein 2 (HTATIP2),* and *tumor susceptibility 101 (TSG101).* However, TP53 mutations are usually rare, though they were found in two out of six cases studied in one series. Uncontrolled neoplastic cell proliferation may be related to failure of the mechanisms that regulate cellular aging. A study of lactotroph carcinomas revealed 61 differentially expressed genes in invasive lesions and 89 of these genes in aggressive-invasive lesions. The genes *chromogranin B (CHGB), importin 7 (IPO7), solute carrier family 2 member 11 (SLC2A11),* and *teneurin transmembrane protein 1 (TENM1)* were cosecreted, which indicated a possible link with malignancy. Risks for aggressive adenoma and pituitary carcinoma have been documented as being linked to MEN1. However, there is not a proven link between pituitary carcinoma and MEN1. Though there has been one case of an SDH-mutated metastatic gonadotroph carcinoma, the AIP and GPR101 genes are not associated.

Pathology

Primary pituitary carcinomas are located in the sellar region as invasive macroadenomas, and only two cases of ectopic adenomas have resulted in multiple metastases to the brain and subarachnoid space. Only a few recorded tumors have remained confined to the sella turcica. Metastasis often happens in the craniospinal axis, due to dissemination within the subarachnoid space. Deeper deposits within the parenchyma usually affect the cerebral cortex and cerebellum. Intraaxial and systemic metastases appear similar to deposits from other solid cancers. Systemic metastases are usually in the bones, liver, lungs, and lymph nodes. Rare sites include the heart, eye orbits, pancreas, middle ears, ovaries and myometrium, and the endolymphatic sac. Via the petrosal sinus, hematogenous dissemination may occur. Since the pituitary gland lacks lymphatic vessels, metastasis to the lymph nodes can occur because of tumor invasion into the soft tissues and skull. Tissue damage during surgery probably does not result in dissemination. Dural metastases, especially when growing slowly, may appear identical to meningiomas in imaging studies. The primary lesions and their metastases are nearly the same as common adenomas. There is no way to determine if an invasive adenoma will become a carcinoma.

Clinical manifestations

Pituitary carcinomas may be hormonally functional or nonfunctional prior to metastasis. However, endocrinologically active tumors are much more common. No clinical or biochemical features exist that are specific for an adenoma that will undergo metastasis. If there is an active primary tumor, metastases usually remain active, often maintaining the endocrine syndrome once the sellar lesion is surgically resected. Diabetes insipidus is rare. Progression to carcinoma has been seen along with Nelson's syndrome. After surgery, continually elevated hormone concentrations should indicate the need to check for metastases. If metastases are hormonally inactive, manifestations are based on the location. If bones are involved, pain and fractures are common. The patient may be asymptomatic for a long time, or the metastasis may only be found during autopsy. Lactotroph carcinomas usually are resistant to dopamine agonists, before and after metastases. Resistance may be present when they form, or can develop over the course of treatment.

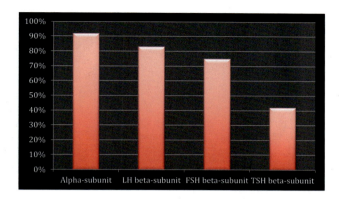

FIGURE 15.11 Common symptoms of pituitary carcinoma.

Though most patients with pituitary carcinoma are symptomatic, two studies have revealed that symptoms from pituitary metastasis were the initial manifestation of metastatic disease in over 50% of patients. The most common symptoms are diabetes insipidus (45% of cases), optic nerve dysfunction (28%), anterior pituitary dysfunction (24%), palsies of cranial nerves III, IV, or VI (22%), and headache (16%)—see Fig. 15.11. Primary pituitary carcinomas usually do not impair pituitary endocrine function. Patients with prolactin-secreting carcinomas and elevated serum prolactin have typical PRL-mediated symptoms of amenorrhea and galactorrhea in females, and erectile dysfunction in males. Serum PRL levels in carcinomas are often similar to values measured in PRL-secreting macroadenomas. For corticotrophic pituitary carcinomas, patients show typical features of hypercortisolism. GH-secreting pituitary carcinomas have symptoms that closely resemble those of GH-secreting adenomas. There are only extremely rare reports of gonadotrophic or thyrotrophic pituitary carcinomas, which usually cause male impotence and oligomenorrhea in females.

Diagnosis

Pituitary carcinoma is only diagnosed when there is a tumor that is noncontiguous with the sellar region (see Fig. 15.12). Diagnosis is difficult, and causes delays that can adversely affect treatment and prognosis. There is often **drop metastasis** of CSF, and CSF or systemic dissemination is required for diagnosis. Four diagnostic requirements are now used to help. These include the following:

- The primary lesion must be an adenohypophyseal tumor, proven by histology.
- Alternative primary lesions must be excluded.
- There must be discontinuous spreading within the craniospinal axis.
- Pathological metastatic features must be similar to the primary pituitary tumor.

Neuronal differentiation is rare with these tumors. About 75% of cases are diagnosed only at autopsy. Clues that increase clinical suspicion of aggressive tumor phenotypes include unresponsiveness or escalating serum prolactin levels, and/or tumor growth, despite adequate dopamine agonist treatment. Pituitary carcinoma is often a diagnosis of exclusion, because of its rarity. Differential diagnosis is *metastasis to the pituitary gland*.

Fine-needle aspiration may be helpful to diagnose extra-CNS metastases—nearly as good as frozen sectioning. Features seen in smears and imprints are of neuroendocrine carcinomas, varying from bland to obviously malignant. In cytology, the appearance is usually similar to the primary or recurrent sellar lesions. Nearly 60% of primary tumors resemble conventional adenomas, but with more proliferation when they recur. These are usually related to longer survival times. However, metastases often have greater abnormality in their histologies than primary tumors, such as high mitotic figures and Ki-67 proliferation indexes, plus vascular invasion. Less often, the metastatic deposits have the original appearance of a benign tumor.

Primary and metastatic carcinomas secrete neuroendocrine differentiation markers, including synaptophysin and chromogranin A. There is variable expression of pituitary hormones. The most common subtype is lactotroph carcinoma. The next most common is corticotroph carcinoma, which includes Crooke cell carcinoma, the carcinoma that occurs with Nelson's syndrome, and silent tumors. Making up less than 30% of overall cases are the gonadotroph, null cell, somatotroph, and thyrotroph subtypes. In rare cases, corticotroph carcinomas cosecrete ACTH, proopiomelanocortin, and CRH.

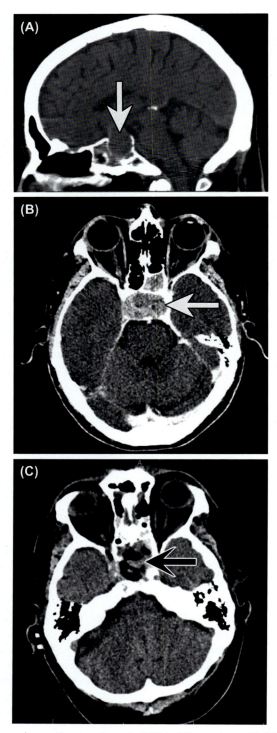

FIGURE 15.12 Three views of a pituitary carcinoma. *Courtesy:* Journal of Clinical Neuroscience, 41, *2017, 75—77.*

Pituitary carcinomas secrete some proteins in greater amounts, including BCL2, cyclooxygenase-2, galectin-3, topoisomerase 2-alpha, and *vascular endothelial growth factor*. They secrete other proteins in lower amounts, including beta-catenin and p27. There has been one documented case of a pituitary carcinoma totally lacking *menin*. The Ki-67 proliferation index is higher in these tumors than in adenomas. There have been indices of 45% in metastatic deposits.

Within a primary lesion, values higher than 10% are linked with aggressive tumor activity, possibly suggesting the primary tumor's diagnosis as a carcinoma. There is much overlap between metastatic lesions and slowly growing adenomas, however. Intense nuclear expression of p53 is often present in primary lesions and metastases, with the overexpression of p53 possibly indicating malignancy. It should be understood that p53 may be undetectable in some of these lesions and their metastases.

There is usually not increased microvascular density unless areas of higher vascular density are the only ones measured. Matrix metalloproteinases are usually increased, resulting in degrading of the extracellular matrix, with angiogenesis and invasion. Secretion of MGMT is easily assessed via immunohistochemistry. This enzyme is part of the body's DNA repair system, which helps resist *temozolomide* through repair of the alkylated guanine. Most carcinomas have less than 10% MGMT-positive cells, with secretion levels often being similar in lesions that recur. In lactotroph carcinomas, low MGMT secretion is common. In corticotroph carcinomas, there may be moderate to high MGMT secretion. Low immunosecretion may predict responses to temozolomide, often resulting in favorable outcomes, but this does not always occur. The response is not always the same in relation to methylation of the MGMT gene promoter.

Nuclear immunosecretion of *mutS homolog 6 (MSH6)*, a DNA mismatch repair protein, is highly related to temozolomide effects. Base mismatches are revealed by heterodimers of MSH6 and MSH2. These assist the **heterodimeric complex** of *mutL homolog 1 (MLH1)* and *PMS1 homolog 2 (PMS2)*, activating the G2-M cell phase checkpoint for DNA damage. This results in apoptosis during the synthesis of DNA. Pituitary carcinomas are able to secrete somatostatin receptors. The type and subtype of the primary lesion and its metastases can be determined via electron microscopy, but this does not reveal tumor aggressiveness or likelihood of spread. Positron emission tomography (PET) with radiolabeled octreotide may aid in detecting and monitoring metastases, as well as *fluorodeoxyglucose (FDG)-PET* and *gallium 68 dotatate (^{68}Ga-DOTATATE)*-PET—especially with nonfunctional metastases.

Treatment

Surgical resection of pituitary carcinomas is rarely curative, since the tumors are locally invasive into the sellar floor, clivus, and cavernous sinus. Sometimes, there are intracranial metastatic deposits within the third or fourth ventricles, associated with high morbidity and mortality. However, it is often possible to surgically debulk, and sometimes, obtain gross total or subtotal resection of tumor tissues. Repeated surgeries can also be performed to remove secondary deposits when they emerge. Recent advances in endoscopic techniques may offer possible advantages over craniotomy. Radiation therapy prevents regrowth in subtotally resected pituitary carcinomas, and slows growth. The two primary methods are stereotactic radiosurgery in a single procedure, or fractionated radiation therapy over 5–6 weeks. Types of radiation therapy include gamma knife, linear accelerator, *cyberknife*, and proton beam therapy. Medications used for pituitary carcinomas are similar to those used for pituitary adenomas, but in higher doses and combinations of agents. These include dopamine agonists, somatostatins, somatostatin analogs, and less often, antiestrogens. Chemotherapy for pituitary carcinomas has shown highly varied results, though cyclo-hexyl-chloroethyl-nitrosourea has been combined with 5-fluorouracil with some success. However, temozolomide has become the first-line chemotherapeutic agent. It was first used to treat a prolactin-secreting carcinoma that had progressed despite many surgeries, radiation therapy, and high-dose dopamine agonist therapy.

Prognosis

Prognosis of pituitary carcinomas is generally poor, and patient often dies from complications of hormonal excess instead of the mass effect of metastases. Approximately 80% of patients die of causes related to the cancer. The SEER database reports overall survival rates as follows:

- 1 year—57.1%,
- 2 years—28.6%,
- 3 years—28.6%, and
- 10 years—28.6%.

There are very different survival rates between invasive carcinomas and adenomas at 1, 2, and 5 years, but this is not true at 10 years. Before metastasis, multiple local recurrences are common. The time between onset and first recurrence varies between several weeks to nearly 30 years. There is also a varied latency period between primary sellar tumor diagnosis and metastasis, from several weeks to 38 years. However, dissemination usually happens in the first 10 years after diagnosis. In early stages, metastases are rare, and nearly never occur at the first disease manifestation. Once they develop, survival time averages less than 4 years. Systemic metastases appear to be more rapidly fatal than

CNS metastases. Corticotroph carcinomas give the shortest survival time. Prognosis for silent corticotroph carcinomas is similar to hormonally active corticotroph carcinomas. There have been some long-term survivors after treatment for pituitary carcinomas.

> **Point to remember**
> Data on pathology and prognosis suggest two distinct types of pituitary carcinoma. The first type makes up most cases, behaving like an invasive adenoma, except for metastatic spread happening after several sellar recurrences. Survival time is longer with this type. The second type presents as a malignant-appearing tumor, with multiple recurrences over a short time period. Metastases occur sooner, and the patient often dies in less than 1 year.

Pituicytoma

A **pituicytoma** is a solid, circumscribed, and low-grade glial neoplasm of the posterior pituitary gland. It forms in the neurohypophysis or infundibulum. The tumor is made up of bipolar-spindled cells, in a storiform or fascicular pattern. Similar to spindle cell oncocytomas and granular cell tumors in the sellar region, pituicytomas have nuclear secretion of *thyroid transcription factor 1 (TTF1)*. This may mean that pituicytomas may make up a spectrum of a single tumor classification. These tumors are considered to be grade I neoplasms by the World Health Organization. As of 2007, pituicytomas have been reclassified as low-grade glial neoplasms originating in the neurohypophysis or infundibulum, with histologically distinct characteristics. Previously, they were considered to be sellar or suprasellar tumors such as granular cell tumors and pilocytic astrocytomas. They were also called *posterior pituitary astrocytomas* and *infundibulomas*.

Epidemiology

Pituicytomas are rare tumors, with only 70 documented cases since 2000. Almost all of them have occurred in adults. The average age at diagnoses was 50 years, in a range between 7 and 83 years. Almost 66% of patients are between the ages of 40 and 60 years, with males affected 1.5 times as often as females.

Etiology and risk factors

The causes and risk factors of pituicytomas are not fully understood, but are believed to be related to those of pituitary adenomas.

Pathology

Pituicytomas form along the neurohypophysis, including the infundibulum (pituitary stalk) and posterior pituitary (see Fig. 15.13). They may be located in the sella, suprasellar region, or in the intrasellar as well as suprasellar regions. The least common pathology is in only the intrasellar region. Pituicytomas are low-grade gliomas. Proliferation of these tumors is low.

Clinical manifestations

Most commonly, pituicytomas have similar features to those of other slow-growth nonhormonally active tumors of the sellar or suprasellar region, compressing the infundibulum, optic chiasm, or pituitary gland itself. Manifestations include headache, visual disturbances, and features related to hypopituitarism, including amenorrhea, fatigue, decreased libido, and mildly elevated serum prolactin—known as the *stalk effect*. In rare cases, asymptomatic pituicytomas are found upon autopsy.

Diagnosis

When viewed in imaging, pituicytomas usually appear solid and circumscribed. They are mostly isointense to gray matter in T1-weighted imaging, hyperintense in T2-weighted imaging, and uniform in their contrast enhancement (see Fig. 15.14). Sometimes, the tumor has heterogeneous contrast enhancement. In rare cases, there is a cystic component. Macroscopically, pituicytomas may be up to several centimeters in diameter, with a rubbery but firm texture. Though radiographical imaging may indicate a smooth tumor contour, it can be adhered firmly to nearby structures in the suprasellar space.

Microscopically, pituicytomas are solid and compact, and almost totally made up of elongated bipolar spindle cells, with a storiform or fascicular pattern. Infiltrative patterns are usually not seen. Each tumor cell has a large amount of eosinophilic cytoplasm. The cell shape can be short and plump, elongated and angulated, or anywhere in between (see

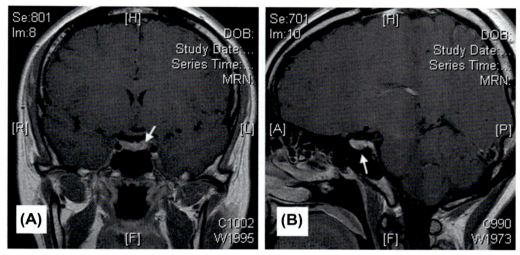

FIGURE 15.13 T1-weighted MRI of a pituicytoma. (A) Coronal view. (B) Sagittal view. *Courtesy:* Journal of Cancer Research and Practice, *5(2), 2018, 47–52.*

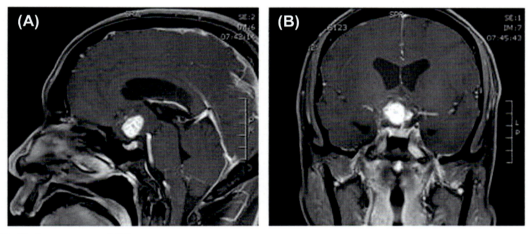

FIGURE 15.14 Pituicytoma seen on MRI.

Fig. 15.15). There is an easily seen cell border, primarily on fascicular cross sections. When the tumor is highly defined, there is no large amount of granularity or vacuolization of the cytoplasm. PAS staining reveals only slight reaction. Also, a great oncocytic change is not usually apparent. Nuclei are of moderate size, oval to elongated, and only have slight nuclear border irregularity. Rarely, there are mitotic figures. The reticulin fibers have a perivascular distribution, while there is only sparse intercellular reticulin. Pilocytic astrocytoma and normal neurohypophysis differ from a pituicytoma, which has no Rosenthal fibers, eosinophilic granular bodies, or Herring bodies. These are axonal dilatations for storage of neuropeptides in the neurohypophysis.

Pituicytomas are positive via immunohistochemistry for vimentin and the S100 protein. They show nuclear staining for TTF1. However, *glial fibrillary acidic protein (GFAP)* staining is highly varied. It may be faint and focal, up to moderate and patchy. It is only, in rare cases, highly positive. Strong and diffuse staining is more common for S100 and vimentin. *B-cell lymphoma 2 (BCL2)* staining is varied, but may be intense. Stains for cytokeratins are negative. Stains for *epithelial membrane antigen (EMA)* may reveal a patchy pattern, which is cytoplasmic, and not membranous. Pituicytomas do not have immunostaining for pituitary hormones, or the neuroendocrine or neuronal makers *aptophysin* and *chromogranin*. Immunoreactivity of *200 kDa neurofilament (NFP)* is only in axons of peritumoral neurohypophyseal tissue. It is not present inside the tumor. The low tumor proliferation is revealed by immunoreactivity for the monoclonal antibody called *MIB1*. The reported Ki-67 proliferation index is between 0.5% and 2.0%.

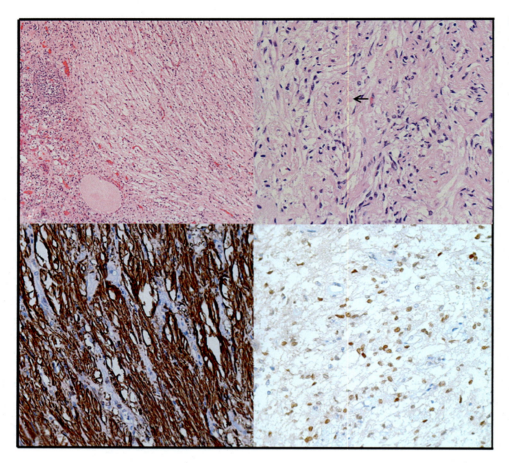

FIGURE 15.15 Cell shapes in a pituicytoma.

Pituicytomas arise from specialized *pituicytes*, which are glial cells of the neurohypophysis. This explains the tumor's anatomical distribution, which is consistent with the morphological and **immunophenotypic** features. Pituicytomas strongly secrete TTF1, similar to the pituicytes of the developing and mature neurohypophysis. This identifies the derivation form pituicytes. Since granular cell tumors of the sellar region and *spindle cell oncocytomas* also secrete nuclear TTF1, their histogenesis is also likely from pituicytes. These other tumor types may in fact be pituicytomas that have lysosome-rich or mitochondrion-rich cytoplasm, inducing their morphologies. These individual connections with pituicytes may occur because of the multiple subtypes of pituicytes within the normal neurohypophysis. There is no identified genetic profile for pituicytomas. In the relatively few documented cases, there has been no *R132H-mutant IDH1*, *KIAA1549-BRAF*, or *V600E-mutant BRAF* fusion. In one particular case only, comparative genomic hybridization analysis, based on microarray, showed losses of the chromosome arms 1p, 14q, and 22q. There were also chromosome arm gains on 5p.

Treatment

The current treatment for pituicytomas is surgical resection. Complete resection may be prevented by the local adherence of the tumor to nearby structures. After surgery, the tumors may slowly regrow over several years.

Prognosis

There is no link between tumor proliferation and clinical outcome. There have been no cases of malignant transformation or distant metastasis.

Spindle cell oncocytoma

A **spindle cell oncocytoma** is an oncocytic, nonneuroendocrine posterior pituitary gland neoplasm. It may be spindled to epithelioid. These tumors develop in adults, and are usually benign. Similar to pituicytomas and granular cell tumors in the sellar region, they have nuclear secretion of TTF. This indicates that these oncocytomas may also be of a unique classification. They are also grade I tumors according to the World Health Organization. In previous times, they were described as *spindle cell oncocytomas of the adenohypophysis*. Based on similarities of ultrastructural and immunohistochemical features, their derivation from **folliculostellate cells** of the anterior pituitary was considered. Today, the TTF1 immunoreactivity of these and the previously discussed tumors, plus nonneoplastic pituicytes, could suggest a similar tumor origin and a link to the posterior pituitary gland.

Epidemiology

Spindle cell oncocytomas are rare. The true incidence is not fully understood. In one study, these oncocytomas made up 0.4% of all types of sellar tumors. Only about 25 cases have ever been reported. The reported patient age range was 24–76 years, with a mean age of 56 years. Distribution has been equal between males and females.

Etiology and risk factors

The causes and risk factors of spindle cell oncocytomas are generally not understood, but are likely to be similar to those of the previously discussed pituitary tumors.

Pathology

Spindle cell oncocytomas occur in intrasellar and suprasellar locations (see Fig. 15.16). Only rarely have they manifested only in the intrasellar region. There has been extension into the cavernous sinus, and invasion into the sellar floor.

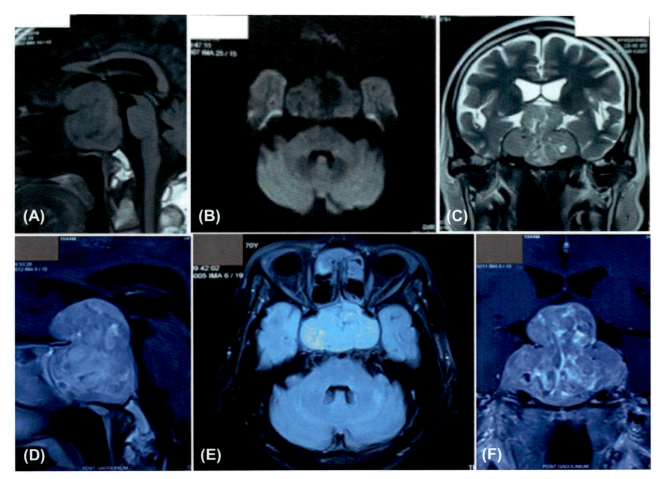

FIGURE 15.16 Six views of a spindle cell oncocytoma. *Courtesy:* Pathology-Research and Practice, 212(3), 2016, 222–225.

Clinical manifestations

Spindle cell oncocytomas have the same features as nonfunctioning pituitary adenomas. The most common presenting symptom is visual disturbances. Other symptoms include pituitary hypofunction, followed by headache, nausea, and vomiting. There have been symptoms of decreased libido, sexual dysfunction, and oligomenorrhea or amenorrhea due to pituitary stalk effect, along with slight hyperprolactinemia. No patients have had diabetes insipidus at onset. One patient with an aggressive tumor that involved the base of the skull also experienced epistaxis.

Diagnosis

In imaging, spindle cell oncocytomas resemble pituitary adenomas. However, they often have avid enhancement after a paramagnetic contrast agent is administered, due to greater vascular supply compared to adenomas. Enhancement may be heterogeneous. A CT scan may reveal calcification. Recurrent intratumoral bleed has been seen, which leads to an incorrect preoperative diagnosis of craniopharyngioma. In magnetic resonance angiography, the rich tumoral blood supply is seen. In one study, there was predominant feeding from the internal carotid arteries on both sides of the neck.

Macroscopically, these tumors often look exactly like pituitary adenomas. They are usually large, having a craniocaudal dimension of 2.5–3 cm. The maximum size reported was 6.5 cm. The tumor can be soft and creamy, resulting in easy resection; but can be firm, adhering to nearby structures. Less often, there is destruction of the sellar floor. Intratumoral hemorrhage may occur. Severe intraoperative bleeding has been documented. There is usually no clear margin or **pseudocapsule**.

Microscopically, spindle cell oncocytomas are usually made up of interlaced fascicles. There are poorly defined lobules of spindle to epithelioid cells, and eosinophilic plus variably oncocytic cytoplasm. The oncocytic changes may be focal or over a wide area. A broad spectrum of morphology exists, including whorl formation, clear cells, focal stromal myxoid changes, follicle-like structures, and osteoclastic-like giant cells. There may be mild-to-moderate nuclear atypia or significant pleomorphism. Many lesions have focal infiltrates of mature lymphocytes. There is usually a low mitotic count, and often, less than 1 mitotic figure out of every 10 high-power fields. There may or may not be increased mitotic activity in recurrent lesions. Ultrastructural examination aids diagnosis. Neoplastic cells are spindled or polygonal, often containing mitochondria that can appear abnormal. The tumor cells do not have secretory granules, which is a key factor of pituitary adenomas. There are often intermediate-type junctions and well-formed desmosomes. Sometimes, there are intracytoplasmic lumina with projecting microvilli.

There is no immunoreactivity for neuroendocrine markers or pituitary hormones. These oncocytomas usually secrete vimentin, S100 protein, BCL2, and EMA. They show staining with the antimitochondrial antibody *MU213-UC clone 131-1*, and nuclear staining for TTF1. The EMA secretion may be weak and focal to diffuse. There is only focal positivity for GFAP. In one case, the *CD44 antigen* and **nestin** were present, which suggests a possible neuronal precursor component. **Alpha-crystallin B chain** was also present in one case. Commonly secreted factors also include **galectin-3** and **annexin A1**, but these markers are not specific, and previously were suggestive of a folliculostellate cell link. The Ki-67 proliferation is usually only within 1%–8%, having an average of 3%. In one example of tumor recurrence, there was a 20% proliferation index, but there was no data on the primary tumor's proliferation rate. Just how these tumors form is uncertain. Pituicytes are believed to be implicated. In one study of seven cases, none of these tumors had positive immunostaining for R132H-mutatnt IDH1. There was no evidence of BRAFV600E mutation or KIAA-BRAF fusion.

Treatment

Spindle cell oncocytomas are usually treated with transsphenoidal surgery for as complete resection as possible. Some surgeries involve alternating central debulking and mobilization of the capsule off the optic chiasm median eminence, hypothalamus, and internal carotid artery branches. After surgery, some patients have required hormone substitution to treat panhypopituitarism.

Prognosis

Most spindle cell oncocytomas are benign, though about 33% of all cases have recurred after 3–15 years. In a small number of cases of recurrent, the first neoplasm had an increased Ki-67 proliferation index between 10% and 20%, or had significant cellular atypia. Incomplete resection is a risk factor for recurrence. Hypervascularity may make complete resection difficult. Recurrent tumors may be more aggressive, having a higher Ki-67 proliferation index with necrosis.

Secondary tumors

Secondary tumors of the pituitary gland develop as metastases from tumors of other body sites. They are also called *metastatic pituitary tumors*. Prior to surgery, these tumors are often misdiagnosed as pituitary adenomas.

Epidemiology

Pituitary metastasis is present in 1%−3.6% of malignant tumors, as found in autopsies. Clinically, metastasis to the pituitary is uncommon, making up 0.14%−28.1% of all brain metastases discovered during autopsy. However, pituitary metastases appear to be increasing in frequency, though this may be due to better cancer survival and better imaging procedures. A study of 190 patients with symptoms of pituitary metastasis revealed that the primary tumors were latent upon presentation in 43.7% of cases. Between 7% and 57.1% of cases involve the pituitary as the single site of metastasis. Approximately 1% of pituitary tumors are metastatic. Most often, breast and lung cancers metastasize to the pituitary gland. There is usually a generalized spread of metastasis, happening with other metastases—usually to the bones. Pituitary metastases usually occur during the sixth or seventh decades of life, with no predilection for males or females.

Etiology and risk factors

Tumors that are metastatic to the pituitary gland are unusual complications of systemic cancer, in elderly patients with diffuse malignant disease. The most common causes of pituitary metastases are breast cancer in women and lung cancer in men. There are no known risk factors except for those that are linked to the primary tumors prior to metastasis.

Pathology

About 57% of metastatic tumors are localized only to the posterior pituitary, followed by 13% only to the anterior pituitary, and 12% to both lobes of the gland. The remaining amount lies within the capsule or infundibulum. The reason that the posterior pituitary receives the majority of metastases may be because of its direct arterial supply of blood. The anterior pituitary receives most of its blood from the hypophyseal portal system. There are rare primary tumors that metastasize to the pituitary. These include differentiated (papillary) thyroid carcinoma, hepatocellular carcinoma, lung neuroendocrine carcinoma, and colonic adenocarcinoma (see Fig. 15.17). Extremely rarely, lung and other cancers may metastasize to pituitary adenomas. In one case, a metastatic GHRH-producing neuroendocrine tumor of the pancreas caused somatotroph hyperplasia within the pituitary gland, which resulted in acromegaly. Also, some metastatic tumors mimic the effects of pituitary adenomas.

Clinical manifestations

Metastatic tumors of the infundibulum may cause visual disturbances. However, the majority of most pituitary metastases are asymptomatic, with only about 7% causing symptoms. The most common symptoms include diabetes insipidus, headache and other pain, ophthalmoplegia, defects of the visual fields, and anterior pituitary dysfunction. Diabetes insipidus occurs more often when the patient has symptomatic pituitary metastases than with symptomatic adenomas,

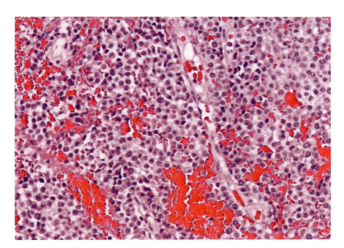

FIGURE 15.17 Histopathology, via H&E stain, of a secondary pituitary tumor that metastasized from a colonic adenocarcinoma.

which only rarely cause the condition. About 60% of patients with pituitary metastases have diabetes insipidus, with less than 1% of patients with adenomas having it. Also, 14%−20% of patients with diabetes insipidus also have pituitary metastasis. Invasive metastatic tumors often cause visual deficits because of suprasellar extension. They cause painful ophthalmoplegia because of invasion into the cavernous sinus. Metastatic tumors can cause dysfunctions of anterior pituitary hormone. In 56%−64% of symptomatic patients, pituitary symptoms are the first presentation signaling malignancy.

Diagnosis

For diagnosis, imaging studies are not extremely useful to distinguish between pituitary metastases and adenomas. However, MRI is better than CT for assessment of the pituitary region. Suspicion of pituitary metastasis is based on a fast increase in a sellar tumor's size, plus aggressive tissue infiltration, and symptoms of diabetes insipidus. Since the majority of metastatic tumors are within the posterior lobe, sections must be made with enough posterior lobe tissue as well as anterior lobe tissue. Pituitary glands have been biopsied during hypophysectomy for palliation in end-stage breast cancer, as well as during autopsy. This examination has revealed pituitary metastasis in 6%−29% of breast cancer patients.

Treatment

Pituitary metastases may be treated with surgical resection, radiation therapy, and chemotherapy. When metastasis is only within the sella turcica, transsphenoidal surgery may allow improvement of visual defects. Pituitary metastases may be suppressed by molecular targeted cancer treatments. Appropriate target therapies may be based on biomarkers, including *EGFR* mutation, oversecretion of *ERBB2 (HER2)*, and *RAS* gene mutations.

Prognosis

The prognosis of pituitary metastases is poor, primarily due to the aggressiveness of primary neoplasms. Outlook depends on the type of primary tumor and the extent of disease spread throughout the body.

Various parasellar masses

There are various parasellar masses that may develop, which include cysts (such as *Rathke's cyst*) and tumors. *Granular cell tumors* include **pituitary choristomas**, which usually occur after age 20. They have abundant cytoplasmic granules without pituitary hormones, and can present with diabetes insipidus. Sometimes, pituitary adenomas are coincidental with these tumors. **Chordomas** are slowly growing, cartilaginous tumors arising from midline remnants of the notochord. They are locally invasive, and can metastasize, usually arising from the vertebrae. About 33% involve the **clivus** region. Chordomas hold a matrix that is rich in mucin, allowing for diagnosis via fine-needle aspiration. They cause headaches, asymmetric visual disturbances, hormone deficiencies, and sometimes, nasopharyngeal obstruction. The tumor is linked with calcification and osteolytic bony erosion. An MRI may distinguish between the mass and the normal pituitary tissue. During surgery, these tumors are lobular, heterogeneous, and rough. Epithelial cell markers are present, which include cytokeratin and vimentin. After surgical excision, recurrences are common. The mean patient survival time is approximately 5 years. In rare cases, these tumors undergo sarcomatous transformation. There is an aggressive nature to them that requires significant surgical dissection. The endoscopic endonasal approach is usually preferred.

Meningiomas (discussed in detail in Chapter 12: Meningioma) may occur in the sellar and parasellar region (about 20% of all meningiomas). Sellar meningiomas are often well circumscribed and smaller than craniopharyngiomas. Suprasellar meningiomas can invade the pituitary gland ventrally. Intrasellar origins of these tumors only occur rarely. There may be coexisting functioning pituitary adenomas with parasellar meningiomas. About 50% of these patients have secondary hyperprolactinemia, headache, progressive visual disturbances, and optic atrophy. The tumors can be hard to distinguish from pituitary adenomas. They have a higher surgical mortality rate than pituitary adenomas.

Gliomas and low-grade **astrocytomas** (discussed in detail in Chapters 6−8) arise from inside the optic chiasm or tract, often infiltrating the optic nerve. Less than 33% are intraorbital, and the same percentage is caused by **Von Recklinghausen disease**. In rare cases, these tumors arise in the sella, associated with hyperprolactinemia. They should be considered for the rare differential diagnosis of a prolactin-secreting pituitary adenoma. About 80% of patients are under 10 years old. **Primary CNS lymphomas** are usually beta-cell non-Hodgkin types. Less than 20 cases have been documented. An MRI will reveal invasion of the cavernous sinus. Also, acute lymphoblastic leukemia may be linked with **periglandular** pituitary infiltrates, and slight pituitary dysfunction.

Metastases to the pituitary gland occur in 3.5% of cancer patients—primarily older adults with diffuse malignancies. The posterior pituitary is the primary site for blood-borne metastatic spread, due to its rich vascular supply that is directly derived from the systemic circulation, through the internal carotid arteries. Carcinomas that metastasize to the pituitary come from the breasts (37.2% of cases), lungs (24.2%), prostate gland (5.2%), and kidneys (4.9%), with the remainder divided over 28 additional sites. Pituitary imaging may not clearly show metastatic deposits from a pituitary adenoma. Metastatic lesions may closely resemble an adenoma. Diagnosis is only via histologic study of resected specimens. Once diagnosed, low-dose pituitary radiation may be able to shrink the metastasis and improve symptoms.

Clinical cases

Clinical case 1

1. How do pituitary adenomas become symptomatic?
2. What do optic chiasm compression and cavernous sinus invasion result in?
3. What are some of the factors concerning recurrent pituitary adenomas?

A 61-year-old man with type 2 diabetes mellitus was referred to a neurosurgeon for consideration of minimally invasive endoscopic surgery. Fourteen years previously, he had undergone a subtotal resection of a large nonsecreting macroadenoma via the transsphenoidal approach. Prior to the surgery, the patient had severe bitemporal hemianopsia, but this improved greatly after the procedure. Four years later, he developed recurrent visual field deficits and progressively worsening headaches. An MRI revealed a large, recurrent tumor with marked compression of the optic chiasm, clival erosion, and invasion of the right cavernous sinus. Using an endoscopic endonasal approach with image guidance, surgical resection of a multicompartmentalized, fibrous adenoma was achieved. The tumor was removed from all regions it affected. The normal pituitary gland was visible toward the left area of the sella, and remained intact. After surgery, the patient's vision was greatly improved, the headaches resolved, and there were no additional endocrine problems. An immediate postoperative MRI revealed total resection of the recurrent tumor.

Answers:

1. **About 30% of pituitary adenomas are nonsecretory and become symptomatic by compressing surrounding structures. Pituitary compression can result in hypopituitarism.**
2. **Optic chiasm compression can result in visual field disturbance, typically bitemporal hemianopia. Cavernous sinus invasion can result in extraocular muscle palsies. Also, headaches and CSF leakage can occur.**
3. **Recurrent or progressive residual pituitary adenomas often invade the cavernous sinuses and other supra- and parasellar structures. They may be scarred, fibrous, and multicompartmentalized, possibly leading to more surgical complications such as CSF leakage, neuroendocrine dysfunction, and visual impairment.**

Clinical case 2

1. What is the linkage between pituitary tumors and acromegaly?
2. How do pituitary tumors such as these form?
3. What are the complications of acromegaly due to these tumors?

A 35-year-old woman was hospitalized after a 1-year history of dull, generalized headache, limb enlargement, weight gain, and changes of her facial features. She had no menstrual periods for nearly 12 months. Clinical examination revealed signs of acromegaly. A skull radiograph revealed a normal-sized sella, with erosion of the sellar floor. An MRI revealed an isointense mass lesion of the clivus, below the sellar floor, which was connected to an intrasellar lesion by a narrow stalk. Her basal growth hormone level was six times higher than normal, but other hormone levels were normal. The tumor was approached via the transsphenoidal route. The diagnosis was pituitary adenoma that cosecreted growth hormone and prolactin.

Answers:

1. **About 95% of cases of acromegaly are caused by a benign pituitary tumor. The other 5% are caused by nonpituitary tumors, usually located in other parts of the brain or in the body. The root cause of acromegaly is excessive secretion of growth hormone.**

2. All pituitary tumors form the same way. A pituitary cell spontaneously grows and multiplies abnormally, due to a genetic mutation in the cell, which is usually sporadic and not inherited. The cells multiply, and the tumor grows, either slowly or quickly.
3. The most serious health consequences of acromegaly include diabetes mellitus, hypertension, cardiovascular disease, colon polyps, sleep apnea, and increased mortality.

Clinical case 3

1. How common are lactotroph adenomas?
2. How does medication therapy work with these tumors?
3. When is surgery considered for lactotroph adenomas?

A 42-year-old obese woman with diabetic retinopathy, type 2 diabetes, hirsutism, and hypertension went to her physician to discuss an elevated prolactin level discovered by her gynecologist. She complained of galactorrhea and no menstruation for 1 year. A pituitary MRI with contrast revealed a subtle area of delayed enhancement on the right side, consistent with a 5-mm microadenoma. The patient was given the dopamine agonist cabergoline, with a follow-up planned for 3 months.

Answers:

1. Lactotroph adenomas (prolactinomas) are the most common type of pituitary adenoma. They are usually benign, with malignant tumors being very rare. The larger the size of the tumor, the greater the prolactin level, and the more likelihood of mass-effect symptoms.
2. Medication therapy involves dopamine agonists that directly inhibit prolactin secretion by the tumor that suppresses tumor growth. The goal of therapy is to suppress the prolactin level to normal range, and restore gradual function. The two dopamine agonists used are bromocriptine and cabergoline.
3. Surgery should be considered if medications cannot be tolerated, or if they do not reduce prolactin levels, restore normal reproduction and pituitary function, or reduce tumor size. Sometimes, medications are continued with surgery or radiation treatment. Usually however, surgery corrects prolactin levels in patients.

Clinical case 4

1. What is the definition of pituitary apoplexy?
2. Which hormone-related problems occur with this condition?
3. What are the predominant causes of pituitary apoplexy?

A 55-year-old man suddenly developed intense headaches that occurred nearly every morning. They became more severe than he could tolerate, and he also began to vomit and experience reduced vision in his right eye. A CT scan was ordered, revealing a pituitary tumor that had undergone apoplexy, leading to the patient's headaches and compression of the optic nerve. Emergency surgery, following an MRI to visualize his brain even more extensively, resulted in successful tumor removal without complications. The patient made an excellent recovery.

Answers:

1. Pituitary apoplexy is bleeding into the pituitary gland, or impaired blood supply of the gland. It usually occurs because of a pituitary tumor. The most common initial symptom is a sudden headache, often with worsening visual field defects.
2. In pituitary apoplexy, the primary initial problem is lack of secretion of adrenocorticotropic hormone. This occurs in about 70% of patients with pituitary apoplexy. The sudden lack of cortisol leads to adrenal crisis, with hypotension, hypoglycemia, and abdominal pain developing.
3. Nearly all cases arise from a pituitary adenoma. In 80% of cases, the patient is unaware of the tumor. Overall, only a small proportion of pituitary tumors will undergo apoplexy. Risks are higher, however, for macroadenomas and tumors that grow more quickly.

Clinical case 5

1. Why is the differentiation between atypical and malignant pituitary tumors difficult?
2. What are the treatment benefits, compared between cabergoline and bromocriptine?
3. What is the outlook for pituitary carcinomas?

A 19-year-old woman presented with primary amenorrhea, galactorrhea, and left-sided hemianopsia. She was examined and tested, and hyperprolactinemia was present, plus a pituitary macroadenoma with invasion of the cavernous sinus, with suprasellar growth. Cabergoline and bromocriptine were used, but these medications did not improve her condition. Surgical resection of the pituitary lesion was performed, and during this procedure, a noncontinuous lesion of the nasal mucosa was discovered. This metastasis resulted in diagnosis of a prolactin-producing pituitary carcinoma. After partial resection, radiotherapy was used, and the patient developed GH deficiency, central hypothyroidism, hypogonadism, and permanent diabetes insipidus. After 6 years, she had developed Graves' disease and hypocortisolism, for which she was treated, though hypothyroidism recurred when the antithyroid drug was withdrawn. Nine years later, the patient remains on bromocriptine, and has localized, stable disease without metastases.

Answers:

1. **Differentiating atypical from malignant pituitary tumors may be difficult since metastases upon initial presentation are extremely uncommon. There are no reliable markers or tumor-specific factors that predict potential malignancy of a prolactin-producing pituitary tumor.**
2. **Cabergoline is more effective and better tolerated compared to bromocriptine. While most patients have a good response to dopamine receptor agonists, based on dosage, others do not achieve normal prolactin levels or a 50% reduction in size of the tumor, even at maximally tolerable doses. Nearly 25% of patients are resistant to bromocriptine and 10% are resistant to cabergoline.**
3. **With the extremely rare pituitary carcinomas, metastases may be found incidentally, or can become symptomatic. There is a worse overall survival than for invasive adenoma. When metastases develop, the mean survival is less than 5 years.**

Key terms

18F-labeled deoxyglucose
Acidophilic
Acromegaly
Adamantinomatous
Addison's disease
Alpha-crystallin B chain
Amenorrhea
Angiofibromas
Annexin A1
Astrocytomas
Bitemporal hemianopsia
Bitemporal superior quadrantanopia
Carney complex
Chordomas
Chromophobic
Clival
Clivus
Collagenomas
Corticotrophic adenomas
Cosyntropism
Craniopharyngioma
Cushing's adenomas
Cushing's disease
Cushing's syndrome
Cyberknife
Diabetes insipidus
Drop metastasis
Endonasal
Exophthalmos
Familial isolated pituitary adenoma
Folliculostellate cells
Gadolinium/pentetic acid
Galactorrhea
Galectin-3
Gamma knife radiosurgery
Gigantism
Gliomas
Gonadotrophic adenomas
Gonadotrophic cell tumors
Graves' disease
Hemicrania continua
Hemihypophysectomy
Heterodimeric complex
Hyperpituitarism
Hypervascularity
Hypopituitarism
Immunophenotypic
Immunostain
Lactotrophic adenomas
Lateral rectus palsy
Leiomyomas
Lentiginosis
Macroadenomas
Meningiomas
Microadenomas
Multiple endocrine neoplasia type 1
Myxomas
Nelson's syndrome
Nestin
Neuralgiform headache
Null cell adenomas
Optic chiasma
Periglandular
Pituicytes
Pituicytoma
Pituitary adenoma
Pituitary apoplexy
Pituitary carcinoma
Pituitary choristomas
Pituitary tuberculoma
Polyethylene glycol precipitation
Primary CNS lymphomas
Prolactinomas
Proton therapy
Pseudocapsule
Rathke's pouch
Sella
Somatotrophic adenomas
Spindle cell oncocytoma
Stereotactic radiosurgery
Thyrotrophic adenomas
Transsphenoidal
Von Recklinghausen disease
Xenoestrogenic

Further reading

Aghi, M. K., & Blevins, L. S. (2012). *Management of pituitary tumors, An issue of neurosurgery clinics of North America: Surgery.* Saunders.
Aghi, M. K., & Blevins, L. S. (2019). *Pituitary adenoma, an issue of neurosurgery clinics of North America (The clinics: Surgery).* Elsevier.
Asa, S. L. (2011). *Tumors of the pituitary gland (Atlas of tumor pathology series 4).* American Registry of Pathology.
Bonneville, J. F., Bonneville, F., Cattin, F., & Nagi, S. (2016). *MRI of the pituitary gland.* Springer.
Casanueva, F. F., & Ghigo, E. (2018). *Hypothalamic-pituitary diseases (Endocrinology).* Springer.
Dubey, S. P., & Schick, B. (2017). *Juvenile angiofibroma.* Springer.
Evans, J. J., & Kenning, T. J. (2015). *Craniopharyngiomas: Comprehensive diagnosis, treatment and outcome.* Academic Press.
Ganz, J. C. (2011). *Gamma knife neurosurgery.* Springer Wien.
Geer, E. B. (2017). *The hypothalamic-pituitary-adrenal axis in health and disease: Cushing's syndrome and beyond.* Springer.
Hayat, M. A. (2012). *Tumors of the central nervous system, Volume 8: Astrocytomas, medulloblastoma, retinoblastoma, chordoma, craniopharyngioma, oligodendroglioma, and ependymoma.* Springer.
Hayat, M. A. (2013a). *Tumors of the central nervous system, Volume 10: Pineal, pituitary, and spinal tumors.* Springer.
Hayat, M. A. (2013b). *Tumors of the central nervous system, Volume 11: Imaging, glioma and glioblastoma, stereotactic radiotherapy, spinal cord tumors, meningioma, and schwannomas.* Springer.
Heron, D. E., Huq, M. S., & Herman, J. M. (2018). *Stereotactic radiosurgery and stereotactic body radiation therapy (SBRT).* Demos Medical.
Kanakis, D. (2016). *Pituitary adenoma: Pathophysiology, diagnosis and treatment options.* Nova Science Publishers Inc.
Kohn, B. (2019). *Pituitary disorders of childhood: Diagnosis and clinical management.* Humana Press.
Lania, A., Spada, A., & Lasio, G. (2016). *Diagnosis and management of craniopharyngiomas (Key current topics).* Springer.
Laws, E. R., Jr., Cohen-Gadol, A. A., Schwartz, T. H., & Sheehan, J. P. (2017). *Transsphenoidal surgery: Complication avoidance and management techniques.* Springer.
Laws, E. R., Jr., & Pace, L. (2016). *Cushing's disease: An often misdiagnosed and not so rare disorder.* Academic Press.
Laws, E. R., Jr., & Sheehan, J. P. (2011). *Sellar and parasellar tumors: Diagnosis, treatments, and outcomes.* Thieme.
Lunsford, L. D. (2015). *Intracranial stereotactic radiosurgery* (2nd ed.). Thieme.
Martinez-Barbera, J. P., & Andoniadou, C. L. (2017). *Basic research and clinical aspects of adamantinomatous craniopharyngioma.* Springer.
Melmed, S. (2010). *The pituitary* (3rd ed.). Academic Press.
Nachtigall, L. B. (2018). *Pituitary tumors: A clinical casebook.* Springer.
Naliato, E. C. O. (2010). *Prolactinomas, Prolactin and Weight Gain (Cancer Etiology, Diagnosis and Treatments).* Nova Novinka.
Norden, A. D., Reardon, D. A., & Wen, P. C. Y. (2011). *Primary central nervous system tumors: Pathogenesis and therapy.* Humana Press.
Pamir, M. N., Al-Mefty, O., & Borba, L. (2017). *Chordomas: Technologies, techniques, and treatment strategies.* Thieme.
Pituitary Network Association. (2013). *Pituitary patient resource guide* (5th ed.). Pituitary Network Association.
Qi, S. (2017). *Craniopharyngiomas — Classification and surgical treatment (Frontiers in neurosurgery), Volume 4.* Bentham Science Publishers.
Qi, S. (2019). *Atlas of craniopharyngioma (Pathology, classification and surgery).* Springer.
Sindwani, R., Recinos, P. F., & Woodard, T. D. (2015). *Endoscopic cranial base and pituitary surgery, an issue of otolaryngologic clinics of North America (The clinics: Surgery).* Elsevier.
Sperring, J. P., & Smith, M. J. (2012). *Acromegaly: Causes, tests and treatments.* CreateSpace Independent Publishing Platform.
Tanase, C., Ogrezeanu, I., & Badiu, C. (2011). *Molecular pathology of pituitary adenomas (Elsevier insights).* Elsevier.
Wongsirisuwan, M. (2015). *Minimally invasive surgery for pituitary adenoma.* Nova Science Publishers, Inc.

Chapter 16

Pineal parenchymal tumors

Chapter Outline

Overview 323
Pineocytoma 323
Pineal parenchymal tumor with intermediate differentiation 326
Papillary tumor of the pineal region 328
Pineoblastoma 330
Clinical cases 333
Key terms 335
Further reading 335

Overview

Pineal parenchymal tumors are generally rare in comparison to other brain tumors. Pineocytomas are well-differentiated grade I tumors that make up about 20% of all pineal parenchymal tumors, with a mean patient age of 42.8 years. Most pineal parenchymal tumors are slightly more common in women than in men. Pineal parenchymal tumors with intermediate differentiation may be either grade II or grade III tumors, make up about 45% of all pineal parenchymal tumors, with a mean patient age of 41 years. Papillary tumors of the pineal region may also be grade II or III, have a mean patient age of 35 years, are the rarest of all pineal parenchymal tumors, and are the only type that affects slightly more men than women. Pineoblastomas are grade IV tumors that have a mean patient age of 17.8 years, and make up about 35% of all pineal parenchymal tumors, also being slightly more common in women.

Pineocytoma

A **pineocytoma** is a rare pineal parenchymal tumor that is well differentiated. It is made up of cells in a uniform pattern that may form large **pineocytomatous** rosettes, pleomorphic cells with **gangliocytic** differentiation, or both. The World Health Organization ranks pineocytomas as grade I tumors. Pineocytomas are sometimes called *pinealomas*.

Epidemiology

Pineocytoma makes up between 17% and 30% (with a mean percentage of 20%) of all pineal parenchymal tumors. It usually affects adults between 20 and 60 years of age, and has a mean patient age of 42.8 years. Females develop pineocytomas slightly more than males (1:0.6). Pineocytomas make up less than 1% of all intracranial neoplasms. About 27% of these tumors are from pineal parenchymal origin.

Etiology and risk factors

No specific genetic alterations have been found regarding pineocytomas. The actual cause of all pineal tumors is unknown, though research is ongoing. There have been some links associated with chromosomal abnormalities, but there are also no identified risk factors.

Pathology

Pineocytomas only develop in the pineal region as well-demarcated, solid masses, and no dissemination or infiltrative growth. They compress nearby brain structures, including the brain stem, cerebellum, and cerebral aqueduct. These tumors also often protrude into the posterior third ventricle. Clearly defined pineocytomas grow locally. They are not linked to CSF seeding. Macroscopically, pineocytomas are a gray-to-tan color, with a homogeneous or granular cut surface. There may be degenerative cyst formation and hemorrhage foci (see Fig. 16.1).

Microscopically, pineocytomas are moderately cellular, made up of small and uniform mature cells that appear similar to pinealocytes (see Fig. 16.2). Growth is usually in sheets, often with many delicate tumor cell processes. Poorly

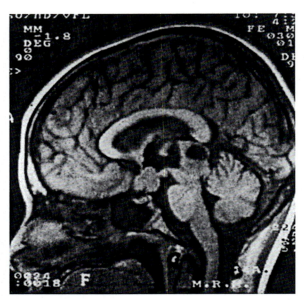

FIGURE 16.1 Graphic showing pineocytoma. The cystic center is lined with a rim of solid and partially calcified tumor (black arrow). Hemorrhage (white arrow) is not uncommon.

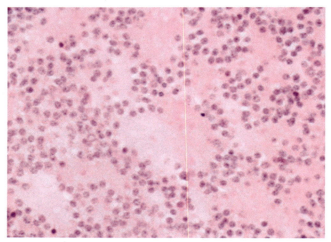

FIGURE 16.2 Cellular appearance of pineocytoma.

defined lobules can be seen, though an obvious lobule-like structure is a feature of the normal pineal gland. Most of the nuclei are round to oval, having small nucleoli and finely dispersed chromatin. There is a moderate amount of cytoplasm that is homogeneously eosinophilic. The tumor processes are obvious, but short and usually end in club-shaped structures able to be easily seen in neurofilament immunostaining or silver impregnation. There are no mitotic figures, of less than 1 mitosis per 10 high-power fields, any except the rarer large tumors. The pineocytomatous rosettes are of differing sizes and amounts. Their centers lack nuclei, and consist of thin cytoplasmic processes that look like neuropil. The nuclei around the edges of each rosette are not regular. Some pineocytomas have a pleomorphic cytological form, characterized by multinucleated giant cells with bizarre nuclei, with or without large ganglion cells. There is still a low mitotic activity, even though the tumor has a semi-malignant appearance to the nuclei. The stroma is a delicate vascular channel network, with a single layer of endothelial cells, and only few reticulin fibers. Microcalcifications are rare, but usually related to calcifications of the rest of the pineal gland.

Pineocytomas have clear and varied amounts of dark cells connected with zonulae adherentes. These cells extend tapered processes that sometimes terminate in bulbous ends. There is fairly abundant cytoplasm, with well-developed organelles. Pineocytoma cells have similar ultrastructural components of normal pinealocytes. These include **annulate**

lamellae, paired and twisted filaments, cilia with a microtubular pattern, fibrous bodies, microtubular sheaves, heterogeneous cytoplasm, small rod-like structures topped with vesicles, membrane whorls, and clusters of mitochondria and centrioles. The cytoplasm and cellular processes have membrane-bound clear vesicles and densely cored granules. Sometimes, the processes have synapse-like junctions.

Pineocytoma cells often have strong immunoreactivity for the protein-coding gene called *synaptophysin, neurofilament protein (NFP)*, and *neuron-specific enolase*, a **metalloenzyme**. Variable staining has occurred for the microtubule element called *class III beta-tubulin*, the *ubiquitin carboxyl-terminal esterase L1 (UCHL1)*, the microtubule-associated protein *tau*, chromogranin-A (parathyroid secretory protein 1), and serotonin receptor *5-HT*. There is **photosensory** differentiation with rhodopsin (visual purple) and the protein called **S-arrestin**. In the pleomorphic variant types, **ganglioid** cells often express NFP and other neuronal markers. Usually, mitotic figures are rare or absent, and the mean Ki-67 proliferation index is less than 1%.

Histogenesis of pineal parenchymal tumors is related to the pinealocytes, which have neuroendocrine and photosensory functions. The developing pineal gland's cells resemble the cells of the developing retinas. They have significant melanin pigment and microtubular cilia. By age 3 months, the pigmented cells decrease in number slowly, and the pigment cannot be detected in **histochemistry**. With continued differentiation, strongly immunoreactive cells (for neuron-specific enolase) accumulate. By age 12 months, the pinealocytes predominate. Tissue cultures reveal that pineocytomas cells also can synthesize melatonin and 5-HT. In a pineocytoma, **immunoexpression** of ASMT and CRX helps indicate a link between tumors and pinealocytes. The transcription factor CRX is involved in development and differentiation of pineal cell families, while ASMT is an important enzyme needed for melatonin synthesis.

Though studies are insufficient, the genetic profile of pineocytomas shows a **hypotriploid** or **pseudodiploid** tendency, with numerical and structural changes. These include loss of some or all of chromosome 22, loss of chromosome 14, loss or partial deletion of chromosome 11, and a gain of chromosomes 5 and 19. In a microarray analysis, pineocytomas have high-level expression of genes that code for enzymes involved in melatonin synthesis and in retinal phototransduction. This means there is bidirectional photosensory and neurosecretory differentiation. There are no syndromes or genetic susceptibilities related to pineocytomas. In one family, this tumor has occurred in siblings.

Clinical manifestations

Due to their expansive growth within the pineal region, pineocytomas cause signs and symptoms that are related to increased intracranial pressure, because of aqueductal obstruction, brain stem or cerebellar dysfunction, and **Parinaud syndrome**, which involves neuro-ophthalmological dysfunction. Signs and symptoms include headache, ataxia, papilledema, impaired vision and ambulation, nausea, vomiting, dizziness, loss of upward gaze, and tremor.

Diagnosis

Pineocytomas usually present as well delineated and round masses less than 3 cm in diameter during a CT scan. They are hypodense and homogeneous, and may have peripheral or central calcification. Sometimes, cystic changes are present that are usually distinct from common pineal cysts. The majority of pineocytomas have heterogeneous contrast enhancement. Occasionally, isodense to partially hyperdense images with homogeneous contrast enhancement have been seen during CT. Hydrocephalus is often present. An MRI usually shows these tumors to be hypointense or isointense in T1-weighted images, but hyperintense on T2-weighted images, with strong and homogeneous contrast enhancement (see Fig. 16.3). Margins are usually very clear, and MRI affords the best method of imaging for this. Confirmation of diagnosis is via brain biopsy.

Treatment

Pineocytomas are surgically resected, and complete resection is usually able to be accomplished since these tumors are usually very well circumscribed. Local recurrence and CSF metastases are very rare. Some patients undergo radiation treatments or chemotherapy, determined individually per patient.

Prognosis

The prognosis of pineocytomas is good following surgical removal. There may be a long interval, of years, between symptoms onset and surgery. No metastasis has been identified, and 5-year survival rates range between 86% and 91%. There has been a documented case study of patients with 100% survival over 5 years. The major prognostic factor is the extent of possible surgical resection.

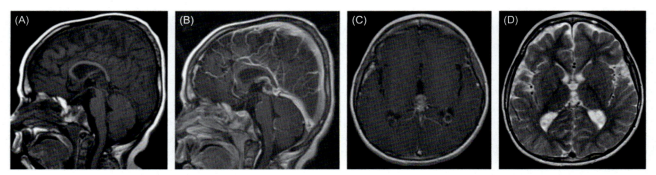

FIGURE 16.3 Pineocytoma on sagittal T1-weighted postcontrast (A) and T2-weighted (B) MRI showing a cystic lesion of the pineal gland with peripheral solid contrast-enhancing components and an intralesional fluid—fluid level after hemorrhage. (C) Top view of skull with T1-weighted imaging. (D) Top view of skull with T2-weighted imaging.

> **Point to remember**
> According to the American Brain Tumor Association, tumor types occurring in the pineal region often involve the pineal gland, but not always. True pineal cell tumors include pineocytomas, pineoblastomas, and mixed pineal tumors. Other tumors that may occur in this region include: germinomas, teratomas, endodermal sinus tumors, embryonal cell tumors, choriocarcinomas, meningiomas, astrocytomas, gangliogliomas, and dermoid cysts.

Pineal parenchymal tumor with intermediate differentiation

A **pineal parenchymal tumor with intermediate differentiation** (PPTID) is of intermediate malignancy, between pineocytoma and pineoblastoma. These tumors are made up of diffuse sheets or large lobules of round cells that have a single structural pattern. They are more differentiated, however, than the cellular patterns of pineoblastomas. The World Health Organization ranks these tumors as either grade II or grade III, and definitive histological grading criteria must still be defined. The relatively new PPTID category was only established in 1993, and these tumors have also been called *malignant pineocytomas, pineocytomas with anaplasia,* and *pineoblastoma with lobules*.

Epidemiology

The PPTIDs occur mostly in adults, with a mean patient age of 41 years. These tumors account for about 45% of all pineal parenchymal tumors, in a total range of 21%—54%. There is a slight female preponderance, and the male-to-female ratio is 0.8:1.

Etiology and risk factors

The PPTIDs arise from pinealocytes or their precursor cells. Immunoexpression of **cone-rod homeobox** (CRX) and ASMT proteins shows that these tumors are biologically related to pinealocytes. There is no proven cause or identified risk factors for the PPTIDs, but frequent chromosomal changes have been identified, with an average of 3.3 gains and 2 losses. The most common are gains in 4q and 12q, and a loss in chromosome 22. No syndromic associations or genetic susceptibilities have been linked.

Pathology

There may be pleomorphic cytology in these intermediately differentiated tumors. They have varied biological and clinical pathologies, ranging from low-grade tumors with common local and delayed recurrences, to high-grade tumors that carry a risk of craniospinal spread. The mitotic activity, neuronal and neuroendocrine differentiation, and Ki-67 proliferation index are all variable. The PPTIDs only occur in the pineal region, with a clinical presentation similar to other pineal parenchymal tumors. Extension of the tumor compresses the corpora quadrigemina, compromising CSF flow through the aqueduct. If the mass compresses the superior colliculus, there may be eye movement abnormalities such as in Parinaud syndrome, paralysis of upward gaze, pupil problems such as slightly dilated pupils with reaction to accommodation but not light, and **nystagmus retractorius**. One case of apoplectic hemorrhage of a PPTID involved sudden symptom onset.

The PPTIDs have local recurrence in about 22% of cases. Though craniospinal dissemination is observed in 10% of cases during diagnosis, it occurs during the disease course in 15% of cases. Macroscopically, PPTIDs resemble pineocytomas. They are soft, may be circumscribed, and lack necrosis. In one case with spinal metastasis, the tumor surface was irregular. Microscopically, PPTIDs may be diffuse, lobulated, or a mixture of both. Diffuse PPTIDs resemble neurocytomas or oligodendrogliomas. Lobulated PPTIDs have vessels that delineate poorly defined lobules. There are also transitional cases, in which the common pineocytomatous areas are linked to a diffuse or lobulated pattern that is more similar to a true PPTID. There is moderate to high cellularity in PPTID, usually with round nuclei that have mild to moderate atypia and a "salt-and-pepper" appearance of the chromatin. The cytoplasm is easier to distinguish than that of pineoblastoma. There is also a pleomorphic cytological variant, with bizarre ganglioid cells that have single or multiple atypical nuclei and significant amounts of cytoplasm. The mitotic activity is low to moderate.

The PPTIDs can be aggressive, and a wide range of mitotic counts complicates diagnosis. In one large study, there were zero mitoses per 10 high-power fields in 54% of patients, one to two mitoses in 28%, three to six mitoses in 15%, and only a few mitoses of more than 6. The mean Ki-67 proliferation index is often much different than that of the pineocytomas and pineoblastomas. It ranges between 3.5% and 16.1%. There is synaptophysin positivity, and variable labeling with the antibodies to chromogranin-A and NFP. Astrocytic interstitial cells usually express GFAP and S-100. In the pleomorphic variants, ganglion cells may express S-100 as well, and the ganglioid cells usually express NFP. There are many more ASMT-positive cells in PPTIDs than in pineoblastomas.

Clinical manifestations

The primary symptoms of PPTIDs are headaches (which are often chronic) and vomiting. This is linked to increased intracranial pressure, which is due to obstructive hydrocephalus caused by tumor extension of the pineal gland, into the posterior third ventricle. Additional symptoms may include dizziness, lethargy, blurred vision, and confusion.

Diagnosis

In imaging studies, PPTIDs are usually bulky masses with localized invasion and are less often circumscribed. Ct scans may show peripheral "exploded" calcifications. An MRI reveals that the tumors are heterogeneous, mostly hypointense on T1-weighted images, and hyperintense on T2-eighted images (see Fig. 16.4A and B). In CT as well as MRI, postcontrast enhancement is usually significant and heterogeneous.

Treatment

Though the goal of surgery is gross total resection, the tumor location and invasive qualities often make this difficult if not impossible. External beam radiation therapy, craniospinal radiation therapy, or whole-ventricular radiation therapy have been used successfully. Chemotherapeutic agents include vincristine, nimustine, carboplatin, and interferon-beta. Overall, the combination of surgery with radiation and chemotherapy is usually able to provide a better outlook.

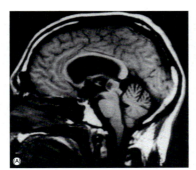 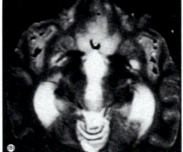 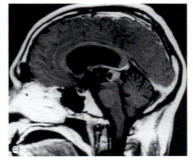

FIGURE 16.4 (A) T1-weighted MRI image of PPTID; (B) T2-weighted MRI image of PPTID; (C) MRI of pineoblastoma.

Prognosis

For PPTIDs, the 5-year overall survival rates range between 39% and 74%. These tumors have a better prognosis than for pineoblastoma, with median overall survival of 13.7 years. The PPTIDs also have a median progression-free survival of 7.7 years. Low-grade PPTIDs have an overall 5-year survival rate of 74%. Recurrence occur in 26% of patients, but are usually local and delayed. High-grade PPTIDs have an overall 5-year survival rate of 39%, and risk of recurrence is 56%, plus risk of spinal spread was 28%. Transformation of PPTID into pineoblastoma is rare.

> **Point to remember**
> Unlike the other histological subtypes of pineal parenchymal tumors, the best treatment for those with intermediate differentiation has not been determined. Surgery has been successful on its own, but the seeding potential of these tumors may indicate postoperative treatments that are similar to those for pineoblastomas. In the limited studies that have been conducted, it appears that surgery, radiation therapy, and chemotherapy together provide the best outcomes.

Papillary tumor of the pineal region

In the pineal region a **papillary tumor** is characterized as developing from neuroepithelial tissues, and combines papillary as well as solid areas. These tumors have epithelial-like cells and are immunoreactive for cytokeratins. The course of these tumors is variable, and may be grade II or grade III, though definitive grading criteria have still not been established.

Epidemiology

In the pineal region, a papillary tumor affects both children and adults, with a mean patient age of 35 years. Actual incidence of these tumors is not available due to how rare they are. To date, less than 200 have been documented globally. There is a male-to-female ratio of 1.06:1.

Etiology and risk factors

The pineal papillary tumors may originate from remnants of the specialized ependymal cells of the **subcommissural organ**. This is based on cytokeratin positivity and ultrastructural visualization of ependymal, neuroendocrine, and secretory organelles. The causes of these tumors are believed to be linked to their high levels of expression of subcommissural genes. This is still under study, however, and there are also no specific risk factors. There have been losses of chromosome 10, and gains of chromosomes 4 and 9. Distinct DNA methylation profiles differentiate these tumors from ependymomas. There are also genetic alterations of the *phosphatase and tensin homolog (PTEN)* protein. No documented syndrome associations or proof of genetic susceptibility exist.

Pathology

Pineal papillary tumors present as large, well-circumscribed masses that often have T1-hyperintensity. They are often gray in color. The tumors recur in 58% of cases within 5 years. However, spinal spread is rare (only about 7% of cases). Macroscopically, these tumors are soft, relatively large—between 20 and 54 mm—and cannot easily be distinguished from a pineocytoma. Microscopically, they have papillary features and more densely cellular areas, often with true rosettes and tubes. In the papillary areas, the vessels are obscured by layers of large columnar cells that are pale to eosinophilic. In the cellular areas, the cells are somewhat clear or vacuolated in regards to the cytoplasm. Sometimes, the cytoplasmic mass may be eosinophilic periodic acid-Schiff-positive. There are round to oval nuclei, which are often irregular, and have **stippled** chromatin. Pleomorphic nuclei may be seen. Mitotic counts range from zero to 13 mitoses per 10 high-power fields, and necrotic foci may be visible. The vessels are hyalinized and have multiple lumina with a **pseudoangiomatous** morphology. There will be a clear demarcation between the tumor and the pineal gland. The Ki-67 proliferation index may range from 1% to 29.7%, and a high index is seen in 39%—40% of cases. High proliferative activity is related to younger patient age.

Electron microscopy reveals ependymal, neuroendocrine, and secretory features. The cells express cytokeratin, synaptophysin, chromogranin, and neurofilament protein (see Fig. 16.5). The tumors are usually made up of alternating clear and dark epithelioid cells connected at the apical region by well formed, intercellular junctions. There are many microvilli and occasional cilia. The nuclei are indented, irregular, or oval, often located at one cellular pole. There may be zonation of the organelles. The cytoplasm usually has many organelles, including clear coated vesicles, mitochondria, and less often, dense-core vesicles. There is abundant rough endoplasmic reticulum, and sometimes, dilated cisternae containing a granular secretory product. Some cells have perinuclear intermediate filaments.

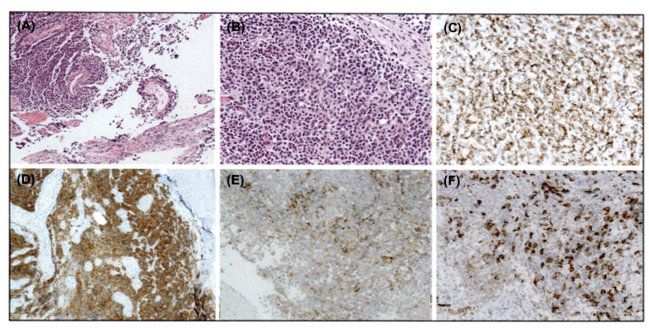

FIGURE 16.5 (A) Histology of the pineal region tumor depicting focal papillary architecture in low magnification. (B) Intermediate magnification shows a solid epithelial tumor with minimal nuclear atypia and mitotic activity. The cells express (C) cytokeratin, (D) synaptophysin, (E) chromogranin, and (F) neurofilament protein.

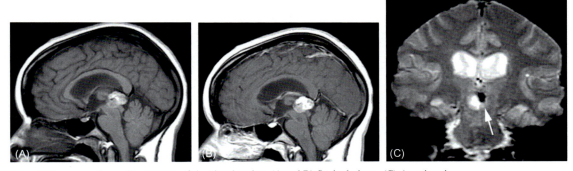

FIGURE 16.6 MRI images of a papillary tumor of the pineal region. (A and B) Sagittal views. (C) Anterior view.

Pineal papillary tumors, distinctively, are reactive for keratins—especially in the papillary regions. The tumors also stain for S-100, vimentin, *microtubule-associated protein 2 (MAP2)*, neuron-specific enolase, *neural cell adhesion molecule 1 (NCAM1)*, and the transthyretin protein. Most of these tumors do not stain for cytoplasmic stanniocalcin-1 (STC1), membranous KIR7.1, the cadherin-1 gene, and the claudin-2 protein, which helps identify them in comparison to choroid plexus tumors.

Clinical manifestations

The symptoms of pineal papillary tumors are nonspecific and may only last a short time. They include headache caused by obstructive hydrocephalus and Parinaud syndrome. Additional symptoms may include strabismus, progressive double vision, nausea, and vomiting.

Diagnosis

Pineal papillary tumors are well-circumscribed, heterogeneous masses made up of limited cystic and solid portions. They are centered on the posterior commissure or in the pineal region itself (see Fig. 16.6). Aqueductal obstruction is common. These tumors may have intrinsic T1-hyperintensity. This may be related—when calcification, hemorrhage, melanin, or fat are absent—to secretory materials high in protein and glycoprotein. In some cases, however, these features have not occurred. The postcontrast enhancement is usually heterogeneous.

Treatment

The treatment of pineal papillary tumors includes the same options as for other pineal tumors, and involves surgery followed by radiation therapy. However, radiation is of disputed value in relation to disease progression of these tumors. Incomplete resection and marked mitotic activity are usually linked to recurrence, and potentially, a worsened outlook.

Prognosis

Overall survival of pineal papillary tumors is 73% at 5 years and 71.6% at 10 years. However, tumor progression has occurred in 72% of cases. When this happens, progression-free survival is at 27% over 5 years. Incomplete resection is linked to decreased survival and tumor recurrence. Increased mitotic activity is associated with shorter progression-free survival of a mean time of 4.3 years. Patients with a Ki-67 proliferation index of 10% or higher have a median progression-free survival time of only 2.4 years.

> **Point to remember**
>
> The 2007 definition of pineal papillary tumors by the World Health Organization is as follows: "A rare neuroepithelial tumor of the pineal region in adults, characterized by papillary architecture and epithelial cytology, immunopositivity for cytokeratin and ultrastructural features suggesting ependymal differentiation."

Pineoblastoma

A **pineoblastoma** is a rare, poorly differentiated malignant tumor that arises in the pineal gland. It is highly cellular and ranked by the World Health Organization as a grade IV tumor. Pineoblastomas are localized in the pineal region, and are also described as *supratentorial midline primitive neuroectodermal tumors*.

Epidemiology

Pineoblastoma usually develops within the first two decades of life, with a mean patient age of 17.8 years. These tumors make up about 35% of all pineal parenchymal tumors, ranging between 24% and 61% in various studies. They occur at any age, but most often in children. Like other pineal parenchymal tumors, the male-to-female ratio is 0.7:1.

Etiology and risk factors

The etiology of pineoblastomas reveals shared features between cells of the pineal gland as well as the retinas of the eyes. There is an occasional association with bilateral (familial) retinoblastoma, in a condition termed *trilateral retinoblastoma syndrome*. This is supported by the occasional progression from a low-grade tumor to a pineoblastoma. Various numerical and structural gene abnormalities are linked to pineoblastomas. There have been common structural alterations of chromosome 1, and losses of all or part of chromosomes 9, 13, and 16. Pineoblastomas occur along with RB1 gene abnormalities, with worsened prognoses. These tumors have also been linked to patients with familial adenomatous polyposis and DICER1 germline mutations. The DICER1 gene provides instructions for building a protein that helps regulate expression of other genes.

Pathology

Patternless sheets of small, immature neuroepithelial cells characterize Pineoblastoma. They have a high nuclear-to-cytoplasmic ratio, small amounts of cytoplasm, and hyperchromatic nuclei, sometimes with small nucleoli. There are frequent mitoses and a Ki-67 proliferation index of always more than 20%. The mean proliferation index ranges widely, however, usually between 23.5% and 50.1%. The retention of SMARCB1 nuclear expression allows for distinction from atypical teratoid or rhabdoid tumors. These tumors usually spread through CSF pathways, and are often aggressive.

Pineoblastomas directly invade nearby brain structures such as the leptomeninges, third ventricle, and **tectal plate**. Macroscopically, pineoblastomas are poorly demarcated and invasive, with a soft texture, friability (an increased likelihood to crumble), and pink to gray color. There may be hemorrhaging, with or without necrosis. Microscopically, these tumors have densely packed small cells with slightly irregular nuclear shapes and hyperchromatic nuclei. Cell borders are usually indistinct. Using hematoxylin and eosin staining, a small amount of rosettes may interrupt the overall diffuse growth pattern. These may be **Homer-Wright rosettes** or **Flexner-Wintersteiner rosettes**, the latter of which reveal *retinoblastic* differentiation (see Fig. 16.7). Mitotic activity is usually high, but can be varied. Necrosis is common.

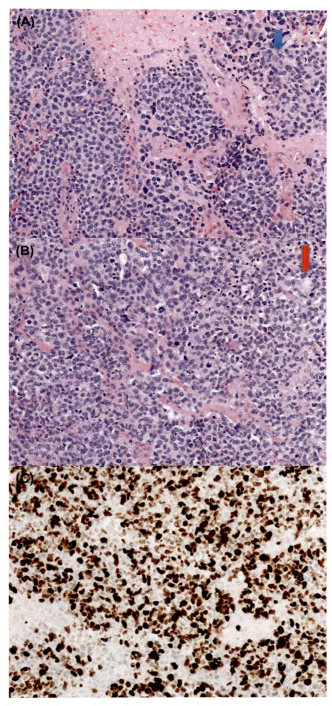

FIGURE 16.7 Microphotographs of a surgically treated pineoblastoma. (A and B) Hematoxylin and eosin stain with high cellularity, numerous mitotic figures (38 mitoses/10 High-Power Fields) and necrosis. Homer-Wright and Flexner-Wintersteiner rosettes (red arrow) and flower-shaped structures called fleurettes (blue arrow) are also seen. (C) Ki67 proliferation index of 60%.

Electron microscopy reveals abundant **euchromatin** and heterochromatin. The cytoplasm has polyribosomes, small amounts of rough endoplasmic reticulum, small mitochondria, and occasionally, intermediate filaments, lysosomes, and microtubules. Poorly formed, short cell processes may have microtubules and small amounts of dense-core granules. Junctions of **zonula adherens** and **zonula occludens** may be present between the cells and processes. Microtubular cilia are sometimes seen. In rare cases, cells are radially arranged around a small central lumen. With pineoblastomas, there is reactivity for glial, neuronal, and photoreceptor markers. There may be positivity for neuro-specific enolase, synaptophysin, class III beta-tubulin, NFP, and chromogranin-A. Also, S-arrestin staining may occur. The reactivity for GFAP should encourage exclusion of entrapped and reactive astrocytes. Also, SMARCB1 is usually expressed.

There is also rare pathological variant called a *mixed pineocytoma-pineoblastoma*, with a biphasic pattern of alternating areas of development. It is important to distinguish the areas that resemble pineocytomas from the normal parenchyma. Another rare variant, *pineal anlage tumors*, resemble pineoblastomas but have a different morphology. They have both neuroectodermal and heterologous **ectomesenchymal** components. The neuroepithelial component has sheets or nests of small, blue, round cells, plus neuronal ganglionic or glial differentiation, with or without melanin-containing epithelioid cells. The ectomesenchymal component has **rhabdomyoblasts** and striated muscle cells, with or without islands of cartilaginous tissues. It is likely that a pineal anlage tumor is totally distinct from pineoblastomas, but this is not totally proven.

Clinical manifestations

The primary symptoms of pineoblastomas are like those of other pineal region tumors. These are related to increased intracranial pressure, and include headaches, nausea, vomiting, plus ocular symptoms such as Parinaud syndrome, diplopia, hearing changes, and reduced visual activity. The time period between initial symptoms may only be 1 month or less. After surgical excision, some patients develop ocular problems such as vertical gaze paresis.

Diagnosis

In imaging studies, pineoblastomas appear as large, multilobulated masses, with frequent invasion of the tectum, thalamus, and splenium of the corpus callosum. There may be small cystic or necrotic areas, with edema. Upon CT imaging, the tumors are usually slightly hyperdense and have postcontrast enhancement. Calcifications are sometimes present, and almost all cases involve obstructive hydrocephalus. A T1-weighted MRI shows the tumors to be usually hypointense to isointense, with heterogeneous contrast enhancement (see Fig. 16.4C). In T2-weighted imaging, they are isointense to slightly hyperintense. Upon diagnosis, craniospinal dissemination is seen in 25%−33% of cases (see Fig. 16.8).

Treatment

Surgical resection is the treatment of choice for pineoblastoma, and is usually performed with minimally invasive techniques that are very difficult but effective. Hydrocephalus may be managed via a ventriculostomy or by shunt placement. Radiation therapy is used as long as the patient is over 3 years of age, due to the possibility of long-term cognitive damage that it can cause in young patients. Chemotherapy may be performed in combination with surgery and radiation therapy.

Prognosis

Pineoblastoma is the most aggressive type of pineal parenchymal tumor, shown by its craniospinal seeding, and rarely, by extracranial drop metastasis that may include the spine. If this occurs, it can lead to radiculopathy in the spine. Recent studies report median overall survival between 4.1 and 8.7 years. Also, 5-year overall survival rates have been reported as being between 10% and 81%. Poor prognoses occur with disseminated disease, partial surgical resection, and younger patient age. Radiation therapy has improved prognoses. Also, because of better chemotherapies and earlier disease detection, 5-year survival of patients with trilateral retinoblastoma syndrome has improved, from 6% of cases to 44%.

> **Point to remember**
>
> Pineoblastomas resemble medulloblastomas and retinoblastomas. They are the most aggressive and highest grade tumor among the pineal parenchymal tumors. They are most common in younger children, and have a well-established association with hereditary retinoblastomas. Pineoblastomas are almost always associated with obstructive hydrocephalus, because of compression of the cerebral aqueduct.

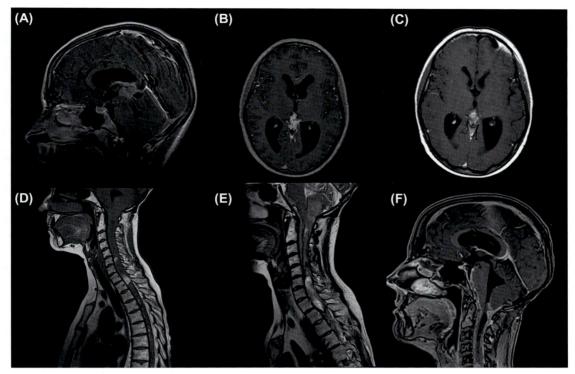

FIGURE 16.8 Preoperative image of a WHO grade IV pineoblastoma (A). (B) Postoperative cranial MRI showing some contrast enhancement at the level of the internal cerebral veins, which disappeared in the following MRI study (C). (D−F) Last craniospinal MRI revealing tumoral infiltration of the midbrain and spinal cord without pineal recurrence.

Clinical cases

Clinical case 1

1. In histopathological examination, how to pineocytoma cells appear?
2. How common are pineocytomas?
3. How successful is surgical resection, in most cases, for pineocytomas?

A 44-year-old woman was admitted to the hospital with headache, vomiting, and Parinaud syndrome. On CT scan, a pineal mass with calcification was found in the caudal area of the tumor, with heterogeneous enhancement after contrast. There was also supratentorial hydrocephalus. The T1-weighted MRI confirmed the heterogeneous enhancement after contrast. An external shunt was inserted, and a stereotactic biopsy was carried out. The specimens revealed the tumor to be a pineocytoma. The tumor was surgically removed one week later. Because of persistent ventricular dilation, an internal shunt was also implanted. The patient, it was determined, did not need radiation therapy after surgery, and over follow-ups for the next 3 years, remained in good health.

Answers:

1. **Pineocytoma cells are often well differentiated, somewhat lobulated, grow in sheets, and have pineocytomatous rosettes and/or pleomorphic cells with gangliocytic differentiation. The cells are uniform, with round to oval nuclei and eosinophilic cytoplasm. They also often have many delicate processes.**
2. **Pineocytomas make up 17%−30% of all pineal parenchymal tumors, usually occur in adults, and have a slightly female preponderance. Overall they make up less than 1% of all intracranial neoplasms.**
3. **Complete surgical resection of pineocytomas is usually possible because of their well circumscribed structure. Local recurrence and CSF metastases is very rare.**

Clinical case 2

1. What is the cellular appearance of these tumors?
2. What is the epidemiology of PPTIDs?
3. What are the differences between the diffuse, lobulated, and transitional subtypes of PPTIDs?

A 37-year-old man presented with a brain tumor that during imaging was found to be 2.5 cm in size. It was had caused obstructive hydrocephalus, resulting in headache and dizziness. An MRI showed intermediate enhancement on T1-weighted imaging and a high intensity of enhancement in T2-weighted imaging. Surgical resection was undertaken, but only subtotal removal was possible. The diagnosis was a pineal parenchymal tumor with intermediate differentiation. One month after surgery, the patient underwent radiation therapy. Four months later, follow-up imaging revealed a remaining mass that had extended to the hypothalamus and third ventricle. The patient was treated with six rounds of chemotherapy. After this, a brain MRI revealed no evidence of any remaining tumor, and all initial symptoms had resolved.

Answers:

1. **The pineal parenchymal tumors with intermediate differentiation consist of diffuse sheets or large lobules of round cells, with a single structural pattern. They are more differentiated than the cell patterns of pineoblastomas.**
2. **The PPTIDs mostly occur in adults, with 41 years being the mean age, and make up 21%−54% of all pineal parenchymal tumors, with a slight female preponderance.**
3. **Diffuse PPTIDs resemble neurocytomas or oligodendrogliomas. Lobulated PPTIDs have vessels that delineate poorly defined lobules. Transitional PPTIDs have pineocytomatous areas linked to a diffuse or lobulated pattern that is more similar to a true PPTID.**

Clinical case 3

1. What is the epidemiology of papillary tumors of the pineal region?
2. What is the distinctive pathological reaction of pineal papillary tumors that identifies them in comparison to other pineal parenchymal tumors?
3. What is the primary difference in treatment of these tumors from other pineal tumors?

A 10-year-old girl presented with a 12 month history of right eye strabismus and diplopia. About 1 month before her visit, she had suffered from an irregular intermittent headache, mostly in the lateral and top areas of her forehead. She also had intermittent nausea and vomiting. An MRI revealed a heterogeneously enhanced, well-defined lesion with limited cystic components, located in the pineal region. The tumor was mostly removed surgically, and appeared gray in color, soft, and well circumscribed. Microscopic examination revealed that the tumor had papillary structures, and the nuclei were slightly irregular. It was diagnosed as a papillary tumor of the pineal region. Postoperatively, the patient did very well, with no tumor recurrence over 15-months of follow-up.

Answers:

1. **These tumors affect children as well as adults, with a mean patient age of 35 years. However, less than 200 cases have been documented globally, with a male-to-female ratio of only 1.06:1.**
2. **The pineal papillary tumors, distinctively, are reactive for keratins—especially in the papillary regions. They also stain for S-100, vimentin, microtubule-associated protein 2 (MAP2), neuron-specific enolase, neural cell adhesion molecule 1 (NCAM1), and the transthyretin protein.**
3. **Radiation therapy is of disputed value with pineal parenchymal tumors in relation to their disease progression. This differs from other pineal tumors.**

Clinical case 4

1. How common are pineoblastomas?
2. With what other structures are the cells of pineoblastomas related?
3. How do pineoblastomas usually spread?

A 23-year-old woman presented with symptoms of hydrocephalus that included headache, nausea, vomiting, diplopia, and blurred vision. An MRI revealed a pineal region tumor that also affected the posterior part of the third ventricle, with secondary obstructive hydrocephalus. The tumor was surgically excised. Postoperatively, the patient developed transient vertical gaze paresis. Pathological studies confirmed a grade IV pineoblastoma with high cellularity and numerous mitotic figures, hyperchromatic nuclei with occasional small nucleoli, little cytoplasm, and Homer-Wright rosettes. The patient underwent craniospinal radiation therapy and five cycles of chemotherapy. However, 7 months later, an MRI revealed an anterior skull base tumor, which was surgically removed and revealed to be another pineoblastoma, with penetration of the bone, dura, and invasion into the mucosal tissues. Stereotactic radiotherapy was able to eradicate all metastatic tumor tissues. Unfortunately, additional tumors developed over time, in the leptomeningeal tissues, and despite repeated therapies, the patient eventually died.

Answers:

1. Pineoblastomas make up about 35% of all pineal parenchymal tumors. They usually occur within the first two decades of life, but can occur at any age. Females have a slight preponderance of these tumors.
2. The etiology of pineoblastomas reveals shared features between cells of the pineal gland as well as the retinas. There is an occasional association with bilateral (familial) retinoblastoma (trilateral retinoblastoma syndrome), supported by occasional progression from a low-grade tumor to a pineoblastoma.
3. Pineoblastomas usually spread through the cerebrospinal fluid pathways, and are often aggressive. They directly invade nearby brain structures such as the leptomeninges, third ventricle, and tectal plate.

Clinical case 5

1. What are the cellular characteristics of pineoblastomas?
2. How do these tumors appear in imaging studies?
3. After surgery for pineoblastoma, how is hydrocephalus managed?

A 16-year-old girl presented with complaints of a chronic headache, diplopia, tinnitus, and recently, a decrease in hearing within her right ear. A CT scan revealed a hyperdense mass in the posterior third ventricular region, with obstructive hydrocephalus. The patient underwent a left ventriculoperitoneal shunt and gross total tumor resection. Histopathology revealed a pineoblastoma. The patient underwent postoperative radiation therapy. Follow-up over 6 years was good, but then the patient developed radiculopathy of the first lumbar to first sacral vertebrae. An MRI was suggestive of drop metastasis and a CSF cytological study revealed clusters of neoplastic cells. She was again treated with radiation, followed up by chemotherapy. At 18 months of follow-up after these procedures, there was no indication of cancer recurrence.

Answers:

1. Pineoblastoma is characterized by patternless sheets of small, immature neuroepithelial cells, which are densely packed and irregular, with indistinct borders. A small amount of rosettes may interrupt their overall diffuse growth pattern.
2. In imaging, pineoblastomas appear as large, multilobulated masses. There is often invasion of the tectum, thalamus, and splenium of the corpus callosum, and there may be small cystic or necrotic areas, with edema.
3. After surgery, hydrocephalus caused by a pineoblastoma may be managed via a ventriculostomy or by shunt placement.

Key terms

Annulate lamellae
Conerod homeobox
Ectomesenchymal
Euchromatin
Flexner-Wintersteiner rosettes
Gangliocytic
Ganglioid
Histochemistry
Homer-Wright rosettes
Hypotriploid
Immunoexpression

Metalloenzyme
Nystagmus retractorius
Papillary tumor
Parinaud syndrome
Photosensory
Pineal parenchymal tumor with intermediate differentiation
Pineoblastoma
Pineocytoma
Pineocytomatous
Pseudoangiomatous

Pseudodiploid
Retinoblastic
Rhabdomyoblasts
S-arrestin
Stippled
Subcommissural organ
Tectal plate
Zonula adherens
Zonula occludens

Further reading

Adesina, A. M., Tihan, T., Fuller, C. E., & Young Poussaint, T. (2016). *Atlas of Pediatric Brain Tumors* (2nd ed.). Springer.
Angelos, P., & Grogan, R. H. (2018). *Difficult Decisions in Endocrine Surgery: An Evidence-Based Approach*. Springer.
Bansal, D. D., Mehra, P., & Kardori, R. (2017). *Introductory Endocrinology: A Concise and Applied Digest*. Jaypee Brothers Medical Publishers Pvt. Ltd.
Bartsch, C., Bartsch, H., Blask, D. E., Cardinali, D. P., Hrushesky, W. J. M., & Mecke, D. (2003). *The Pineal Gland and Cancer: Neuroimmunoendocrine Mechanisms in Malignancy*. Springer.

Berry, J. L., Kim, J. W., Damato, B. E., & Singh, A. D. (2019). *Clinical Ophthalmic Oncology: Retinoblastoma* (3rd ed.). Springer.
Bruce, J. N., & Parsa, A. T. (2011). *Pineal Region Tumors, An Issue of Neurosurgery Clinics (The Clinics: Surgery)*. Saunders.
Chiocca, E. A., & Breakefield, X. O. (2013). *Gene Therapy for Neurological Disorders and Brain Tumors (Contemporary Neuroscience)*. Humana Press.
Drevelegas, A. (2002). *Imaging of Brain Tumors with Histological Correlations*. Springer.
Ellenbogen, R. G., & Sekhar, L. N. (2012). *Principles of Neurological Surgery: Expert Consult (Principles of Neurosurgery)* (3rd ed.). Saunders.
Fountas, K., & Kapsalaki, E. Z. (2019). *Epilepsy Surgery and Intrinsic Brain Tumor Surgery: A Practical Atlas*. Springer.
Francis, J. H., & Abramson, D. H. (2015). *Recent Advances in Retinoblastoma Treatment (Essentials in Ophthalmology)*. Springer.
Gajjar, A., Reaman, G. H., Racadio, J. M., & Smith, F. O. (2018). *Brain Tumors in Children*. Springer.
Goldblum, J. R., Weiss, S. W., & Folpe, A. L. (2019). *Enzinger and Weiss' Soft Tissue Tumors* (7th ed.). Elsevier Science/Health Science.
Gondor, A. (2016). *Chromatin Regulation and Dynamics*. Academic Press.
Greenberg, H. S., Chandler, W. F., & Sandler, H. M. (2000). *Brain Tumors (Contemporary Neurology Series)*. Oxford University Press.
Gursoy-Ozdemir, Y., Bozdag-Pehlivan, S., & Sekerdag, E. (2017). *Nanotechnology Methods for Neurological Diseases and Brain Tumors: Drug Delivery Across the Blood-Brain Barrier*. Academic Press.
Hattingen, E., & Pilatus, U. (2016). *Brain Tumor Imaging (Medical Radiology)*. Springer.
Hayat, M. A. (2013a). *Tumors of the Central Nervous System, Volume 10: Pineal, Pituitary, and Spinal Tumors*. Springer.
Hayat, M. A. (2013b). *Tumors of the Central Nervous System, Volume 11: Imaging, Glioma and Glioblastoma, Stereotactic Radiotherapy, Spinal Cord Tumors, Meningioma, and Schwannomas*. Springer.
Jain, R., & Essig, M. (2015). *Brain Tumor Imaging*. Thieme.
Kobayashi, T., & Lunsford, L. D. (2009). *Pineal Region Tumors: Diagnosis and Treatment Options (Progress in Neurological Surgery, Volume 23)*. S. Karger.
Kornienko, V. N., & Pronin, I. N. (2009). *Diagnostic Neuroradiology*. Springer.
Morita, S. Y., Dackiw, A. P. B., & Zeiger, M. A. (2009). *McGraw-Hill Manual: Endocrine Surgery*. McGraw-Hill Education/Medical.
Norden, A. D., Reardon, D. A., & Wen, P. C. Y. (2010). *Primary Central Nervous System Tumors: Pathogenesis and Therapy (Current Clinical Oncology)*. Humana Press.
Quinones-Hinojosa, A., Raza, S. M., & Laws, E. R. (2013). *Controversies in Neuro-Oncology: Best Evidence Medicine for Brain Tumor Surgery*. Thieme.
Rajendran, J., & Manchanda, V. (2010). *Nuclear Medicine Cases (McGraw-Hill Radiology Series)*. McGraw-Hill Education/Medical.
Ramasubramanian, A., Shields, C. L., Tasman, W., Meadows, A. T., & Knudson, A. G. (2012). *Retinoblastoma*. Jaypee Brothers Medical Publishers Pvt. Ltd.
Thomas, P. (2013). *Endocrine Gland Development and Disease, Volume 106 (Current Topics in Development Biology)*. Academic Press.
Turgut, M., & Kumar, R. (2011). *The Pineal Gland and Melatonin: Recent Advances in Development, Imaging, Disease and Treatment (Endocrinology Research and Developments)*. Nova Science Publishers Inc.
Warmuth-Metz, M. (2017). *Imaging and Diagnosis in Pediatric Brain Tumor Studies*. Springer.
Wick, M. R. (2008). *Diagnostic Histochemistry*. Cambridge University Press.
Wilson, M. W. (2010). *Retinoblastoma (Pediatric Oncology)*. Springer.
Zhou, J. (2017). *Histochemistry (De Gruyter Textbook)*. De Gruyter.

Chapter 17

Melanocytic tumors

Chapter Outline

Overview	337	Meningeal melanocytosis	342
Meningeal melanoma	337	Clinical cases	344
Meningeal melanocytoma	339	Key terms	346
Meningeal melanomatosis	340	Further reading	346

Overview

Melanocytic tumors in the central nervous system are believed to form from leptomeningeal melanocytes derived from the neural crest. Primary melanocytic tumors may be diffuse or localized. Diagnosis is based on recognizing tumor cells with melanocytic differentiation, usually via histopathological examination. The majority of these tumors have finely distributed melanin pigment in their cytoplasm, but coarsely distributed melanin in the tumor stroma and the cytoplasm of macrophages that are referred to as melanophages. Primary CNS melanocytic tumors must be distinguished from other melanotic tumors, including metastatic melanoma, and tumors undergoing melanization, such as schwannomas, gliomas, medulloblastomas, and paragangliomas. Though there have been rare cases of melanocytic colonization in meningiomas, there is only slight evidence of any true melanotic meningioma. Most melanocytic tumors react with anti-melanosomal antibodies such as human melanoma black 45 (HMB45), melan-A, and microphthalmia-associated transcription factor (MITF), while also expressing the S-100 protein. Staining for neuron-specific enolase and vimentin are variable. Normally, CNS melanocytes are localized at the base of the brain, around the ventral medulla, and along the upper cervical portion of the spinal cord.

Meningeal melanoma

Meningeal melanoma is a malignant melanocytic tumor of the central nervous system (CNS) arising from the leptomeningeal melanocytes. These tumors do not form in groups. Sometimes, primary melanomas do not appear to have any melanin pigment. When melanin is lacking in any melanocytic tumor, it is accurately identified by its immunohistochemistry and mutation profile. The melanocytes in this type of tumor, and other melanocytic tumors, are closer in appearance to melanocytes of the **uveal tract** than of the skin or mucosal membranes. Like most melanocytic tumors, there is finely distributed melanin in the cytoplasm, but coarsely distributed melanin in the stroma and cytoplasm of macrophages called **melanophages**.

Epidemiology

Primary meningeal melanomas have been documented in patients between the ages of 15 and 73 years, with a mean incidence of 43 years. The annual incidence is 0.5 cases per 10 million people, globally.

Etiology and risk factors

Like all melanocytic tumors, meningeal melanomas are not fully understood regarding their likelihood of being inherited. Similar to skin melanomas, these tumors are linked with activating somatic mutations of *neuroblastoma rat sarcoma (NRAS)* that usually involve codon 61. *BRAF V600E* gene mutations have not been identified in CNS melanocytic tumors as they have been in cases of neurocutaneous melanosis. For childhood melanomas, there is also a strong connection with NRAS mutations—primarily *codon 61*. Primary CNS melanomas also have *guanine nucleotide-binding protein G(q) subunit alpha (GNAQ)* or *subunit alpha-11 (GNA 11)* mutations, but less often than melanocytomas have these mutations.

The tumors progress to melanoma, similar to uveal melanoma with early GNAQ or GNA11 mutations, followed by *breast cancer type 1 (BRCA1*, also called *BAP1)* activation or *splicing factor 3B subunit 1 (SF3B1)* or *eukaryotic translation initiation factor 1A, X-chromosomal (EIF1AX)* mutations, then becoming malignant. Rarely, there are mutations of the *telomerase reverse transcriptase (TERT) promoter*, NRAS, BRAF, and *proto-oncogene c-KIT*, but when these occur, they increase suspicion of a metastasis. Risk factors for meningeal melanoma may include previous radiation exposure, excess body fat, and neurofibromatosis type 2.

Pathology

Meningeal melanomas usually present as solitary mass lesions, with aggressive growth. They are dural-based, occurring through the neuroaxis, with a slightly more common occurrence in the spinal cord and posterior fossa. Expression of collagen IV and **reticulin** does not occur around individual melanoma cells, but does occur around blood vessels and large tumor nests. Rarely, cytokeratins, epithelial membrane antigen (EMA), glial fibrillary acidic protein (GFAP), or neurofilament proteins (NFPs) are expressed. In primary meningeal melanomas, the Ki-67 proliferation index is about 8%.

Tumors related to **neurocutaneous melanosis** are believed to form from melanocyte precursor cells that first acquire somatic mutations—mostly of NRAS—and then reach the CNS. Though NRAS mutations are not inherited, they probably occur early during embryogenesis, before they migrate to the leptomeninges. In cases of neurocutaneous melanosis, it is believed that just one postzygotic NRAS mutation is causative of multiple lesions. Most malignant melanomas are single, with an extra-axial location. They may range from red-brown to blue to black, or even be nonpigmented.

The microscopic view of melanomas is similar to the appearance of melanomas in other body sites. There are anaplastic epithelioid or spindled cells, arranged in loose fascicles, nests, or sheets. There are variable amounts of cytoplasmic melanin. In some cases, there are large tumor cells with bizarrely appearing nuclei. There may be many typical and atypical mitotic figures, large nucleoli, significant pleomorphism, or densely cellular, less pleomorphic features. These usually involve tightly packed spindle cells and a high nuclear to cytoplasmic ratio. **Fontana−Masson stains** may reveal a pigmented basal layer in the tumor tissue. HMB45 staining shows large amounts of HMB45-positive, dark red stained cells as well as being cytoplasmic-positive for antimelanosomal antibody called *melanoma antigen recognized by T cells 1 (MART1)*, also known as *Melan A* (see Fig. 17.1). Compared to melanocytomas, melanomas are more pleomorphic. They are also more anaplastic and mitotically active, with a higher cell density. They may have obvious tissue invasion or coagulative necrosis. Meningeal melanomatosis can develop from diffuse spread of a melanoma through the subarachnoid space.

Clinical manifestations

The clinical features of meningeal melanomas, like melanocytomas, involve focal neurological signs that are based on tumor location. Documented symptoms have included limb weakness, altered consciousness, confusion, agitation, and foot drop.

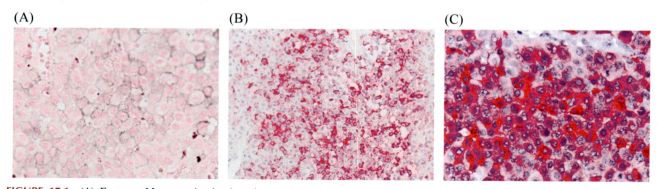

FIGURE 17.1 (A) Fontana−Masson stain showing pigmented basal layer in tumor tissue. (B) HMB45 staining showing a significant amount of HMB45-positive (dark red stained) cells. (C) Stain revealing that the tumor is cytoplasmic-positive for antimelanosomal antibody MART-1 (Melan A).

Diagnosis

In imaging studies, CNS structures close to a meningeal melanoma are often T2-hyperintense. This indicates vasogenic edema, generated because of rapid tumor growth. DNA methylation patterns help identify melanotic tumors, along with their mutations and histological classes.

Treatment

Treatment of meningeal melanoma is based on tumor location and metastasis. Intrathecal therapy has been used, with the agents known as methotrexate, liposomal cytarabine, and thiotepa. Radiation therapy is of uncertain effectiveness and is still under study. However, new advances involve systemic targeted therapy and immune checkpoint inhibitors.

Prognosis

Malignant meningeal melanomas have a poor prognosis since they are extremely aggressive and radiation-resistant, and often metastasize widely. However, prognosis is better if there is no metastasis and total surgical resection is possible. For melanomas not linked with congenital nevi, BAP1 loss is believed to be related to a poorer prognosis.

> **Point to remember**
>
> Primary CNS melanomas are much less common than melanomas that affect the skin. They are primary melanocytic tumors of the leptomeningeal melanocytes, derived from neural crest cells, and present as solitary masses. They are most common in the spinal cord and posterior fossa, but can appear anywhere in the CNS. These tumors are identified by their cellular pleomorphism, atypia, necrosis, and mitotic activity.

Meningeal melanocytoma

Meningeal melanocytoma is a term that describes a benign or intermediate-grade melanocytic tumor. It is a well differentiated and solid, noninfiltrative tumor arising from the leptomeningeal melanocytes. In rare cases, these tumors do not have any visible melanin pigment. The tumors are most commonly found near the foramen magnum, posterior cranial fossa, **Meckel's cave**, or near the cranial nerve nuclei.

Epidemiology

Meningeal melanocytoma make up only 0.06% − 0.1% of all brain tumors. Annual incidence, globally, is estimated at 1 case per 10 million people. It may occur in patients of all ages, with the youngest documented case being a 9-year-old child, and the oldest being a 73-year-old. However, they are most common during the fifth decade, usually between 45 and 50 years of age. The female predominance is 1.5 to 1.

Etiology and risk factors

Though the exact causes and risk factors are unknown, hotspot mutations of *GNAQ* or *GNA 11* have been seen, usually involving *codon 209*. These mutations are similar to those found in uveal melanoma and blue nevus. Cytogenetic losses of *chromosome 3* and the long arm of *chromosome 6* have also been documented.

Pathology

Meningeal melanocytoma is characterized by epithelioid, fusiform, **polyhedral**, or spindled melanocytes. They are often very dark and even black in color, though less often they are lighter in color or nonpigmented. The cells show no evidence of anaplasia, elevated mitotic activity, or necrosis. Expression of collagen IV and reticulin does not occur around individual melanocytoma cells, but is present around blood vessels and larger tumor nests.

In rare cases, cytokeratins, EMA, GFAP, or NFPs are expressed. The Ki-67 proliferation index is usually less than 1% − 2%. Meningeal melanocytomas arise mostly in the extramedullary, intradural compartment, at the level of the cervical and thoracic spine. They may be dural-based, or involve the nerve roots or spinal foramina. Less often, they develop from the leptomeninges of the posterior fossa or supratentorial compartments. The trigeminal cave is a location in which primary melanocytic tumors occur more often. Tumors in this area are linked to *ipsilateral nevus of Ota*.

Clinical manifestations

Meningeal melanocytomas cause symptoms that are related to compression of the spinal cord, cerebellum, or cerebrum. Focal neurological signs are based on tumor location. Some patients are asymptomatic. Symptoms may include progressive pain, weakness, gait disturbances, hemiparesthesia, and sensory deficits. Rarely, subarachnoid hemorrhage has occurred.

Diagnosis

Generally, imaging studies show homogeneous enhancement, postcontrast. Tumors with large amounts of melanin have a characteristic pattern of precontrast T1 hypersensitivity and T2 hypointensity. Melanocytomas are single mass lesions, usually extra-axial. Under a microscope, melanocytomas show no invasion of surrounding structures. There may be slightly spindled or oval tumor cells, with various amounts of melanin, forming tight groupings that superficially resemble whorls, as in a meningioma. At the periphery of these tight "nests," there are heavily pigmented tumor cells and tumoral macrophages.

There are other variants of melanocytomas, which include cells of sheet-like, storiform, or **vasocentric** arrangements. Nuclei may be bean- or oval-shaped, sometimes with grooves, and small eosinophilic nucleoli. Generally, cytologic atypia and mitoses are not present. On average, there is less than 1 mitosis per 10 high-power fields. The melanocytomas with bland cytology, but with CNS invasion or increased mitotic activity, are classified as intermediate-grade melanocytic neoplasms. Rarely, **amelanotic** melanocytomas have been documented.

At the cellular level, there are no junctions, but melanosomes are contained, in various developmental stages. Different from a schwannoma, there is no pericellular basal lamina of any significant form, yet groupings of melanocytoma cells may be within sheaths. No desmosomes or interdigitating cytoplasmic processes are present, which is unlike the structures of a meningioma.

Treatment

Complete resection is usually indicated. If it is only partial, postoperative radiation therapy is used.

Prognosis

Meningeal melanocytomas do not have anaplastic features, but sometimes locally recur (in about 26% of cases). One study reported a fatality rate of 10.5% over 46 months. Intermediate-grade tumors often invade the CNS but insufficient data exists to predict their outlook. There have been rare cases of malignant transformation, possibly worsening prognosis.

> **Point to remember**
> Meningeal melanocytomas are most common in the spinal canal, near the foramen magnum, as well as the posterior cranial fossa, Meckel's cave, or close to the cranial nerve nuclei. The upper cervical region is a common site because melanocytes are highly concentrated in this location. Melanocytomas usually cause progressive pain, weakness, and sensory deficits.

Meningeal melanomatosis

Meningeal melanomatosis is a term describing primary malignant melanocytic tumors that are either diffuse or multifocal. They arise from leptomeningeal melanocytes and may spread through the subarachnoid space and the **Virchow – Robin spaces**. Often, the brain and/or spinal cord are invaded (see Fig. 17.2). This condition is much more aggressive than meningeal melanocytosis.

Epidemiology

Diffuse cases of meningeal melanomatosis are rare. There is no incidence based on population available. There has been one study of 39 patients who had meningeal melanomatosis as well as **giant congenital nevi** earlier in life. Patients ranged in age from stillborn infants to the second decade of life. There was an even distribution between genders and no racial or ethnic predisposition.

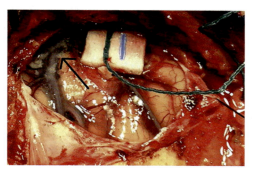

FIGURE 17.2 Meningeal melanomatosis as seen during surgical resection.

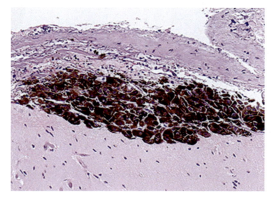

FIGURE 17.3 The appearance of tumor cells in meningeal melanomatosis.

Etiology and risk factors

Meningeal melanomatosis is strongly linked with neurocutaneous melanosis, a rare type of **phacomatosis** that is usually associated with giant congenital nevi, presenting prior to 2 years of age. However, the exact causes and risk factors are unknown.

Pathology

Meningeal melanomatosis involve the infratentorial and supratentorial leptomeninges. They may involve the superficial brain parenchyma via extension into the perivascular Virchow − Robin spaces. Generally, these tumors also involve large areas of the subarachnoid space. Sometimes, focal or multifocal nodules are seen. In most cases, the tumors occur in the cerebellum, pons, medulla oblongata, and temporal lobes. Tumor cells that diffusely involve the leptomeninges may be cuboidal, oval, round, or spindled (see Fig. 17.3).

Clinical manifestations

Meningeal melanomatosis may cause neurological symptoms linked to hydrocephalus or localized effects upon the CNS parenchyma. Chronic headache is common. These symptoms include bowel and bladder dysfunction, neuropsychiatric symptoms, and sensory or motor disturbances. Symptoms progress rapidly with malignant transformation. There is increased intracranial pressure, causing irritability, lethargy, seizures, and vomiting.

This disease is often discovered in childhood because of hydrocephalus from blockage of the subarachnoid and perivascular spaces. Abnormal CSF flow may also cause syringomyelia. If there is concurrent giant or numerous congenital melanocytic nevi, usually of the skin of the trunk, head, or neck, the disorder is referred to as *neurocutaneous melanosis*. There may also be other malformations, such as lipomas or **Dandy − Walker syndrome**.

Diagnosis

Diagnosis of meningeal melanomatosis is usually via histopathological examination, but has sometimes been done via cerebrospinal fluid cytology. Extensive invasion of the subarachnoid space with perivascular infiltration of the cerebral cortex. In CT and MRI, there is diffuse thickening and enhancement of the leptomeninges (see Fig. 17.4). There is often

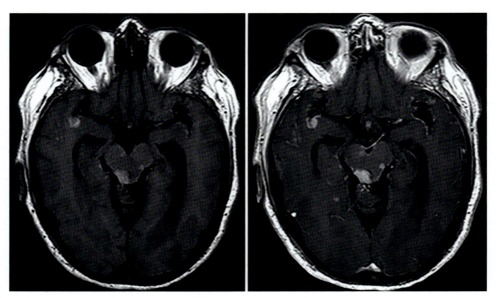

FIGURE 17.4 Meningeal melanomatosis.

focal or multifocal nodularity. In CT scans, hyperdensity is due to melanin outlining the sulci. In MRI, melanin may be seen on unenhanced T1-weighted imagine as hyperintensity, and in 20% of cases, there is diffuse enhancement of the meninges of the brain and spine.

Pathologic proliferation of leptomeningeal melanocytes and their production of melanin provide the primary microscopic appearance. If there is obvious parenchymal invasion, this is a diagnostic factor that differentiates melanomatosis from melanocytosis. Diagnosis of melanomatosis is also based on the presence of mitotic activity, severe cytological atypia, or necrosis. Differential diagnoses include subarachnoid hemorrhage, other melanocytic tumors, leptomeningeal carcinomatosis, meningitis, and diffuse leptomeningeal **glioneuronal tumors**.

Treatment

Treatment of meningeal melanomatosis must occur as early as possible because once there is malignancy and metastasis, surgical resection is not viable and both chemotherapy and radiation therapy are relatively ineffective. Experimental treatments are ongoing, using the *mitogen-activated protein kinase-kinase (MEK)* inhibitor called *MEK162*, which may inhibit tumor growth.

Prognosis

The prognosis of symptomatic meningeal melanomatosis, especially when it is part of neurocutaneous melanosis, is poor. Those with malignancy have a mortality rate of 77%. If a Dandy − Walker malformation is also present, the prognosis is even worse, and the median survival time is 6.5 months after becoming symptomatic.

> **Point to remember**
>
> Meningeal melanomatosis is an aggressive condition that is strongly linked to cutaneous melanocytic lesions (neurocutaneous syndrome). It is usually discovered in childhood, because of hydrocephalus. The cells often involve the perivascular spaces, and commonly directly invade the brain—a feature that aids in correct diagnosis.

Meningeal melanocytosis

Meningeal melanocytosis is a term that usually describes diffuse, benign melanocytic tumors that do not form macroscopic masses. There also can be multifocal proliferation of **bland cells**, arising from the leptomeninges, and involving the subarachnoid space.

Epidemiology

There are no accurate statistics on incidence and prevalence of meningeal melanocytosis, but experts have estimated it to occur in 1 of every million people, with females affected 1.5 times more often than males. The condition has occurred in all age groups, but primarily during the fifth decade of life.

Etiology and risk factors

Like meningeal melanomatosis, there is a strong link between meningeal melanocytosis and neurocutaneous melanosis. However, the exact causes and risk factors are unknown.

Pathology

Meningeal melanocytosis cells are able to spread into the perivascular spaces without clinically evident invasion of the brain. Like meningeal melanomatosis, these tumors involve the infratentorial and supratentorial leptomeninges, sometimes occur in the superficial brain parenchyma and perivascular Virchow − Robin spaces, and generally exist in large areas of the subarachnoid space. There is also occasional focal or multifocal nodule formation, and the cerebellum, pons, medulla, and temporal lobes are most often affected by meningeal melanocytosis. Tumor cells that diffusely involve the leptomeninges may be cuboidal, oval, round, or spindled, similar to those of melanomatosis. However, in melanocytosis, individual cells are bland, accumulate in the subarachnoid and Virchow − Robin spaces, but do not have extreme CNS invasion (see Fig. 17.5). Obvious parenchymal invasion is not seen.

Clinical manifestations

Like meningeal melanomatosis, the symptoms of meningeal melanocytosis may include bowel and bladder dysfunction, low back and leg pain, neuropsychiatric symptoms, and sensory or motor disturbances, including vision disturbances and hallucinations. Symptoms also progress rapidly with malignant transformation, including increased intracranial pressure, causing headache, irritability, lethargy, seizures, and vomiting. Neurocutaneous melanosis, lipomas, syringomyelia, or Dandy − Walker syndrome can also manifest. About 25% of meningeal melanomatosis cases involve significant cutaneous lesions. Oppositely, 10% − 15% of patients with large congenital melanocytic skin nevi develop clinical symptoms of CNS melanocytosis. Also, melanocytosis may be linked to *congenital nevus of Ota*.

Diagnosis

Diagnosis of meningeal melanocytosis has sometimes been made by cerebrospinal fluid cytology, but usually is done via histopathological examination. In up to 23% of children with giant congenital nevi but no symptoms, there has been radiological evidence of CNS involvement. In CT or MRI, there is diffuse thickening and enhancement of the leptomeninges, often accompanied by focal or multifocal nodularity (see Fig. 17.6). Based on the amount of melanin, diffuse and circumscribed melanocytic tumors may have a unique appearance on MRI. This is because of the paramagnetic properties of melanin. These result in an isodense or hyperintense signal on T1-weighted images, and a hypointense signal on T2-weighted images. Diffuse melanocytic tumors present as a dense, black replacement of the subarachnoid

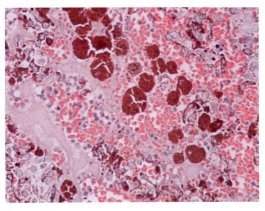

FIGURE 17.5 Tumor cells of meningeal melanocytosis.

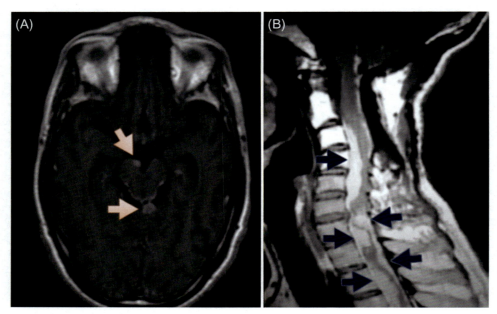

FIGURE 17.6 Meningeal melanocytosis.

space, or as dark clouding of the meninges. The primary microscopic findings are due to pathological proliferation of leptomeningeal melanocytes and their production of melanin.

Treatment

Due to the limited amount of cases of meningeal melanocytosis, there is not a large amount of treatment data about this condition. In the past, treatments have included surgeries involving craniotomy, biopsy, and tumor debulking. Temozolomide has been used for a patient to control seizures and reduce edema.

Prognosis

The prognosis is worse for the diffuse form of melanocytosis in comparison to the multifocal form, even without a histological malignancy. However, the lack of cases has not provided accurate prognostic statistics.

> **Point to remember**
>
> Meningeal melanocytosis is a benign primary melanocytic CNS tumor that may also be associated with neurocutaneous syndrome. It is less aggressive than meningeal melanomatosis, from which it differs because the cells are bland, without atypia or necrosis, and lack significant mitotic activity.

Clinical cases

Clinical case 1

1. Is this patient within the common age range for primary meningeal melanomas?
2. How do meningeal melanomas usually present?
3. What is the outlook for this type of melanocytic tumor?

A 65-year-old man presented with weakness in his right hand. This was followed by a brief period of unresponsiveness. At the time of hospitalization, these symptoms had resolved, but the patient was confused and agitated. His medical history included hyperlipidemia, hypertension, asthma, chronic otitis media, and nasal polyps. Examination revealed chronic left foot drop, but normal reflexes. There were no focal sensory deficits or focal cranial nerve deficits. A CT scan raised suspicions for a subarachnoid hemorrhage within the precentral gyrus.

An MRI showed abnormal leptomeningeal enhancement of the left parietal, left frontal, and right frontal lobes. The left central sulcus signal in the left primary sensory and motor cortices showed the most enhancement. An electroencephalogram suggested mild encephalopathy with no epileptiform discharges. When his CSF was sampled, there were occasional atypical mononuclear cells, and normal protein and glucose. Brain biopsy revealed a nested melanocytic neoplasm with extension into the perivascular spaces. Diagnosis was of a primary meningeal melanoma.

Answers:

1. **Yes, since primary meningeal melanomas have been documented between the ages of 15 and 73, with a mean incidence of 43 years.**
2. **Meningeal melanomas usually present as solitary mass lesions, with aggressive growth. They are dural-based, occurring through the neuroaxis.**
3. **Malignant meningeal melanomas have a poor prognosis since they are extremely aggressive and resistant to radiation, and often metastasize widely. Prognosis is better without metastasis, and total surgical resection is possible.**

Clinical case 2

1. Where do meningeal melanocytomas usually develop?
2. Is this patient's age at diagnosis common?
3. Is this patient likely to have a good prognosis?

A 45-year-old man started having increasing pain in the neck region, which resulted in hospitalization about a year later. Over the following months, his symptoms intensified, and included gait disturbance and hemiparesthesia. An MRI of the patient's spine revealed a cervical mass. CT scan revealed a homogenously enhancing, dura-attached mass at the C1 level. An additional MRI revealed a contrast-enhancing mass in the dura that was about 25 mm in diameter. It was hyperintense on T1-weighted imaging and hypointense on T2-weighted imaging. The presurgical neuroradiological diagnosis was a meningioma. During surgery, a black discoloration of the dura was seen, and the tumor was also black, multilobulated, and well circumscribed. Gross total section was successful. Diagnosis was confirmed by immunohistochemical analysis, and the final diagnosis was of a meningeal melanocytoma.

Answers:

1. **Meningeal melanocytomas are most commonly found near the foramen magnum, posterior cranial fossa, Meckel's cave, or near the cranial nerve nuclei.**
2. **Yes, meningeal melanocytomas are most common during the fifth decade, usually between 45 and 50 years of age.**
3. **Since meningeal melanocytomas do not have anaplastic features, in most cases the prognosis, as for this patient, is good. They locally recur in about 26% of cases, and one study reported a relatively low fatality rate of 10.5% over 46 months. There have been rare cases of malignant transformation.**

Clinical case 3

1. What is the aggressiveness of meningeal melanomatosis in comparison to meningeal melanocytosis?
2. Which parts of the brain are usually the site of meningeal melanomatosis?
3. How is diagnosis usually accomplished?

A 66-year-old female presented with a chronic headache that had persisted over a few months. A neurological examination was normal, but CT of her brain revealed diffuse, high density lesions with multiple branched linear enhancements in the base of the right temporal lobe. They appeared similar to subarachnoid hemorrhages. An MRI revealed high signal intensity lesions in T1-weighted imaging, low signal intensity lesions in T2-weighted imaging, and strong enhancements with the same pattern. The CSF fluid was sampled and showed high cellularity, pleomorphic cells with abundant cytoplasm and a black pigment, nuclear pleomorphism, prominent nucleoli, and 7 mitoses per 10 high power fields. The diagnosis was meningeal melanomatosis.

Answers:

1. **Meningeal melanomatosis is much more aggressive than meningeal melanocytosis. Also, this patient's diffuse form is relatively rare. If malignancy is present, the mortality rate is 77%.**
2. **In most cases, meningeal melanomatosis develops in the cerebellum, pons, medulla oblongata, and temporal lobes (as in this patient).**

3. **Diagnosis of meningeal melanomatosis is usually via histopathological examination, but has sometimes been done via CSF cytology. In CT and MRI, there is diffuse thickening and enhancement of the leptomeninges, and often, focal or multifocal nodularity.**

Clinical case 4

1. Is meningeal melanocytosis usually diffuse or multifocal?
2. How to the tumor cells appear in the leptomeninges?
3. What is the prognosis for the diffuse form of meningeal melanocytosis, as in this patient's case?

A 61-year-old woman presented with headache, vision loss, and hallucinations that had persisted for about 2 months. She also complained of low back pain, and pain in her right leg Analysis of her CSF revealed a low glucose level and a high protein level. CT and MRI showed diffuse and small shaded areas of the basal leptomeninges and the presence of hydrocephalus. Tumor cells appeared round and bland. Prior to any surgery, the patient fell into a coma and developed an acid-base imbalance, with significant tachycardia. This resulted in cardio-respiratory failure. The patient unfortunately did not survive, and autopsy revealed a dark brown mass on the basal leptomeninges that had blurred boundaries. Histopathological analysis revealed a primary leptomeningeal melanocytosis, a relatively uncommon diagnosis.

Answers:

1. **Meningeal melanocytosis usually describes diffuse, benign melanocytic tumors that do not form macroscopic masses. There can also be multifocal proliferation of bland cells that involve the subarachnoid space.**
2. **Tumor cells that diffusely involve the leptomeninges may be cuboidal, oval, round, or spindled—similar to those of melanomatosis—but with a bland appearance, and a lack of extreme CNS invasion.**
3. **The prognosis is worse for the diffuse form of meningeal melanocytosis in comparison to the multifocal form, even without a histological malignancy.**

Key terms

Amelanotic
Bland cells
Dandy-Walker syndrome
Fontana — Masson stains
Giant congenital nevi
Glioneuronal tumors
Meckel's cave

Melanophages
Meningeal melanocytoma
Meningeal melanocytosis
Meningeal melanoma
Meningeal melanomatosis
Neurocutaneous melanosis
Phacomatosis

Polyhedral
Reticulin
Uveal tract
Vasocentric
Virchow — Robin spaces

Further reading

Arnautovic, K. I., & Gokaslan, Z. L. (2019). *Spinal cord tumors.* Springer.
Barnhill, R. L., Piepkorn, M. W., & Busam, K. J. (2014). *Pathology of melanocytic nevi and melanoma* (3rd Edition). Springer.
Busam, K. J., Gerami, P., & Scolyer, R. A. (2018). *Pathology of melanocytic tumors.* Elsevier.
Chang, E. L., Brown, P. D., Lo, S. S., Sahgal, A., & Suh, J. H. (2018). *Adult CNS radiation oncology: Principles and practice.* Springer.
Crowson, A. N., Magro, C. M., & Mihm, M. C., Jr. (2014). *The Melanocytic proliferations: A comprehensive textbook of pigmented lesions* (2nd Edition). Wiley-Blackwell.
Desai, M. C. (2013). *Annual reports in medicinal chemistry (Volume 48).* Academic Press.
Dicato, M. A., & Van Cutsem, E. (2018). *Side effects of medical cancer therapy: Prevention and treatment* (2nd Edition). Springer.
Fawdington, T. (2015). *A case of melanosis: With general observations on the pathology of this interesting disease.* Palala Press.
Fountas, K., & Kapsalaki, E. Z. (2019). *Epilepsy surgery and intrinsic brain tumor surgery: A practical atlas.* Springer.
Goldblum, J. R., Weiss, S. W., & Folpe, A. L. (2019). *Enzinger and Weiss's soft tissue tumors* (7th Edition). Elsevier.
Greenfield, J. P., & Long, C. B. (2017). *Common neurosurgical conditions in the pediatric practice: Recognition and management.* Springer.
Gupta, N., Banerjee, A., & Haas-Kogan, D. A. (2017). *Pediatric CNS tumors (Pediatric oncology)* (3rd Edition). Springer.
Gursoy Ozdemir, Y., Bozdag Pehlivan, S., & Sekerdag, E. (2017). *Nanotechnology methods for neurological diseases and brain tumors — Drug delivery across the blood-brain barrier.* Academic Press.
Hayat, M. A. (2012a). *Tumors of the central nervous system, volume 10: Pineal, pituitary, and spinal tumors.* Springer.
Hayat, M. A. (2012b). *Tumors of the central nervous system, volume 9: Lymphoma, supratentorial tumors, glioneuronal tumors, gangliogliomas, neuroblastoma in adults, astrocytomas, ependymomas, hemangiomas, and craniopharyngiomas.* Springer.

Hayat, M. A. (2014). *Tumors of the central nervous system, volume 13: Types of tumors, diagnosis, ultrasonography, surgery, brain metastasis, and general CNS diseases.* Springer.

Hayat, M. A. (2015). *Tumors of the central nervous system, volume 14: Glioma, meningioma, neuroblastoma, and spinal tumors.* Springer.

Hayat, M. A. (2016). *Brain metastases from primary tumors, volume 3: Epidemiology, biology, and therapy of melanoma and other cancers.* Academic Press.

Hoang, M. P., & Mihm, M. C., Jr. (2014). *Melanocytic lesions: A case based approach.* Springer.

Islam, M. P., & Roach, E. S. (2015). *Neurocutaneous syndromes (Volume 132) (Handbook of clinical neurology).* Elsevier.

Jain, R., & Essig, M. (2015). *Brain tumor imaging.* Thieme.

Kesharwani, P., & Gupta, U. (2018). *Nanotechnology-based targeted drug delivery systems for brain tumors.* Academic Press.

Kirshblum, S., & Lin, V. W. (2018). *Spinal cord medicine, 3rd edition — comprehensive evidence-based clinical reference for diagnosis and treatment.* Demos Medical.

Longo, C., Argenziano, G., Lallas, A., Moscarella, E., & Piana, S. (2017). *Atlas of diagnostically challenging melanocytic neoplasms.* Springer.

Mahajan, A., & Paulino, A. (2017). *Radiation oncology for pediatric CNS tumors.* Springer.

Massi, G., & LeBoit, P. E. (2014). *Histological diagnosis of nevi and melanoma* (2nd Edition). Springer.

McKee, P. H., & Calonje, J. E. (2009). *Diagnostic atlas of melanocytic pathology: Expert consult.* Mosby.

National Comprehensive Cancer Network. (2017). *NCCN guidelines for patients: Melanoma 2018.* National Comprehensive Cancer Network (NCCN).

Norden, A. D., Reardon, D. A., & Wen, P. C. Y. (2011). *Primary central nervous system tumors: Pathogenesis and therapy (Current clinical oncology).* Humana Press.

Panteliadis, C. P., Hagel, C., & Benjamin, R. (2016). *Neurocutaneous disorders: A clinical, diagnostic and therapeutic approach.* Urban & Fischer/Elsevier.

Riker, A. I. (2018). *Melanoma: A modern multidisciplinary approach.* Springer.

Ruggieri, M., Pascual Castroviejo, I., & Di Rocco, C. (2008). *Neurocutaneous disorders: Phakomatoses & hamartoneoplastic syndromes.* SpringerWien.

Scheinemann, K., & Bouffet, E. (2015). *Pediatric neuro-oncology.* Springer.

Sciubba, D. M. (2019). *Spinal tumor surgery: A case-based approach.* Springer.

Sepehr, A. (2019). *Melanocytic proliferations: A case-based approach to melanoma diagnosis.* JP Medical Ltd.

Shea, C. R., Reed, J. A., & Prieto, V. G. (2015). *Pathology of challenging melanocytic neoplasms: diagnosis and management.* Springer.

Sik Kang, H., Woo Lee, J., & Lee, E. (2017). *Oncologic imaging: Spine and spinal cord tumors.* Springer.

Tate, M., Cooper-Knock, J., Hunter, Z., & Wood, E. (2014). *Neurology and clinical neuroanatomy on the move (Medicine on the move).* CRC Press.

Tonn, J. C., Reardon, D. A., Rutka, J. T., & Westphal, M. (2019). *Oncology of CNS tumors* (3rd Edition). Springer.

Warmuth-Metz, M. (2017). *Imaging and diagnosis in pediatric brain tumor studies.* Springer.

Watts, C. (2013). *Emerging concepts in neuro-oncology.* Springer.

Chapter 18

Primary lymphoma of the brain

Chapter Outline

Overview	349	T-cell and NK/T-cell lymphomas	357
Diffuse large B-cell lymphoma of the central nervous system	349	MALT lymphoma of the dura	359
Intravascular large B-cell lymphoma	353	Anaplastic large cell lymphoma	360
Low-grade B-cell lymphomas	355	Clinical cases	361
Immunodeficiency-associated central nervous system lymphomas	356	Key terms	365
		References	365

Overview

Primary lymphomas of the brain are of generally unknown causes, but people with weakened immune systems are of higher risk for developing them. These lymphomas may be linked to the Epstein−Barr virus (EBV), especially in people with the human immunodeficiency virus (HIV). Primary brain lymphomas are more common in middle-aged to elderly adults, but overall, they are still rare. General signs and symptoms include changes in speech or vision, headaches, nausea, vomiting, difficulty with body movements, seizures, and hemiparesis. Without treatment, primary cerebral lymphomas can be fatal within a few months, but with proper and early diagnosis and treatment, many of these lymphomas offer good survival rates.

Diffuse large B-cell lymphoma of the central nervous system

A **diffuse large B-cell lymphoma** (DLBCL) is a tumor that is confined to the central nervous system (CNS). It does not include lymphomas that develop in the dura, lymphomas of T-cell or NK-cell origin, intravascular large B-cell lymphomas, and any lymphomas with systemic effects, or with secondary CNS involvement. The tumors can be benign or malignant.

Epidemiology

Diffuse large B-cell lymphoma of the CNS is the most common form of non-Hodgkin's lymphoma in adults (30%−58% of cases). There is an annual incidence of 7−20 cases per 100,000 population in both the United States and the United Kingdom. In Europe, incidence is 3.8 of every 100,000 population. It mostly occurs in older people, with a mean age of diagnosis of 70 years. Rarely, it occurs in children and young adults.

Etiology and risk factors

The causes of diffuse large B-cell lymphomas are generally unknown. However, it is known that the *Epstein−Barr virus (EBV), human herpesviruses 6 and 8 (HHV6 and HHV8)*, and **polyomaviruses** (*simian vacuolating virus 40 or SV40*, as well as the *BK virus*) are *unrelated* to the development of these lymphomas. The BK virus was named from the initials of the first patient ever diagnosed. Risk factors for diffuse large B-cell lymphomas are unknown.

However, approximately 8% of patients with a primary CNS lymphoma have had a previous extracranial tumor, usually within the hematopoietic system. In a patient with a primary CNS lymphoma preceded by an extraneural lymphoma, there may be a common clonal origin. This distinguishes a CNS relapse from an unrelated, secondary cerebral lymphoma. However, analyses of these factors are not common, so the associations between them are often misunderstood. Folate and methionine metabolism may be linked to susceptibility for primary CNS lymphomas.

Pathology

Most primary CNS lymphomas (about 60%), including DLBCLs, involve the supratentorial space. Their development is in the frontal lobe (in 15% of cases), posterior fossa (13%), basal ganglia (BG) and periventricular parenchyma (PP) (10%), temporal lobe (8%), parietal lobe (7%), corpus callosum (5%), occipital lobe (3%), and spinal cord (1%)—see Fig. 18.1. In 60%−70% of cases, there is a single tumor. While the leptomeninges are often involved, manifestation only within the meninges is rare. Sometimes, intracranial development follows ocular development. Extraneural manifestations are extremely rare.

Diffuse large B-cell lymphomas have large central areas of necrosis in most cases, and some have viable perivascular "islands" (see Fig. 18.2). At the edges, there is usually an angiocentric infiltration pattern. Fragmentation of the **argyrophilic** fiber network occurs due to infiltration of cerebral blood vessels. Tumor cells from these perivascular "cuffs" invade the neural parenchyma. There is either a well-demarcated invasion border with small clusters or with individual tumor cells that diffusely infiltrate. This shows a significant microglial and astrocytic activation, with reactive inflammatory infiltrates of mature B and T cells. There are large and atypical cells with large irregular, oval, pleomorphic or round nuclei. The nucleoli are distinct and correspond to **immunoblasts** or **centroblasts**. Some patients' histopathologies show monomorphic cells mixed with macrophages, a factor that resembles Burkitt lymphoma.

The tumor cells are mature B cells, positive for *cluster of differentiation 19 (CD19)*, CD20, CD22, CD79a, and *paired box 5 (PAX5)*. On their surfaces, IgM and IgD are expressed, but not IgG. There is either lambda or kappa **light chain restriction**. Most cells (60%−80%) express *B-cell lymphoma 6 (BCL6) protein*, and the *interferon regulatory factor protein (MUM1/IRF4)* in 90% of cases. However, plasma cell markers such as CD38 and CD138 are usually negative. CD10 is expressed more often in diffuse large B-cell lymphomas than in other primary CNS lymphomas. It is important, when a tumor has diffuse large B-cell lymphoma and CD10 positivity, that a thorough investigation is performed for metastasis into the CNS from another body site. *Human leukocyte antigens* HLA-A/B/C and HLA-DR may

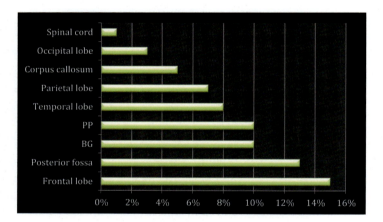

FIGURE 18.1 Locations of diffuse large B-cell lymphomas.

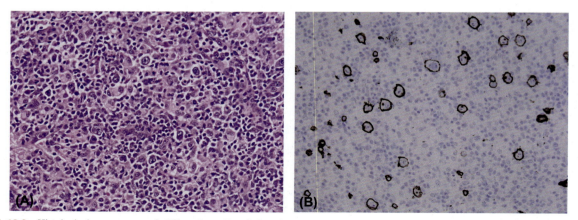

FIGURE 18.2 Histological appearance of diffuse large B-cell lymphoma. (A) The neoplasm involves the lymph node in a diffuse pattern. (B) Scattered neoplastic large B-cells positive for CD20 are within a background of reactive small T-cells, histiocytes, and rare small B-cells.

or may not be expressed. About half of all primary CNS lymphomas have lost expression of HLA class I, HLA class II, or both. BCL2 expression is common, and 82% of primary CNS lymphomas have a phenotype with high BCL2 and the regulator gene/proto-oncogene called *MYC*. There is fast mitotic activity. Therefore, the Ki-67 proliferation index is usually between 70% and 90%. There may be frequent apoptotic cells. Except for isolated cases, there are no signs of *EBV* infection. If this virus is present, there should be an immediate evaluation of immunodeficiency.

The lymphoma's cells correspond to late germinal center exit B cells, with blocked terminal B-cell differentiation. This means that they carry somatically mutated and rearranged immunoglobulin (IG) genes. There are signs of continuing somatic **hypermutation**, and persistent activity of BCL6. This hypermutation extends to other genes linked to tumorigenesis, including BCL2, MYC, *oxysterol binding protein-like 10 (OSBPL10)*, PAX5, *proto-oncogene serine/threonine-protein kinase (PIM1)*, *Ras homolog member H (RHOH)*, and *Sushi domain containing 2 (SUSD2)*. This means that unusual somatic hypermutation has a great effect on pathogenesis of primary CNS lymphomas. The tumor cells' fixed IgM/IgD phenotype is partly due to the ending of IG class switch rearrangements, when the **Smu region** is deleted. Mutations of *PR domain zinc finger protein 1 (PRDM1)* also add to impaired IG class switch recombination.

In 38% of cases, translocations affect the IG genes, and in 17%−47% of cases, they affect BCL6. However, MYC translocations rarely occur and there are no translocations of BCL2. There are recurrent gains of genetic material that usually affect, in 43% of cases, chromosome 18q21.33-q23, which includes the BCL2 and *mucosa-associated lymphoid tissue lymphoma translocation protein 1 (MALT1)* genes. In 26% of cases, chromosomes 12 and 10q23.21 are affected. In 52% of cases, there are losses of genetic material involving 6q21. In 37% of cases, this affects 6p21, and in 32% of cases, 8q12.1-q12.2 and 10q23.21. About 73% of primary CNS lymphomas are affected by homozygous or heterozygous loss, or partial uniparental **disomies** of 6p21.32. This chromosomal region contains the HLA class II-encoding genes called HLA-DQA, HLA-DQB, and HLA-DRB. In relation, 55% of primary CNS lymphomas lose expression of HLA class I gene products, and 46% lose expression of HLA class II gene products.

Due to genetic alterations, some important pathways are activated. The pathways include the B-cell receptor, the NF-κB pathway, and the toll-like receptor. The affected genes include:

- *Myeloid differentiation primary response 88 (MYD88)—over 50% of cases*;
- *MALT1—43% of cases*;
- *BCL2—43% of cases*;
- *Inositol polyphosphate 5-phosphatase1 (INPP5D)—25% of cases*;
- *CD79B—20% of cases*;
- *Caspase recruitment domain-containing protein 11 (CARD11)—16% of cases*;
- *Casitas B-lineage (CBL)—4% of cases*;
- *B-cell linker (BLNK)—4% of cases*.

The activation of the pathways due to these genetic alterations may help tumor cells proliferate, and also prevent apoptosis. Epigenetic alterations may contribute to tumor pathogenesis. This includes epigenetic silencing from DNA methylation. In 84% of cases, there is DNA hypermethylation of *death-associated protein kinase 1 (DAPK1)*, followed by hypermethylation of *cyclin-dependent kinase inhibitor 2A (CDKN2A, 75%)*, *O-6-methylguanine-DNA methyltransferase (MGMT, 52%)*, and *replication factor C (RFC, 30%)*.

Clinical manifestations

Diffuse large B-cell lymphomas cause slowed psychomotor and speech functions, cognitive dysfunction, memory loss, receptive aphasia, and focal neurological symptoms in most cases. Less often, there are headaches, bilateral or unilateral numbness or weakness, cranial nerve palsies, facial paralysis and seizures. When there is also ocular involvement, blurred vision and **eye floaters** may develop.

Diagnosis

The most sensitive diagnostic method is MRI, which is hypointense on T1-weigted imaging, isointense to hyperintense on T2-weighted imaging, and in postcontrast imaging, densely enhanced (see Fig. 18.3). Compared to malignant gliomas and metastases, peritumoral edema is usually not severe. Hyperintense enhancement may signal meningeal involvement. Meningeal dissemination is found in 15.7% of patients. Most of these diagnoses (12.2%) are via CSF studies, with 10.5% by *polymerase chain reaction (PCR)*, and 4.1% by MRI. Pleocytosis is related to meningeal dissemination, and is found in 35%−60% of patients. Cell counts may be normal in some cases. Magnetic resonance spectroscopy (MRS) helps to differentiate malignant from benign forms of these tumors, and PET/CT scans may be done to detect highly metabolic nodules.

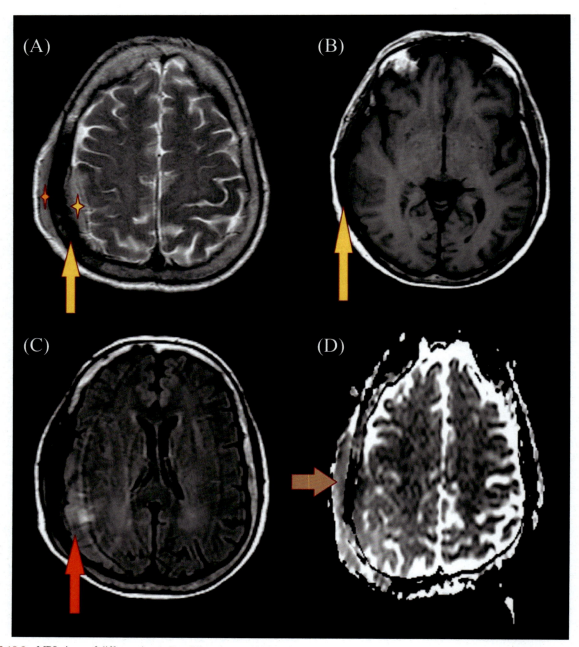

FIGURE 18.3 MRI views of diffuse primary B-cell lymphoma. (A) hypointense diploic signal of right parietal bone on T2 (long yellow arrow) with an isointense dural mass (yellow star) and swelling of the scalp (orange star); (B) the extended hypointense diploic signal on T1 (long yellow arrow); (C) the hyperintense edema on fluid-attenuated inversion recovery (FLAIR) MRI (long red arrow); (D) the low apparent diffusion coefficient of soft tissue tumefaction (short orange arrow).

In a few cases, the CSF has neoplastic cells when there is leptomeningeal involvement. Detection of these cells may require several lumbar punctures. Detection of tumor cells may be enhanced by combining cytological and immunohistochemical analysis with flow cytometry covering a variety of parameters. When PCR analysis is conducted of the *complementarity-determining region 3 (CDR3)* of the *immunoglobulin heavy (IGH)* gene area, that the PCR products are sequenced, a clonal B-cell population may be identified in the CSF. However, this does not allow for classification of the lymphoma. Differentiation of large B-cell lymphomas and similar tumors can occur via elevated CSF levels of *microRNA* subtypes *miR-19*, *miR-21*, and *miR-92a*.

Macroscopically, these lymphomas may be single or multiple in the parenchyma, and usually in the cerebral hemispheres. They are often deeply located and close to the ventricles. The tumors may be grayish-tan or yellow in color, and also firm, friable, granular, hemorrhagic, and have central necrosis. Sometimes they almost exactly resemble the nearby neuropil. There may be varied demarcation from the surrounding parenchyma. Some tumors resemble metastases in that they are well delineated. If diffuse borders and architectural effacement exist, the tumors look similar to gliomas, and similarly to those tumors, may have diffuse infiltration of much of the hemispheres, with no distinctly formed mass. This occurrence is called **lymphomatosis cerebri**, but this term is not a replacement for the accurate lymphoma diagnosis. When the meninges are involved, it may appear similar to meningioma or meningitis, but may be hard to visualize.

Microscopically, there are three distinct pathological factors, which involve stereotactic biopsy, histopathology, and immunophenotype. Stereotactic biopsy is the primary method to establish diagnosis and classification of all CNS lymphomas. Corticosteroids must be withheld before it is done, since they cause the tumors to rapidly decrease in size, and may prevent diagnosis in up to 50% of patients. Histopathology need or put in pathology

Treatment

Steroid therapy may cause diffuse large B-cell lymphomas to disappear within hours. However, the current treatment of choice is high-dose methotrexate-based poly chemotherapy. In elderly patients with primary CNS lymphoma and methylated MGMT, treatment with only temozolomide has been effective. Additional medications include mannitol and dexamethasone. While whole-brain radiation therapy may improve outcomes, there is a risk of neurotoxicity, and severe autonomic, cognitive, and motor dysfunction—especially in older adults. Gamma knife treatment is another option. Total surgical resection is often possible.

Prognosis

Diffuse large B-cell lymphomas have a much better prognosis than most primary CNS lymphomas. Patients with a worsened prognosis, however, are those over age 65. There is shorter survival time and higher risk for neurotoxicity if the patient has the *missense variant Tc2c.776C to G mutation* of **transcobalamin C**. Usually, there is an average progression-free survival time of about 1 year, and overall survival of about 3 years. Improved survival is linked to the presence of reactive perivascular CD3 T-cell infiltrates, and with *LIM domain only 2 (LMO2)* protein expression by tumor cells.

> **Point to remember**
> Diffuse large B-cell lymphomas are much more common than T-cell lymphomas, and they are also the most common form of non-Hodgkin's lymphomas. Incidence increases with age, and they are most common in older adults. These are aggressive lymphomas that can also develop in many other body locations besides the CNS. Tissue biopsy is required for the definitive diagnosis of these lymphomas.

Intravascular large B-cell lymphoma

Intravascular large B-cell lymphoma is distinctively characterized by only intravascular growth. It is a rare form of non-Hodgkin's lymphoma that has a variety of other names. These include: *angiotropic large-cell lymphoma, intralymphatic lymphomatosis, intravascular lymphomatosis*, and *intravascular lymphoma*. In previous decades, these lymphomas were incorrectly believed to arise from the endothelium, hence the incorrect name, *malignant angioendotheliomatosis*.

Epidemiology

Intravascular large B-cell lymphoma involves the CNS in 75%−85% of all cases. The global prevalence is only 1 case out of every 1 million population. Most intravascular lymphomas are of this type, arising from the B cells. The true incidence is unknown. Median age at diagnosis is in the sixth to seventh decades, with an average age of 70 years, and there is no gender predilection. All races and ethnic groups can be affected.

Etiology and risk factors

Most cases of these lymphomas are linked to infiltration of B cells into the lumens of small blood vessels. However, the cause is generally not understood, and risk factors are also unknown. Certain genetic mutations are suspected, but not proven. However, general contributing factors in relation to lymphomas include family history of immune disease, older age, decreased immunity, systemic diseases, smoking, exposure to radiation and industrial chemicals, chemotherapy, EBV infection, X-rays, CT scan exposure, professions involving radiation exposure, and certain medications and drugs.

Pathology

The brain is usually invaded by these lymphomas, with spinal cord involvement being less common. There is a lack of CD29 and ICAM1 (CD54) expression, which is believed to cause the tumor cells to be unable to migrate across the walls of blood vessels. Therefore, they remain in the lumina. The cells are positive for CD20. The lymphomas have a large B cell morphology, in which tumor cells are twice the size of normal lymphocytes (or even larger), usually having a prominent nucleolus. These atypical lymphoma cells especially proliferate within the lumina of capillaries and postcapillary venules. There is no obvious extravascular tumor mass or easily seen lymphoma cells in the peripheral blood. Though these lymphomas are common in the brain, they usually spare the lymph nodes.

Clinical manifestations

The intravascular growth of these lymphomas causes symptoms that resemble those of cerebral infarction or subacute encephalopathy. The most common symptoms include rash, neurologic problems, fever, weight loss, and night sweats. Other symptoms include fatigue, headache, and frequent infections. Some patients have also developed changes in appetite, limb weakness, anemia, difficulty breathing, hypotension, back pain, leg swelling, abdominal pain and swelling, constipation, and frequent urination or urinary retention. Enlargement of a lymphoma within the brain may cause progressive dementia, facial numbness, vision changes, and potentially fatal inflammation of the meninges or brain itself.

Diagnosis

Macroscopically, there is a large, necrotizing diffuse appearance of intravascular large B-cell lymphomas, with or without hemorrhage (see Fig. 18.4). In some cases, abnormalities are not easily visualized. Microscopic diagnosis is based on large, atypical B cells that occlude the cerebral blood vessels. It may be difficult to diagnose these tumors quickly. Immunohistochemical staining is helpful for diagnosis. Imaging studies include X-rays, ultrasound, CT, MRI, PET

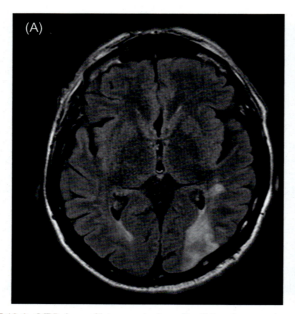 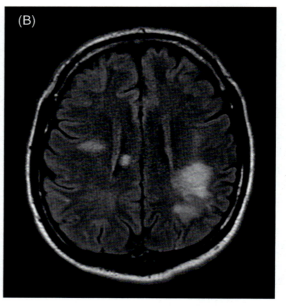

FIGURE 18.4 MRI views of intravascular large B-cell lymphoma.

scans, and vascular studies. Lumbar puncture is used as a method of determining CNS involvement. Often, enhancement of the cauda equina is seen.

Treatment

Without treatment, intravascular large B-cell lymphomas are quickly fatal. However, the tumors may respond well to combination chemotherapy. This usually involves the combination known as *R-CHOP*, which consists of rituximab, cyclophosphamide, adriamycin, oncovin, and prednisone. For brain lymphomas, intrathecal administration may be used to administer additional medications such as methotrexate, cytarabine, and prednisolone. Radiation therapy may also be administered, mostly for early stage lymphomas and along with chemotherapy. Since lymphomas are systemic, surgery is uncommon. Additional treatment options include bone marrow transplantation and stem cell transplantation.

Prognosis

Prognosis of intravascular large B-cell lymphomas is based on tumor progression and staging, responses to treatment, and overall patient health. Generally, the prognosis is poor in comparison to other CNS lymphomas. Prognosis is also worse with increased age, vital organ involvement, recurrent tumors, and late diagnosis or treatment. Progression to bone marrow failure is usually associated with shorter survival.

> **Point to remember**
>
> Intravascular large B-cell lymphoma is a rare type of extranodal large B-cell lymphoma, characterized by selective growth of lymphoma cells within the microvasculature. It is very rare, but aggressive, and most common middle-aged or elderly adults. The lymphoma cells circulate in the blood and can block small blood vessels of the brain, spinal cord, and other body areas.

Low-grade B-cell lymphomas

Low-grade B-cell lymphomas have a variety of subclassifications. These include small lymphocytic lymphomas that are positive for CD5 and CD23, lymphoplasmacytic lymphomas, extranodal **MALT lymphomas**, and follicular lymphomas. In *small lymphocytic lymphoma*, there are sometimes enlarged lymph nodes, and the disease is a form of *chronic lymphocytic leukemia*. In *lymphoplasmacytic lymphoma*, also known as **Waldenström's macroglobulinemia**, both **lymphoplasmacytoid cells** and **plasma cells** are affected; there are high levels of circulating IgM, and the condition is considered a type of *plasma cell dyscrasia*. In *extranodal MALT lymphomas*, there is malignant transformation of *marginal zone B-cells*. In *follicular lymphomas*, there is uncontrolled division of the **centrocytes** and centroblasts that normally occupy the follicles of lymphocytes in germinal centers of lymphoid tissues.

Epidemiology

Low-grade B-cell lymphomas almost always affect adults and not children. About 40% of non-Hodgkin's lymphomas are classified as "low-grade." The low-grade B-cell lymphomas are more common in older people, and affect both genders nearly equally.

Etiology and risk factors

Low-grade B-cell lymphomas develop when abnormal B cells become cancerous and multiply. There is no link between these lymphomas and immunodeficiency. However, there was one documented case of a MALT lymphoma, which involves the mucosa-associated lymphoid tissue, in which the patient had a chronic history of white matter disease and features of multiple sclerosis. In another case of MALT lymphoma, the patient also had *Chlamydophila psittaci* infection, a form of respiratory **psittacosis**. Other than these rare cases, there are no known risk factors.

Pathology

Low-grade B-cell lymphomas develop more slowly than primary CNS diffuse large B-cell lymphomas. The lymphocytes are mostly small in size, with varied amounts of admixed plasma cells. Usually, the lymphomas are initially localized, but in severe cases, can spread to different body sites. They commonly occur in the white matter in or near the subcortical parieto-insular lobes and BG. The disease is usually limited to one or two groups of lymph nodes.

Clinical manifestations

Low-grade B-cell lymphomas may cause focal neurological findings, memory impairment, seizures, and visual defects. The lymph nodes may become enlarged, though not painful. Other symptoms include night sweats, fever, extreme fatigue, hemiparesis, weight loss, nausea, vomiting, shortness of breath, coughing, back pain, and headache.

Diagnosis

The diagnosis of low-grade B-cell lymphomas requires a stereotactic biopsy and microscopic examination of the cells, revealing a diffuse, dense, or perivascular infiltrate. An MRI is usually easily able to reveal these lesions. They are usually hypointense on T1-weighted imaging, with slight enhancement after contrast administration. On T2-weighted imaging, they show moderate hyperintensity. Immunophenotyping shows a large amount of CD20-positive B cells. Usually they are negative for CD5 and CD10. Sometimes, there are monotypic plasma cells that have a low proliferation index. About one of every five patients is diagnosed with these lymphomas while they are still in their early stages and the disease is localized. Differential diagnosis is of an atypical colloidal cyst.

Treatment

Treatment of low-grade B-cell lymphomas is widely varied. It includes complete or partial surgical resection, corticosteroids, radiation therapy, and chemotherapy. For some patients, several of these treatments are combined. Chemotherapeutic agents include the R-CHOP regimen, as well as bendamustine and fludarabine. For more severe cases, monoclonal antibodies such as ibritumomab and tositumomab can be administered as second-line therapies. For early stage, slowly growing B-cell lymphomas, radiation therapy may be sufficient and often provides long-term nonrecurrence. Stem cell transplantation is indicated for some patients.

Prognosis

For most patients with low-grade B-cell lymphomas, the prognosis is good, unless the disease has become severe and spread widely.

> **Point to remember**
> Low-grade B-cell lymphomas are different than other forms of lymphoma since they are caused by abnormal B-cells. They often spread unnoticed because symptoms may not be highly significant. These lymphomas grow slowly and are usually incurably, though many patients respond well to treatments. Like other lymphomas, they are more common in older adults.

Immunodeficiency-associated central nervous system lymphomas

Immunodeficiency-associated CNS lymphomas are primary types of CNS lymphomas that have four subtypes. These include: AIDS-related diffuse large B-cell lymphoma, EBV-positive diffuse large-cell lymphoma that is not otherwise specified, lymphomatoid granulomatosis, and primary CNS posttransplant lymphoproliferative disorder.

Epidemiology

Immunodeficiency-associated CNS lymphomas occur, in 90% of cases, in patients who have the EBV. There is no predilection for any particular age group, though slightly more cases have occurred in patients who are in their 50s or 60s. However, the incidence of primary CNS lymphomas of this classification has increased by more than 10 times in immunocompromised individuals, from about 2.5 cases per 10 million population, globally, to about 30 cases per 10 million population.

Etiology and risk factors

Though exact causes are not known, patients may be predisposed to these lymphomas because of inherited or acquired immunodeficiency. Causative syndromes may include **ataxia-telangiectasia**, *IgA deficiency*, and **Wiskott – Aldrich syndrome**. There is an increased risk for CNS lymphomas because of autoimmune disorders, iatrogenic immunosuppression, and gradual deterioration of the immune system due to aging. Autoimmune disorders include **Sjögren syndrome** and *systemic lupus erythematosus*. Iatrogenic immunosuppression may be instituted either for organ transplantation or because of treatment for other conditions with drugs such as azathioprine, methotrexate, or mycophenolate. In the 1980s, HIV/AIDS infection was highly responsible for the large increase in CNS lymphoma cases that occurred. Overall, immunosuppression-related CNS lymphomas are related to the EBV.

Pathology

In relation to the EBV, immunodeficiency-associated CNS lymphomas express Epstein − Barr early ribonucleoprotein 1 (EBER1), EBER2, Epstein − Barr nuclear antigens 1 to 6 (EBNA1−6), and latent membrane protein 1 (LMP1). Regarding the four subtypes of these lymphomas, there are some different pathological factors. In AIDS-related diffuse large B-cell lymphoma, there may be a more frequent multifocal presentation than in other many CNS lymphomas, and more necrosis with larger necrotic areas. However, this subtype has largely been eradicated by the use of HAART therapy. In EBV-positive diffuse large-cell lymphoma that is not otherwise specified, the elderly are often affected with no known immunodeficiency, and **lymphomagenesis** is linked to reduced immune system function from aging. In **lymphomatoid granulomatosis**, the brain is affected in about 1 of 4 cases, with angiocentric and angiodestructive lymphoid infiltrates being present. In primary CNS posttransplant lymphoproliferative disorder, CNS involvement is not frequent, but may be the only area of the body affected.

Clinical manifestations

Signs and symptoms include seizures, headache, cranial nerve abnormalities, altered mental status, other focal neurological deficits, fever, night sweats, weight loss, diplopia, dysphagia, orbital cellulitis, vertigo, eye bulging, abnormalities of the irises, monocular vision loss, sinus abnormalities, nasal cavity abnormalities and infections, progressive dementia or stupor, and facial hypoesthesia.

Diagnosis

On radiographic imaging, immunodeficiency-associated CNS lymphomas are more likely than other CNS lymphomas to be heterogeneous peripherally enhancing, with central nonenhancement, due to necrosis. This is different from immunocompetent lymphomas, which show solid homogeneous enhancement. There is a multifocal presentation surrounded by a higher degree of vasogenic edema. An MRI or contrast enhanced CT scan will show ring-enhancing lesions in the deep white matter of the brain. Differential diagnosis is cerebral **toxoplasmosis**. Brain biopsy is confirmative.

Treatment

Treatments are essential the same for these lymphomas as for the other types previously discussed, bearing in mind that an immunocompromised patient is in more danger of adverse treatment outcomes. Addition of methotrexate and leucovorin has extended survival time. For AIDS patients, the use of HAART therapy may be the most important treatment consideration.

Prognosis

The prognosis for immunocompromised patients with CNS lymphomas is usually poor. Median survival is 10−18 months, and less if the patient has AIDS (only about 2.5 months). Survival time may be extended to a median of 3.5 years by adding methotrexate and leucovorin to the treatment regimen. Radiation therapy may increase median survival time, if added to methotrexate treatment, beyond 4 years, but is not recommended due to increased risk of leukoencephalopathy and dementia if the patient is older than 60 years.

> **Point to remember**
>
> Patients with primary immunodeficiency diseases are at increased risks for developing lymphomas, as well as other cancers. However, immunotherapies are available to target lymphomas. Examples include monoclonal antibodies, which recognize antigens present on surfaces of cancer cells; antibody-drug conjugates, which also attach to cancer cells but then enter them and kill them by targeting critical cell functions. There is also radioimmunotherapy, consisting of monoclonal antibodies attached to a source of radiation.

T-cell and NK/T-cell lymphomas

Primary T-cell lymphoma of the CNS is extremely rare. There is only one reported subclassification, which is called *peripheral T-cell lymphoma*. These tumors make up about 10% of all non-Hodgkin's lymphomas. Subtypes of

T-cell lymphomas are the extranodal, cutaneous, anaplastic large cell, and angioimmunoblastic forms. **Primary NK/T-cell lymphoma** of the CNS is similar to other examples of this type of lymphoma, which occur in the nasal area and other sites.

Epidemiology

Primary T-cell lymphoma of the CNS makes up only 2% of all primary CNS lymphomas. It is more common in Asia than anywhere else in the world, mostly affecting young to middle-aged adults. Primary NK/T-cell lymphomas usually affect young to middle-aged adult men. They also occur mostly in Asia, and to a lesser degree in Mexico, Central America, and South America. For some reason, they are rare in Europe and North America. In about 80% of cases, the median age of affected patients is between 50 and 60 years, with a slight male predominance.

Etiology and risk factors

T-cell lymphomas are linked to the EBV and the human T-cell leukemia virus-1. Chronic inflammation and lymphadenopathy are risk factors. NK/T-cell lymphomas are linked to the EBV as well, and usually develop because of an NK cell malignancy much more often than a cytotoxic T cell malignancy. Risk factors include increasing age, lymph node disease, and excessive EBV levels in the patient's DNA.

Pathology

Primary T-cell lymphomas occur as single or multiple tumors with medium-sized lymphoid cells in a perivascular pattern. They are most common in the cerebral hemispheres (64% of cases), corpus callosum (13%), BG (11%), brain stem (9%), cerebellum (7%), spinal cord (4%), and meninges (2%)—see Fig. 18.5. Malignant T cells express CD45 and the T-cell antigens known as *CD2, CD3, CD4 or CD8, CD5, CD7*, and *CD56*. Loss of T-cell antigens can occur. Sometimes, there is positivity for **granzyme B** and **perforin**, which are cytotoxic granule proteins.

Clinical manifestations

Signs and symptoms of primary T-cell lymphomas include fever, headache, nausea, vomiting, confusion, unexplained weight loss, coughing, shortness of breath, difficulty initiating body movements, postural problems, night sweats, chronic fatigue, lethargy, a feeling of tiredness, pain, and inflammation of the lymph nodes. Signs and symptoms of NK/T-cell lymphomas are similar, but based on the affected region of the head and neck, can also include facial inflammation, diplopia, decreased visual acuity, impaired hearing, dysphagia, and hoarseness.

Diagnosis

To distinguish T-cell or NK/T-cell lymphomas from T-cell-rich large B-cell lymphoma with inflammation, molecular genetic demonstration of T-cell or NK cell **monoclonality** is needed. While their immunophenotypic and histological features are varied, a low-grade appearance is common, often with edema. Diagnostic methods include blood tests, EBV titers, CT, MRI, stereotactic biopsy, and lumbar puncture. On CT, these lymphomas often appear hyperdense. Diffusion-weighted images reveal hyperintensity of lesions.

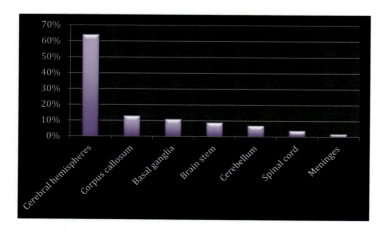

FIGURE 18.5 Locations of primary T-cell lymphomas.

Treatment

Treatments of T-cell or NK/T-cell lymphomas include corticosteroids, chemotherapy, radiation therapy, bone marrow transplantation, and stem cell transplantation. Multiple therapies are usually needed since monotherapy has resulted in rates of local and distance recurrence as high as 49%. CHOP therapy is usually used, combining cyclophosphamide, doxorubicin, vincristine, and prednisone, accompanied by radiation therapy. For some patients, methotrexate is used as a single chemotherapeutic agent. Disease progression commonly occurs despite treatment, however. A newer therapy called the **SMILE protocol** requires additional study because of significant **myelotoxicity**. This protocol combines dexamethasone, methotrexate, ifosfamide, L-asparaginase, and etoposide.

Prognosis

The prognosis of primary T-cell lymphoma is similar to or slightly better than for that of diffuse large B-cell lymphoma. The prognosis of NK/T-cell lymphomas of the CNS has a poorer prognosis. Combination therapies for these lymphomas have yielded 5-year survival rates between 20% and 80%.

> **Point to remember**
>
> T-cell lymphomas can develop in lymphoid tissues, or outside of them. T-cell, non-Hodgkin's lymphoma accounts for about 10% of all cases. Natural killer (NK) cell lymphoma is diagnosed in less than 1% of cases, but may be extremely aggressive. NK cells are hybrid immune cells that are similar but not identical to T cells.

MALT lymphoma of the dura

MALT lymphoma of the dura is also known as *extranodal marginal zone lymphoma of mucosa-associated lymphoid tissue of the dura*. These lymphomas are much less common than primary lymphomas of the brain. Also, in very rare cases, MALT lymphomas have formed within the brain itself.

Epidemiology

Dural MALT lymphomas affect middle-aged adults, with women developing them about five times more often than men. Though overall uncommon in relation to other CNS lymphomas (less than 1% of all CNS lymphomas), they are the most common form of *primary dural lymphoma*.

Etiology and risk factors

The cause of dural MALT lymphoma is not well understood since the dura does not have any lymphoid tissue. It has been suggested that a benign inflammatory condition of the dura might attract polyclonal lymphocytes from which a monoclonal lymphoma could form. Chronic infections and autoimmune diseases may be implicated and can be considered potential risk factors.

Pathology

Most MALT lymphomas arise within the cranial dura as solid, localized, extra-axial masses. MALT lymphoma developing in the dura over the spinal cord is much less common, and may be related to compression of the spinal cord. Histology and immunophenotype are similar to those of MALT lymphomas that develop in other areas of the body. They are made up of small lymphocytes, a few large cells, marginal zone cells that have slightly irregular and pale cytoplasm, and often, large amounts of plasma cells. These sometimes have remnants of reactive follicles or amyloid. Dural MALT lymphomas can invade adjacent brain tissues, usually in the Virchow − Robin spaces. The cells are CD5- and CD-10 negative B-cells. There are often monotypic plasma cells, which suggest plasmacytic differentiation. While there is no evidence of systemic IgG4-related disease, the monotypic plasma cells are sometimes IgG4-positive. Trisomies—usually of chromosome 3—are sometimes found, but MALT lymphoma-related translocations are rare. MALT lymphomas can recur in other locations, such as the subcutaneous tissues of the back.

Clinical manifestations

Signs and symptoms of dural MALT lymphomas include headache, a feeling of pressure, dizziness, seizures, focal neurological defects, and visual abnormalities. Nausea, vomiting, and ataxia have also been documented. Rarely, there is progressive cranial nerve dysfunction, causing bilateral visual and hearing loss. In one case, the patient developed acute hemiparesis. Clinical presentation is usually indolent, however, and symptoms are related to mass effect.

Diagnosis

Imaging studies reveal a mass, or thickening of the dura that resembles a plaque, and may be similar to the appearance of a meningioma. MALT lymphomas are usually localized when found, but in a small amount of cases, there is extradural disease. The characteristic feature of MALT lymphoma is the presence of neoplastic cells in epithelial structures, and the formation of **lymphoepithelial** lesions. Immunohistochemistry is used to diagnose MALT lymphomas from other small B-cell non-Hodgkin's lymphomas. On CT scans, they appear somewhat hyperdense compared to the brain and generally enhance greatly. On MRI, they appear isodense to the gray matter in T1-weighted imaging. When contrast is used, this imaging method shows vivid enhancement that is usually homogeneous, sometimes with an indistinct appearance from other brain tissues. In T2-weighted imaging, MALT lymphomas are isodense to hypodense to the gray matter, and edema is common in adjacent brain tissues. Diffusion weighted imaging (DWI)/apparent diffusion coefficient (ADC) shows restricted diffusion. Differential diagnoses include meningioma, dural metastases, **Erdheim − Chester disease**, and **Rosai − Dorfman disease**.

Treatment

Treatments for MALT lymphomas are varied. They can be surgically resected and may or may not also require chemotherapy. This usually involves multiple cycles of fludarabine, mitoxantrone, and rituximab to achieve complete remission. Also, rituximab is sometimes used as a single chemotherapeutic agent. Radiation therapy can be deferred as long as there is close clinical and neuroimaging follow-ups.

Prognosis

Fortunately, most patients with MALT lymphomas achieve complete remission and remain disease-free. Since these are mucosa-associated lymphoid tissue-related tumors, they have an indolent course and a good prognosis.

> **Point to remember**
> MALT lymphomas involve the mucosa-associated lymphoid tissues, and originate from B cells. They usually grow slowly, but have the potential to become aggressive. MALT lymphomas are usually low-grade, marginal zone lesions, with a monomorphous, diffuse infiltrate of small lymphocytes and plasma cells, plus lymph follicle formation that resembles chronic inflammation.

Anaplastic large cell lymphoma

Anaplastic large cell lymphoma is a primary CNS tumor that is *anaplastic lymphoma kinase (ALK)* positive. It is a rare type of lymphoma. Of all cases, less than 1% occur within the central nervous system. Most cases of anaplastic large cell lymphoma are of the common type, but there have also been documented cases of *primary CNS anaplastic large cell lymphoma of the lymphohistiocytic variant*, and combined *lymphohistiocytic and small-cell* variants. Also, there is *primary CNS anaplastic large cell lymphoma* that is ALK-negative.

Epidemiology

For anaplastic large cell lymphoma, most affected patients are children and young adults. Ages range from 23 months to 31 years, with a slightly male preponderance. The rare ALK-negative form of these lymphomas affects adults of both genders near equally.

Etiology and risk factors

A translocation of the ALK and *nucleophosmin 1 (NPM1)* genes is the causative genetic abnormality. There are no known risk factors, but chronic inflammation may be related.

Pathology

In most cases, there are one or more intracerebral masses, often with involvement of the leptomeninges, and sometimes, of the dura or skull. The parieto-occipital area is a common location. The lymphoma is rarely confined only to the leptomeninges, and develops aggressively. Of the ALK-negative subtype, the tumor is usually supratentorial, with diffuse white matter involvement.

Clinical manifestations

Anaplastic large cell lymphoma may cause headache, nausea, fever, seizures, or combinations of these. Additional signs and symptoms may include weight loss and night sweats.

Diagnosis

Often, the patient is at first believed to have an infection before correct diagnosis is achieved. Biopsy will reveal large atypical cells that include hallmark pleomorphic cells, and a variable admixture of reactive, nonneoplastic cells. They are positive for ALK, the cell membrane protein *CD30*, and in most cases, *epithelial membrane antigen (EMA)*, with variable expression of T-lineage antigens. The cytoplasm may be amphiphilic, with bizarre nuclei. The chromatin is finely dispersed and the nucleoli are prominent. The mitotic activity is prominent, and the Ki-67 proliferation index is usually high. Imaging studies commonly include MRI and CT scans. The lymphoma usually appears as a well enhancing mass that is hypointense on T1-weighted imaging, hyperintense on T2-weighted imaging, and homogeneously enhanced with contrast. There is often edema of the adjacent parenchyma.

Treatment

Treatments for anaplastic large cell lymphomas include anthracycline-based chemotherapy initially, with the most common regimen being CHOP, sometimes along with intrathecal methotrexate. Second-line therapy is with the antibody-drug conjugate medication called brentuximab vedotin. Other second-line therapies include the following regimens:

- GDP—gemcitabine, dexamethasone, cisplatin;
- DHAP—dexamethasone, high-dose cytarabine, cisplatin;
- ICE—ifosfamide, carboplatin, etoposide.

Total surgical resection can often be successfully performed. Radiation therapy may be used when indicated.

Prognosis

Fortunately, with prompt diagnosis and treatment, complete remission and long-term survival is often possible. Prognosis of the rare ALK-negative form of these lymphomas is poorer than for the ALK-positive form.

> **Point to remember**
> Anaplastic large cell lymphomas, if positive for the ALK protein, are more common in young people and usually respond well to chemotherapy. These lymphomas, when negative for the ALK protein, are more common in older people and are also more likely to recur. The anaplastic large cell lymphomas are rare types of non-Hodgkin's lymphomas.

Clinical cases

Clinical case 1

1. How common are diffuse large B-cell lymphomas of the CNS?
2. Where do most of these tumors develop?
3. What is the prognostic outlook for these tumors?

A 58-year-old man presented to his local hospital with facial paralysis and slurred speech. He had no history of genetic or immunodeficiency disorders. Lumbar puncture revealed that his CSF was normal and contained no tumor cells. The patient was negative for HIV and EBV. An MRI revealed a lesion of 1.8 cm at the right BG region and corona radiata. The patient refused a stereotactic biopsy. Treatment with mannitol and dexamethasone improved his symptoms. Two weeks later, he underwent gamma knife treatment and remained in the hospital with a headache and left-sided numbness. Another MRI revealed high and low signals in the right BG and radial area, with a lesion of 1.6 cm in diameter. MRS revealed that the tumor was malignant. He underwent three cycles of temozolomide therapy, which greatly decreased the tumor size. However, a PET/CT scan showed highly metabolic nodules in the right BG and left temporal lobe. Four months later, another MRI revealed that the left temporal lesion was resolved, and the right-sided lesion was in full remission. However, the next year, the patient developed slurred speech, memory loss, receptive aphasia, and right-sided weakness. An MRI revealed a right-sided lesion of 5.5 cm in diameter within the left frontal lobe. It was successfully resected during surgery, and the patient's symptoms improved greatly. The diagnosis was of a diffuse large B-cell lymphoma. Over 20 months of follow-up, the patient had no tumor recurrence.

Answers:

1. Diffuse large B-cell lymphomas of the CNS are the most common of all non-Hodgkin's lymphomas in adults (30% − 58% of cases). Incidence ranges from 7 to 20 cases per 100,000 populations, annually, in the United States and United Kingdom. In Europe, incidence is lower, at 3.8 out of every 100,000 population.
2. About 60% of diffuse large B-cell lymphomas of the CNS develop in the supratentorial space. They are in the frontal lobe in 15% of cases, followed by the posterior fossa (13%), BG and PP (10%), temporal lobe (8%), parietal lobe (7%), corpus callosum (5%), occipital lobe (3%), and spinal cord (1%).
3. Diffuse large B-cell lymphomas have a much better prognosis than primary CNS lymphomas. Patients with the worst prognosis are those over age 65 or if there is a mutation of transcobalamin C.

Clinical case 2

1. How often does intravascular large B-cell lymphoma involve the CNS, and what is its global prevalence?
2. What is the appearance of the lymphoma cells?
3. What are the common treatments for these lymphomas?

A 64-year-old woman was assessed at a hospital because of general fatigue and fever, plus leg swelling and limb weakness. There was no lymphadenopathy. Large, atypical lymphoid cells were found in the lumina of her small blood vessels. The cells were positive for CD20 and negative for CD29 and CD54. Her symptoms gradually worsened, and she developed urinary retention. Head and spinal MRI revealed enhancement of the cauda equina, and CSF examination showed infiltration with abnormal lymphocytes. Diagnosis was of an intravascular large B-cell lymphoma. The patient was immediately started on an R-CHOP regimen, combined with intrathecal injection of methotrexate, cytarabine, and prednisolone. Urinary retention gradually improved. After eight courses of R-CHOP and four courses of intrathecal therapy, there was complete tumor remission. The limb weakness remained, but the patient became able, over time, to walk with a cane. Over 28 months of follow-up, the patient has had no recurrence.

Answers:

1. Intravascular large B-cell lymphoma involves the CNS in 75%−85% of all cases. The global prevalence is only 1case out of every 1 million population.
2. Intravascular large B-cell lymphoma cells are twice the size of normal lymphocytes, or larger, usually with a prominent nucleolus.
3. Treatments include R-CHOP combination chemotherapy; intrathecal methotrexate, cytarabine, and prednisolone; radiation therapy; surgery in rare cases only; bone marrow transplantation; and stem cell transplantation.

Clinical case 3

1. Which gender is affected more often by low-grade B-cell lymphomas?
2. What are the subtypes of these lymphomas?
3. Are these lymphomas usually diagnosed when they are localized or widespread?

A 61-year-old woman presented with hemiparesis that had gradually been worsening. She complained of increased fatigue, but had not experienced weight loss, night sweats, fever, or headache. An MRI scan revealed a lesion of 2.8-cm in diameter, within the white matter at the left subcortical parieto-insular lobe and BG. It was hypointense on T1-weighted imaging, with some areas showing slight enhancement after contrast administration, and moderate hyperintensity on T2-weighted scans. A stereotactic biopsy was performed, and the lesion was first diagnosed as an atypical colloidal cyst. However, a follow-up MRI showed that the lesion had grown to 3.6-cm in diameter, and the patient's symptoms had worsened. Another stereotactic biopsy revealed a low-grade B-cell lymphoma. Complete surgical resection was performed successfully. Over time, her hemiparesis was nearly totally gone and no further manifestations of B-cell lymphoma could be found.

Answers:

1. **Low-grade B-cell lymphomas affect both genders nearly equally, and usually affect adults, not children.**
2. **Subtypes of low-grade B-cell lymphomas include: small lymphocytic lymphomas that are positive for CD5 and CD23, lymphoplasmacytic lymphomas, extranodal MALT lymphomas, and follicular lymphomas.**
3. **About one of every five patients with low-grade B-cell lymphomas is diagnosed while they are still in their early stages and the disease is localized.**

Clinical case 4

1. What are the four subtypes of immunodeficiency-associated CNS lymphomas?
2. Is the incidence of these lymphomas increasing or decreasing?
3. What are the risk factors for these lymphomas?

A 49-year-old female was hospitalized because of orbital cellulitis. She had previously been diagnosed with HIV and a CNS lymphoma, for which she was taking methotrexate and rituximab. Physical examination revealed bulging of her right eye and nonreactivity of the iris. An infection was suspected, and the patient was started on vancomycin and piperacillin-tazobactam. A CT scan revealed right periorbital cellulitis and diffuse sinus disease. There were extensive adhesions and black, necrotic tissue in her nasal cavity. Biopsies were taken and treatment began for a concurrent rhino-cerebral fungal infection. Such infections typically can affect the brain and eyes in patients with HIV.

Answers:

1. **The four subtypes include: AIDS-related diffuse large B-cell lymphoma, EBV-positive diffuse large-cell lymphoma that is not otherwise specified, lymphomatoid granulomatosis, and primary CNS posttransplant lymphoproliferative disorder.**
2. **The incidence of primary CNS lymphomas of this classification has increased by more than 10 times in immunocompromised individuals, from about 2.5 cases per 10 million population (globally) to about 30 cases per 10 million population.**
3. **There is an increased risk for CNS lymphomas because of autoimmune disorders, iatrogenic immunosuppression, and gradual deterioration of the immune system due to aging. Overall, immunosuppression-related CNS lymphomas are related to the EBV.**

Clinical case 5

1. What are the subtypes of T-cell lymphomas?
2. What is the common epidemiology of T-cell and NK/T-cell lymphomas?
3. In which parts of the brain are T-cell lymphomas most likely to develop?

A 23-year-old man was hospitalized because of nausea, vomiting, and acute confusion. Neurological examination revealed slow responses and postural instability. A CT scan revealed a 6-cm in diameter mild hyperdense mass in the right frontotemporal region, with related edema. There was also a similar lesion in the right cerebellar lobe, of 3.5-cm in diameter. Diffusion-weighted imaging showed hyperintense lesions. Stereotactic biopsy was performed, revealing medium-sized lymphoid cells with a perivascular pattern. Immunohistochemistry was positive for CD3, CD8, and CD56, indicating that the tumor cells were of T-cell origin. The diagnosis was primary T-cell CNS lymphoma. Chemotherapy was started using high-dose methotrexate, and a specialist determined that radiation therapy could be used as well.

Answers:

1. **Subtypes of T-cell lymphomas include the extranodal, cutaneous, anaplastic large cell, and angioimmunoblastic forms.**

2. T-cell lymphomas mostly affect young to middle-aged adults, while NK/T-cell lymphomas usually affect young to middle-aged men. Both lymphomas occur mostly in Asia.
3. T-cell lymphomas usually develop in the cerebral hemispheres, in 64% of cases. Less commonly, they develop in the corpus callosum, followed in decreasing order by the BG, brain stem, cerebellum, spinal cord, and meninges.

Clinical case 6

1. How common are MALT lymphomas of the dura in women as compared to men?
2. If a dural MALT lymphoma invades nearby brain tissues, where does this most often occur?
3. What is the characteristic feature of these lymphomas, and what other brain tumors may they resemble?

A 46-year-old woman complained of pain and pressure in the right frontal region of her head, with blurring of her right eye's vision. An MRI revealed a 6-cm extra-axial enhancing mass along the right frontal convexity that extended anteriorly to the superior ridge of the eye orbit. There was thickening and enhancement of the left frontal dura mater. The dural mass was resected, and its pathology was consistent with a MALT lymphoma. The patient was treated with fludarabine, mitoxantrone, and rituximab in four cycles. She was closely monitored with several MRIs to monitor for intracranial recurrence. Within 1 year, she relapsed with another MALT lymphoma, but it was located in the subcutaneous tissues of her back. She was again treated with chemotherapy, but this time only with rituximab, and 33 months later was in total remission, with no CNS or other disease.

Answers:

1. MALT lymphomas affect middle-aged adults, with women developing them about five times more often than men. They are the most common form of primary dural lymphoma.
2. Dural MALT lymphomas usually invade nearby brain tissues within the Virchow−Robin spaces. These spaces are perivascular, and accompany the lenticulostriate, perforating branches of the middle cerebral arteries in the BG.
3. The characteristic feature of MALT lymphoma is the presence of neoplastic cells in epithelial structures, and the formation of lymphoepithelial lesions. They may resemble meningiomas in appearance.

Clinical case 7

1. What is the epidemiology of the ALK-positive form of anaplastic large cell lymphomas?
2. What will a biopsy reveal?
3. What is the most common initial chemotherapy and the second-line chemotherapies?

A 30-year-old man presented with a progressive headache of the left parietal area that had bothered him over 6 weeks. He was alert with no neurological deficits, fever, recent illnesses, weight loss, night sweats, fatigue, or any lymph node enlargement. An MRI revealed a 5-mm in diameter mass in his left parieto-occipital dura, with edema of the adjacent parenchyma. The mass was hypointense on T1-weighted imaging, hyperintense on T2-weighted imaging, and homogeneously enhanced with contrast. His CSF was negative for malignancy. He underwent total surgical resection of the tumor. Microscopic examination revealed a pleomorphic neoplasm with large lymphoid cells and a moderate amount of amphiphilic cytoplasm, plus bizarre nuclei. There was finely dispersed chromatin and prominent nucleoli. The cells were admixed with reactive nonneoplastic cells. Prominent mitotic activity was noted and the Ki-67 proliferation index was 75%. Diagnosis was of an anaplastic large cell lymphoma. The patient was treated with a CHOP regimen, intrathecal methotrexate, and radiation therapy. He remained disease-free over 16 months of follow-up.

Answers:

1. Most affected patients are children and young adults. Ages range from 23 months to 31 years, with a slightly male preponderance.
2. Biopsy will reveal large atypical cells that include hallmark pleomorphic cells, and a variable admixture of reactive, nonneoplastic cells.
3. The most common initial chemotherapy regimen is CHOP, sometimes along with intrathecal methotrexate. The second-line chemotherapies include brentuximab vedotin, GDP, DHAP, and ICE.

Key terms

Anaplastic large cell lymphoma
Argyrophilic
Ataxia-telangiectasia
Centroblasts
Centrocytes
Diffuse large B-cell lymphoma
Disomies
Erdheim − Chester disease
Eye floaters
Granzyme B
Hypermutation
Immunoblasts
Immunodeficiency-associated CNS lymphomas
Intravascular large B-cell lymphoma
Light chain restriction
Low-grade B-cell lymphomas
Lymphoepithelial
Lymphomagenesis
Lymphomatoid granulomatosis
Lymphomatosis cerebri
Lymphoplasmacytoid cells
MALT lymphoma of the dura
MALT lymphomas
Monoclonality
Myelotoxicity
Perforin
Plasma cells
Polyomaviruses
Primary NK/T-cell lymphoma
Primary T-cell lymphoma
Psittacosis
Rosai − Dorfman disease
Sjögren syndrome
SMILE protocol
Smu region
Toxoplasmosis
Transcobalamin C
Waldenström's macroglobulinemia
Wiskott − Aldrich syndrome

References

Abla, O., & Attarbaschi, A. (2019). *Non-Hodgkin's lymphoma in childhood and adolescence.* Springer.
Abla, O., & Janka, G. (2018). *Histiocytic disorders.* Springer.
Abutalib, S. A., & Hari, P. (2017). *Clinical manual of blood and bone marrow transplantation.* Wiley-Blackwell.
Adler, E. M., Bishop, M. R., & Baker, W. J. (2016). *Living with lymphoma: A patient's guide* (2nd Edition). Johns Hopkins University Press.
Andreou, J. A., Kosmidis, P. A., Gouliamos, A. D., Vrakidou, E. P., Prassopoulos, V. K., & Vassilakopoulos, T. P. (2016). *PET/CT in lymphomas: A case-based atlas.* Springer.
Aster, J. C., Pozdnyakova, O., & Kutok, J. L. (2013). *Hematopathology: A volume in the high yield pathology series (Expert consult).* Saunders.
Cai, Q., Yuan, Z., & Lan, K. (2017). *Infectious agents associated cancers: Epidemiology and molecular biology (Advances in experimental medicine and biology).* Springer.
Carver, A. R. (2018). *Primary CNS lymphoma: Patient Care Journal.* Carver.
Cerci, J., Fanti, S., & Delbeke, D. (2016). *Oncological PET/CT with histological confirmation.* Springer.
Crawford, D. H., Johannessen, I., & Rickinson, A. B. (2014). *Cancer virus: The discovery of the Epstein-Barr virus.* Oxford University Press.
Ferry, J. A. (2011). *Extranodal lymphomas: expert consult.* Saunders.
Foss, F. (2013). *T-cell lymphomas (Contemporary hematology).* Humana Press.
Frappier, L. (2013). *EBNA1 and Epstein-Barr Virus associated tumours (Springer briefs in cancer research).* Springer.
Gobina, R. (2011). *Lymphomas in HIV patients: A hospital-based study.* Lap Lambert Academic Publishing.
Graeber, C. (2018). *The breakthrough: immunotherapy and the race to cure cancer.* Twelve.
Hayat, M. A. (2012). *Tumors of the central nervous system, volume 9: Lymphoma, supratentorial tumors, glioneuronal tumors, gangliogliomas, neuroblastoma in adults, astrocytomas, ependymomas, hemangiomas, and craniopharyngiomas.* Springer.
Hentrich, M., & Barta, S. K. (2016). *HIV-associated hematological malignancies.* Springer.
Hudnall, S. D., Much, M. A., & Siddon, A. J. (2019). *Pocket guide to diagnostic hematopathology.* Springer.
Kelly, W. K., & Halabi, S. (2018). *Oncology clinical trials: Successful design, conduct, and analysis, 2nd edition − oncology clinical trials book for designing, conducting and analyzing clinical trials.* Demos Medical.
Macklis, R. M., & Conti, P. S. (2010). *Image-guided radiation therapy in lymphoma management: The increasing role of functional imaging.* CRC Press.
Marcus, R., Sweetenham, J. W., & Williams, M. E. (2014). *Lymphoma: Pathology, diagnosis, and treatment, 2nd edition.* Cambridge University Press.
Munz, C. (2015). *Epstein Barr Virus volumes 1 and 2: One herpes virus: many diseases (Current topics in microbiology and immunology).* Springer.
National Comprehensive Cancer Network. (2019). *NCCN guidelines for patients: Follicular lymphoma.* National Comprehensive Cancer Network (NCCN).
National Comprehensive Cancer Network. (2017). *NCCN guidelines for patients: Diffuse large B-cell lymphoma.* National Comprehensive Cancer Network (NCCN).
Norden, A. D., Reardon, D. A., & Wen, P. C. Y. (2011). *Primary central nervous system tumors: Pathogenesis and therapy.* Humana Press.
Olch, A. J. (2013). *Pediatric radiotherapy: Planning and treatment.* CRC Press.
Querfeld, C., Zain, J., & Rosen, S. T. (2019). *T-cell and NK-cell lymphomas: From biology to novel therapies (Cancer treatment and research).* Springer.
Shao, H. (2016). *Diffuse large B-cell lymphoma: Symptoms, treatment and prognosis.* Nova Science Publishers Inc.
Sugita, Y. (2019). *Primary central nervous system lymphomas and related diseases: Biology, pathology, and treatment.* Amazon Services LLC.
Trajkova, S. (2013). *Diffuse large B-cell lymphoma.* Lap Lambert Academic Publishing.
Yarchoan, R. (2014). *Cancers in people with HIV and AIDS: progress and challenges.* Springer.

Chapter 19

Li–Fraumeni syndrome

Chapter Outline

Overview	367	Key terms	376
Li–Fraumeni syndrome	367	Further reading	376
Clinical cases	373		

Overview

Li–Fraumeni syndrome is a rare autosomal disorder characterized by a familial clustering of tumors. There is a predominance of sarcomas, breast cancers, brain tumors, and adrenocortical carcinomas diagnosed before the age of 45 years. Other cancers are also present in excess in certain families, including leukemia, lung cancer, skin melanoma, gastrointestinal cancer, and prostate cancer. Also, germ cell tumors, choroid plexus papilloma, and Wilms' tumor have been reported as part of the syndrome. Most affected families have tumor suppressor gene TP53 mutations. Loss of p53 protein function is believed to suppress a mechanism of protection against the accumulation of gene alterations. Despite much research, it is widely believed that there are possibly many more undiscovered genetic defects responsible for Li–Fraumeni syndrome.

Li–Fraumeni syndrome

Li–Fraumeni syndrome is a rare form of familial tumor syndrome, and is an autosomal dominant disorder. It involves germline mutations that can be inherited, or can arise from mutations early during embryogenesis, or within one of a parent's germ cells. This syndrome involves multiple primary tumors that affect children as well as younger adults. The most common tumors that appear include brain tumors, adrenocortical carcinoma, breast cancer, osteosarcomas, and soft tissue sarcomas. The syndrome was named after the two American physicians who first identified it in 1969: Dr. Frederick Pei Li and Dr. Joseph F. Fraumeni, Jr. At the time, they were studying pediatric and familial cancers at the National Cancer Institute. However, the syndrome was actually first reported under the name "Li–Fraumeni syndrome" in 1982, by researchers in the United Kingdom, who described two families with multiple forms of cancer in their younger members. Another name for the syndrome is *sarcoma, breast, leukemia, and adrenal gland syndrome*.

Epidemiology

The TP53 germline mutation linked to Li–Fraumeni syndrome is believed to occur at a rate of approximately 1 in every 5000–20,000 births. It accounts for up to 17% of all cases of familial cancers. Individuals with Li–Fraumeni syndrome have about a 50% chance of developing cancer by age 40, and up to 90% of a chance by age 60. Women with the syndrome have a lifetime breast cancer risk of 49% by age 60. The International Agency for Research on Cancer has a TP53 mutation database that contains the genetic and pedigree information of nearly 800 families that have carriers of this mutation. This database can be accessed at http://p53.iarc.fr. The updated version of this database was released in 2019, and includes a major update of data on germline variations and the functional assessment of mutant proteins.

Approximately 13% of patients with TP53 germline mutations develop central nervous system (CNS) tumors, and the male-to-female ratio of patients with brain tumors related to the mutation is 1.5:1. Like sporadic brain tumors, patient age with CNS tumors linked to the mutation has a **bimodal distribution**. The first peak in incidence is in children—mostly medulloblastomas or related primitive neuroectodermal tumors, tumors of the choroid plexus, and ependymomas. The second peak is in the third and fourth decades of life—mostly astrocytic brain tumors. Of 139 families with at least one brain tumor, the mean number of CNS tumors per family was 1.55. There are several documented families with a high clustering of brain tumors. This may indicate that some mutations have a cell-specific or organ-specific risk.

In the United States alone, there have been about 400 documented families affected by Li−Fraumeni syndrome since its discovery. Actual population incidence is unknown. Every year, about 5−10 cases of soft tissue sarcoma occur out of every 1 million children younger than 15 years of age. In a study of sarcoma patients, 10% of families with an osteosarcoma before age 20 or a soft tissue sarcoma before age 16 were found to have a germline TP53 mutation. There is no evidence of any ethnic or racial predisposition for the syndrome. However, cancer **penetrance** is 93% for female carriers and 73% for male carriers. The prevalence and penetrance of Li−Fraumeni syndrome is shown in Fig. 19.1.

> **Point to remember**
>
> Though the documented number of cases is relatively low, there are more than 1000 multigenerational families believed to have Li−Fraumeni syndrome globally. Until today, people have contacted the *Li-Fraumeni Syndrome Association* from 172 different countries. Their website is https://www.lfsassociation.org.

Etiology and risk factors

Li−Fraumeni syndrome is usually caused by a germline mutation in the *TP53 (tumor suppressor gene)* located on chromosome 17p13. Brain tumors are more likely linked with **missense mutations** in the DNA-binding surface of the p53 protein that contacts the minor groove of DNA. Mutation types are related to patient age at the onset of a brain tumor. **Truncating mutations**, which result in incomplete protein synthesis, are linked to early-onset brain tumors. Gene−environment interactions may cause familial clustering of brain tumors. Therefore risk factors include environmental carcinogens and harmful lifestyle factors. Not everyone with a TP53 gene mutation will develop cancer, but risks are much higher than in the general population. The risk of developing a second cancer increases with younger age at diagnosis of the first cancer. A second cancer usually occurs 6−12 years after the first cancer. The probability of

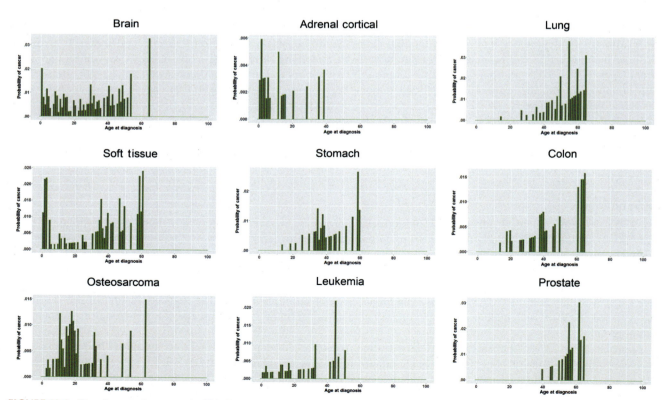

FIGURE 19.1 Prevalence and penetrance of Li−Fraumeni syndrome.

a person with Li—Fraumeni syndrome developing a second cancer is 57% within 30 years of the first cancer. Risks increase with exposure to radiation. Subsequent primary tumors, especially sarcomas, are more likely to develop in body areas that were previously exposed to radiation.

Pathology

About 80% of all tumors in individuals carrying the TP53 germline mutation occur as breast cancer, bone tumors, adrenal gland tumors, CNS tumors, and soft tissue sarcomas. Sporadic forms of these tumors also have a high amount of TP53 mutations. This may mean that the mutations can begin the process of malignant transformation. Generally, the tumors linked to a TP53 germline mutation occur earlier than in the sporadic form, yet there are significant organ-specific variances. For example, adrenocortical carcinoma linked to a TP53 germline mutation develops in children nearly always. This is different from sporadic adrenocortical carcinoma, which has a wide age distribution and a peak in incidence in individuals aged 40 or older. The TP53 gene on chromosome 17p13 has 11 **exons** that span 20 **kilobases**. A "base" is also referred to as a "base pair," and is equivalent to the number of **nucleotides** in the length of a strand of DNA. Exon 1 is noncoding, while exons 5—8 are highly conserved. Table 19.1 illustrates where tumors usually develop in individuals carrying the TP53 germline mutation.

Most TP53 germline mutations occur over exons 5—8, and also at **codons** 133, 175, 245, 248, and 273 *or* 337. Missense mutations are the most common mutation subtype. They result in mutant proteins with a total loss of function, dominant-negative phenotypes, and oncogenic activities. However, **nonsense mutations, deletion/insertion mutations**, and **splice-site mutations** also happen. Codons that have never been reported as germline mutations include *arginine 249 (Arg249)* and *cysteine 176 (Cys176)*, which are usually somatically mutated in sporadic tumors. *Residue 176* (cysteine) aids in coordinating a zinc atom, which creates a bridge between *domains 1 and 3*. This is essential for stabilizing the entire DNA-binding domain architecture. *Residue 249* (arginine) forms essential contacts with various residues of the scaffold via hydrogen bridges. Residues from codons 133 and 337 are **hotspots** for TP53 germline mutations, yet less common in sporadic cancers. In families with clusters of early-onset breast cancers, meaning 3—6 cases per family and an average patient age at onset of 34 years, a mutation at codon 133 (or M133T) has been found. For some reason, there is also a mutation at codon 337 (or R337H) that is common only in Brazilian children with adrenocortical carcinomas and in Brazilian families with Li—Fraumeni syndrome.

The proportion of guanine:cytosine to adenine:thymine (G:C $\Rightarrow$ A:T) transitions at cytosine-plus-guanine (CpG) sites is larger in TP53 germline mutations compared to somatic mutations. However, proportions of G:C $\Rightarrow$ A:T transitions at non-CpG sites, as well as G:C $\Rightarrow$ A:T **transversions** are lower. The G:C $\Rightarrow$ A:T transitions at CpG sites are believed to be endogenous, such as occurring because of deamination of 5-methylcytosine. This happens spontaneously in nearly all types of cells, but is usually corrected by DNA repair activities. The difference between these various proportions could be because non-CpG G:C $\Rightarrow$ A:T and G:C $\Rightarrow$ A:T mutations are linked to exogenous carcinogen exposures, while germline mutations appear to mostly occur from endogenous processes. An illustration of a CpG site is shown in Fig. 19.2.

TABLE 19.1 Percentages of tumor development in TP53 germline mutation carriers.

Organ or tissue	Usual histology	% of all tumors	Peak age at onset (years)
Breast	Carcinoma	31	26—30
Soft tissues	Sarcoma	14	0—5
Central nervous system	Astrocytoma, choroid plexus tumor, glioblastoma, medulloblastoma	13	0—5
Adrenal cortex	Carcinoma	12	0—5
Bones	Osteosarcoma	9	11—15

FIGURE 19.2 A CpG (cytosine-plus-guanine) site.

TABLE 19.2 Tumor sites in 1350 carriers of a TP53 germline mutation.

Site	%	Female cases	Male cases	Average age (years)
Breast	30.6	412	0	34
Soft tissues	13.8	101	84	22
Brain	13	62	114	18
Adrenal cortex	11.7	119	39	7
Bones	8.6	70	46	17
Hematopoietic system	2.7	23	13	24
Skin	2.5	26	8	45
Lungs	2.5	17	17	44
Colon	2.1	14	14	39
Ovaries	1.9	26	0	40
Stomach	1.4	6	13	39
Other	9.3	82	44	34

In a study of 1350 patients who carried a TP53 germline mutation, the tumor locations were assessed, as shown in Table 19.2.

In a relatively recent study, a family with Li–Fraumeni syndrome, having a TP53 germline mutation (codon 236 deletion), plus multiple CNS tumors, also had additional germline mutations. Missense mutations of the *MutS protein homolog 4 (MSH4)* DNA repair gene were found in three patients who had gliomas—a total of two anaplastic astrocytomas and two glioblastomas. Two other family members developed peripheral schwannomas but lacked a TP53 germline mutation. However, they did have the MSH4 germline mutation and a germline mutation of the *large tumor suppressor kinase 1 (LATS1)* gene, which is a downstream mediator of the *neurofibromatosis 2 (NF2)* gene.

The TP53 tumor suppressor gene encodes a 2.8 kb transcript, which encodes a 393 amino acid protein that is expressed, at low levels, very widely. This amino acid protein is a transcription factor with many functions, and helps control cell cycle progression, survival of cells that are exposed to agents that damage DNA, **nongenotoxic** stimuli (including hypoxia), and the actual integrity of the DNA. Should hypoxia or DNA damage occur, it encourages a transient nuclear accumulation and activation of the p53 protein. There is also transcriptional activation of target genes responsible for inducing apoptosis or arrest of the cell cycle.

The p53 protein has many functions also, but is mostly involved in control of the cell cycle, DNA integrity, and cell survival following exposure to DNA-damaging agents. It is also believed to regulate various other important processes. These may include cell oxidative metabolism, the cells' response when deprived of nutrients, division and renewal of stem cells, and fertility. Stress and cell type affect how much of a biological response will be influenced by p53, as well as the results from this. Functions of p53 are mostly based on its transcriptional activity. However, it can also function through interactions with different proteins. In the majority of cancers, TP53 is inactivated by gene mutations, which bring about loss of the tumor suppressor role. The p53-mutant proteins are different from each other in regards to the amount of lost suppressor function, and their ability to inhibit **wild-type p53** in a dominant-negative fashion. Also, some p53-mutants appear to have their own oncogenic activity. However, this gain-of-function phenotype's molecular basis is not fully understood. Each p53-mutant protein's functionality may be based, partly or totally, on the amount of protein structural alteration resulting from the mutation.

In addition, a missense germline mutation of *glycine-245-serine (Gly245Ser)* has been identified as a mutation hotspot of the TP53 gene in two affected members of a Li–Fraumeni syndrome family. In this study, a liposarcoma and a colorectal carcinoma both showed p53 protein accumulation, yet there was no loss of wild-type p53 (see Fig. 19.3). In a review of germline TP53 mutation carriers, the majority had a loss of wild-type p53. Some had retention of heterozygosity, a small amount had loss of the allele that harbored the TP53 germline mutation.

> **Point to remember**
> A person with Li–Fraumeni syndrome is 25 times more susceptible of developing several cancer types when compared to a non-affected individual. This is because patients inherit a mutated copy of the p53 gene from one parent, and the other copy becomes mutated by environmental factors, leading to collapse of natural defense mechanisms against cancer.

Clinical manifestations

The manifestations of Li–Fraumeni syndrome include development of several multiple cancers. Usually, this means soft tissue and bone sarcomas, breast cancer, brain tumors, adrenocortical carcinoma, and acute leukemia. Other cancers that have developed include gastrointestinal cancers (of the colon or pancreas) and cancers of the kidneys, lungs (adenocarcinoma), skin (melanoma), thyroid, ovaries, testicles, and prostate gland.

Diagnosis

Screening tests for cancer are very important for Li–Fraumeni syndrome's early detection. They involve detailed physical examinations, routine blood tests, screening for endocrine cancer—specifically for adrenal gland function, yearly brain and body MRI scans, and intermittent ultrasonography. If an affected individual is in his or her early 20s, a colonoscopy may be required every 2–5 years. Women in their early 20s should have mammographies every 6 months and yearly breast MRI scans. The classic Li–Fraumeni syndrome's clinically diagnostic criteria include three major areas, as follows:

- The first sarcoma must develop before age 45.
- At least one first-degree relative must have any type of tumor before age 45.
- A first- or second-degree relative must have cancer before age 45, or must have a sarcoma at any age.

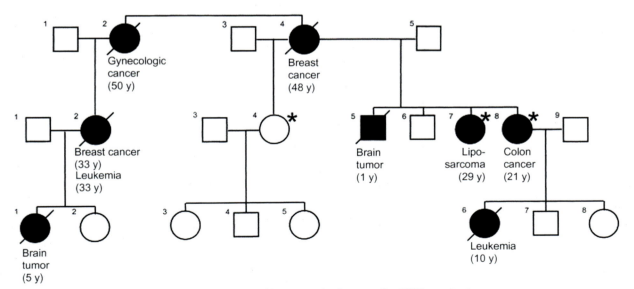

FIGURE 19.3 A Li–Fraumeni syndrome family with retained heterozygosity for a germline TP53 mutation in two tumors.

There is also a variant form called **Li—Fraumeni-like syndrome**, which also has three main sets of criteria. The three sets of criteria are as follows:

- The *Li—Fraumeni-like (LFL)-E2 definition*—established by British geneticist Rosalind A. Eeles. It was her second definition, hence the abbreviation "E2." In this definition, there must be sarcoma at any age in the **proband**, plus any two of the following tumors within the same family, which includes in a single individual:
  - Brain tumor
  - Adrenocortical tumor
  - Breast cancer at any age under 50 years
  - Leukemia
  - Melanoma
  - Pancreatic cancer at any age under 60 years
  - Prostate cancer
  - Sarcoma at any age
- The *LFL-B definition*—established by British epidemiologist Jillian M. Birch. The first letter of her last name signifies the "B" in the name of this definition. In this definition, there must be any childhood cancer or sarcoma, brain tumor, or adrenocortical carcinoma at an age less than 45 years in the proband, plus:
  - One first- or second-degree relative with a cancer that is commonly associated with Li—Fraumeni syndrome (such as adrenocortical carcinoma, brain tumor, breast cancer, leukemia, sarcoma) at any age, plus
  - One first- or second-degree relative of the same lineage, with any cancer diagnosed at any age below 60 years
- The *Chompret definition*—established by French researcher Agnes Chompret. This definition was updated in 2009, and has three components as follows:
  - A tumor of the Li—Fraumeni tumor spectrum, such as adrenocortical carcinoma, brain tumor, premenopausal breast cancer, leukemia, lung adenocarcinoma in situ, osteosarcoma, or soft tissue sarcoma, at any age below 46 years in the proband, and at least one first- or second-degree relative either with a Li—Fraumeni-type tumor (other than breast cancer if the proband is affected by breast cancer) at any age below 56 years or with multiple tumors, or
  - Multiple tumors (except for multiple breast cancers) in the proband, with the first occurring before age 46, and with two or more belonging to the Li—Fraumeni spectrum, or
  - Adrenocortical carcinoma or choroid plexus carcinoma in the proband, regardless of any family history.

The genomic profile of a Li—Fraumeni-like syndrome patient presenting without TP53 mutation is shown in Fig. 19.4.

There are many differential diagnoses for Li—Fraumeni syndrome itself. These include adrenal carcinoma, astrocytoma, breast cancer, choroid plexus papilloma, chromosomal breakage syndromes, colon cancer, cutaneous melanoma, gastrointestinal cancers, intestinal polyposis syndromes, nonrhabdomyosarcoma soft tissue sarcomas, retinoblastoma, and Wilms tumor-aniridia-genitourinary anomalies-mental retardation syndrome. Also, for children, differential diagnoses include acute lymphoblastic leukemia, acute myelocytic leukemia, osteosarcoma, and rhabdomyosarcoma.

Treatment

There is no clear evidence that people with Li—Fraumeni syndrome who develop cancers should be treated any differently from other patients with cancer, in regards to surgery or chemotherapy. However, radiation therapy should be used cautiously because of concerns of increased risks for radiation-induced secondary tumors. The specifics of therapy are related to the actual type of cancer that develops. The National Comprehensive Cancer Network (NCCN) has created an algorithm for the testing and treatment of Li—Fraumeni syndrome in adults. Physicians must discuss prophylactic mastectomy for women with the syndrome, in order to reduce risks for developing breast cancer. This must be considered on an individual basis depending on the degree of cancer risk and the available reconstructive options. In one study, the doxorubicin to breast cancer dose response was shown to be stronger in survivors of Li—Fraumeni syndrome—related childhood cancers. Genetic counseling is essential for affected families to provide appropriate understanding of risks and to evaluate genetic predisposition markers. Patients with Li—Fraumeni syndrome may develop anxiety, depression, and serious distress, so psychological monitoring is important.

> **Point to remember**
>
> Many strategies using small molecule drugs to reactivate or modify dysfunctional TP53 protein are being studied today. Clinical trials with Li—Fraumeni syndrome patients have been conducted, and some are ongoing, at the Garvan Institute of Medical Research, Peter MacCallum Cancer Centre, St. Jude Children's Research Hospital, Dana-Faber Cancer Institute, and University of Texas' M.D. Anderson Cancer Center.

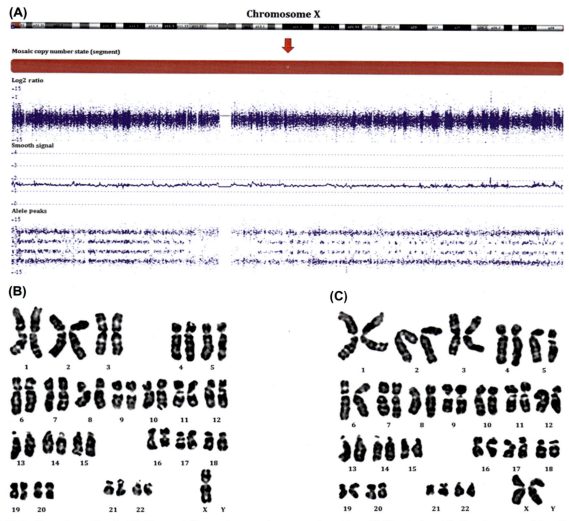

FIGURE 19.4 Genomic profile of a Li–Fraumeni-like syndrome patient presenting without TP53 mutation. (A) Genetic screen of the affected individual. (B) Note that chromosome 17 appears mutated. (C) Note that chromosome 17 appears normal.

Prognosis

Improvements in treatments for various childhood cancers have led to longer survival rates for many children diagnosed with Li–Fraumeni syndrome, though possible late effects still include other primary malignancies. It is important to test children in families affected by Li–Fraumeni syndrome for their likelihood to develop the same condition. Children in families with this syndrome who survive an initial cancer have a relative risk of developing a second cancer that is 83 times greater than in the general population. Prognosis may be better when affected individuals are advised about potential risks, available screening methods, and family members are also evaluated. Outlook may be improved by having physicians and other healthcare professionals who are aware of Li–Fraumeni syndrome become involved with care.

Clinical cases

Clinical case 1

1. What are the most common types of tumors that occur as part of Li–Fraumeni syndrome?
2. What is the other name for this syndrome?
3. What are the chances for developing cancer in relation to this syndrome?

A 5-year-old girl was diagnosed with a tumor in the left cerebellar hemisphere. It was surgically removed and diagnosed as a classic type of medulloblastoma. Radiation therapy was followed, and after 18 months of follow-up, no recurrence was detected. Soon, the patient's 3-year-old brother was diagnosed with a mass in the posterior areas of the left lateral ventricle, which was resected and diagnosed as a choroid plexus papilloma. Again, after 18 months of follow-up, there was no recurrence. As a result of these two siblings having brain tumors, the family history was carefully evaluated. Their mother had been diagnosed with breast cancer in situ previously, at the age of 34 years. She had a radical mastectomy and had no recurrence over 2 years. Previously, the childrens' aunt had a modified radical mastectomy and was diagnosed with occult breast cancer, which caused her death due to distant metastasis. The aunt's own daughter was diagnosed with adrenal pheochromocytoma at 3 years of age, and then a renal cyst at 12 years of age. The grandfather of the proband (the 5-year-old girl) had died from a liver mass when he was in his 40s. As a result, the tumors in this family were aggregated, likely to be Li–Fraumeni syndrome.

Answers:

1. **The most common tumors that appear include brain tumors, adrenocortical carcinoma, breast cancer, osteosarcomas, and soft tissue sarcomas.**
2. **The other name for Li–Fraumeni syndrome is** *sarcoma, breast, leukemia, and adrenal gland syndrome*.
3. **Individuals with Li–Fraumeni syndrome have about a 50% chance of developing cancer by age 40, and up to 90% of a chance by age 60. Women with the syndrome have a lifetime breast cancer risk of 49% by age 60.**

Clinical case 2

1. How many cases of Li–Fraumeni syndrome have been documented in the United States?
2. With this syndrome, what are the risks of developing second cancers, and in what time frame?
3. In relation to this syndrome, what are the average ages of patients who develop osteosarcomas, brain tumors, and breast cancer?

Three members of one family were diagnosed with malignancies. The first was a 24-year-old man, diagnosed and treated for osteosarcoma of the maxilla, which proved to be fatal. His younger brother developed osteosarcoma of the mandible later, was treated, and had no signs of recurrence over 12 months of follow-up. The men's mother developed glioblastoma multiforme brain cancer along with ductal carcinoma of the breast, and died from her malignancies. This family highlights the need for careful examination, inspection, and notification of the risks of family members diagnosed with Li–Fraumeni syndrome–related tumors.

Answers:

1. **In the United States, there have been about 400 documented families affected by Li–Fraumeni syndrome since its discovery. Actual population incidence is unknown.**
2. **With Li–Fraumeni syndrome, the risk of developing a second cancer increases with younger age at diagnosis of the first cancer, usually within 6–12 years after the first cancer. The probability of a person with the syndrome developing a second cancer is 57% within 30 years of the first cancer.**
3. **In relation to Li–Fraumeni syndrome, the average age of patients to develop osteosarcomas is 17 years, with a female predominance. For brain tumors, the average age is 18 years, with a male predominance. For breast cancer, the average age of the female patient is 34 years.**

Clinical case 3

1. How often is the TP53 germline mutation linked to Li–Fraumeni syndrome?
2. How many patients with the TP53 germline mutation develop CNS tumors and what is the male-to-female ratio?
3. Where does the TP53 germline mutation occur?

The proband in this case is one of five siblings who was examined at 2½ years of age because of headaches and repeating vomiting. An MRI revealed a large, heterogeneously enhancing right lateral intraventricular mass, with calcification and hemorrhaging. It was confirmed to be a choroid plexus tumor. Since this type of tumor is often found in families with Li–Fraumeni syndrome, immunohistochemistry staining for p53 was performed. The clinician found strong positive nuclear accumulation, suggestive of p53 dysfunction, and consistent with Li–Fraumeni-related tumors. There was a loss of TP53 heterozygosity, which is also found in the syndrome's related tumors. After surgical resection and chemotherapy, the patient remained without any recurrence. The patient's older sister developed a choroid plexus tumor as well, and immunohistochemistry staining for p53 showed strong positive nuclear accumulation, plus loss of TP53 heterozygosity. Unfortunately, her tumor recurred and was eventually fatal, at the age of 7 years. A thorough review of the family history revealed a great-grandmother and great-aunt who also had brain tumors, another aunt had liver cancer, and two uncles had colorectal cancer.

Answers:

1. **The TP53 germline mutation linked to Li–Fraumeni syndrome is believed to occur at a rate of about 1 in every 5000–20,000 births. It accounts for up to 17% of all cases of familial cancers.**
2. **About 13% of patients with TP53 germline mutations develop CNS tumors, and the male-to-female ratio of patients with brain tumors related to the mutation is 1.5:1.**
3. **The TP53 germline mutation occurs on chromosome 17p13.**

Clinical case 4

1. For people with the TP53 mutation, where do most tumors develop?
2. What do screening tests for cancer, in relation to Li–Fraumeni syndrome, involve?
3. What are the three clinically diagnostic criteria for Li–Fraumeni syndrome?

A 37-year-old woman, known to be a carrier of the TP53 mutation, was diagnosed with breast cancer. She was treated with a variety of medications, and underwent a bilateral mastectomy of the right breast. Her family history was assessed, and it was found that her father and half-brother were diagnosed with brain tumors. In her father's case, it was fatal. Several aunts, cousins, and a half-sister also had various malignancies including colorectal cancer, and another half-sister developed stage IV breast cancer with metastasis to her bones.

Answers:

1. **About 80% of all tumors in individuals carrying the TP53 germline mutation occur as breast cancer, bone tumors, adrenal gland tumors, CNS tumors, and soft tissue sarcomas.**
2. **Screening tests involve detailed physical examinations, routine blood tests, screening for endocrine cancer—specially for adrenal gland function, yearly brain and body MRI scans, and intermittent ultrasonography.**
3. **The three major criteria areas include the first sarcoma must develop before age 45; at least one first-degree relative must have any type of tumor before age 45; and a first- or second-degree relative must have cancer before age 45, or must have a sarcoma at any age.**

Clinical case 5

1. For Li–Fraumeni-like syndrome, what is the LFL-E2 definition established by British geneticist Rosalind A. Eeles?
2. What is the LFL-B definition as established by Jillian M. Birch?
3. What is the Chompret definition?

A 47-year-old woman went to her local hospital with a painless lump in her right breast. There was no family history of breast or ovarian cancer, but two of her family members had died from late-onset cancers. Mammography and histological examination confirmed invasive ductal carcinoma of the breast. There was no metastasis. A partial mastectomy was performed. The tumor cells were positive for p53 and results for Ki67 were positive in 5% of the cells. After the surgery, she underwent four cycles of adjuvant chemotherapy, and had no recurrence. Because of family history of cancer, samples of her peripheral blood lymphocytes and tumor tissue were assessed for the TP53 mutation. Later, her three children underwent genetic screening. Her 20- and 16-year-old daughters also had the mutation, but no signs or symptoms of cancer. It was determined that this patient's case and family data were an example of Li–Fraumeni-like syndrome.

Answers:

1. **In the LFL-E2 definition, there must be sarcoma at any age in the proband (initial family member involved), plus any two of the following tumors within the family, or within one single individual: brain tumors, adrenocortical tumor, breast cancer at any age under 50 years, leukemia, melanoma, pancreatic cancer at any age under 60 years, prostate cancer, and sarcoma at any age.**

2. The LFL-B definition requires that there is any childhood cancer or sarcoma, brain tumor, or adrenocortical carcinoma at an age less than 45 years in the proband, plus: one first- or second-degree relative with a cancer commonly associated with Li–Fraumeni syndrome at any age, plus one first- or second-degree relative of the same lineage, with any cancer diagnosed below 60 years of age.
3. The Chompret definition, established by Agnes Chompret, has three components: a tumor of the Li–Fraumeni spectrum at any age below 46 years in the proband, and at least one first- or second-degree relative with a Li–Fraumeni-type tumor (other than breast cancer if the proband is affected by breast cancer) at any age below 56 years or with multiple tumors; or multiple tumors (except for multiple breast cancers) in the proband, with the first occurring before age 46, with two or more of the Li–Fraumeni spectrum; or adrenocortical carcinoma or choroid plexus carcinoma in the proband, regardless of family history.

Key terms

Bimodal distribution
Codons
Deletion/insertion mutations
Exons
Hotspots
Kilobases
Li–Fraumeni syndrome
Li–Fraumeni-like syndrome
Missense mutations
Nongenotoxic
Nonsense mutations
Nucleotides
Penetrance
Proband
Splice-site mutations
Transversions
Truncating mutations
Wild-type p53

Further reading

Armstrong, S. (2016). *p53: The gene that cracked the cancer code*. Bloomsbury Sigma.
Berliner, J. (2014). *Ethical dilemmas in genetics and genetic counseling: Principles through case scenarios*. Oxford University Press.
Bettinger, B. T. (2019). *The family tree guide to DNA testing and genetic genealogy* (2nd ed.). Family Tree Books.
Bettinger, B. T., & Parker Wayne, D. (2016). *Genetic genealogy in practice*. National Genealogical Society.
Biesecker, B. B., Peters, K. F., & Resta, R. (2019). *Advanced genetic counseling: Theory and practice*. Oxford University Press.
Boslaugh, S. (2019). *Genetic testing (Health and medical issues today)*. Greenwood.
Chung, D. C., & Haber, D. A. (2010). *Principles of clinical cancer genetics: A handbook from Massachusetts General Hospital*. Springer.
Clarke, A. (2019). *Harper's practical genetic counselling* (8th ed.). CRC Press.
Coleman, W. B., & Tsongalis, G. J. (2016). *Diagnostic molecular pathology: A guide to applied molecular testing*. Academic Press.
Edwards, Q. T. (2017). *Genetics and genomics in nursing: Guidelines for conducting a risk assessment*. Springer.
Ellis, C. N. C. (2010). *Inherited cancer syndromes: Current clinical management* (2nd ed.). Springer.
Ellis, N. C., Jr. (2003). *Inherited cancer syndromes: Current clinical management*. Springer.
Farkas Patenaude, A. (2004). *Genetic testing for cancer: Psychological approaches for patients and families*. American Psychological Association.
Friedman, T., Dunlap, J. C., & Goodwin, S. F. (2015). *Advances in genetics* (Vol. 91). Academic Press.
Goodenberger, M. L., Thomas, B. C., & Kruisselbrink, T. (2017). *Practical genetic counseling for the laboratory*. Oxford University Press.
Hodgson, S. V., Foulkes, W. D., Eng, C., & Maher, E. R. (2014). *A practical guide to human cancer genetics* (4th ed.). Springer.
Israeli, M., & Schwartz, L. (2005). *Cancer: A dysmethylation syndrome*. John Libbey Eurotext Ltd.
Jorde, L. B., Carey, J. C., & Bamshad, M. J. (2019). *Medical genetics* (6th ed.). Elsevier.
Kadhum, R. F., & Al-Dragi, W. A. (2019). *Gene toxicity of ionizing radiation*. Lap Lambert Academic Publishing.
Kasper, C., Schneidereith, T. A., & Lashley, F. R. (2015). *Lashley's essentials of clinical genetics in nursing practice* (2nd ed.). Springer.
Lejeune, F., Benhabiles, H., & Jia, J. (2016). *Nonsense mutation correction in human diseases: An approach for personalized medicine*. Academic Press.
LeRoy, B. S., Veach, P. M., & Bartels, D. M. (2010). *Genetic counseling practice: Advanced concepts and skills*. Wiley-Blackwell.
MacFarlane, I., McCarthy Veach, P., & LeRoy, B. (2014). *Genetic counseling research: A practical guide (Genetic counselling in practice)*. Oxford University Press.
Mathiesen, A., & Roy, K. (2018). *Foundations of perinatal genetic counseling (Genetic counseling in practice): A guide for counselors*. Oxford University Press.
Matloff, E. (2013). *Cancer principles and practice of oncology: Handbook of clinical cancer genetics*. LWW.
McCarthy Veach, P., LeRoy, B. S., & Callanan, N. P. (2018). *Facilitating the genetic counseling process: Practice-based skills* (2nd ed.). Springer.
McKinlay Gardner, R. J., & Amor, D. J. (2018). *Gardner and Sutherland's chromosome abnormalities and counseling (Oxford monographs on medical genetics)* (5th ed.). Oxford University Press.
Minna Stern, A. (2012). *Telling genes: The story of genetic counseling in America*. Johns Hopkins University Press.
Nose, V. (2019). *Diagnostic pathology: Familial cancer syndromes* (2nd ed.). Elsevier.

Parker, P. M. (2007). *Li-Fraumeni syndrome – A bibliography and dictionary for physicians, patients, and genome researchers*. ICON Group International, Inc.

Rajasekaran, R., Sethumadhavan, R., & Chandrasekaran, P. (2012). *Cancer informatics: Computational analysis of significant missense mutations in various cancer genes and their drug targets*. Lap Lambert Academic Publishing.

Robin, N. H., & Farmer, M. (2017). *Pediatric cancer genetics*. Elsevier.

Schneider, K. A. (2011). *Counseling about cancer: Strategies for genetic counseling* (3rd ed.). Wiley-Blackwell.

Stanford, J. (2016a). *M.D. Anderson cancer genetic testing: A guide for patients to know how M.D. Anderson Cancer Center treats cancer*. M.D. Anderson.

Stanford, J. (2016b). *Cancer: Genetic testing for patients: A guide to fight cancer for a cure in genetics way*. Stanford.

Stevenson, F. K., di Genova, G., Ottensmeier, C. H., & Savelyeva, N. (2013). *Cancer immunotherapy: Chapter 15. Genetic vaccines against cancer: Design, testing and clinical performance*. Academic Press.

Sutton, C. L. (2017). *Genetic testing: Defining your path to a personalized health plan: An integrative approach to optimize health*. Dallas Chiropractic & Kinesiology, PLLC.

Uhlmann, W. R., Schuette, J. L., & Yashar, B. (2009). *A guide to genetic counseling* (2nd ed.). Wiley-Blackwell.

Williams, J. R. (2019). *The immunotherapy revolution: The best new hope for saving cancer patients' lives*. Williams Cancer Institute.

Yousef, G. M., & Jothy, S. (2014). *Molecular testing in cancer*. Springer.

Chapter 20

Turcot syndrome

Chapter Outline

Overview	379	Clinical cases	384
Turcot syndrome	379	Key terms	386
Brain tumor-polyposis syndrome 1	379	Further reading	386
Brain tumor-polyposis syndrome 2	381		

Overview

Turcot syndrome is also described as brain tumor-polyposis syndrome. It comprises two forms of disease that are characterized by the manifestations of colorectal and central nervous system neoplasms. Type 1, also called mismatch repair cancer syndrome, has features of hereditary nonpolyposis colorectal cancer combined mostly with glioblastomas. It was the original type described by Jacques Turcot. Type 2, also called familial adenomatous polyposis (FAP), mostly manifests with medulloblastomas. In general, the prognosis is worse for Type 1 than for Type 2.

Turcot syndrome

Turcot syndrome was the original term used to describe cases of brain tumors occurring with gastrointestinal polyps and actual cancerous tumors. However, today, it is understood that there are two extremely different cancer syndromes that involve these manifestations. The inheritance patterns and factors involved are also distinct. Turcot syndrome was first described in 1959 by the Canadian surgeon *Jacques Turcot*. In addition, the early development of colon cancer and gliomas has been documented as occurring in relation to Li–Fraumeni syndrome, which was discussed in Chapter 19, Li–Fraumeni Syndrome.

> **Point to remember**
> Turcot syndrome is a variation of a polyposis syndrome. It is characterized by multiple colonic polyps, with an increased risk of colon cancer and primary brain tumors.

Brain tumor-polyposis syndrome 1

Brain tumor-polyposis syndrome 1 (BTP1) is also known as *mismatch repair cancer syndrome*. It is an autosomal dominant syndrome that has decreased penetrance. The syndrome is associated with biallelic DNA mismatch repair mutations. Neoplasia usually occurs in the central nervous system and the gastrointestinal system.

Epidemiology

There have been more than 200 cases of BTP1 reported, but the syndrome is underdiagnosed. It is most prevalent in Middle Eastern and South Asian countries, where **consanguinity** is high. Brain tumors—usually malignant gliomas—occur in the first two decades of life, making up 25%–40% of all BTP1 cancers. In one specific study, median patient age at development of glioblastoma in BTP1 was 18 years, while the peak incidence in the general population is between 40 and 70 years.

Etiology and risk factors

BTP1 is caused by biallelic mutations in one of the four **mismatch repair genes**, which include *MLH1, MSH2, MSH6, and PMS2*. MLH1 is also known as *MutL homolog1*. MSH2 and MSH6 are also known as *MutS homolog 2* and *MutS homolog 6*, respectively. PMS2 is also known as *PMS1 protein homologue 2*. Family history of **Lynch syndrome** is usually not a significant factor for BTP1 or Lynch-related cancers. The genetic defect that underlies BTP1 is the inability to recognize or repair DNA mismatches during replication. The genes causing BTP1 are as follows:

- *MLH1*—at chromosome 3p21.3;
- *MSH2*—at chromosome 2p16;
- *MSH3*—at chromosome 5q11-q13;
- *MSH6*—at chromosome 2p16;
- *PMS1*—at chromosome 2q32; and
- *PMS2*—at chromosome 7p22; the **C-terminus** of PMS2 interacts with MLH1, and the PMS2–MLH1 complex binds to MSH2/MSH6 **heterodimers**, forming a functional mismatch recognition complex that is specific to these DNA strands.

The recognition and repair of base-pair mismatches are based on heterodimers of MSH2 and MSH6, since they form a *sliding clamp* on human DNA. Cells deficient in any of the above genes that cause BTP1 become defective in the repair of mismatched bases. They are also defective in repair of insertions/deletions of single nucleotides. This results in high rates of mutation, and also in microsatellite instability, though cancers of the BTP1 syndrome often do not have microsatellite instability. Instead, they are characterized by very high rates of single-nucleotide mutations. This is different from the microsatellite instability seen in all cancers developing in heterozygous carriers. Other than gene mutations, there are known no risk factors.

Pathology

The pathology of BTP1 is different from that of Lynch syndrome, which involves development of primarily colon and genitourinary cancers in adults. Instead, BTP1 features development of multiple brain tumors and other malignancies in childhood. Most gene mutations in this syndrome cause loss of gene expression. Many brain gliomas that occur as part of BTP1 have prominent nuclear pleomorphism, plus **multinucleation** that is similar to that of pleomorphic xanthoastrocytoma or giant cell glioblastoma. Recognizing these factors may result in immunohistochemical testing for the loss of mismatch repair proteins. Other types of tumors that have been seen include low-grade gliomas such as oligodendrogliomas and pleomorphic astrocytomas. Additionally, medulloblastomas and primitive neuroectodermal tumors have been documented, sometimes including glial features.

At the molecular levels, the tumors have a specific ultra-hypermutation phenotype. This makes them stand out as different from other childhood tumors. The genotype and phenotype are difficult to determine because of the rarity of BTP1. Mutations of PMS2 and MSH6 may predominate, while MSH2 germline mutations are rare. This is different from Lynch syndrome, in which MLH1 and MSH2 germline mutations are the most common. Heterozygous carriers are usually not affected by BTP1.

Clinical manifestations

Over 90% of patients have café-au-lait macules, plus other skin abnormalities such as axillary freckles or hypopigmentation. It is important to distinguish these from the skin lesions that are related to neurofibromatosis type 1. In up to 30% of patients, hematological malignancies occur in the first decade of life—usually T-cell lymphomas. However, gastrointestinal polyposis and cancers—often adenocarcinomas—manifest in nearly all patients by the second decade of life. Abdominal pain, blood in the stool, diarrhea, and melena are common manifestations. Sarcomas and urinary tract cancers have also developed. Gliomas including glioblastomas and astrocytomas may cause limping, gait abnormalities, and hemiparesis. There is often perifocal edema in the brain, and tumors may have intense ring-enhancing rims. Neurological symptoms include headache, nausea, vomiting, and seizures.

Diagnosis

The diagnosis of BTP1 is based on the presence of café-au-lait macules, consanguinity, and certain tumors of the brain, gastrointestinal tract, and blood—during childhood. In recent years, a diagnostic scoring system has been developed in relation to genetic testing for BTP1. The diagnosis is based on detection of a germline biallelic mutation in a mismatch repair gene. However, the *PMS2* gene has many variants that are not understood, and also has technical problems in its sequencing due to many pseudogenes. Because of this, several functional assays have been developed to provide rapid detection of mismatch repair gene deficiencies. Microsatellite instability is not a reliable test for BTP1, even though it is reliable for Lynch syndrome. Immunohistochemical staining reveals loss of expression of the protein that is encoded by the gene, in the tumor and in normal tissue, in more than 90% of patients. Cell-based assays on normal lymphoblasts and fibroblasts are able to detect microsatellite instability, the failure to repair guanine—thymine (G-T) mismatches, and the resistance to a variety of compounds.

Treatment

Patients with BTP1, and their families, may benefit from genetic counseling. There are surveillance protocols in existence, so early detection may increase survival, regardless of whether the carrier is biallelic or heterozygous. The mismatch repair-deficient cells have a high resistance to chemotherapies such as temozolomide. This must be evaluated in the treatment of gliomas that are related to BTP1. Oppositely, the ultra-hypermutation phenotype of cancers related to BTP1 may respond to immune checkpoint blockade. Treatment options include surgical resection, chemotherapy, and radiation therapy.

Prognosis

Patients with BTP1-related glioblastomas have an average survival of more than 27 months. This is much longer than in sporadic cases of glioblastoma (1 year). Many long-term survivors have a biallelic germline PMS2 mutation, and have lived more than 10 years after treatment for their gliomas.

> **Point to remember**
>
> With brain tumor-polyposis syndrome 1, children are likely to also developing cancers of the blood and lymphatic system, uterine or ovarian cancer, and various rare pediatric cancers. The syndrome can cause more than one type of cancer to develop at a time, or at different ages throughout a patient's lifetime.

Brain tumor-polyposis syndrome 2

Brain tumor-polyposis syndrome 2 (BTP2) is also an autosomal dominant syndrome. Like BTP1, this syndrome is also associated with multiple tumors of the GI tract, but the primary type of brain tumor is usually a medulloblastoma, and not a glioma. In the GI tract, tumors usually form mainly in the epithelium of the large intestine. They can undergo malignant transformation into colon cancer. There are three variants that have been identified, as follows:

- *Familial adenomatous polyposis (FAP)*—the most severe and common form, with hundreds or thousands of polyps and extremely high likelihood of developing colon cancer.
- *Attenuated familial adenomatous polyposis*—the APC gene is functional but slightly impaired; this form carries a likely 70% lifetime risk of cancer, but there are usually about 30 colon polyps instead of hundreds or thousands; it arises at an age when FAP is not likely to develop—usually at an average of 55 years.
- *Autosomal recessive familial adenomatous polyposis*—also known as *MYH-associated polyposis* since it is caused by defects in the *MutY homolog (MUTYH) gene* on chromosome 1; this is a milder form of the disease, and only occurs when both parents are carriers.

The MUTYH gene encodes the DNA repair enzyme called **MYH glycosylase**. When this enzyme does not function normally, DNA errors may increase, initiating the development of tumors. Mutations of the gene are inherited in an autosomal recessive pattern.

Epidemiology

BTP2 causes about 1% of all colon cancers. However, brain tumors are actually rare in BTP2, and make up less than 1% of all malignancies of patients with the syndrome. The median patient age of BTP2-related medulloblastomas is 15 years, which is the same as for sporadic medulloblastomas. The incidence of the causative gene mutation is between 1 in 10,000 and 15,000 births. By age 35, about 95% of patients will have colon polyps, with colon cancer being almost certain to develop. In untreated individuals, the mean age of colon cancer is 39 years.

Etiology and risk factors

BTP2 is caused by heterozygous mutations of tumor suppressor gene called *adenomatous polyposis coli (APC)*. The gene is located on chromosome 5q21, and is a major tumor suppressor of the **WNT pathway**, which was named from the terms "Wingless" and "Int-1." Wnt signaling pathways use nearby cell-to-cell communication (paracrine) or same-cell communication (autocrine)—see Fig. 20.1. The three Wnt signaling pathways are known as the following:

- *Canonical Wnt pathway*—leads to regulation of gene transcription.
- *Noncanonical planar cell polarity pathway*—regulates the cytoskeleton, which is responsible for cell shape.
- *Noncanonical Wnt/calcium pathway*—regulates calcium inside cells.

The activation of a WNT pathway usually occurs because of beta-catenin alterations. This is seen in 10%—15% of all medulloblastomas. However, the link between WNT activation and BTP2-related medulloblastomas is not fully understood. Other than gene mutations, there are no known risk factors.

Pathology

The only clearly associated type of brain tumor with BTP2 is medulloblastoma. This type of tumor is 90 times more likely to develop along with BTP2 than in people of the general population. Though the APC gene regulates beta-catenin, which is important for cell communication, growth, signaling, and controlled destruction, when the gene is uncontrolled, it can allow numerous cancers to form. In the attenuated form of BTP2, the polyps are sometimes flat instead of polypoid in their morphology. They are more proximal to the splenic flexure. Gastric fundic polyps and duodenal adenomas are also seen. This means that polyps and cancers can manifest in the upper portion of the colon or the upper gastrointestinal tract rather than the usual locations.

Clinical manifestations

People with BTP2 are at an extremely high risk for colorectal cancer, with an incidence of malignancy being close to 100%. They may also develop osteosarcomas (an average of 70% of patients), aggressive fibromatosis (average 12.5%), thyroid cancer (average 2.5%), and much less commonly, hepatoblastoma (1%). Fig. 20.2 illustrates these percentages. Patients with BTP2 are often symptomatic, but signs and symptoms can include blood in the stool, iron deficiency anemia, weight loss, alterations of bowel habits, and metastasis to the liver or elsewhere in the body. Polyps can also form

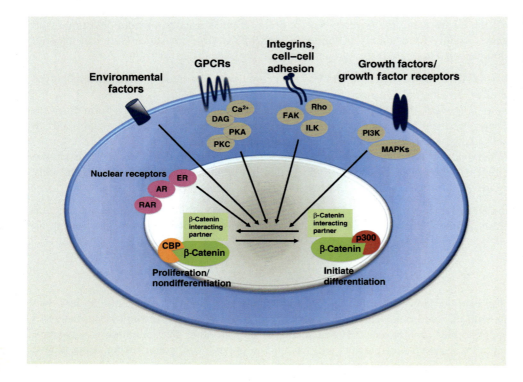

FIGURE 20.1 The WNT pathway. Courtesy: Jia-Ling Teo, Michael Kahn (Advanced Drug Delivery Reviews Volume 62, Issue 12).

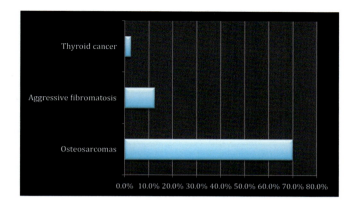

FIGURE 20.2 Average percentages of the likelihood of BTP2 developing into other tumors.

in the duodenal tract. Unfortunately, BTP2 may be silent in its development, until it has become advanced colorectal cancer. Other common cancers that may develop include ampullary adenocarcinoma, jaw cysts, sebaceous cysts, and **osteomata**. When BTP2 develops along with osteomas, fibromas, and sebaceous cysts, it is referred to as **Gardner's syndrome**, and may or may not involve abnormal scarring.

Diagnosis

The hallmark of BTP2 diagnosis is the development of hundreds to thousands of colon polyps relatively early in life. The syndrome is well characterized and has clinical and molecular diagnostic criteria. There are surveillance protocols in place, with most patients requiring a preventive colectomy. To assess more deeply into the gastrointestinal tract, colonoscopy is preferred over sigmoidoscopy since it provides better observation of the common right-sided location of the polyps. Because brain tumors are rare, surveillance does not include brain MRI. Genetic testing provides the ultimate diagnosis, in 95% of cases. Individuals may be diagnosed as being "at risk" for BTP2, requiring routine monitoring to assess for development of the syndrome. Monitoring may involve providing outpatient colonoscopy, and sometimes, upper gastric tract esophagogastroduodenoscopy, usually every 1–3 years.

The typical core diagnostic criteria for FAP is 100 or more polyps when the patient is under age 40, or the patient is of any age, has polyps, and the condition is also present in a family member. For the attenuated form, diagnostic criteria are indefinite. Basically, there are two general sets of criteria, as follows:

- No family history of 100 or more polyps before age 30, plus one of the following:
  - 10–99 polyps,
  - 100 or more polyps with patient age of 35–40, or
  - colorectal cancer before age 60 and relatives with multiple adenomatous polyps.
- Family history of 10–99 adenomas diagnosed after age 30.

Treatment

There are no specific therapies recommended for BTP2-related medulloblastomas. Treatment for the other symptoms of BTP2 is based on the genotype. Prophylactic surgery may be recommended before age 25, or upon detection of polyps or tumors. Prophylactic colectomy is indicated when there are more than 100 polyps, extremely dysplastic polyps, or there are multiple polyps larger than 1 cm in diameter. There are also several surgical options that involve removal of the colon or the rectum as well. Medications available that can significantly decrease the number of polyps include the nonsteroidal antiinflammatory drugs.

Prognosis

The prognosis for BTP2-related medulloblastomas is generally favorable, but it's not known if this is also true for WNT-activated medulloblastomas. A poorer prognosis may be indicated by the appearance of medulloblastoma in a young patient who lacks evidence of gastrointestinal polyps. When BTP2 is detected and treated early, or the polyps are only inside the gastrointestinal tract, there is a high rate of success regarding prevention or removal of cancer without recurrence.

> **Point to remember**
> Brain tumor-polyposis syndrome 2 results in the likely development of colorectal cancer as well as medulloblastomas of the brain. Most people with this syndrome have mutations of the APC gene on chromosome 5q21, increasing the risk for medulloblastomas. Other types of brain tumors that may develop, but less often, include astrocytomas and ependymomas.

Clinical cases

Clinical case 1

1. What is the other name for brain tumor-polyposis syndrome 1?
2. When do malignant brain tumors usually occur as part of this syndrome?
3. What genetic defect underlies BTP1?

A family had two children affected by what was initially described as Turcot syndrome. The son was referred for a genetic evaluation because he had developed lymphoblastic lymphoma at age 5 and invasive adenocarcinoma of the colon at age 8. There was no evidence of familial adenomatous polyposis. However, café-au-lait macules and axillary freckles were noticed during examination. There were no neurofibromas or Lisch nodules seen. Unfortunately, the boy died at the age of 9 years. His sister was referred for evaluation of a brain tumor. At age 8, she developed a right-sided limp and gait abnormalities. An MRI revealed a mass in the left temporal parietal region, as well as lesions in the left frontal lobe. A needle biopsy was consistent with an infiltrative astrocytic neoplasm. However, after resection, the final diagnosis was glioblastoma multiforme. The girl unfortunately also died at the age of 10, of progressive disease from the glioblastoma. The other two siblings of these children had no incidence of cancer, nor did their parents. Based on the brain tumors, colon adenocarcinoma, a clinical diagnosis of brain tumor-polyposis syndrome 1 (BTP1) was made.

Answers:

1. **This syndrome is also known as mismatch repair cancer syndrome. It is associated with biallelic DNA mismatch repair mutations.**
2. **Brain tumors—usually malignant gliomas—usually occur in the first two decades of life with BTP1. They make up 25%–40% of all BTP1 cancers.**
3. **The genetic defect that underlies BTP1 is the inability to recognize or repair DNA mismatches during replication. The causative genes are MLH1, MSH2, MSH3, MSH6, PMS1, and PMS2.**

Clinical case 2

1. What is unique about the gene mutations seen in BTP1?
2. How does BTP1 differ from Lynch syndrome?
3. What types of brain tumors are most common in BTP1?

An 11-year-old boy presented with headache over a week, vomiting, and history of a single tonic–clonic seizures. He also had vague abdominal pain over about 1 month prior to evaluation. His family history was positive for colon cancer. His sister had multiple café-au-lait spots and colon polyps. During examination, his skin was found to have multiple café-au-lait spots and areas of hypopigmentation. A brain CT scan revealed a 4-cm lesion in the left frontal lobe, with perifocal edema. The tumor had an intense ring-enhancing rim. Gross total resection was performed, and the tumor was diagnosed as glioblastoma multiforme. The patient developed diarrhea, melena, and intermittent bleeding from the rectum. Colonoscopy revealed multiple colonic polyps of various sizes. A punch biopsy of the polyps showed grade II adenocarcinoma. A molecular study revealed mutation of the DNA mismatch repair gene. He was diagnosed with BTP1. For the adenocarcinoma, he underwent a total colectomy procedure. One month later, rectal bleeding recurred and new rectal polyps developed, which were treated by additional surgery and chemotherapy. He also underwent radiation therapy for the brain tumor. Within 9 months, the patient developed chronic headache and right hemiparesis. There was now a left temporal ring-enhancing lesion of 5 cm in size, revealed by CT and MRI. The lesion was surgically resected and diagnosed again as a glioblastoma. The patient was started on temozolomide as palliative therapy, but additional tumors developed, and there was a massive progression of colonic adenocarcinoma. The boy died within 13 months of the initial diagnosis.

Answers:

1. The gene mutations seen in BTP1 are characterized by very high rates of single-nucleotide mutations. This is different from the microsatellite instability seen in all cancers developing in heterozygous carriers.
2. BTP features development of multiple brain tumors and other malignancies in childhood. Lynch syndrome involves development of primarily colon and genitourinary cancers in adults.
3. Gliomas of various forms are common with BTP1. Many gliomas have prominent nuclear pleomorphism plus multinucleation that is similar to that of pleomorphic xanthoastrocytoma or giant cell glioblastoma. Other types of tumors include oligodendrogliomas, pleomorphic astrocytomas, medulloblastomas, and primitive neuroectodermal tumors.

Clinical case 3

1. How early do gastrointestinal polyposis and cancers such as adenocarcinomas usually manifest in BTP1?
2. What does immunohistochemical staining reveal?
3. How do the cells react to various treatment methods?

A 12-year-old boy developed headache and nausea, accompanied by bright red blood in his stool. Left-side hemiparesis and a more intense headache had developed by the time of examination. A CT scan revealed a cystic tumor in the right parietal region. Surgical resection was performed, with the diagnosis of a glioblastoma. The boy received radiation therapy for 2 months, and the rectal bleeding continued. Through biopsy, a sigmoid colon polyp was revealed, further diagnosed as a proliferating tubular adenoma. A recurrent brain tumor developed within a few months, and during evaluation of this tumor, the patient's family discussed that they had no knowledge of any brain or colon cancer occurring in their relatives. An MRI revealed a cystic tumor in the same area as the previous tumor. An additional examination of his colon revealed a giant polyp that completely overlapped the lumen of the sigmoid colon. The polyp was able to be totally removed, again diagnosed as a tubular adenoma. There were no signs of malignancy. Polypoid formations up to 15 mm in size were defined throughout the sigmoid colon, numbering about 30. The recurrent brain tumor was also removed successfully, again diagnosed as a glioblastoma. The postoperative period was uncomplicated, and the boy was discharged, with regular follow-ups planned.

Answers:

1. Gastrointestinal polyposis and cancers—often adenocarcinomas—manifest in nearly all patients with BTP1 by the second decade of life.
2. Immunohistochemical staining reveals loss of expression of the protein encoded by the PMS2 gene, in the tumor and in normal tissue, in more than 90% of patients.
3. The cells of patients with BTP1 have a high resistance to chemotherapies such as temozolomide. However, the ultra-hypermutation phenotype in relation to BTP1 may respond to immune checkpoint blockade. Treatment options include surgical resection, chemotherapy, and radiation therapy.

Clinical case 4

1. What is the primary type of brain tumor associated with BTP2?
2. What are the three variants of BTP2, and which is most severe and common?
3. How common is BTP2 in comparison to all types of colon cancers?

An 11-year-old girl was confirmed to have multiple colorectal polyps, and to have a positive family history of these cancers. She underwent subtotal colectomy. At 39 years of age, she was found to have hundreds of polyps, with two adenocarcinomas. At age 41, she was referred for genetic testing after developing recurrent rectal cancer. The 2-year follow-up CT scan revealed multiple lung and liver metastases. The patient was treated with combination chemotherapy. A deeper study into the patient's family history showed that her mother had first been diagnosed with rectal cancer at age 24, and by middle age, she had multiple colectomies due to many polyps, one of which was an adenocarcinoma. Additional family members were also screened. The 13-year-old daughter of the original patient eventually developed multiple polyps and adenocarcinoma as well. Additional family members had medulloblastomas along with colorectal cancers. It was determined that this family was an example of having BTP2.

Answers:

1. The primary type of brain tumor associated with BTP2 is usually a medulloblastoma, and not a glioma such as glioblastoma multiforme.
2. The three variants of BTP2 are familial adenomatous polyposis (the most severe and common, with hundreds or thousands of polyps and extremely high likelihood of developing colon cancer); attenuated familial adenomatous polyposis (with a 70% lifetime risk of cancer); and autosomal recessive familial adenomatous polyposis (which only occurs when both parents are carriers).
3. BTP2 causes only about 1% of all colon cancers.

Clinical case 5

1. What is the median patient age of BTP2-related medulloblastomas, and what percentage of patients will have colon polyps?
2. Regarding BTP2, what are the three Wnt signaling pathways?
3. How common are medulloblastomas with BTP2 in comparison to the general population?

A 32-year-old man was diagnosed with the classic form of BTP2. He denied any knowledge of his relatives having colorectal cancer or brain tumors. However, two of his children developed multiple colorectal polyps as well. Signs and symptoms included diarrhea, blood in the stool, weight loss, and weakness. Genetic studies revealed that they all had the APC gene mutation. Over time, the man and one of the children developed a brain tumor. Imaging studies, biopsies, and surgical resection revealed them to both have medulloblastomas.

Answers:

1. The median patient age of BTP2-related medulloblastomas is 15 years. By age 35, about 95% of patients will have colon polyps, with colon cancer being almost certain to develop.
2. The three Wnt signaling pathways related to BTP2 are the canonical Wnt pathway that leads to regulation of gene transcription; the noncanonical planar cell polarity pathway that regulates the cytoskeleton and cell shape; and the noncanonical Wnt/calcium pathway that regulates calcium inside cells.
3. Medulloblastomas are 90 times more likely to develop along with BTP2 than in people of the general population.

Key terms

Consanguinity
C-terminus
Gardner's syndrome
Heterodimers
Lynch syndrome
Mismatch repair genes
Multinucleation
MYH glycosylase
Osteomata
Turcot syndrome
WNT pathway

Further reading

Arnan, M. (2011). *Cancer biology, a study of cancer pathogenesis: How to prevent cancer and diseases*. Xlibris.
Barrett, Q., & Lum, L. (2016). *Wnt signaling: Methods and protocols (Methods in molecular biology)*. Humana Press.
Boardman, L. A. (2018). *Intestinal polyposis syndromes: Diagnosis and management*. Springer.
Bonavida, B. (2015). *Nitric oxide and cancer: Pathogenesis and therapy*. Springer.
Chang, E. L., Brown, P. D., Lo, S. S., Sahgal, A., & Suh, J. H. (2018). *Adult CNS radiation oncology: Principles and practice*. Springer.
Ellis, C. N. (2010). *Inherited cancer syndromes: Current clinical management* (2nd ed.). Springer.
Frank, D. A. (2012). *Signaling pathways in cancer pathogenesis and therapy*. Springer.
Goss, K. H., & Kahn, M. (2011). *Targeting the Wnt pathway in cancer*. Springer.
Gregory, C. D. (2016). *Apoptosis in cancer pathogenesis and anti-cancer therapy: Perspectives and opportunities (Advances in experimental medicine and biology)*. Springer.
Gregory, J. E. (2012). *Pathogenesis of cancer*. Literary Licensing, LLC.
Gupta, N., Banerjee, A., & Hass-Kogan, D. A. (2017). *Pediatric CNS tumors (Pediatric oncology)* (3rd ed.). Springer.
Hayat, M. A. (2013). *Tumors of the central nervous system, Volume 13: Types of tumors, diagnosis, ultrasonography, surgery, brain metastasis, and general CNS diseases*. Springer.
Hoppler, S. P., & Moon, R. T. (2014). *Wnt signaling in development and disease: Molecular mechanisms and biological functions*. Wiley-Blackwell.

Kelley, M. R., & Fishel, M. L. (2016). *DNA repair in cancer therapy: Molecular targets and clinical applications* (2nd ed.). Academic Press.

Lindon, J. C., Nicholson, J. K., & Holmes, E. (2018). *The handbook of metabolic phenotyping*. Elsevier.

Litchman, C. (2012). *Desmoid tumors*. Springer.

Low, V. H. S. (2012). *Gastrointestinal imaging: Case review series* (3rd ed.). Saunders.

Mahajan, A., & Paulino, A. (2018). *Radiation oncology for pediatric CNS tumors*. Springer.

Mercier, I., Jasmin, J. F., & Lisanti, M. P. (2012). *Caveolins in cancer pathogenesis, prevention and therapy (Current cancer research)*. Springer.

Musella, A. (2014). *Brain tumor guide for the newly diagnosed* (9th ed.). Musella Foundation for Brain Tumor Research & Information, Inc.

National Comprehensive Cancer Network. (2018). *NCCN guidelines for patients: Colon cancer*. National Comprehensive Cancer Network (NCCN).

National Comprehensive Cancer Network. (2019). *NCCN guidelines for patients: Rectal cancer*. National Comprehensive Cancer Network (NCCN).

Nose, V. (2019). *Diagnostic pathology: Familial cancer syndromes* (2nd ed.). Elsevier.

Nose, V., Greenson, J. K., & Paner, G. P. (2013). *Diagnostic pathology: Familial cancer syndromes*. Lippincott Williams & Wilkins.

Pennisi, C. P., Prasad, M. S., & Rameshwar, P. (2017). *The stem cell microenvironment and its role in regenerative medicine and cancer pathogenesis (Series in research and business chronicles: Biotechnology and medicine)*. River Publishers.

Rajendran, J., & Manchanda, V. (2010). *Nuclear medicine cases (McGraw-Hill Radiology series)*. McGraw-Hill Education/Medical.

Tonn, J. C., Reardon, D. A., Rutka, J. T., & Westphal, M. (2019). *Oncology of CNS tumors* (3rd ed.). Springer.

Strickland, J., & Green, E. (2011). *Lynch syndrome: Tests, causes and treatments*. CreateSpace Independent Publishing Platform.

Vincan, E. (2008). *Wnt signaling: Volume 1: Pathway methods and mammalian models (Methods in molecular biology)*. Humana Press.

Vogelsang, M. (2013). *DNA alterations in Lynch syndrome: Advances in molecular diagnosis and genetic counseling*. Springer.

Wahlsten, D. (2019). *Genes, brain function, and behavior: What genes do, how they malfunction, and ways to repair damage*. Academic Press.

Williams, C. K. O. (2019). *Cancer and AIDS: Part II: Cancer pathogenesis and epidemiology*. Springer.

Chapter 21

Von Hippel–Lindau disease

Chapter Outline

Von Hippel–Lindau disease 389
Clinical cases 394
Key terms 396
Further reading 396

Von Hippel–Lindau disease

Von Hippel–Lindau disease is an autosomal dominant disorder. The disease is characterized by the following: clear cell renal cell carcinoma (RCC), capillary **hemangioblastoma** of the *central nervous system* (CNS) and retinas, inner ear tumors, adrenal gland pheochromocytoma, and pancreatic tumors. Swedish pathologist and bacteriologist *Arvid Lindau* described capillary hemangioblastomas, linking them to retinal vascular tumors, which were described earlier by German ophthalmologist *Eugen von Hippel*. Lindau also pointed out the relationship between capillary hemangioblastomas and tumors of the kidneys and other visceral organs. The von Hippel–Lindau (VHL) tumor suppressor gene, which is implicated in the disease, was first identified in 1993.

Epidemiology

VHL disease is estimated to affect 1 out of every 36,000 to 45,500 people, on an annual basis, globally. There is over 90% penetrance by age 65. The comparisons between VHL-related hemangioblastomas and sporadic hemangioblastomas are illustrated in Table 21.1.

Etiology and risk factors

The VHL disease is caused by germline mutations of the VHL tumor suppressor gene. This gene is located on chromosome 3p25–26, and has three exons, plus a coding sequence of 639 nucleotides. The VHL germline mutations are spread over the three exons. The most common type of mutation is a *missense mutation*. However, other mutations include microdeletions/insertions, nonsense mutations, large deletions, and splice-site mutation*s*. The VHL gene mutations are also common in sporadic hemangioblastomas and RCCs. Also, the VHL tumor suppressor protein is highly important for cellular oxygen sensing (see Fig. 21.1). The VHL gene is expressed in many tissues, but mostly in the epithelial skin cells. It is also expressed in the endocrine and exocrine organs, and the gastrointestinal, respiratory, and urogenital tracts. Within the CNS, VHL protein immunoreactivity is strong in the neurons, including the Purkinje cells within the cerebellum.

TABLE 21.1 Comparisons between VHL-related and sporadic hemangioblastomas.

Factors	VHL-related	Sporadic
Patient age	23 years (7-64)	44 years (7-82)
Female	56%	41%
Multiple	65%	5%
Intracranial	73%	79%
Spinal	75%	11%

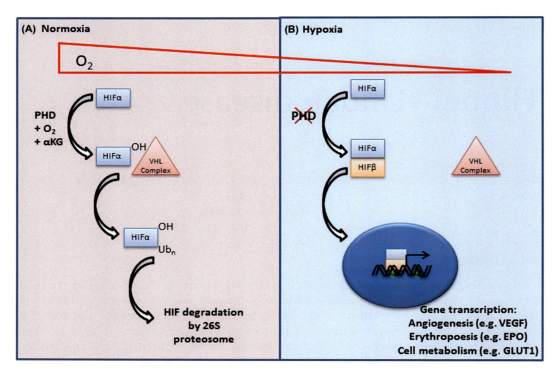

FIGURE 21.1 The VHL tumor suppressor gene in relation to cellular oxygen sensing. (A) Normoxia. (B) Hypoxia.

Mutational inactivation of the VHL gene in family members who are affected makes them genetically susceptible to development of tumors in various organs. However, there is only partial understanding of how inactivation or loss of the VHL protein causes neoplastic transformation. The role of the VHL protein in actual protein degradation and angiogenesis is linked to a specific signaling pathway. The alpha domain of the protein creates a complex with the hydrophobic scaffold protein *cullin-2, RING-box protein 1*, and *transcription elongation factor B polypeptides* 1 and 2. This complex has ubiquitin ligase activity. It targets cellular proteins for ubiquitination as well as proteasome-regulated degradation. The VHL gene's domain that is involved in binding to transcription elongation factor B (also known as **elongin**) is often mutated in VHL-related neoplasms.

The VHL protein plays an important role in cellular oxygen sensing by targeting factors that bring about hypoxia. These factors mediate the cellular responses to hypoxia. Their targeting is linked to ubiquitination and **proteasomal degradation**. The VHL protein's beta-domain interacts with *hypoxia-inducible factor 1-alpha (HIF1A)*. As the protein's hydroxylated subunit binds, this causes **polyubiquitination**, targeting HIF1A for proteasome degradation. If hypoxia is present, or there is a lack of functional VHL, HIF1A will accumulate, activating transcription of hypoxia-inducible genes (see Fig. 21.2). These include:

- *EPO*—the erythropoietin gene
- *PDGFB*—platelet-derived growth factor subunit B
- *TGFA*—transforming growth factor alpha
- *VEGF-A*—vascular endothelial growth factor A

Overexpression of VEGF explains why VHL-related neoplasms have such a large capillary component. This growth factor may be therapeutically relevant by using neutralizing antiVEGF antibody. The induction of erythropoietin sometimes causes paraneoplastic erythrocytosis in people with CNS hemangioblastoma or kidney cancer.

Study of RCC cell lines indicate that the VHL protein affects control of cell cycle exit, such as the transition from the *gap 2 (G_2) phase* to the inactive G_0 phase. This may occur by prevention of the accumulation of the cyclin-dependent kinase inhibitor known as *cyclin-dependent kinase inhibitor 1B (CDKN1B)*. Tumor-derived VHL protein does not bind to fibronectin, but *wildtype* **VHL protein** does. Accordingly, the carcinoma cells have a poorly assembled extracellular fibronectin matrix. The VHL protein-deficient tumor cells have a much higher chance of being invaded.

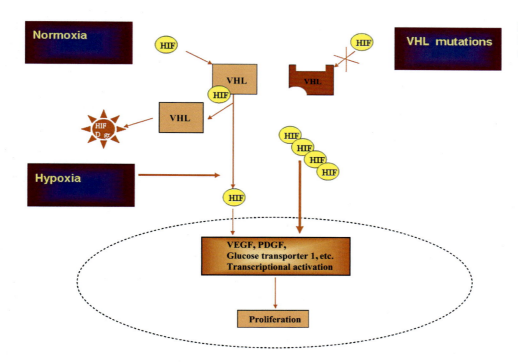

FIGURE 21.2 Causative links between the VHL protein, hypoxia, and development of VHL disease.

This is through down-regulation of the cells' response to **hepatocyte growth factor or scatter factor**, and reduced levels of *tissue inhibitor of metalloproteinases 2*. The causative germline mutations of VHL include the following:

- *Type 1*—commonly, there are hemangioblastomas and RCCs; pheochromocytomas are rare or absent; this type is usually caused by deletions, missense mutations, or truncations
- *Type 2A*—has a high risk for hemangioblastomas and pheochromocytomas; RCCs are rare; this type is caused by missense mutations
- *Type 2B*—has a high frequency of hemangioblastomas, pheochromocytomas, and RCCs; this type is usually caused by missense mutations
- *Type 2C*—has frequent pheochromocytomas, but a lack of hemangioblastomas and RCCs; this type is caused by VHL missense mutations, but uniquely, show no hypoxia-inducible factor dysregulation; along with VHL functioning as a tumor suppressor gene, VHL gene mutations are common in up to 78% of sporadic hemangioblastomas, and are widespread in clear cell RCCs.

Since VHL disease is genetic, there are no other identified risk factors.

Pathology

In VHL disease, while nonneoplastic cysts often form in the kidneys or pancreas, a variety of different tumors occur in these organs, as well as in a large variety of other areas. Kidney lesions and tumors occur in carriers of VHL germline mutations, and are usually multifocal and bilateral. They usually manifest at a mean patient age of 37 years, with onset between 16 and 67 years. This is different from the sporadic clear cell RCCs, which usually manifest at approximately 61 years. By age 70, there is a 70% chance of developing clear cell RCCs. Hemangioblastomas of the retinas manifest earlier, at a mean age of 25 years. Therefore, they are more likely to be diagnosed earlier as well. In the CNS, hemangioblastomas develop at a mean age of 29 years, and are mostly located in the cerebellum. They are also seen in the brain stem and spinal cord. About 25% of all cases are linked to VHL disease.

In the adrenal glands, pheochromocytomas may be difficult to manage, especially in families with VHL that are predisposed to these types of tumors. Often, adrenal pheochromocytomas are linked to pancreatic cysts. Additional extrarenal manifestations include:

- Broad ligament cystadenomas
- Epididymal cystadenomas

TABLE 21.2 Areas of tumors and lesions in VHL disease.

Area	Tumors	Lesions (nonneoplastic)
Adrenal glands	Pheochromocytoma	N/A
Central nervous system	Hemangioblastoma	N/A
Epididymis	Papillary cystadenoma	N/A
Eyes (retinas)	Hemangioblastoma	N/A
Inner ears	Endolymphatic sac tumors	N/A
Kidneys	Clear cell renal cell carcinoma	Cysts
Pancreas	Neuroendocrine islet cell tumors	Cysts

- Inner ear endolymphatic sac tumors
- Neuroendocrine tumors

Areas where tumors and lesions linked to VHL commonly develop are summarized in Table 21.2.

Clinical manifestations

Signs and symptoms associated with VHL disease include headaches (including migraines), balance and walking problems, dizziness, limb weakness, vision problems, seizures, and hypertension. Related conditions include angiomatosis, hemangioblastomas, pheochromocytoma, RCC, pancreatic serous cystadenoma, bilateral papillary cystadenomas of the epididymis, bilateral papillary cystadenomas of the broad ligament of the uterus, and endolymphatic sac tumors. Angiomatosis occurs in about 37% of affected patients, usually within the retina. Loss of vision is common. Common additional symptoms include strokes, heart attack, and cardiovascular disease. About 40% of VHL disease patients present with CNS hemangioblastomas, but they are actually present in between 60% and 80% of cases. Spinal hemangioblastomas are found in 13%-59% of patients. Though all of these tumors are common as part of VHL disease, about 50% of cases have only one type of tumor.

Diagnosis

VHL disease is clinically diagnosed on the basis of capillary hemangioblastoma in the CNS or retina, plus the presence of a typical VHL-related extraneural tumor, or a positive family history. Germline VHL mutations are nearly always seen as part of this disease. Detection of tumors specific to VHL disease is diagnostically important, and mostly involves CT and MRI. If there is a positive family history of the disease, just a single hemangioblastoma, pheochromocytoma, or RCC may be enough to make the diagnosis. If there is no family history, at least two tumors must be identified to make the diagnosis. Imaging often reveals significantly enhancing lesions that are hypointense to isointense on T1-weighted imaging (see Fig. 21.2). They are often hyperintense on T2-weighted imaging, with intense contrast enhancement. For abdominal lesions, ultrasound may be used for diagnosis. For retinal tumors, funduscopy is often used. In hereditary VHL disease, **Southern blot** and **gene sequencing** techniques can be used, to screen family members of diagnosed patients. *De novo* cases producing genetic mosaicism are harder to detect since mutations are not found in white blood cells used for analysis (see Fig. 21.3).

> **Point to remember**
> Ongoing studies of VHL disease are focused on many different areas. These include visualizing vascular endothelial growth factor-producing lesions, the effects of various drugs upon hemangioblastomas as part of the disease, use of contrast-enhanced ultrasound as a screening tool, deeper genetic epidemiology, and assessment of residual VHL function that may predict individual patients' disease courses.

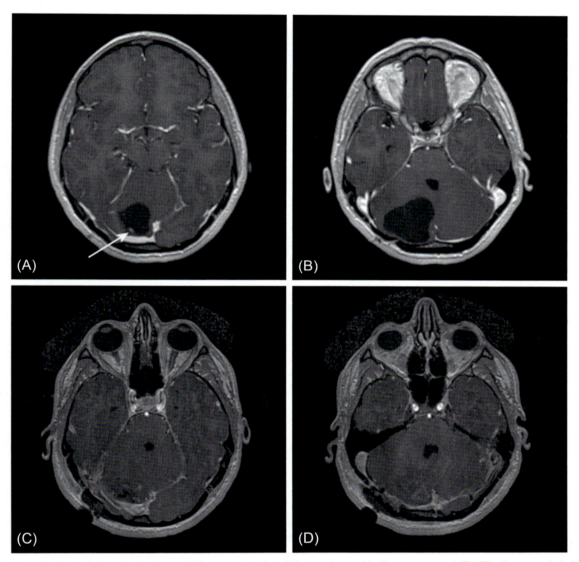

FIGURE 21.3 Dural-based frontal lobe hemangioblastoma, showing different views with T1, contrast, and FLAIR. *Courtesy: Rafael Antonio Vincente Lacerda, Antonio Gilvan Teixeria Jr., Euler, Nicolau Sauaia Filho, et al. (World Neurosurgery Volume 129).*

Treatment

The patient with a VHL germline mutation must have continued medical and genetic counseling. If a retinal or CNS hemangioblastoma has developed, especially in younger patients or when the lesions are multiple, analyses for germline mutations of the VHL gene are important. This allows for prompt detection of VHL-related tumors. Periodic screening of VHL patients is mandatory. This begins with retinoscopy at age 5, and with MRI of the CNS and abdomen beginning at age 10. The earlier the diagnosis and treatment, the better. Hemangioblastomas of the CNS that are symptomatic are usually surgically removed partially or totally. Propranolol has been used for relief of migraine headaches, and has been shown to prevent additional tumor growth, but adverse effects such as orthostatic hypotension may limit its use. For symptomatic retinal angiomas, laser **photocoagulation** therapy and cryotherapy are commonly used, and sometimes, antiangiogenic treatments are helpful. Kidney tumors can be removed by partial nephrectomy or radiofrequency ablation.

Prognosis

Metastatic RCC is the leading cause of death, in up to 75% of cases, from VHL disease. About 40% of patients develop this carcinoma, and about 10% of patients have this carcinoma upon diagnosis of VHL disease. By age 60, the risk for RCC is about 70% in VHL patients. Median life expectancy for VHL patients is 49 years. Diligent surveillance helps to increase life expectancy. Morbidity varies, based on the organ system involved and the extent of progression. CNS hemangioblastomas are the second most common cause of morbidly and mortality in VHL patients. About 70% of patients develop these tumors.

> **Point to remember**
> Parents with VHL disease have a 50% chance of passing it on to their children. It is a lifelong disease that results in the body's cells mutating and acting as if they have an insufficient amount of oxygen. The latest information on VHL can be found on *MyVHL: The Patient Natural History Study*, which is located online at: http://www.vhl.org/MyVHL.

Clinical cases

Clinical case 1

1. What are the clinical hallmarks of VHL disease?
2. Which gene is implicated in this disease?
3. How common is this disease, globally?

A 24-year-old man presented with weakness in his right lower leg that had persisted for about 1 month. Neurological evaluation revealed compressive myelopathy at cervical vertebral levels C6-C7. An MRI revealed a cystic mass in the left cerebellar hemisphere, and a peripheral nodule showing significant contrast enhancement. There were also enhancing intramedullary mass lesions in the spine at the C2-C3, D1-D2, and D9-D10 vertebral levels. The lesions ranged in size between 5 mm and 13 mm. They were hypointense to isointense on T1-weighted imaging, and hyperintense on T2-weighted imaging with intense contrast enhancement. An initial diagnosis of cerebellar and spinal hemangioblastomas was made. Further imaging, using ultrasound, revealed multiple cysts in both kidneys, of between 3 mm and 13 mm in diameter. A diagnosis of VHL disease was then made, and funduscopy also revealed multiple angiomas in the right eye.

Answers:

1. **VHL disease is signified by retinal and CNS hemangioblastomas, pheochromocytomas, cysts in the pancreas and kidneys, and a high risk for malignant transformation of kidney cysts into RCC. There are also cases involving tumors of the inner ears.**
2. **The VHL tumor suppressor gene is implicated in this disease. It was first identified in 1993.**
3. **VHL disease is estimated to affect 1 out of every 36,000-45,500, on an annual basis, globally.**

Clinical case 2

1. Are VHL-related hemangioblastomas more common in younger or older patients in comparison to sporadic hemangioblastomas?
2. Are VHL-related hemangioblastomas more likely to be multiple in number than sporadic hemangioblastomas, and where do they occur most often?
3. What germline mutations are implicated in VHL?

A 32-year-old man presented with sudden onset of blurred vision in his left eye. Examination also revealed bilateral retinal angiomas and renal tumors. He was treated with laser therapy for the retinal angiomas. A CT scan of his abdomen revealed the renal tumors to be between 1 cm and 4 cm in diameter, and there were also multiple pancreatic cysts. The patient underwent a right partial nephrectomy, since the tumor in the right kidney was the largest one. He was closely followed-up. Then, his twin brother was found to also have similar tumors, and underwent surgery for bleeding from a cerebellar hemangioblastoma. Genetic testing revealed both brothers to be positive for VHL disease. Five years later, the initial patient was found to have a left renal tumor of about 4 cm. A brain CT scan was normal. A left partial nephrectomy was performed, and a histological study revealed that the margins were tumor-free. The tumor was a clear cell variant of a low-grade RCC. The patient recovered well from the surgery and continued with follow-up.

Answers:

1. **VHL-related hemangioblastomas have a mean patient age of 23 years (out of a total range of 7-64 years), while sporadic hemangioblastomas have a mean patient age of 44 years (out of a total range of 7-82 years).**
2. **VHL-related hemangioblastomas are multiple in 65% of cases while sporadic hemangioblastomas are multiple only in 5% of cases. VHL-related hemangioblastomas occur in the spinal cord in 75% of cases, and are intracranial in 73% of cases. The reason for these percentages to be similar and nonexclusive of each other is because of the multiple location occurrence.**
3. **The VHL disease is caused by germline mutations of the VHL tumor suppressor gene, which is located on chromosome 3p25–26. The most common type of mutation is a missense mutation.**

Clinical case 3

1. What is the cause of family members developing tumors in various organs in relation to VHL?
2. What are the four subtypes of VHL germline mutations?
3. What is the mean age of development of CNS hemangioblastomas, and where are they mostly located?

A 33-year-old woman was diagnosed with VHL disease, just prior to the death of her mother. An autopsy of her mother revealed that she also had VHL disease. The daughter was examined by a neurologist. A single hemangioblastoma of 2 cm in diameter was found at the medulla oblongata during MRI. The patient was asymptomatic, but underwent surgery because of the risk of complications due to the size of the tumor. She recovered well, and continued to have annual MRIs as follow-up. Seven years later, two hemangioblastomas in the medulla were found after the patient began complaining of migraine headaches. Propranolol was used, in increasing doses, and the migraines began to reduce in severity. After 9 months of treatment, she had significant improvement. During the treatments, the patient underwent a cerebral and spinal MRI that showed no tumor changes from the previous MRI performed 1 year before. Adverse effects from the propranolol appeared, which included orthostatic hypotension. Over time, her doses were reduced and eventually stopped, with the migraines not worsening. However, in few years, clear growth of the tumors was seen in MRI, and additional surgery was performed to remove them. Since that time, the patient remained asymptomatic.

Answer:

1. **Mutational inactivation of the VHL gene in family members who are affected, makes them genetically susceptible to development of tumors in various organs. However, there is only partial understanding of how inactivation or loss of the VHL protein causes neoplastic transformation.**
2. **The four subtypes of VHL germline mutations, include Type 1, 2A, 2B, and 2C. Type 1 usually involves hemangioblastomas and RCCs. Type 2A has a high risk for hemangioblastomas and pheochromocytomas. Type 2B has a high frequency of hemangioblastomas, pheochromocytomas, and renal carcinomas. Type 2C has frequent pheochromocytomas.**
3. **In the CNS, hemangioblastomas develop at a mean age of 29 years, and are mostly located in the cerebellum. They are also seen in the brain stem and spinal cord.**

Clinical case 4

1. What are the four additional extrarenal manifestations of VHL disease, not including CNS tumors, retinal tumors, adrenal tumors, and pancreatic tumors?
2. What are the common signs and symptoms of VHL?
3. How is VHL clinically diagnosed?

A 55-year-old woman had surgical resection of a left cerebellar hemisphere hemangioblastoma. Also, an abdominal CT scan revealed a pheochromocytoma, which was also resected. Genetic analysis revealed a VHL gene mutation, and evaluation of the family history showed VHL disease in two of the patient's siblings. As a result, the patient was diagnosed with VHL disease. After 1 year, an MRI showed no evidence of recurrence or abnormal findings. However, 2 years later, another MRI showed a solid mass with strong enhancement, located in the right cerebellar hemisphere, plus a hyperintense solid mass at the posterior falx. Both lesions were growing very slowly, causing no symptoms, and the patient was followed via MRI for about 4 years. Then, a stereotactic biopsy was performed of the posterior falx lesion, which was diagnosed as a grade I meningothelial meningioma. The lesion was removed, with no postoperative complications. The patient decided on clinical and radiological follow-up.

Answers:

1. **The four additional extrarenal manifestations of VHL include broad ligament cystadenomas, epididymal cystadenomas, inner ear endolymphatic sac tumors, and neuroendocrine tumors.**
2. **The common signs and symptoms of VHL include headaches (including migraines), balance and walking problems, dizziness, limb weakness, vision problems, seizures, and hypertension.**
3. **VHL disease is clinically diagnosed on the basis of capillary hemangioblastoma in the CNS or retina, plus the presence of a typical VHL-related extraneural tumor, or a positive family history.**

Clinical case 5

1. When there is a positive family history of VHL disease, what may be sufficient for diagnosis?
2. Why are analyses for VHL germline mutations important?
3. What is the leading cause of death from VHL disease?

A 40-year-old man presented with a headache, left sided body weakness that had persisted for about 6 months, and seizures that had persisted for about 2 months. He had three relatives who had developed brain tumors, two of which had also developed pheochromocytomas. Neurological examination revealed weakness in his left arm and leg, with no sensory involvement. A contrast-enhanced CT revealed a well-defined, oval, hypodense lesion of about 6 cm in diameter, in the right frontoparietal region. It was highly cystic. The tumor was surgically resected, and histopathology revealed it to be a hemangioblastoma. Follow-up after 3 months showed no recurrence. However, after 2 years, the patient developed tumors in his spinal cord and brain stem. These were also surgically resected, and the patient continued to be tumor-free after 1 year of follow-up.

Answers:

1. **With a positive family history of VHL disease, just a single hemangioblastoma, pheochromocytoma, or RCC may be sufficient to make the diagnosis.**
2. **Analyses for VHL germline mutations are important if a retinal or CNS hemangioblastoma has developed, especially in younger patients or when the lesions are multiple. This allows for prompt detection of VHL-related tumors. Periodic screening of VHL patients is mandatory.**
3. **Metastatic RCC is the leading cause of death, in up to 75% of cases, from VHL disease. About 40% of patients develop this carcinoma, and about 10% of patients have this carcinoma upon diagnosis.**

Key terms

Elongin
Gene sequencing
Hemangioblastoma
Hepatocyte growth factor
Photocoagulation
Polyubiquitination
Proteasomal degradation
Scatter factor
Southern blot
Von Hippel—Lindau disease
Wildtype VHL protein

Further reading

Bunz, F. (2016). *Principles of cancer genetics* (2nd edition). Springer.
Eckerman, A., Kruger, M., Doyle, C., & Chan-Smutko, G. (2009). *VHL handbook kids' edition: A handbook for parents and kids living with von Hippel—Lindau*. VHL Family Alliance.
Ellis, C. N. (2011). *Inherited cancer syndromes: Current clinical management* (2nd edition). Springer.
Fior, R., & Zilhao, R. (2019). *Molecular and cell biology of cancer: When cells break the rules and hijack their own planet (learning materials in biosciences)*. Springer.
Furtado, L. V. (2018). *Precision molecular pathology of neoplastic pediatric diseases (molecular pathology library)*. Springer.
Gorczyca, W. (2010). *Flow cytometry in neoplastic hematology: Morphologic-immunophenotypic correlation* (2nd edition). Informa Healthcare.
Govindan, R., & Devarakonda, S. (2019). *Cancer genomics for the clinician*. Demos Medical.
Hayat, M. A. (2011). *Tumors of the central nervous system, . Astrocytomas, hemangioblastomas, and gangliogliomas* (volume 5). Springer.

Heim, S., & Mitelman, F. (2015). *Cancer cytogenetics: Chromosomal and molecular genetic aberrations of tumor cells (4th edition)*. Wiley-Blackwell.

Hodgson, S. V., Foulkes, W. D., Eng, C., & Maher, E. R. (2014). *A practical guide to human cancer genetics (4th edition)*. Springer.

Kim, I. J. (2017). *Cancer genetics and genomics for personalized medicine*. Jenny Stanford Publishing.

Matloff, E. (2013). *Cancer principles and practice of oncology: Handbook of clinical cancer genetics*. LWW.

Parker, P. M. (2007). *Von Hippel–Lindau syndrome—A bibliography and dictionary for physicians, patients, and genome researchers*. ICON Group International, Inc.

Pichert, G., & Jacobs, C. (2016). *Rare hereditary cancers: Diagnosis and management (recent results in cancer research)*. Springer.

Ruggieri, M., Pascual Castroviejo, I., & Di Rocco, C. (2008). *Neurocutaneous disorders: Phakomatoses & hamartoneoplastic syndromes*. SpringerWien.

Stratakis, C. A., Ghigo, E., & Guaraldi, F. (2013). *Endocrine tumor syndromes and their genetics (frontiers of research, volume 41)*. S. Karger.

VHL Alliance. (2015). *VHL handbook: What you need to know about VHL, edition 5*. CreateSpace Independent Publishing Platform.

VHL Alliance. (2015). *VHL patient vignettes: Personal stories, thoughts, and tips submitted by people dealing with the diagnosis of von Hippel–Lindau*. VHL Alliance.

VHL Family Alliance, & Wilcox Graff, J. (2009). *What you need to know about VHL: A reference handbook for people with von Hippel–Lindau, their families, and support personnel*. VHL Family Alliance.

Weinberg, R. A. (2013). *The biology of cancer (2nd edition)*. W.W. Norton & Company.

Chapter 22

Cowden syndrome

Chapter Outline

Cowden syndrome 399
Clinical cases 403
Key terms 405
Further reading 405

Cowden syndrome

Cowden syndrome is characterized by more than one hamartoma, involving tissues that form from all three of the germ cell layers. It is an autosomal dominant disorder. The syndrome carries with it significant risks of breast, colon, endometrial, renal, and epithelial thyroid cancer. There are also less common developments of lung cancer. Facial **trichilemmomas** are often present. Cowden syndrome was named after *Rachel Cowden*, the first patient who was ever documented with its characteristic manifestations. She died at age 31 of breast cancer. The syndrome was first described in 1963 and is sometimes described as *multiple hamartoma syndrome*. It is a form of **genodermatosis**.

Epidemiology

After identification of the causative gene, prevalence of Cowden syndrome, based on molecular studies, has been estimated to occur in one of every 200,000 people. Before identification of the gene, the estimate was one of every million people. However, this syndrome is difficult to identify, meaning that prevalence may be underestimated. In a recent study, it was estimated that *de novo* mutations of the *phosphatase and tensin homolog (PTEN)* protein may occur in between 11% and 48% of individuals. Cowden syndrome usually manifests during the third decade of life. By this time, about 99% of patients develop mucocutaneous signs, even though the other features of the syndrome may be present. The mean patient age at diagnosis of breast cancer with Cowden syndrome is about 10 years younger than for breast cancer in the general population. The total age range for Cowden syndrome is between 13 and 75 years. Lifetime risks of epithelial thyroid cancer with Cowden syndrome are as high as 10%.

Etiology and risk factors

Cowden syndrome is mainly caused by germline mutations in the PTEN protein. These are usually found to be present in about 85% of Cowden syndrome cases, but less so in a few newer studies. The mutations include intragenic and promoter mutations, and large deletions/rearrangements. The presence of a germline mutation is linked to a familial risk for malignant breast cancer. The PTEN susceptibility gene is located on chromosome 10q23.3. It consists of nine exons that span 120–150 kb of genomic distance. The gene encodes a 1.2 kb transcript, plus a 403-amino acid lipid dual-specificity phosphatase. This dephosphorylates protein as well as lipid substrates. The substrate is homologous to the focal adhesion molecules known as **tensin** and **auxilin**. The amino acid sequence related to these adhesion molecules is encoded by exons 1 to 6. There is a classic phosphatase core recurrence encoded in exon 5, the largest of the exons. The exon makes up 20% of the coding region. There is also a longer isoform of PTEN. This is believed to interact with the mitochondrion, but its effects are unclear.

PTEN is nearly totally expressed, and a few studies have shown that high levels of PTEN protein expression in the CNS, skin, and thyroid are affected by the neoplasias of Cowden syndrome. PTEN is also prominently expressed as the gastrointestinal (GI) tract and autonomic nervous system are developing. The primary substrate of PTEN is called *phosphatidylinositol (3,4,5)-trisphosphate (PIP$_3$)*, a part of the **PI3K pathway**. PI3K activation phosphorylates and activates *protein kinase B (AKT)* to localize it in the plasma membrane, with many downstream effects. When PTEN is in good supply and is functional, PIP3 is converted to *phosphatidylinositol 4,5-bisphosphate (PIP$_2$)*, resulting in **hypophosphorylated** and apoptotic AKT, which is a well-understood factor of cell survival. PTEN in the cytoplasm is mostly

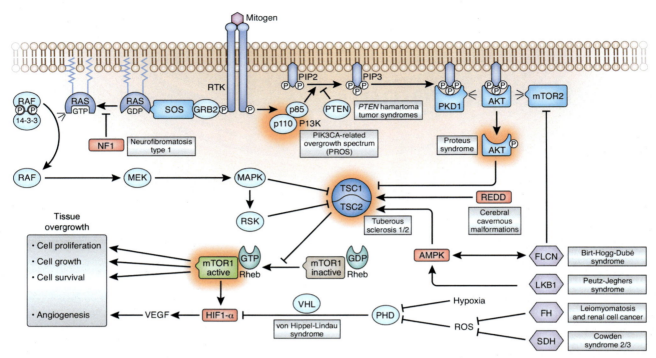

FIGURE 22.1 The complexity of the PI3K-AKT-mTOR and interacting signaling pathways, with associated heritable diseases with skin involvement as a result of mutations in the corresponding genes. The key molecules in the pathway are highlighted by colored overlay. *AMPK*, AMP kinase; *NF1*, neurofibromatosis type 1; *VEGF*, vascular endothelial growth factor.

able to send signals by using its lipid phosphatase activity down the PI3K and AKT pathways (see Fig. 22.1). However, when PTEN is within the nucleus, it mostly signals using protein phosphatase activity down the **cyclin-D1** / *mitogen-activated protein kinase (MAPK)* pathway. This causes a *phase G1* arrest in breast and glioma cells, and possibly other cells. It is believed that PTEN is able to dephosphorylate *focal adhesion kinase (FAK)*, and also inhibit signaling of integrin and MAPK.

There is also a causative link to adult-onset dysplastic cerebellar **gangliocytoma**, which is also known as **Lhermitte−Duclos disease**. In recent studies, additional genes have been identified as predisposing individuals to Cowden syndrome. These include the *succinate dehydrogenase (SDH)* genes, *p53-regulated DNA replication inhibitor killin (KLLN), phosphatidylinositol-4,5-bisphosphate 3-kinase catalytic subunit alpha (PIK3CA)*, and *RAC-alpha serine/threonine-protein kinase (AKT1)*. In one small study of children with *juvenile polyposis of infancy*, deletion of *bone morphogenetic protein receptor type 1A (BM-PR1A)*, the upstream of PTEN, was present. Germline deletion of both of these characterized a subset of juvenile polyposis of infancy.

Predictive risk factors for Cowden syndrome are linked to activation of *mammalian target of rapamycin (mTOR)* signaling. This is an important downstream response to dysfunction or deficiency of PTEN. This mTOR inhibition is effective in both vitro and animal studies. Another syndrome that shares PTEN gene mutations is called *segmental overgrowth lipomatosis arteriovenous malformation epidermal nevus syndrome* or *SOLAMEN syndrome*.

Pathology

Cowden syndrome has varied, broad expression, plus penetrance that is related to age. Dysplastic cerebellar gangliocytoma is an unusual CNS tumor closely linked to Cowden syndrome. Adult-onset Lhermitte−Duclos disease is extremely predictive of a PTEN germline mutation, even without any other features or family history. In patients or families with the syndrome, other benign tumors and malignancies have occurred. It is uncertain whether lymphomas, sarcomas, leukemias, and meningiomas are actual components of Cowden syndrome.

In one study, just 78% of patients with Cowden syndrome had a PTEN germline mutation, but 100% had **hamartomatous** intestinal polyps. These polyps may be of several types. Some are hamartomas that are extremely similar to juvenile polyps made up of mixed connective tissues, which are normally present in the smooth muscle, continuous with the muscularis mucosae. Other types include **ganglioneuromatous** and lipomatous lesions, along with lymphoid hyperplasia. The polyps may develop in the colon, rectum, small intestine, or stomach. Inside the colon or rectum, they are usually between 3 and 10 mm in diameter, but some can be larger than 2 cm. While some polyps have clearly defined structures, others appear like small tags growing out of the mucosa. In some cases, the polyps contain adipose tissue. Inside the lesions, the mucosal glands are normal or elongated, with irregular formations. However, the epithelium above is normal, containing columnar and goblet cells.

In juvenile-like polyps, there may be some ganglion tissue. There have been cases with lesions containing predominant autonomic nerves. This is rare, but results in a **ganglioneuroma**-like appearance. It is likely that GI malignancies are linked to Cowden syndrome, but unproven. In one study, about 86% of cases showed glycogenic acanthosis of the esophagus, and all of these cases had the PTEN mutation.

Clinical manifestations

The most common pathological manifestations include mucocutaneous lesions, pallor, fibrocystic disease and carcinoma of the breasts, thyroid changes, gastrointestinal hamartomas, macrocephaly—especially **megalencephaly**, mental retardation, headaches, cognition and memory problems, and multiple early onset uterine leiomyomas. Regarding gastrointestinal polyps, most are asymptomatic, but in young patients, adenomatous polyps and colon cancers have occurred. It is believed that in the future, many more Cowden syndrome patients will be identified via colon cancer screening.

Carcinomas of the breasts and thyroid gland are the most common cancers related to Cowden syndrome. Breast cancer has also been seen in male patients with the syndrome. For women with the syndrome, lifetime risk of breast cancer ranges between 25% and 50%, while in the general population, this is only about 11%. The predominant histology is ductal adenocarcinoma. Most related breast carcinomas occur as ductal carcinoma in situ, adenosis, atypical ductal hyperplasia, and sclerosis. Thyroid cancer is histologically usually follicular carcinoma, though papillary histologies have been seen in rare cases. In patients with Cowden syndrome, medullary thyroid carcinoma has never been documented.

Diagnosis

The diagnosis of Cowden syndrome involves a large variety of pathognomonic, major, and minor criteria. It is based on the fact that PTEN mutations exist in most patients diagnosed with dysplastic cerebellar gangliocytoma. As a result of this, adult-onset Lhermitte − Duclos disease was changed, from a major criterion to a pathognomonic criterion. It was given the highest weighting amount (10) in the Cleveland Clinic scoring system, which is considered to be the most accurate method of diagnosis. Their website is located at: https://www.lerner.ccf.org/gmi/ccscore.

For women, breast awareness must start at 18 years of age. Clinical breast examinations, every 6–12 months should begin at age 25, or at 5 to 10 years before the earliest known familial breast cancer patient's age of diagnosis. Annual mammography and breast MRI screening should start at age 30–35, or be individualize based on the earliest age of familial onset. For endometrial cancer screening, patient education is vital. There must be a prompt response to any symptoms, and participation in clinical trials is advised to determine effectiveness or the need for various types of screening. The patient should be counseled about risk-reducing mastectomy and hysterectomy as well as the degree of protection, cancer risks, and reconstruction options. Psychosocial, social, and quality-of-life factors must be addressed.

For both men and women, there must be an annual comprehensive physical examination starting at 18 years of age, or 5 years before the youngest diagnosis age of cancer in a family member, paying special attention to the breasts and thyroid gland. There should be an annual thyroid evaluation at age 18 or 5–10 years before the earliest known diagnosis of familial thyroid cancer. Colonoscopy should start at age 35, then every 5 years or more often, if the patient has polyps or is symptomatic. Renal ultrasound should be considered beginning at 40 years of age, and then every 1–2 years. Some patients will require dermatological evaluation. Psychomotor assessment may be needed for children, plus brain MRI, if there are any symptoms. All patients must be educated about cancer signs and symptoms. Diagnostic imaging methods include MRI, CT, and ultrasonography.

Before 1996, diagnosis of Cowden syndrome was difficult because of a lack of uniform diagnostic criteria. Therefore, the *International Cowden Consortium (ICC)* developed criteria for the syndrome that was based on existing

TABLE 22.1 Diagnostic criteria developed by the International Cowden Consortium.

Characteristic (pathognomonic) criteria	Major criteria	Minor criteria
Adult Lhermitte–Duclos disease (LDD)	Breast cancer	Fibrocystic breast disease
Mucocutaneous lesions: Facial trichilemmomas Papillomatous papules Distal extremity (acral) keratoses Mucosal lesions	Endometrial carcinoma	Fibromas
	Macrocephaly (above the 97th percentile)	Genitourinary tumors or malformations (such as renal cell carcinoma, uterine fibroids)
	Thyroid cancer, especially follicular	Hamartoma-like intestinal polyps
		Lipomas
		Mental retardation
		Other thyroid lesions, such as goiter or nodules

literature and actual clinical cases. It is extremely important to identify trichilemmomas and **papillomatous** papules. The ICC diagnostic criteria are summarized in Table 22.1.

The ICC's requirements for diagnosis include the following factors:

- Mucocutaneous lesions:
  - If there are six or more facial papules (with three or more being trichilemmomas), OR
  - If there are cutaneous facial papules with oral mucosal **papillomatosis**, OR
  - If there are oral mucosal papillomatosis with acral keratoses, OR
  - If there are six or more **palmoplantar** keratoses,
  - If two or more major criteria met, with one being macrocephaly or LDD.

Overall, one major criteria plus three minor criteria must exist, or four minor criteria for diagnosis of Cowden syndrome. The requirements for diagnosis of a patient with a family member who has the syndrome include:

- one pathognomonic criterion;
- any one major criterion, with or without minor criteria;
- two minor criteria;
- a history of **Bannayan-Riley-Revalcaba syndrome**—this is characterized by lipomatosis, macrocephaly, hemangiomatosis, and speckled penis, and is believed to be allelic to Cowden syndrome; in one study about 60% of patients had a PTEN germline mutation and another 10% had larger PTEN germline deletions; it is believed that this syndrome and Cowden syndrome are allelic, and part of a single spectrum, molecularly.

Point to remember

The *National Comprehensive Cancer Network* has proposed operational clinical criteria for diagnosis of Cowden syndrome, but this is not considered to be as accurate. Their website is https://www.nccn.org.

Treatment

Malignancies related to Cowden syndrome are usually treated the same as for sporadic cancers, except for breast and thyroid cancer. With Cowden syndrome and a first-time breast cancer diagnosis, there should be mastectomy of the involved breast, plus prophylactic mastectomy of the uninvolved breast. If there is thyroid cancer or follicular adenoma, a total thyroidectomy is recommended, even when one lobe of the gland is affected. This is because of the high likelihood of recurrence, plus difficulty distinguishing benign from malignant growths via hemithyroidectomy alone. Benign mucocutaneous lesions are not usually treated until they become disfiguring or symptomatic. Options include topical agents, curettage, cryosurgery, excision, and laser ablation.

In one study of a child with Proteus syndrome plus a PTEN germline mutation, the mTOR inhibitor called *sirolimus* (also called *rapamycin*) was effective for treatment. In a 2009 study, sirolimus was evaluated to treat Cowden syndrome and other syndromes with PTEN germline mutations. Sirolimus was used over different time periods, based on whether each Cowden syndrome patient did or did not have cancer. The patients in the study were between the ages of 18 and 65 years, except for three who were over age 65. Half of the participants were male and half were female, and all were Caucasian, except for one person of Asian heritage. The results were that skin and GI lesions regressed, and neurological evaluations revealed improvement in the patients' cerebellar function scores within 1 month. There were also decreases in mTOR signaling.

Prognosis

There is a lack of study on whether the prognosis of Cowden syndrome patients with cancer is any different from cases of sporadic cancers. Mortality rates are linked to the types of cancer present. Benign hamartomas may also cause major debilitation.

> **Point to remember**
> There is also a condition called *Cowden-like syndrome*, in which few patients have mutations of PTEN or variants in the SDHB gene, which is another gene related to Cowden syndrome. However, some of the cancers involved with Cowden-like syndrome are similar. Additionally, there is *type 2 segmental Cowden syndrome*, which is associated with a Cowden-type nevus and considered a type of epidermal nevus syndrome.

Clinical cases

Clinical case 1

1. When do the characteristic dermatologic lesions of Cowden syndrome usually develop?
2. What are the lifetime risks of epithelial thyroid cancer with this syndrome?
3. What is papillomatosis, as seen in this patient?

A 31-year-old woman presented with many raised, solid lesions on her neck, axillae, groin, and forehead. Two years previously, she had undergone subtotal thyroidectomy for a multinodular goiter. Physical examination revealed macrocephaly, pallor, and nodules on her left middle finger and right ring finger. Cowden syndrome was considered. Biopsy of a groin lesion revealed hyperkeratosis, acanthosis, and papillomatosis. The patient underwent electrocauterization of the neck and forehead lesions, and the axillae and groin lesions were excised. The patient was scheduled for a 1-year follow-up.

Answers:

1. **Usually with Cowden syndrome, multiple hamartomas develop, often within the third decade of life and about 80% of patients present with facial trichilemmomas, acral keratoses, papillomatous papules, or mucosal lesions.**
2. **The lifetime risks of epithelial thyroid cancer with Cowden syndrome are as high as 10%.**
3. **Papillomatosis is the development of numerous papillomas, which are benign tumors derived from epithelium. They may arise from the skin, mucous membranes, or glandular ducts.**

Clinical case 2

1. What is the term "polyposis" used to identify, as in this case study?
2. What are the major criteria for Cowden syndrome?
3. What is the overall age range for this syndrome?

A 73-year-old man was evaluated by a cancer genetics clinic because of extensive upper and lower gastrointestinal polyposis. A previous biopsy of two gastric polyps revealed gastric carcinomas. One was a moderately differentiated adenocarcinoma, and the other was a poorly differentiated adenocarcinoma arising within a gastric adenoma. Both carcinomas were stage 1 lesions, and the patient underwent a total removal of the stomach, abdominal lining, and gallbladder. The patient had previously undergone excision of an inflamed fibroma on his right hand and the removal of a ganglioneuroma during a colonoscopy procedure. Also, two tubular adenomas had been excised from his cecum and rectum. The patient's mother had died of a gastric adenocarcinoma. His head was also larger than the 97th percentile for his age, and he had hyperpigmented macules of the lower lip, plus mucosal swellings of the inner lip. His upper limbs and trunk had multiple skin tags, and there was a lipoma on the right side of his abdomen. He was clinically diagnosed with Cowden syndrome.

Answers:

1. **Polyposis is a condition characterized by the presence of numerous internal polyps. In conditions such as familial adenomatous polyposis, there is a high potential of these polyps to become malignant.**
2. **The major criteria for Cowden syndrome include breast cancer, endometrial carcinoma, macrocephaly above the 97th percentile, and thyroid cancer (especially follicular).**
3. **The total age range for Cowden syndrome has been documented as being between 13 and 75 years.**

Clinical case 3

1. What is the name of the linked condition between Cowden syndrome that involves adult-onset dysplastic cerebellar gangliocytoma?
2. Aside from PTEN and KLLN, what other genes predispose individuals to Cowden syndrome?
3. How common are hamartomatous intestinal polyps with Cowden syndrome?

A 64-year-old man complained of progressive memory loss over about 2 weeks, and a neurological examination revealed moderate declines in cognition. Initially, it was believed that he had some form of paraneoplastic encephalitis. However, over time, it was discovered that this patient had a gangliocytoma. Additional studies found two malignant tumors: a small cell lung carcinoma and a renal clear cell carcinoma. He also had many benign tumors, which included an abdominal wall hamartoma, colon adenomas, and multiple kidney and liver cysts. Further examination revealed macrocephaly, and multiple mucocutaneous lesions. Genetic sequencing revealed a pathogenic, heterozygous mutation of the KLLN gene, and Cowden syndrome was diagnosed.

Answers:

1. **The causative link between adult-onset dysplastic cerebellar gangliocytoma and Cowden disease is also known as Lhermitte − Duclos disease.**
2. **Aside from PTEN and KLLN, the other genes that predispose individuals to Cowden syndrome include the SDH genes, PIK3CA, and AKT1.**
3. **In one study, 100% of patients with Cowden syndrome had hamartomatous intestinal polyps, while just 78% had a PTEN germline mutation.**

Clinical case 4

1. Are thyroid changes common in relation to Cowden syndrome?
2. Which type of thyroid cancer is most common in relation to the syndrome?
3. What are the pathognomonic criteria for Cowden syndrome?

A 17-year-old boy was brought in for evaluation for symptoms that appeared related to a thyroid disorder: tremor, palpitations, and anxiety. His father had previously undergone a total thyroidectomy for multinodular goiter, and later died of an anaplastic oligodendroglioma. The boy's paternal grandfather also had developed thyroid cancer but it was not fatal. The boy had macrocephaly, a hemangioma on his left leg, problems with cognition, facial trichilemmomas, and thyroid gland enlargement. Thyroid ultrasonography showed a heterogeneous goiter with no visible nodules. He was treated with methimazole and propranolol, but did not achieve a euthyroid state. Eventually, fine needle aspiration revealed a benign, hyperplastic thyroid nodule. Genetic testing revealed the boy to have a PTEN germline mutation. A total thyroidectomy was performed. However, final diagnosis was Cowden syndrome.

Answers:

1. Yes, thyroid changes are amongst the most common pathological manifestations of Cowden syndrome, which also include mucocutaneous lesions, pallor, fibrocystic disease and carcinoma of the breasts, GI hamartomas, macrocephaly, mental retardation, headaches, cognition and memory problems, and multiple early-onset uterine leiomyomas.
2. Thyroid cancer is usually found to be follicular carcinoma, though in rare cases, papillary histologies have been seen.
3. The pathognomonic criteria of Cowden syndrome include adult Lhermitte − Duclos disease, facial trichilemmomas, papillomatous papules, distal extremity (acral) keratoses, and mucosal lesions.

Clinical case 5

1. Is gangliocytoma a common type of CNS tumor?
2. What is the predominant histology for breast carcinomas in relation to this syndrome?
3. What is the link between diagnosis of Cowden syndrome and dysplastic cerebellar gangliocytoma?

A 64-year-old woman presented with a two-year history of a left cerebellar hemisphere lesion and chronic headaches. Enlargement of the lesion prompted neurosurgical examination. A brain MRI revealed a well delineated lesion with a thickened appearance on T2-weighted imaging. T1-weighted imaging revealed a small enhancing lesion on the left side, near the mastoid, which was believed to be a hemangioma. Lhermitte − Duclos disease was considered, and immunostaining appeared to be confirmative. However, the lesion was a dysplastic cerebellar gangliocytoma and not a hemangioma. The patient had a history of multiple thyroid colloid nodules, but no thyroid malignancy. She was also found to have a ductal adenocarcinoma of the right breast. Final diagnosis was Lhermitte − Duclos disease, a close relation to Cowden syndrome.

Answers:

1. Actually, dysplastic cerebellar gangliocytoma is an unusual CNS tumor, even though it is closely linked to Cowden syndrome and Lhermitte − Duclos disease.
2. The predominant histology is ductal adenocarcinoma of the breast. Most related breast carcinomas occur as ductal carcinoma in situ, adenosis, atypical ductal hyperplasia, and sclerosis.
3. The diagnosis of Cowden syndrome involves many pathognomonic, major, and minor criteria, and is based on the fact that PTEN mutations exist in most patients diagnosed with dysplastic cerebellar gangliocytoma. As a result, adult-onset Lhermitte − Duclos disease was changed from a major criterion to a pathognomonic criterion for Cowden syndrome.

Key terms

Auxilin
Bannayan-Riley-Revalcaba syndrome
Cowden syndrome
Cyclin-D1
Gangliocytoma
Ganglioneuroma
Ganglioneuromatous
Genodermatosis
Hamartomatous
Hypophosphorylated
Lhermitte − Duclos disease
Megalencephaly
Palmoplantar
Papillomatosis
Papillomatous
PI3K pathway
Trichilemmomas

Further reading

Abbas, A. K., Lichtman, A. H., & Pillai, S. (2017). *Cellular and molecular immunology* (9th Edition). Elsevier.
Adesina, A. M., Tihan, T., Fuller, C. E., & Poussaint, T. Y. (2010). *Atlas of pediatric brain tumors.* Springer.
Argenyi, Z., & Jokinen, C. H. (2011). *Cutaneous neural neoplasms: A practical guide (Current clinical pathology).* Humana Press.
Boardman, L. A. (2018). *Intestinal polyposis syndromes: Diagnosis and management.* Springer.
Bunz, F. (2016). *Principles of cancer genetics* (2nd Edition). Springer.
Butterfield, L. H., Kaufman, H. L., & Marincola, F. M. (2017). *Cancer Immunotherapy principles and practice.* Demos Medical.
Cassidy, S. B., & Allanson, J. E. (2010). *Management of genetic syndromes* (3rd Edition). Wiley-Blackwell.
Delaini, G. G., Skricka, T., Colucci, G., & Nicholls, J. (2009). *Intestinal polyps and polyposis: From genetics to treatment to follow-up.* Springer.

Dong, H., & Markovic, S. N. (2018). *The basics of cancer immunotherapy*. Springer.
El-Darouti, M. A., & Al-Ali, F. M. (2019). *Challenging cases in dermatology volume 2: Advanced diagnoses and management tactics*. Springer.
Ellis, C. N. (2010). *Inherited cancer syndromes: Current clinical management* (2nd Edition). Springer.
Gupta, N., Banerjee, A., & Haas-Kogan, D. A. (2017). *Pediatric CNS tumors (Pediatric oncology)* (3rd Edition). Springer.
Icon Group International. (2010). *Hamartoma: Webster's Timeline History, 1935-2007*. Icon Group International, Inc.
Icon Group International. (2010). *Papillomatosis: Webster's Timeline History, 1929-2007*. Icon Group International, Inc.
Landsberg, L. (2018). *Pheochromocytomas, paragangliomas and disorders of the sympathoadrenal system: Clinical features, diagnosis and management (Contemporary endocrinology)*. Humana Press.
Micali, G., Lacarrubba, F., Stinco, G., Argenziano, G., & Neri, I. (2018). *Atlas of pediatric dermatoscopy*. Springer.
Nose, V., Greenson, J. K., & Paner, G. P. (2013). *Diagnostic pathology: familial cancer syndromes*. Lippincott Williams & Wilkins.
Parker, P. M. (2007). *Cowden syndrome – A bibliography and dictionary for physicians, patients, and genome researchers*. ICON Group International, Inc.
Philipone, E., & Yoon, A. J. (2017). *Oral pathology in the pediatric patient: A clinical guide to the diagnosis and treatment of mucosal lesions*. Springer.
Starr, T. K. (2019). *Cancer driver genes: Methods and protocols (Methods in molecular biology)*. Humana Press.
Stratakis, C. A., Ghigo, E., & Guaraldi, F. (2013). *Endocrine tumor syndromes and their genetics (Frontiers of hormone research, volume 41)*. S. Karger.
Suster, S. (2015). *Atlas of mediastinal pathology (Atlas of anatomic pathology)*. Springer.
Tan, A. C., & Huang, P. H. (2017). *Kinase signaling networks (Methods in molecular biology)*. Humana Press.
Weinberg, R. A. (2013). *The biology of cancer* (2nd Edition). W.W. Norton & Company.
Xu, K. (2013). *PTEN: Structure, mechanisms-of-action, role in cell signaling and regulation (Protein science and engineering)*. Nova Biomedical.

Chapter 23

Tuberous sclerosis

Chapter Outline

Tuberous sclerosis 407
Clinical cases 413
Key terms 416
Further reading 416

Tuberous sclerosis

Tuberous sclerosis is actually a collection of autosomal dominant phacomatosis disorders, involving benign tumors of the CNS and different nonneural tissues, plus hamartomas. Abbreviated as "TSC", which stands for *tuberous sclerosis complex*, tuberous sclerosis is a rare, multisystem disease-involving noncancerous tumors to develop mostly in the brain, eyes, skin, heart, lungs, liver, and kidneys. Though overall being a rare condition, tuberous sclerosis is the second most common neurocutaneous syndrome, after neurofibromatosis type 1. The usual appearance of this disease combines epilepsy, mental retardation, and facial angiofibromas, referred to as *Vogt's triad*. However, this only occurs in 30%-40% of patients.

Epidemiology

In previous years, the many different clinical manifestations of tuberous sclerosis resulted in the condition being under-diagnosed. Today, data shows that it affects between 25,000 and 40,000 people in the United States, and about 1-2 million people globally. The estimated prevalence is 1 case per 6000-10,000 live births. The disease usually manifests before the age of 10 years, but it can manifest much later. Most cases of tuberous sclerosis (60%) occur sporadically, with no family history. There is no predilection for gender, race, or ethnic group.

Etiology and risk factors

Tuberous sclerosis is caused by a mutation of the *tuberous sclerosis 1 (TSC1) gene* on chromosome 9q, or a mutation of *tuberous sclerosis 2 (TSC2)* on chromosome 16p. The TSC1 gene contains 23 **exons**, with 21 of them carrying coding information. Both of these genes are large and complex. Genetic abnormalities can vary from point mutations to deletions. Via immunohistochemistry and western blotting, the proteins **tuberin** and **hamartin** can be identified throughout many body organs and tissues. Tuberin is produced by the TSC2 gene, while hamartin is produced by the TSC1 gene. Hamartin has a molecular weight of 130 kDa and is strongly expressed in the brain, heart, and kidneys. Its pattern of expression overlaps that of tuberin. Small deletions and nonsense mutations of the TSC1 gene make up about 30% of all mutations of that gene. Nearly all of them form a truncated gene, with more than 50% of the changes affecting exons 15 and 17. The TSC2 gene contains 40 exons and encodes a large area of 5.5 kb. There is widespread expression in the brain and the other organs that are affected by tuberous sclerosis. Alternative splicing of various mRNAs has been seen. Some of the 180 kDA protein product tuberin has close homology with the active catalytic site of *Rap1GTPase-activating protein 1 (RAP1GAP)*, which is part of the *RAS* signaling protein family.

The gene mutation spectrum of TSC2 is larger than that of TSC1, including large deletions and missense mutations, and less often, splice junction mutations. The highest number of mutations occur in exons 16, 33, and 40. Large deletions may extend into the nearby *polycystic kidney disease 1* gene, causing a phenotype of tuberous sclerosis and polycystic kidney disease. TSC2 mutations have been well documented as being linked with a more severe phenotype. There is usually earlier onset of seizures, larger amounts of tubers, and a lower **cognition index**. However, TSC2 missense mutations are also linked with milder phenotypes. Somatic inactivation of the **wildtype allele** has been found in cardiac and kidney lesions, as well as in subependymal giant cell astrocytomas (SEGAs). Some lesions in tuberous sclerosis are believed to be caused by **haploinsufficiency**. This means a situation in which the total level of a gene product—a certain protein—produced by the cell is about 50% of the normal level, and is not sufficient to permit normal cell function to continue. There is also no evidence of the inactivation of TSC2 (or TSC1) in the histologically similar focal cortical dysplasias. Additionally, loss of TSC1 in periventricular zone neuronal stem cells may be enough to cause abnormal migration and a giant cell phenotype. This supports the "two-hit hypothesis" of tuber formation.

Hamartin and tuberin interact within the cells, forming a complex. Mutation of either gene causes disruption of the complex function, and then, similar disease phenotypes. When tuberous sclerosis is sporadic, mutations occur five times more often in TSC2 than in TSC1. However, in families with more than one affected member, the mutation ratio between the two genes is exactly the same. The TSC1 or TSC2 mutations are found in approximately 85% of patients with tuberous sclerosis. The remaining 15% of patients could be mosaics, or could have a mutation in a noncoding gene area that has not been identified. Mosaicism has been seen in parents of some patients with sporadic cases, as well as those with tuberous sclerosis. Patients with tuberous sclerosis and no gene mutations are found have a milder phenotype.

The tuberin-hamartin complex is a signaling node. It integrates stress signals and growth factor from the upstream phosphoinositide 3-kinase/protein kinase B pathway. The signals are transmitted downstream, coordinating many cellular processes that include cell proliferation and size. The mTOR pathway is negatively regulated by the tuberin-hamartin complex. Disruption of the complex results in upregulation of the mTOR pathway, plus increased proliferation and cell growth via the effector molecules called *4E-binding protein 1* and *S6 kinase beta-1*. There are no proven risk factors for tuberous sclerosis.

> **Point to remember**
>
> TSC gene mutations often initially cause noticeable skin changes. These include angiofibromas, periungual fibromas, hypomelanotic macules, and shagreen patches. Other major features of tuberous sclerosis include tubers and nodules in the brain, SEGA, and various tumors of the heart, kidneys, eyes, and liver.

Pathology

In the central nervous system, lesions related to tuberous sclerosis may include cerebral cortical tubers, subependymal hamartomatous nodules, and white matter heterotopia. The tubers consist of giant cells, similar to SEGAs, and dysmorphic neurons. They have disruptions of cortical lamination; calcification of blood vessel walls, parenchyma, or both; gliosis; and loss of myelin. The surrounding cortex often has what appears to be normal **cytoarchitecture**, meaning arrangement of cells in the tissue, but newer and more detailed immunohistochemical and morphometric studies are finding more abnormalities. In all cortical layers and the underlying white matter, dysmorphic and giant cells may be present. The dysmorphic neurons have changes in their cortical radial orientation, **perikaryal fibril** accumulation, and atypical dendritic branching. Deep sequencing of the TSC1, TSC2, and *Kirsten rat sarcoma* genes shows that small, second-hit mutations are rare within tubers. Insulin signal pathways are normally affected via inhibition by tuberin and hamartin, but can show slight yet definitive differences in tubers, in comparison with severe cortical dysplasia.

Clinical manifestations

The primary CNS manifestations of tuberous sclerosis include **tubers**, also known as cortical hamartomas; subependymal glial nodules; SEGAs; and subcortical glioneuronal hamartomas. Additional manifestations include seizures, intellectual disability, and developmental delay. Children and adolescents may also have learning difficulties, mental retardation, behavioral problems, autism spectrum disorder, and attention-deficit/hyperactivity disorder (ADHD). The major extraneural manifestations include **peau chagrin**, cutaneous angiofibromas (also known as **adenoma sebaceum**), cardiac rhabdomyomas, subungual fibromas, intestinal polyps, pulmonary lymphangioleiomyomatosis, visceral cysts, and renal **angiomyolipomas**. In newborns and infants under 2 years of age, cardiac rhabdomyomas are often presenting features. These tumors contain striated muscle fibers. More than 50% of young patients with cardiac rhabdomyomas have tuberous sclerosis. Cutaneous signs include facial angiofibromas, hypomelanotic nodules, and **shagreen patches**, which are ovoid and elevated plaques. They may be skin colored or pigmented, wrinkled or smooth, and appear on the trunk or lower back in most cases. Subungual fibromas are most common in childhood, not in older patients. Renal angiomyolipomas occur in up to 80% of patients by the age of 10 years. Though polycystic kidney disease only appears in 3%-5% of cases, renal cysts manifest in up to 20% of cases. Lymphangioleiomyomatosis development is not understood, but can be fatal because of severe lung function impairment. It exists in up to 40% of adult females who have tuberous sclerosis.

Every phenotypic feature of tuberous sclerosis may also occur sporadically when the genetic condition is not present. About half of patients with lymphangioleiomyomatosis do not have tuberous sclerosis. In these patients, when sporadic angiomyolipomas occur, they are usually solitary. In tuberous sclerosis cases, related angiomyolipomas are usually multiple or bilateral. Neurological symptoms are extremely common with tuberous sclerosis, and are often serious or even life threatening. Most often, initial neurological signs, of various causes, include the following (also, see Fig. 23.1):

- Untreatable epilepsies that include infantile spasms—80%-90% of cases
- Neurobehavioral disorders—in more than 60% of cases
- Cognitive impairment—50% of cases
- Autism spectrum disorder—in up to 40% of cases

These signs are related to structural changes of the cortex and subcortical white matter, usually as tubers. However, autopsies on a small amount of patients have indicated that cortical and white matter disorganization may be less severe. The primary manifestations of tuberous sclerosis are further summarized in Table 23.1.

> **Point to remember**
>
> A large variety of symptoms of tuberous sclerosis have been reported. These include developmental delays, intellectual disabilities, arrhythmias, benign brain tumors, calcium deposits on the brain, benign heart and kidney tumors, growths around or under the nails, growths on the retinas, pale patches on the eyes, growths on the gums or tongue, pitting of the teeth, areas of decreased skin pigmentation, red patches of skin on the face, raised skin with an orange peel-like texture, and seizures.

Diagnosis

The diagnosis of tuberous sclerosis is mostly based on clinical features. It can be difficult because of wide variations in phenotype, penetrance in those carrying the gene mutation, and the age of each patient when symptoms begin. The patient should be assessed with a personal history plus a family history that goes back for three generations. Genetic counseling helps determine others who may be at risk. In 2012, the diagnostic criteria were revised at the *International Tuberous Sclerosis Complex Consensus Conference* (https://www.tscinternational.org). The criteria are highly important given the fact that genetic testing is only available in a few locations, and may not be able to be accessed by clinicians. The clinical manifestations of tuberous sclerosis are categorized as major or minor. Diagnostic categories are created by the numbers of major or minor manifestations present in each patient. These categories define the likelihood of tuberous sclerosis as either definite or possible. Testing for the causative gene mutations may be confirmative if a patient does not have the clinical criteria for an accurate diagnosis, yet his or her phenotype is suggestive of likelihood of developing the disease. Genetic testing for the disease, however, is not widely available. For prenatal patients, diagnosis is via mutation analysis, but can only occur when the mutation is proven to exist in other family members. The clinical diagnostic criteria for tuberous sclerosis is summarized in Table 23.2.

In imaging studies, CNS subependymal hamartomatous nodules appear similar to "candle dripping", like wax running down from a burning candle. Cortical tubers are commonly detected via CT or MRI, but can also be seen via

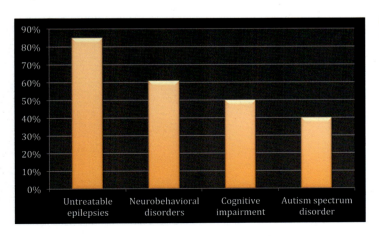

FIGURE 23.1 Initial neurological signs of various causes, in tuberous sclerosis.

TABLE 23.1 Primary manifestations of tuberous sclerosis.

Manifestation	Occurrence
Central nervous system	
Cortical tuber	90%-100%
Subependymal nodule	90%-100%
White matter hamartoma, white matter heterotopia	90%-100%
Subependymal giant cell astrocytoma (SEGA)	6%-16%
Eyes	
Retinal hamartoma	50%
Retinal giant cell astrocytoma	20%-30%
Hypopigmented iris spot	10%-20%
Skin	
Facial angiofibroma (adenoma sebaceum)	80%-90%
Hypomelanotic macule	80%-90%
Shagreen patch	20%-40%
Forehead plaque	20%-30%
Periungual and subungual fibroma	20%-30%
Heart	
Cardiac rhabdomyoma	50%
Lungs	
Pulmonary cyst	40%
Lymphangioleiomyomatosis	1%-2.3%
Micronodular pulmonary hyperplasia of type II pneumocytes	Rare
Digestive system	
Microhamartomatous rectal polyp	70%-80%
Liver hamartoma	40%-50%
Hepatic cyst	24%
Adenomatous polyp of the duodenum and small intestine	Rare
Kidneys	
Multiple, bilateral angiomyolipoma	50%
Isolated renal cyst	10%-20%
Polycystic kidney disease	2%-3%
Renal cell carcinoma	1.2%
Other	
Gingival fibroma	50%-70%
Bone cyst	40%
Pitting of tooth enamel	30%
Arterial aneurysm, in the aorta, axillary artery, or intracranial arteries	Rare

TABLE 23.2 Clinical diagnostic criteria for tuberous sclerosis.

Major features	Minor features	Definitive diagnosis	Possible diagnosis
Three or more angiofibromas or fibrous cephalic plaques	Four or more dental enamel pits	Two major features, or one major feature with two or more minor features	One major feature, or Two or more minor features
Three or more hypomelanotic macules 5 mm or more in diameter	Two or more intraoral fibromas		
Two or more subungual fibromas	Confetti skin lesions		
Multiple retinal hamartomas	Multiple renal cysts		
A "shagreen patch" (an oval-shaped, elevated nevoid plaque on the trunk or lower back in early childhood)	Nonrenal hamartomas		
Cortical dysplasias, which include tubers and cerebral white matter radial migration lines	Retinal achromic (colorless) patch		
SEGA			
Subependymal nodules			
Cardiac rhabdomyoma			
Lymphangioleiomyomatosis			
Two or more angiomyolipomas			

X-rays (see Fig. 23.2). Tubers have low signal on T1-weighted images and high signal on both T2-weighted and fluid-attenuated inversion recovery (FLAIR) images, unless they are calcified. The white matter lesions are hypointense on T1-weighted images, and hyperintense on T2-weighted and FLAIR images (see Fig. 23.3). Structural abnormalities that often represent tubers may be found through metabolic brain studies, such as *fluorodeoxyglucose-positron emission tomography*. Along with intraoperative **electrocorticography**, many tubers linked to epilepsy may be identified. These malformations are strongly related to infantile spasms and generalized tonic-clonic seizures. The lesions resemble sporadic cortical malformations that are not related to tuberous sclerosis that are classified by the *International League Against Epilepsy (ILAE)* as *cortical dysplasia Type IIb*. The ILAE's website is located at: https://www.ilae.org.

Perikaryal fibrils that are present can be highlighted with silver impregnation techniques. These show many neurons that have a neurofibrillary tangle-like morphology. In the tubers and nearby cortex and white matter, characteristic **balloon cells** may be seen. These cells look like gemistocytic astrocytes, because there is eosinophilic glassy cytoplasm that may be clustered in small groupings. However, the cells often have prominently **nucleolated** nuclei, which is unlike gemistocytes. Cellular elements may have both neurons and astrocytes. Though the neurons express proteins that are neuronal-associated, they have cytoarchitectural features of immature neurons, or of poorly differentiated neurons. These include reduced spine density and axonal projections. In cortical tubers, giant cells have a molecular and cellular heterogeneity that is similar to that of SEGA. A mixed glioneuronal origin of these cells is suggested by immunohistochemical markers that are characteristic of both glial and neuronal phenotypes.

Many giant cells in the tubers express **nestin protein** and mRNA. Some cells have immunoreactivity for *glial fibrillary acidic protein*, while others that have the same morphological phenotype express neuronal markers. These markers include *gap junction beta-1 protein* or *connexin 32*, and *gap junction beta-2 protein* or *connexin 26*. Other expressed neuronal markers include *class III beta-tubulin, neurofilaments, alpha-internexin*, and *microtubule-associated protein 2*. Formation of well-formed synapses between giant cells and nearby neurons, however, does not occur consistently. Cortical hamartomas that cannot be morphologically distinguished from tubers can be present with chronic focal epilepsies, with no evidence of any underlying tuberous sclerosis. Pathogenesis of such sporadic lesions is not fully understood. Subependymal hamartomas are elevated and often calcified nodules made up of cells highly similar to those in cortical tubers, yet smaller in size.

After the TSC1 and TSC2 genes were cloned in the 1990s, there was development of probes for the gene transcripts as well as the translated proteins. Immunostaining of a tuber with anti-hamartin or anti-tuberin antibodies does not

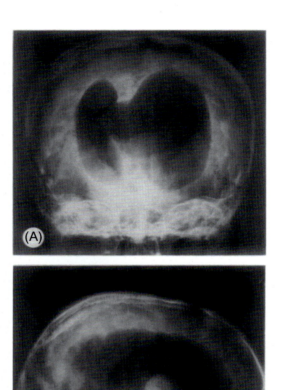

FIGURE 23.2 (A) Posterior and (B) lateral X-rays showing tuberous sclerosis. Courtesy: *Benjamin E. Medley MD, Richard A. McLeod MD, and O.Wayne Houser MD, Seminars in Roentgenology Volume 11.*

prove which mutation may be present, so is not helpful diagnostically. The proteins are widely expressed in the CNS when brain development is normal. Surgically resected tubers have been studied, but because autopsies of patients with tuberous sclerosis are rare, it has been difficult. This is because proteins, DNA, and mRNA are preserved better in samples from living patients. Studies of how hamartin and tuberin may regulate cell adhesion through *ezrin, radixin, and moestrin proteins*, plus the *guanosine triphosphate hydrolase enzyme (GTPase) Ras homolog (Rho) family* have been conducted. Electrophysiological studies have revealed differences in synaptic and neurophysiological abnormalities, from surgically resected brain tissue, compared between tuberous sclerosis and severe cortical dysplasia.

For adult women, it is important to test pulmonary function and perform a high-resolution CT of the chest. Use of a **Wood's lamp** can aid in examination of the face and mouth. Echocardiograms are used to find rhabdomyomas, and electrocardiograms are used to assess arrhythmias. Electroencephalographs are used for patients with seizure activity. Funduscopes are used to spot retinal hamartomas or achromic patches.

> **Point to remember**
>
> Tuberous sclerosis is diagnosed via genetic testing, brain MRI, head CT scans, electrocardiogram, echocardiogram, kidney ultrasound, eye examinations, and use of a Wood's lamp to examine the skin.

Treatment

Tuberectomy is a good surgical approach in the treatment of intractable seizures related to tuberous sclerosis. This is the resection of an epileptogenic tuber only, without additional tissue resection. Also, mTOR inhibitors such as

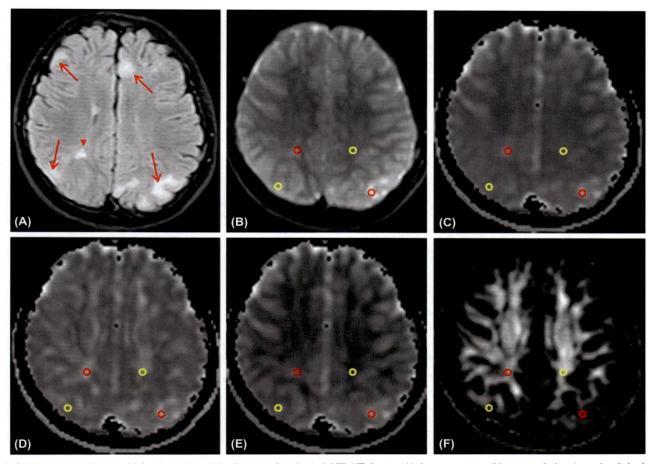

FIGURE 23.3 A 11-year-old female patient with tuberous sclerosis. Axial FLAIR image (A) demonstrates a white matter lesion (arrowhead) in the right parietal lobe, and multiple tubers (arrows) on both hemispheres. Placement of the reactive oxygen intermediates within a tuber (red circle), a white matter lesion (red circle), and corresponding contralateral normal regions (yellow circles) are shown in (B), apparent diffusion coefficient (C), axial diffusivity (D), radial diffusivity (E), and fractional anisotropy (F) maps. Courtesy: *M.S. Dogan, K.Gumus, G.Koc, et. al., Diagnostic and Interventional Imaging Volume 97.*

everolimus have been used to treat the disease. They cause significant tumor size reduction for renal angiomyolipomas, **lymphangioleiomyomas**, and SEGAs. When treatment with mTOR inhibitors is stopped, the tumors regrow. The effects of these agents are being evaluated for clinical management of epilepsy and other neurological manifestations of the disease. Other mTOR inhibitors include: temsirolimus and ridaforolimus. The first generation mTOR inhibitors are known as *rapalogs*, with *rapamycin* being one of the examples of a *small molecule inhibitor*. Infantile spasms are best treated with vigabatrin or adrenocorticotropic hormone.

Prognosis

Tuberous sclerosis usually shortens the lifespan of patients, but usually only slightly. The most common causes of death, in the second decade of life, include brain tumors and status epilepticus, and less often, renal abnormalities. If the patient is 40 years or older, death is usually related to kidney problems such as cystic diseases or neoplasms, as well the uncommon proliferative lung condition known as lymphangioleiomyomatosis. Earlier ages of seizure development are linked to worsened prognoses.

Clinical cases

Clinical case 1

1. Which gene mutations are involved with tuberous sclerosis?

2. How common is tuberous sclerosis?
3. When does tuberous sclerosis usually manifest?

A 26-year-old man presented with multiple growths in his upper and lower gums that had developed over 5 years. The man also had experienced seizures since he was 10 years old. Additionally, he had slurred speech and problems understanding others. Physical examination revealed many nodular growths on his forehead, nose, cheeks, and angiofibromas on his chest and back. He also had a shagreen patch on the right lumbosacral region of this back. The patient's teeth were somewhat pitted. A brain CT revealed hypodense areas of 5 mm in diameter in the subependymal regions of both ventricles, indicating multiple calcified tuberous nodules. The patient was diagnosed with tuberous sclerosis.

Answers:

1. **Tuberous sclerosis is a neurodevelopmental disease-involving mutations of the TSC1 or TSC2 genes, which code for inhibitors of central cell growth, and control the mTOR pathway.**
2. **Tuberous sclerosis affects between 25,000 and 40,000 people in the United States, and about 1-2 million people globally. The estimated prevalence is 1 case per 6000-10,000 live births.**
3. **Tuberous sclerosis usually manifests before the age of 10 years, but may manifest much later.**

Clinical case 2

1. How is tuberous sclerosis inherited?
2. What types of gene mutations occur with tuberous sclerosis?
3. What are the differences in manifestations of tuberous sclerosis between the causative TSC1 and TSC2 genes?

A 11-year-old boy was brought to his local emergency department because of experiencing multiple generalized tonic-clonic seizures for the previous 10 days. He had been having seizures since the age of 1 year, and was taking antiepileptic seizures, but these provided little control. There was no family history of seizures. The boy's father had skin nodules on his face and neck, and a hypopigmented macule on his chest. The boy also had multiple hyperpigmented papules in the nasolabial region, and five hypopigmented macules on his legs. There was a shagreen patch on his left buttock. A CT scan of the boy's head revealed subependymal nodules and tuberous sclerosis was diagnosed.

Answers:

1. **Tuberous sclerosis is inherited as an autosomal dominant disease, but sporadic mutations are found most often. It is actually a collection of autosomal dominant disorders, involving benign tumors of the CNS and different nonneural tissues, plus hamartomas.**
2. **The TSC1 gene mutations can vary from point mutations to deletions. Small deletions and nonsense mutations of the TSC1 gene make up about 30% of all mutations of this gene. The gene mutation spectrum of TSC2 is larger than that of TSC1, including large deletions and missense mutations, and less often, splice junction mutations.**
3. **TSC2 mutations have been well documented as being linked with a more severe phenotype. There is usually earlier onset of seizures, larger amounts of tubers, and a lower cognition index. However, TSC2 missense mutations are also linked with milder phenotypes.**

Clinical case 3

1. What types of lesions related to tuberous sclerosis may develop in the central nervous system?
2. What is the pathology of CNS tubers?
3. What are the major extraneural manifestations of tuberous sclerosis?

A 35-year-old woman presented with papules on her cheeks, which she said had been there since childhood. She had previously been diagnosed with tuberous sclerosis, and had experienced seizures for the past 3 years. Echocardiography results were normal, but a brain CT revealed calcification in the subependymal region of the bilateral ventricles. A CT scan of her abdomen showed angiomyolipomas in both kidneys. She was prescribed increased doses of phenytoin to manage her seizures.

Answers:

1. **In the CNS, lesions related to tuberous sclerosis may include cerebral cortical tubers, subependymal hamartomatous nodules, and white matter heterotopia.**

2. Tubers in the CNS consist of giant cells, similar to SEGAs, and dysmorphic neurons. They have disruptions of cortical lamination; calcification of blood vessel walls, parenchyma, or both; gliosis; and loss of myelin. Tubers are also known as cortical hamartomas.
3. The major extraneural manifestations of tuberous sclerosis include peau chagrin, cutaneous angiofibromas (or adenoma sebaceum), cardiac rhabdomyomas, subungual fibromas, intestinal polyps, pulmonary lymphangioleiomyomatosis, visceral cysts, and renal angiomyolipomas.

Clinical case 4

1. How common are renal angiomyolipomas in patients with tuberous sclerosis?
2. What are the differences in development of sporadic angiomyolipomas and angiomyolipomas linked to tuberous sclerosis?
3. What are the overall most common initial neurological signs of tuberous sclerosis.

A 28-year-old man presented with multiple facial eruptions, and told the examining physician that he had them since he was 12 years old. For about 2 years, he had experienced headaches and seizures. The patient also was mentally retarded. There was no family history of tuberous sclerosis. On physical examination, four macules were found on his right arm, and there were multiple shagreen patches on his back, abdomen, buttocks, thighs, and knees. His fingernails and toenails had small benign growths. A CT scan of the patient's head showed subependymal calcified tubers in the left parietal region, with right-sided cerebral atrophy. Echocardiography and electrocardiography showed left ventricular hypertrophy. Funduscopy revealed chorioretinal atrophy and bilateral retinal hematoma, suggesting tuberous sclerosis. Abdominal ultrasonography showed multiple echogenic angiomyolipomas in both kidneys, cysts in the left kidney, and an angiomyolipoma in the liver.

Answers:

1. **Renal angiomyolipomas occur in up to 80% of patients by the age of 10 years.**
2. **When sporadic angiomyolipomas occur, they are usually solitary. In tuberous sclerosis cases, related angiomyolipomas are usually multiple or bilateral.**
3. **Overall, untreatable epilepsies, including infantile spasms are the initial signs of tuberous sclerosis. They appear in 80%-90% of cases.**

Clinical case 5

1. What manifestations of tuberous sclerosis are common in children and adolescents?
2. What are the primary manifestations of tuberous sclerosis in the CNS?
3. What are the major clinically diagnostic criteria for tuberous sclerosis?

A 27-year-old woman, who had experienced seizures since childhood, presented with worsening of her seizures. The patient was mentally retarded and had acute visual loss. She had facial angiofibromas, hypomelanotic macules in the lower abdomen, and both periungual fibromas and subungual fibrosis of her left thumb. There was a shagreen patch in the lower right lumbar area, and most of her teeth were either missing or extremely pitted. Her left kidney had multiple cysts and two angiolipomas. A CT scan of her brain revealed small, nodular protrusions into the lateral ventricles with calcified foci and a "candle dripping" appearance. A final diagnosis of tuberous sclerosis was made.

Answers:

1. **Children and adolescents with tuberous sclerosis may have learning difficulties, mental retardation, behavioral problems, autism spectrum disorder, and ADHD.**
2. **The primary manifestations of tuberous sclerosis in the CNS include cortical tuber (90%-100% of cases), subependymal nodules (90%-100% of cases), white matter hamartoma and/or white matter heterotopia (90%-100% of cases), and SEGA (6%-16% of cases).**
3. **The major features of clinically diagnostic criteria for tuberous sclerosis include: three or more angiofibromas or fibrous cephalic plaques; three or more hypomelanotic macules that are 5 mm in diameter; two or more subungual fibromas; multiple retinal hamartomas; a shagreen patch; cortical dysplasias (including tubers and cerebral white matter radial migration lines); SEGA; subependymal nodules; cardiac rhabdomyoma; lymphangioleiomyomatosis; and two or more angiomyolipomas.**

Key terms

Adenoma sebaceum
Angiomyolipomas
Balloon cells
Cognition index
Cytoarchitecture
Electrocorticography
Exons
Hamartin
Haploinsufficiency
Lymphangioleiomyomas
Nestin protein
Nucleolated
Peau chagrin
Perikaryal fibril
Shagreen patches
Tuberectomy
Tuberin
Tuberous sclerosis
Tubers
Wildtype allele
Wood's lamp

Further reading

Abouelmagd, A., & Ageely, H. M. (2013). *Basic genetics: A primer covering molecular composition of genetic material, gene expression and genetic engineering, mutations, and human genetic disorders*. Universal Publishers.

Anderson, L. (2019). *Living with tuberous sclerosis, a congenital neurological chromosomal genetic disorder*. Amazon.com Services LLC.

Azmi, A. (2016). *Conquering RAS: From biology to cancer therapy*. Academic Press.

Baumgartner, C. (2003). *Clinical electrophysiology of the somatosensory cortex: A combination study using electrocorticography, scalp-EEG, and magnetoencephalography*. Springer.

Cho, Z. H., Calamante, F., & Chi, J. G. (2015). *7.0 Tesla MRI brain white matter atlas* (2nd ed.). Springer.

Dabbs, D. J. (2018). *Diagnostic immunohistochemistry: Theranostic and genomic applications* (5th ed.). Elsevier.

Delaini, G. G., Skricka, T., Colucci, G., & Nicholls, J. (2009). *Intestinal polyps and polyposis: From genetics to treatment and follow-up*. Springer.

Eckdahl, T. T. (2018). *Newborn screening for genetic disorders*. Momentum Press Health.

Fernandez, I., & Jackson, L. (2018). *mRNA: Molecular biology, processing and function (cell biology research progress)*. Nova Science Publishers Inc.

Feuerstein, G. Z., Hunter, A. J., Metcalf, B. W., Poste, G., & Ruffolo, R. R., Jr. (2014). *Inflammatory cells and mediators in CNS disease (new horizons in therapeutics, volume 2)*. CRC Press.

Hayat, M. A. (2012). *Tumors of the central nervous system, volume 9: Lymphoma, supratentorial tumors, glioneuronal tumors, gangliogliomas, neuroblastoma in adults, astrocytomas, ependymomas, hemangiomas, and craniopharyngiomas*. Springer.

Icon Group International. (2010a). *Angiomyolipoma: Webster's timeline history, 1955-2007*. ICON Group International, Inc.

Icon Group International. (2010b). *Haploinsufficiency: Webster's timeline history, 1995-2007*. ICON Group International, Inc.

Icon Group International. (2010c). *Leiomyoma: Webster's timeline history, 1932-2007*. ICON Group International, Inc.

Kee, K. (2019). *A simple guide to tuberous sclerosis, diagnosis, treatment and related conditions*. Amazon.com Services LLC.

Keegan, B. M. (2016). *Common pitfalls in multiple sclerosis and CNS demyelinating diseases, case-based learning*. Cambridge University Press.

Koziol, L. F., & Ely Budding, D. (2009). *Subcortical structures and cognition: Implications for neuropsychological assessment*. Springer.

Kurien, B. T., & Scofield, R. H. (2015). *Western blotting: Methods and protocols (methods in molecular biology)*. Humana Press.

Kwiatkowski, D. J., Holets Whittemore, V., & Thiele, E. A. (2010). *Tuberous sclerosis complex: genes, clinical features, and therapeutics*. Wiley-Blackwell.

Lisak, R. P., Truong, D. D., Carroll, W. M., & Bhidayasiri, R. (2016). *International neurology* (2nd ed.). Wiley-Blackwell.

McDermott, M. W., & Larson, D. (2006). *Radiosurgery for benign CNS tumors, an issue of neurosurgery clinics of North American (volume 17, number 2) (The clinics: Surgery)*. Saunders.

Mita, M., Mita, A., & Rowinsky, E. K. (2015). *mTOR inhibition for cancer therapy: Past, present and future*. Springer.

Parker, P. M. (2007). *Tuberous sclerosis—A bibliography and dictionary for physicians, patients, and genome researchers*. ICON Group International, Inc.

Paxinos, G., Furlong, T., & Watson, C. (2019). *Human brainstem: Cytoarchitecture, chemoarchitecture, myeloarchitecture*. Academic Press.

Petrides, M. (2018). *Atlas of the morphology of the human cerebral cortex on the average MNI brain*. Academic Press.

Ruggieri, M., Castroviejo, I. P., & Di Rocco, C. (2008). *Neurocutaneous disorders: Phakomatoses & hamartoneoplastic syndromes*. SpringerWien.

Rust, C. (2017). *It's nice to meet me too: The laughter and lessons learned from raising randy, a young man with tuberous sclerosis*. Beaver's Pond Press.

Samkoff, L. M., & Goodman, A. D. (2014). *Multiple sclerosis and CNS inflammatory disorders (neurology in practice)*. Wiley-Blackwell.

Shettleworth, S. J., Bloom, P., & Nadel, L. (2012). *Fundamentals of comparative cognition (fundamentals in cognition)*. Oxford University Press.

Singh, N. N. (2016). *Handbook of evidence-based practices in intellectual and developmental disabilities (evidence-based practices in behavioral health)*. Springer.

Singh, P. (2019). *Tuberous sclerosis: Benign burden of brain*. Amazon.com Services LLC.

Vargo, F. E. (2015). *Neurodevelopmental disorders: A definitive guide for educators*. W.W. Norton & Company.

Weichhart, T. (2012). *mTOR: Methods and protocols (methods in molecular biology)*. Humana Press.

Yokota, T., & Maruyama, R. (2018). *Exon skipping and inclusion therapies: Methods and protocols (methods in molecular biology)*. Humana Press.

Chapter 24

Brain tumors in children

Chapter Outline

Overview 417
Germ cell tumors 417
    Germinoma 421
    Teratoma 423
    Yolk sac tumor 426
    Embryonal carcinoma 428
    Choriocarcinoma 429
    Mixed germ cell tumors 431
    Medulloblastoma 432
Clinical cases 433
Key terms 435
Further reading 436

Overview

The majority of brain tumors in children, except for medulloblastomas, are germ cell tumors. Overall, pediatric brain tumors are rare, yet they are the second most common cancers in children after leukemia. In children, the most common type of brain tumor is malignant medulloblastoma. Pediatric brain tumors in younger children often cause changes in skull diameter and bulging of the fontanelles. For treatment, chemotherapy is often used in younger children instead of radiation therapy, since radiation may have negative effects upon the brain as it is developing. Fortunately, the majority of children are able to survive most forms of primary brain tumors.

Germ cell tumors

The major germ cell tumors include germinoma, teratoma, yolk sac tumor, embryonal carcinoma, choriocarcinoma, and mixed germ cell tumors that include various types of tumors. Pure germinomas that contain **syncytiotrophoblastic giant cells** are a distinct variant of germ cell tumors. Teratomas may be classified as mature, immature, or tumors that show signs of malignant transformation.

Epidemiology

Germ cell tumors of the central nervous system (CNS) mostly occur in children and adolescents. They are most prevalent in eastern Asia in comparison with the United States and Europe, making up 2%–3% of all primary intracranial neoplasms. In a study conducted in Japan, China, the Republic of Korea, and Taiwan, CNS germ cell tumors accounted for 8%–15% of pediatric brain tumors. However, in North America and Europe, they made up only 0.3%–0.6% of all primary intracranial tumors, and 3%–4% of pediatric brain tumors. Highest incidence is in Japan, with 0.45 cases of every 100,000 population under the age of 15. This is more than twice the incidence rate in the United States and Germany.

In a Japanese study, 70% of patients with germ cell tumors were between 10 and 24 years of age, and 73% of these were males. For congenital germ cell tumors—mostly teratomas—only 2.9% of patients were under 5 years of age, and 6.2% were above 35 years of age. In males, most cases occur in the pineal region, while in females, most cases occur in the suprasellar region. In this study, male patients greatly predominated, having 89% of the teratomas documented, as well as 78% of germinomas, and 75% of other types of germ cell tumors. While pure germinomas are most common, mixed lesions and teratomas are also seen significantly. In another Japanese study, the following results were obtained (also shown in Fig. 24.1):

- Germinomas (including those with syncytiotrophoblastic features)—41% of patients (of this percentage, the syncytiotrophoblastic features made up 5.2%);
- Mixed germ cell tumors—32%;

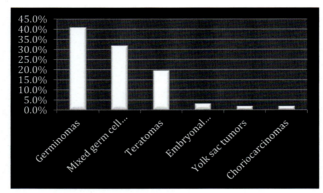

FIGURE 24.1 Prevalence of various germ cell tumors in a Japanese study.

- Teratomas—19.6% (of this percentage, 63.3% were mature teratomas, while 23.3% were immature and 13.3% showed malignant development);
- Embryonal carcinomas—3.3%;
- Yolk sac tumors—2%;
- Choriocarcinomas—2%.

Peak incidence in children is between 10 and 14 years of age, with males affected between 2 and 3 times as often as females. It is important to understand that relative incidence rates of each tumor type vary based on the CNS location in which they develop.

Etiology and risk factors

The majority of CNS germ cell tumors occur in peripubertal patients. Circulating **gonadotropin** levels are involved in their development. This is proven by their development within diencephalic centers that regulate gonadal activity, as well as their increased incidence along with **Klinefelter syndrome**. The link with this syndrome may show an X chromosome overdosage, which is a commonly seen genetic component of these lesions. Risk factors for CNS germ cell tumors are unknown.

Pathology

Approximately 80% of CNS germ cell tumors develop along a midline axis that extends from the pineal gland to the suprasellar compartment (the second most common site). Suprasellar tumors originate within the neurohypophyseal-infundibular stalk. Also, there are diffuse periventricular, intraventricular, cerebral hemispheric, thalamostriate, bulbar, cerebellar, intramedullary, and intrasellar variants. Teratomas are often congenital **holocranial** variants. While germinomas are most common in the suprasellar area and the basal ganglionic-thalamic regions, the nongerminomatous forms are not as common in these sites. When they are multifocal, CNS germ cell tumors are usually in the suprasellar compartment and pineal region. This can occur in both areas at the same time, or in one area, followed by the other. Bilateral basal ganglionic and thalamic lesions are also well documented.

The exact histogenesis of germ cell tumors is controversial. Gonadal and **neuraxial** germ cell tumors have similar histological, immunophenotypic, and genetic features. Neuraxial germ cell tumors are believed to be derived from primordial germ cells. This cell may migrate to the developing CNS, or may actually target it. Studies of fetal pineal glands with antibodies to *placental alkaline phosphatase (PLAP*, a primordial germ cell marker) have not proven primitive germ cell elements to be present. Germ cells may differentiate into multiple forms once they enter the CNS. Skeletal muscle-like cells from the developing pineal gland may descend from primitive germinal elements that migrate as part of **neuroembryogenesis**. Striated muscle-type cells that have unknown activities are also within the thymus, which basically lacks germ cells, and even so, this organ is a common site of germ cell tumorigenesis.

Another possibility implicates native cells that are embryonic (pluripotent) or neural in type. If this is true, there must be selective genetic programming of these precursors over the germ cell differentiation pathway, and they must undergo neoplastic transformation. This may be supported by the fact that the cells have **hypomethylation** of the imprinted *small nuclear ribonucleoprotein polypeptide N (SNRPN)* gene, along with primordial germ cell elements, such as in intracranial germ cell tumors, and with gonadal and other extraneuraxial germ cell tumors. Overexpression of transcription factor *OCT4* in the cells is linked to formation of teratomas. Some researchers believe that pure spinal cord teratomas are actual neoplasms of germ cell origin, but other experts believe they are complex malformations.

Pure intracranial teratomas, whether congenital or infantile, are much different from CNS germ cell tumors that form after early childhood. They resemble teratomas that grow in the testes of infants, usually being diploid, with general chromosomal integrity. After early childhood, CNS germ cell tumors, regardless of histology, have aneuploid profiles, overlapped patterns of genetic imbalance, and complicated chromosomal anomalies. The imbalances are mostly gains of chromosomes 1q, 2p, 7q, 8q, 10q, 12p, and X. There are usually losses of chromosomes 5q, 10q, 11q, 13, and 18q. There is debate as to whether gain of 12p and formation of isochromosome 12p occur in the CNS in high frequencies, along with the prevalence of X duplication. Most studies suggest that isochromosome 12p is found in fewer CNS germ cell tumors than in those in other body locations. Areas of higher gains involve *cyclin D2 (CCND2)* and *PR/SET domain 14 (PRDM14)*, which is a regulator of primordial germ cell specification. Losses of the *RB transcriptional corepressor 1 (RB1)* locus may implicate the cyclin-*cyclin dependent kinase (CDK)*-retinoblastoma protein-*eukaryote 2 factor (E2F)* pathway.

Upregulated *tyrosine-protein kinase (KIT)/rat sarcoma (RAS)* signaling is a component of gonadal seminomas as well as mediastinal **seminomas**. It is also present in most intracranial germinomas. The majority of these tumors either activate KIT family or *RAS* family mutations linked to severe chromosomal instability. These alterations are less often found in other germ cell tumor types. In CNS germinomas, an inactivation mutation of *Casitas B-lineage lymphoma (CBL)* has been documented as encoding a negative KIT regulator. Less common abnormalities in many intracranial germ cell tumors include inactivating mutations of tumor suppressor gene *BCL6 corepressor-like 1 (BCORL1)* and activating *protein kinase B (AKT)-mammalian target of rapamycin (mTOR)* lesions. Epigenetic alterations in gonadal and some intracranial germ cell tumors involve hypomethylation of the imprinted *SNRPN* gene and the *insulin-like growth factor 2 (IGF2)-H19 gene* imprinting control area. However, amplification of chromosome 19's *microRNA cluster*, known as *C19MC*, has not been seen in immature teratomas, even though neural tube-like rosettes are common in these tumors.

The CNS germ cell tumors usually are sporadic in occurrence. Klinefelter syndrome increases risks for intracranial and mediastinal germ cell tumorigenesis. This is characterized by a 47 XXY genotype and may be related to increased dosage of an X chromosome-related gene, since CNS and other germ cell tumors usually have additional X chromosomes, along with potentiating effects of chronically raised levels of gonadotropin. Down syndrome is linked to increased risk of testicular germ cell tumorigenesis, and may also be linked with intracranial germ cell tumors. These tumors are also related to neurofibromatosis type 1 (NF1) between siblings, between parents and offspring, and in the fetus of a woman with independent ovarian teratoma. In rare cases, CNS germ cell tumors have developed secondary gonadal or mediastinal germ cell neoplasms. One of these patients had Down syndrome. Germline variants of the chromatin-modifying gene called *Jumonji domain-containing 1C (JMJD1C)* may be linked to increased risks for intracranial germ cell tumors, according to a Japanese study. The risk factors for germ cell tumors are discussed individually for each subtype below.

Clinical manifestations

Tumor histologies and locations influence clinical manifestations. Germinomas usually cause symptoms more slowly than the other forms. If a lesion is within the pineal region, the cerebral aqueduct becomes compresses and obstructed, causing progressive hydrocephalus and intracranial hypertension. They often compress and invade the **tectal plate**, resulting in paralysis of upwards gaze and convergence, known as **Parinaud syndrome**. The optic chiasm is impinged upon by neurohypophyseal-suprasellar germ cell tumors. This results in visual field defects, and often, disruption of the **hypothalamohypophyseal axis**. This occurs prior to development of diabetes insipidus and signs of pituitary failure, such as delays in sexual maturation and growth. In males, precocious puberty may develop because of secretion of *human chorionic gonadotropin (hCG)* from the neoplastic syncytiotrophoblasts, which stimulates production of testosterone. Rarer instances of precocious puberty occur in females with hCG-producing tumors, due to additional expression of cytochrome P450 aromatase, catalyzing the conversion of C19 steroids to estrogen. In this example, hCG may have some additional intrinsic follicle stimulating hormone-like activity. Local recurrence of some CNS germ cell tumors, along with cerebrospinal fluid-borne dissemination is usually how they progress. However, spread to the abdomen through ventriculoperitoneal shunts, plus hematogenous spread mostly to the lungs and bones is possible.

Diagnosis

Except for teratomas, germ cell tumors usually present as solid. Germ cell tumors are contrast enhancing on CT and MRI scans. Fig. 24.2 shows an MRI of a suprasellar germ cell tumor. Germinomas often enhance more

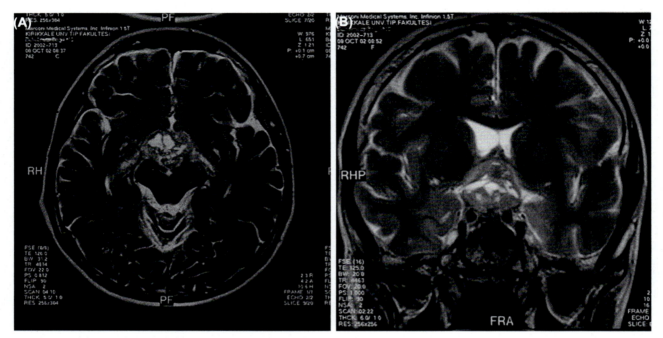

FIGURE 24.2 MRI of a suprasellar germ cell tumor. (A) Without contrast. (B) With contrast.

homogeneously than other types. The tumor tissue is usually hypointense to isointense on T1-weighted imaging, and isointense to hyperintense on T2-weighted imaging. Thalamic and basal ganglia germinomas may be more likely to become calcified or cystic than tumors within the pineal or suprasellar areas. They can show only slight T1-weighted signal abnormalities, and very poor definition of T2-hyperintensity, with no enhancement or only very slight enhancement. Calcified regions, intratumoral cysts, and features having low-signal-attenuation characteristics similar to those of fat suggest teratomas. Choriocarcinomas are often hemorrhagic. Tumor spread, along or beneath the ependymal lining of a brain ventricle, may be evaluated by **neuroendoscopic** procedures, yet not be visible on MRI. Teratomas and some other congenital germ cell neoplasms may be detected in utero by ultrasonography or **fast MRI**.

Treatment

Treatments are quite varied based on histological subtypes of these tumors. Mature teratomas may be cured via surgical excision. Surgery is commonly performed for teratomas, and may be used for recurrent germ cell tumors. Pure germinomas are highly sensitive to radiation therapy. Options include external radiation therapy, stereotactic radiosurgery, and internal radiation. External radiation therapy is commonly used to treat CNS germ cell tumors in children. Chemotherapy is another treatment option, and high-dose chemotherapy with **stem cell rescue** is often chosen. Systemic chemotherapy is commonly used to treat CNS germ cell tumors. New types of treatment are being tested in clinical trials, with include **targeted therapy**. Patients may be enrolled into clinical trials before, during, or after starting treatment.

Prognosis

The histological subtype of a germ cell tumor is the most prognostic factor. Long-term survival rates of more than 90% have been achieved after only craniospinal irradiation of pure germinomas. When chemotherapy is added, there may be comparable control using lower doses of radiation and smaller field volumes. A more recent study showed that beta-hCG mRNA expression over various CNS germ cell tumors was not related to recurrence of pure germinomas. Tumors with the worst prognosis include yolk sac tumors, embryonal carcinomas, choriocarcinomas, and mixed lesions involving these subtypes. Treatment of the more aggressive subtypes has resulted in survival between 60% and 70% of patients, when chemotherapy is combined with radiation therapy.

> **Point to remember**
> According to the Pediatric Brain Tumor Foundation, more children die of brain tumors than from any other form of cancer. In the United States, over 28,000 children and adolescents are living with the diagnosis of a primary brain tumor. Approximately 4600 children and adolescents are diagnosed with a primary brain or CNS tumor every year—that is about 13 cases *per day*. There are more than 100 different types of brain tumors, making diagnosis and treatment challenging.

Germinoma

A *germinoma* is a malignant germ cell tumor with large primordial germ cells that have prominent nucleoli and various amounts of cytoplasmic clearing. These tumors are not differentiated upon discovery, and may be benign or malignant.

Epidemiology

Germinomas are the most common CNS germ cell tumors. Males are about twice as likely to develop these tumors than females. They are most common between the ages of 10 and 21 years. Intracranial germinomas occur in only 0.7 of every 1 million children, mostly on or near the midline of the pineal or suprasellar areas, but also within the medulla oblongata. In 5%–10% of patients, the tumor infiltrates both the pineal and suprasellar areas. In some cases, the tumor has extended into the fourth ventricle.

Etiology and risk factors

Germinomas are believed to originate from developmental errors when primordial germ cells fail to migrate normally. However the causes and risk factors for germinomas are not fully understood.

Pathology

Almost all germinomas have a large population of reactive lymphoid cells. These tumors can also occur as part of mixed germ cell tumors, in combination with other types of germ cell tumors. They are composed of soft, friable tissue of a pink to tan to white color. While generally solid, there may be focal cystic changes. Hemorrhage and necrosis only rarely occur. The tumor cells are large, undifferentiated, and often irregular. They resemble primordial germinal elements, arranged in lobules, sheets, or when there is **stromal desmoplasia**, regular cords and trabeculae. Cytological features include nuclei that are centrally position, round, and vesicular. They have prominent nucleoli and discrete cell membranes. There is usually a fairly large amount of cytoplasm that is often clear because of accumulation of glycogen. Mitotic activity is seen and can be conspicuous. Thin fibrovascular septa, often infiltrated with small lymphocytes, are common. Some of these tumors have such a strong **lymphoplasmacellular** reaction that neoplasia is obscured. When a germinomas provokes intense granulomatous responses, it may appear like sarcoidosis or tuberculosis. Examples of germinoma cells from a pineal gland tumor are shown in Fig. 24.3.

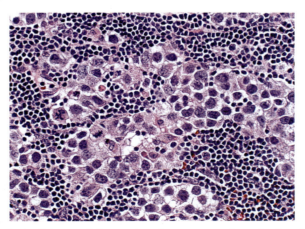

FIGURE 24.3 Pineal germinoma. The dual cellular population consists of numerous small reactive lymphocytes and large clear tumor cells that resemble primordial germ cells (hematoxylin–eosin stain, ×400).

Germinomas are distinguished from solid types of embryonal carcinomas and yolk sac tumors by consistent cell membrane and Golgi region immunoreactivity for KIT and membranous *monoclonal antibody D2−40* labeling. Cytoplasmic or membranous PLAP expression is nonspecific and less consistent. Germinomas often have immunoreactivity for the RNA-binding *LIN28A homolog protein*, along with nuclear expression of transcription factors that include: *estrogen receptor gamma (ESRG), homeobox protein NANOG, OCT4, Sal-like protein 4 (SALL4)*, and *undifferentiated embryonic cell transcription factor 1 (UTF1)*. They are usually nonreactive for *cluster of differentiation 30 (CD30)* and **alpha-fetoprotein**. A small amount show cytoplasmic labeling by the *AE1-AE3* and *CAM5.2* cytokeratin antibodies. This occurrence, along with ultrastructural proof of lumen and intercellular junction formation, may indicate an ability to differentiate towards embryonal carcinoma, or along epithelial lines. Pure germinomas may have syncytiotrophoblastic components expressing beta-HCG and human placental lactogen. The presence of such cells should not result in a diagnosis of choriocarcinoma. T cells often dominate reactive lymphoid elements in germinomas. This includes *CD4-expressing helper-inducer* and *CD8-expressing cytotoxic-suppressor* elements. However, CD20-labeling B cells and CD138-labeling plasma cells may be obvious, proving humoral immune responses to the tumor.

Clinical manifestations

Signs and symptoms of CNS germinomas include headache, hydrocephalus, nausea, vomiting, behavior or cognitive changes, fatigue, ataxia, vision changes, facial numbness, cranial nerve deficits, delayed or precocious puberty, diabetes insipidus, stunted growth, constipation, incontinence, leg weakness, gait instability, shortness of breath, and wheezing. These manifestations are based on the actual location of the tumor within the CNS.

Diagnosis

At the time of diagnosis, metastasis is seen in about 22% of patients. Often, serum and spinal fluid tumor markers of alpha-fetoprotein and beta-HCG are tested. Pure germinomas are not linked with these markers. In 1%−15% of cases, a low level of beta-HCG is produced. Though CT and MRI are used for imaging, the neuroimaging features of germinomas are similar to other tumors (see Fig. 24.4). Therefore, biopsy is the preferred method of diagnosis. Cytology of the CSF is often studied to detect spinal metastasis, which is important for staging and plans for radiation therapy.

Treatment

Germinomas are extremely sensitive to radiation therapy as well as chemotherapy, and combination treatments offer the potential for a cure. Chemotherapy is not usually given alone unless radiation therapy has contraindications. Chemotherapies often include combinations of bleomycin, cisplatin, and etoposide. Gross total surgical resection is not usually done since it can increase risks of complications. Complications from surgical resection include dysphagia. The

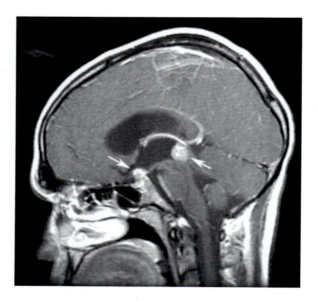

FIGURE 24.4 MRI of a germinoma.

best treatment results come from craniospinal radiation therapy with local tumor boost of more than 4000 **centigray** (cGy). Gamma knife radiation treatment is often used.

Prognosis

Fortunately, because of their sensitivity to radiation therapy, cure rates are high for germinomas. Intracranial germinomas have an approximate 90% survival to 5 years after diagnosis. The cure rate does not appear to be influenced by subtotal resection, so gross total resection is not usually necessary.

> **Point to remember**
>
> The prognoses for germinomas are based on their histologic variants, according to the Japanese Pediatric Brain Tumor Study Group. Pure, mature germinomas have a good prognosis, while a poor prognosis is given for histologies resembling yolk sac tumors, choriocarcinomas, embryonal carcinomas, and mixed tumors (of yolk sac tumors, choriocarcinomas, or embryonal carcinomas).

Teratoma

A **teratoma** is a germ cell tumor made up of **somatic tissue**, from two or three germ layers (out of the ectoderm, endoderm, and mesoderm). They are subclassified as the following:

- *Mature*—only containing adult-type tissues such as mature skin or skin appendages, adipose tissue, neural tissue, cartilage, smooth muscle, bone, minor salivary glands, gastrointestinal epithelium, and respiratory epithelium;
- *Immature*—containing immature, fetal, or embryonic tissues alone, or along with mature tissues;
- *Malignant transformative*—a rare form, containing a component that results from malignant transformation of a somatic tissue — usually a sarcoma or carcinoma; however, embryonal tumors with features of primitive neuroectodermal tumors can form in the CNS as well.

Epidemiology

Teratomas are generally uncommon, yet make up the largest proportion of fetal intracranial neoplasms. They constitute between 26% and 50% of all types of fetal brain tumors, and occur in a wide range of ages in children.

Etiology and risk factors

The causes of teratomas involve abnormal development of **pluripotent cells**, which include germ cells and embryonal cells. The teratomas of embryonic origin are congenital, while the teratomas of germ cell origin may or may not be congenital. However, teratomas in the CNS are usually derived from embryonic cells. Risk factors for teratomas are not understood.

Pathology

Mature teratomas consist only of completely differentiated adult-type tissues, with little or no mitotic activity (see Fig. 24.5). The ectodermal components often have epidermis and skin appendages, choroid plexus, and central nervous tissue. Common mesodermal components are smooth and striated muscle, bone, adipose tissue, and bone. The common endodermal components are glands that are often dilated as cysts, and lined with enteric-type or respiratory-like epithelia. However, hepatic-like and pancreatic-like tissue may be found. Also, gastrointestinal-like (with muscular coats) and bronchus-like structures (with cartilaginous rings), along with mucosa, may be found.

Immature teratomas are made up of incompletely differentiated elements that resemble fetal tissues (see Fig. 24.6). If they are admixed with mature tissues, any immature teratoma components mean that the tumor must be classified as immature. This is true even if incompletely differentiated components make up only a small part of the neoplasm. Often, there are compact, mitotically active blastema-like **stromas** resembling embryonic mesenchyme that have concentric bands of spindle cells encircling glandular epithelium. The spindle cells have hyperchromatic nuclei and significant mitoses and apoptosis. There are islands of primitive neuroectodermal elements that can form tubules and rosettes (neuroepithelial and multilayered). The rosettes may have a central lumen or **canalicular** structures resembling a developing neural tube. The primitive neuroepithelium has small cells and little cytoplasm, with hyperchromatic oval or

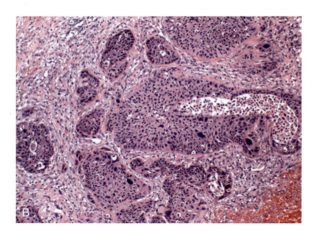

FIGURE 24.5 Mature teratoma cells.

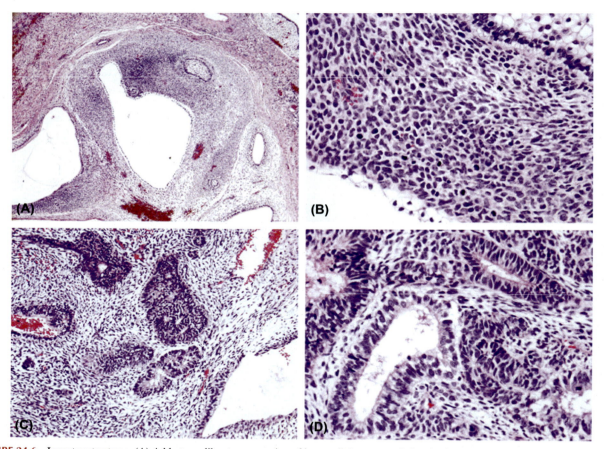

FIGURE 24.6 Immature teratoma. (A) A blastema-like stroma consists of hypercellular concentric bands of spindle cells encircling glandular epithelium. (B) On close examination, the spindle cells show hyperchromatic nuclei with ample mitoses and apoptosis. (C) Islands of primitive (embryonic-appearing) neuroepithelium including tubules and rosettes. (D) The primitive neuroepithelium is made up of small cells with scant cytoplasm and hyperchromatic oval or carrot-shaped nuclei with mitoses and apoptosis.

carrot-shaped nuclei, plus mitoses and apoptosis. Often, clefts with melanotic neuroepithelial lining are seen. These form when differentiation of the retinal pigment epithelium stops.

For malignant transformative teratomas (see Fig. 24.7), intracranial germ cell tumors may involve different somatic-type cancers. Usually, these are rhabdomyosarcomas or undifferentiated sarcomas. Less often, they are enteric-type adenocarcinomas, primitive neuroectodermal tumors, and squamous carcinomas. **Erythroleukemia** and leiomyosarcoma

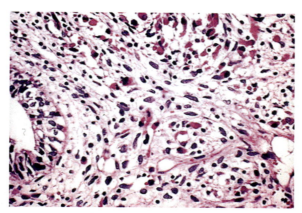

FIGURE 24.7 Teratoma with an additional malignant component (type III) consisting of mature teratomatous elements admixed with rhabdomyosarcoma.

has also occurred, as well as carcinoid tumors linked with intradural spinal teratomas. Yolk sac tumor components may be progenitors of enteric-type adenocarcinomas developing in intracranial germ cell tumors.

Clinical Manifestations

The clinical manifestations of teratomas are based on whether the tumors are intra-axial or extra-axial. Intra-axial teratomas usually present before birth or in the newborn period. They are large in size, increasing head circumference and often causing difficulty during delivery. They mostly occur supratentorially. Extra-axial teratomas usually present in childhood or early adulthood and are smaller in most cases. They usually arise in the pineal or suprasellar regions, presenting due to mass effect, with headache, vomiting, papilledema, obstructive hydrocephalus, Parinaud syndrome, or optic chiasm compression.

Diagnosis

Careful evaluation is required for the malignant transformative variant of teratomas, such as for pineal teratomas. The pathologist must specify the type of secondary cancer that is prevented. The tumor must never simply be diagnosed as a *malignant teratoma*. These tumors may be associated with elevated levels of serum alpha-fetoprotein or serum carcinoembryonic antigen (CEA). Imaging is usually heterogeneous, with a mixture of tissue densities and signal intensity. If fat is present, it helps in narrowing the differential diagnosis. CT scans are often performed, often without enhancement. Most intracranial teratomas have some amount of fat and calcification, which is usually solid and appears like "clumps." There are usually cystic and solid components that give the tumor and irregular outline. Solid components show variable enhancement. On T1-weighted images, the hyperintense components are due to fat and proteinaceous/lipid-rich fluid. The intermediate components are of soft tissue, and the hypointense components are due to calcification and blood products. Enhanced T1-weighted imaging offers enhancement of solid soft tissue components). T2-weighted imaging shows mixed signals from differing components (see Fig. 24.8). Based on tumor location, differential diagnoses are quite varied. For intra-axial tumors, they include *supratentorial primitive neuroectodermal tumor (sPNET)*, atypical rhabdoid/teratoid tumor (ATRT), and choroid plexus carcinoma. For extra-axial tumors, they include intracranial lipoma or dermoid, craniopharyngioma, and other pineal region tumors. Funduscopy is often performed to evaluate papilledema.

Treatment

The treatment for mature or immature teratomas begins with surgical resection. After surgery, if hydrocephalus develops, it will require proper treatment. Because teratomas are usually well encapsulated and noninvasive, they are relatively easy to resect, though this may be more difficult within the brain itself. For malignant teratomas, surgery is usually followed with chemotherapy. Teratomas that are in surgically inaccessible locations, or are extremely complex or likely to be malignant, are sometimes treated first with chemotherapy. Some patients experience hemiparesis following surgical resection.

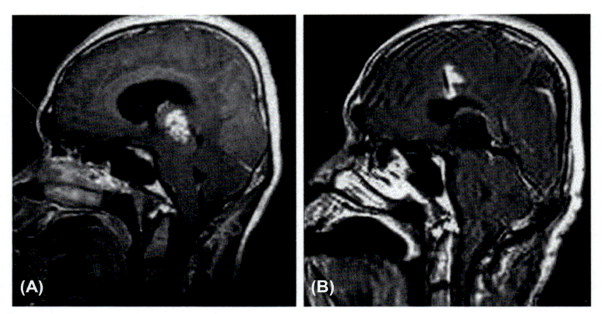

FIGURE 24.8 Different MRI views of a pineal teratoma.

Prognosis

Location of the tumor and its size determine prognosis. For intra-axial tumors, stillbirth is relatively common. Smaller extra-axial lesions have a good prognosis if they can be successfully resected.

Yolk sac tumor

A **yolk sac tumor** is an aggressive, nongerminomatous malignant germ cell tumor. It is made up of primitive germ cells that are arranged in a variety of patterns. These patterns may repeat those of the embryonic yolk sac, allantois, and the extra-embryonic mesenchyme, producing alpha-fetoprotein. Yolk sac tumors may occur as components of mixed germ cell tumor, along with other germ cell tumors. These tumors are also referred to as *endodermal sinus tumors* and *infantile embryonal carcinoma*. Yolk sac tumors often occur in non-CNS locations such as the chest, abdomen, pelvis, and testes.

Epidemiology

Though overall rare, intracranial yolk sac tumors mostly develop in the pineal or suprasellar regions, but have been documented in various lobes of the brain. The actual incidence and prevalence rates are not documented.

Etiology and risk factors

The cause of yolk sac tumors is not understood, but it is most likely linked to gene mutations. There are no known risk factors for CNS yolk sac tumors.

Pathology

Yolk sac tumors are usually gray to tan in color and solid. They are most often friable or, because of significant myxoid changes, gelatinous. There may be obvious focal hemorrhage. The cells are primitive-appearing epithelial cells in a loose, often myxoid matrix that resembles the extraembryonic **mesoblast**. There is variable cellularity. The cells are believed to differentiate towards the yolk sac endoderm. Solid sheets may be formed by the epithelial cellular elements, but more often, are arranged loosely within a network of irregular tissue spaces, called a **reticular pattern** (see Fig. 24.9). They may also be located around anastomosing sinusoidal channels as cuboidal epithelia. Sometimes, the cells form drape-like structures over fibrovascular projections, forming papillae, which are known as **Schiller—Duval bodies** (see Fig. 24.10). There may be oddly constricted microcysts with flat epithelia, known as **polyvesicular vitelline patterns**, along with entire-type glands that have goblet cells, and foci of hepatocellular

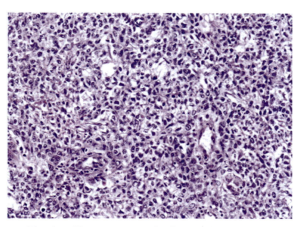

FIGURE 24.9 Yolk sac tumor, solid pattern. The sheet-like arrangement of cells may mimic a seminoma.

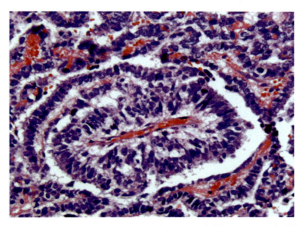

FIGURE 24.10 Yolk sac tumor. On transverse section, the festoons resemble glomeruloid structures (Schiller − Duval bodies).

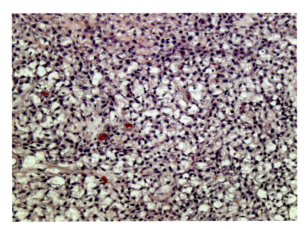

FIGURE 24.11 Yolk sac tumor, reticular microcystic pattern. The tumor cells have characteristic prominent cytoplasmic vacuoles creating a sieve-like appearance and microcysts.

differentiation known as the **hepatoid variant**. Characteristic appearances include brightly eosinophilic, periodic acid-Schiff-positive, and diastase-resistant hyaline globules or vacuoles with a sieve-like appearance (see Fig. 24.11). These may be varied, or occupy epithelial cell cytoplasm, or lie within extracellular spaces. There is widely varying mitotic activity, but necrosis is rare.

Yolk sac tumors are distinguished by cytoplasmic immunoreactivity of their epithelial elements for alpha-fetoprotein. The hyaline globules are also reactive. The epithelial components regularly label for cytokeratins, and are often positive for the protein-coding gene called *glypican-3*. They may sometimes be positive for *placental alkaline phosphatase (PLAP)*, and usually label for *Lin-28 homolog A (LIN28A)*, with nuclear expression of *Sal-like protein 4 (SALL4)*. However, *octamer-binding transcription factor 4 (OCT4)* expression is unusual, and *tyrosine-protein kinase KIT* reactivity is rare, but if is it present, is usually focal and cytoplasmic instead of membranes, and lacks Golgi area accentuation. There is no expression of beta-hCG or human placental lactogen.

Clinical manifestations

Yolk sac tumors may cause seizures, nausea, vomiting, headache, coughing, fever, changes in bowel function or habits, fatigue, breathing difficulties, weight loss, and night sweats. They can also cause hypotension, back pain, weakness, dizziness, papilledema, cranial nerve palsies, and ataxia. A positive Babinski sign is often seen.

Diagnosis

Diagnosis of CNS yolk sac tumors is based on histology, which usually includes malignant endodermal cells that secrete alpha-fetoprotein. The cells can be detected in tumor tissue, serum, cerebrospinal fluid, and even the urine and amniotic fluid. These tumors may occur as small malignant foci within a larger teratoma or other tumor. Biopsy of the tumor may reveal only the larger tumor, while elevate alpha-fetoprotein reveals the presence of the yolk sac tumor. The transcription factor *GATA-4* may also aid in diagnosis. However, the extremely high levels of alpha-fetoprotein found in these groups complicate diagnosis in pregnant and women. Tumor surveillance via monitoring of alpha-fetoprotein requires accurate corrections for gestational age in pregnant women, and the age of the infant. In infants, tumor marker tests are interpreted with a reference table or graph of normal alpha-fetoprotein. Imaging studies for yolk sac tumors include CT and MRI scans, usually with contrast. A *pure primary intracranial yolk sac tumor* is rare, but has been documented, and does not involve other areas of the body.

Treatment

Treatments for yolk sac tumors usually involve surgical resection combined with chemotherapies such as bleomycin, cisplatin, and etoposide. If promptly treated, death from yolk sac tumors is rare.

Prognosis

The prognosis for intracranial yolk sac tumors is poorer than for germinoma, with a median survival of 2 years or less. In previous decades, yolk sac tumors were extremely lethal, but today's advances have greatly improved prognosis.

Embryonal carcinoma

Embryonal carcinoma is also an aggressive, nongerminomatous malignant germ cell tumor. It features large epithelioid cells that are similar to those of the embryonic germ disc. There is usually geographical necrosis and a high mitotic count. Structures that are present include pseudopapillary and pseudoglandular formations. This type of tumor may also develop as part of a mixed germ cell tumor, combined with other germ cell tumors.

Epidemiology

Embryonal carcinoma is relative rare, and most often occurs outside the CNS, in the ovaries and testes. However, these tumors are least common in prepubertal children. Intracranial embryonal carcinomas make up 10% of all intracranial germ cell tumors. The majority of these tumors in children are malignant. Very little is known about the incidence of embryonal carcinomas in the CNS. Patients have ranged from birth to 16 years.

Etiology and risk factors

CNS embryonal carcinomas may begin in embryonic cells that remain in the brain after birth. Though risk factors are uncertain, these tumors may be related to genetic conditions such as Turcot syndrome, **Rubinstein−Taybi syndrome**, Gorlin syndrome, Li−Fraumeni syndrome, and **Fanconi anemia**. The tumors may spread via the CSF to other areas of the brain and spinal cord.

Pathology

Embryonal carcinomas are solid, and made up of friable grey to white tissues that may have focal hemorrhage and necrosis. They have large cells proliferating in sheets and nests. The cells form premature papillae, or line spaces resembling glands. The tumor cells form embryoid bodies with germ discs and tiny amniotic cavities. There may be abundant clear to violet-colored cytoplasm, macronucleoli, a high mitotic count, and areas of coagulative necrosis. Embryonal carcinomas are distinguished from other germ cell tumors by cytoplasmic immunoreactivity for *cluster of differentiation 30 (CD30)*. They are strongly and uniformly reactive for cytokeratins, and in many cases, positive for PLAP. There is labeling for LIN28A, plus nuclear expression of *embryonic stem cell related gene (ESRG)*, OCT4, SALL4, *Sry-Box transcription factor 2 (SOX2)*, and *undifferentiated transcription factor 1 (UTF1)*. There may be KIT expression that is mostly focal and nonmembranous. Usually, embryonal carcinomas are negative for alpha-fetoprotein, beta-hCG, and human placental lactogen.

Clinical manifestations

Signs and symptoms of embryonal carcinomas include eyelid retraction, vertical gaze paresis, convergence or retraction nystagmus, and pupils that respond poorly to light. There may be oculomotor difficulties, symptoms of increased intracranial pressure, and hemiparesis. In a few cases, Parinaud syndrome and precocious puberty developed.

Diagnosis

It is important to distinguish embryonal carcinomas, via biopsy, from other intracranial tumors. From the limited data collected, it is believed that embryonal carcinomas are more common than initially documented, and that current treatments for them may be insufficient. Diagnostic methods include angiography, CT scan with or without contrast, MRI, and arteriography. Usually, imaging reveals displacement of the deep venous system and the origin of feeding tumor vessels. If the tumor is in the pineal region, there may be a heterogenous mass with scattered calcification, which enhances with contrast. Suprasellar masses may be isodense and enhance homogenously with contrast.

Treatment

Treatments for intracranial embryonal carcinomas include radiation therapy and chemotherapy, as well as partial or total surgical resection. Chemotherapies have included methotrexate, citrovorum, and cisplatin.

Prognosis

Prognosis for embryonal carcinomas is varied. Patients have survived for over 2 years following treatment, but the majority of children survive less than 1 year.

> **Point to remember**
>
> The four types of nonmedulloblastoma embryonal tumors include: embryonal tumors with multilayered rosettes, medulloepitheliomas, CNS neuroblastomas, and CNS ganglioneuroblastomas. Embryonal tumors with multilayered rosettes most commonly occur in young children, and are fast growing tumors. Medulloepitheliomas occur most often in infants and young children. CNS neuroblastomas may be large, and spread to other areas of the brain or spinal cord.

Choriocarcinoma

Choriocarcinoma, like the other tumors in this chapter, is an aggressive, nongerminomatous malignant germ cell tumor. However, it is imposed of **cytotrophoblasts** and syncytiotrophoblasts. Sometimes, it is made up of intermediate trophoblasts. Choriocarcinoma belongs to the spectrum known as *gestational trophoblastic disease*.

Epidemiology

Overall, choriocarcinoma is a rare CNS tumor. The documented cases of choriocarcinoma in children have occurred between the ages of 1 month and 8 years. However, these tumors do occur more often in young adults, in their 20s.

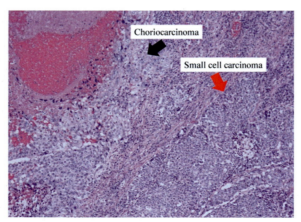

FIGURE 24.12 Choriocarcinoma occurring along with a small cell carcinoma.

Etiology and risk factors

It is believed that choriocarcinomas originate from retained primordial germ cells that migrate abnormally during embryogenesis. There are no known risk factors.

Pathology

With choriocarcinoma, necrosis and hemorrhage are commonly present. There is usually a significant elevation of hCG in the CSF or blood. These tumors sometimes occur as components of mixed germ cell tumors, or with other germ cell tumors (see Fig. 24.12). They are solid tumors. Neoplastic syncytiotrophoblast surrounds or drapes the cytotrophoblastic components, which are made up of masses of large, mononucleated cells having vesicular nuclei. Their cytoplasm is clear or acidophilic. Characteristic appearances include pools of blood, ectatic vascular channels, and a lot of hemorrhagic necrosis. The syncytiotrophoblasts have diffuse cytoplasmic immunoreactivity for beta-hCG and human placental lactogen. Cytokeratin labeling is seen, with some of these tumors expressing PLAP. However, there is no KIT or OCT4 labeling.

Clinical manifestations

Signs and symptoms of choriocarcinoma are varied based on location. They may include increased quantitative chorionic gonadotropin, and symptoms related to increased intracranial pressure, including headache and dizziness.

Diagnosis

Histologic diagnosis requires the presence of cytotrophoblasts and syncytiotrophoblasts. These are both forms of giant cells usually containing multiple hyperchromatic or vesicular nuclei. The nuclei are often seen in knot-like clusters, in large amounts of purple-colored or basophilic cytoplasm. Imaging studies include CT, MRI, and X-rays.

Treatment

Treatments for choriocarcinoma include surgical resection, chemotherapy, and radiation therapy. Choriocarcinoma is extremely responsive to chemotherapies such as methotrexate, with up to 95% of patients responding favorably. For intermediate and high-risk patients, **EMACO therapy** is given, which consists of etoposide, methotrexate, actinomycin, cyclosporine, and vincristine.

Prognosis

Children with choriocarcinomas have a better prognosis compared to adults who develop these tumors. Patients without metastases have a high recovery rate, but the tumor is usually fatal once metastasis has occurred.

> **Point to remember**
>
> Intracranial choriocarcinomas are highly malignant tumors arising from the embryonal chorion. These tumors have been documented with disseminated brain metastases as well as being primary tumors within the intracranial third ventricle. Most documented cases have been in patients in their 20s, but these tumors are also prevalent in children.

Mixed germ cell tumors

A **mixed germ tells tumor** is a rare type of cancer made up of at least 2 different types of germ cell tumors. These include choriocarcinoma, embryonal carcinoma, yolk sac tumor, teratoma, and seminoma. These tumors occur in the brain, chest, and abdomen, but more often in the ovaries or testicles.

Epidemiology

About 60% of all germ cell tumors are mixed. The three most common combinations of tumors are as follows:

- Teratoma plus embryonal carcinoma plus yolk sac tumor;
- Seminoma plus embryonal carcinoma;
- Teratoma plus embryonal carcinoma.

There has been an increase in incidence of mixed germ cell tumors over the past 20 years, but this is probably due to better methods of classifying them.

Etiology and risk factors

The causes and risk factors for mixed germ cell tumors are based on the tumor subtypes that are combined, as previously discussed individually.

Pathology

Mixed germ cell tumors show the macroscopic features of the germ cell tumors that comprise them, which have already been described previously. Any combination of germ cell tumors can be seen in combination. Individual tumor components have the same antigenic profiles as already discussed. The heterogeneous appearance of mixed germ cell tumors includes distinctive regions that look different from one another, encompassing neuroectodermal tissue, squamous epithelium, and even cartilage and skeletal muscle. The neuroectodermal tissue may consist of neuroepithelial elements with anaplastic or mature giant cells. Mixed germ cell tumors are usually solid, encapsulated, and cystic, and often white in color. The tumor mass often has many low-density areas that represent necrosis and hemorrhage. The microscopic features are based on the component tumors. There are documented cases of pituitary mixed germ cell tumors, with parasellar and suprasellar extension.

Clinical manifestations

Signs and symptoms of mixed germ cell tumors of the CNS are based on the individual subtypes of tumors combined, as discussed individually above. Most often, headache and visual impairment are seen. There may be bony erosion within the skull in some cases.

Diagnosis

The pathologist reporting a mixed germ cell tumor must specific which of the subtypes of tumors are present, and document the relative proportions of each of them. Immunostains are useful for differentiating the component tumors. The individual subtypes of tumors are diagnosed as previously discussed for each of them. A variety of imaging studies are used, including MRI, CT, and arteriography.

Treatment

Treatments of mixed germ cell tumors of the CNS are based on the individual subtypes of tumors combined, as already discussed individually above. After surgical resection, some patients experience transient diabetes insipidus. Chemotherapies include cisplatin-etoposide-bleomycin combination therapy. Radiation therapy is used on an individualized basis.

Prognosis

The general prognosis of mixed germ cell tumors is based on the combinations of the tumor subtypes that are present, as previously discussed. Generally, the worst prognosis exists for the teratoma + embryonal carcinoma + yolk sac tumor subtype.

Medulloblastoma

Medulloblastoma is a significant CNS tumor in children. Medulloblastomas are most common in the cerebellum and dorsal brainstem. They are embryonal neuroepithelial tumors presenting mainly in childhood, and consisting of densely packed small, round, undifferentiated cells that have mild to moderate nuclear pleomorphism, with a high mitotic count. Most medulloblastomas are grade IV tumors. There are five subtypes of medulloblastomas, which include:

- WNT-activated—anaplastic (very rare);
- Sonic hedgehog (SHH)-activated tumor protein 53 (TP53)-mutant—anaplastic or desmoplastic/nodular (very rare);
- SHH-activated TP53-wildtype—anaplastic or desmoplastic/nodular, with extensive nodularity;
- Non-WNT/non-SHH, group 3—anaplastic;
- Non-WNT/non-SHH, group 4—anaplastic, but more rare.

Epidemiology

Medulloblastomas are the most common CNS embryonal tumors and the most common malignant tumors of childhood. The *SHH-activated, TP53-wildtype* form of medulloblastoma is prevalent in infants, but when it has extensive nodularity, is a tumor of low-risk. The SHH-activated, TP53-mutant subtype is a high-risk tumor prevalent in children between 7 and 17 years of age. Also, the WNT pathway-activated subtype is more common in children than in adults, and the non-WNT/non-SHH subtype is more common in infants and children. The WNT pathway-activated subtype usually presents in children between 7 and 14 years of age. The likelihood of developing a medulloblastoma during childhood increases with age—children approaching age 18 have the highest incidence. Annual incidence of all medulloblastomas during childhood is six cases per 1 million children. In the United States, overall incidence is highest in Caucasian non-Hispanics (2.2 cases per 1 million), Hispanics (2.1 cases per 1 million), and African Americans (1.5 cases per 1 million). The median patient age at diagnosis is 9 years of age, though there are peaks in incidence at 3 and 7 years of age. About 77% of patients with medulloblastomas are under 19 years of age. There is an overall male-to-female ratio of 1.7:1, though incidence between the genders in patients 3 years of age or less is basically equal.

Etiology and risk factors

There is genetic susceptibility to medulloblastomas in monozygotic twins, siblings, and relatives. The exact underlying cause of these tumors, however, is unknown. Most cases occur sporadically, but there may be links to Li—Fraumeni syndrome, Gorlin syndrome, and Turcot syndrome. There also may be a specific chromosomal abnormality, known as *isochromosome 17q*, with associated loss or inactivation of genetic information. Additional chromosomal abnormalities have been identified, on chromosomes 1, 7, 8, 9, 10q, 11, and 16.

Pathology

Medulloblastomas usually spread through CSF fluid pathways, seeding the neuroaxis with metastatic tumor deposits. In rare cases, they reach the lymphatic system and bones. Most medulloblastomas form in the cerebellar vermis, appearing as pink to gray, often-friable masses, filling the fourth ventricle. In the cerebellar hemispheres, the tumors are usually firm and well circumscribed. Extensive necrosis is rare. If disseminated, discrete tumor nodules may be found in the craniospinal leptomeninges of CSF pathways. In extremely rare cases, medulloblastomas have occurred along with Rubinstein-Taybi syndrome, **Nijmegen breakage syndrome**, neurofibromatosis, and **ataxia-telangiectasia**. Risk factors may include ionizing radiation, male gender (which has a slight prevalence), younger age, and genetic conditions such as nevoid basal cell carcinoma syndrome (NBCCS) or *breast cancer type 1 (BRCA1)* gene mutations.

Clinical manifestations

Medulloblastomas may cause increased intracranial pressure, headaches (most often in the morning), nausea, vomiting, and ataxia. Symptoms often manifest 1–5 months before diagnosis is made. The child often shows no interest in any stimuli, and later manifestations include a stumbling gait, truncal ataxia, falling, diplopia, papilledema, and sixth cranial nerve palsy. Dizziness and nystagmus, with or without facial sensory loss or motor weakness may develop.

Diagnosis

Medulloblastoma is distinctive on T1-weighted and T2-weighted MRI imaging, with heterogeneous enhancement. A medulloblastoma using T2-weighted MRI and T2-fluid-attenuated inversion recovery (FLAIR) MRI imaging is shown

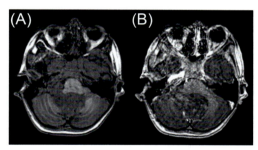

FIGURE 24.13 Medulloblastoma. (A) T2-weighted MRI image, and (B) T2 FLAIR MRI image. *Courtesy: Elizabeth Presutto BS, Matthew Chappell MD, Joseph Fullmer MD, and Sajeev Ezhapilli MBBS, DNB, DABR; Radiology Case Reports Volume 13, Issue 2.*

in Fig. 24.13. The tumors appear well circumscribed and solid, with extreme cellularity. They have high mitotic activity, small amounts of cytoplasm, and often form clusters and rosettes. Diagnosis may be made via the **Chang staging system**. The correct diagnosis of medulloblastoma may require the ruling out of *atypical teratoid rhabdoid tumor*. Also, lumbar puncture to check the CSF is used to determine tumor metastasis.

Treatment

The first step in treatment is surgical resection of the tumor. Adding radiation therapy to the entire neuroaxis, plus chemotherapy may increase disease-free survival. Different chemotherapeutic regimens may be used. Most of these involve a combination of cisplatin, lomustine, carboplatin, cyclophosphamide, or vincristine. In patients less than age 4, chemotherapy can delay or eliminate the need for radiation therapy, but both therapies can have long-term toxicities. Proton beam irradiation may reduce the impact of radiation, and can reduce the cognitive late effects of cranial irradiation. The adverse effects of radiation therapy include cognitive impairment, bone growth retardation, psychiatric illness, endocrine disruption, and hearing loss. Increased intracranial pressure may require use of corticosteroids or a ventriculoperitoneal shunt. Some patients require physical therapy rehabilitation after treatment.

Prognosis

In children, the WNT-pathway-activated subtype of medulloblastoma has an excellent prognosis today. Prognosis approaches 100% because of available surgical techniques and adjuvant therapies. Unfortunately, the prognosis is poor for the SHH-activated, TP-53 mutant subtype. The combination of proton beam irradiation with surgical resection and chemotherapy allows for 5-year survival in more than 80% of cases. Prognosis is worse if the child is under 3 years of age, the degree of surgical resection is inadequate, or if there is spread to the CSF, spine, supratentorial region, or other body systems.

> **Point to remember**
> The National Cancer Institute describes medulloblastoma as a primary CNS tumor of grade IV. They are malignant and grow quickly. In the diagnostic Chang staging system, an M0 score means "no evidence of gross subarachnoid metastasis." An M1 score means "microscopic tumor cells in CSF," and an M2 score means "gross nodules in the cerebellum, cerebral subarachnoid space or the third or lateral ventricles." An M3 score means "gross nodules in the spinal subarachnoid space," and an M4 score means "extraneural metastasis."

Clinical cases

Clinical case 1

1. What is the description of the cells of a germinoma?
2. What is the basic epidemiology of germinomas?
3. How sensitive are germinomas to various treatments?

A 14-year-old boy presented with right facial numbness and gait instability that had been present for 6 months. Neurological examination revealed right-sided deficits in cranial nerves V, VIII, and IX. An MRI showed a lesion within the dorsal region of the medulla oblongata. A subtotal resection was performed. The tumor was pink, soft, and irregular. It originated from the medulla oblongata and extended into the fourth ventricle. After surgery, the patient had transient dysphagia. Histopathological analysis confirmed diagnosis of a germinoma. The patient was treated via gamma knife, plus two rounds of bleomycin-cisplatin-etoposide chemotherapy. He returned to school within 3 months, and after 4 years of follow-up had no tumor recurrence.

Answers:

1. **A germinoma has large primordial germ cells with prominent nucleoli and various amounts of cytoplasmic clearing. The cells are large, undifferentiated, and often irregular. They resemble primordial germinal elements, arranged in lobules, sheets, or when there is stromal desmoplasia, regular cords and trabeculae.**
2. **Germinomas are the most common CNS germ cell tumors. Males are about twice as likely to develop them, and they are most common between the ages of 10 and 21 years. Intracranial germinomas occur in only 0.7 of every 1 million children.**
3. **Germinomas are extremely sensitive to radiation therapy as well as chemotherapy, and combination treatments offer the potential for a cure. The best treatment results come from craniospinal radiation therapy, and gamma knife radiation treatment is often used.**

Clinical case 2

1. What are the three types of teratomas?
2. How common are teratomas in children?
3. What is the difference between intra-axial and extra-axial teratomas?

An 8-year-old boy was brought to the emergency department with headache and vomiting that had persisted for 2 days. Bilateral papilledema was detected via funduscopy. A nonenhanced CT scan revealed a midline lesion, with hydrocephalus. An MRI revealed a large, intraventricular, heterogeneous mass that partly compressed the third ventricle. A subtotal resection of the mass was performed, and histopathology revealed it to be a teratoma. After surgery, the patient developed left arm and left leg hemiparesis, and left facial paralysis. A follow-up MRI revealed small tumor residue in the right thalamus, but his symptoms resolved over time and there was no tumor recurrence.

Answers:

1. **Teratomas are subclassified as mature, immature, and malignant transformative. The mature subtype only contains adult-type tissues. The immature subtype contains immature, fetal, or embryonic tissues alone, or along with mature tissues. The malignant transformative subtype is rare, containing a component resulting from malignant transformation of usually a sarcoma or carcinoma.**
2. **Teratomas make up the largest proportion of fetal intracranial neoplasms. They constitute 26%–50% of all types of fetal brain tumors, and occur in a wide range of ages in children.**
3. **Intra-axial teratomas are usually present before birth or in the newborn period. They are large in size and mostly occur supratentorially. Extra-axial teratomas usually present in childhood or early adulthood, and are usually smaller. They usually arise in the pineal or suprasellar regions.**

Clinical case 3

1. What are the other names for yolk sac tumors?
2. Regarding immunoreactivity, what distinguishes these tumors from others?
3. Upon what is the diagnosis of yolk sac tumors based?

A 2-year-old boy was hospitalized because of focal epileptic seizures and recurrent vomiting. Physical examination revealed papilledema, partial third nerve palsy, and a positive Babinski sign. CT and MRI scans, with contrast, showed a tumor within the left temporoparietal lobe. The tumor was surgically resected, and histopathologic examination showed it to be a yolk sac tumor. Thorough imaging studies of the boy's chest, abdomen, pelvis, and testes failed to detect any primary tumor. Therefore, the diagnosis was adjusted to *pure primary intracranial yolk sac tumor*. The boy survived without tumor recurrence.

Answers:

1. **Yolk sac tumors are also referred to as endodermal sinus tumors and infantile embryonal carcinoma.**
2. **Yolk sac tumors are distinguished by cytoplasmic immunoreactivity of their epithelial elements for alpha-fetoprotein. The hyaline globules are also reactive. The epithelial components regularly label for cytokeratins, and are often positive for the protein-coding gene called glypican-3.**

3. **Diagnosis of yolk sac tumors is based on histology, which usually includes malignant endodermal cells that secrete alpha-fetoprotein. The cells can be detected in tumor tissue, serum, cerebrospinal fluid, and even the urine and amniotic fluid.**

Clinical case 4

1. What are the three most common combinations of tumors within mixed germ cell tumors?
2. What is the common appearance of mixed germ cell tumors?
3. Which mixture of tumors in a mixed germ cell tumor has the worse prognosis?

A 16-year-old girl presented with headache, impairment of visual acuity, and diplopia. There was an impairment of upward gaze in the left eye. Skull X-rays revealed a marked bony erosion of the dorsum of the sella. Cerebral arteriography showed a large pituitary tumor with parasellar and suprasellar extension that was greater on the left side. There were multiple low-density areas in the tumor mass, likely representing tumor necrosis and hemorrhage. Surgical resection was performed, and the tumor was solid, encapsulated, and mostly white in color. Microscopic examination showed 80% neuroectodermal tissue, squamous epithelium, cartilage, and skeletal muscle. The neuroectodermal tissue consisted of neuroepithelial elements and mature or anaplastic giant cells. The diagnosis of a mixed germ cell tumor was made. After surgery, the patient experienced transient diabetes insipidus. Chemotherapy was started, using cisplatin-etoposide-bleomycin combination therapy. Over time, her symptoms improved greatly. Radiation therapy was also given once, and since then, the patient has had no tumor-related symptoms.

Answers:

1. **The three most combination combinations of tumors include: teratoma + embryonal carcinoma + yolk sac tumor; seminoma + embryonal carcinoma; and teratoma + embryonal carcinoma.**
2. **Mixed germ cell tumors are usually solid, encapsulated, and cystic, and often white in color. The tumor mass often has many low-density areas that represent necrosis and hemorrhage.**
3. **The worst prognosis is for a mixture of teratoma with embryonal carcinoma and yolk sac tumor.**

Clinical case 5

1. Where do medulloblastomas usually form, and what is their most common tumor grade?
2. How do medulloblastomas usually spread?
3. In younger children, what are the adverse effects of radiation therapy for medulloblastomas?

A 9-year-old girl presented with morning headaches, vomiting, loss of balance, and double vision. An MRI revealed a large left cerebellar brain tumor that extended to the fourth ventricle, but no evidence of metastasis. The tumor was surgically resected, but since it was partially adhered to the ventricles, a follow-up surgery was required. At that time, total resection was successful, and a chemotherapy and radiation therapy was implemented per the National Cancer Institute's protocol for standard risk medulloblastoma. The chemotherapy combined cisplatin, vincristine, cyclophosphamide, and lomustine. The patient recovered, but required physical therapy rehabilitation. However, brain and spinal MRIs and lumbar puncture to check the CSF were negative for tumor recurrence.

Answers:

1. **Medulloblastomas are most common in the cerebellum and dorsal brainstem, and most of them are grade IV tumors.**
2. **Medulloblastomas usually spread through the CSF fluid pathways, seeding the neuroaxis with metastatic tumor deposits. In rare cases, they reach the lymphatic system and bones.**
3. **In younger patients, the adverse effects of radiation therapy include cognitive impairment, bone growth retardation, psychiatric illness, endocrine disruption, and hearing loss.**

Key terms

Alpha-fetoprotein
Ataxia-telangiectasia
Canalicular
Centigray
Chang staging system

Choriocarcinoma
Cytotrophoblasts
EMACO therapy
Embryonal carcinoma
Erythroleukemia

Fanconi anemia
Fast MRI
Germinoma
Gonadotropin
Hepatoid variant

Holocranial
Hypomethylation
Hypothalamohypophyseal axis
Klinefelter syndrome
Lymphoplasmacellular
Medulloblastoma
Mesoblast
Mixed germ cell tumor
Neuraxial
Neuroembryogenesis
Neuroendoscopic
Nijmegen breakage syndrome
Parinaud syndrome
Pluripotent cells
Polyvesicular vitelline patterns
Reticular pattern
Rubinstein-Taybi syndrome
Schiller-Duval bodies
Seminomas
Somatic tissue
Stem cell rescue
Stromal desmoplasia
Stromas
Syncytiotrophoblastic giant cells
Targeted therapy
Tectal plate
Teratoma
Yolk sac tumor

Further reading

Abbas, A. K., Lichtman, A. H., & Pillai, S. (2017). *Cellular and molecular immunology* (9th Edition). Elsevier.
Adesina, A. M., Tihan, T., Fuller, C. E., & Young Poussaint, T. (2016). *Atlas of pediatric brain tumors* (2nd Edition). Springer.
Ahluwalia, M., Metellus, P., & Soffietti, R. (2019). *Central nervous system metastases*. Springer.
Amin, M. B., Edge, S. B., Greene, F. L., Byrd, D. R., Brookland, R. K., et al. (2017). *AJCC cancer staging manual* (8th Edition). Springer.
Baehring, J. M., & Piepmeier, J. M. (2006). *Brain tumors: practical guide to diagnosis and treatment (Neurological disease and therapy)*. CRC Press.
Brandao, L. A. (2016). *Pediatric brain tumors update, an issue of neuroimaging clinics of North America (volume 27-1) (The clinics: radiology)*. Elsevier.
Burger, P. C., & Scheithauer, B. W. (2007). *Tumors of the central nervous system (AFIP atlas of tumor pathology, series 4)*. American Registry of Pathology.
Chitale, A. R. (2014). *Diagnostic problems in tumors of central nervous system: Selected topics (Diagnostic problems in tumor pathology series, volume 4)*. Chitale Publications.
Compton, C. C., Byrd, D. R., Garcia-Aguilar, J., Kurtzman, S. H., Olawaiye, A., & Washington, M. K. (2012). *AJCC cancer staging Atlas: A companion to the seventh editions of the AJCC cancer staging manual and handbook* (2nd Edition). Springer.
DeMonte, F., Gilbert, M. R., Mahajan, A., McCutcheon, I. E., Buzdar, A. U., & Freedman, R. S. (2007). *Tumors of the brain and spine (MD Anderson cancer care series)*. Springer.
DeVita, V. T., Rosenberg, S. A., & Lawrence, T. S. (2018). *DeVita, Hellman, and Rosenberg's cancer: Principles & practice of oncology* (11th Edition). LWW.
Fountas, K., & Kapsalaki, E. Z. (2019). *Epilepsy surgery and intrinsic brain tumor surgery: A practical atlas*. Springer.
Gajjar, A., Reaman, G. H., Racadio, J. M., & Smith, F. O. (2018). *Brain tumors in children*. Springer.
Gupta, N., Banerjee, A., & Haas-Kogan, D. A. (2017). *Pediatric CNS tumors (Pediatric oncology)* (3rd Edition). Springer.
Gursoy Ozdemir, Y., Bozdag Pehlivan, S., & Sekerdag, E. (2017). *Nanotechnology methods for neurological diseases and brain tumors: Drug delivery across the blood-brain barrier*. Academic Press.
Hattingen, E., & Pilatus, U. (2015). *Brain tumor imaging (Medical radiology)*. Springer.
Hayat, M. A. (2012). *Pediatric cancer, volume 3: Diagnosis, therapy, and prognosis*. Springer.
Hayat, M. A. (2013). *Tumors of the central nervous system, volume 10: Pineal, pituitary, and spinal tumors*. Springer.
Hayat, M. A. (2014). *Tumors of the central nervous system, volume 12: Molecular mechanisms, children's cancer, treatments, and radiosurgery*. Springer.
Hayat, M. A. (2014). *Tumors of the central nervous system, volume 13: Types of tumors, diagnosis, ultrasonography, surgery, brain metastasis, and general CNS diseases*. Springer.
Hayat, M. A. (2011). *Tumors of the central nervous system, volume 3: Brain tumors (Part 1)*. Springer.
Hayat, M. A. (2011). *Tumors of the central nervous system, volume 4: Brain tumors (Part 2)*. Springer.
Jain, R., & Essig, M. (2015). *Brain tumor imaging*. Thieme.
Keating, R. F., Goodrich, J. T., & Packer, R. (2013). *Tumors of the pediatric central nervous system* (2nd Edition). Thieme.
Lacruz, C. R., Saenz de Santamaria, J., & Bardales, R. H. (2018). *Central nervous system intraoperative cytopathology (Essentials in cytopathology)* (2nd Edition). Springer.
Lakhi, N., & Moretti, M. (2016). *Alpha-fetoprotein: Functions and clinical applications*. Nova Science Publishers Inc.
Mahajan, A., & Paulino, A. (2018). *Radiation oncology for pediatric CNS tumors*. Springer.
McLendon, R. E., Rosenblum, M. K., & Bigner, D. B. (2006). *Russell & Rubinstein's pathology of tumors of the nervous system (Contemporary neurology)* (7th Edition). CRC Press.
Moliterno Gunel, J., Piepmeier, J. M., & Baehring, J. M. (2017). *Malignant brain tumors: State-of-the-art treatment*. Springer.
Mucci, G. A., & Torno, L. R. (2015). *Handbook of long term care of the childhood cancer survivor (Specialty topics in pediatric neuropsychology)*. Springer.
Newton, H. B. (2018). *Handbook of brain tumor chemotherapy, molecular therapeutics, and immunotherapy* (2nd Edition). Academic Press.

Newton, H. B., & Maschio, M. (2015). *Epilepsy and brain tumors*. Academic Press.
Nogales, F. F., & Jimenez, R. E. (2017). *Pathology and biology of human germ cell tumors*. Springer.
Norden, A. D., Reardon, D. A., & Wen, P. Y. C. (2010). *Primary central nervous system tumors: Pathogenesis and therapy (Current clinical oncology)*. Humana Press.
Provenzale, J., Bigner, D., Bigner, S., & McLendon, R. (2000). *Pathology of tumors of the central nervous system: A guide to histologic diagnosis*. CRC press.
Robin, N. H., & Farmer, M. (2017). *Pediatric cancer genetics*. Elsevier.
Sampson, J. H. (2017). *Translational immunotherapy of brain tumors*. Academic Press.
Terezakis, S. A., & MacDonald, S. M. (2018). *Target volume delineation for pediatric cancers (Practical guides in radiation oncology)*. Springer.
Tonn, J. C., Reardon, D. A., Rutka, J. T., & Westphal, M. (2019). *Oncology of CNS tumors* (3rd Edition). Springer.
Warmuth-Metz, M. (2016). *Imaging and diagnosis in pediatric brain tumor studies*. Springer.

Chapter 25

Peripheral nerve sheath tumors

Chapter Outline

Hybrid nerve sheath tumors	439	Malignant peripheral nerve sheath tumor with perineurial differentiation	449
Malignant peripheral nerve sheath tumors	442	**Perineurioma**	450
Malignant peripheral nerve sheath tumor with divergent differentiation	444	**Clinical cases**	451
Epithelioid malignant peripheral nerve sheath tumor	446	**Key terms**	453
		Further reading	454

Hybrid nerve sheath tumors

Hybrid nerve sheath tumors are benign types of peripheral nerve sheath tumors (PNSTs). They have features of more than one common type of these tumors, which include schwannomas, perineuriomas, and neurofibromas. The two most common forms of hybrid nerve sheath tumors are the schwannoma/perineuriomas and the neurofibroma/schwannomas.

Epidemiology

The epidemiology of hybrid nerve sheath tumors is varied based upon the actual subtypes of the tumors involved. In general, these tumors are most prevalent in younger adults, and slightly more common in males than in females. There is no predilection for racial or ethnic groups.

Etiology and risk factors

The schwannoma/perineurioma form is usually sporadic in occurrence, while the neurofibroma/schwannoma is usually linked to schwannomatosis, neurofibromatosis type 1 (NF1) or neurofibromatosis type 2 (NF2). There are also rare cases of neurofibroma/perineurioma, which is usually related to NF1. Other than these related factors, there are no other proven risk factors.

Pathology

Hybrid nerve sheath tumors may be widely distributed over the body, often in the dermis and subcutaneous tissue. In rare cases, they are associated with the cranial or spinal nerves. The majority of biphasic schwannoma or reticular perineurioma cases occur on the fingers or toes. Macroscopically, hybrid tumors are almost exactly the same in appearance as neurofibromas or schwannomas. They generally are pink to tan in color, and soft. Microscopically, hybrid schwannoma/perineurioma tumors usually have a Schwannian **cytomorphology**. However, their structures are more like perineuriomas. They are often well circumscribed and lack encapsulation. There are spindle cells with fat yet tapered nuclei. The cytoplasm is pale but eosinophilic, with indistinct cell borders that have a storiform, lamellar, and/or whorled architecture. There may be myxoid stromal changes, in about 50% of all cases. Degenerative cytological atypia is often seen, which is similar to classic changes in schwannomas.

Hybrid neurofibromas/schwannomas have two primary components, as follows:

- Schwannoma-like component—nodular Schwann cell proliferation, sometimes with Verocay bodies; mainly composed of cellular Antoni A areas, often with the **Verocay bodies** and Schwann cells having nuclear palisading
- Neurofibroma-like component—with a mixed cell population, collagen, and myxoid changes; may have large amounts of collagen, fibroblasts, and myxoid changes; the Schwannian cells have unique wavy and elongated appearances

Often, there is a plexiform architecture to these tumors. The less common hybrid neurofibromas/perineuriomas have plexiform neurofibromas and large areas of **perineuriomatous** differentiation when the patient has NF1. With immunohistochemistry, biphasic differentiation is apparent, which is both Schwannian and perineuriomatous. In rare cases, the neurofibromatous and perineuriomatous areas can be seen in routine *hematoxylin & eosin* stains.

The schwannoma/perineurioma hybrid tumors have dual differentiation via immunohistochemistry (see Fig. 25.1). The Schwannian plump-spindled cells are positive for *S100 protein*. The perineurial slender-spindled cells, however,

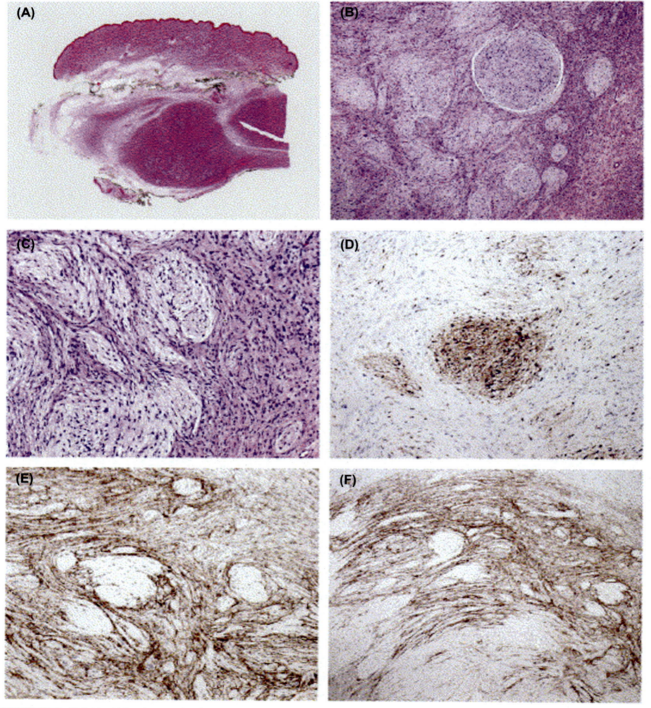

FIGURE 25.1 Hybrid peripheral nerve sheath tumors. (A) Schwannoma. (B) Perineurioma. (C) Tumor cells forming giant collagenous rosettes. (D) Malignant peripheral nerve sheath tumor. (E) Immunohistochemical analysis positive for EMA. (F) Different magnification with EMA staining, revealing stellate cells.

have variable immunoreactivity for *claudin-1, epithelial membrane antigen (EMA)*, and *glucose transporter-1 (GLUT1)*. Via **double staining** for EMA and S100, there are parallel layers of alternating EMA-positive and S100-positive cells. These cells do not show **coexpression** of antigens. Based on these results, the majority of the tumors are made up of about 66% Schwann cells and 33% perineurial cells.

The best staining methods for neurofibromas/schwannomas reveal the presence of monomorphic Schwann cells within the schwannoma component, such as S100 and *transcription factor SOX10*, or polymorphic cells in the neurofibroma component, which includes S100 and SOX10, plus perineurial (EMA and GLUT1) cells and fibroblasts. In NF-related schwannomas, there may be entrapped axons. However, entrapped, large axon bundles are more common in neurofibromas. In neurofibromas/perineuriomas, biphasic Schwannian and perineuriomatous differentiation is revealed in immunohistochemistry. The perineuriomatous areas stain positively for claudin-1, EMA, and GLUT1, but stain negatively for S100. Intraneural perineurial proliferations have been identified during screening NF lesions and normal serves in patients with NF1. This is supportive of the fact that both pure and hybrid perineuriomatous lesions occur in PNSTs of NF1.

Since more than 50% of patients with hybrid PNSTs have multiple tumors, a tumor syndrome is suggested. In schwannomatosis, hybrid neurofibromas/schwannomas are common, and occur in 71% of patients. A significant link to neurofibromatosis exists, in which this hybrid lesion occurs in 26% of NF2 cases and in 9% of NF1 cases. For patients with schwannomatosis, 61% of the tumors appear as schwannoma-like nodules in a neurofibroma-like tumor. This is related to hybrid neurofibromas/schwannomas. When there is a hybrid morphology with or without mosaic SWItch/sucrose non-fermentable *(SWI/SNF)-related matrix-associated actin-dependent regulator of chromatin subfamily B member 1* (SMARCB1, or *integrase interactor 1 [INI1]*) expression in immunohistochemistry studies, a schwannoma may be related to neurofibromatosis—especially NF2—and schwannomatosis. Also, hybrid neurofibromas/perineuriomas occur mostly with NF1.

> **Point to remember**
> The nerve sheath is the tissue covering and protecting the nerves. Nerve sheath tumors grow directly from nerves themselves, usually sporadically, but sometimes due to a condition such as neurofibromatosis. Nerve sheath tumors are either benign, such as neurofibromas and schwannomas, or malignant, such as sarcomas.

Clinical manifestations

Peripheral nerve sheath hybrid tumors are either similar or nearly identical to those of other benign PNSTs. The manifestations are mostly based on the site of origin. Hybrid nerve sheath tumors, such as those of the spine, can cause pain or neurological deficits if they involve large peripheral nerves. Signs and symptoms are quite varied based on the actual tumor location, but include thoracic limb lameness, pain, numbness, burning sensations, pins and needles sensations, weakness, muscle degeneration, vertigo, difficulty with balance, restricted movements, ataxia, lack of reflexes, symptoms of Horner syndrome, paresis, and ptosis.

Diagnosis

Diagnosis of hybrid nerve sheath tumors involves MRI, CT, and PET scans. A biopsy is required to diagnose the nature of the tumor.

Treatment

Surgical resection is the primary treatment method for hybrid nerve sheath tumors, to help restrict tumor spread. In some cases, an entire limb must be amputated, or nearby nerves must be removed. For malignant tumors, radiation therapy and chemotherapy may be indicated. Radiation therapy can be both preoperative and postoperative. Chemotherapy is mainly applied when the tumor size is too small for surgical resection, or when available localized treatment options will not be beneficial. For some patients with spinal hybrid tumors, intrathecal drug delivery pumps are used to deliver pain relief.

Prognosis

Prognosis for hybrid nerve sheath tumors is based on their type. Benign tumors usually have a good prognosis, but malignant tumors are usually highly metastatic, and are often quickly fatal. Life span is reduced also in the case of tumors associated with NF1.

Malignant peripheral nerve sheath tumors

Malignant peripheral nerve sheath tumors (MPNSTs) develop in the protective lining that covers nerves. They have Schwann cell or perineurial cell differentiation, and often arise within a peripheral nerve or in extraneural soft tissue. These tumors are considered to be aggressive since there is up to 65% chance of recurrence after surgical resection.

Epidemiology

MPNSTs mostly affect young to middle-aged adults but have occurred up to the sixth decade of life. However, they also occur in adolescents. Approximately 50% of these tumors are linked to NF1, where they often develop from a pre-existing plexiform or intraneural neurofibroma, affecting patients of younger age. Overall, the MPNSTs make up 5% or less of all malignant soft tissue tumors. However, the mean patient age when associated with NF1 is 28-36 years. For sporadic cases, the mean patient age is 40-44 years. There is no gender predilection. Between 10% and 20% of these tumors develop in children and adolescents.

According to the *Surveillance, Epidemiology, and End Results Program* of the *National Cancer Institute* in the United States, a study of 1711 cases of MPNSTs revealed differences in the age and sex distribution of these tumors. In relation to this study alone, in males, they were most common between the ages of 30 and 39, and least common in children from birth to age 9. In females, the tumors were most common between ages 50 and 59, and least common from birth to age 9.

Etiology and risk factors

Most MPNSTs that occur sporadically arise from large peripheral nerves, with no known benign precursor. The majority of these tumors have combined inactivation of the *cyclin-dependent kinase inhibitor 2A, protein complex 2,* and NF1 component genes. About half of all MPNSTs are linked to NF1 and usually develop from deep plexiform neurofibromas or intraneural neurofibromas that are large in size. Approximately 40% of these tumors occur in patients with no identified predisposition. About 10% are linked to previous treatment with radiation. The tumors may also develop because of surgical trauma. In rare cases, MPNSTs develop from conventional schwannomas, pheochromocytomas, or ganglioneuroblastomas/ganglioneuromas.

Pathology

In general, there is a lack of any accurate grading system for MPNSTs. Though there is not any well documented or validated criteria, one method is to divide the tumors into high-grade and low-grade types. The high-grade tumors make up about 85% of cases while the low-grade tumors make up about 15% of the remaining cases. Examples of high-grade MPNSTs include:

- *Conventional monomorphous spindle cell MPNSTs*
- *Highly pleomorphic MPNSTs*
- *MPNSTs with divergent differentiation*—such as:
  - *Glandular MPNSTs*
  - *Malignant triton tumors*
  - *Tumors of* **angiosarcomatous**, **chondrosarcomatous**, *and* **osteosarcomatous** *differentiation.*

The low-grade MPNSTs are well differentiated, and usually develop transitionally from neurofibromas. Though not needed for diagnosis, they usually have an increased mitotic rate.

MPNSTs usually affect large and medium-sized nerves instead of small nerves. The most common sites include the buttocks, thighs, brachial plexus, upper arms, and paraspinal area. The most commonly affected nerve is the sciatic nerve. When MPNSTs affected the cranial nerves, which is rare, they usually arise from schwannomas. They only rarely develop as primary intraparenchymal tumors. The tumors often infiltrate nearby soft tissues, and can spread through hematogenous and intraneural routes. Between 20% and 25% of cases involve metastasis, which is usually to the lungs.

Macroscopically, these tumors have varied appearances. A large amount of them arise in neurofibromas, sometimes as focal transformations. Therefore, the growth process may be only slightly visible in gross examination. Oppositely, larger and usually high-grade tumors that originate in or are not associated with a nerve are fusiform and expansive masses, or are globular and unencapsulated soft tissue tumors. Both of these forms invade surrounding tissues. Almost all of the tumors are larger than 5 cm in size. There have even been cases in which the tumors were larger than 10 cm.

MPNSTs may be soft to hard. The cut surface is usually gray or cream-colored. There is usually necrosis and hemorrhage, both of which may be extreme.

Microscopically, there are also wide variances. Many tumors resemble fibrosarcomas, with a herringbone pattern. Others have interwoven fasciculated cell growth patterns. Both patterns have extremely packed spindle cells, and varied amounts of eosinophilic cytoplasm (see Fig. 25.2). The nuclei are usually wavy and elongated. Unlike cells of smooth muscle, these cells have tapered ends. There is either a diffuse growth pattern, or alternating loose and dense cellular areas. Often, there is perivascular hypercellularity and tumor components that appear to herniate into the vascular lumina. Unusual growth patterns may include hemangiopericytoma-appearing areas. Rarely, there is nuclear palisading. The tumors grow in nerve fascicles, yet often invade nearby soft tissues. There is often a pseudocapsule of varying thickness. About 75% of these tumors have mitotic activity and geographical necrosis. Often, high-grade MPNSTs show more than four mitoses per high-power field.

Clinical manifestations

In most cases, MPNSTs appear as progressively enlarging masses in the extremities. There may or may not be any neurological symptoms. When these tumors develop in the spine, there is usually radicular pain. Additional signs and symptoms include peripheral edema, difficulty with movements, soreness, discomfort, numbness, burning, a prickling sensation, paraparesis, bowel/bladder involvement, and dizziness or loss of balance.

> **Point to remember**
>
> MPNST often initially presents as a growing lump or mass that sometimes causes pain or tingling sensations. The tumors may also cause weakness when the patient tries to move the affected body part. It is important to make an appointment with a physician if signs and symptoms are persistent or worsening.

Diagnosis

In imaging studies, MPNSTs resemble soft tissue sarcomas. Often, there is inhomogeneous contrast enhancement, and irregular tumor contours, which is suggestive of invasion. For detecting MPNSTs in patients with NF1, fluorodeoxyglucose-positron emission tomography is sensitive. Other imaging studies may include MRI, X-rays, CT scan, and bone scans. These tumors often appear hypointense on T1-weighted imaging and hyperintense on T2-weighted imaging (see Fig. 25.3). With axial contrast-enhanced MRI, there may be **inhomogeneous** enhancement of the lesions, with some nonenhancing areas of necrosis. However, the most conclusive test is a tumor biopsy.

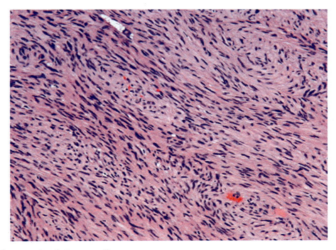

FIGURE 25.2 Malignant peripheral nerve sheath tumor cells.

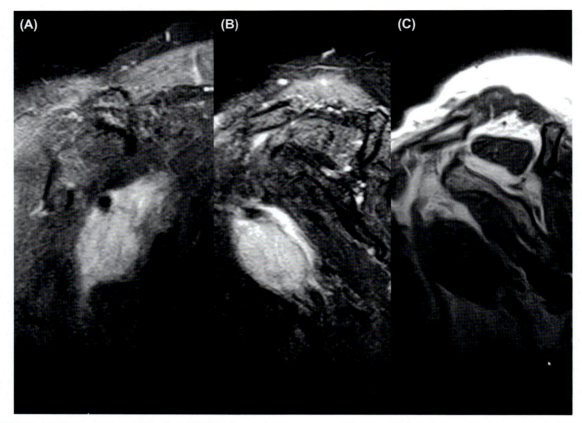

FIGURE 25.3 (A) The noncontrast MRI of an MPNST shows a heterogeneous, solid mass with hyperintense signal in a sagittal view, with T2-weighting. (B) A coronal fat-saturated T2-weighted view of the same tumor. (C) Hypointensity of the same tumor is shown in a T1-weighted image.

Treatment

Treatment for MPNSTs includes surgical resection of the tumors plus as much surrounding tissue as needed, radiation therapy, bone removal and replacement, and chemotherapy. In severe cases, amputation is suggested in order to save the patient's life, though limb-saving treatment is always attempted.

Prognosis

The prognosis for MPNSTs is quite varied, based on patient age, health, and tolerance to treatments. Poorer prognoses are given for tumors that are larger than 5 cm, high-grade disease, presence of neurofibromatosis, and if there is metastasis. There is a relatively poor prognosis in pediatric patients. With NF1, these tumors are extremely life threatening.

Malignant peripheral nerve sheath tumor with divergent differentiation

A MPNST *with divergent differentiation* is also known as a **malignant triton tumor** or a *glandular* MPNST. Tumors with glandular components were first reported in 1892, but tumors with rhabdomyoblastic differentiation were first identified in 1938. There have been about six documented cases with both glandular and rhabdomyoblastic elements.

Epidemiology

This subtype of MPNSTs make up about 15% of all cases of MPNSTs. There are no accurate epidemiological statistics on incidence or prevalence, however.

Etiology and risk factors

Between 60% and 75% of these tumors are linked to the presence of NF1. The link with NF1 and spindle cells that cannot be distinguished from the common form of MPNSTs reveals a clear relationship between high-grade MPNSTs with divergent differentiation, and common high-grade MPNSTs.

Pathology

Many different mesenchymal tissues, including bone, cartilage, skeletal muscle, smooth muscle, and angiosarcoma-like tissues may be present in these tumors. They show rhabdomyosarcomatous differentiation. The tumors may contain glandular epithelium resembling the epithelium of the intestines. Neuroendocrine differentiation is common, while squamous epithelium is less common. The tumors are generally firm and tan-colored, with areas of hemorrhage and necrosis. They are usually well-encapsulated, with a smooth surface (see Fig. 25.4). The spindle cells are usually arranged in interlacing fascicles. The nuclei are pleomorphic, hyperchromatic, and mitotically active, with eosinophilic cytoplasm. There may be areas of palisading necrosis, perivascular proliferations of "plump" tumor cells, amianthoid fibers, and myxoid stroma (see Fig. 25.5).

Clinical manifestations

The clinical manifestations of MPNSTs with divergent differentiation are not different than for the common form of these tumors.

Diagnosis

The same diagnostic methods exist for these tumors as for common MPNSTs. In MRI, these tumors usually appear with some amount of calcification (see Fig. 25.6).

Treatment

The same treatment methods also exist for these tumors as for common MPNSTs.

Prognosis

The prognosis for these MPNSTs is the same as for the common MPNSTs.

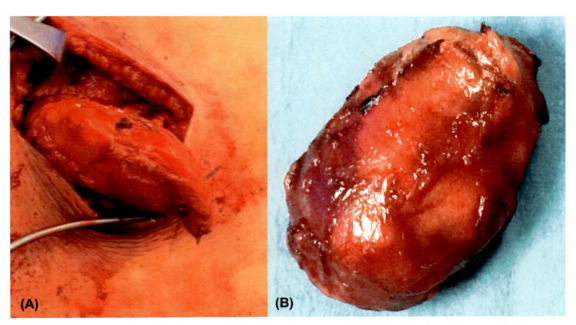

FIGURE 25.4 (A) An operative view showing a malignant peripheral nerve sheath tumor with divergent differentiation, as it was dissected from the surrounding structures. (B) The excised mass is shown, which is well-encapsulated, with a smooth surface.

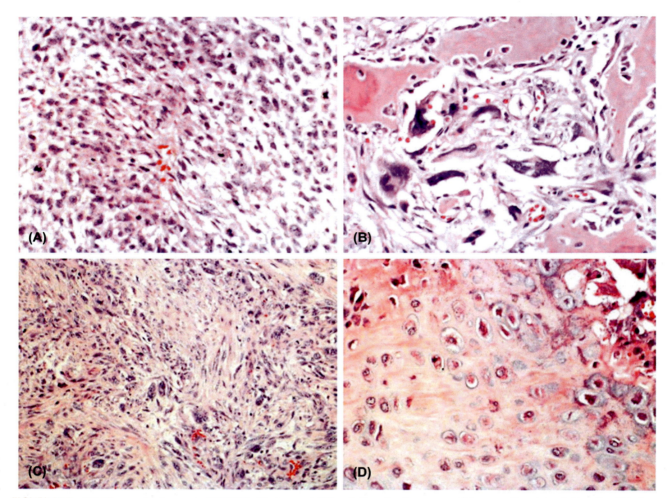

FIGURE 25.5 (A) Micrograph of a resected MPNST with divergent differentiation, showing an area of the tumor featuring spindle cells within loose stroma, with associated nuclear atypia and scattered mitotic figures. (B) The presence of bony trabeculae surrounded by plump malignant hyperchromatic, multinucleated bizarre cells. (C) The tumor cells have a pleomorphic undifferentiated (malignant fibrous histiocytoma-like) pattern. (D) An area with chondrosarcomatous differentiation.

Epithelioid malignant peripheral nerve sheath tumor

Epithelioid MPNSTs are rare, and make up less than 5% of all MPNSTs. They are not related to NF1, and may arise from a schwannoma that is malignantly transformed. Adults are affected more than children.

Epidemiology

Epithelioid MPNSTs have been seen in patients between 6 and 80 years of age, with a mean age of 44 years. There is a nearly equal incidence between males and females.

Etiology and risk factors

There are no specific causes or risk factors that have been identified for this subtype of MPNSTs that are any different than for the common MPNSTs.

Pathology

These tumors have occurred above the fascia (superficial) and in deeper locations. They show diffuse S100 protein positivity, and some cases have had a loss of INI1 expression. The tumors have ranged from 0.4 to 20 cm in size. The epithelioid cells are relatively uniform but clearly atypical. They can be round, polygonal, or ovoid in shape, with round,

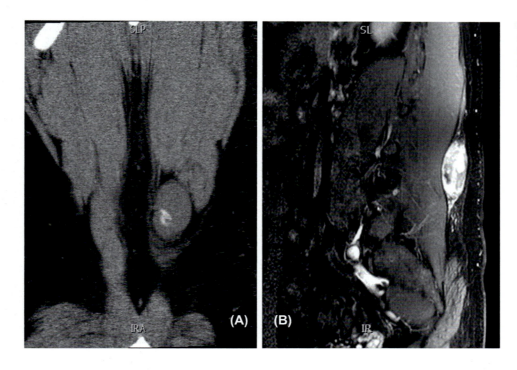

FIGURE 25.6 MRI of MPNST with divergent differentiation. (A) Coronal view. (B) Sagittal view. Both views clearly show the lesion with some amount of calcification.

vesicular nuclei. The cytoplasm can range from abundant amphiphilic to palely eosinophilic. About 33% of cases show focal spindled morphology. The median mitotic rate is 5 mitoses per 10 high-power fields. Most of the tumors have a multilobular growth pattern, with lobules and nests that are surrounded by myxoid stroma, fibroid stroma, or both. An epithelioid MPNST, in this example a schwannoma, is shown in Fig. 25.7.

Clinical manifestations

The signs and symptoms of epithelioid MPNSTs are the same as for common MPNSTs.

Diagnosis

Diagnosis of epithelioid MPNSTs uses the same methods as for common MPNSTs. During imaging studies, this subtype appears hypointense on T1-weighted MRI imaging, and as a heterogeneous hyperintense mass on T2-weighted MRI imaging, with peripheral hyperintensity and central hypointensity. This likely reflects elements of necrosis with or without hemorrhage. When enhanced T2-weighted imaging is used, peripheral enhancement is seen.

Treatment

Most epithelioid MPNSTs are treated by surgical resection, but some cases require chemotherapy and/or radiation therapy. These tumors are slightly less sensitive to chemotherapy, however, than other forms.

Prognosis

Risks of metastasis, disease-related death, and recurrence appear to be lower than for common MPNSTs. About 66% of patients will not have tumor recurrence following treatment, though when recurrence has happened, it has been fatal in a small number of cases.

> **Point to remember**
> Epithelioid MPNSTs are rare sarcomas that are relatively less sensitivity to chemotherapy than other forms. They have diffuse S100 protein positivity, infrequent associations with NF1, and occasional origins in schwannomas.

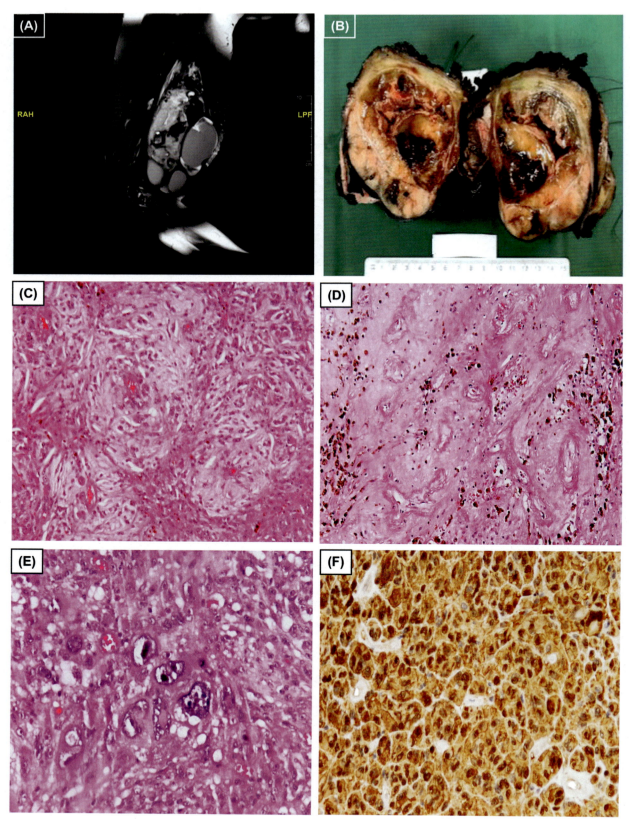

FIGURE 25.7 Epithelioid MPNST (schwannoma).

Malignant peripheral nerve sheath tumor with perineurial differentiation

MPNSTs with *perineurial differentiation* are also rare. Similar to benign perineuriomas, these tumors are S100-negative but EMA-positive. They can metastasize, but are not as aggressive than the common MPNSTs. These tumors are also referred to as **malignant perineuriomas**, and therefore, have a tumor grade of III.

Epidemiology
The epidemiology of these tumors varies widely, based upon their location. There are no accurate epidemiological statistics available.

Etiology and risk factors
The causes and risk factors of MPNSTs with perineurial differentiation are not clear but appear to be the same as for common MPNSTs.

Pathology
MPNSTs with perineurial differentiation show hypercellularity, atypical nuclei, and increased mitotic activity. Diffuse S100 protein expression is common. SMARCB1 (INI1) is lost in half of all cases. These tumors show expression of related differentiation markers. They are composed of neoplastic perineural cells, hence the name. They demonstrate extension of these cells around axons, largely contained by endoneurium.

Clinical manifestations
There are no distinct clinical manifestations of these tumors in comparison with the common MPNSTs.

Diagnosis
MPNSTs with perineurial differentiation such as these are distinguished from other high-grade sarcomas mostly by determining that they originate from either a peripheral nerve, a benign precursor, or because of genetic or immunohistochemical features. Malignant spindle cell tumors, when occurring along with NF1, are considered to be MPNSTs until they are proven otherwise. Synovial sarcoma of the nerves must be considered as a differential diagnosis. Additional differential diagnoses include neurofibromas, schwannomas, chronic inflammatory polyneuropathy, **Dejerine-Sottas disease**, and **Charcot-Marie-Tooth disease**. Synovial sarcomas are common in soft tissues, but also occur as unique and rare primary nerve tumors. They are very similar, morphologically and in immunohistochemistry, with common MPNSTs. However, they carry an *SS18-SSX2* or an *SS18-SSX1 fusion gene*. MPNSTs with perineurial differentiation may be distinguished from malignant melanomas from a lack of immunostaining for HMB45 and melan-A, together with an origin from a peripheral nerve or a benign nerve sheath tumor. On MRI, these tumors appear as fusiform enlargements of nerves, with increased T2-weighted signals and contrast enhancement.

Treatment
After histological confirmation via biopsy, MPNSTs with perineurial differentiation can be surgically resected, with or without nerve grafting. Radiation therapy and chemotherapy may also be helpful.

Prognosis
Fortunately, the MPNSTs with perineurial differentiation have a better prognosis than other MPNSTs since they are less aggressive. The worst prognosis for these tumors involves a tumor size of 5 cm or more, a truncal location, local recurrence, and a high-grade designation. Survival is decreased if NF1 is also present, and gains at chromosome 16p, losses from 10q or Xq, and CDK4 amplifications also worsen prognosis.

> **Point to remember**
>
> MPNST with perineurial differentiation is a rare subtype. They are usually not related to NF1, and have no encapsulation or neurofibroma components. Nerve involvement is infrequent. Generally, they are of better prognosis than for the common form of MPNSTs.

Perineurioma

A **perineurioma** is a tumor that is entirely made up of neoplastic perineurial cells. Intraneural perineuriomas are benign, with proliferating perineurial cells in the endoneurium that form characteristic **pseudo-onion bulbs**, which are rounded expansions that have the appearance of actual onions. Soft tissue perineuriomas are usually not associated with nerves. They have variably whorled patterns and are usually benign. Malignant soft tissue perineurioma is a rare type of MPNST, with perineurial differentiation, and may be grade I to III. Intraneural perineurioma was previously considered to be a type of hypertrophic neuropathy, and is a grade I tumor.

Epidemiology

Intraneural perineuriomas usually develop in early adulthood or adolescence. There is no difference in prevalence between the genders. Soft tissue perineuriomas mostly occur in adult females, about twice as often as in males. Intraneural and soft tissue perineuriomas are both rare. The intraneural perineuriomas make up about 1% of nerve sheath tumors. With a similar epidemiology, the soft tissue perineuriomas make up about 1% of soft tissue tumors. Over 50 cases of intraneural perineuriomas have been documented, including cranial nerve tumors. Over 100 cases of soft tissue perineuriomas have been seen. There has been one documented case of a perineurioma arising within a lateral ventricle.

Etiology and risk factors

It was not until 2017 that any cause of intraneural perineuriomas was known. In that year, the *Mayo Clinic* identified a common cause for these tumors and an unexpected shared pathogenesis with intracranial meningiomas. There are three recurrent mutations in the WD40 domain of the TRAF7 protein. These mutations also frequently occur as a genetic cause of meningioma. This may mean that both types of tumors originate from the same cell. There are no identified risk factors for intraneural perineuriomas.

Pathology

Intraneural perineuriomas mostly affect peripheral nerves of the arms and legs. Cranial nerve lesions are rarely seen. A segmental, tubular enlargement of the affected nerve occurs, in which the nerve can become several times larger than normal. The nerve fascicles are pale and coarse. The majority of tumors are less than 10 cm in length, though there has been one case of a 40 cm long sciatic nerve tumor. A "bag-of-worms" plexiform growth pattern is not seen even though many fascicles are usually involved. In one case, there was involvement of two adjacent spinal nerves.

Neoplastic perineurial cells proliferate through the endoneurium, forming concentric layers around axons. The fascicles become enlarged, and the pseudo-onion bulbs form (see Fig. 25.8). Cross-sectioning reveals variable cellularity of the fascicles. Normal-appearing perineurial cells mostly occur within the endoneurium, but also in the perineurium.

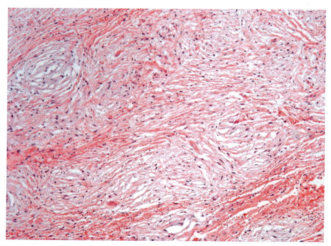

FIGURE 25.8 Perineurioma, with whirling growth pattern of bland spindled cells that have elongated cytoplasmic processes. Myxoid changes in the stromal background are seen, along with "onion bulb" formations.

Very large whorls may surround many nerve fibers. Sometimes, perineurial cells around one or more axons add to a nearby onion bulb as well. Therefore, the bulbs anastomose to form a complicated endoneurial network. In just one fascicle, cell density and lesion complexity may be varied. Mitotic activity is rare. In a newly formed lesion, axonal density and myelination may be nearly normal. In older lesions, however, with most fibers surrounded by perineurial cells and wide separation, myelin may be absent or only present in small amounts. In later stages, the centers of the perineurial whorls may only contain Schwann cells, and no accompanying axons. There may be prominent hyalinization. Myxoid changes in the stromal background are common.

All intraneural perineuriomas are immunoreactive for vimentin. There is a membranous pattern of EMA staining, along with the patterns of collagen IV and laminin. Axons centralized in the pseudo-onion bulbs stain for *neurofilament protein*, while those in the residual Schwann cells stain for S100 protein. There has also been staining for p53 protein. Intraneural perineuriomas may have a Ki-67 proliferation index of 5%-15%. There is a monosomy of chromosome 22.

Clinical manifestations

Intraneural perineuriomas cause progressive muscle weakness, with or without atrophy, more often than they cause sensory disturbances. There may be pain along the affected nerve.

Diagnosis

With electron microscopy, intraneural perineuriomas have myelinated nerve fibers that are surrounded by normal-appearing perineurial cells. The cells have cytoplasmic processes that are long and thin, with many pinocytotic vesicles. The cells are lined by a patchy surface basement membrane, and there may be abundant stromal collagen. Biopsy is all that is needed for a confirmative diagnosis. Differential diagnoses include neurofibroma, schwannoma, intraneural MPNST, neural lipofibroma, Pacinian neuroma, low-grade fibromyxoid sarcoma, ectopic meningioma, desmoid fibromatosis, smooth muscle tumor, and solitary fibrous tumor.

Treatment

The treatment for intraneural perineurioma may involve surgical resection of the entire tumor—but this is reserved for severe symptoms since risks for surgery may outweigh the benefits. Since the tumor runs along the length of a nerve, surgery is determined on a case-by-case basis. Postoperative care is important, and the patient must maintain only minimum activity levels until the surgical wound heals. Regular screening and check-ups are important.

Prognosis

Intraneural perineuriomas are benign, and usually do not recur or metastasize. Prognosis is generally excellent, but especially when the tumors are small and completely excised.

> **Point to remember**
>
> In most cases, intraneural perineuriomas present with muscle weakness. They are encountered mostly in young adults and adolescents, with no gender predilection. The tumors present as a fusiform expansion of part of a peripheral nerve, ranging greatly in length, but usually longer than 10 cm.

Clinical cases

Clinical case 1

1. What types of tumors may be involved in hybrid nerve sheath tumors?
2. Is this patient's diagnosis inconsistent with the usual etiology of neurofibroma/schwannomas.
3. What are the two primary components of hybrid neurofibromas/schwannomas?

A 47-year-old man presented with multiple painful and enlarging right-sided nerve sheath tumors on his hip, pubic area, paraspinal, and spinal foraminal regions. He had no diagnostic features consistent with any type of neurofibromatosis. His CT scans revealed multiple cutaneous and subcostal lesions. While a brain MRI showed no tumors, a spinal MRI revealed heterogeneously enhancing right-side masses at multiple thoracic and lumbar nerve roots. The lesions were painful, and OTC pain medications were not effective. Therefore, an intrathecal drug delivery pump was needed for implantation, to deliver analgesia as required. The patient underwent surgical resection of some of the most serious tumors, which all were pathologically similar, appearing as tan-to-pink soft tissue nodules. Though features of plexiform neurofibroma predominated, there were focal regions of Schwann cell proliferation resembling schwannoma. The final diagnosis was of multiple hybrid nerve sheath neurofibroma/schwannomas.

Answers:

1. Hybrid nerve sheath tumors may include neurofibromas, perineuriomas, and schwannomas. The two most common forms are the schwannoma/perineuriomas and the neurofibroma/schwannomas.
2. This patient's diagnosis is somewhat inconsistent because neurofibroma/schwannomas are usually (but not always) linked to schwannomatosis, NF1, or NF2. He had no diagnostic features consistent with any type of neurofibromatosis, yet developed the neurofibroma/schwannomas.
3. The two primary components of these hybrid tumors are the schwannoma-like component and the neurofibroma-like component. The first involves nodular Schwann cell proliferation, sometimes with Verocay bodies, mainly composed of cellular Antoni A areas, often with the Verocay bodies and Schwann cells having nuclear palisading. The second component involves a mixed cell population, collagen, and myxoid changes. There may be large amount of these as well as fibroblasts. The Schwannian cells have unique wavy and elongated appearances.

Clinical case 2

1. Are MPNSTs considered aggressive, and why?
2. In which age range are MPNSTs most common?
3. Is there any accurate grading system for MPNSTs?

A 67-year-old woman with a history of breast cancer presented with a soft tissue mass located where she had previously had lumbar surgery. An excisional biopsy revealed a traumatic neuroma. Five years later, the patient returned with a quickly growing and tender nodule in the same location. Another excisional biopsy was performed, showing a tumor made of malignant epithelioid spindle cells that merged with the residual traumatic neuroma. The malignant cells were positive for S100 protein. They were negative for cytokeratins, melan-A, HMB45, and GFAP. Electron microscopy revealed no melanosomes. The diagnosis was of a MPNST.

Answers:

1. MPNSTs are considered to be aggressive since there is up to a 65% chance of recurrence after surgical resection.
2. MPNSTs mostly affect young to middle-aged adults, but have occurred up to the sixth decade of life. They also occur in adolescents.
3. In general, there is a lack of any accurate grading system for MPNSTs. One method is to divide the tumors into high-grade and low-grade types. The high-grade tumors make up about 85% of cases while the low-grade tumors make up about 15% of the remaining cases. Examples of high-grade MPNSTs include: conventional monomorphous spindle cell MPNSTs, highly pleomorphic MPNSTs, and MPNSTs with divergent differentiation.

Clinical case 3

1. Which nerves are most commonly affected by MPNSTs?
2. How common is tumor metastasis with these types of tumors?
3. What do these tumors resemble in imaging studies?

A 35-year-old woman presented with a history of severe lower back pain that radiated to both thighs, paraparesis, and bowel/bladder involvement over the past 3 months. An MRI of the lumbar spine revealed an expansive soft tissue mass destroying the L5 vertebral body and surrounding structures. There was paraspinal extension and epidural cord compression. The mass was hypointense on T1-weighted imaging and hyperintense on T2-weighted imaging. An axial contrast-enhanced MRI revealed an inhomogeneous enhanced lesion and some nonenhancing areas of necrosis within. The paraspinal and intraspinal tumors were almost totally excised during surgery, and local radiation therapy was then administered. Over time, the patient experienced lung metastasis and died from related complications.

Answers:

1. **MPNSTs** usually affect large and medium-sized nerves instead of small nerves. The most common sites include the buttocks, thighs, brachial plexus, upper arms, and paraspinal area. The most commonly affected nerve is the sciatic nerve.
2. **Between 20% and 25%** of cases of MPNSTs involve metastasis, which is usually to the lungs.
3. MPNSTs **resemble soft tissue sarcomas** in imaging studies. Often, there is inhomogeneous contrast enhancement, and irregular tumor contours, which is suggestive of invasion.

Clinical case 4

1. What is an MPNST with divergent differentiation also called?
2. How common are these tumors in comparison with all cases of MPNSTs?
3. When resected, how do these tumors appear?

A 23-year-old man, with no history of neurofibromatosis, presented with a swelling on his back that had increased in size and become uncomfortable. The tumor was surgically resected, and histopathological examination revealed it to be a MPNST. It had extensive osseous and cartilaginous differentiation. Within 1 year, the patient developed pulmonary metastasis, and this tumor was also resected. Its histopathology revealed it to also be a metastatic MPNST, this time without any osseous or cartilaginous differentiation. The patient had no additional recurrence at 9 months of follow-up.

Answers:

1. **An MPNST with divergent differentiation is also known as a malignant triton tumor or a glandular** MPNST.
2. **MPNSTs with divergent differentiation make up about 15% of all cases of MPNSTs.**
3. **These tumors are generally firm and tan-colored, with areas of hemorrhage and necrosis. There may be areas of palisading necrosis, perivascular proliferations of plump tumor cells, amianthoid fibers, and myxoid stroma.**

Clinical case 5

1. What is the characteristic formation that occurs in perineuriomas?
2. When do intraneural perineuriomas usually develop, and what is their basic epidemiology?
3. What is the recently identified cause of intraneural perineuriomas?

A 22-year-old woman presented with a chronic history of paroxysmal numbness in both legs, and also an aching sensation of pain. She had no family history of neuromuscular symptoms. Clinical examination revealed weakness and hypoesthesia of the limbs. She also had weakness of the ulna muscle of the left hand. An MRI showed nerve enlargement and an abnormal hyperintense signal on T2-weighted imaging, including the cervical plexus, brachial plexus, intercostal nerves, spinal roots, and sciatic nerve. Nerve biopsy revealed circumscribed lesions surrounded by fibrous connective tissue made up of layers of spindle-shaped cells around a central axon, and the presence of pseudo-onion bulbs. The perineurial cells were positive for EMA, collagen IV, and CD34. The Schwann cells were positive for S100 protein. The final diagnosis was intraneural perineurioma.

Answers:

1. **Proliferating perineurial cells in the endoneurium of a perineurioma form characteristic pseudo-onion bulbs, which are rounded expansions that have the appearance of actual onions.**
2. **Intraneural perineuriomas usually develop in adolescence or early adulthood. There is no difference in prevalence between the genders. The intraneural perineuriomas make up about 1% of nerve sheath tumors.**
3. **In 2017, the Mayo Clinic identified a common cause for intraneural perineuriomas, which shares an unexpected pathogenesis with intracranial meningiomas. There are recurrent mutations in the WD40 domain of the TRAF7 protein, meaning that both types of tumors originate from the same cell.**

Key terms

Angiosarcomatous
Charcot-Marie-Tooth disease
Chondrosarcomatous
Coexpression
Cytomorphology
Dejerine-Sottas disease
Double staining
Hybrid nerve sheath tumors
Inhomogeneous
Malignant perineuriomas
Malignant triton tumor
Osteosarcomatous
Perineurioma
Perineuriomatous
Pseudo-onion bulbs
Verocay bodies

Further reading

Ambrosini, V., & Fanti, S. (2016). *PET/CT in neuroendocrine tumors (clinicians' guides to radionuclide hybrid imaging)*. Springer.
Argenyi, Z., & Jokinen, C. H. (2011). *Cutaneous neural neoplasms: A practical guide (current clinical pathology)*. Humana Press.
Batchelor, T., Nishikawa, R., Tarbell, N., & Weller, M. (2017). *Oxford textbook of neuro-oncology (Oxford textbooks in clinical neurology)*. OUP Oxford.
Bernstein, M., & Berger, M. S. (2014). *Neuro-oncology: The essentials (3rd ed.)*. Thieme.
Bignold, L. P. (2019). *Principles of tumors: A translational approach to foundations (2nd ed.)*. Academic Press.
Billings, S. D., Patel, R. M., & Buehler, D. (2019). *Soft tissue tumors of the skin*. Springer.
Binder, D. K., Sonne, D. C., & Fischbein, N. J. (2010). *Cranial nerves: Anatomy, pathology, imaging*. Thieme.
Brennan, M. F., Antonescu, C. R., & Maki, R. G. (2013). *Management of soft tissue sarcoma*. Springer.
Chaichana, K., & Quinones-Hinojosa, A. (2019). *Comprehensive overview of modern surgical approaches to intrinsic brain tumors*. Academic Press.
Delbeke, D., & Israel, O. (2010). *Hybrid PET/CT and SPECT/CT imaging: A teaching file*. Springer.
Earlstein, F. (2016). *Charcot Marie Tooth disease: Diagnosis, symptoms, treatment, causes, doctors, nervous disorders, prognosis, research, history, surgery, and more! facts & information*. NRB Publishing.
Franco, R., Marino, F. Z., & Giordano, A. (2018). *The mediastinal mass: A multidisciplinary approach (current clinical pathology)*. Humana Press.
Gram Hansen, C. (2019). *Disease and the Hippo pathway: Cellular and molecular mechanisms*. Mdpi AG.
Harsh, G. R., & Vaz-Guimaraes, F. (2017). *Chordomas and chondrosarcomas of the skull base and spine (2nd ed.)*. Academic Press.
Henshaw, R. M. (2017). *Sarcoma: A multidisciplinary approach to treatment*. Springer.
Khalbuss, W. E., & Li, Q. K. (2015). *Diagnostic cytopathology*. Springer.
Khatua, S., & Pillay Smiley, N. (2019). *Update in pediatric neuro-oncology*. Mdpi AG.
Kim, D. H., Chang, U. K., Kim, S. H., & Bilsky, M. H. (2008). *Tumors of the spine*. Saunders.
Lloyd, S. K. W., & Evans, D. G. R. (2013). *Peripheral nerve disorders: Chapter 54. neurofibromatosis type 2 (NF2): Diagnosis and management (Handbook of clinical neurology 115)*. Elsevier.
National Comprehensive Cancer Network (NCCN). *NCCN guidelines for patients: Soft tissue sarcoma.* (2014) National Comprehensive Cancer Network (NCCN).
Nayar, R., Lin, X., Paintal, A. S., Gupta, R., & Nemcek, A. A., Jr. (2019). *Atlas of cytopathology and radiology*. Springer.
Newton, H. B., & Malkin, M. G. (2010). *Neurological complications of systemic cancer and antineoplastic therapy (neurological disease and therapy)*. CRC Press.
Penkert, G., & Fansa, H. (2004). *Peripheral nerve lesions: Nerve surgery and secondary reconstructive repair*. Springer.
Pollock, R., Randall, R. L., O'Sullivan, B., & Pollock, R. E. (2019). *Sarcoma oncology: A multidisciplinary approach*. PMPH USA, Ltd.
Seyfried, T. (2012). *Cancer as a metabolic disease: On the origin, management, and prevention of cancer*. Wiley.
Singh, A., Yadav, D., & Kant Singh, K. (2019). *Brain tumor detection from MRI images of brain: Using hybrid genetic FCM*. Lap Lambert Academic Publishing.
Suster, S. (2015). *Atlas of mediastinal pathology (Atlas of anatomic pathology)*. Springer.
Tonn, J. C., Reardon, D. A., Rutka, J. T., & Westphal, M. (2019). *Oncology of CNS tumors (3rd ed.)*. Springer.
Ueda, T., & Kawai, A. (2016). *Osteosarcoma*. Springer.
Vallat, J. M., & Weis, J. (2014). *Peripheral nerve disorders: Pathology and genetics*. Wiley-Blackwell.
Watts, C. (2013). *Emerging concepts in neuro-oncology*. Springer.
Winfree, C., & Spinner, R. J. (2008). *Peripheral nerves: Tumors and entrapments, an issue of neurosurgery clinics of North America, volume 19, number 4 (The clinics: Surgery)*. Saunders.
Wu, M. H. (2017). *Metabolic factor ET-1 imbalance enhanced condrosarcoma malignant*. Jinlang Academic Press.
Zhou, L., Burns, D. K., & Cai, C. (2019). *A case-based guide to neuromuscular pathology*. Springer.

Chapter 26

Metastatic tumors

Chapter Outline

Metastatic tumors	455	Key terms	463
Clinical cases	461	Further reading	463

Metastatic tumors

Metastatic tumors are those originating outside the central nervous system (CNS). They spread through the hematogenous route to the CNS, in most cases. Less often, they directly invade the CNS from nearby body structures.

Epidemiology

The incidence rates for brain metastases has probably been underreported and underdiagnosed. Incidence of brain metastases in Sweden along, between 1987 and 2006, was documented at 14 cases out of every 100,000 people. This was double the levels seen in earlier years, likely because of the development of better imaging techniques. Generally, CNS metastases are seen in 25% of people who die of cancer, according to autopsy studies. In patients with solid tumors, leptomeningeal metastases occur in 4% − 15% of cases. In patients with advanced cancer, 8% − 9% of cases showed dural metastases. Regarding spinal tumors, spinal epidural metastases occur in 5% − 10% of cases, which is much more common than spinal leptomeningeal or intramedullary metastases (see Fig. 26.1).

In adults, CNS metastatic tumors are the most common neoplasms. However, metastases make up only 2% of all pediatric CNS tumors. Up to 30% of adults and 6% − 10% of children with cancer will develop metastases to the brain. Various primary tumors occur in different proportions between the two genders. However, gender has no great independent effect upon occurrence of CNS metastasis, for the majority of tumor types. Brain metastases have been of highest incidence in patients with primary lung cancer between the ages of 40 and 49 years. It has also been high in patients with colorectal cancer, primary melanoma, or renal cancer between ages 50 and 59, and with breast cancer between ages 20 and 29 years. Incidence of CNS metastases appears to be increasing, partly because of increased detection from better imaging, but also increased tumor incidence with predilections for involving the brain—such as lung cancer. Also, new therapeutic agents that may be effective overall and prolong life may also be ineffective for preventing or

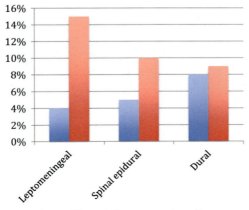

FIGURE 26.1 Lowest to highest ranges of metastases to specific central nervous system sites.

treatment CNS metastatic disease. Table 26.1 compares the relative frequencies of primary tumors and their brain metastases in males. Table 26.2 compares the same information for females.

Etiology and risk factors

The causes of changes from normal to cancer cells, in metastatic brain tumors as well as primary brain tumors, are not fully understood. There is a large range of tumors that cause brain metastases. Some subtypes are more likely to metastasize to the CNS. Metastatic brain tumors are most commonly caused by lung cancer (in both genders). For women, the second most common cause involves breast cancers described as **epidermal growth factor receptor-related (ERBB2)-positive** and **triple-negative**. Any breast cancer that tests negative for three primary factors—estrogen, progesterone, and the **HER2 protein**—is referred to as "triple-negative." The third most common cause is gastrointestinal cancer. For men, the second most common cause is gastrointestinal cancer, and the third is melanoma. Aside from this, actual risk factors for CNS metastases are unknown.

Pathology

The most common source of brain metastasis, in adults, is lung cancer. This is predominantly either adenocarcinoma or small cell carcinoma. The next most common sources are breast cancer, melanoma, renal cell carcinoma, and colorectal cancer. The most common origins of spinal epidural metastasis are prostate, breast, and lung cancers, followed by non-Hodgkin's lymphoma, multiple myeloma, and renal cancer. Tumors and their subtypes have large variance in their likelihood of metastasizing to the CNS. In pediatric patients, the most common sources of CNS metastases are leukemias and lymphomas. This is followed by nonhematopoietic CNS neoplasms, which include germ cell tumors, neuroblastoma, osteosarcoma, Ewing sarcoma, and rhabdomyosarcoma. Sometimes, primary head and neck neoplasms will extend intracranially via direct invasion. This may occur along the cranial nerves, presenting as intracranial tumors.

About 80% of brain metastases occur in the central hemispheres—especially in arterial border zones, and where the cerebral cortex and white matter join. About 15% occur in the cerebellum (see Fig. 26.2). Another 5% occur in the brainstem. Less than 50% of all brain metastases are single, being the only metastatic lesion in the brain. Also, only a few are solitary—the only metastasis within the body. Additional intracranial locations include the leptomeninges and dura mater, where extension from or to other brain compartments is seen more often. A large majority of metastases that reach the spinal cord expand from the paravertebral tissues or vertebral body into the epidural space (see Fig. 26.3). Sometimes, metastatic tumors seed in the walls of the brain ventricles or are found in the pituitary gland or choroid plexus.

There is only rare occurrence of tumor-to-tumor metastasis. Lung and breast cancers are the most common tumors related to this. Meningioma is the type of tumor that usually receives tumor cells from these cancers. Dural metastases are fairly common with cancers of the prostate, breasts, and lungs, as well as in hematological malignancies. Leptomeningeal metastases are seen more often with lung and breast cancers, melanoma, and hematopoietic tumors. Spinal epidural metastases are more often linked to cancers of the breasts, prostate, kidneys, lungs, and with

TABLE 26.1 Primary tumors and their brain metastases, in males.

Primary tumor	Related brain metastases
Prostate 30%	0.6%
Gastrointestinal 30%	11.4%
Lungs 15%	50%
Other areas 9%	5.6%
Bladder 8%	1.4%
Kidneys 5%	6.9%
Melanoma 3%	10%
N/A	Carcinoma of unknown primary (CUP) 14%

TABLE 26.2 Primary tumors and their brain metastases, in females.

Primary tumor	Related brain metastases
Breasts 30%	28.4%
Gastrointestinal 25%	8.3%
Other areas 15%	3.5%
Lungs 8%	31.6%
Uterus 8%	3.1%
Ovaries 5%	3.5%
Melanoma 5%	7.1%
Kidneys 4%	6.1%
N/A	Carcinoma of unknown primary (CUP) 8.4%

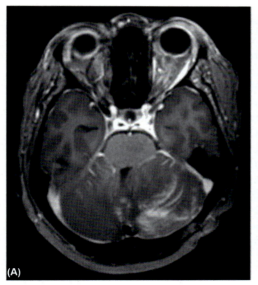
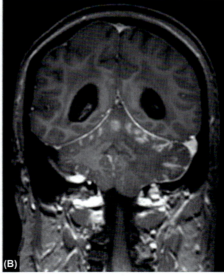

FIGURE 26.2 Metastatic tumor of the cerebellum shown (A) without contrast and (B) with contrast.

non-Hodgkins lymphoma or multiple myeloma. Small cell lung carcinoma is more often implicated in intramedullary spinal cord metastases.

In the CNS, parenchymal metastases are often grossly rounded and circumscribed, with grey to white to tan masses and varying amounts of peritumoral edema and central necrosis. When an adenocarcinoma metastasizes, there may be groupings of mucoid material. With metastases of clear cell renal cell carcinoma, choriocarcinoma, and melanoma, hemorrhage is relatively common. With melanoma metastases having a large amount of melanin, the metastasis appears brown to black in color. Diffuse opacity of membranes or multiple nodules may be produced by leptomeningeal metastasis of a non-Hodgkin lymphoma. Dural metastases can form localized nodules or plaques and also diffuse lesions. Primary head and neck area neoplasms can extend intracranially via direct invasion, potentially destroying the skull bones to a large degree. Sometimes, the skull is penetrated, without major bone destruction, by less severe perineural or perivascular invasion.

Under a microscope, histology and immunohistochemistry of secondary CNS tumors may be as wide-ranging as the primary tumors they form from. The majority of brain metastases are well demarcated and have varied perivascular growth in adjacent CNS tissue. This growth is referred to as **vascular cooption**. Sometimes, there is more diffuse, infiltrative **pseudogliomatous** growth with lymphomas and small cell lung carcinomas. There may be extensive tumor

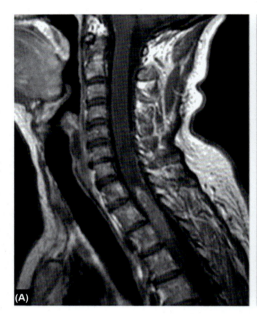 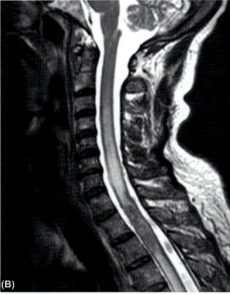

FIGURE 26.3 Intraspinal metastasis from a breast cancer. (A) There is a "flame-like" appearance of contrast enhancement in the upper pole of the lesion. (B) There is considerable edema around the lesion on this T2-weighted image.

necrosis, with identifiable tumor tissue only found around blood vessels and at the lesion edges. With leptomeningeal metastasis of a non-Hodgkin lymphoma, tumors cells are dispersed through the Virchow-Robin and subarachnoid spaces. The cells may invade nearby nerve roots and CNS parenchyma. Infiltration patterns may help to distinguish metastatic tumors from primary CNS tumors such as diffuse gliomas. However, sometimes glioblastomas may have a pushing margin with or without invasion via the Virchow − Robin spaces, instead of a single-cell type of infiltration. This is especially true in focal areas. The tumors can sometimes be confused for metastatic tumors because of similar morphology. Metastatic CNS tumors have variable pathologies, often with significant mitotic activity. The proliferation index may be much higher than that of the primary neoplasm.

Before presenting as hematogenous metastases within the CNS, a tumor cell must first escape from the primary tumor. It enters the blood stream and survives, and then enters and survives in the microenvironment of the CNS. Even today, how a tumor cell is able to survive all of these factors is not fully understood. Secondary CNS tumors may develop via direct extension from primary tumors in the paranasal sinuses, bones, and other nearby structures. These tumors are not actually considered to be metastases since they are still in contact with the primary neoplasm. When they contact the CSF-containing compartments, the tumor cells may seed throughout the CNS.

> **Point to remember**
> Metastatic CNS cancer is one that has spread from an earlier tumor in another body location. Under a microscope, metastatic cancer cells usually look the same as cells of the original type of cancer. Most types of metastatic cancer cannot be cured with the currently available treatments.

Clinical manifestations

Intracranial metastases generally cause neurological signs and symptoms from increased intracranial pressure, or local effect of a tumor upon nearby brain tissue. Symptoms may progress slowly. They include altered mental status, headache, ataxia, paresis, nausea, visual changes, and sensory abnormalities. Sometimes, there is an acute presentation, with hemorrhage, infarct, or seizures.

Diagnosis

No primary tumor is found at presentation in up to 10% of patients with brain metastases. A serious diagnostic challenge is the metastasis of renal cell carcinoma to a hemangioblastoma, accompanying von Hippel − Lindau disease. It usually takes up to 1 year for a diagnosed primary lung carcinoma to metastasize to the CNS. However, breast cancer and melanoma can take multiple years for CNS metastasis to occur. Often, patients with leptomeningeal metastasis

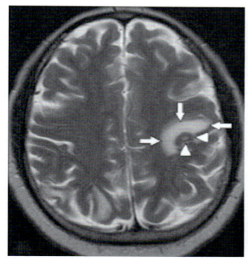
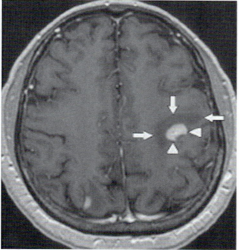

FIGURE 26.4 Axial cut in a contrast MRI of the brain. The left image, with T2-weighting, shows a hyperintense left frontal region, marked with arrows. The right image, with T1-weighting, shows perilesional hypointensity and central hyperintensity of the left frontal region, marked with arrows.

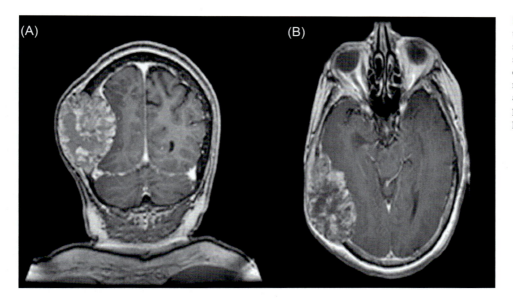

FIGURE 26.5 MRI of the brain, using T1-weighted imaging. (A) shows the coronal cut while (B) shows the axial cut, with the presence of a hyperintense image in the right parietal-temporal-occipital area, with bone infiltration and displacement of the brain parenchyma.

have many different neurological signs and symptoms upon presentation. These may include headache, ataxia, mental changes, cranial nerve abnormalities, and radiculopathy. Diagnostic examination reveals malignant cells in the first cerebrospinal fluid sample in approximately 50% of cases. This can be 80% or more if CSF sampling is repeated, and when 10 mL or more is available for analysis. Spinal metastases usually compress the spinal cord or nerve roots, causing back pain, arm or leg weakness, sensory abnormalities, and incontinence. These symptoms can develop over hours, days, or even weeks.

Via MRI, intraparenchymal metastases are usually circumscribed, with mild T1-hypointensity and T2-hyperintensity. There may be ring-like or diffuse contrast enhancement, with a surrounding area of parenchymal edema (see Fig. 26.4). Hemorrhagic metastases and metastatic melanomas that contain melanin may show hyperintensity on noncontrast CT or MRI. With leptomeningeal metastasis, an MRI may reveal diffuse or focal leptomeningeal thickening and contrast enhancement. There are sometimes tumor nodules dispersed through the subarachnoid space. Also, enlargement and enhancement of the cranial nerves, with communicating hydrocephalus, may be present. An MRI can show dural metastases as nodular masses, or thickening of the dura along bone structures (see Fig. 26.5). Metastases within the vertebral bodies appear as confluent, diffuse, or discrete areas of low signal intensity. A CT scan may be helpful in detecting bone involvement.

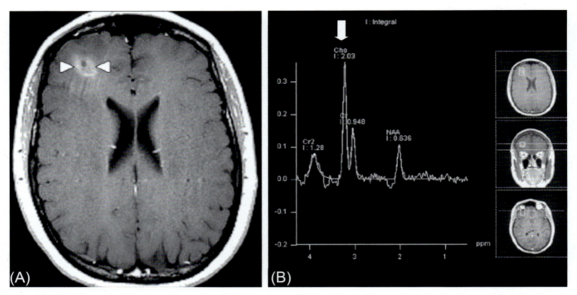

FIGURE 26.6 (A) X-ray spectroscopy of a right frontal lesion, marked with arrows. (B) A graph with an arrow indicating an increase in choline, which represents accelerated formation of cell membranes. This is compatible with a metastatic brain tumor.

Since immunohistochemical characteristics of secondary CNS tumors are usually similar to the tumors from which they originate, immunohistochemical analysis is often helpful in distinguishing primary from secondary CNS tumors. It is also good for assessment of natures and origins of metastatic neoplasms—especially for unknown primary tumors. Additionally, X-ray **spectroscopy** may be used for imaging, helping to identify increases in choline, which are compatible with metastatic brain tumors (see Fig. 26.6).

Treatment

Systemic therapy with targeted chemotherapies is used more than ever before for patients with CNS metastases. However, in previous years, it was less common since chemotherapeutic agents penetrate the blood-brain barrier poorly. Biomarker tests considered for these therapies include **EGFRALK rearrangement** for nonsmall cell lung cancer. Tests for **ERBB2 amplification** and estrogen and progesterone receptor expression are done for breast cancer. For melanoma, tests for **BRAF mutations** are performed. For colorectal cancer, tests are done for **RAF mutations**, and for gastroesophageal cancer, tests are done for ERBB2 amplification. Some of these markers show large differences between primary tumors and brain metises. This may result in analyzing biomarkers in brain metastasis tissue samples in order to plan treatments.

Treatment is often primarily palliative, to reduce symptoms and prolong life. However, in younger, healthier patients, aggressive treatments may include maximal surgical resection, chemotherapy, and radiosurgical (Gamma knife) therapy. Symptomatic care is given to all patients, which includes use of corticosteroids and anticonvulsants. Radiation therapy may also include whole-brain irradiation and fractionated radiothrearpy. Chemotherapeutic agents include alectinib and crizotinib. Intrathecal chemotherapy has been used experimentally, in which a chemotherapeutic drug is delivered via **intralumbar** injection into the CSR. Also, immunotherapy appears to be effective in some patients when they are asymptomatic, stable, and have not been previously treated. The agents *antiprogrammed cell death protein 1 (PD-1)* and *anticytotoxic T-lymphocyte-associated protein 4 (CTLA-4)* may be used alone, or in combination.

Prognosis

Prognosis for CNS metastases is based on the age of the patient, the **Karnofsky performance status**, the amount of metastases that exist, and the severity of the extracranial disease. Other prognostic factors include specific tumor type, and which molecular drivers are involved. Prognosis may also be affected by the presence of peritumoral brain edema. Recent improvements in overall survival may be linked to better focal and systemic therapies, plus earlier detection of CNS metastases.

> **Point to remember**
> The majority of brain metastases occur in the cerebral hemispheres, followed by the cerebellum and brainstem. Today, there is a rapid pace of development of treatments for metastasis of other cancers to the brain, with continuous clinical trials being conducted. There is significant evidence that we are getting closer to being able to manage tumor metastasis to the CNS than ever before.

Clinical cases

Clinical case 1

1. How do most tumors originating in other body sites spread to the central nervous system?
2. How often are CNS metastases seen in people who die of cancer?
3. What percentage of adults with cancer will develop metastases to the brain?

A 53-year-old woman, who had smoked for most of her life, was diagnosed with nonsmall cell lung cancer. She also had abdominal pain, and a CT scan revealed a 7-cm adrenal mass on the right adrenal gland. A PET scan revealed a right-sided intra-pulmonary lesion. The adrenal gland tumor was found to be an epidermal growth factor receptor (EGFR) negative adenocarcinoma of nonsmall cell lung cancer origin. A brain MRI revealed three small tumors, and the patient was treated with whole brain radiation therapy. She next received three regimens of systemic chemotherapy and external beam radiation to the primary lung tumor. Within a year, she had a good systemic clinical response to these treatments. In another year, the patient report vision changes, and another brain MRI revealed new metastatic tumors. She was treated with stereotactic radiosurgery. Chemotherapy continued until the following year, when imaging revealed stable systemic disease.

Answers:

1. **Most metastatic tumors spread from other body sites through the hematogenous route to the central nervous system. Less often, they directly invade the CNS from nearby body structures.**
2. **Generally, CNS metastases are seen in 25% of people who die of cancer, according to autopsy studies.**
3. **Up to 30% of adults with cancer will develop metastases to the brain. Also, brain metastases has been of highest incidence in patients with primary lung cancer between the ages of 40 and 49.**

Clinical case 2

1. Brain metastases are of highest incidence in breast cancer patients of what age range?
2. What are the top three most common sources of tumors that metastasize to the brain, in women?
3. Which type of breast cancers are most likely to metastasize to the brain?

A 52-year-old woman who had previously had breast cancer was hospitalized because for progressive right-sided body weakness and inability to speak. A brain MRI revealed a metastatic tumor in the deep frontal region with a large amount of edema. The tumor was removed using minimally invasive surgery, and the patient's ability to speak returned shortly after recovery. She regained strength on the right side of her body, and her speech was totally normal again after a few months.

Answers:

1. **Brain metastases are of highest incidence in breast cancer patients between ages 20 and 29.**
2. **The top three most common sources of tumors metastasizing to the brain, in women, are the lungs (31.6%), breasts (28.4%), and gastrointestinal system (8.3%).**
3. **Breast cancers described as "epidermal growth factor receptor-related (ERBB2)-positive" and "triple-negative" are most likely to metastasize to the brain. Any breast cancer testing negative for estrogen, progesterone, and the HER2 protein is referred to as triple-negative.**

Clinical case 3

1. What are the top three tumor sources of metastases to the brain, in men?
2. Where do the vast majority of brain metastases occur?
3. Are most brain metastases single or multiple?

A 55-year-old man presented with headaches, and both memory and visual changes. Neuro-imaging revealed a right temporal parietal arteriovenous malformation and an adjoining, hyper-enhancing occipito-temporal lobe lesion. The patient was treated with radiosurgery for the arteriovenous malformation. One month later, his symptoms worsened, and more imaging revealed two small, enhancing lesions of the left temporal lobe. After microsurgical resection of the temporal lobe lesions, the patient was also found to have an adenocarcinoma of his left lung.

Answers:

1. The top three tumor sources of brain metastases, in men, are the lungs (50%), gastrointestinal system (11.4%), and from melanoma (10%).
2. About 80% of brain metastases occur in the central hemispheres—especially in arterial border zones, and where the cerebral cortex and white matter join.
3. Less than 50% of all brain metastases are single, being the only metastatic lesion in the brain. Also, only a few are solitary — the only metastasis within the body.

Clinical case 4

1. How common are cerebellar metastases from other sites?
2. How do most brain metastases appear?
3. In what manner do intracranial metastasis cause neurological signs and symptoms?

A 78-year-old man had a history of gastrointestinal cancer. He presented to his local hospital with nausea and a severe headache. A head CT scan revealed a cerebellar mass. He also had a 5-cm abdominal mass and a 3-cm right hepatic lobe mass. A brain MRI confirmed a mixed solid and cystic mass in the medial and inferior right cerebellar hemisphere with marked mass effect upon the fourth ventricle. There was also an enhancing mass in the left occipital lobe. The right cerebellar tumor was surgically resected and found to be a metastatic adenocarcinoma. He was scheduled for additional surgeries and radiation therapy, but his overall poor health required time for him to be stabilized prior to these being carried out.

Answers:

1. Only about 15% of all brain metastases occur in the cerebellum, and another 5% occur in the brainstem.
2. Most brain metastases are well demarcated and have varied perivascular growth in adjacent CNS tissue. This growth is referred to as vascular cooption.
3. Intracranial metastases generally cause neurological signs and symptoms from increased intracranial pressure, or local effect of a tumor upon nearby brain tissue. Symptoms may progress slowly. They include altered mental status, headache, ataxia, paresis, nausea, visual changes, and sensory abnormalities. Sometimes there is an acute presentation, with hemorrhage, infarct, or seizures.

Clinical case 5

1. How do intraparenchymal metastases appear via MRI?
2. Why is it often helpful to use immunohistochemical analysis to distinguish between primary and secondary CNS tumors?
3. What is the prognosis for CNS metastases based on?

An 86-year-old woman with a history of kidney cancer presented with a right temporo-parietal tumor and symptoms of headache, altered mental status, weakness, and seizures. The tumor, a hemangioblastoma, was treated with surgery and stereotactic radiosurgery. The patient was also diagnosed with von Hippel − Lindau disease. Five months later, the patient returned with a second solitary right lesion, and again received surgery and radiosurgery, successfully. Over six months of follow-up there had been no tumor recurrence.

Answers:

1. Using MRI, intraparenchymal metastases are usually circumscribed, with mild T1-hypointensity and T2-hyperintensity. There may be ring-like or diffuse contrast enhancement, with a surrounding area of parenchymal edema.
2. Since immunohistochemical characteristics of secondary CNS tumors are usually similar to the tumors from which they originate, immunohistochemical analysis is often helpful in distinguishing primary from secondary CNS tumors. It is also good for assessment of natures and origins of metastatic neoplasms—especially for unknown primary tumors.

3. **Prognosis for CNS metastases is based on the patient's age, the Karnofsky performance status, the amount of existing metastases, and the severity of the extracranial disease. Other prognostic factors include the specific tumor type, which molecular drivers are involved, and the presence of peritumoral brain edema**.

Key terms

BRAF mutations
EGFRALK rearrangement
Epidermal growth factor receptor-related (ERBB2)-positive
ERBB2 amplification
HER2 protein
Intralumbar
Karnofsky performance status
Metastatic tumors
Pseudogliomatous
RAF mutations
Spectroscopy
Triple-negative
Vascular cooption

Further reading

Ahluwalia, M., Metellus, P., & Soffietti, R. (2019). *Central nervous system metastases*. Springer.
Amiji, M. M., & Ramesh, R. (2018). *Diagnostic and therapeutic applications of exosomes in cancer*. Academic Press.
Bilsky, M. H., & Laufer, I. (2019). *Spinal oncology: An issue of neurosurgery clinics of North America (Volume 31-2) (The clinics: Surgery)*. Elsevier.
Birbrair, A. (2019). *Tumor microenvironment: Recent advances (Advances in experimental medicine and biology, Book 1225)*. Springer.
Chang, E. L., Brown, P. D., Lo, S. S., Sahgal, A., & Suh, J. H. (2018). *Adult CNS radiation oncology: Principles and practice*. Springer.
Cid, R. P. (2019). *Circulating tumor cells in breast cancer metastatic disease (Advances in experimental medicine and biology)*. Springer.
Cristofanilli, M. (2017). *Liquid biopsies in solid tumors (Cancer drug discovery and development)*. Humana Press.
Dolgushin, M., Kornienko, V., & Pronin, I. (2018). *Brain metastases: advanced neuroimaging*. Springer.
Ferguson, C. (2019). *Current research in brain cancer*. Foster Academics.
Gokaslan, Z. L., Boriani, S., Fisher, C. G., & Gomes Vialle, L. R. (2014). *AOSpine masters series volume 1: Metastatic spinal tumors*. Thieme.
Hayat, M. A. (2014). *Brain metastases from primary tumors, volume 1: Epidemiology, biology, and therapy*. Academic Press.
Hayat, M. A. (2015). *Brain metastases from primary tumors, volume 2: Epidemiology, biology, and therapy*. Academic Press.
Hayat, M. A. (2016). *Brain metastases from primary tumors, volume 3: Epidemiology, biology, and therapy of melanoma and other cancers*. Academic Press.
Hayat, M. A. (2014). *Tumors of the central nervous system, volume 13: Types of tumors, diagnosis, ultrasonography, surgery, brain metastasis, and general CNS diseases*. Springer.
Ironside, J., Nicoll, J., & Moss, T. (2007). *Intra-operative diagnosis of CNS tumors*. CRC Press.
Kantarjian, H., & Wolff, R. (2016). *The MD Anderson manual of medical oncology* (3rd Edition). McGraw-Hill Education / Medical.
Kim, D. G., & Lunsford, L. D. (2012). *Current and future management of brain metastasis (Progress in neurological surgery, volume 25)*. S. Karger.
Peabody, T. D., & Attar, S. (2014). *Orthopedic oncology: Primary and metastatic tumors of the skeletal system (Cancer treatment and research)*. Springer.
Piotrowski, W., Brock, M., & Klinger, M. (2004). *CNS metastases — neurosurgery in the aged (Advances in neurosurgery 12)*. Springer.
Raizer, J. J., & Abrey, L. E. (2007). *Brain metastases (Cancer treatment and research)*. Springer.
Rajendran, J., & Manchanda, V. (2010). *Nuclear medicine cases (McGraw-hill radiology series)*. McGraw-Hill Education / Medical.
Ramakrishna, R., Magge, R. S., Baaj, A. A., & Knisely, J. P. S. (2019). *Central nervous system metastases: Diagnosis and treatment*. Springer.
Randall, R. L. (2016). *Metastatic bone disease: An integrated approach to patient care*. Springer.
Sawaya, R. (2004). *Intracranial metastases: Current management strategies*. Blackwell Futura.
Schiff, D., & Van den Bent, M. J. (2018). *Metastatic disease of the nervous system (Volume 149) (Handbook of clinical neurology)*. Elsevier.
Tonn, J. C., Reardon, D. A., Rutka, J. T., & Westphal, M. (2019). *Oncology of CNS tumors* (3rd Edition). Springer.
Van Meir, E. G. (2009). *CNS cancer: Models, markers, prognostic factors, targets, and therapeutic approaches (Cancer drug discovery and development)*. Humana Press.
Vilensky, J. A., Weber, E. C., Sarosi, T., & Charmichael, S. W. (2010). *Medical imaging of normal and pathologic anatomy*. Saunders.
Wagman, L. (2014). *Hepatocellular cancer, cholangiocarcinoma, and metastatic tumors of the liver, an issue of surgical oncology clinics of North America (Volume 24-1) (The Clinics: Surgery)*. Elsevier.
Wang, Y., & Crea, F. (2017). *Tumor dormancy and recurrence (Cancer Drug discovery and development)*. Humana Press.

Chapter 27

Spinal cord tumors

Chapter Outline

Spinal cord tumors	465	**Clinical cases**	474
Extradural tumors	469	**Key terms**	476
Intradural tumors	471	**Further reading**	476

Spinal cord tumors

Spinal cord tumors can develop within the spinal cord parenchyma, where they are known as *intramedullary tumors*. They may also develop outside of the cord parenchyma, known as *extramedullary tumors*, and often compress the spinal cord or its nerve roots. Of the intramedullary tumors, the most common forms are gliomas, which include ependymomas and low-grade astrocytomas. These can infiltrate and destroy the parenchyma, and extend over many segments, or cause a syrinx. Extramedullary tumors may be *intradural* or *extradural*.

Epidemiology

Primary spinal cord tumors are less common than primary brain tumors. They make up only about 5.3% of all primary central nervous system tumors. In the United States, just over 2500 new cases are diagnosed annually, occurring most often in people between 20 and 60 years of age. Spinal cord tumors occur with a 1:1 ratio between the genders—except for meningiomas, which occur slightly more often in women. Primary spinal cord tumors are much less common than tumors that metastasize to the spine, which affect between 5% and 10% of people with systemic cancer.

Etiology and risk factors

The causes of most primary spinal cord tumors are unknown. Patients with neurofibromatosis type 1 (NF1) may develop spinal astrocytomas or neurofibromas. In patients with NF2, ependymomas or schwannomas (spinal nerve root tumors) may be present. Patients with von Hippel−Lindau syndrome are at risk for hemangioblastomas of the spine. Additional risk factors are unknown.

Pathology

Primary spinal cord tumors rarely metastasize outside of the central nervous system. The clinical manifestations will be based on the rate of tumor growth which part of the spinal cord is affected. The tumors are classified by the cell of origin and their location (see Fig. 27.1). The primary anatomic consideration is where the tumor is located in relation to the spinal dura mater. The most common primary spinal cord tumors are listed in Table 27.1.

Clinical manifestations

Signs and symptoms of spinal cord tumors usually evolve because of compression of the cord. For primary spinal cord tumors, symptom duration is often 3−4 years prior to diagnosis. Symptoms evolve more slowly when the tumor is slow-growing and benign. The spinal cord can even be compressed so that it appears like a thin ribbon, without serious neurologic deficits manifesting. However, a quickly-growing tumor causes a much faster manifestations of symptoms, due to extensive edema and increased pressure. Generally, signs and symptoms of spinal cord tumors include pain, motor weakness, sensory impairment, and bowel and bladder dysfunction. Pain is the most common symptom. For epidural metastases, neck or back pain can be present for weeks to months. Intradural tumors may cause pain for years

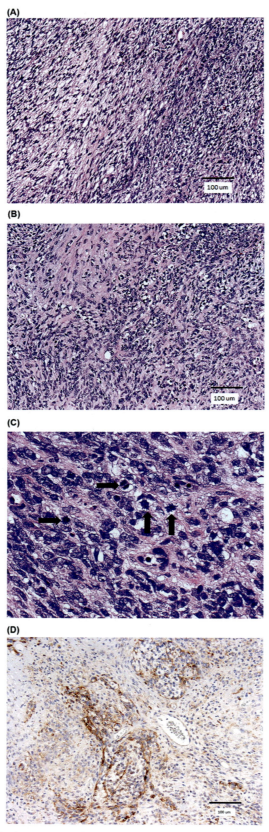

FIGURE 27.1 The tumors at L4 (A) and in the conus medullaris (B) both show interweaving cellular fascicles of spindle cells (original magnification ×200). (C) Over 20 mitotic figures per 10 high-power fields were identified focally. Four mitotic figures (arrows) are identified in this tumor at the level of the conus medullaris. (D) Immunohistochemistry for epithelial membrane antigen highlights epithelioid areas in the tumor at the level of the conus medullaris.

TABLE 27.1 Common primary spinal cord tumors.

Benign tumors	Malignant tumors
Aneurysmal bone cyst	Chondrosarcoma
Giant cell tumors	Chordoma
Hemangiomas	Ewing's sarcoma
Osteoid osteoma/osteoblastoma	Osteosarcoma

prior to diagnosis. The pain is often misdiagnosed as arthritis, intervertebral disc disease, or back strain. In general, the signs and symptoms of spinal cord tumors may be summarized as three syndromes:

- **Compressive (sensorimotor) syndrome**—mostly due to compression, but sometimes due to invasion and destruction of spinal tracts;
- **Irritative (radicular) syndrome**—a combination of spinal cord compression and its manifestations with radiating pain;
- **Syringomyelic syndrome**—due to actual inflammation of the spinal cord.

Neck or back pain may be localized or radicular. Especially with epidural metastases, localized pain and tenderness over the involved region are common. Radicular pain is often described as "band-like." It follows the distribution of the dermatomes (spinal nerve roots), varying from mild to severe. It may be described as burning, dull, or sharp. In nearly every case, it becomes more severe as time passes. Many patients complain of pain that worsens at night, and it is often aggravated by laying down. Spinal instability may be indicated by pain that is aggravated by movement, but relieved by remaining immobile. Spinal pressure and more intense pain may be the result of any activity that produces a Valsalva maneuver. These include coughing, sneezing, and straining during defecation.

Weakness can follow sensory symptoms and is the most commonly identified objective factor, occurring in 35% − 75% of patients. The **myotomes** (muscle groups) that are involved determine the amount of impairment. Weakness is often related to a positive Babinski sign, hyperreflexia, and spasticity, and will progress to total paraplegia without treatment. Between 50% and 68% of patients cannot walk at the time of initial diagnosis of epidural spinal cord compression. Individual motor symptoms vary based on tumor location. A lateral tumor affects voluntary arm and leg movements, muscle tone, posture, and coordination. Tumors of the anterior cord affect equilibrium, posture, and voluntary trunk muscle movements.

Between 50% and 70% of patients have sensory deficits, and 60% have bowel and bladder dysfunction. Sensory deficits vary based on tumor location, with lateral tumors affecting pain and temperature, resulting in numbness, tingling, and cold sensations. If the posterior aspect of the spinal cord is affected, the awareness of proprioception and vibration of body areas is affected. If the tumor is anterior, touch and pressure on the opposite side of the body is affected. Spinal cord compression affects function below the tumor, so it is important to determine the highest functional level. In about 20% of patients, **Brown-Séquard syndrome** is present. This involves a loss of position, touch, and vibration sense, plus motor ability on the same side as the lesion, with contralateral loss of temperature and pain sensation.

Diagnosis

Assessment of a patient with a known or suspected spinal cord tumor starts with a thorough history. Symptom description and duration must be recorded, along with any factors that improve or worsen them, and the order in which they appeared. A neurologic examination is performed to assess motor and sensory function, gait, and reflexes. Presence of bowel or bladder dysfunction is evaluated, and a pain assessment is done. This creates a baseline to compare to all future evaluations. The neurological evaluations should attempt to determine the likely tumor location.

Diagnosis of spinal cord tumors is via MRI of the affected area, with and without contrast. It provides detailed views of the spinal cord (see Fig. 27.2). Use of contrast is common since many tumors show enhancement regardless of their histologic grade. Also, MRI assists in planning treatment. CT with myelography is less accurate but can assist when vertebral metastases is suspected. Bone scans help to identify vertebral disease but are not always specific. Unfortunately, a spinal cord tumor is often not even considered until extremely obvious neurologic manifestations appear. For segmental neurologic deficits or possible spinal cord compression, emergency diagnosis and treatment is needed.

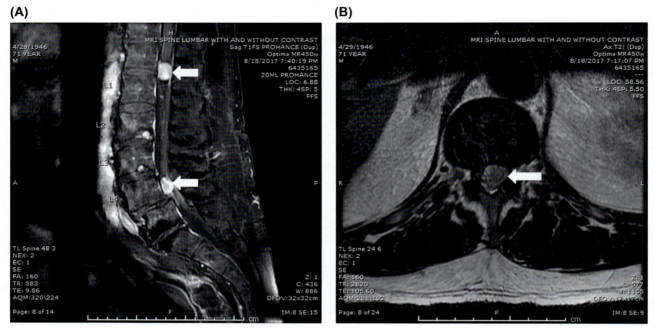

FIGURE 27.2 MRI of the lumbar spine. (A) Sagittal T2-weighted imaging shows multiple rounded enhancing nodules. The largest measures 15 × 15 mm in size. It is homogeneously enhancing. A second large nodule is seen at L4 measuring 1.7 × 1.4 cm in size. Numerous other small nodules are seen along the cauda equina. (B) Axial T2-weighted imaging shows severe canal narrowing at L4 due to compression by the large nodule.

Spinal tumors are suggested by back or radicular pain that is progressive, nocturnal, or unexplained. Segmental neurologic deficits and unexplained neurologic deficits referred to the spinal cord or nerve roots are also diagnostic. Unexplained back pain in patients who have primary lung, breast, prostate, kidney, thyroid, or lymphoma-related tumors may also indicate a spinal cord tumor. Sensory assessment of each patient starts at the toes, moving upward, to determine the spinal level at which function remains, which is usually the location of the tumor. Sometimes this is not entirely true, such as when the lesion is located one or two vertebrae higher than the level of spinal cord compression. The tumor level is often accompanied by a narrow band of hyperesthesia directly above it. Sensory effects can be bilateral and symmetrical, asymmetrical, and even unilateral. There may be a combination of sensory and motor deficits that are discovered.

> **Point to remember**
> It is extremely important for patients who have cancer in another body location, and begin to experience neck or back pain, to be evaluated for cancer metastasis to the spinal cord.

Treatment

With neurologic deficits, corticosteroids are started immediately to reduce cord edema and preserve functionality, especially in the setting of **cauda equina syndrome**. Tumors compressing the cord are treated quickly. Some tumors, if well-localized and primary, can be excised surgically. Surgery is able to provide rapid spinal cord decompression. It is also indicated for spinal instability or bone collapse into the spinal canal, tumor recurrence that is untreatable with additional radiation therapy, for radio-resistant tumors, or when the patient is having fast neurological deterioration. Aside from common surgical risks, complications from surgery also include cerebrospinal fluid leakage, wound dehiscence, and the development of neurologic deficits. Patients with extreme or long-term neurologic deficits often are not improved after surgery. The most important complication that requires treatment is a new neurologic deficit when neurologic function is likely not to return. New deficits are usually related to spinal cord vascular insults or manipulation during surgery. CSF leaks may develop when the dura is not totally sealed, or a tear is left unrepaired. Lumbar drainage for several days is usually enough to repair this type of leak, though surgery may also be needed.

While corticosteroids are often the first line of treatment for spinal cord tumors, optimal loading and maintenance doses have not been determined. Since spinal cord damage may become irreversible without relief of compression, high-dose dexamethasone is often prescribed and then tapered off. Chemotherapy is indicated for epidural tumors that are likely to respond, and for patients with recurrent tumors, if they have received radiation therapy. There have been no trials of chemotherapy for primary spinal cord tumors. Even so, drugs that are effective against intracranial gliomas may be effective for the same types of tumors within the spinal cord. Radiation therapy may be used with or without surgical decompression in some cases, but is generally not recommended for totally resected intradural, low-grade spinal cord tumors.

Prognosis

Prognosis for spinal cord tumors is based on whether they are extradural or intradural, the patient's age and overall health, if the tumor is benign or malignant, and if it is primary or metastatic. Prognosis for intramedullary tumors is worse since they may not be able to be removed without causing neural damage. The outlook for metastatic tumors is varied since treatments are mostly palliative.

> **Point to Remember**
> A spinal cord tumor is an abnormal mass of tissue within or surrounding the spinal cord, spinal column, or both. Spinal tumors can be benign or malignant. Spinal tumors are referred to in one of two ways: by the region of the spine in which they occur and by their location within the spine (intradural-extramedullary, intradural-intramedullary, or extradural).

Extradural tumors

Extradural tumors are located outside of the spinal cord, yet outside of the dura that contains the cord, nerve roots, and spinal fluid. Subtypes are referred to as *primary* and *secondary* tumors. The primary tumors are the least common subtype and are mostly seen in children and young adults. Most extradural tumors represent metastases, and include masses in the bones, disks, and paraspinal soft tissues.

Epidemiology

Extradural tumors are the most common type of spinal tumors. They make up 65% of all cases. Secondary extradural tumors are 40 times more common than primary extradural tumors. Extradural metastatic disease, which may seed the epidural space or vertebra, makes up 95% of spinal metastases. There are no accurate statistics on the overall incidence or prevalence of extradural spinal tumors.

Etiology and risk factors

Secondary extradural tumors are caused by metastatic spread of cancer from a primary tumor somewhere else in the body. Most secondary extradural tumors are caused by metastasis to the vertebral column, which is a common site of bone metastasis. They may occur at various spinal levels, or continuously throughout the spine. Metastatic spinal cord tumors usually originate from breast cancer, lung cancer, prostate cancer, renal cancer, and from multiple myeloma. Less common metastases come from gastrointestinal tract cancers, melanoma, lymphoma, sarcoma, and thyroid cancer.

Pathology

Most extradural tumors, before compressing the spinal cord, will invade and destroy bones. Primary extradural tumors may be benign or malignant, and originate within the spine. For secondary extradural tumors, epidural metastatic tumors that compress the spinal cord are common complications and are considered to be neurologic emergencies. Compression of the spinal cord, and not actual invasion of the cord, is the usual cause of neurologic symptoms. The cord is most often compressed anteriorly, which distorts and damages the nervous tissue. Spinal cord compression may occur from direct tumor extension into the epidural space, from direct extension through the intervertebral foramina, or by collapse of the vertebrae with displacement of bone into the epidural space. The most common location of epidural spinal cord compression, in 70% of cases, is the thoracic spine. This is followed by 20% of cases in the lumbosacral spine and 10% of cases in the cervical spine. However, in about 30% of all cases, there are multiple spinal levels involved (see Fig. 27.3) Thoracic compression is usually due to metastases from breast or lung cancer. Lower thoracic or lumbosacral compression is usually due to metastases from gastrointestinal, renal, or prostate tumors.

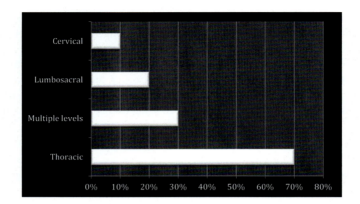

FIGURE 27.3 The most common locations of epidural spinal cord compression.

Spinal cord compression may be due to lymphomas since they may extend directly, through the intervertebral foramina. Primary extradural tumors can develop, usually chordomas or sarcomas. Chordomas grow slowly, but are extremely invasive. They are common in the sacrum, but also develop in the cervical spine or at the base of the skull. Chordomas erode bone and soft tissue extensively, and though histologically benign, are difficult to remove completely. Extradural spinal metastases are believed to occur via hematogenous arterial spread. Other method may be direct invasion through the intervertebral foramina by a paravertebral mass, or retrograde venous spread, from a primary site via **Batson's plexus**.

Clinical manifestations

Anterior compression of the spinal cord causes edema and ischemia. Especially for extradural tumors, early pain is common. It will be progressive and unrelated to physical activity, but worsened by lying down or leaning against an object. There may be back pain that radiates down the sensory distribution of a certain dermatome. Neurologic deficits are usually referable to the spinal cord, developing over time. These commonly include incontinence, spastic weakness, and dysfunction of sensory tracts in a certain cord region and below. Deficits are commonly bilateral.

Diagnosis

For most spinal tumors, the amount of deficits may be reduced and recovery may be improved if they can be diagnosed early. Comprehensive medical and neurological examinations are usually performed. Diagnosis involves a complete medical evaluation, and MRI or CT scans. Fig. 27.4 shows an extradural tumor that metastasized from lung cancer. Biopsies are obtained to help categorize tumors. Additional imaging studies may involve X-rays, myelography, and bone scans.

Treatment

For extradural tumors, radiation therapy and corticosteroids are the most widely used treatments. Higher doses of radiation are often initially given, especially with evidence of neurologic dysfunction. Spinal cord radiation does not cause any acute clinical symptoms, but the major complication is radiation myelopathy. This results from demyelination and white matter necrosis, or from intramedullary microvascular injury. Myelopathy may appear as a subacute or a more extreme, delayed reaction.

Prognosis

Fortunately, the prognosis for many patients with extradural spinal tumors is good, especially with adequate surgery, radiation therapy, and chemotherapy. Documentation exists of patients surviving with no signs of disease recurrence for years. A worse outlook is given when the tumor has metastasized from another site, meaning it is often advanced when discovered. Many metastatic cases have a survival of only 3−6 months. Improved survival is linked to the patient's ability to walk, a single site of cord compression, no brain or visceral metastases, and a tumor that is sensitive to radiation therapy. For neurologic recovery, the severity of weakness is most significant. About 80% of patients who can

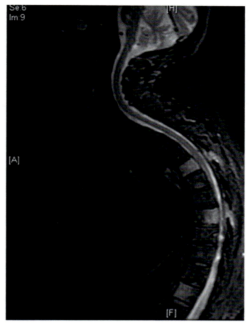

FIGURE 27.4 Extradural tumor metastases from lung cancer.

walk when they are diagnosed will remain able to do so after treatment. Likelihood of recovery is less with increased neurologic dysfunction. Between 30% and 45% of initially *paraparetic* and nonambulatory patients will become able to walk. Those who are paraplegic at diagnosis probably will remain in this state, and only 10% will become able to walk again. A better prognosis is also given for patients whose symptoms develop slowly rather than quickly.

> **Point to remember**
> Primary extradural tumors of the spine comprise only a small percentage of all spinal tumors. They may pose significant challenges to surgeons, but improvements in radiation therapy and chemotherapy have increased survival rates and improved overall outcomes.

Intradural tumors

Intradural tumors are located within the spinal cord. **Intradural-extramedullary tumors** are within the dura, but not within the parenchyma of the spinal cord. **Intradural-intramedullary tumors** actually occur totally within the spinal cord, inside the dura and parenchyma. The intradural tumors are linked to NF1, NF2, and von Hippel – Lindau syndrome.

Epidemiology

Intradural-extramedullary tumors make up nearly 90% of all primary spinal cord tumors, plus 30% of all spinal tumors. The most common subtypes are meningiomas and nerve sheath tumors in the form of neurofibromas and schwannomas. Meningiomas usually form in the thoracic spine, in 80% of cases. Schwannomas are common in the lumbar spine, located on any one of the multiple nerve roots of the cauda equina. The less common intradural-intramedullary tumors include chordomas, epidermoid tumors, and vascular tumors. Intradural-intramedullary tumors only account for about 2%–4% of all CNS tumors.

Etiology and risk factors

There are many genetic factors related to intradural spinal tumors. Most often, these include NF1, NF2, and von Hippel – Lindau syndrome. However, the actual cause of most tumors is not fully understood. Other than the genetic factors listed, there are no actual identified risk factors for intradural spinal tumors.

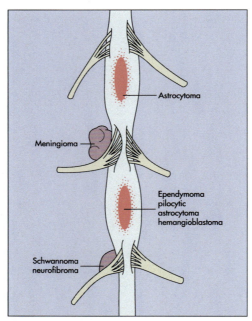

FIGURE 27.5 Intradural-extramedullary and intradural-intramedullary spinal cord tumors.

Pathology

Both intradural and extradural tumors cause neurologic damage via compressing the spinal cord or the nerve roots. Intradural-extramedullary tumors arise from the nerve roots of spinal cord coverings, while intradural-intramedullary tumors develop within the cord itself. Fig. 27.5 shows intradural-extramedullary tumors (meningioma and schwannoma) along with intradural-intramedullary tumors (astrocytoma, ependymoma, pilocytic astrocytoma, and hemangioblastoma). Spinal meningiomas and schwannomas are both usually benign and slow-growing. Sarcomas may arise as extramedullary tumors. Intradural-intramedullary tumors form from the same cells as intracranial-intramedullary tumors. However, they are usually less malignant, resulting in most primary spinal cord tumors to be benign. Between 40% and 60% of intramedullary tumors are ependymomas, and less often astrocytomas, which are more common in children. About 67% of thoracic tumors are astrocytomas, while about 49% of cervical tumors involve this type of tumor

The least common histologies include gangliogliomas, hemangioblastomas, hemangiomas, medulloblastomas, and oligodendrogliomas. Most ependymomas form in the cervical area, while myxopapillary ependymomas are usually located in the lumbar region. Hemangioblastomas are benign vascular tumors occurring about 50% of the time in the thoracic region. About 66% are sporadic, with the remainder linked to von Hippel−Lindau syndrome. Since 80% of these are solitary lesions, when multiple lesions are present, von Hippel−Lindau syndrome may be indicated. Intradural-intramedullary metastases can occur via cerebrospinal fluid pathways.

Clinical manifestations

Intradural-extramedullary tumors often cause pain because of traction upon or irritation of the spinal nerve roots, displacement of the cord, alteration of blood supply, or CSF circulation obstruction. Sometimes, distal lower extremity sensory deficits, segmental neurologic deficits, and symptoms of spinal cord compression appear quickly, resulting in paraplegia and bowel and bladder incontinence. Pain is usually followed by paresthesias, then sensory loss, and muscular weakness. If the compression is chronic, there will be muscle wasting along the distribution of the affected roots. Intradural-intramedullary tumors invade and destroy the spinal cord, altering blood supply, affecting cellular membrane stability, and disrupting afferent and efferent impulses. As a result, there are sensory, motor, and reflex deficits based on which spinal nerves are compromised. Resultant edema causes more deficits.

Diagnosis

As with extradural spinal tumors, diagnosis of intradural spinal tumors involves medical and neurological examination, plus imaging studies that include X-rays, CT, MRI, myelography, and bone scans; along with biopsies. In most cases,

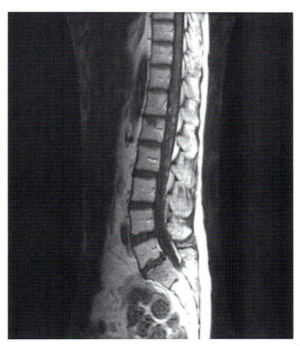

FIGURE 27.6 Intradural extramedullary metastases from breast cancer.

MRI is the imaging technique of choice because of its high soft tissue contrast potential. Fig. 27.6 shows intradural-extramedullary metastases from breast cancer. Intramedullary lesions are characterized by abnormal spinal cord signal and expansion. Extramedullary lesions displace instead of expand the spinal cord, and a CSF cleft may be seen separating the tumor from the cord. Differential diagnoses for intramedullary lesions include infections, demyelinating disease, vascular malformations, and metabolic abnormalities. Differential diagnoses for extramedullary lesions may be of much wider variance since the imaging features of these lesions are not as specific.

Treatment

Both spinal meningiomas and schwannomas can often be totally removed surgically, and recurrence is rare with complete resection. Treatment of spinal ependymomas is usually maximal surgical resection, and radiation therapy is used if only incomplete resection is possible, or a higher-grade tumor exists. When an ependymoma involves the CSF, it is treated with craniospinal radiation therapy. Chemotherapy is usually only used for tumor recurrence. While surgical resection is the treatment of choice for intradural tumors, their infiltrative nature complicates surgery. This is why radiation therapy is often needed. Since up to 20% of spinal cord astrocytomas are malignant, they may only be treatable via biopsy and radiation therapy.

For most intradural-extramedullary tumors, surgery is the first treatment. A posterior surgical approach, known as a laminectomy, is usually performed. Recurrence risk is about 10% for complete resections, but rates increase to 20% for incompletely resected lesions. If possible, recurrences are usually also surgically resected. For intradural-intramedullary tumors such as ependymomas and astrocytomas—except for malignant astrocytomas, surgery is also the first treatment. The most important factor here is the amount of tumor infiltration of the surrounding cord. Indistinct tumor margins and microscopic infiltration may prevent total surgical resection, which may also cause loss of neurological function. High-grade astrocytomas are treated with biopsy only, followed by radiation therapy. However, for well-delineated astrocytomas, surgical resection may give long-term control and even be curative. While astrocytomas usually recur within 3 years, for ependymomas, recurrence is often delayed as long as 12 years.

Both intramedullary and extramedullary tumors can be treated with radiation therapy when incompletely resected, or if their recurrence should not be treated with additional surgery. Ependymomas are sensitive to radiation, so even with subtotal resection; postoperative radiation therapy lends a 15-year survival rate of 75%. For patients with high-grade astrocytomas, radiation therapy is the standard treatment, but still does not greatly improve the outcome. Also, survival is not affected by radiation therapy with low-grade astrocytomas.

Prognosis

For intradural tumors, a better prognosis is given for larger ability for surgical resection, lower tumor grade, younger patient age, and slower onset of neurological dysfunction. When completely resected, schwannomas, ependymomas, and meningiomas are not likely to recur. This is not true for astrocytomas, however, which carry a worse prognosis. The patient's level of neurological function before surgery is the most significant prognostic factor for the outcome. However, prognosis for intradural astrocytomas is poor, with a median survival of only 10−15 months. The 10-year survival rate for low-grade astrocytomas is 78%, but for infiltrative astrocytomas, it is only 17%. Generally, the intradural-intramedullary tumors have a worse prognosis since they usually present as astrocytomas or gliomas. The intradural-extramedullary tumors have a generally better prognosis since they present as meningiomas or neurofibromas.

> **Point to remember**
>
> An intradural-extramedullary tumor is a spinal cord tumor that causes spinal cord compression. Nerves in and around the spinal cord are compressed as the tumor grows in size. The vast majority of these tumors are fortunately benign—usually schwannomas or meningiomas. For the intradural-intramedullary tumors, malignant examples are usually various types of gliomas, while benign examples include epidermoid cysts and lipomas.

Clinical cases

Clinical case 1

1. What are the descriptions of intramedullary and extramedullary spinal cord tumors?
2. How common are primary spinal cord tumors?
3. What are the general signs and symptoms of spinal cord tumors?

A 37-year-old woman presented complaining of back pain that she initially thought had been caused by lifting heavy boxes at her house. Over weeks, the pain became worse and worse. Initial X-rays and a CT scan revealed large lesions of the right iliac bone, T11 vertebra and left ninth rib. Her spinal cord was severely compressed by the thoracic lesion, and the tumor had totally replaced the spine's structural bony support. The iliac lesion was biopsied and revealed to be a metastatic adenocarcinoma. The patient underwent resection of the spinal tumor and fusion of spinal segments T9 through L2. She soon underwent stereotactic radiosurgery for the surgical site's tumor bed, left rib mass, and iliac mass. The final pathology was triple-negative breast cancer, which was biopsied and revealed to be a metastatic carcinoma.

Answers:

1. Spinal cord tumors can develop within the spinal cord parenchyma, where they are known as intramedullary tumors. They may also develop outside of the cord parenchyma, known as extramedullary tumors, and often compress the spinal cord or its nerve roots. Extramedullary tumors may be intradural or extradural.
2. Primary spinal cord tumors are less common than primary brain tumors. They make up only about 5.3% of all primary CNS tumors. In the United States, just over 2500 new cases are diagnosed annually. Primary spinal cord tumors are much less common than tumors that metastasize to the spine, which affect 5%−10% of people with systemic cancer.
3. Generally, signs and symptoms of spinal cord tumors include pain (most common), motor weakness, sensory impairment, and bowel and bladder dysfunction.

Clinical case 2

1. With spinal cord tumors, how is neck or back pain often described?
2. What does weakness usually follow, and to what is it related?
3. How common is it for spinal cord tumors to cause patients to lose the ability to walk?

A 60-year-old man with a history of prostate cancer had been experiencing back pain for a few months, and then suddenly lost his ability to walk. He also developed urinary incontinence. Examination revealed severe weakness in the patient's legs and numbness from the torso down to the feet. X-rays and MRIs were ordered, and imaging confirmed vertebral column tumors that had metastasized from the prostate, to multiple thoracic spinal vertebrae. There was a pathologic fracture of the T4 vertebra and spinal cord compression. During laminectomy, it was found that the cancer cells had also reached the T3 and T5 vertebra. The laminae from T3 to T5 was removed, and the T4 vertebrae was removed. The spinal tumor was then excised. Spinal fusion of T3 to T5 using a bone graft to replace T4 was performed, and additional repairs were made to T3 and T5. The patient recovered from all of this with no tumor recurrence.

Answers:

1. **Neck or back pain from spinal cord tumors may be localized or radicular. Especially with epidural metastases, localized pain and tenderness over the involved region are common. Radicular pain is often described as "band-like." It follows the distribution of the spinal nerve roots, varying from mild to severe, and may be described as burning, dull, or sharp.**
2. **Weakness can follow sensory symptoms and is the most commonly identified objective factor (35%−75% of patients). Weakness is often related to a positive Babinski sign, hyperreflexia, and spasticity. It will progress to total paraplegia without treatment.**
3. **Between 50% and 68% of patients cannot walk at the time of initial diagnosis of epidural spinal cord compression.**

Clinical case 3

1. How do posterior and anterior spinal cord tumors differ?
2. What is the primary method of diagnosis for spinal cord tumors?
3. Why are corticosteroids started immediately for spinal cord tumors?

A 68-year-old man presented with recurring low back pain that had been getting worse. He stated that the pain got worse at night when he went to bed. The patient also had leg weakness, foot numbness, and cold sensations in the feet and legs. He previously had been diagnosed and treated for a kidney tumor. An MRI was ordered, and the L5 vertebra was compressed by a tumor mass. The patient received corticosteroids and soon after, radiation treatment. The tumor mass was successfully resected during surgery. Spinal cord compression was fortunately discovered early, and the patient responded well to all of the treatments. Over 1-year of follow-up, there had been no recurrence.

Answers:

1. **If the posterior aspect of the spinal cord is affected, the awareness of proprioception and vibration of body areas is affected. If the tumor is anterior, touch and pressure on the opposite side of the body is affected.**
2. **Diagnosis of spinal cord tumors is via MRI of the affected area, with and without contrast. It provides detailed views of the spinal cord. Use of contrast is common since many tumors show enhancement regardless of their histologic grade.**
3. **With neurologic deficits, corticosteroids are started immediately to reduce cord edema and preserve functionality, especially in the setting of cauda equina syndrome. While corticosteroids are often the first line of treatment for spinal cord tumors, optimal loading and maintenance doses have not been determined.**

Clinical case 4

1. How common are intradural-intramedullary spinal cord tumors?
2. Which type of tumor is most common out of all intradural-intramedullary spinal cord tumors?
3. What are the signs and symptoms of intradural-intramedullary tumors?

A 76-year-old woman complained of progressive lower extremity numbness and subjective weakness. Imaging revealed an intramedullary lesion in her thoracic spine at the T4 level. The patient had no focal weakness or myelopathy. A sagittal T2-weighted image revealed extensive spinal cord edema. Sagittal T1-contrasted imaging showed a discrete enhancing lesion at the T4 level. The diagnosis was of an intradural-intramedullary spinal cord tumor, which was partially resected. The patient began additional treatments to address residual tumor tissue.

Answers:

1. **Intradural-intramedullary tumors are much less common than intradural-extramedullary tumors, only accounting for about 2%−4% of all CNS tumors.**

2. Between 40% and 60% of intradural-intramedullary tumors are ependymomas, and less often astrocytomas, which are more common in children. About 67% of thoracic tumors are astrocytomas, while about 49% of cervical tumors involve this type of tumor.
3. Intradural-intramedullary tumors invade and destroy the spinal cord, altering blood supply, affecting cellular membrane stability, and disrupting afferent and efferent impulses. As a result, there are sensory, motor, and reflex deficits based on which spinal nerves are compromised. Resultant edema causes more deficits.

Clinical case 5

1. What is the epidemiology of intradural-extramedullary spinal cord tumors?
2. Why do intradural-extramedullary spinal cord tumors cause pain?
3. Do intradural-extramedullary spinal cord tumors carry a better or worse prognosis than intradural-intramedullary tumors?

A 30-year-old man presented with complaints of pain and stiffness in his neck, and numbness and weakness of his arms and legs. The symptoms had been present for three months and were intensifying. An MRI of the cervical spine showed an intradural extramedullary enhancing mass lesion at the second and third cervical vertebral level. It compressed the spinal cord with focal cord edema and anterior-right lateral displacement of the spinal cord. The lesion was hyperintense on T1-imaging and hypointense on T2-weighted imaging. Another intradural extramedullary mass lesion was found at the cranio-vertebral level. It was hyperintense to hypointense on T1 and isointense on T2. The patient underwent laminectomy and surgical resection with decompression. No residual lesions were found on a postoperative MRI scan.

Answers:

1. Intradural-extramedullary tumors make up nearly 90% of all primary spinal cord tumors, plus 30% of all spinal tumors.
2. Intradural-extramedullary tumors often cause pain because of traction upon or irritation of the spinal nerve roots, displacement of the cord, alteration of blood supply, or CSF circulation obstruction.
3. The intradural-extramedullary tumors have a generally better prognosis since they present as meningiomas or neurofibromas. Generally, the intradural-intramedullary tumors have a worse prognosis since they usually present as astrocytomas or gliomas.

Key terms

Batson's plexus
Brown-Séquard syndrome
Cauda equina syndrome
Compressive (sensorimotor) syndrome
Extradural tumors
Intradural-extramedullary tumors
Intradural-intramedullary tumors
Intradural tumors
Irritative (radicular) syndrome
Myotomes
Paraparetic
Syringomyelic syndrome

Further reading

Ames, C., Boriani, S., & Jandial, R. (2013). *Spine and spinal cord tumors: Advanced management and operative techniques*. Thieme.
Arnautovic, K. I., & Gokaslan, Z. L. (2019). *Spinal cord tumors*. Springer.
Aryan, H. (2009). *Spinal tumors: A treatment guide for patients and family*. Jones & Bartlett Learning.
Bilsky, M. H., & Laufer, I. (2019). *Spinal oncology: An issue of neurosurgery clinics of North America (Volume 31-2) (The clinics: Surgery)*. Elsevier.
Chang, E. L., Brown, P. D., Lo, S. S., Sahgal, A., & Suh, J. H. (2018). *Adult CNS radiation oncology: Principles and practice*. Springer.
Dickman, C. A., Fehlings, M. G., & Gokaslan, Z. L. (2006). *Spinal cord and spinal column tumors: Principles and practice*. Thieme.
Gokaslan, Z. L., Boriani, S., Fisher, C. G., & Gomes Vialle, L. R. (2014). *AOSpine masters series volume 1: Metastatic spinal tumors*. Thieme.
Gokaslan, Z. L., Boriani, S., Fisher, C. G., & Gomes Vialle, L. R. (2014). *AOSpine masters series volume 2: Primary spinal tumors*. Thieme.
Gunzburg, R., Szpalski, M., & Aebi, M. (2007). *Vertebral tumors*. LWW.
Gupta, N., Banerjee, A., & Haas-Kogan, D. A. (2017). *Pediatric CNS tumors (Pediatric Oncology)* (3rd Edition). Springer.
Hayat, M. A. (2012). *Tumors of the central nervous system, volume 6: Spinal tumors (Part 1)*. Springer.

Hayat, M. A. (2012). *Tumors of the central nervous system, volume 7: Spinal tumors (Part 2)*. Springer.
Hayat, M. A. (2013). *Tumors of the central nervous system, Volume 10: Pineal, pituitary, and spinal tumors*. Springer.
Hayat, M. A. (2013). *Tumors of the central nervous system, volume 11: Imaging, glioma and glioblastoma, stereotactic radiotherapy, spinal cord tumors, meningioma, and schwannomas*. Springer.
Hayat, M. A. (2015). *Tumors of the central nervous system, volume 14: Glioma, meningioma, neuroblastoma, and spinal tumors*. Springer.
Hsu, W., & Jallo, G. I. (2013). *Pediatric neurology: Pediatric spinal tumors (Handbook of clinical neurology #112)*. Elsevier.
Klekamp, J., & Samii, M. (2007). *Surgery of spinal tumors*. Springer.
Koller, H., Robinson, Y., Demetriades, A., & Himmelhan, R. (2019). *Cervical spine surgery: Standard and advanced techniques: Spine research society — Europe instructional surgical atlas*. Springer.
Landi, A., Gregori, F., & Delfini, R. (2019). *Spinal cord and spinal column tumors*. Nova Science Publishers Inc.
Manfre, L. (2017). *Vertebral lesions (New procedures in spinal interventional neuroradiology)*. Springer.
Meyer, B., & Rauschmann, M. (2019). *Spine surgery: A case-based approach*. Springer.
Sahgal, A., Lo, S. S., Ma, L., & Sheehan, J. P. (2016). *Image-guided hypofractionated stereotactic radiosurgery: A practical approach to guide treatment of brain and spine tumors*. CRC Press.
Sciubba, D. M. (2018). *Spinal tumor surgery: A case-based approach*. Springer.
Sik Kang, H., Woo Lee, J., & Lee, E. (2017). *Oncologic imaging: Spine and spinal cord tumors*. Springer.
Van Gotham, J. W. M., van den Hauwe, L., Parizel, P. M., & Baert, A. L. (2007). *Spinal imaging: Diagnostic imaging of the spine and spinal cord (Medical radiology)*. Springer.
Walker, D., Perilongo, G., Taylor, R., & Punt, J. (2004). *Brain and spinal tumors of childhood*. Hodder Arnold Publications.

Chapter 28

Rare brain tumors

Chapter Outline

Overview	479	Anaplastic pleomorphic xanthoastrocytoma	491
Rare brain tumors	479	Primitive neuro-ectodermal tumors	493
Atypical teratoid-rhabdoid tumor	479	**Clinical cases**	495
Diffuse midline gliomas	483	**Key terms**	497
Gliomatosis cerebri	487	**Further reading**	497
Pleomorphic xanthoastrocytoma	488		

Overview

Rare brain tumors are those diagnosed in less than 10% of patients for which there is very little available information. They include tumors collectively referred to as embryonal brain tumors. Examples include atypical teratoid-rhabdoid tumors, diffuse midline gliomas, gliomatosis cerebri, pleomorphic xanthoastrocytoma (both nonanaplastic and anaplastic in form), and primitive neuro-ectodermal tumors. Recent research has revealed that some rare brain tumors with different names actually share similar genetic makeups and may actually represent the same disease. Physicians and researchers are hoping to develop new diagnostic methods to improve recognition and to devise more specific and effective treatments.

Rare brain tumors

There are other rare brain tumors that affect patients of different ages. The patterns of tumor growth are also widely varied for these tumors. According to the *National Cancer Institute's Center for Cancer Research*, some grow very quickly, others very slowly, and still others in between these growth rates. Some often spread to the spinal cord. Various tumors are likely to recur, while others recur only rarely.

Atypical teratoid-rhabdoid tumor

An **atypical teratoid-rhabdoid tumor** (AT-RT) is a malignant embryonal tumor of the central nervous system (CNS). It is mostly made up of poorly differentiated elements. These often include rhabdoid cells, with inactivation of *SWItch/ Sucrose non-fermentable-related, matric associated, actin dependent regulator of chromatin, subfamily B, member 1 (SMARCB1)* (also known as INI1). In extremely rare cases, there is inactivation of *SMARCA4* (also known as *transcription activator BRG1*). The AT-RTs are abbreviated as *AT-RTs*. They are classified as grade IV tumors by the World Health Organization.

Epidemiology

AT-RTs are most common in young children. In several studies, AT-RTs make up 1% − 2% of all pediatric brain tumors, rarely occur in children over 6 years of age, and are extremely rare in adults. Because of the overwhelming incidence of cases in children below 3 years of age, these tumors are believed to make up 10% or more of all CNS tumors in infants. There is a male predominance, of 1.6 to 2 cases, to every 1 female case. In two large studies of children with AT-RTs, the ratio of supratentorial to infratentorial tumors was 4:3. The supratentorial tumors are most common in the cerebral hemispheres and less common in the pineal gland, suprasellar region, and ventricular system. Infratentorial tumors occur in the cerebellar hemispheres, cerebellopointine angle, and brainstem, being comparatively prevalent during the first 2 years of life. Less often, these tumors develop in the spinal cord. Tumor seeding often occurs via the cerebrospinal fluid pathways. It is revealed in up to 25% of all patients at presentation. In the uncommon adult cases, infratentorial tumors are extremely rare.

Etiology and risk factors

With familial rhabdoid tumors, such as in rhabdoid tumor predisposition syndrome 1, which involves the SMARCB1 gene, and in rhabdoid tumor predisposition syndrome 2, which involves SMARCA4, there have been documented adult carriers who were unaffected, and the presence of gonadal mosaicism. With a risk of germline mutations of 33% or more in the SMARCB1-deficient AT-RTs, plus an even higher risk with SMARCA4-deficient tumors, it is crucial to perform molecular genetic studies for all newly diagnosed patients. Though actual causes and other risk factors are not known, the SMARCB1 and SMARCA4 deficiencies have definitely been proven to be related to development of AT-RTs.

Pathology

The neoplastic cells show histological and immunohistochemical **polyphenotypic differentiation**. This occurs along epithelial, mesenchymal, and neuroectodermal lines. These tumors, plus their deposits through cerebrospinal fluid pathways, usually appear similar to medulloblastomas and other CNS embryonal tumors. They are mostly soft, pink to red in color, and demarcated from nearby parenchyma. They usually contain necrotic foci and are sometimes hemorrhagic. If these tumors have large amounts of mesenchymal tissue, they can be firm and tan to white in color in some areas. Tumors forming in the cerebellopontine angle wrap around the cranial nerves and vessels, invading the brainstem and cerebellum. The tumors have also been seen in the temporal lobe, midbrain, hypothalamus, thalamus, diencephalon, optic pathways, and fourth ventricle (see Fig. 28.1). In rare cases, there is involvement of the nearby bones.

The AT-RTs are heterogeneous lesions that may be hard to recognize histopathologically. In many cases, the cells have classic rhabdoid features, with oddly located nuclei that contain vesicular chromatin, obvious eosinophilic nucleoli, and abundant cytoplasm. There is an obvious eosinophilic globular cytoplasmic inclusion, with cell borders that are well-defined. The cells usually range from the classic rhabdoid phenotype to those of less obvious nuclear atypia, and significant amounts of pale eosinophilic cytoplasm. The cytoplasm has a finely grained and homogeneous character, but may contain a dense pink body that resembles an inclusion and is poorly defined. In the rhabdoid cells, there are usually whorled bundles of intermediate filaments that nearly fill the perikaryon. Cytoplasmic vacuolation is common. The rhabdoid cells may be structured in sheets or nests that are often "jumbled." The cells, only in a small amount of cases, are the predominant or exclusive histopathological factor.

The majority of AT-RTs contain various components that have primitive mesenchymal, neuroectodermal, and epithelial features. In about 66% of cases, a small-cell embryonal component is found. Less common, there is mesenchymal differentiation, which usually appears as areas having spindle cell features and a background of basophils or mucopolysaccharides. The least common histopathology is epithelial differentiation, which can form adenomatous areas, papillary structures, or cords and ribbons with poor differentiation. Rarely there is a myxoid matrix, and when this occurs, it may be difficult to distinguish these tumors from choroid plexus carcinomas. There are usually abundant mitotic figures. Often, there are broad areas of widespread necrosis and hemorrhage.

With AT-RTs, there is also a wide spectrum of immunohistochemical reactivities. The rhabdoid cells usually show expression of *epithelial membrane antigen (EMA), smooth muscle antibody (SMA),* and *vimentin*. There are also often immunoreactivities for cytokeratins, *glial fibrillary acidic protein (GFAP), neurofilament protein (NFP),* and *synaptophysin* (see Fig. 28.2). Usually, there is no expression of germ cell markers or markers of skeletal muscle differentiation. In AT-RTs, there is a loss of nuclear expression of SMARCB1. Pediatric CNS embryonal tumors that lack rhabdoid features but show loss of SMARCB1 expression, are also classified as AT-RTs. Rarely, these tumors form

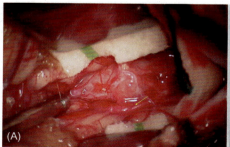
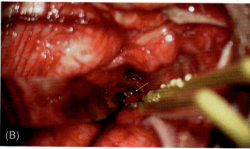

FIGURE 28.1 Intraoperative photographs of a pediatric atypical teratoid-rhabdoid tumor. (A) Tumor extruding from the fourth ventricle upon opening of the dura and arachnoid. (B) Following tumor resection and hematoma evacuation, a rostrally oriented view of the fourth ventricle and aqueduct.

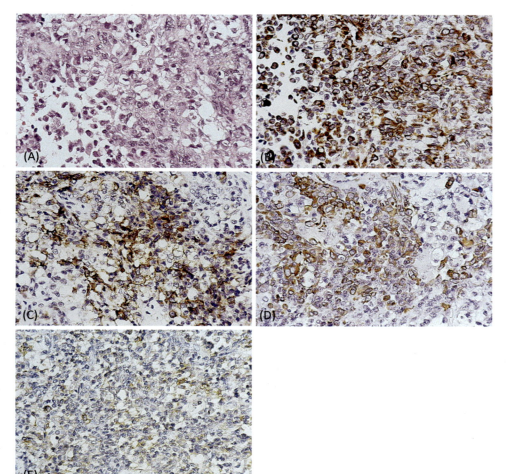

FIGURE 28.2
Immunohistochemistry distinguishes the highly malignant atypical teratoid-rhabdoid tumor from medulloblastoma. (A) This posterior fossa tumor in an infant, using hematoxylin and eosin staining, contains vimentin (B), and shows focal immunoreactivity to epithelial membrane antigen (C) and glial fibrillary acidic protein (D) and a focally weak response to synaptophysin (E). *Courtesy: McKeever, P. E. (2004). New methods of brain tumor analysis. In: H. Mena & G. Sandberg (Eds.) Dr. Kenneth M. Earle memorial neuropathology review. Washington, DC: Armed Forces Institute of Pathology.*

cribriform strands, well-defined surfaces, and trabeculae. In these cases, the tumors are called *cribriform neuroepithelial tumors*, which is probably an epithelioid AT-RT variant that may be less aggressive. Also rare are tumors having the clinical and morphological features of AT-RT but which retain SMARCB1 expression. In these cases, loss of nuclear expression of SMARCA4 (BRG1) is rare. In children, there is significant proliferative activity, often with Ki-67 proliferation indexes of 50% or higher. In adult patients, though data is limited, the proliferation index has been much lower.

Histogenesis of rhabdoid tumors is not known. Since epithelial, mesenchymal, and neural markers can be expressed and these tumors are most common in young children, they may form from **pluripotent fetal cells**. It is possible that AT-RTs arise from other tumor types, in the settings of ganglioglioma and other low-grade CNS tumors. The tumors can develop sporadically or along with a rhabdoid tumor predisposition syndrome. The genetic hallmark is mutation or loss of the SMARCB1 locus at chromosome 22q11.2. The genomes of AT-RTs are very simple, with very low rates of mutations, and loss of SMARCB1 being the main recurrence. Loss of SMARCB1 expression at the protein level occurs in nearly all cases. Most AT-RTs have detectable mutations or deletions of SMARCB1. Other cases show loss of SMARCB1 function because of reduced RNA or protein expression.

In 20%−24% of cases, homozygous deletions of the SMARCB1 locus are found. In other cases, a single SMARCB1 allele becomes mutated, with the second allele being lost by deletion or mitotic recombination. Two coding sequence mutations are found in rare cases. In the majority, nonsense and frameshift mutations may lead to protein truncations. In the SMARCB1 gene, localization of mutations appears to vary between rhabdoid tumors in various body sites. Exons 5 and 9 contain "hotspots" in AT-RT tumors of the CNS. No evidence of hypermethylation has been detected, and there is no established connection between the type of SMARCB1 alteration and disease outcomes.

In only a few cases, there may be tumors with AT-RT features and intact SMARCB1 protein expression. Instead, these cases have shown mutation and inactivation of SMARCA4, and are associated with very young patients, and a poor prognosis. There has been one case in which a link existed between ovarian cancer (small cell carcinoma) in a mother and an AT-RT in her newborn. Loss of SMARCB1 generally appears to cause widespread yet specific deregulation of genes and pathways related to the cell cycle, its survival, and its differentiation. Overexpressed cell cycle regulatory genes include *aurora kinase A (AURKA)* and *cyclin D1 (CCND1)*. A loss of SMARCB1 results in transcription activation of *enhancer of zeste homolog 2 (EZH2)*, and repression plus increased *histone methylation on tail of histone H3 (H3K27me3)* of **polycomb gene targets**, to maintain the **epigenome**. The harmful effects of SMARCB1 deficiency involve the hippo signaling pathway and its main effector, *yes associated protein 1 (YAP1)*, which is overexpressed in AT-RTs.

There may be two or more different molecular classes of AT-RTs, with distinctly different outcomes. Tumors with enrichment of neurogenic or forebrain markers are linked to a supratentorial location. These respond better to therapies and have better long-term survival. The tumors with mesenchymal lineage markers are usually infratentorial, with worse outcomes.

Clinical manifestations

The signs and symptoms of AT-RTs are based on patient age, tumor size and tumor location. Infants usually present with nonspecific signs, including failure to thrive, lethargy, and vomiting. Head tilt and cranial nerve palsy, usually of the sixth and seventh nerves, are specific manifestations. In children age 3 or older, headache, hemiplegia, and nystagmus are commonly reported.

Diagnosis

Correct diagnosis of AT-RTs requires demonstration of inactivation of the SMARCB1 or SMARCA4 genes. This is usually done by standard immunohistochemical staining for their proteins. Tumors having this morphology that do not have this molecular/genetic confirmation are classified as *CNS embryonal tumors with rhabdoid features*. Like other embryonal tumors, CT and MRIs usually show these tumors to be isodense to hyperintense in FLAIR imaging, with restricted diffusion (see Fig. 28.3). Cystic and necrotic regions are obvious, as zones of heterogeneous signal intensity. Nearly every AT-RT is contrast-enhancing to some degree. Leptomeningeal disseminations are seen in up to 25% of cases upon presentation. For diagnosis, immunohistochemical staining for expression of the SMARCB1 protein (INI1) can be sensitive and specific for AT-RTs.

Treatment

The treatment of AT-RTs involves surgery, chemotherapy, radiation therapy, and chromatin remodeling agents. Surgical resection may be difficult for tumors of the cerebellopontine angle. Total or near-total resections are often not possible. No standard treatment for AT-RT is known, and about half of all tumors transiently respond to chemotherapy. Agents include carboplatin, cisplatin, cyclophosphamide, etoposide, and vincristine. Combination therapies have been shown to improve outcomes for many patients. For some patients, intrathecal chemotherapy has been effective. Also,

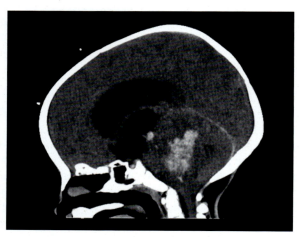

FIGURE 28.3 Sagittal CT scan of a child's head demonstrating a large, hemorrhagic posterior fossa atypical teratoid-rhabdoid tumor filling the fourth ventricle, with associated hydrocephalus and brainstem compression.

high-dose chemotherapy with stem cell rescue completely suppresses the bone marrow activities, and then the stem cells are returned to the patient's body to regrow the bone marrow.

Radiation therapy is usually deferred until the patient is at least 3 years of age, though sometimes it is necessary for very young patients. External beam (conformal) radiation uses several beams intersecting at the tumor location so that normal brain tissue receives less radiation and cognitive function is not as significantly affected. Proton beam radiation, though only available on a limited basis, has been effective against AT-RTs. Though still in preclinical evaluation, chromatin remodeling agents such as **histone deacetylase inhibitors** are being studied. They appear to be effective against some but not all of the AT-RT cell lines.

Prognosis

Overall prognosis for AT-RTs is poor. However, new data shows that not all of these tumors have the same prognosis. In one German study of children with AT-RTs, there was a 3-year overall survival rate of 22%, and an event-free survival rate of 13%. One subset (14%) survived long-term with no events. In another study that utilized chemotherapy along with other treatments, there was a 2-year disease progression-free survival rate of 53%, plus or minus 13%, and an overall survival rate of 70%, plus or minus 10%. A Canadian study reviewed prognosis based on combined use of high-dose chemotherapy with radiation, with children having a 2-year overall survival rate of 60%, plus or minus 12.6%. This was compared to children having on standard-dose chemotherapy, leading a 2-year overall survival rate of 21.7%, plus or minus 8.5%. However, results from various studies have not been consistent. For example, in the *Children's Cancer Group CCG-9921 Study*, with only chemotherapy, there was a 5-year event-free survival rate of 14%, plus or minus 7%, and an overall survival rate of 29%, plus or minus 9%. Even for the subgroup of tumors with the best prognosis, AT-RTs are still aggressive, having a 5-year progression-free and overall survival rate of 60%, with recurrence happening in about 33% of cases.

> **Point to remember**
> There is a subtype of tumor known as *CNS embryonal tumor with rhabdoid features*, also highly malignant. The cells show evidence of polyphenotypic differentiation, but insufficient data exists to determine if they are significantly different from classic AT-RTs. These tumors are also grade IV, and made up mostly of poorly differentiated elements that include rhabdoid cells.

Diffuse midline gliomas

Diffuse midline gliomas, such as the type known as the *H3 K27M-mutant*, are infiltrative, midline high-grade anaplastic gliomas. They have mostly astrocytic differentiation, plus a K27M gene mutation in either chromosome *H3F3A* or *HIST1H3B/C*. This particular subtype was added in 2016 by the World Health Organization to their classifications of CNS tumors. They represent the majority of diffuse intrinsic pontine gliomas (DIPGs), though identical tumors can be found in the brainstem, thalamus, and spinal cord. These tumors are classified as grade IV.

Epidemiology

Diffuse midline glioma is a tumor most common in children, but can also occur in adults. Brainstem and pontine examples were previously identified as *brain stem gliomas* and *DIPGs*. Incidence data of these tumors arising only in midline structures are not available, primarily because many of the world's brain tumor registries have not yet included them as their own distinct category. Children between 5 and 11 years are primarily diagnosed. The pontine tumors usually arise approximately at age 7, while the thalamus tumors usually arise approximately at age 11. There is no clear gender predilection.

Etiology and risk factors

For high-grade diffuse midline gliomas, gene sequencing has revealed recurrent heterozygous mutations at position K27 of the histone coding genes *H3F3A, HIST1H3B*, and *HIST1H3C* from the following:

- Pons—80% of cases;
- Spinal cord—60% of cases;
- Thalamus—50% of cases (see Fig. 28.4).

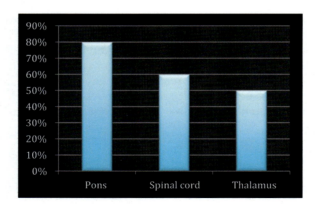

FIGURE 28.4 Locations, by percentage, of histone coding gene mutations related to diffuse midline gliomas.

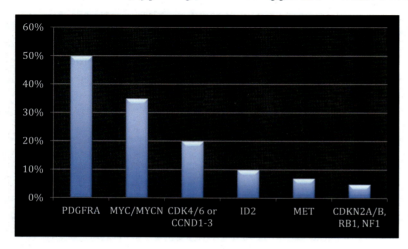

FIGURE 28.5 High-level focal amplifications in diffuse midline gliomas.

Within the brain, the mutations happen exclusively in diffuse midline gliomas. The K27M mutations that affect H3.3, which is encoded by H3F3A, occur about 3 times as much as the same mutation within histone variant H3.1, which occurs in HIST1H3B or HIST1H3C. The K27M substitution involves lysine being replaced by methionine at amino acid 27. This causes a decrease in H3K27me3, possible due to inhibition of PRC2 activity. Chromatin regulation is also targeted by nonrecurrent mutations of chromatin readers and writers. These include types of *mixed-lineage leukemia (MLL)*, **chromodomain** *helicase DNA-binding protein families*, and lysine-specific demethylase.

Additional gene mutations in many cancer pathways often target the receptor tyrosine kinase/RAS/PI13K pathway, such as mutations in platelet-derived growth factor receptor A (PDGFRA), phosphatidylinositol-4,5-bisphosphate 3-kinase catalytic subunit alpha (PIK3CA), phosphatidylinositol 3-kinase regulatory subunit alpha (PIK3R1), or phosphatase and tensin homolog (PTEN). They occur in about 50% of all cases. Other targets are the p53 pathway, such as mutations in ataxia-telangiectasia mutated (ATM), checkpoint kinase 2 (CHEK2), protein phosphatase 1D (PPM1D), or tumor protein P53 (TP53), which occur in about 70% of cases. To a lesser degree, the retinoblastoma protein pathway is targeted. In about 10% of thalamic high-grade gliomas, activating mutations or fusions that targeted fibroblast growth factor receptor 1 (FGFR1) were found. Oppositely, recurrent mutations of the gene encoding the BMP receptor activin A receptor type 1 (ACVR1) were found in about 20% of DIPGs. These appear to be related to H3.1 mutations.

High-level focal amplifications in diffuse midline gliomas include the following:

- *PDGFRA*—in up to 50% of cases;
- *Myelocytomatosis/N-myc proto-oncogene protein (MYC/MYCN)*—in up to 35% of cases;
- *Cyclin-dependent kinase 4/6 (CDK4/6)* or *cyclin D1–3 (CCND1–3)*—in 20% of cases;
- *DNA-binding protein inhibitor* ID2—in 10% of cases;
- *Mesenchymal-epithelial transition (MET)*—in 7% of cases;
- *Cycline-dependent kinase inhibitor 2A/B (CDKN2A/B), retinoblastoma 1 (RB1)*, or *neurofibromatosis 1 (NF1)*—in less than 5% of cases (see Fig. 28.5).

The fusion events involving the tyrosine kinase receptor gene *FGFR1* happen in thalamic diffuse gliomas. Also, about 4% of pontine gliomas carry **neurotrophin receptor** (NTRK) fusion genes. Common, widespread chromosome changes include single copy gains of chromosomes 1q and 2. There may also be up to 20% of pontine gliomas with few copy number changes. In rare cases, patients with Li−Fraumeni syndrome or NF1 develop midline infiltrating gliomas. However, there is no identified specific genetic susceptibility for the tumors. Additional risk factors are unknown.

Pathology

Mitotic activity is present in most of these tumors. Microvascular proliferation and necrosis may also exist. The tumor cells diffusely infiltrate nearby and distant brain structures. Autopsy studies have shown that DIPGs become disseminated into the leptomeninges in approximately 40% of cases. With this subtype, diffuse tumor invasion of the brainstem is common. About 25% spread to also involve the supper cervical cord and thalamus. Distal spread may reach as far as the frontal lobe. Rarely, the occipital lobe is affected, causing an amount of phenotypic overlapping with gliomatosis cerebri. These tumors have also been seen in the third ventricle. Some cases of gliomatosis cerebri show the H3F3A K27M mutation. This may indicate that the thalamic tumors also have a tendency for diffuse invasion.

Diffuse infiltration of the CNS parenchyma distorts and enlarges various structures. There is often fusiform enlargement of the pons, which may be symmetrical or asymmetrical (see Fig. 28.6). If the tumor has areas of hemorrhage or necrosis, there will be focal softening and discoloration. The tumors infiltrate both gray and white matter, with the cells being mostly small and monomorphic. Sometimes, however, the cells are large and pleomorphic. There is usually an astrocytic morphology, or sometimes, an oligodendroglial morphology. About 1 in 10 DIPGs do not have mitotic figures, microvascular proliferation, or necrosis—these are grade II tumors instead of grade IV. The rest of the cases are high-grade, and 25% have mitotic figures. The remaining cases have mitotic figures plus foci of necrosis and microvascular proliferation. For some reason with DIPGs, tumor grade is not predictive of outcome in the genetically classic cases.

Nearly all tumor cells express neural cell adhesion molecule 1 (NCAM1), oligodendrocyte transcription factor 2 (OLIG2), and the S100 protein. Yet, immunoreactivity for GFAP is variable. Expression of microtubule-associated protein 2 (MAP2) is common, with synaptophysin immunoreactivity often focal, though chromogranin-A and neuronal nuclear protein (NeuN) are not usually expressed. Nuclear p53 immunopositivity can be present, which suggests a TP53 mutation. This is present in about half of all cases. An alpha thalassemia retardation X-linked (ATRX) mutation leads to loss of nuclear ATRX expression in 10%−15% of cases. Examination of the spatiotemporal distribution of neural precursor cells in the brainstem shows a nestin-like and OLIG2-expressing neural precursor-like cell population within the ventral pons. This may be the cell of origin of diffuse pontine gliomas. The exact cell of origin of diffuse midline gliomas is still, however, unknown.

Clinical manifestations

Majority of patients develop with brainstem dysfunction or CSF obstruction that often presents within only 1−2 months, resulting in obstructive hydrocephalus. Classic symptoms include multiple cranial neuropathies, with long tract signs and ataxia. There may also be headaches, papilledema, nausea, and vomiting. For thalamic gliomas, initial symptoms often include signs of increased intracranial pressure, gait disturbances, and motor weakness or hemiparesis.

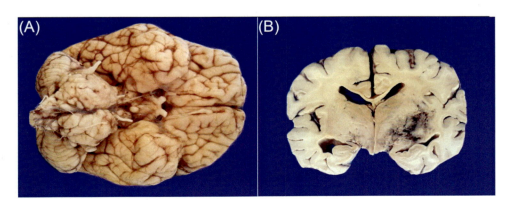

FIGURE 28.6 Diffuse midline glioma. (A) The base shows a tumor that bilaterally increases the volume of the pons. (B) Coronal sections of the brain show a poorly delimited neoplasm with areas of necrosis and hemorrhage, and loss of parenchyma affecting the gray nuclei of the right side base.

Diagnosis

Though usually present, mitotic activity is not required for diagnosis. A CT scan is often performed first to assess obstructive hydrocephalus. In MRI imaging, these tumors are usually T1-weighted hypointense and T2-weighted hyperintense (see Fig. 28.7). There may be contrast enhancement, hemorrhage, or necrosis. The tumors are usually large and expansive, but asymmetrical. For brainstem tumors, more than 66% of the pons may be occupied. There can be an exophytic component that encases the basilar artery or protrudes into the fourth ventricle. Often, the tumors infiltrate into the cerebellar peduncles or hemispheres, the midbrain, and the medulla oblongata. Contrast enhancement only rarely involves more than 25% of the tumor volume. Immunohistochemistry that uses a mutation-specific antibody can detect the *H3F3A K27M* if it is present.

Treatment

Treatment options for these tumors include radiation therapy, neurosurgery, chemotherapy, and other drug therapies. Standard treatment for DIPGs is 6 weeks of radiation therapy, which can greatly improve symptoms, though they usually recur after 6 − 9 months and then progress rapidly. For DIPGs, neurosurgery is often not possible since the tumors invade diffusely throughout the brainstem. Neurosurgically performed brainstem biopsies for **immunotyping** are also potentially dangerous. Pontine biopsies are similarly dangerous. Conventional radiation therapy, therefore, is the mainstay of treatment for DIPGs. Gamma knife or cyberknife radiosurgery may be selected for certain patients. The role of chemotherapy is unclear, but **radiosensitizers**, which increase effects of radiation therapy are under investigation. Future clinical trials may involve signal transfer inhibitors, which interfere with cellular pathways.

Prognosis

The prognosis for diffuse midline glioma, H3 K27M-mutant is poor even with today's better treatments. There is a 2-year survival rate of less than 10%. Finding an H3 K27M mutation means a worse prognosis than with the wildtype cases. Within the thalamus, high-grade tumors are related to short overall survival, no matter what the histone gene status may be. Though this is not apparently a factor for pontine tumors with classic radiological and clinical features, less than 10% of patients will survive past 2 years.

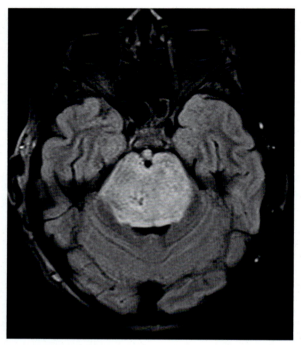

FIGURE 28.7 Diffuse midline glioma on T2-weighted magnetic resonance imaging showing a hyperintense mass within the pons.

> **Point to remember**
> Diffuse midline gliomas are primary CNS tumors and rare subtype of glial tumors. They are all classified as grade IV tumors and occur more often in children. Unfortunately, these tumors have a poor prognosis and are usually fatal.

Gliomatosis cerebri

Gliomatosis cerebri is a rare primary brain tumor often characterized by diffuse infiltration of the brain. There are neoplastic glial cells affecting various areas of the cerebral lobes, with infiltrative threads that spread quickly and deeply. They may spread into surrounding brain tissues or into multiple parts of the brain at the same time. This makes them very difficult to remove or treat. The tumors behave very similarly to glioblastomas (see Chapter 7: Glioblastoma). The first international registry for gliomatosis cerebri was established in 2014 by the *Weill Cornell Brain and Spine Center*.

Epidemiology

Gliomatosis cerebri may occur at any age, but usually develop within the third and fourth decades, between ages 46 and 53. The tumors occur just slightly more often in males in comparison to females.

Etiology and risk factors

The causes of gliomatosis cerebri, and risk factors for the disease, are unknown, but the tumors usually arise from the glial cells of the brain.

Pathology

Grade II gliomatosis cerebri are mid-grade tumors, with a high chance of recurrence after surgical removal. There is usually a genetic change involving *isocitrate dehydrogenase (IDH)*. Grade III and IV gliomatosis cerebri are malignant and fast-growing, often becoming resistant to treatment. The tumor cells resemble glial cells and involve many areas of the cerebrum. They can spread to other areas via the CSF, but do not spread outside the CNS. The tumor cells may be extremely similar to those of an anaplastic oligoastrocytoma.

Clinical manifestations

Gliomatosis cerebri may affect any part of the brain or spinal cord, as well as the optic nerve and compact white matter. The signs and symptoms are varied, including fatigue, headache, mood changes, seizures, corticospinal tract deficits, visual disturbances, dementia, memory changes, and lethargy. The disease has also been known to cause rapidly progressing symptoms that resemble those of Parkinson's disease.

Diagnosis

With today's medical advances, diagnosis of gliomatosis cerebri occurs, but with difficulty. While not considered a *formal diagnosis*, gliomatosis cerebri refers to a special pattern of diffuse and extensive growth of glioma cells. In previous years, the lack of MRI meant that most people were diagnosed after death during an autopsy. In most cases, gliomatosis cerebri is diagnosed as a diffuse and poorly circumscribed, infiltrative, nonenhancing lesion. It is hyperintense on T2-weighted imaging and causes expansion of the cerebral white matter. It is hard to distinguish from anaplastic astrocytomas or glioblastoma multiforme. For the grade II tumors, on imaging there is no obvious mass, but a widespread tumor pattern or a "fluffy"-looking abnormality. For the grade III and IV tumors, there is a tumor mass accompanying the "fluffy" appearance. Therefore, diagnosis is more often based on imaging instead of on pathology (see Fig. 28.8). Other imaging methods include CT and PET.

Treatment

If possible, the first treatment is surgery to determine the tumor type and remove as much of it as possible without causing additional symptoms. Surgery is usually limited to a biopsy since there is no central mass for removal. Other treatments include radiation therapy, chemotherapy, and newer methods that are part of clinical trials. Often, gliomatosis cerebri responds well to radiation therapy. Clinical trials are ongoing for newer chemotherapeutic drugs, targeted therapy, and immunotherapy.

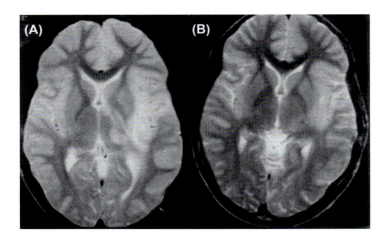

FIGURE 28.8 Axial (A and B) images, showing increased signal of the left frontal and bilateral parietal lobes with extension into the overlying cortex.

Prognosis

Gliomatosis cerebri generally carries a poor prognosis with an average survival of about 50% at 1 year, and of about 25% at 3 years. Prognosis is worsened if the disease transforms into glioblastoma multiforme.

> **Point to remember**
> Gliomatosis cerebri refers to patterns of diffuse and extensive growth of glioma cells, invading multiple lobes of the brain. Under a microscope, the tumor cells resemble glial cells, involving many cerebral areas. They can spread via the cerebrospinal fluid.

Pleomorphic xanthoastrocytoma

Pleomorphic xanthoastrocytoma is an astrocytic glioma that has large, pleomorphic, and often multinucleated cells, plus spindle and lipidized cells, many eosinophilic granular bodies, and a dense pericellular reticulin network. There is often neuronal differentiation. It is a grade II tumor, according to the World Health Organization.

Epidemiology

Pleomorphic xanthoastrocytoma is rare, making up less than 1% of all astrocytic neoplasms. It usually affects children and younger adults. The median patient age at diagnosis is 22 years. Pleomorphic xanthoastrocytoma is listed by the *Central Brain Tumor Registry of the United States (CBTRUS)* as a unique astrocytoma variant, with an annual incidence of 0.3 cases per 100,000 people. Occurrence in older adults, even in the seventh and eight decades, has been documented. The gender predilection is 1:1. There was one United States study that showed this tumor to be slightly more common in African-Americans than in other racial or ethnic groups, but data is limited.

Etiology and risk factors

A *BRAF V600E gene mutation* is common with this tumor. If present without an IDH mutation, the tumor is most likely a pleomorphic xanthoastrocytoma. Otherwise, no specific etiologies have been discovered. Formation of these tumors may be related to malformations since they are sometimes seen along with cortical dysplasia or ganglion cell lesions. There may also be a relation to defective NF1 function. There are no proven links with other hereditary tumor syndromes. This is not a surprise considered the high frequency of MAPK pathway alterations seen in sporadic pleomorphic xanthoastrocytomas. There has been no documentation of familial clustering of pleomorphic xanthoastrocytomas. There has been one case of this tumor, having the classic histology and molecular features, developing along with **DiGeorge syndrome**. However, there are no identified risk factors.

Pathology

The mitotic activity of pleomorphic xanthoastrocytoma is low, with less than 5 mitoses per 10 high-power fields. The tumor, in 98% of cases, is located superficially in the cerebral hemispheres—usually the temporal lobe. There is often involvement of the nearby leptomeninges and formation of cysts. The tumors have also developed in the cerebellum and spinal cord. In only two cases, there was a primary pleomorphic xanthoastrocytoma of the retina in pediatric patients. The tumors are often soft in texture and yellowish in color.

Most of these tumors are superficial and extend to the leptomeninges, often accompanied by a cyst and sometimes forming a **mural nodule** in the cyst wall. Less often, there is invasion of the dura, leptomeningeal dissemination, multifocality, and predominantly exophytic growth. Spindled cells are mixed with mononucleated or multinucleated giant astrocytes. Their nuclei have wide variances in both size and staining. Intranuclear inclusions and prominent nucleoli are common. Sometimes, the astrocytes are packed tightly together to create what appears to be an epithelioid pattern. Also, there may be sheets of fusiform cells. The often multinucleated xanthomatous cells have intracellular lipid accumulation, usually as droplets that occupy a large part of the cell body. The droplets push cytoplasmic organelles and glial filaments to the cell edges. This usually results in the astrocytic component easy to visualize using GFAP or *hematoxylin & eosin (H&E)* stains. Intensely eosinophilic or pale granular bodies are almost always present. Also common are focal groupings of small lymphocytes, sometimes with plasma cells.

Another hallmark is the existence of reticulin fibers that can best be visualized by using silver impregnation. Though the fibers may be due to reactive meningeal changes, tumor cells can be surrounded by basement membranes that stain positively for reticulin. These can be seen as pericellular basal laminae. Though the mitotic count is low, and necrosis is rarely present, necrosis itself is not enough to achieve a grade III classification.

Pleomorphic xanthoastrocytomas, while almost always having immunoreactivity for GFAP and S100 protein, tend to show neuronal differentiation. In tumors that have the usual histological features of these tumors, expression of neuronal markers has been reported in different amounts. These markers include neurofilament, synaptophysin, class III beta-tubulin, and MAP2. Sometimes, the biphenotypic glioneuronal appearance has been ultrastructurally identified. The cells often express *cluster of differentiation 34 (CD34)*. BRAF V600E mutations can be detected via immunohistochemistry using antibodies, and the immunophenotype with V600E-mutant BRAF expression plus loss of CDKN2A is secondary to the homozygous deletion of CDKN2A. This is often seen in pleomorphic xanthoastrocytomas. Usually, mitotic figures are rare or absent, with the Ki-67 proliferation index less than 1%.

Complicated karyotypes have been identified along with gains of chromosomes 3 and 7, plus alterations of the long arm of chromosome 1. However, these are not specific only to these tumors. Most of the tumors are diploid and sometimes polyploid. This may be linked to subgroups of very bizarre and multinucleated cells. In a genomic hybridization study, there was a loss of chromosome 9. Also, homozygous chromosome 9p21.3 deletions of the CDKN2A/CDKN2B loci were seen in 60% of cases. Therefore, a loss of CDKN2A protein expression has been shown via immunohistochemistry in 61% of these tumors.

Between 50% and 78% of cases involve BRAF point mutations (see Fig. 28.9). The majority involve V600E. Yet, BRAF V600E mutations are not specific to these tumors. They are also seen in gangliogliomas, pilocytic astrocytoma, and less often in other primary CNS tumors. BRAF mutations seem not to be related to the present anaplastic factors that are present. Rare examples of other gene mutations have included the *tuberous sclerosis complex subunit 2 (TSC2)*, NF1, and *ETS variant transcription factor 6 (ETV6)-neurotrophic receptor tyrosine kinase 3 (NTRK3)* fusion genes. The TP53 gene was mutated in 6% of cases as well, but this was not related to present anaplastic features. In another study, amplifications of the CDK4, *epidermal growth factor receptor (EGFR)*, and *mouse double minute 2 (MDM2)* genes did not occur, and in a different study, no IDH1 mutations were found using immunohistochemistry for R132H-mutant IDH1. This distinguishes pleomorphic xanthoastrocytoma from *diffusely infiltrating cerebral astrocytoma*. There also has been no evidence of chromosome 7q34 duplication, which is often linked to BRAF fusion, analyzed by comparative genomic hybridization based on microarray.

Clinical manifestations

Because of the common tumor location, may patients first present with a long history of seizures. These occur in approximately 75% of cases. Headaches are another common complaint. Dizziness may occur, while some patients are asymptomatic.

Diagnosis

On CT and MRI scans, pleomorphic xanthoastrocytomas often involve the cortex and overlying leptomeninges. On CT, the tumor may appear hypodense, hyperdense, or mixed, having strong and occasionally heterogeneous contrast enhancement. The tumor cysts appear hypodense. On MRI, the tumor's solid area is either hypointense or isointense to

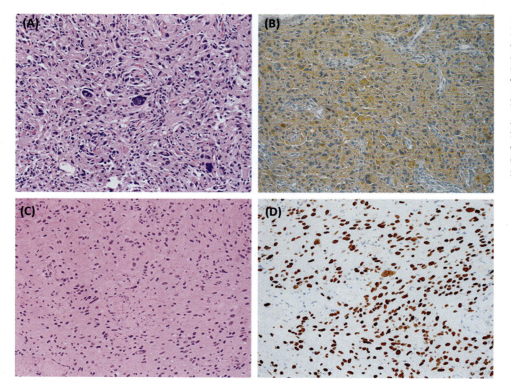

FIGURE 28.9 Immunohistochemical assessment of BRAF expression in gliomas. (A and B) Pleomorphic xanthoastrocytoma, BRAF V600E mutated. Tumor cells show cytoplasmic granular expression of BRAF V600E. Pleomorphic mononucleated or multinucleated cells with frequent nuclear inclusions are seen on hematoxylin & eosin staining. (C and D) Glioblastoma, H3K27M mutated, IDH1 wild type.

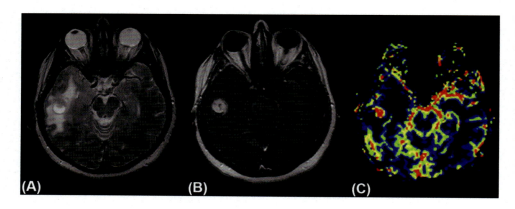

FIGURE 28.10 Pleomorphic xanthoastrocytoma. T2-weighted (A) and postcontrast T1-weighted (B) MRI showing a well-circumscribed, partially cystic, focally enhancing right medial temporal lobe mass with minimal surrounding edema and mass effect. (C) T2-weighted MRI from a right parietal view showing superficial cortical localization and effects upon the adjacent skull.

gray matter, in T1-weighted images, and hyperintense or mixed on T2-weighted and FLAIR imaging (see Fig. 28.10). However, the cystic area is isointense to the CSF. Postcontrast enhancement will be moderate or strong. There is usually not significant peritumoral edema because of slow tumor growth.

Diffuse astrocytoma is a common differential diagnosis because pleomorphic xanthoastrocytoma often has a diffusely infiltrative, nonpleomorphic component. Along with more typical pleomorphic components, correct diagnosis is supported by a BRAF V600E gene mutation and the lack of an IDH mutation. Less often, ganglioglioma may present with a glial component that resembles pleomorphic xanthoastrocytoma. In rare cases, composite tumors that have features of pleomorphic xanthoastrocytomas and gangliogliomas, yet with only slight intermingling, have been documented. Areas that show eosinophilic granular bodies and spindle cells may resemble pilocytic astrocytoma. However, areas with a more typical histological appearance, with no BRAF translocations or factors resulting in *mitogen-activated protein kinase (MAPK)* activation, are supportive of pleomorphic xanthoastrocytoma. Differential diagnoses include mesenchymal tumors. However, this is usually avoided because of positivity for GFAP in nonreactive cells, though this positivity can be focal or absent in small biopsy specimens.

Treatment

For pleomorphic xanthoastrocytoma, surgery is often the treatment of choice. Total resection is often possible. Additional treatment is sometimes not required, but repeated MRIs are suggested to monitor tumor recurrence. When a tumor cannot be totally resected, radiation therapy may be recommended. Steroids may be used to control tissue swelling. Chemotherapy, targeted therapies, and treatments under clinical study may be determined on an individual basis.

Prognosis

Pleomorphic xanthoastrocytoma as a generally favorable prognosis in comparison to diffusely infiltrative astrocytoma. It does not recur in nearly 71% of cases, and there is a 90.4% overall survival rate at 5 years. The extent of surgical resection influences prognosis. In patients with tumors having low mitotic activity, the estimated recurrence-free survival rate is 73.7%. Breaking this down further, 73.5% of children and 74.4% of adults survive without recurrence. The estimated 5-year overall survival rate for these patients is 89.4%. This is broken down as 92.9% of children and 86.8% of adults. The BRAF V600E mutation is more common in grade II tumors than in grade III anaplastic tumors. The status of this mutation is not much different between tumors in children and adults, and it is not certain whether the mutation affects prognosis.

> **Point to remember**
> Pleomorphic xanthoastrocytomas usually form in the cerebral hemispheres and are most common in children and younger adults. They are slow-growing tumors that usually do not spread. However, the anaplastic form of these tumors is fast-growing, usually invading brain tissue throughout the brain lobe where they formed.

Anaplastic pleomorphic xanthoastrocytoma

Anaplastic pleomorphic xanthoastrocytomas have 5 or more mitoses per 10 high-power fields and are much more aggressive than the nonanaplastic form. The anaplastic form is a grade III tumor, according to the World Health Organization. The tumors are usually supratentorial, most often involving the temporal lobe.

Epidemiology

There is no proven epidemiological data available on anaplastic pleomorphic xanthoastrocytomas. There has been a study of 74 cases, in which anaplasia was present in 31% of the patients at initial diagnosis. This can be broken down further, as 37% of adult cases, and 23% of pediatric cases.

Etiology and risk factors

The frequency of a BRAF V600E mutation is not as common in these tumors than in the nonanaplastic grade II form. There are no distinct links between anaplastic pleomorphic xanthoastrocytomas and hereditary tumor syndromes, and no risk factors have been identified.

Pathology

In anaplastic pleomorphic xanthoastrocytomas, necrosis may be present, but it is not understood if necrosis is significant along with lack of increased mitotic activity. They can occur anywhere in the cerebral hemispheres and may cross to the opposite hemisphere through the corpus callosum, resulting in perilesional edema. Anaplasia usually appears with brisk mitotic activity, while the tumor otherwise appears similar to the nonanaplastic form. The high levels of mitotic activity can be diffuse or focal. Necrosis is usually present along with this mitotic activity. Though microvascular proliferation is not common, it is usually linked to necrosis and brisk mitotic activity. Anaplasia may be seen upon first diagnosis or when a tumor recurs. This may indicate progression for a grade II tumor to a grade III tumor. There have been tumors with features of anaplasia upon initial resection that have shown features considered to be in the upper limits of grade II tumors upon recurrence. This probably indicates tumor heterogeneity. The grade III tumors, at first diagnosis and in recurrence, may show less pleomorphism, plus more diffuse infiltration than the common grade II form. Histology of anaplasia has not been formally studied for these tumors. However, transformations have been small-cell, fibrillary, and epithelioid-rhabdoid.

Anaplastic pleomorphic xanthoastrocytomas may closely resemble epithelioid glioblastomas, histologically and molecularly. Both tumor types often have BRAF V600E mutations. There has been a case of a pleomorphic xanthoastrocytoma recurring as an epithelioid glioblastoma. The anaplastic tumors, however, do not have the uniform cells that are common with epithelioid glioblastomas, and also have eosinophilic granular bodies, which are unique. The cells are often large, with bizarre nuclei and abundant eosinophilic cytoplasm (see Fig. 28.11). Often, a lower-grade component resembling a nonanaplastic tumor is present, at least on a focal basis. Rarely, a SMARCB1-deficient and AT-RT-like lesion has appeared in pleomorphic xanthoastrocytoma. This is linked to a loss of SMARCB1 expression in the high-grade rhabdoid component, which is morphologically unique. This clonal evolution may be related to an extremely aggressive biology, even in a low-grade original tumor. Therefore, such lesions have been described as *SMARCB1-deficient anaplastic pleomorphic xanthoastrocytomas*. The phenotype of the anaplastic form is similar to the nonanaplastic form.

Genetic alterations specific to the anaplastic form are not understood. Some of these tumors develop via malignant progression from a nonanaplastic tumor, but the sequence of genetic events is not known. BRAF V600E mutations are less common (47.4% in one study) compared to being present in about 75% of cases of the nonanaplastic form. The BRAF V600E mutation status is very similar between adult and pediatric cases.

Clinical manifestations

Seizures are the most common presenting symptom of anaplastic pleomorphic xanthoastrocytomas. Like the nonanaplastic form, they may also cause headaches and dizziness. Vomiting is also a component of these tumors.

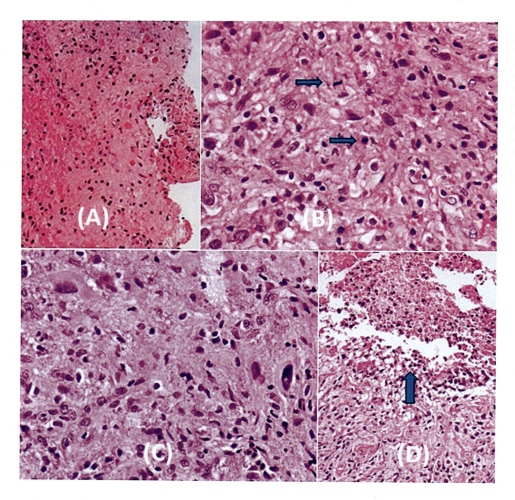

FIGURE 28.11 (A) Anaplastic pleomorphic xanthoastrocytoma exhibiting numerous eosinophilic hyaline globules. (B) Sheets of astrocytic cells having pleomorphic nuclei with some cells having xanthomatous cytoplasm. Presence of frequent mitoses (arrows showing two mitotic figures). (C) Sheets of spindly to large polygonal astrocytic cells having pleomorphic nuclei with nuclear inclusions. (D) Necrotic tissue.

Diagnosis

In imaging studies, these tumors are often seen as a circumscribed supratentorial mass that is peripherally located. The tumor is often cystic, affecting the cerebral cortex and leptomeninges above it. An MRI with contrast usually reveals a heterogeneous intensely enhancing tumor (see Fig. 28.12).

Treatment

Like the nonanaplastic form, treatment options include surgical resection and radiation therapy, plus steroids to reduce tissue swelling. Similarly, chemotherapy, targeted therapies, and treatments under clinical study may be given, determined on an individual basis.

Prognosis

Anaplastic pleomorphic xanthoastrocytomas have a much worse survival rate than the nonanaplastic form. Also, the prognostic significance of a BRAF V600E mutation with these tumors is not known. Shorter survival time is likely when there is a high mitotic count and necrosis. Survival rates are much worse when the mitotic count is 5 or more per 10 high-power fields (55.6% of patients) compared to a mitotic count of less than 5 per 10 high-power fields (89.4% of patients). The 5-year survival rate for patients without necrosis is 90.2% while for those with necrosis it is only 42.2%. In general, children have a 67.9% 5-year recurrence-free survival rate, and adults have a 62.4% rate. The overall 5-year survival rate for children is 87.4% and for adults is 76.3%.

Primitive neuro-ectodermal tumors

Primitive neuro-ectodermal tumors, also known as *PNETs*, are primary CNS tumors. Today, they are being reclassified and given other names, based upon their molecular features. However, they are all classified as grade IV tumors, since they are malignant and grow quickly. The subtypes of tumors that belong to this group include *CNS ganglioneuroblastomas, CNS neuroblastomas, embryonal tumors with multilayered rosettes* (and *other unspecified embryonal tumors*), and *medulloepitheliomas*.

Epidemiology

PNETs are most common in children, but may also occur in adults. They are extremely rare, with no accurate data on their incidence or prevalence. They are most common in Caucasians and slightly more common in males than in females. There have been estimations that in the United States, about 129 people are diagnosed per year, with 72 cases being children under age 15, and 57 cases being children age 15 or older, or adults. There are an estimated 950 people living with PNETs in the United States. There are no global estimates.

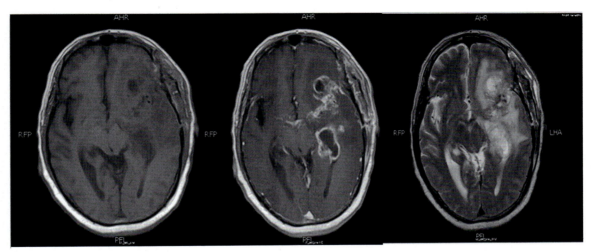

FIGURE 28.12 MRI showing significant overall increase in the size of the anaplastic tumor involving the left frontal, temporal, and parietal lobes, with perilesional edema and postsurgical changes in T1, T1 postcontrast and T2-weighted images, respectively.

Etiology and risk factors

The cause of PNETs is unknown, but some of the subtypes are believed to be linked to genetic changes. There are no identified risk factors.

Pathology

PNETs usually originate from anywhere within the brain, and less often in the brainstem or spinal cord. They form from the ectoderm and can become larger than 5 cm in diameter. The tumors can spread to other areas of the CNS and various body organs. The center of these tumors is often necrotic, and there is often high mitotic activity. In many cases, the tumor cells show immunohistochemical expression of synaptophysin (see Fig. 28.13).

Clinical manifestations

Symptoms related to PNETs are based on tumor location and may include headaches, seizures, nausea, vomiting, problems with cognition and memory, weakness, numbness, vision abnormalities, and problems with balance or movement. If the tumor is in the spine, there may be bowel or bladder incontinence, and pain in the back and legs.

Diagnosis

Upon initial diagnosis, most PNETs have already spread. Diagnosis requires MRI, in which the tumor usually appears as a single heterogeneous mass in the cerebral cortex. PNETs often enhance with contrast, and multiple tumors can be identified. There may be cysts or a collection of fluid within the mass, and swelling may be present around the tumor.

Treatment

The first treatment option for PNETs is surgical resection, when it is possible. Additional treatments may include radiation therapy, chemotherapy, or treatments under clinical trials—such as new chemotherapeutic agents, targeted therapies, or immunotherapies. In most cases, surgery is followed by radiation when the patient is at least 3 years old, and the radiation may be targeted to the brain and spine.

Prognosis

Due to the small numbers of people with PNETs, the 5-year survival rate has not been able to be calculated.

> **Point to remember**
>
> PNETs are highly malignant, and most develop in young children. Children below the age of 3 have a lower survival rate since the use of radiation therapy is generally contraindicated. Relapses of PNETs are associated with high fatalities.

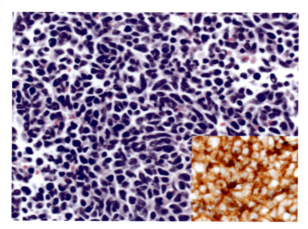

FIGURE 28.13 Primitive neuro-ectodermal tumor cells forming a 2-centimeter mass within an 8-cm retroperitoneal metastasis. The inset shows diffuse immunohistochemical expression of synaptophysin, whereas all germ cell immunostains were negative.

Clinical cases

Clinical case 1

1. In which age group are AT-RTs most common?
2. How do these tumors usually appear?
3. What does correct diagnosis of these tumors require?

A 6-year-old boy had developed nystagmus when he was only 18 months old. An MRI at that time revealed a tumor that extended from the right temporal lobe into the right midbrain, hypothalamus, thalamus, and the right optic pathway. Heterogeneous enhancement and intratumoral cysts were seen. There was no family history of neurofibromatosis, but one of the boy's uncles had died from a rhabdomyosarcoma. The initial tumor had been partially resected, diagnosed as a pilocytic astrocytoma, and throughout his life the boy was treated with various chemotherapy regimens. At age 5, he had developed left hemiparesis, which was linked to growth of intratumoral cysts in the right diencephalon. He underwent radiation therapy and surgical debulking of the cysts and tumor tissue. Histological study with immunohistochemical staining revealed the presence of an AT-RT, based on inactivation of the SMARCB1 gene. Unfortunately, the patient died within 8 months of completing adjuvant chemotherapy.

Answers:

1. **AT-RTs are most common in young children and have been documented as making up 1%−2% of all pediatric brain tumors. They rarely occur in children over 6 years of age, and are extremely rare in adults. They are believed to make up 10% of more of all CNS tumors in infants.**
2. **The AT-RTs usually appear similar to medulloblastomas and other CNS embryonal tumors. They are mostly soft, pink to red in color, and demarcated from nearby parenchyma.**
3. **Correct diagnosis of AT-RTs requires demonstration of inactivation of the SMARCB1 or SMARCA4 genes, usually via standard immunohistochemical staining for their proteins.**

Clinical case 2

1. What ages of children are most often diagnosed with diffuse midline gliomas?
2. Do these tumors infiltrate gray or white matter, and how can the cells appear?
3. How do most patients present?

A 10-year-old girl presented with a headache, nausea, and vomiting. The headache had become a chronic problem for her over about 6 months, but her parents were not greatly worried until the nausea and vomiting began. Examination revealed severe bilateral papilledema. A CT scan showed obstructive hydrocephalus with a mass in the posterior aspect of the third ventricle. An MRI was performed, with and without contrast, revealing a midline neoplasm that originated from the thalamic-hypothalamic area. It was slightly hypointense on T1-weighted imaging with minimal enhancement. In size, the mass was just less than 2 cm in diameter, and located anterior to the pineal region, above the dorsal midbrain. The tumor was biopsied and a ventriculoperitoneal shunt was inserted to manage the hydrocephalus. Markedly atypical and pleomorphic astrocytes were seen, with occasional mitotic figures identified. An H3F3A K27M gene mutation was found via molecular testing, and a diffuse midline glioma with an anaplastic astrocytoma subtype was diagnosed. Total surgical resection was not possible due to the tumor location, so the girl was treated with chemotherapy and radiation therapy, which stabilized her condition with no tumor progression for 5 months after the biopsy.

Answers:

1. **Children between 5 and 11 years of age are primarily diagnosed with diffuse midline gliomas. The pontine tumors usually arise at about age 7, and the thalamus tumors usually arise at about age 11.**
2. **These tumors infiltrate both gray and white matter. The cells may be mostly small and monomorphic or sometimes, large and pleomorphic. There is usually an astrocytic morphology, or sometimes, an oligodendroglial morphology.**
3. **Most patients present with brainstem dysfunction or CSF obstruction that often develop within only 1−2 months, resulting in obstructive hydrocephalus.**

Clinical case 3

1. Why is gliomatosis cerebri difficult to treat?

2. What genetic change is usually involved with this disease?
3. Which signs and symptoms signify gliomatosis cerebri?

A 52-year-old man presented with recurrent seizures that had been occurring over the past 2 months. An MRI revealed multiple regions of hyperintense T2-signals with positive mass effect, involving the temporal lobe on the right side. The signals extended up into the corpus callosum, occipital lobe, and posterior thalamus. There was additional hyperintense T2-signal and thickening of the cerebral cortex. Postcontrast imaging revealed subtle enhancement of the corpus callosum. A biopsy was conducted, and gliomatosis cerebri was diagnosed. The tumor cells, however, were similar to those of an anaplastic oligoastrocytoma. The patient began treatment with radiation.

Answers:

1. **In gliomatosis cerebri, neoplastic glial cells have infiltrative threads that spread quickly and deeply. They may spread into surrounding brain tissues, or into multiple parts of the brain at the same time. This makes them very difficult to remove or treat.**
2. **There is usually a genetic change involving IDH.**
3. **Signs and symptoms of gliomatosis cerebri include fatigue, headache, mood changes, seizures, corticospinal tract deficits, visual disturbances, dementia, memory changes, lethargy, and symptoms that resemble those of Parkinson's disease.**

Clinical case 4

1. What cell types are involved in pleomorphic xanthoastrocytoma?
2. Which gene mutation is common with these tumors?
3. Where are pleomorphic xanthoastrocytomas usually located?

A 24-year-old man was hospitalized because of recurrent seizures. A CT scan revealed a cystic mass in the right temporal lobe, of about 3 cm in diameter. After administration of contrast, the outer and anterior portions of the lesion were enhanced. The tumor was hypointense on precontrast T1-weighted MRI imaging, and enhanced intensely anterolaterally on postcontrast T1-weighted imaging. On T2-weighted imaging, it was hyperintense. There was no peritumoral edema. Surgical resection was performed, and the tumor was soft and yellow in color. The patient recovered and was discharged within 13 days. Histological examination of the tumor revealed astrocytic proliferation with GFAP-positive multinuclear large pleomorphic cells. There was no mitosis or necrosis, and pale granular bodies were present. The diagnosis was of a pleomorphic xanthoastrocytoma.

Answers:

1. **Pleomorphic xanthoastrocytoma has large, pleomorphic, and often multinucleated cells, plus spindle and lipidized cells.**
2. **A BRAF V600E gene mutation is common with these tumors. If present without an IDH mutation, the tumor is most likely a pleomorphic xanthoastrocytoma.**
3. **Pleomorphic xanthoastrocytomas are located, in 98% of cases, superficially in the cerebral hemispheres—usually the temporal lobe.**

Clinical case 5

1. What is the mitotic activity in anaplastic pleomorphic xanthoastrocytomas?
2. Where to these tumors occur?
3. Which type of tumor does an anaplastic pleomorphic xanthoastrocytoma closely resemble?

A 35-year-old man was examined after presenting with early morning headaches and vomiting over the previous 10 days. An MRI with contrast revealed a heterogeneous intensely enhancing right frontal mass lesion that crossed to the opposite hemisphere through the corpus callosum, with perilesional edema. Subtotal surgical resection was able to be performed, and histological examination revealed the tumor to have large pleomorphic cells with bizarre nuclei and abundant eosinophilic cytoplasm. Eosinophilic granular bodies were present. There was mitosis with atypical mitotic figures, and large foci of necrosis. The diagnosis was of an anaplastic pleomorphic xanthoastrocytoma. The patient underwent radiation therapy and over 6 months of follow-up, did not experience tumor progression.

Answers:

1. **Anaplastic pleomorphic xanthoastrocytomas have 5 or more mitosis per 10 high-power fields, and are much more aggressive than the nonanaplastic form.**
2. **These tumors can occur anywhere in the cerebral hemispheres, and may cross to the opposite hemisphere through the corpus callosum, resulting in perilesional edema.**
3. **Anaplastic pleomorphic xanthoastrocytomas may closely resemble epithelioid glioblastomas, histologically and molecularly. Both tumor types often have BRAF V600E mutations.**

Clinical case 6

1. What are the subtypes of PNETs?
2. From where do PNETs originate, and what do they form from?
3. How do PNETs appear in imaging studies?

A 22-year-old woman presented with headaches, blurred and double vision, and vomiting that had been present for almost a month, but which became much worse in the past week. An MRI revealed a left posterior parietal heterogeneously enhancing mass. The tumor was close to 7 cm in diameter. The patient underwent surgical resection, and the lesion was found to have a necrotic center. It was diagnosed as a primitive neuroectodermal tumor (PNET) and had 5 − 10 mitoses per high-power field. The patient recovered with no postoperative complications or tumor recurrence.

Answers:

1. **The PNET subtypes include: CNS ganglioneuroblastomas, CNS neuroblastomas, embryonal tumors with multilayered rosettes (and other unspecified embryonal tumors), and medulloepitheliomas.**
2. **PNETs usually originate from anywhere within the brain, and less often in the brainstem or spinal cord. They form from the ectoderm.**
3. **PNETs are diagnosed by MRI, in which they usually appear as a single heterogeneous mass in the cerebral cortex. They often enhance with contrast, and multiple tumors can be identified. There may be cysts or a collection of fluid within the mass, and swelling may be present around the tumor.**

Key terms

Atypical teratoid-rhabdoid tumor
Chromodomain
Diffuse midline gliomas
DiGeorge syndrome
Epigenome
Gliomatosis cerebri
Histone deacetylase inhibitors
Immunotyping
Mural nodule
Neurotrophin receptor
Pleomorphic xanthoastrocytoma
Pluripotent fetal cells
Polycomb gene targets
Polyphenotypic differentiation
Primitive neuro-ectodermal tumors
Radiosensitizers

Further reading

Adesina, A. M., Tihan, T., Fuller, C. E., & Young Poussaint, T. (2016). *Atlas of pediatric brain tumors* (2nd Edition). Springer.
Ahluwalia, M., Metellus, P., & Soffietti, R. (2019). *Central nervous system metastases*. Springer.
Baehring, J. M., & Piepmeier, J. M. (2006). *Brain tumors: Practical guide to diagnosis and treatment (Neurological disease and therapy)*. CRC Press.
Bigner, D. D., Friedman, A. H., Friedman, H. S., McLendon, R., & Sampson, J. H. (2016). *The duke glioma handbook: Pathology, diagnosis, and management*. Cambridge University Press.
Brandao, L. A. (2017). *Pediatric brain tumors update, an issue of neuroimaging clinics of North America (Volume 27-1) (The clinics: Radiology)*. Elsevier.
Chaichana, K., & Quinones-Hinojosa, A. (2019). *Comprehensive overview of modern surgical approaches to intrinsic brain tumors*. Academic Press.
Chang, E. L., Brown, P. D., Lo, S. S., Sahgal, A., & Suh, J. H. (2018). *Adult CNS radiation oncology: Principles and practice*. Springer.
Duffau, H. (2017). *Diffuse low-grade gliomas in adults* (2nd Edition). Springer.
Fountas, K., & Kapsalaki, E. Z. (2019). *Epilepsy surgery and intrinsic brain tumor surgery: A practical atlas*. Springer.
Furtado, L. V., & Husain, A. N. (2018). *Precision molecular pathology of neoplastic pediatric diseases (Molecular pathology library)*. Springer.
Gajjar, A., Reaman, G. H., Racadio, J. M., & Smith, F. O. (2018). *Brain tumors in children*. Springer.
Gursoy Ozdemir, Y., Bozdag Pehlivan, S., & Sekerdag, E. (2017). *Nanotechnology methods for neurological diseases and brain tumors: Drug delivery across the blood-brain barrier*. Academic Press.

Hattingen, E., & Pilatus, U. (2016). *Brain tumor imaging (Medical radiology)*. Springer.
Hayat, M. A. (2014). *Tumors of the central nervous system, volume 13: Types of tumors, diagnosis, ultrasonography, surgery, brain metastasis, and general CNS diseases*. Springer.
Jain, R., & Essig, M. (2015). *Brain tumor imaging*. Thieme.
Jeltsch, A., & Rots, M. G. (2018). *Epigenome editing: Methods and protocols (Methods in molecular biology)*. Humana Press.
Kaye, A. H., & Laws, E. R., Jr. (2011). *Brain tumors: An encyclopedic approach, expert consult* (3rd Edition). Saunders.
Ladyani, M., & Gerald, W. L. (2003). *Expression profiling of human tumors: Diagnostic and research applications*. Humana Press.
Moliterno Gunel, J., Piepmeier, J. M., & Baehring, J. M. (2016). *Malignant brain tumors: State-of-the-art treatment*. Springer.
Newton, H. B. (2018). *Handbook of brain tumor chemotherapy, molecular therapeutics, and immunotherapy* (2nd Edition). Academic Press.
Newton, H. B., & Maschio, M. (2015). *Epilepsy and brain tumors*. Academic Press.
Ozsunar, Y., & Senol, U. (2019). *Atlas of clinical cases on brain tumor imaging*. Springer.
Pope, W. (2019). *Glioma imaging: Physiologic, metabolic, and molecular approaches*. Springer.
Quinones-Hinojosa, A., Raza, S. M., & Laws, E. R. (2013). *Controversies in neuro-oncology: Best evidence medicine for brain tumor surgery*. Thieme.
Rosenfeld, C. S. (2015). *The epigenome and developmental origins of health and disease*. Academic Press.
Sahgal, A., Lo, S. S., Ma, L., & Sheehan, J. P. (2016). *Image-guided hypofractionated stereotactic radiosurgery: A practical approach to guide treatment of brain and spine tumors*. CRC Press.
Sampson, J. H. (2017). *Translational immunotherapy of brain tumors*. Academic Press.
Sughrue, M. E. (2019). *The glioma book*. Thieme.
Sughrue, M. E., & Yang, I. (2019). *New techniques for management of 'inoperable' gliomas*. Academic Press.
Warmuth-Metz, M. (2016). *Imaging and diagnosis in pediatric brain tumor studies*. Springer.
Weis, S., Sonnberger, M., Dunzinger, A., Voglmayr, E., Aichholzer, M., Kleiser, R., & Strasser, P. (2019). *Imaging brain diseases: A neuroradiology, nuclear medicine, neurosurgery, neuropathology and molecular biology-based approach*. Springer.

Glossary

18F-labeled deoxyglucose A radiopharmaceutical used in PET scans.
Absolute refractory period Depolarization and repolarization of an organ or cell.
Acidophilic Referring to an organism that thrives under highly acidic conditions.
Acrochordon A skin tag.
Acromegaly A disorder caused by excess growth hormone after growth plates have closed.
Action potentials Rapid rising and falling of membrane potentials in specific cells, causing nearby locations to depolarize.
Acute lymphoblastic leukemia A cancer of the lymphoid blood cells involving large numbers of immature lymphocytes.
Acute myelogenous leukemia A cancer of the myeloid blood cells involving rapid growth of abnormal cells in the bone marrow and blood.
Adamantinomatous Referring to the presence of calcifications.
Addison's disease An endocrine disorder in which the adrenal glands do not produce enough steroid hormones.
Adenocarcinomas Neoplasia of epithelial tissue that has glandular characteristics or origin.
Adenohypophyseal placode A plate-like structure of the adenohypophysis.
Adenoma sebaceum A cutaneous disorder with angiofibromas starting in childhood.
Adrenal chromaffin cells Neuroendocrine cells, mostly in the medulla of the adrenal glands.
Age-specific death rates The amount of deaths based on certain age groups.
Aicardi syndrome A genetic malformation involving absence of the corpus callosum, with retinal abnormalities and infantile spasms.
Alar plate An embryonic neural structure involved in general somatic and visceral sensory impulses.
Alcian blue A dye used to stain polysaccharides for examination.
Alpha-crystallin B chain A protein that helps to bind misfolded proteins to prevent protein aggregation.
Alpha-fetoprotein A major plasma protein produced by the yolk sac and fetal liver during development.
Alternating electric field therapy An electromagnetic therapy that uses low-intensity, intermediate frequency electrical fields to treat cancer.
Amelanotic Lacking the pigment melanin.
Amenorrhea Absence of a menstrual period in a woman of reproductive age.
Amianthoid Having a crystalline appearance.
Amygdaloid body A portion of the amygdala involved in processing memory, decision-making, and emotional responses.
Anaplasia Loss of characteristics of mature cells and their orientation, compared to each other and to endothelial cells.
Anaplastic astrocytoma A rare grade III brain tumor that is usually fatal within 5 years.
Anaplastic ependymoma A rare type of brain tumor that is often fatal.
Anaplastic large-cell lymphoma A tumor that involves the T-cells that have the best prognosis of all T-cell lymphomas.
Anaplastic meningiomas Grade III tumors that are the most aggressive meningiomas.
Anaplastic oligodendroglioma A grade III neuroepithelial tumor believed to originate from oligodendrocytes.
Ancient schwannoma A benign peripheral nerve tumor, usually in the retroperitoneal space.
Aneurysmal bone cyst An osteolytic bone neoplasm with sponge-like, blood- or serum-filled nonendothelialized spaces.
Angiocentric Perivascular.
Angiofibromas Small, smooth papules on the sides of the nose and cheeks.
Angiolipoma A subcutaneous nodule with vascular structure that is often painful.
Angiomatous meningiomas Angioma-like meningeal tumors of the CNS.
Angiomyolipomas The most common benign tumor of the kidneys.
Angiopoietin/Tie2 receptor A receptor that binds the protein growth factor angiopoietin.
Angiosarcoma A cancer of the endothelial cells lining the walls of blood or lymphatic vessels.
Angiosarcomatous Referring to an angiosarcoma.
Annexin A1 A cellular protein found in the blood or elsewhere outside the cellular environment.
Annulate lamellae Cell membranes that occur as parallel elements with double, identical membranes, like the nuclear envelope.
Anosmia Loss of the ability to detect one or more odors.
Anterior (ventral) horn A structure of either the lateral ventricle or the spinal cord located toward the front of the body.
Anterior cerebral artery One of two brain arteries that supplies oxygenated blood to most midline areas of the frontal and superior/medial/parietal lobes.

Anterior commissure A white matter tract connecting the temporal lobes across the midline, located in front of the columns of the fornix.
Anterior corticospinal tract A small bundle of descending fibers connecting the cerebral cortex to the spinal cord.
Anterior horns See "anterior (ventral) horn."
Anterior median fissure A fissure containing a double fold of pia mater, with the floor formed by the anterior white commissure; located in both the spinal cord and medulla oblongata.
Anterior roots The ventral roots of a spinal nerve with efferent motor functions.
Anterograde flow The forward movement of oxygenated blood in the vertebral arteries, toward the brain.
Anterolateral sulcus In the spinal cord, the location where ventral fibers exit; in the medulla, the location where the rootlets of the hypoglossal nerve emerge.
Antoni A pattern A highly cellular pattern within a schwannoma.
Antoni B pattern A loose microcystic pattern within a schwannoma.
Aptamers Oligonucleotide or peptide molecules that bind to target molecules.
Arachnoid mater One of the three meninges derived from the neural crest mesectoderm in the embryo.
Arachnoiditis ossificans A form of an inflammatory condition of the arachnoid mater, in which ossification occurs.
Archicortex The oldest region of the cerebral cortex, making up the three cortical layers of the hippocampus.
Arcuate fibers Internal, anterior external, or posterior external fibers, from second-order sensory neurons; they are present in the medulla, carrying proprioceptive information.
Argyrophilic Referring to grains and coiled bodies in brain tissue, such as in the form of dementia known as argyrophilic grain disease (AGD).
Association areas The parts of the cerebral cortex that function to produce meaningful perceptions, interactions, abstract thinking, and language.
Association fibers The fibers located in the association areas of the cerebral cortex.
Astrocytes Star-shaped glial cells in the CNS.
Astrocytomas CNS tumors that originate from astrocytes.
Ataxia-telangiectasia A rare neurodegenerative, autosomal recessive disease that impairs movement and coordination, weakens the immune system, and predisposes to cancer.
Atonia A condition in which muscles lose strength or become either paralyzed or extremely relaxed.
ATP transcriptional regulator, x-linked A protein involved in chromatin remodeling; when the related ATRX gene is mutated, mental retardation and alpha-thalassemia may develop.
Atypia A condition of being irregular.
Atypical choroid plexus papilloma A form of rare, benign neuroepithelial intraventricular tumor of the choroid plexus; it has irregular architecture, density, cytology, proliferation, and necrosis.
Atypical meningiomas Irregular grade II meningeal tumors that involve brain invasion.
Atypical neurofibroma An irregular type of benign nerve-sheath tumor in the peripheral nervous system.
Atypical teratoid-rhabdoid tumor A rare tumor usually diagnosed in childhood, usually in the cerebellum, with irregular features.
Auditory association area A portion of the auditory cortex in the temporal lobe, which processes hearing, and interacts with language switching.
Auditory cortex The part of the temporal lobe that processes auditory information.
Autonomic (visceral) reflexes Those that involve the response of a visceral effector, such as cardiac or smooth muscles, or glands.
Auxilin An enzyme that helps regulate molecular chaperone activities.
Axolemma The cell membrane of an axon.
Axon hillock A specialized part of the cell body of a neuron, where membrane potentials are summated before transmission to the axon.
Axon terminals Distal terminations of the telodendria of an axon.
Axonal transport A cell process responsible for movement of organelles to and from a neuron's cell body.
Axons Long, slender projections of neurons that usually conduct action potentials away from the nerve cell body.
Axoplasm The cytoplasm within the axon of a neuron.
Balloon cells Large degenerated cells with pale-staining cytoplasm; or, a large form of nevus cell with abundant nonstaining cytoplasm.
Band of Bungner One of the linear bands of interdigitating Schwann cells.
Bannayan-Riley-Ruvalcaba syndrome Multiple subcutaneous lipomas, macrocephaly, and hemangiomas.
Basal nuclei The basal ganglia; subcortical nuclei in the brain, associated with motor movements, learning, eye movements, cognition, and emotion.
Basal plate In the neural tube, the region that is ventral to the sulcus limitans.
Basal vein A vein in the brain formed at the anterior perforated substance.
Basilar artery One of the arteries that supplies the brain with oxygenated blood.
Batson's plexus A network of valveless veins that connect the deep pelvic and thoracic veins to the internal vertebral venous plexuses.
Beckwith-Wiedemann syndrome An overgrowth disorder with an increased risk of childhood cancer and congenital abnormalities.
Benign fibrous histiocytoma Dermatofibromas, a benign skin growth.
Biallelic inactivation A dominant genomic change in which both alleles of a gene are inactivated.
Bimodal distribution A statistical pattern in which frequencies of values in a sample have two distinct peaks, with some amount of overlap.
Biphenotypic sinonasal sarcoma A rare, low-grade malignant tumor of the nasal cavity; it was only recognized recently, in 2017.

Bitemporal hemianopsia Partial blindness in which vision is missing in the outer half of the right and left visual fields.
Bitemporal superior quadrantanopia Partial blindness in which the upper, outer visual quadrants of both eyes are affected.
Bizarre neurofibroma A benign nerve-sheath tumor of the peripheral nervous system in which there are bizarre, multinucleated cells.
Bland cells Those that appear nonmalignant, lack atypical mitoses, are not hyperchromatic, and are not bizarre in appearance.
Blepharoblasts Basal bodies, especially of flagellated cells.
Blister cells Red blood cells containing peripherally located vacuoles.
Blood–brain barrier (BBB) A CNS structure that keeps various substances in the bloodstream out of the brain, while allowing in substances essential to metabolic function.
Blooming Having the characteristics of an autosomal recessive genetic disorder that predispose to a wide variety of cancers.
Bodian silver A stain used to examine neurofilament subunits.
Botryoid Resembling a bunch of grapes in appearance.
Brachytherapy A form of radiotherapy where a sealed radiation source is placed inside or next to the area requiring treatment.
Brachyury A protein that functions as a transcription factor within the T-box family of genes.
BRAF mutations Those affecting the proto-oncogene B-Raf gene, resulting in birth defects or cancer.
Broca's area A region in the frontal lobe of the dominant brain hemisphere, with functions linked to speech production.
Brown-Séquard syndrome Spinal hemiparaplegia, caused by damage to one-half of the spinal cord.
Bulbous corpuscles Ruffini endings that are slowly adapting mechanoreceptors in the cutaneous tissue, between the dermal papillae and hypodermis.
Bumpy cell A cell that has an irregular and uneven appearance.
Burden of disease The impact of a health problem as measured by financial cost, mortality, and morbidity.
Burkitt lymphoma A cancer of the lymphatic system, mostly the lymphocytes in the germinal center; mostly occurring in children.
Butterfly glioma A bilateral glioma.
C-terminus The carboxyl-terminus; the end of an amino acid chain terminated by a free carboxyl group.
Caldesmon A calmodulin-binding protein, involved in regulation of smooth muscle and nonmuscle contraction.
Calretinin A calcium-binding protein involved in calcium signaling.
Canalicular Related to small passageways known as canaliculi.
Candidate population A chosen group of people for a specific medical study.
Capicua A conserved transcription factor; deficiency disrupts bile acid homeostasis.
Cardiac myxoma A rare benign tumor of the heart.
Cardiac nerves Autonomic nerves that supply the heart.
Cardiac plexus A nerve plexus at the base of the heart, which innervates it.
Carney complex An autosomal dominant condition, with myxomas, skin hyperpigmentation, and endocrine overactivity.
Castleman's disease A group of uncommon lymphoproliferative disorders, with lymph node enlargement and unique microscopic features.
Cathepsins Enzymes that degrade proteins.
Cauda equina A bundle of spinal nerves and rootlets that arise from the lumbar enlargement and conus medullaris of the spinal cord.
Cauda equina syndrome A condition that occurs when the cauda equina is damaged, resulting in back and leg pain, anal numbness, and loss of bowel of bladder control.
Caudal A term describing how close a structure is to the trailing end of an organism, such as the back or spine.
Caudate nucleus One of the structures that makes up the corpus striatum, involved in Parkinson's disease, learning, and inhibition of actions.
Cavernous hemangiomas Benign vascular tumors in which collections of dilated blood vessels form lesions.
Celiac ganglia Two large, irregular masses of nerve tissue in the upper abdomen; part of the sympathetic prevertebral chain processing.
Celiac plexus The solar plexus, located in the abdomen; if it receives trauma, it can cause the diaphragm to spasm.
Cellular schwannoma A type of usually benign nerve-sheath tumor composed of Schwann cells, in which the patterns are extremely cellular in appearance.
Centigray 1/100 of a gray, which is a derived unit of ionizing radiation dose.
Central canal The cerebrospinal fluid-filled space running through the spinal cord.
Central neurocytomas Very rare, usually benign intraventricular brain tumors that mostly form from the neuronal cells of the septum pellucidum.
Central sulcus A fold in the cerebral cortex, separating the parietal and frontal lobes, and the primary motor and primary somatosensory cortices.
Centrioles Cylindrical organelles mostly made up of the protein called tubulin.
Centroblasts Activated B-cells that are enlarged, and rapidly proliferate in the germinal center of lymphoid follicles.
Centrocytes Generally, B-cells with cleaved nuclei, which may appear in follicular lymphomas.
Cerebellar hemispheres The one median and two lateral portions of the cerebellum.
Cerebellar tonsils Rounded lobules on the underside of the cerebellar hemispheres.
Cerebiform Also spelled "cerebriform"; similar to the brain in structure.
Cerebral aqueduct The aqueduct of Sylvius in the midbrain; it connects the third and fourth ventricles.
Cerebral arterial circle The circle of Willis, a circulatory anastomosis that supplies blood to the brain and surrounding structures.

Cerebral cavernous malformations Cerebral cavernomas; benign vascular tumors or hemangiomas in which collections of dilated blood lesions form lesions.
Cerebral cortex The outer layer of neural tissue of the cerebrum.
Cerebral hemispheres The right and left halves of the cerebrum.
Cerebrospinal fluid The clear, colorless fluid of the CNS.
Cervical enlargements The attachments of the large nerves that supply the limbs.
Chang staging system An older method of diagnosis for medulloblastomas, based on tumor diameter, extension, seeding, and metastasis.
Charcot-Marie-Tooth disease A peripheral nervous system disease, with progressive loss of muscle tissue and touch sensation.
Chemical synapse A biological junction through which neuronal signals can be shared, and also send to cells in muscles or glands.
Chemoattractants Chemical substances that influence the recruitment and migration of cells.
Chemoreceptors Chemosensors; specialized sensory receptor cells that transduce chemical substances to generate biological signals.
Childhood ependymoma A tumor that arises from the ependymal tissues; in children, it is usually intracranial.
Cholesterotic Related to cholesterol.
Cholinergic synapses Those of the neuromuscular junctions, handling compounds that mimic the actions of acetylcholine or butyrylcholine.
Chondroma A benign cartilaginous tumor that is encapsulated, with a lobular growth pattern.
Chondromyxoid Related to a cartilaginous tumor.
Chondrosarcoma A bone sarcoma made up of cells derived from those that produce cartilage.
Chondrosarcomatous Related to chondrosarcomas.
Chordoid meningiomas Rare forms of atypical meningiomas, with a greater risk of recurrence than typical meningiomas, but less than malignant meningiomas.
Chordomas Rare, slow-growing neoplasms that may arise from cellular remnants of the notochord.
Choriocarcinoma A malignant, trophoblastic cancer, usually of the placenta.
Choroid plexus A cell plexus that arises from the tela choroidea of the brain ventricles.
Choroid plexus carcinoma A malignant cancer of the choroid plexus.
Choroid plexus papilloma A rare, benign neuroepithelial intraventricular grade I lesion of the choroid plexus.
Chromodomain A protein structural domain of 40–50 amino acid residues usually found in proteins related to alteration of chromatin.
Chromophobic Related to a chromophobe cell, a histological structure that does not easily stain, and appears relatively pale under a microscope.
Chromothripsis A mutational process by which hundreds or thousands of clustered chromosomal rearrangements occur at once; related to cancer and congenital diseases.
Chronic lymphocytic leukemia A type of cancer in which the bone marrow makes too many lymphocytes, usually worsening slowly over years.
Chronic myelogenous leukemia A cancer of the white blood cells, with increased, unregulated growth of myeloid cells in the bone marrow, and their accumulation in the blood.
Cingulate gyrus The brain area within the medial aspect of the cerebral cortex that receives inputs from the thalamus and neocortex.
Circle of Willis See "cerebral arterial circle."
Circular sulcus In the insula, a semicircular fissure that separates the insula from the gyri of the operculum.
Classic ependymoma The most common form of ependymal tumor.
Claudin One of the family of proteins that are important in the tight junctions.
Claustrum A thin bilateral structure that connects to cortical and subcortical brain regions; it allows for integration of color, sound, touch, and other cortical inputs allowing the experience of multiple sensory events.
Clear cell ependymoma A form of ependymal tumor with cells that have clear cytoplasm when stained with hematoxylin and eosin (H&E).
Clear cell meningiomas A form of meningeal tumor with cells that have clear cytoplasm when stained with hematoxylin and eosin (H&E).
Clival Related to the clivus.
Clivus A bony part of the cranium at the skull base, a shallow depression behind the dorsum sellae that slopes obliquely backward.
Codman's triangle An area of new subperiosteal bone created when a tumor or other lesion raises the periosteum away from the bone.
Codons Collections of three DNA bases.
Coexpression The simultaneous expression of two or more genes.
Cognition index A method of measuring an individual's cognitive abilities.
Collagenomas Connective tissue lesions made up mostly of collagen, which may be present at birth or appear within the first few years of life.
Collaterals Small blood vessels or nerves that supply or innervate various areas.
Compressive (sensorimotor) syndrome A medical condition caused by direct pressure upon a nerve, involving pain, tingling, numbness, and weakness.
Cone-rod homeobox A protein that is a photoreceptor-specific transcription factor, involved in differentiation of photoreceptor cells.
Conglomeration A joining together of structures from various sources or types.
Connexons Assemblies of six proteins called connexins that form pores for gap junctions between cytoplasm of two adjacent cells.
Consanguinity The property as being from the same bloodline as another person.
Continuous propagation The process in which an action potential in an axon spreads from a nearby region of its membrane, via a series of small steps.
Conus medullaris The tapered, lower end of the spinal cord.

Corpus callosum The callosal commissure, a thick nerve tract made up of a flat bundle of commissural fibers under the cerebral cortex.
Cortical dysplasias Common types of childhood epilepsies that are usually difficult to control with medications.
Corticospinal system The functional system related to the spinal cord.
Corticotrophic adenomas Benign pituitary adenomas made up mostly of corticotrophs, which may lead to Cushing's disease.
Costello syndrome A rare genetic disorder, with delayed development, intellectual disabilities, unique facial features, highly flexible joints, and loose folds of extra skin.
Cosyntropism A condition related to cosyntropin and adrenal insufficiency.
Cowden syndrome A condition of benign hamartomas and increased risk of breast, thyroid, uterine, and other cancers.
Cranial meninges The three membranes that envelop the brain.
Cranial nerves Those that emerge directly from the brain and relay information, mostly to and from the head and neck, including vision, taste, smell, and hearing.
Cranial placodes Neurogenic thickenings of the epithelium in the embryonic head ectoderm layer, giving rise to neurons and other sensory structures.
Cranioarchischisis The most severe neural tube defect in which the brain and spinal cord both remain open, and anencephaly and spina bifida are present.
Craniopharyngioma A rare brain tumor of pituitary gland embryonic tissue that is most common in children.
Crista galli The upper part of the perpendicular plate of the ethmoid bone.
Cumulative incidence A measure of disease frequency during a period of time.
Curvilinear Having a formation of curved lines.
Cushing's adenomas Benign adrenal tumors that may cause Cushing's syndrome.
Cushing's disease A cause of Cushing's syndrome that involves increased secretion of adrenocorticotropic hormone (ACTH) from the anterior pituitary.
Cushing's syndrome Signs and symptoms caused by prolonged exposure to glucocorticoids such as cortisol; also called hypercortisolism.
Cyberknife A radiation therapy that delivers radiosurgery for tumors.
Cyclin-D1 A cyclin protein that is involved in regulating cell cycle progression.
Cytoarchitecture The study of cellular composition of CNS tissues.
Cytokeratin A keratin protein in the intracytoplasmic cytoskeleton of epithelial tissues.
Cytokines Small proteins that are important in cell signaling.
Cytomorphology The study of the structure of cells.
Cytotrophoblasts Stem cells of the inner layer of trophoblasts.
Dandy-Walker syndrome A rare congenital brain malformation in which the cerebellar vermis does not fully form, and CSF enlarges the fourth ventricle and posterior fossa.
de novo From the beginning; or, in gene mutation, an alteration present for the first time in one family member.
Deep tendon reflexes Those that demonstrate the homeostasis between the cerebral cortex and spinal cord.
Deep veins Those that are deep in the body, below the superficial veins.
Dejerine-Sottas disease A hereditary neurological disorder, with damage to the peripheral nerves and resulting progressive muscle wasting.
Deletion/insertion mutations Gene mutations due to deletions or insertions of short regions by strand slippage, or of longer regions via homologous recombination.
Dendrites Branched, protoplasmic extensions of a nerve cell that propagate electrochemical stimulation received from other neural cells to the cell body.
Dendritic zone An area of branched, tree-like structures.
Dentate gyrus Part of the hippocampal formation involved in memory.
Denticulate ligaments Triangular ligaments that anchor the spinal cord.
Depolarization A shift in electric charge distribution, resulting in less negative charge inside the cell.
Dermatofibromas Benign skin growths; benign fibrous histiocytomas.
Desmin A protein filament that integrates sarcomere structures and regulates their architecture.
Desmoid-type fibromatosis A group of soft tissue tumors that are firm, white, and whorled.
Desmosomal Referring to a desmosome, a structure specialized for cell-to-cell adhesion.
Diabetes insipidus A condition characterized by large amounts of dilute urine and increased thirst; complications include dehydration and seizures.
Diencephalon A division of the forebrain situated between the telencephalon and midbrain, which forms the thalamus, hypothalamus, posterior pituitary, and pineal gland.
Diff-Quik stain A type of stain used to quickly differentiate pathology specimens.
Diffuse astrocytoma A brain tumor with diffuse zones of infiltration, usually within the cerebral hemispheres of adults.
Diffuse large B-cell lymphoma The most common of the large-cell lymphomas.
Diffuse midline gliomas Rare astrocytomas that usually occur in children, most often in the brainstem.
Diffusion-weighted imaging A type of magnetic resonance imaging (MRI) that uses computer software to generate images by diffusion of water molecules, creating a contrast appearance.

DiGeorge syndrome A condition caused by deletion of a small segment of chromosome 22, with heart problems, distinct facial features, infections, developmental delay, learning problems, and cleft palate.
Disability-adjusted life years Measures of overall disease burden, as the number of years lost due to ill-health, disability, or early death.
Disomies Also called aneuploidies; presence of abnormal numbers of chromosomes in cells.
Dorsal median septum Also called the posterior median septum; a sheet of glial tissue in the midsagittal plane of the spinal cord that separates it into halves.
Dorsal median sulcus Also called the posterior median sulcus; a shallow groove along the midline of the posterior spinal cord, separating the posterior funiculi.
Dorsolateral placodes Thickened areas that give rise to neurons; the trigeminal and otic placodes.
Double staining Mixing two dyes that each stain different portions of tissues or cells.
Drop metastasis When a tumor metastasizes down to the cauda equina.
Dura mater A thick membrane of dense irregular connective tissue surrounding the CNS; it is the outer layer of the meninges.
Dural tail A common but unspecific feature of meningioma that is due to dural hypertrophy.
Dysembryoplastic Referring to tissues, as in neuroepithelial tumors, that comprise glial and neuron cells, and may be linked to focal cortical dysplasia.
Dysmetria A lack of coordination of movement that is a form of ataxia.
Dysplasia epiphysealis hemimelica Also called Trevor's disease; a congenital bone developmental disorder that is more common in males.
Dystrophin A rod-shaped cytoplasmic protein that helps connect the cytoskeleton of a muscle fiber to the extracellular matrix, through the cell membrane.
Ectoderm The most exterior of the three primary germ layers in a new embryo.
Ectomesenchymal Referring to the ectomesenchyme, which is important for the formation of tissues of the head and neck.
Ectomesenchymoma A rare, fast-growing tumor of the nervous system or soft tissue that occurs mainly in children.
Edinger-Westphal nucleus The accessory oculomotor nucleus that innervates the iris sphincter muscle and ciliary muscle.
EGFRALK rearrangement Epidermal growth factor (EGF) and anaplastic lymphoma kinase (ALK) changes, associated with cancer.
Ehlers-Danlos syndrome (EDS) A group of rare genetic connective tissue disorders, with loose and painful joints, soft stretchy skin, and abnormal scarring.
Electrochemical gradient A gradient of electrochemical potential, usually for an ion that can move across a membrane.
Electrocorticography A type of electrophysiological monitoring that uses electrodes placed directly on the exposed brain surface to record cerebral cortex electrical activity.
Elephantiasis neuromatosa Enlargement of a limb due to diffuse neurofibromatosis of the skin and subcutaneous tissue.
Elongin A protein that is a regulatory subunit of the transcription factor B complex, involved in controlling transcription of information from DNA to messenger RNA.
EMACO therapy A combination of etoposide, methotrexate, and dactinomycin.
Embryonal carcinoma An uncommon germ cell tumor of the ovaries and testes.
En bloc resection Removal of a large bulky tumor virtually without dissection.
Encapsulated nerve endings Those that have either a brush border encapsulation, or fluid-filled sacs at the ends.
Endoderm The inner of the three primary germ layers of a new embryo.
Endonasal Within the nose.
Endothelial meningioma Also called meningothelial meningioma, the most common histological subtype, often combined with fibrous meningioma.
Enostoses Several areas of compact bones within cancellous bones.
Enteric nervous system (ENS) A mesh-like system of neurons controlling gastrointestinal tract functions.
Enteroceptors Those that respond to stimuli inside the body.
Eosinophilic granuloma A form of Langerhans cell histiocytosis.
Ependymal cells Ependymocytes; glial cells that make up the ependyma.
Ependymoblastoma A primitive neuroectodermal tumor that is a malignant neural crest tumor, usually occurring in children and younger adults.
Epibranchial placodes Thickened areas of the segment between the ceratobranchial and pharyngobranchial in a branchial arch.
Epidermal growth factor receptor (EGFR) A receptor for EGF that stimulates cell growth, proliferation, and differentiation.
Epidural space The potential space between the two layers of the dura mater.
Epigenome A record of chemical changes to DNA and histone proteins.
Epiretinal On the surface of the retina.
Epithalamus A posterior segment of the diencephalon.
Epithelioid hemangioendothelioma An uncommon vascular tumor of intermediate malignancy that resembles a proliferation of epithelioid cells.
Equilibrium potential Also called reversal potential; the membrane potential of an ion, at which there is no overall flow of the ion from one side of a membrane to the other.
ERBB2 amplification Making multiple copies of the ERBB2 gene, which commonly occurs in breast and ovarian cancers.
Erdheim-Chester disease A rare systemic disorder with proliferation of histiocytes, via symmetric sclerosis in portions of the lower limbs, and by extraskeletal involvement.

Erythroleukemia A malignant blood dyscrasia, with atypical erythroblasts and myeloblasts in the peripheral blood.
Erythropoietin A glycoprotein hormone that acts on stem cells of the bone marrow to stimulate red blood cell production.
Ethylnitrosourea A nitrosourea with potential antineoplastic activity.
Euchromatin A tightly packed form of chromatin that is the most active portion of the genome within the cell nucleus.
Ewing's sarcoma A type of cancer that may be a bone sarcoma or a soft tissue sarcoma.
Excitatory postsynaptic potential (EPSP) A postsynaptic potential that makes the postsynaptic neuron more likely to fire an action potential.
Exons Parts of genes that encode part of the final mature RNA produced, after introns have been removed via RNA splicing.
Exophthalmos Also called proptosis; a bulging of the eye anteriorly out of the orbit.
Exophytic Outward growing.
Exostoses Bone spurs; the formation of new bone on the surface of a bone.
Extradural tumors Those that are situated or occur outside the dura mater.
Extraneural Located outside of a nerve.
Extraosseous Occurring outside a bone or bones.
Extraskeletal Located in the soft tissues.
Eye floaters Small spots that drift through the field of vision.
Factor VIII-related antigen Also called von Willebrand factor; a glycoprotein involved in hemostasis.
Falx cerebri A large, crescent-shaped fold of meningeal layer of dura mater.
Familial isolated pituitary adenoma A condition of autosomal dominant inheritance characterized by two or more relatives affected by pituitary adenomas and no other associated symptoms.
Fanconi anemia A rare genetic disease resulting in impaired response to DNA damage.
Fasciculus cuneatus The larger lateral subdivision of the posterior funiculus.
Fasciculus gracilis The smaller medial subdivision of the posterior funiculus.
Fast MRI A supplemental breast cancer screening method, used for women with dense breast tissue.
Fetal hydantoin syndrome A group of fetal defects caused by exposure to teratogenic effects of phenytoin, commonly used to treat epilepsy.
Fetal lipomas Hibernomas; benign neoplasms of brown fat in a fetus.
Fibrolipomatous stalk An attachment between a lipomyelomeningocele's subcutaneous fat component and the underlying spinal cord.
Fibrosarcoma A malignant mesenchymal tumor characterized by immature proliferating fibroblasts or undifferentiated anaplastic spindle cells in a storiform pattern.
Fibroxanthoma A low-grade skin malignancy related to malignant fibrous histiocytoma.
Filum terminale A delicate strand of fibrous tissue proceeding down from the conus medullaris.
Fissure A groove, division, deep furrow, elongated cleft, or tear; also generally referred to as a sulcus.
Flaccid paralysis A neurological condition of weakness or paralysis, with reduced muscle tone, and no other obvious cause.
Flexner-Wintersteiner rosettes Palisading in a halo or spoke-and-wheel arrangement, seen in retinoblastoma and other ophthalmic tumors.
Flocculonodular lobes Lobes of the cerebellum consisting of nodules and flocculi.
Flow voids Areas of normal active CSF flow at the cerebral aqueduct and fourth ventricle.
Folia Ridges or gyri in the cerebellar cortex; they resemble leaves in shape.
Folliculostellate cells Nonendocrine cells found in the anterior lobe of the pituitary gland.
Fontana-Masson stains Silver stains used to identify dematiaceous fungi by marking the pigment melanin.
Foramen cecum A small opening converted from a notch, via articulation with the ethmoid, where the frontal crest of the frontal bone ends.
Foramen magnum A large oval opening in the occipital bone through which the spinal cord passes.
Foramen of Luschka The lateral aperture, a paired structure that is an opening in each lateral recess of the fourth ventricle, allowing for CSF flow.
Foramen of Magendie The median aperture that drains CSF from the fourth ventricle into the cisterna magna.
Foramen of Monro The interventricular foramina that connect the paired lateral ventricles with the third ventricle at the midline of the brain.
Forme fruste An atypical or attenuated manifestation of a disease or syndrome that seems incomplete, partial, or aborted.
Fourth ventricle The brain ventricle that extends from the cerebral aqueduct to the obex, and is filled with CSF.
Frameshift mutations A genetic mutation caused by insertions or deletions of nucleotides in a DNA sequence that cannot be divided by three.
Frameshifts Additions or deletions of nucleotides in a strand of DNA that result in aberrant proteins and gene mutations.
Frond A leaf-like structure.
Funduscopy Ophthalmoscopy; a test that allows visualization of the fundus of the eye and other structures using a funduscope (ophthalmoscope).
Funiculi Cord-like structures; axon bundles within nerve fascicles; or, paired white matter regions of the spinal cord.
Gadolinium A metal or salt that absorbs neutrons and is sometimes used for shielding in neutron radiography and as a contrast agent.
Gadolinium/pentetic acid A combination of agents used for contrast in an MRI, to aid in imaging of blood vessels, tissues, and intracranial lesions.

Galactorrhea Spontaneous flow of milk from the breast.
Galectin-3 A protein that plays an important role in cell−cell adhesion, cell−matrix interactions, macrophage activation, angiogenesis, metastasis, and apoptosis.
Gamma knife radiosurgery A treatment that uses gamma radiation to treat tumor cells, especially in the brain.
Gangliocytic Referring to gangliocytes (ganglion cells).
Gangliocytoma A rare, slow-growing primary CNS tumor that usually occurs in the temporal lobe of children and young adults.
Ganglioid Referring to a ganglion, a group of neuron cell bodies of the PNS.
Ganglion A group of neuron cell bodies in the peripheral nervous system.
Ganglioneuroma A rare, benign tumor of autonomic nerve fibers arising from neural crest sympathogonia.
Ganglioneuromatous Referring to a ganglioneuroma.
Gardner's syndrome Also called familial polyposis of the colon; characterized by multiple colon polyps along with osteomas of the skull, thyroid cancer, epidermoid cysts, fibromas, and desmoid tumors.
Gemistocytes Swollen, reactive astrocytes that usually appear during acute injury.
Gemistocytic astrocytes See "gemistocytes."
Gene sequencing DNA sequencing or whole genome sequencing.
General senses Those that are perceived due to receptors of the body, such as touch, temperature, and hunger.
Genodermatosis An inherited genetic skin condition, including epidermolysis bullosa, ichthyosis, palmoplantar keratoderma, and neurofibromatosis.
Germinoma A type of germ cell tumor, not differentiated, which may be benign or malignant.
Giant cell tumor A relatively uncommon bone tumor characterized by multinucleated osteoclast-like cells.
Giant congenital nevi Also called congenital melanocytic nevus, found in infants at birth, usually on the head or neck.
Gigantism A condition of excessive growth and height caused by overproduction of growth hormone in childhood.
Glial cells Neuroglia; nonneuronal cells of the CNS and PNS, which do not produce electrical impulses, with wider ranging activities than neurons.
Glial septum A structure in the midline raphe of the hindbrain and spinal cord that may be a physical barrier during outgrowth of the corticospinal tract.
Glioblastoma The most aggressive type of cancer that begins in the brain, representing about 15% of all brain tumors.
Glioblastoma multiforme See "glioblastoma."
Gliofibrillary Relating to, or composed of glial fibrils.
Glioma A malignant tumor that starts in the glial cells of the CNS.
Glioma susceptibility 1 The first type of susceptibility to developing a glioma, linked to a genetic defect on chromosome 3p25.
Gliomagenesis The formation and development of gliomas.
Gliomatosis cerebri A rare primary brain tumor characterized by diffuse infiltration with neoplastic glial cells affecting areas of the cerebral lobes.
Gliomatous Referring to a glioma.
Glioneuronal tumors Rare, histologically mixed neuronal and glial tumors.
Gliosarcoma A rare, malignant glioma with gliomatous and sarcomatous components.
Globus pallidus The dorsal pallidum of the brain, and a major component of the basal nuclei.
Glomeruloid Referring to a glomerulus.
Golgi apparatus An organelle in most eukaryotic cells that packages proteins into membrane-bound vesicles.
Gonadotrophic adenomas Those that express and secrete biologically active gonadotropins and cause reproductive system abnormalities.
Gonadotropin Glycoprotein polypeptide hormones secreted by gonadotrope cells of the anterior pituitary gland.
Gorlin-Koutlas syndrome A complex genetic disorder or multiple tumors, including multiple schwannomas.
Graded potential A change in membrane potential that may vary in size, and may be excitatory or inhibitory.
Gradient echo sequence A technique used to obtain images very quickly; it is characterized by a single excitation followed by a gradient applied along the reading axis.
Granule cells Neurons with very small cell bodies, which interact with mossy fibers, Purkinje cells, thalamus, and cerebral cortex.
Granzyme B A serine protease that activates apoptosis by activating caspases.
Graves' disease Toxic diffuse goiter, an autoimmune disease of the thyroid that is the most common cause of hyperthyroidism.
Gray commissure A thin strip of gray matter around the central canal of the spinal cord; with the anterior white commissure, it connects the two halves of the cord.
Gustatory cortex A brain structure responsible for the perception of taste.
Gyri Ridges on the cerebral cortex usually surrounded by sulci.
Gyriform Having a ridge-like structure.
Habenula The habenular nucleus; involved in nociception, sleep−wake cycles, reproduction, and mood.
Hair follicle receptors Networks of hair plexuses around hair follicles that send and receive nerve impulses to and from the brain when the hair moves.
Hairy cell leukemia An uncommon hematological malignancy, with accumulation of abnormal B-lymphocytes.
Hamartin Tuberous sclerosis 1, a protein encoded by the TSC1 gene that, when defective, causes tuberous sclerosis.

Hamartoma A mostly benign, local malformation of cells that resembles a neoplasm of local tissue; they may predispose the individual toward malignancy.
Haploinsufficiency Dominant gene action, with a single copy of a wild-type allele at a locus in heterozygous combination with a variant allele being insufficient to produce the standard phenotype.
Hemangioblastoma A vascular tumor of the CNS originating from the vascular system, usually in middle age.
Hemangioma A usually benign vascular tumor derived from blood vessel cell types; usually called a "strawberry mark," seen soon after birth.
Hemangiopericytoma A soft tissue sarcoma originating in the pericytes in capillary walls; in the nervous system, it is often aggressive.
Hemicrania continua A persisting unilateral headache, considered to be a primary headache disorder.
Hemihypophysectomy Surgical removal of half of the pituitary gland.
Hemispheric lateralization Functional differences between the left and right cerebral hemispheres.
Hemosiderotic Referring to hemosiderosis, an iron overload disorder caused by the accumulation of hemosiderin.
Hepatocyte growth factor Scatter factor that acts mostly on epithelial and endothelial cells, but also hemopoietic progenitor cells and T-cells.
Hepatoid variant A form of tumor with cells resembling liver cells.
HER2 protein Human epidermal growth factor receptor 2 protein; amplification or overexpression is implicated in aggressive breast cancer.
Herringbone Scintillating scotoma, a common visual aura; or, pattern that appears to be "interwoven."
Heterodimeric complex A complex composed of two different subunits that differ in composition.
Heterodimers Proteins composed of two polypeptide chains differing in composition in the order, number, or type of their amino acid residues.
Heterozygosity The state of being heterozygous, which is having two alleles at corresponding loci on homologous chromosomes different for one or more loci.
Hexaribonucleotide A six-sided form of a ribonucleotide, which is a nucleotide with ribose as its pentose component.
Hibernoma A benign neoplasm of vestigial brown fat.
Hippocampal formation A compound structure in the medial temporal lobe implicated in memory, spatial navigation, and attention.
Hippocampus Part of the limbic system, playing important roles in memory.
Histiocytoid Referring to histiocytes, which are nonmotile macrophages of extravascular tissues, and primarily, connective tissues.
Histiocytoma A tumor consisting of histiocytes; examples include myxofibrosarcoma, benign fibrous histiocytoma, and malignant fibrous histiocytoma.
Histochemistry Also called immunohistochemistry, the most common application of immunostaining, to diagnose cancer and other abnormal cells.
Histone deacetylase inhibitors Chemical compounds that inhibit histone deacetylases; used as mood stabilizers, anticonvulsants, and to treat cancer.
Histopathological Referring to histopathology, the microscopic examination of tissue to study manifestations of disease.
Holocranial Encompassing the entire brain or head.
Homer-Wright rosettes Those in which differentiated tumor cells surround the neuropil, such as in neuroblastoma or medulloblastoma.
Hotspots Areas with a high level of radioactive uptake or contamination.
Hyalinized Having a glassy appearance, with transparency.
Hybrid nerve-sheath tumors Uncommon tumors usually seen with neurofibromatosis 1 or 2, such as schwannoma+perineurioma.
Hyperchromasia Developing excess chromatin or nuclear staining, especially as part of a pathological process.
Hyperdiploidy A condition in which there are slightly more than the diploid number of chromosomes.
Hyperglobulinemia The presence of excess globulins in the blood.
Hypermutation The process of producing an unusually high number of mutations or changes.
Hyperostosis Excessive growth or thickening of bone tissue.
Hyperpolarization An increase in potential differences across a membrane.
Hypertonic Having a higher osmotic pressure than a surrounding medium or fluid.
Hypodense Being less dense than normal.
Hypodiploidy A condition in which there are slightly less than the diploid number of chromosomes.
Hypogastric plexus The superior or inferior plexuses of nerves on the vertebral bodies, anterior to the bifurcation of the abdominal aorta.
Hypoglossal trigone A structure above the area where the hypoglossal nucleus approaches the rhomboid fossa.
Hypointense Being less than the normal level of intensity.
Hypomelanosis of Ito Incontinentia pigmenti achromians, a skin condition with patterns of bilateral or unilateral hypopigmentation.
Hypophosphorylated Having less than the normal amount of phosphorylation.
Hypothalamic sulcus A groove in the lateral wall or the third ventricle that marks the boundary between the thalamus and hypothalamus.
Hypothalamus The brain structure that links the nervous and endocrine systems via the pituitary gland.
Hypotriploid Containing less than three times the number of chromosomes of a haploid.
Immunoblasts Lymphocytes activated by an antigen; the most immature members of the protective cells involved in an immune response.
Immunoexpression The expression of a protein as a result of an immune response.
Immunonegative Generating a negative response to a test for a specific antigen or antibody.
Immunophenotypic Referring to an immunophenotype, which is identified by studying proteins expressed by cells.
Immunopositivity The quality of being immunopositive, which is positive to a test for a specific antigen or antibody.
Immunoprofile An immunological profile of an individual.

Immunostain The result of immunostaining, which is the use of an antibody-based method to detect a specific protein.
Immunotyping The use of techniques to identify immunotypes.
Incidence density rate The person-time incidence rate, which uses an incidence rate that has a denominator that is the product of the person-time of the at-risk population.
Incidence rate A measure of the frequency with which a disease or other incident occurs over a specified time period.
Infantile hemangiomas Benign vascular tumors in infants that appear as red or blue raised lesions.
Inferior colliculi The principal midbrain nuclei of the auditory pathway.
Inferior sagittal sinus An area below the brain that allows blood to drain outward posteriorly from the center of the head.
Infratentorially Referring to the infratentorial region of the brain, the area below the tentorium cerebelli.
Infundibulum A funnel-shaped cavity or organ, such as the pituitary stalk.
Inhomogeneous Anything that is not uniform in composition or character.
Integrative centers Those that integrate and process sensory information and perform complicated motor activities and analytic functions.
Interdigitations Nipple-like extensions.
Internal capsule A white matter structure in the inferomedial part of each cerebral hemisphere that carries information via ascending and descending axons.
Internal carotid arteries Those located in the inner sides of the neck to supply blood to the brain and eyes.
Internal jugular veins Paired veins that collect blood from the brain and superficial parts of the face and neck.
Interneurons Also called association neurons that are the central nodes of neural circuits, allowing communication between sensory and motor neurons, and the CNS.
Internexin A protein that is a major component of the intermediate filament network in small interneurons and cerebellar granule cells.
Interoceptors Internal receptors that respond to changes inside the body.
Intertumoral Between tumors.
Interventricular foramen One of the channels that connects the paired lateral ventricles with the third ventricle; or, in embryology, the temporary opening between the developing ventricles of the heart.
Intrachromosomal inversion Between several chromosomes, the end-to-end reversal of various segments.
Intracortical Within a cortex.
Intradural tumors Those that are within the dura mater.
Intralabyrinthine Within the labyrinth of the ear.
Intralumbar Within the lumbar region of the spine.
Intraparenchymatous Within a parenchyma.
Intumescences Swollen body structures.
Irritative (radicular) syndrome A pain syndrome usually caused by compressed or irritated nerve roots; it also causes numbness, tingling, and weakness.
Isocitrate dehydrogenase An enzyme that catalyzes oxidative decarboxylation of isocitrate, producing alpha-ketoglutarate and carbon dioxide.
Isointense Having the same intensity as another object.
Joint kinesthetic receptors Those that sense self-movement and body position within the joints.
Juxtallocortex The mesocortex, which is the transitional area of the cerebral cortex, formed at borders between the true isocortex and allocortex.
Kaposi sarcoma A cancer that may form masses in the skin, lymph nodes, and other organs, with the lesions usually being purple in color.
Karnofsky score A ranking method that allows physicians to evaluate a patient's ability to survive chemotherapy for cancer.
Kilobases Units of measure of the length of nucleic acid chains equaling 1000 base pairs.
Kinesin A motor protein found in eukaryotic cells.
Klinefelter syndrome The set of symptoms resulting from two or more X chromosomes in males; primarily, there is infertility and small testicles.
Klippel-Trenaunay syndrome A rare congenital condition in which blood or lymph vessels fail to form normally.
Knee-jerk reflex The patellar reflex, a stretch reflex that tests the L2, L3, and L4 segments of the spinal cord.
Lamellar corpuscles Pacinian corpuscles; one of the four major types of mechanoreceptor cells in hairless skin; they are sensitive to vibration and pressure.
Lamina terminalis A thin lamina in the forebrain that stretches from the interventricular foramen to the recess at the base of the optic stalk.
Large deletions Genetic deletions that, due to the amount of information that is missing, are usually fatal.
Lateral corticospinal tract The largest part of the corticospinal tract, extending throughout the entire length of the spinal cord.
Lateral division The part of the spinal cord that contains almost exclusively unmyelinated and thinly myelinated fibers.
Lateral rectus palsy Abducens nerve palsy, causing contraction of the lateral rectus muscle to abduct the eye.
Lateral sulcus The Sylvian fissure, separating the frontal and parietal lobes from the temporal lobe.
Leak channels Potassium channels that are regulated by signaling lipids, oxygen tension, pH, mechanical stretch, and G-proteins.
Legius syndrome An autosomal dominant condition characterized by café au lait spots, often mistaken for neurofibromatosis type 1.
Leiomyoma Also called fibroid; a benign smooth muscle tumor that usually occurs in the uterus, small intestine, and esophagus.
Leiomyosarcoma A malignant smooth muscle tumor that may remain dormant for a long time, and recur after years.
Lemnisci Bundles of secondary sensory fibers in the brainstem.

Lens placode A thickened portion of ectoderm that serves as the precursor to the lens of the eye.
Lentiform nucleus A structure comprising the putamen and globus pallidus within the basal nuclei.
Lentiginous Having the characteristics of lentigines, which are small pigmented spots on the skin with clearly defined edges.
Leptomeningeal Referring to the arachnoid and pia mater, which are collectively known as the leptomeninges.
Leptomyelolipomas Lumbosacral lipomas.
Lhermitte-Duclos disease Dysplastic gangliocytoma of the cerebellum.
Li-Fraumeni syndrome A rare, autosomal dominant disorder that predisposes carriers to cancer development.
Life expectancy A statistical measure of the average time a person is expected to live, based on birth year, current age, gender, and demographic factors.
Life table A graphical depiction of the probability for a person of a specific age to die before their next birthday.
Lifetime prevalence The proportion of individuals in a population that have experienced a disease, traumatic event, or certain behavior.
Light chain restriction Blocking the function of small polypeptide subunits of protein complexes.
Limbic system Brain structures that support emotions, behaviors, motivation, long-term memory, and olfaction.
Lipidization Reaction of a lipid with (usually) a polypeptide.
Lipoblasts Precursor cells for adipocytes.
Lipofuscin Yellow-brown pigment granules made up of lipid-containing residues of lysosomal digestion.
Lipoma A benign tumor made of fat tissue.
Liposarcoma A rare cancer that arises in fat cells in soft tissues.
Lisch nodules Iris hamartomas; pigmented aggregates of dendritic melanocytes in the irises that are related to neurofibromatosis type 1.
Local potentials Those that are graded, weaken as they spread from the point of stimulation, and reversible.
Longitudinal fasciculi The dorsal, inferior, or medial white matter fiber tracts in the brainstem that convey visceral motor and sensory signals.
Longitudinal fissure The deep grove that separates the cerebral hemispheres.
Lower motor neurons Those that are located in the CNS and required for all voluntary movements.
Lumbar cistern The space that extends from the conus medullaris to the second sacral vertebra; this cistern is used to withdraw CSF during lumbar puncture.
Lumbar enlargements The widened areas of the spinal cord that provide attachments to nerves supplying the lower limbs.
Lumbar puncture Spinal tap; the insertion of a needle into the spinal canal, usually to collect CSF for diagnostic testing.
Luse bodies Collagen fibers with very long spacing between electron-dense bands.
Lymphangioleiomyomas Fluid-filled hypodense structures in the retroperitoneal regions of the abdomen and pelvis in some patients with lymphangioleiomyomatosis.
Lymphoepithelial Referring to lymphoid cells and epithelium.
Lymphomagenesis The formation of a lymphoma.
Lymphomatoid granulomatosis A rare lymphoproliferative disorder, with microscopic characteristics of granulomas with polymorphic lymphoid infiltrates and focal necrosis.
Lymphomatosis cerebri A rare form of a primary CNS non-Hodgkin's lymphoma, with diffuse infiltration of tumor cells throughout the brain parenchyma.
Lymphoplasmacellular Referring to disorders such as plasmacytoma, multiple myeloma, lymphoplasmacytic lymphoma, MALT lymphoma, and amyloidosis.
Lymphoplasmacytoid cells Those that have features of both plasma cells and lymphocytes.
Lynch syndrome Hereditary nonpolyposis colorectal cancer.
Macrocysts Aggregates of cells formed during sexual reproduction, enclosed in cellulose walls; these aggregates engulf other cells.
Macronucleoli Extremely large nucleoli, which are the largest structures in the nuclei of eukaryotic cells.
Maffucci syndrome A rare disorder with multiple benign tumors of cartilage within bones, usually in the hands, feet, and limbs.
Magnetic resonance spectroscopy A spectroscopic technique to observe local magnetic fields around atomic nuclei.
Malignant perineuriomas A malignant tumor composed entirely of neoplastic perineurial cells; they may be intraneural or within soft tissues.
Malignant triton tumor A rare, aggressive tumor made up of malignant schwannoma cells and malignant rhabdomyoblasts.
MALT lymphomas Tumors that involve the mucosa-associated lymphoid tissue (MALT), originating from B-cells.
Mammillary bodies Two small, round bodies on the undersurface of the brain that form part of the limbic system.
Mammillothalamic tract A fasciculus arising from cells in the medial and lateral nuclei of the mammillary body, and by fibers directly continued from the fornix.
Mechanoreceptors Sensory cells that respond to mechanical pressure or distortion.
Meckel's cave The trigeminal cave, a dura mater pouch containing CSF.
Medulla oblongata The lower part of the brainstem responsible for involuntary functions, containing the cardiac, respiratory, vomiting, and vasomotor centers.
Medulla spinalis The spinal cord.
Medullary pyramids Paired white matter structures of the medulla that contain motor fibers of the pyramidal tracts.
Medulloblastoma The most common primary brain cancer in children; it is invasive and fast-growing, spreading through the CSF to metastasize.

Medullomyoblastomas Variants of medulloblastomas with scattered smooth, striated muscle cells.
Megalencephaly A developmental disorder in which the brain is abnormally large.
Melanoma A malignant skin cancer developing from melanocytes, usually caused by ultraviolet light exposure.
Melanophages Tissue-resident macrophages that can absorb pigment from the extracellular space.
Melanotic schwannoma A nerve-sheath tumor with uniform composition of variable melanin-producing Schwann cells and metastatic potential.
Melatonin A hormone mostly released by the pineal gland; it regulates the sleep–wake cycle.
Membrane potential The difference in electric potential between the interior and exterior of a biological cell.
Memory consolidation A category of processes that stabilize a memory trace after its first acquisition.
Mendelian syndrome A group of genetic disorder caused at a single genetic locus.
Meningeal melanocytoma A rare, pigmented tumor found on the leptomeninges of the brain, usually at the base of the brain and brainstem, but in the spine.
Meningeal melanoma A primary melanocytic CNS tumor representing malignant neoplasms of leptomeningeal melanocytes, derived from neural crest cells.
Meningioangiomatosis A rare disease characterized by a benign lesion of the leptomeninges, usually of the cerebral cortex, and by leptomeningeal and meningovascular proliferation.
Meningioma Usually, a slow-growing tumor that forms from the meninges; it is linked to family history, ionizing radiation, and neurofibromatosis type 2.
Meningothelial meningioma The most common subtype of meningioma, often combined with fibrous meningioma.
Merlin A tumor suppressor protein involved in neurofibromatosis type 2.
Mesencephalon The midbrain; the forward-most part of the brainstem, associated with vision, hearing, motor control, sleep, wakefulness, alertness, and temperature regulation.
Mesenchymal neoplasm Usually, abnormal cellular growth in bone, connective tissue, or the lymph or circulatory systems.
Mesenteric ganglia Part of the prevertebral ganglia; they innervate the intestines.
Mesoblast The mesoderm, especially in its early, undifferentiated stages.
Mesoderm The middle layer of the three primary germ layers in a new embryo.
Metachronous Occurring at a different time than a similar event.
Metalloenzyme An enzyme in which a metal ion is bound with one labile coordination site; the ion is usually in a pocket with a suitable shape, and catalyzes reactions that are difficult to achieve in organic chemistry.
Metaplasia The transformation of one differentiated cell type into another.
Metaplastic meningiomas Uncommon variants of benign meningiomas, with tumor cells sharing characteristics of other body tissues.
Metencephalon The embryonic part of the hindbrain that differentiates into the pons and cerebellum.
Methylnitrosourea A carcinogen, mutagen, teratogen, and alkylating agent.
Methylome The set of nucleic acid methylation changes in a genome or cells.
Microcystic meningiomas Rare variants of meningiomas with atypical appearances that are difficult to diagnose.
Microfilaments Actin filaments in the cytoplasm of eukaryotic cells that form part of the cytoskeleton.
Microgemistocytes Smaller forms of gemistocytes, which are swollen, reactive astrocytes.
Microglia Glial cells in the CNS; they make up 10%–15% of all brain cells.
Microrosettes Small forms of rosettes, which are flower-shaped structures.
Microtubules Polymers of tubulin that form part of the cytoskeleton and provide structure and shape to eukaryotic cells.
Midbrain See "mesencephalon."
Middle cerebral artery One of the three major paired arteries that supplies blood to the cerebrum.
Midkine Neurite growth-promoting factor 2, which binds heparin.
Minigemistocytes Tiny forms of gemistocytes.
Mismatch repair cancer syndrome A condition associated with biallelic DNA mismatch repair mutations; also called Turcot syndrome.
Missense mutations Point mutations in which a single nucleotide change results in a codon that codes for a different amino acid.
Mitogens Peptides or small proteins that induce cells to begin mitosis.
Mixed germ cell tumor A teratoma or teratocarcinoma.
Monoclonality The production of cells from one ancestral cell via repeated cellular replication.
Monosynaptic reflex A reflex arc that consists of only one sensory neuron and one motor neuron.
Mucocele A distension of a hollow organ or cavity because of mucus buildup.
Multinucleation The state having more than one nucleus.
Multiple myeloma A cancer of plasma cells that is generally incurable.
Multipotency A condition in which progenitor cells have the gene activation potential to differentiate into discrete cell types.
Multivacuolated Having multiple vacuoles.
Mural nodule A small mass of solid tissue on the inner wall of a cyst.
Muscle spindles Stretch receptors within the body of a muscle that mostly detect changes in the length of the muscle.
Myelencephalon The afterbrain; the most posterior region of the embryonic hindbrain, from which the medulla oblongata develops.

Myelin A lipid-rich substance that surrounds nerve cell axons to insulate them and increase the rate at which action potentials are passed along.
Myelotoxicity Bone marrow suppression; the decrease in production of cells responsible for providing immunity, carrying oxygen, and for normal blood clotting.
Myoepithelioma A salivary gland tumor of the head and neck that is usually benign.
Myofibroblastoma A rare, benign tumor, usually of the breast.
Myogenin Myogenic factor 4, involved in the coordination of skeletal muscle development and repair.
Myotomes Groups of muscles innervated by a single spinal nerve.
Myxoid foci Primary locations of something with mucus-like qualities.
Myxomas Myxoid tumors of primitive connective tissue.
Myxopapillary ependymoma A localized, slow-growing, low-grade tumor that usually originates from the lumbosacral nervous tissue of younger patients.
Myxopapillary variant A form of a tumor that has both mucus-like and papillary-like qualities.
Necrobiotic Relating to necrobiosis, the natural death of cells or tissues through aging.
Nelson's syndrome A rare disorder sometimes occurring when both adrenal glands were removed to treat Cushing's syndrome, resulting in macroadenomas.
Neocortex The isocortex of the cerebral cortex, involved in higher order brain functions.
Neovascularity The state of neovascularization, which is the natural formation of new blood vessels.
Nestin An intermediate filament protein mostly expressed in nerve cells.
Nestin protein See "nestin."
Neural crest cells Temporary cells arising from the embryonic ectoderm, which give rise to melanocytes, craniofacial cartilage, bone, smooth muscle, peripheral and enteric neurons, and glial cells.
Neural fold A structure arising during neurulation in embryonic development.
Neural groove A shallow median groove of the neural plate between the neural folds of an embryo.
Neural plate The developmental structure that is the basis for the nervous system.
Neural tube The embryonic precursor to the CNS.
Neuralgiform headache A stabbing headache with features resembling neuralgia and its related pain.
Neuraxial Related to the neuroaxis; the axial, unpaired part of the CNS, including the spinal cord, rhombencephalon, mesencephalon, and diencephalon.
Neurilemma The outermost nucleated cytoplasmic layer of Schwann cells surrounding the axon of a neuron.
Neurocristopathies Pathologies that may arise from defects in the development of tissue containing cells commonly derived from the embryonic neural crest cell lineage.
Neurocutaneous melanosis A congenital disorder, with congenital melanocytic nevi on the skin, and melanocytic tumors in the leptomeninges of the CNS.
Neuroembryogenesis The formation of neurologic structures during the embryonic period.
Neuroendoscopic Referring to neuroendoscopy, a minimally invasive surgical procedure in which a tumor is removed through small holes in the skull or through the mouth or nose.
Neurofibrils Proteinaceous filaments that extend in the cytoplasm of nerve cells.
Neurofibroma A benign nerve-sheath tumor in the PNS that may or may not be part of neurofibromatosis type 1.
Neurofibromatosis A group of three conditions in which tumors grow in the nervous system: NF1, NF2, and schwannomatosis.
Neurofibromin A tumor suppressor protein.
Neurofibrosarcoma Also called malignant peripheral nerve-sheath tumor, a cancer of the connective tissue surrounding nerves.
Neurofilaments Intermediate filaments in the cytoplasm of neurons.
Neuroglandular junction The site of communication between a nerve fiber and a gland.
Neuroglia Glial cells of the CNS and PNS.
Neuromuscular junction A chemical synapse between a motor neuron and muscle fiber.
Neuron An electrically excitable nerve cell, communicating with other cells via synapses.
Neuron-specific enolase Enolase 2, an enzyme in mature neurons and cells of neuronal origin.
Neuropores Openings formed in various regions, such as the cranial and caudal neuropores.
Neuropril Any area in the nervous system composed mostly of unmyelinated axons, dendrites, and glial cell processes, forming a synaptically dense region with a low number of cell bodies.
Neurosarcoid Also called neurosarcoidosis, a condition with granulomas of the CNS.
Neurotrophin receptor One of the group of growth factor receptors that specifically bind to neurotrophins.
Neurotubules Microtubules in neurons within nervous tissues.
Neurula An embryo at the early stage of development in which neurulation occurs.
Nijmegen breakage syndrome A rare autosomal recessive congenital disorder causing chromosomal instability.
Nissl substance Also called a Nissl body; a large granular material found in neurons; the body changes because of pathological conditions.
Nociceptors Pain receptors.
Nodes of Ranvier Myelin-sheath gaps along myelinated axons where the axolemma is exposed to the extracellular space.

Nodosal placode A thickened area associated with the third branchial cleft that generates the nodose ganglion and distal parts of cranial nerve X.
Nodular fasciitis A benign soft tissue lesion usually found in the superficial fascia.
Non-Hodgkin lymphoma A group of blood cancers that includes all types of lymphomas except Hodgkin lymphomas.
Nonpsammomatous Not having psammoma bodies.
Nonsense mutations Also called point-nonsense mutations; they are point mutations in a sequence of DNA that result in a premature stop codon and in a usually nonfunctional protein product.
Notochord A flexible rod made of material similar to cartilage.
Nuclear pores Openings in a nuclear pore complex, which span the nuclear envelope.
Nuclear pseudopalisading Development of a hypercellular zone, usually surrounding necrotic tissue, in a nucleus.
Nucleolated Containing a nucleolus or nucleoli.
Nucleolin The major nucleolar protein of growing eukaryotic cells.
Nucleotides Organic molecules consisting of a nucleoside and a phosphate.
Nucleus basalis A group of neurons located mainly in the substantia innominata of the basal forebrain.
Null cell adenomas Pituitary adenomas that do not secrete hormones, and often cause compressive effects upon the pituitary stalk.
Nystagmus retractorius A spasmodic backward movement of the eyeball occurring on attempts to move the eye, a sign of midbrain disease.
Obex The ependyma-lined junction of the teniae of the fourth ventricle, at the inferior angle.
Oculomotor nucleus The group of motor neurons innervating all external eye muscles except the rectus lateralis and obliquus superior.
Olfactory cortex The piriform cortex corresponding to the rostral half of the uncus.
Oligodendrocytes Cells of the oligodendroglia.
Oligodendroglioma A neoplasm derived from and composed of oligodendroglia.
Oligosarcomas Gliosarcomas arising from oligodendrogliomas.
Ollier disease A rare sporadic nonhereditary skeletal disorder, with typically benign enchondromas developing near the growth plate cartilage.
Optic chiasma The part of the brain where the optic nerves cross.
Osteoblastoma An uncommon osteoid tissue-forming primary bone neoplasm.
Osteochondroma The most common benign bone tumor.
Osteoid osteoma A benign bone tumor arising from osteoblasts and some components of osteoclasts.
Osteomata The plural term for osteoma; new pieces of bone usually growing on other pieces of bone—mostly in the skull; they are benign tumors.
Osteosarcoma A cancerous bone tumor also known as osteogenic sarcoma.
Osteosarcomatous Referring to an osteosarcoma.
Otic pit The auditory pit; the first rudiment of the internal ear in development.
Otic placode A thickening of the ectoderm on the outer surface of an embryo from which the ear develops.
Oxytocin A peptide hormone and neuropeptide implicated in reproduction, childbirth, and milk production.
Pachymeningeal fibrosis Formation of fibrous tissue within the dura mater.
Paget's disease Any of three diseases that include Paget's disease of bone, Paget's disease of the breast, and extramammary Paget's disease.
Paleocortex The part of the cerebral cortex that, along with the archeocortex, develops along with the olfactory system; it mostly consists of the piriform cortex and parahippocampal gyrus.
Palmoplantar Referring to the palms and soles.
Papillary ependymoma A primary CNS tumor that forms linear, epithelial-like surfaces along areas of CSF exposure.
Papillary meningiomas Rare malignant meningiomas that usually occur in young patients, and are aggressive, with atypical imaging features.
Papillary tumor A papilloma, which is a circumscribed, benign epithelial tumor projecting from a surrounding surface.
Parameningial Also spelled parameningeal; of or related to the structures proximal to the meninges.
Paranuclear inclusions Structures enclosed near but outside of the cell nucleus.
Paraparetic Relating to paraparesis.
Parasubiculum A narrow region of cortex between the entorhinal area and the subiculum.
Paresis Slight or incomplete paralysis.
Parinaud syndrome Paralysis of conjugate upward gaze, with a lesion at the level of the superior colliculi.
Peau chagrin A texture of shagreen patches, with rough, orange-like skin.
Penetrance The frequency with which a heritable trait is manifested by people carrying the principle gene or genes conditioning it.
Perforin A protein in the cytoplasmic granules of T-cytotoxic lymphocytes and natural killer cells, implicated in target cell lysis.
Pericellular Surrounding a cell.
Pericytes Elongated, contractile cells wrapped around precapillary arterioles outside the basement membrane.
Perifocal Surrounding a focus; denoting tissues or contained blood in the area of an infective focus.
Periglandular Surrounding a gland.

Perikaryal fibril A small fiber of or related to the perikaryon of a nerve, involving or occurring within the perikaryon.
Perikaryon The cell body of a neuron, containing the nucleus and organelles.
Perineurioma A tumor composed entirely of neoplastic perineurial cells.
Perineuriomatous Referring to a perineurioma.
Perineuronal satellitosis The microanatomical clustering of glioma cells around neurons in the tumor microenvironment.
Period prevalence The number of people with a specific disease at one point in time, divided by the total number of people in the population.
Peripheral nervous system One of the two components that makes up the nervous system, along with the CNS; the PNS consists of the nerves and ganglia outside of the CNS.
Peritumoral edema Swelling around a tumor location.
Perivascular aggregation Clustering of neoplastic clear cells around vessels.
Petrosal placode A thickened area associated with the second pharyngeal groove that generates the inferior ganglion of the glossopharyngeal nerve and distal parts of cranial nerve IX.
Phacomatosis One of a group of rare syndromes involving structures arising from the embryonic ectoderm, characterized by birthmarks or skin lesions, often involving multiple organ systems.
Pheochromocytomas Rare, chromaffin cell tumors of the adrenal medulla.
Photocoagulation Laser coagulation surgery for eye diseases, to cauterize ocular blood vessels.
Photosensory Relating to the perception of light.
Pia mater The delicate innermost layer of the meninges around the CNS.
Pilocytic astrocytoma A brain tumor occurring more often in children and young adults, usually in the cerebellum.
Piloerection Goose bumps that develop due to tickling, cold, or strong emotions.
Piloid Hair-like.
Pilomotor Causing movement of body hairs.
Pineal gland Also called the pineal body; a small, conical structure attached to the posterior wall of the third ventricle of the cerebrum involved in melatonin biosynthesis.
Pineoblastoma A poorly differentiated tumor of the pineal gland that usually occurs in the first 30 years of life.
Pineocytoma A tumor arising in the pineal gland that resembles normal pineal parenchyma.
Pituicytoma An older term describing a tumor of the neurohypophysis.
Pituitary adenoma A benign neoplasm of the anterior pituitary gland that may be secretory or nonsecretory.
Pituitary apoplexy A stroke occurring in the pituitary area.
Pituitary carcinoma A malignant tumor of the pituitary gland, made of epithelial cells.
Pituitary choristomas Masses formed by maldevelopment of tissues not normally found in the pituitary gland.
Pituitary tuberculoma A tumor-like mass resulting from aggregation or enlargement of broken-down tubercles.
Plasma cells Antibody-secreting cells in lymphoid tissue, derived from B-cells via lymphokines stimulation, reacting with specific antigens.
Plasmocytoma A plasma cell dyscrasia, or a discrete plasma cell tumor mass that is believed to be solitary.
Pleomorphic Referring to a variable appearance or morphology.
Pleomorphic xanthoastrocytoma A rare variant of astrocytoma presenting early in life and causing seizures.
Plexiform neurofibroma An anomaly instead of an actual neoplasm, in which a proliferation of Schwann cells occurs from the inner aspect of a nerve sheath.
Plexus of Auerbach A plexus of unmyelinated fibers and postganglionic autonomic cell bodies in the muscular coat of the esophagus, stomach, and intestines.
Plexus of Meissner An autonomic plexus in the submucosa of the alimentary tube that regulates secretions of the mucosa.
Pluripotent fetal cells Those that have the capacity to affect more than one organ or tissue and are not fixed as to their potential development.
Point prevalence The prevalence of a disease during a specific time period.
Poisson distribution A distribution function used to describe occurrence of rare diseases or events.
Polycomb gene targets Those that are silenced in multiple myeloma.
Polycythemia An increase in the total red blood cell mass of the blood.
Polyhedral Having many sides or surfaces.
Polymorphic variant of Zulch A genetically heterogeneous group of anaplastic gliomas, including tumors of actual oligodendroglial differentiation.
Polyomaviruses A family of viruses containing DNA, with virions about 45 nm in diameter, that infect humans, other mammals, and birds.
Polyphenotypic differentiation An increase in morphological or chemical heterogeneity over multiple phenotypes.
Polyploidy The state of having more than two sets of homologous chromosomes.
Polypoid Resembling a polyp.
Polysomy An excess of a specific chromosome.
Polysynaptic reflex One in which there are one or more interneurons that connect afferent and efferent signals.
Polyubiquitination An enzymatic posttranslational modification in which a ubiquitin protein is attached to a substrate protein, but affects a chain of ubiquitin.
Pons The part of the brainstem inferior to the midbrain and superior to the medulla oblongata; it contains nuclei involved in sleep, respiration, swallowing, bladder control, hearing, equilibrium, taste, eye movement, facial expression and sensation, and posture.

Pontine flexure The rhombic flexure between the metencephalon and myelencephalon.
Posterior commissure The epithalamic commissure, a rounded band of white fibers crossing the middle line on the dorsal aspect of the rostral end of the cerebral aqueduct; important in the bilateral pupillary light reflex.
Posterior funiculus An area between the posterolateral and posterior median sulci.
Posterior horns The occipital horns of the lateral ventricle, impinging into the occipital lobe in a posterior direction.
Posterior intermediate sulcus A longitudinal furrow between the posterior median and posterolateral sulci of the spinal cord, in the cervical region.
Posterior root ganglion The dorsal root ganglion, also known as the spinal ganglion, which is a group of nerve cell bodies.
Posterolateral sulcus A longitudinal furrow on either side of the posterior median sulcus of the spinal cord, marking the line of entrance of the posterior nerve roots.
Postsynaptic neuron A neuron to the cell body or dendrite of which an electrical impulse is transmitted across a synaptic cleft via neurotransmitter release from the axon terminal of a presynaptic neuron.
Postsynaptic potentials Changes in the membrane potential of the postsynaptic terminal of a chemical synapse.
Potential spaces Regions in which two surface membranes adjoin, separated only by a small amount of fluid lubrication.
Prefrontal cortex The anterior part of the frontal lobe.
Premotor cortex An area of the motor cortex within the frontal lobe, just anterior to the primary motor cortex.
Presubiculum Also called "Brodmann area 27"; a rostral part of the parahippocampal gyrus of the guenon.
Presynaptic neuron A neuron from the axon terminal of which an electrical impulse is transmitted across a synaptic cleft to the cell body, or one or more dendrites of a postsynaptic neuron, via neurotransmitter release.
Prevalence Referring to the number of people in a population who have a disease at a given time.
Prevertebral plexuses The three plexuses of autonomic nerve division: the cardiac, celiac, and hypogastric plexuses.
Primary motor cortex The brain region located in the dorsal portion of the frontal lobe that is the main region of the motor system.
Primary neurulation The main part of the formation in the early embryo of the neural plate and neural folds, followed by its closure with development of the neural tube.
Primary somatosensory cortex Brodmann areas 1, 2, and 3 within the postcentral gyrus, involved in body position, movements, distribution of somatosensory information, touch perception, memories, and processing of textures, sizes, and shapes.
Primary T-cell lymphoma A rare form of cancerous lymphoma arising mainly from uncontrolled proliferation of T-cells.
Primary vesicles The three early subdivisions of the embryonic neural tube, including the forebrain, midbrain, and hindbrain.
Proband The patient or family member that brings a family under genetic study.
Prolactinomas Pituitary adenomas made up of lactotrophs that secrete excessive amounts of prolactin.
Propagation Reproduction.
Proprioceptors Sensory nerve endings that give information about body movements and position.
Propriospinal tract Exclusively concerned with the spinal cord.
Proteasomal degradation Conversion of chemical compounds to less complex forms, in this case, related to the cytoplasmic organelles known as proteasomes.
Proteus syndrome A sporadic disorder characterized by gigantism of the hands and feet, abnormal growth, pigmented nevi, thickening of the palms and soles, vascular malformations, and subcutaneous lipomas.
Proton therapy A type of radiotherapy that uses a beam of protons to irradiate cancerous or otherwise diseased tissue.
Psammomatous meningiomas Benign tumors that usually present as heavily calcified intracranial or spinal mass lesions.
Pseudarthrotic Referring to pseudarthrosis, a pathologic entity with deossification of a weight-bearing long bone, followed by bending and fracture.
Pseudo-onion bulbs Circumferential Schwann cell processes often surrounding axons or axon sprouts.
Pseudoangiomatous Referring to having a somewhat angioma-like appearance.
Pseudocapsule A structure, similar to a capsule, that surrounds some carcinomas.
Pseudodiploid Having two RNA genomes per virion, but giving rise to only one DNA copy in infected cells.
Pseudogliomatous Referring to having a somewhat glioma-like appearance.
Pseudohypoxia A phenomenon in which cells and tissues have a hypoxic phenotype even in the presence of oxygen.
Pseudoinclusions Having an apparent inclusion, which in cytology means a nuclear or cytoplasmic aggregate of stainable substances; in histology, it means an object completely inside a tissue.
Pseudopapillary Having the outward appearance of a papilla.
Pseudopsammoma Having a somewhat psammoma-like appearance.
Pseudorosettes Structures resembling rosettes, which are tumor cell formations.
Pseudotumor cerebri A disease with increased intracranial pressure of idiopathic etiology, usually affecting middle-aged women.
Pseudotumoral Relating to a pseudotumor, which is an enlarged portion of tissue that resembles a tumor.
Psittacosis An infection with *Chlamydia* bacteria, transmitted by birds, with fever, headaches, and pneumonia.
Purkinje cells GABAergic neurons within the cerebellar cortex.
Putamen A round structure at the base of forebrain that regulates learning and movement.
Pyramidal cells Triangular cells in the cerebral cortex, hippocampus, and amygdala, involved in motor control and cognition.
Radicular pain Also called radiculitis; pain radiated along the sensory distribution of a nerve due to inflammation or irritation of the nerve root.

Radiculopathy Also called a pinched nerve; conditions in which one or more nerves do not function properly (neuropathy), with radicular pain, weakness, numbness, or difficulty controlling specific muscles.
Radiosensitizers Agents that make tumor cells more sensitive to radiation.
Rathke's pouch An evagination at the roof of the developing mouth that gives rise to the anterior pituitary gland.
Refractory period Recovery time of an excitable membrane to be ready for a second stimulus.
Relative refractory period A form of recovery time that corresponds to hyperpolarization.
Remak bundle A group C nerve fiber of the CNS and PNS that respond to pain, heat, chemicals, mechanical stimuli, and physiological changes.
Resting membrane potential Also called resting voltage; the relatively static membrane potential of resting cells.
Reticular activating system A network of brainstem neurons that projects to the hypothalamus to mediate behavior, and to the thalamus and cortex to activate awake, desynchronized EEG patterns.
Reticular formation Interconnected nuclei in the brainstem involved in behavioral arousal and consciousness.
Reticular pattern A diffuse or patchy appearance.
Reticulin A type of fiber in connective tissue composed of type III collagen, secreted by reticular cells.
Retiform Composed of crossing lines, resembling a net.
Retrograde flow The flow of a body fluid in a direction other than normal.
Rhabdoid meningiomas Rare and aggressive tumors, classified as grade III, which have increased proliferative activity.
Rhabdomyoblastic Referring to a rhabdomyoblast, a cell type that is essential to the diagnosis of a rhabdomyosarcoma.
Rhabdomyoma A benign tumor of striated muscle, common in children.
Rhabdomyosarcoma An aggressive, highly malignant tumor that develops from skeletal muscle cells that have not fully differentiated.
Rhombencephalon The developing hindbrain (medulla, pons, and cerebellum).
Rhombic lips Posterior sections of the developing metencephalon.
Rosai-Dorfman disease Sinus histiocytosis with massive lymphadenopathy.
Rosenthal fibers Thick, elongated, worm-like bundles found on staining of brain tissue, accompanying gliosis, occasional tumors, and some metabolic disorders.
Rosettes Rose-shaped tumor formations.
Rostral Situated toward the oral or nasal region.
Rothmund-Thomson syndrome A rare autosomal recessive skin condition that may be related to osteosarcoma; with poikiloderma, alopecia, and defects of the knees, thumbs, and bones.
Rubinstein-Taybi syndrome A condition characterized by short stature, learning difficulties, distinctive facial features, and broad thumbs and first toes.
Salivatory nuclei The nuclei that innervate the salivary glands.
Saltatory propagation The conduction of action potentials along myelinated axons from one node of Ranvier to the next, increasing conduction velocity.
Satellite cells Muscle stem cells with very little cytoplasm, in mature muscle.
Scatter factor Hepatocyte growth factor secreted by mesenchymal cells.
Schiller-Duval bodies Cellular structures in yolk sac tumors.
Schwann cells Neurolemmocytes, the principle glial cells of the PNS.
Schwannoma A usually benign nerve-sheath tumor made up of Schwann cells.
SCL protein A streptococcal collagen-like protein involved in resistance to phagocytosis, adherence to plasma and extracellular matrix proteins, and degradation of host proteins.
Secondary neurulation The process in which the neural ectoderm and some endoderm cells form the medullary cord.
Secondary vesicles The telencephalon, diencephalon, metencephalon, myelencephalon, and neuroepithelium (ventricular zone).
Secretory meningiomas Uncommon variants of meningiomas that cause significant peritumoral edema.
Sella Also called the sella turcica; a depression in the body of the sphenoid bone where the pituitary gland or hypophysis is located.
Seminomas Germ cell tumors of the testicles, or more rarely of the mediastinum.
Sense organs The eyes, ears, skin, nose, and mouth.
Serotonin A monoamine neurotransmitter involved in regulating mood, cognition, reward, learning, memory, and many physiological processes.
Serpiginous Referring to a chronic condition that is slowly progressive.
Sessile Lacking a stalk; or, unable to move.
Shagreen patches Connective tissue nevi that appear early in life.
Sjögren syndrome A chronic autoimmune disease of the lacrimal and salivary glands, and often, of the lungs, kidneys, and nervous system.
Small-cell lung cancers Highly malignant lung tumors that have faster growth and earlier metastases than non-small-cell lung cancers.
Somatic mosaicism A state in which the somatic cells have more than one genotype.
Somatic motor neurons Those that originate in the CNS and project their axons to skeletal muscles; they are involved in locomotion.
Somatic reflexes Those related to the muscles of the body.
Somatic tissue The tissue of the body cells and not the reproductive cells.

Somatosensory association cortex Brodmann area 5, involved in movement and association, and part of the posterior parietal cortex.
Somatotrophic adenomas Those that secrete growth hormone.
Southern blot A method used in molecular biology to detect a specific DNA sequence; it transfers DNA to a filter membrane to be detected by probe hybridization.
Spastic paralysis Spasticity; reduced skeletal muscle performance combined with paralysis, increased tendon reflex activity, and hypertonia.
Spatial summation Eliciting an action potential in a neuron with input from multiple presynaptic cells.
Special senses Vision, hearing, balance, smell, and taste.
Spectroscopy The study of interactions between matter and electromagnetic radiation, as a function of wavelengths or frequency.
Spin-lattice effect In nuclear magnetic resonance imaging, a mechanism by which the component of total nuclear magnetic moment vector, parallel to the constant magnetic field, relaxes from a higher energy, nonequilibrium state to thermodynamic equilibrium with its surroundings (the "lattice").
Spinal cord A long, thin, tubular structure of nervous tissue, extending from the medulla oblongata to the lumbar region of the vertebral column.
Spinal reflexes Those involving the spinal cord, including the stretch, jaw jerk, biceps, brachioradialis, extensor digitorum, triceps, knee-jerk, and ankle jerk reflexes.
Spindle cell oncocytoma A rare variant of an oncocytoma that is made up of spindle cells; it usually develops in the anterior pituitary gland.
Splanchnic nerves Paired visceral nerves carrying fibers of the autonomic nervous system and sensory fibers from the organs.
Splenium One of the nerve tracts of the corpus callosum.
Splice-site mutations Genetic mutations involving insertion, deletion, or changes in numbers of nucleotides where splicing occurs during the processing of precursor messenger RNA into mature messenger RNA.
Staghorn vessels Those that have a sharply branched, jagged appearance.
Stem cell factor A cytokine that may exist as a transmembrane protein and soluble protein; it is important in hematopoiesis, spermatogenesis, and melanogenesis.
Stem cell rescue During chemotherapy or radiation therapy, the removal of stem cells from the blood before treatment, with reinfusion occurring after treatment.
Stereotactic biopsy A procedure that uses a computer and imaging in at least two planes to localize a target lesion in three-dimensional space, and guide the removal of tissue for examination.
Stereotactic radiosurgery A minimally invasive surgery using a three-dimensional coordinate system to locate small target lesions.
Stereotaxis A technique used to localize breast lesions for biopsy.
Stevens-Johnson syndrome A type of severe skin reaction that includes fever, flu-like symptoms, blistering, and peeling; it can become life threatening.
Straight sinus The tentorial sinus below the brain that receives blood from the superior cerebellar veins and inferior sagittal sinus.
Strap cells Elongated rhabdomyoblasts that are also called tadpole cells.
Stretch reflex The myotatic reflex, a muscle contraction in response to stretching within a muscle.
Stromal cells Connective tissue cells of any organ, such as fibroblasts.
Stromal desmoplasia Growth of fibrous or connective tissue that contains stromal cells.
Stromas Connective, functionally supportive networks of cells, tissues, or organs.
Structures of Scherer Secondary patterns of glioma cell infiltration in a glioma.
Subarachnoid space The space that exists between the arachnoid and pia mater, is filled with CSF, and continues down the spinal cord.
Subcommissural organ A small glandular structure in the third brain ventricle, involved in reabsorption and circulation of CSF, and with functions related to electrolyte and water balance.
Subdural space A cavity that can be opened by separation of the arachnoid mater from the dura mater, due to trauma, pathology, or lack of CSF.
Subependymoma A rare ependymal tumor, usually in the fourth ventricle, that is in most cases benign.
Subicular complex A set of structures in the subiculum of the hippocampal formation; they include the presubiculum, postsubiculum, and parasubiculum.
Subiculum The most inferior part of the hippocampal formation.
Subpial spread The spread of a malignant glioma beneath the pia mater.
Substantia nigra A basal nuclei structure of the midbrain, important for reward and movement.
Sulcus A space, groove, crevice, or furrow.
Sulcus limitans A groove in the fourth brain ventricle that separates the cranial nerve motor nuclei from the sensory nuclei.
Summation Achieving action potential in a neuron.
Superior colliculi Structures on the roof of the midbrain involved in eye movements and distractibility.
Superior mesenteric ganglia The ganglia located in the upper part of the superior mesenteric plexus where pre- and postsynaptic nerves of the sympathetic nervous system synapse.
Superior sagittal sinus The superior longitudinal sinus along the margin of the falx cerebri that allows blood to drain from the anterior cerebral hemispheres to the confluence of sinuses.
Synapse A structure that permits a neuron to pass an electrical or chemical signal to another neuron, or to the target effector cell.

Synaptic cleft A small space between neurons into which neurotransmitter molecules are released.
Synaptic delay A very short time required for chemical transmission to occur.
Synaptic fatigue Short-term synaptic depression, in which neurons cannot fire or transmit input signals.
Synaptic junctions Membranes of the presynaptic neuron and postsynaptic receptor cell, along with the synaptic cleft.
Synaptophysin Major synaptic vesicle protein 38 present in nearly all CNS neurons that participate in synaptic transmission.
Synchronous tumors Two or more malignancies identified simultaneously or within 6 months of initial diagnosis.
Syncytiotrophoblastic giant cells Extremely large cells within a syncytiotrophoblast, the epithelial covering of the embryonic placental villi.
Syringomyelic syndrome Also called syringomyelia, in which a cyst or cavity forms in the spinal cord, causing its eventual destruction.
Syrinx A fluid-filled cavity in the nervous system.
Tactile (Merkel) disks Mechanoreceptors in the basal epidermis and hair follicles that can respond to light touch.
Tactile corpuscles Meissner's corpuscles; mechanoreceptors in the skin that respond to light touch.
Tanycytic ependymoma An uncommon fibrillary variant of ependymoma, with hair-like cells that have ependymal-like nuclei.
Targeted therapy A treatment that blocks the growth of cancer cells by interfering with specific targeted molecules they need for carcinogenesis and tumor growth.
Tectal plate The junction of gray and white matter in an embryo.
Tecum Also spelled "tectum"; a structure that serves as a "roof," such as the dorsal part of the midbrain.
Telencephalon The cerebrum.
Telodendria The many branches that divide off of an axon.
Telomerase reverse transcriptase A catalytic subunit of the enzyme called telomerase; it maintains telomere ends.
Temporal summation A condition occurring when a high frequency of action potentials in the presynaptic neuron elicits postsynaptic potentials that summate with each other.
Tendon organs Also called Golgi tendon organs; proprioceptive sensory receptor organs that sense changes in muscle tension.
Tentorium cerebelli An extension of the dura mater that separates the cerebellum from the inferior portion of the occipital lobes.
Teratoid-rhabdoid tumors Atypical, rare tumors usually diagnosed in childhood, mostly within the brain, and especially in the cerebellum.
Teratoma A tumor made up of different tissues, such as hair, muscle, teeth, or bone.
Tethered spinal cord syndrome A group of neurological disorders related to spinal malformations, including tight filum terminale, lipomeningomyelocele, split cord malformations, dermal sinus tracts, and dermoids.
Thalamus A large mass of gray matter in the dorsal part of the diencephalon; it generally acts as a relay station between subcortical areas and the cerebral cortex.
Thermoreceptors Sensory receptors that detect temperature changes.
Threshold The level that must be reached for an effect to be produced, or the value at which a stimulus produces a sensation.
Thyrotrophic adenomas Rare pituitary tumors that manufacture thyroid-stimulating hormone, leading to hyperthyroidism.
Tight junctions Multiprotein junctional complexes that prevent leakage of transported solutes and water, or can serve as leaky pathways.
Toxoplasmosis A parasitic disease that may cause a flu-like illness.
Trabecular bone A porous bone composed of trabeculated bone tissue.
Transcapsular Across or through a capsule.
Transcobalamin C A carrier protein that binds cobalamin (B_{12}).
Transcriptome The set of all RNA transcripts in a person or a cell population.
Transitional meningiomas Benign tumors with transitional cell characteristics.
Transsphenoidal Through the nose and sphenoid bone.
Transverse fissure A horizontal fissure within the cerebellum.
Transversions Point mutations in DNA, either spontaneous or caused by ionizing radiation or alkylating agents.
Trichilemmomas Benign cutaneous neoplasms, with differentiation toward cells of the outer root sheath.
Triple-negative Referring to an uncommon type of breast cancer, in which cancer cells test negative for three common markers: the estrogen receptor, the progesterone receptor, and the HER2 cell-growth protein.
Truncating mutations Point mutations in genes that generate one of the three known stop codons.
Trunk ganglia Those that lie within one of the trunks of the body, such as in the sympathetic trunk.
Tuberectomy A surgical procedure for the treatment of tuberous sclerosis.
Tuberin Tuberous sclerosis complex 2, a protein; mutations of the gene that encodes this protein lead to tuberous sclerosis.
Tuberous sclerosis A rare multisystem autosomal dominant genetic disease, with noncancerous tumors growing in the brain and other vital organs.
Tubonodular Having a tube-like and nodular appearance.
Tumefactive Swollen.
Tumor protein p53 The protein encoded by the p53 gene, which is mutated in many cancers.
Tumorlets Small benign growths, usually made of smooth muscle cells.
Turcot syndrome A rare form of multiple intestinal polyposis associated with brain tumors.
Ultrastructural Referring to the architecture of cells and biomaterials that is visible at higher magnifications than found on a standard optical light microscope.

Upper motor neurons The cerebral cortex and brainstem neurons that carry information down to activate interneurons and lower motor neurons.
Uveal tract Also called the uvea; the pigmented middle of the three concentric layers of the eye.
Vacuolation The development or formation of vacuoles that are membrane-bound organelles inside cells.
Vascular cooption An alternative mechanism to interact with blood vessels, used by normal cells as well as cancer cells.
Vasocentric Also spelled "vasocentric"; having a vessel at the center.
Vasogenic edema Extracellular brain edema, caused by an increase in the permeability of the blood−brain barrier.
Vasopressin Antidiuretic hormone, synthesized in the hypothalamus and released in response to extracellular fluid hypertonicity.
Vasostimulatory Stimulating vasodilation, the widening of the blood vessels.
VCB complex The VHL-ElonginC-ElonginB complex; it is involved in VHL tumor suppressor functions.
Venous dural sinuses Also called the cerebral sinuses; channels between the endosteal and meningeal layers of the dura matter; they receive blood and CSF.
Ventral median fissure Also called the anterior median fissure; it contains a fold of pia mater and extends along the entire length of the medulla oblongata.
Ventral posteromedial nucleus A nucleus of the thalamus that communicates with the face, oral cavity, postcentral gyrus, and cortical gustatory area.
Ventriculoperitoneal shunt A device placed in the cerebrum to drain CSF into the peritoneal cavity.
Vermis Also called the cerebellar vermis, in the posterior fossa of the cranium; it is associated with body posture and locomotion.
Verocay bodies Components of Antoni A, the dense areas of schwannomas located between palisading spindle cells in neoplasms.
Vimentin A structural protein expressed in mesenchymal cells.
Virchow-Robin spaces Perivascular spaces around blood vessels in the brain and other organs, involved in immunological functions and to disperse neural and blood-derived messengers.
Viscus An internal organ.
Visual association area The area of the brain that contains the visual area V2, also known as the secondary visual cortex or prestriate cortex.
Visual cortex The area of the cerebral cortex that processes visual information.
Von Hippel-Lindau disease A rare genetic disorder, with visceral cysts and benign tumors that can become malignant.
Von Recklinghausen disease Neurofibromatosis type 1.
Waardenburg-Shah syndrome The fourth type of Waardenburg syndrome, signified by congenital hearing loss, pigmentation deficiencies, a white patch in the hair, patches of light skin, and Hirschsprung's disease.
Waldenström's macroglobulinemia A cancer that affects two types of B-cells: the lymphoplasmacytoid and plasma cells.
Wallerian degeneration An active process that results when a nerve fiber is cut or crushed, and the part of the axon distal to the injury degenerates.
Weibel-Palade bodies Storage granules of endothelial cells; they release von Willebrand factor and P-selectin, and are involved in hemostasis and inflammation.
Wernicke's area A part of the cerebral cortex involved in the comprehension of written and spoken language.
White communicating rami The preganglionic sympathetic outflow nerve tract from the spinal cord.
White epidermoid A rare type of epidermoid cyst that does not have the typical near-CSF density on CT scan or intensity on MRI.
White matter Areas of the CNS mainly made up of myelinated axons, also called tracts; involved in learning, modulating distribution of action potentials, and in relaying and coordinating communication between different brain regions.
Wild-type allele An allele considered to be normal for an individual.
Wild-type VHL protein A normally functioning von Hippel-Lindau protein.
Wiskott-Aldrich syndrome A rare X-linked recessive disease, with eczema, thrombocytopenia, immune deficiency, and bloody diarrhea.
Withdrawal reflex A spinal reflex that quickly coordinates contractions of all flexor muscles and relaxations of extensors in a limb, causing sudden withdrawal from a potentially damaging stimulus.
Wood's lamp A black light lamp that emits long-wave ultraviolet light and very little visible light.
Xanthogranulomas Yellowish lesions, as a form of histiocytosis.
Xenoestrogenic Referring to xenoestrogens, which are xenohormones that imitate estrogen.
Years of life lived with a disability The number of years that a person lives with some type of disabling disease; it is a component of the disability-adjusted life year (DALY).
Years of life lost A calculation of the age at which deaths occur by giving more weight to deaths at a younger age and less weight to deaths at an older age.
Years of potential life lost An estimate of the average years a person would have lived if they had not died prematurely.
Yolk sac tumor Also called an endodermal sinus tumor, which is a type of germ cell tumor; the most common testicular tumor in children under age 3.
Zonula adherens Also called adherens junctions; protein complexes at cell−cell junctions in epithelial and endothelial tissues, usually more basal than tight junctions.
Zonula occludens Also called tight junctions, which mainly prevent leakage of transported solutes and water, and seal the paracellular pathway.
Zonulae adherentes See zonula adherens.
Zonulae occludentes See zonula occludens.

Index

Note: Page numbers followed by "*f*" and "*t*" refer to figures and tables, respectively.

A

Abducens nerve, 12
Absolute refractory period, 53
Accessory nerve, 73
Acetylcholine (ACh), 51, 58
Acetylcholinesterase (AChE), 58
Acoustic neuroma. *See* Vestibular schwannoma
Acquired immunodeficiency syndrome (AIDS), 90, 253
Acquired reflex. *See* Learned reflex
Acrochordon, 199
Acromegaly, 287
Action potentials, 43, 46–47, 52. *See also* Graded potentials
   axon diameter, 55–56
   comparison of graded potentials and, 54*t*
   generation, 53–54
   propagation, 54–55
      speed, 55–56
   threshold and all-or-none principle, 55
Activating protein kinase B (AKT), 399–400, 419
Active ion channels, 51
Activin A receptor type 1 (ACVR1), 484
Acute lymphoblastic leukemia (ALL), 89–90
Acute myelogenous leukemia (AML), 89–90
Adamantinomatous form, 302
Addison's disease, 299–300
Adenocarcinomas, 87–88
Adenohypophyseal placode, 70–71
Adenoid glioblastoma, 136
Adenoma sebaceum, 408
Adenomatous polyposis coli (APC), 382
3′Adenomatous polyposis coli mutation (APC mutation), 269
Adenosine diphosphate (ADP), 54
Adenosine triphosphate (ATP), 45
Adrenal chromaffin cells, 36
Adrenal medulla, 34–36
Adrenocorticotropic hormone (ACTH), 286
   adrenocorticotropic hormone–secreting tumors, 299–302
   clinical manifestations, 300
   diagnosis, 300–301
   epidemiology, 299
   etiology and risk factors, 299–300
   pathology, 300
   prognosis, 302
   treatment, 301–302
Afterbrain, 68

Age-specific death rates, 85
Aging of population, 83–85
Aicardi syndrome, 234
Alar plate, 66
Alcian blue, 169–170
All-or-none principle, 55
Alpha thalassemia retardation X-linked (ATRX) mutation, 485
Alpha-crystallin B chain, 316
Alpha-fetoprotein (AFP), 77–78, 422
Alpha-internexin, 411
Alpha-thalassemia/mental retardation syndrome, X-linked (ATRX) loss, 141
Alternating electric field therapy, 138
Alveolar rhabdomyosarcomas, 265
Amelanotic melanocytomas, 340
Amygdala, 9, 14
Amygdaloid body, 9, 22
Amygdaloid nuclear complex, 14
Amyotrophic lateral sclerosis (ALS), 62
Anaplastic astrocytoma, 112, 121–123
   clinical manifestations, 122
   diagnosis, 122
   epidemiology, 121
   etiology and risk factors, 121–122
   pathology, 122
   prognosis, 123
   treatment, 123
Anaplastic ependymoma, 172–173
Anaplastic large cell lymphoma, 360–361
Anaplastic lymphoma kinase (ALK), 270, 360
Anaplastic meningioma, 225–227
Anaplastic oligodendroglioma (AO), 135, 147, 153–156
   clinical manifestations, 153
   diagnosis, 154–155
   epidemiology, 153
   etiology and risk factors, 153
   pathology, 153
   prognosis, 155–156
   treatment, 155
Anaplastic pleomorphic xanthoastrocytoma, 491–493
   clinical manifestations, 492
   diagnosis, 493
   epidemiology, 491
   etiology and risk factors, 491
   pathology, 491–492
   prognosis, 493
   treatment, 493

Anaplastic rhabdomyosarcomas, 265
Anaxonic neurons, 47
Ancient neurofibroma, 198
Ancient schwannoma, 183
Anemia, 222
Aneurysmal bone cyst, 91
Angioblastic meningiomas, 221–222
Angiocentric arrangement, 113
Angiofibromas, 286
Angiolipoma, 268–269
Angiomatous meningioma, 221–222
Angiomyolipomas, 408
Angiopoietin/Tie2 receptor, 137
Angiosarcoma, 253–255
   clinical manifestations, 254
   diagnosis, 254
   epidemiology, 253
   etiology and risk factors, 253
   pathology, 253
   prognosis, 255
   treatment, 254
Angiosarcomatous differentiation, 442
Angiotropic large-cell lymphoma, 353
Annexin A1, 316
Annulate lamellae, 324–325
Anterior cerebral artery, 13
Anterior commissure, 8
Anterior external arcuate fibers, 73
Anterior horns, 26, 66
Anterior median fissure, 26
Anterior roots, 26
Anterograde flow or transport, 47
Anterolateral sulcus, 26
Anticytotoxic T-lymphocyte-associated protein 4 (CTLA-4), 460
Antifibrotic agent, 199
Antiprogrammed cell death protein 1 (PD-1), 460
Antoni A pattern, 181–183
Antoni B pattern, 183
Aperta neural tube defect, 78
Apoptosis, 137
Apoptotic AKT, 399–400
Apparent diffusion coefficient (ADC), 360
Aptamers, 62
Aquaporin-1, 244
Arachnoid mater, 23
Archicortex, 6
Arcuate fibers, 8
Arginine 249 (Arg249), 369

519

Argyrophilic fiber network, 350
Aryl hydrocarbon receptor-interacting protein (AIP), 286
Ascending pathways, 27
Ascending tracts, 26
Association areas, 7–8
Association fibers, 8
Association neurons, 49
Astrocytes, 41–42, 112
Astrocytoma, 111–123. *See also* Diffuse astrocytoma
　anaplastic, 121–123
　clinical cases, 123–126
　diffuse, 119
　infiltrative low-grade, 116–117
　pilocytic, 111–114
　pleomorphic xanthoastrocytoma, 117–121
　SEGA, 111, 115–121
Ataxia-telangiectasia, 356–357, 432
Ataxia-telangiectasia mutated (ATM), 484
ATP transcriptional regulator, x-linked (ATRX), 132–133
Attention-deficit/hyperactivity disorder (ADHD), 408
Attenuated familial adenomatous polyposis, 381
Attributable deaths, 91
Atypia, 233–234, 249
Atypical choroid plexus papilloma, 236
Atypical meningioma, 223–224
Atypical neurofibroma, 200–202
Atypical teratoid-rhabdoid tumor (AT-RT), 425, 479–483
　clinical manifestations, 482
　diagnosis, 482
　epidemiology, 479–480
　etiology and risk factors, 480
　pathology, 480–482
　prognosis, 483
　treatment, 482–483
Auditory association area, 7, 20–21
Auditory cortex, 7
Aurora kinase A (AURKA), 481–482
Autonomic nervous system (ANS), 3, 33–37, 35f, 48. *See also* Peripheral nervous system (PNS)
　parasympathetic division, 37
　　parasympathetic neurons and pathways, 38t
　sympathetic division, 33–37
　　adrenal medulla, 34–36
　　postganglionic sympathetic fibers, 37
　　preganglionic sympathetic fibers, 34–36
　　sympathetic neurons and pathways, 38t
Autonomic reflexes, 29
Autosomal dominant phacomatosis disorders, 407
Autosomal recessive familial adenomatous polyposis, 381
Auxilin, 399
Axoaxonic synapses, 56–57
Axodendritic synapse, 56–57
Axolemma, 46–47
Axon(s), 43
　diameter, 55–56
　hillock, 45
　reaction, 44
　terminals, 47
Axonal transport, 47
Axoplasm, 46–47
Axosomatic synapses, 56–57

B

B-cell linker (BLNK), 351
B-cell lymphoma 2 (Bcl2), 272, 313. *See also* Diffuse large B-cell lymphoma (DLBCL)
B-cell lymphoma 6 protein (BCL6 protein), 350–351
B-Raf, 112
Balloon cells, 411
Band of Bungner, 61–62
Bannayan-Riley-Revalcaba syndrome, 402
Basal ganglia (BG), 8–9, 350
　areas, 15
Basal nuclei, 8–9, 8f
Basal plate, 66
Basal vein, 13
Base pair, 369
Bases, 369
Basilar artery, 13
Batson's plexus, 470
BCL-6 corepressor protein (BCOR), 256–257
BCL6 corepressor-like 1 (BCORL1), 419
Beckwith–Wiedemann syndrome, 265
Benign fibrous histiocytoma, 272–273
Benign schwannoma, 179. *See also* Cellular schwannoma
Benign stromal spindle cell tumors, 271
Benign triton tumors. *See* Neuromuscular choristomas
Beta-catenin, 303–304
Betacatenin, 137
34betaE12, 170
Bevacizumab, 210
Biallelic inactivation, 244–245
Bimodal distribution, 367–368
Biochemical isolation, 23
Biphenotypic sinonasal sarcoma, 258
Bipolar neurons, 47
Bisphenol-A, 295
Bitemporal hemianopsia, 286
Bitemporal superior quadrantanopia, 286
Bizarre neurofibroma. *See* Atypical neurofibroma
BK virus, 349
Bland cells, 342
Blepharoblasts, 163
Blood pressure (BP), 10
Blood supply, 13
　deep arteries of head and neck, 15f
　deep veins of head and neck, 17f
　superficial arteries of head and neck, 14f
　superficial veins of head and neck, 16f
Blood–brain barrier (BBB), 23, 41, 43, 89
Blooming, 268
Bodian silver, 198

Bone morphogenetic protein receptor type 1A (BM-PR1A), 400
Brachytherapy, 254
Brachyury, 244
BRAF mutations, 460
Brain, 3–13, 4f
　blood supply, 13
　cerebral hemispheres, 5–9, 5f
　cerebrospinal fluid, 23–24, 26
　diencephalon, 9–11
　functional brain systems, 13–19
　higher mental functions, 19–23
　language processing, 20–21
　memory, 21–22
　meninges, 23–24
　metastases, 89
　reticular formation, 19, 20f
　sleep and sleep-wake cycles, 22–23
　stem, 5, 12–13
　tumors, 233
　　choriocarcinoma, 429–430
　　clinical cases, 433–435
　　embryonal carcinoma, 428–429
　　germ cell tumors, 417–433
　　germinoma, 421–423
　　medulloblastoma, 432–433
　　mixed germ cell tumors, 431
　　teratoma, 423–426
　　yolk sac tumor, 426–428
Brain and spinal tumors, global epidemiology of
　aging of population, 83–85
　attributable deaths, 91
　burden of brain and spinal tumors in United States, 95–96
　comparison of brain and spinal tumors with other cancers, 97–99
　DALYs, 93–94
　distribution by age, gender, and race, 87–91
　global mortality, 94
　global prevalence, 92–93
　life expectancy, 85–87
　mortality from brain and spinal tumors in United States, 96
　QALY, 94
　world population, 83
Brain tumor-polyposis syndrome 1 (BTP1), 379–381
Brain tumor-polyposis syndrome 2 (BTP2), 381–384
Breast cancer type 1 (BRCA1), 338, 432
Broca's area, 20–21
Bromocriptine, 296
Brown-Séquard syndrome, 467
Bulbous corpuscles, 30–31
Burden of disease, 103–104
Burkitt lymphoma, 89–90
Butterfly glioma, 133

C

C-terminus of PMS2, 380
C-X-C chemokine receptor type 4 (CXCR4), 244

Cabergoline, 289, 294, 296
Caldesmon, 272
Calmodulin-binding transcription activator 1 (CAMTA1), 252
Calretinin, 187
Canalicular structures, 423–424
Cancer, 367
Candidate population, 104
Canonical Wnt pathway, 382
Carbonic anhydrase isozymes, 244
Carcinoembryonic antigen (CEA), 425
Cardiac nerves, 37
Cardiac plexus, 37
Carney complex, 286
Case fatality ratio, 94
Casitas B-lineage lymphoma (CBL), 351, 419
Caspase recruitment domain-containing protein 11 (CARD11), 351
Caspase-8 (CASP8), 292
Castleman's disease, 222
Catenin beta-1 (CTNNB1), 269
Cathepsins, 133–134
Cauda equina syndrome, 33, 34f, 468
Caudate nucleus, 9
Cavernous hemangiomas, 107, 249
Celiac ganglia, 34–36
Cell body, 45–46, 46t
Cell division control protein 42 homolog (CDC42), 133–134
Cellular schwannoma, 184–185
　clinical manifestations, 184
　diagnosis, 184
　epidemiology, 184
　etiology and risk factors, 184
　pathology, 184
　prognosis, 185
　treatment, 184–185
Centigray (cGy), 422–423
Central Brain Tumor Registry of the United States (CBTRUS), 95–96, 148, 246, 488
Central canal, 26
Central nervous system (CNS), 3, 41, 65, 87, 233–234, 243, 337, 367–368, 389, 417, 455, 479
　DLBCL, 349–353
　glial cells of, 41–43, 42t
　malformation, 77–78, 77t
Central sulcus, 5–6
Centrioles, 45–46
Centroblasts, 350
Centrocytes, 355
Cerebellar hemispheres, 12
Cerebellar pilocytic astrocytomas, 113
Cerebellar tonsils, 166–167
Cerebellum, 12–13
Cerebral aqueduct, 76
Cerebral arterial circle, 13
Cerebral cavernous malformations, 249
Cerebral cortex, 5–6
　sensory and motor areas of, 7f
　functions, 21t
Cerebral hemispheres, 5–9, 5f, 162
　association areas, 7–8

　basal nuclei, 8–9, 8f
　cerebral cortex, 6
　cerebral white matter, 8
　higher mental functions, 8
　integrative centers, 8
　lobes, 5–6, 6f
　motor areas, 7
　sensory areas, 7
Cerebral toxoplasmosis, 357
Cerebral white matter, 8
Cerebrospinal fluid (CSF), 23–24, 26, 43, 220, 233, 296
　production, 24f
Cervical enlargements, 25–26
Cervical spinal nerve 5 (C_5), 25–26
Chang staging system, 432–433
Checkpoint kinase 2 (CHEK2), 484
Chemical synapses, 56
　function, 57–58
Chemically gated ion channels, 51
Chemoattractants, 197–198
Chemoreceptors, 30
Childhood ependymoma, 173
Chloride channel capicua mutation (CIC mutation), 149
Cholesterotic variants, 272
Cholinergic synapses, 58
Chondroma, 276–278, 318
Chondromyxoid stroma, 252
Chondrosarcoma, 255–256
Chondrosarcomatous differentiation, 442
Chordoid meningioma, 222
Choriocarcinoma, 429–430
Choroid plexus, 23–25, 76
Choroid plexus carcinoma, 425
Choroid plexus papilloma, 233–238
　clinical cases, 239–241
　clinical manifestations, 234–235
　diagnosis, 235
　epidemiology, 233
　etiology and risk factors, 233
　pathology, 233–234
　prognosis, 235–236
　treatment, 235
Chromatolysis, 44
Chromodomain, 484
Chromogranin B (CHGB), 308
Chromosome, 407
Chronic disease, 84
Chronic lymphocytic leukemia (CLL), 89–90
Chronic myelogenous leukemia (CML), 89–90
Cilia, 43
Cingulate bundle, 18
Cingulate cortex, 15
Cingulate gyrus, 18
Circle of Willis, 13
Circular sulcus, 6
Class III beta-tubulin, 325, 411
Classic ependymoma, 161–165
　cellular appearance, 163f
　clinical manifestations, 164
　diagnosis, 164–165
　epidemiology, 161–162
　etiology and risk factors, 162

　MRI, 164f
　pathology, 162–164
　prognosis, 165
　treatment, 165
Classic oligodendroglioma, 149
Classical glioblastoma, 133
Claudin, 201–202
　claudin-1, 440–441
Clear cell ependymoma, 171
Clear cell meningioma, 225
Cleveland Clinic scoring system, 401
Clival invasion, 289
Clivus region, 318
Cluster of differentiation 19 (CD19), 350–351
Cluster of differentiation 30 (CD30), 422, 429
Cluster of differentiation 31 (CD31), 252
Cluster of differentiation 34 (CD34), 201–202, 489
Cluster of differentiation 44 (CD44), 308
Cochlear nuclei, 74
Codman's triangle, 263
Codons, 369
Coenzyme A (CoA), 58
Cognition index, 407–408
Collagen IV, 183
Collagenomas, 286
Collaterals, 46–47
Colorectal cancer, 379
Columns, 46
Commissural fibers, 8
Complementarity-determining region 3 (CDR3), 352
Compressive (sensorimotor) syndrome, 467
Computed tomography (CT), 89
Cone-rod homeobox (CRX), 326
Confidence interval, 103
Congenital infantile fibrosarcoma, 257
Congenital malformation, 77
Conglomeration, 197
Connective tissue tumors, 249
Connexin 26, 411
Connexin 32, 411
Connexons, 56
Consanguinity, 379
Consciousness, 19–20
Continuous propagation, 55
Conus medullaris, 25–26
Corpus callosum, 8
Corpus callosum lipoma, 266
Corpus striatum. See Striatum
Cortex, 3–5
Cortical dysplasias, 117–118
　Type IIb, 409–411
Corticospinal system, 27–28
Corticotrophic adenomas, 288
Corticotropin-releasing hormone (CRH), 299
Costello syndrome, 265
Cowden syndrome, 399
　clinical cases, 403–405
　clinical manifestations, 401
　diagnosis, 401–402, 402t
　epidemiology, 399
　etiology and risk factors, 399–400
　pathology, 400–401

Cowden syndrome (*Continued*)
  prognosis, 403
  treatment, 403
Cranial meninges, 23
Cranial nerves (CN), 31–33, 32f
  classifications, 32t
  cranial nerve IV, 12
  sensory and motor nuclei, 31–33
Cranial placodes, 70–71
Craniorachischisis, 77
Craniopharyngioma, 134, 302–304
  clinical manifestations, 302–303
  diagnosis, 303–304
  epidemiology, 302
  etiology and risk factors, 302
  pathology, 302
  prognosis, 304
  treatment, 304
Cresyl violet, 45
Cribriform neuroepithelial tumors, 480–481
Crista galli, 13
Crooke cell adenoma, 300
Cullin-2, 389–390
Cumulative incidence, 105–106
Curvilinear types, 267
Cushing's adenomas, 285
Cushing's disease, 287, 299–300
Cushing's syndrome, 287
Cutaneous neurofibroma, 199
Cyclin B3 (CCNB3), 256–257
Cyclin D1 (CCND1), 399–400, 481–482
Cyclin D1–3 (CCND1–3), 484
Cyclin D2 (CCND2), 419
Cyclin-dependent kinase (CDK), 419
  CDK4, 141
  CDK4/6, 484
  CDKN1B, 390–391
  CDKN2A, 133, 292, 351
  CDKN2A/B, 484
Cysteine 176 (Cys176), 369
Cytoarchitecture, 408
Cytokeratin (CK), 247
  CK5/6, 170
  CK7, 234, 303–304
  cytokeratin-8, 170
Cytokines, 112–113
Cytomegalovirus, 130
Cytomorphology, 439
Cytoplasm, 45
Cytosine-plus-guanine sites (CpG sites), 369
Cytotrophoblasts, 429

D
Dandy–Walker syndrome, 341
Death certificates, 94
Death-associated protein kinase 1 (DAPK1), 292, 351
Decussation of the pyramids, 73
Deep arteries, 13
Deep sleep, 22
Deep tendon reflexes, 29, 29t
Deep veins, 13

Dehydroepiandrosterone sulfate (DHEAS), 306–307
Deletion of q arm of chromosome 19 (19q deletion), 147
Deletion/insertion mutations, 369
Dendrites, 43
Dendritic spines, 47
Dendritic zone, 47
Densely granulated corticotroph adenomas, 300
Densely granulated somatotroph adenomas (DGSAs), 291
Dentate gyrus, 16–17
Denticulate ligaments, 25–26
Depolarization, 51
Dermatofibromas, 272
Descending pathways, 27–28
Descending tracts, 26
Desmin, 247
Desmoid-type fibromatosis, 269–270
Diabetes insipidus, 289
Diacylglycerol kinase zeta (DGKZ), 308
Dicer 1
  ribonuclease III (DICER1), 300
Diencephalon, 9–11, 67, 75
  epithalamus, 11
  hypothalamus, 10, 11t
  pineal gland, 11
  thalamus, 9f, 10
Diff-Quik stain, 239
Diffuse astrocytoma, 119, 490. *See also*
  Astrocytoma; Infiltrative low-grade astrocytomas
  clinical manifestations, 120
  diagnosis, 120
  epidemiology, 119
  etiology and risk factors, 119
  pathology, 119
  prognosis, 121
  treatment, 121
Diffuse cutaneous neurofibroma, 199
Diffuse intrinsic pontine gliomas (DIPGs), 483
Diffuse large B-cell lymphoma (DLBCL), 349–353
  clinical manifestations, 351
  diagnosis, 351–353
  epidemiology, 349
  etiology and risk factors, 349
  pathology, 350–351
  prognosis, 353
  treatment, 353
Diffuse midline gliomas, 483–487
  clinical manifestations, 485
  diagnosis, 486
  epidemiology, 483
  etiology and risk factors, 483–485
  pathology, 485
  prognosis, 486–487
  treatment, 486
Diffusion weighted imaging (DWI), 122, 360
DiGeorge syndrome, 488
Disability adjusted life years (DALYs), 93–94
Disease frequency measures, 106
Disomies, 351
Dominant hemisphere, 19

Dopamine, 45–46
Dorsal (posterior) funiculi, 27
Dorsal column nuclei, 74
Dorsal median septum, 71
Dorsal median sulcus, 71
Dorsal nucleus of vagus nerve, 74
Dorsolateral placodes, 71
Double homeobox 4 (DUX4), 256–257
Double staining, 440–441
Dura mater, 23
Dural-based frontal lobe hemangioblastoma, 393f
Dural folds, 23
Dural tail, 217
Dynein, 47
Dysembryoplastic neuroepithelial tumor, 149
Dysmetria, 245
Dysplasia epiphysealis hemimelica.
  See Trevor's disease
Dystrophin, 276

E
4E-binding protein 1, 408
EAAT1 glutamate transporter, 234
Ectoderm, 65
Ectomesenchymal components, 332
Ectomesenchymoma, 265–266
Ectopic pituitary adenoma, 289
Edinger–Westphal nucleus, 37
Effector, 29
Efferent fibers, 48
EGFR/ALK rearrangement, 460
Ehlers–Danlos syndrome (EDS), 22
Electrical synapses, 56
  function, 57
Electrochemical gradient, 52
Electrocorticography, 409–411
Elephantiasis neuromatosa, 199, 201–202
Elongin, 389–390
EMACO therapy, 430
Embryology
  central nervous system malformation, 77–78, 77t
  cranial placodes, 70–71
  diencephalon, 75
  medulla spinalis, 71
  mesencephalon, 75
  metencephalon, 75
  myelencephalon, 71–74
  neural crest, 69–70
  neural tube, 65–68
  telencephalon, 75–76
Embryonal carcinoma, 428–429
Embryonal rhabdomyosarcoma, 265
Embryonic stem cell related gene (ESRG), 429
En plaque meningiomas, 217–218
Encapsulated nerve endings, 30–31
Endoderm, 65
Endodermal sinus tumors, 426
Endonasal surgery, 289
Endothelial meningioma, 219
Enteric nervous system (ENS), 3
Enteroceptors, 30

Entorhinal cortex, 18
Eosinophilic granuloma, 91
Eosinophilic keratinous material. *See* Wet keratin
Ependymal cells, 43, 71
Ependymal tumors, 173. *See also* Germ cell tumors
  anaplastic ependymoma, 172–173
  classic ependymoma, 161–165
  clear cell ependymoma, 171
  clinical cases, 174–176
  ependymoblastoma, 173–174
  myxopapillary ependymoma, 168–171
  papillary ependymoma, 171
  RELA fusion-positive ependymoma, 171–172
  subependymoma, 165–168
  tanycytic ependymoma, 171
Ependymoblastoma, 173–174
Ependymomas, 134, 162–163
Ephrin (Eph), 137
Epibranchial placodes, 71
Epidermal growth factor receptor (EGFR), 132–133, 244, 299, 461, 489
Epidermal growth factor receptor-related-positive (ERBB2-positive), 456
Epidural lipomatosis, 267
Epidural space, 23
Epigenomes, 134, 481–482
Epipharngeal placodes. *See* Epibranchial placodes
Epiphysis cerebri. *See* Pineal gland
Epithalamus, 11, 67
Epithelial membrane antigen (EMA), 163, 200–201, 218, 234, 245–246, 313, 338, 361, 440–441, 480–481
Epithelioid cell glioblastoma, 136
Epithelioid hemangioendothelioma, 252–253
Epithelioid MPNSTs, 446–448
Epstein–Barr early ribonucleoprotein 1 (EBER1), 357
Epstein–Barr nuclear antigens 1 to 6 (EBNA1–6), 357
Epstein–Barr virus (EBV), 260, 349
Equilibrium potential, 52
Erb-b2 receptor tyrosine kinase 2 (ERBB2), 308
ERBB2 amplification, 460
Erdheim–Chester disease, 360
Erlotinib, 138
Erythroblast transformation-specific related gene (ERG), 252
Erythroleukemia, 424–425
Erythropoietin, 245
Erythropoietin gene (EPO gene), 390
Estrogen receptor gamma (ESRG), 422
Ethylnitrosourea, 148–149
ETS variant transcription factor 6 (ETV6), 489
Eukaryote 2 factor (E2F), 419
Eukaryotic translation initiation factor 1A, X-chromosomal (EIF1AX), 338
European Quality of Life with Five Domains (EQ-5D), 94
Everolimus, 412–413

EVI2A gene, 203
EVI2B gene, 203
Ewing sarcoma, 90, 249, 256–257
Ewing sarcoma breakpoint region EWS-1 (EWSR1), 257
Excitable cell membrane, 51
Excitatory postsynaptic potential (EPSP), 59
Exons, 369, 407
Exophthalmos, 295–296
Exostoses, 262
Exteroceptors, 48
Extracellular fluid (ECF), 43
Extracellular space, 43
Extradural tumors, 469–471
Extramedullary tumors, 95–96, 465
Extraneural manifestations, 209
Extranodal MALT lymphomas, 355
Extranodal marginal zone lymphoma of mucosa-associated lymphoid tissue of dura. *See* MALT lymphoma of dura
Extraosseous chondrosarcomas, 255
Extraskeletal myxoid chondrosarcomas, 255
Eye floaters, 351
Ezrin, 411–412

F

F-fluorodeoxyglucose (FDG), 201
Facial nerves, 12, 37
Fact memories, 21
Factor VIII-related antigen, 254
Falx cerebri, 5
Familial adenomatous polyposis (FAP), 269, 379, 381
Familial isolated pituitary adenoma (FIPA), 286
Fanconi anemia, 428
Far upstream binding protein 1 (FUBP1), 152
Fasciculi, 8, 46
Fasciculus cuneatus, 27
Fasciculus gracilis, 27
Fasciculus proprius, 27
Fast MRI, 419–420
Fibroblast growth factor receptors (FGFRs), 292
  FGFR1, 484
Fibrolipomatous stalk, 266
Fibrosarcoma, 257–258
Fibrous meningioma, 220
Fibrous xanthoma, 272
Fibroxanthoma, 272
Filum terminale, 25–26
Fissure, 5
Flaccid paralysis, 27–28
Flexner-Wintersteiner rosettes, 330–332
Flocculonodular lobes, 12–13
Flow voids, 140
Fluid-attenuated inversion recover magnetic resonance imaging (FLAIR MRI), 304
Fluid-attenuated inversion recovery technique (FLAIR technique), 199
Fluorescence in situ hybridization (FISH), 149, 257
Fluorodeoxyglucose-PET (FDG-PET), 311

Fluorodeoxyglucose-positron emission tomography, 409–411
Focal adhesion kinase (FAK), 133–134, 399–400
Focal cortical dysplasia, 117–118
Folia, 12
Follicle-stimulating hormone (FSH), 286
Folliculostellate cells, 314–315
Fontana–Masson stains, 338
Foramen cecum, 73
Foramen magnum, 25–26
Foramen of Monro, 115
Foramina, 31
Forebrain. *See* Diencephalon
Fornix, 18
Fourth ventricle, 67–68
Frameshift mutations, 207
Free nerve endings. *See* Nonencapsulated nerve endings
Friend leukemia integration transcription factor 1 (FLI1), 254
Frond-like structure, 233–234
Frontal lobe, 5–6
Frontonasal dysplasia, 69–70
Functional brain systems, 13–19
  hippocampal formation, 16–18, 18f
  limbic system, 14–18, 18f
  functions, 15–16
Funiculi, 27, 46

G

G-protein-coupled receptor 101 (GPR101), 292
Gadolinium, 149–150
Gadolinium-diethylene triamine pentaacetic acid (DTPA), 272
Galactorrhea, 296
Galectin-3, 316
Gallium 68 dotatate-PET (^{68}Ga-DOTATATE-PET), 311
Gamma knife radiosurgery, 289–291
Gamma-aminobutyric acid (GABA), 60
Ganglia, 46
Gangliocytic differentiation, 323
Gangliocytoma, 400
Ganglioid cells, 325
Ganglion, 48
Ganglioneuroma-like appearance, 401
Ganglioneuromatous lesions, 400–401
Gap junction
  beta-1 protein, 411
  beta-2 protein, 411
Gardner's syndrome, 260, 382–383
Gastrointestinal tract (GI tract), 399–400
Gated ion channels, 51
Gefitinib, 138
Gemistocytes, 115, 136
Gemistocytic astrocytes, 115
Gene, 407
  sequencing technique, 392–393
General senses, 30
General sensory receptors, 30
General transcription factor 2H subunit 1 (GTF2H1), 308

Genodermatosis, 399
Germ cell tumors, 92–93, 417–433
  clinical manifestations, 419
  diagnosis, 419–420
  epidemiology, 417–418
  etiology and risk factors, 418
  pathology, 418–419
  prognosis, 420–421
  treatment, 420
Germinoma, 421–423
  clinical manifestations, 422
  diagnosis, 422
  epidemiology, 421
  etiology and risk factors, 421
  pathology, 421–422
  prognosis, 423
  treatment, 422–423
Germline mutation, 367
Gestational trophoblastic disease, 429
GH adenomas. See Somatotroph adenoma
GH-cell adenomas. See Somatotroph adenoma
GH-producing adenomas. See Somatotroph adenoma
Giant cell
  glioblastoma, 136
  tumor of bone, 90
Giant congenital nevi, 340
Glandular MPNST, 444
Glial cells, 41–44, 66, 167
  of CNS, 41–43
  of PNS, 43–44
  types, 42f, 42t
Glial fibrillary acidic protein (GFAP), 42, 115, 150, 163, 313, 338, 411, 480–481
Glial hamartias, 209
Glial septum, 26
Glioblastoma, 107, 112, 129–141, 131f, 132f
  clinical cases, 141–143
  clinical manifestations, 134
  diagnosis, 134–138
  epidemiology, 129
  etiology and risk factors, 129–130
  pathology, 130–134
  prognosis, 139
  treatment, 138–139
Glioblastoma multiforme (GBM). See Glioblastoma
Gliofibrillary oligodendrocytes, 155
Glioma (cytosine-phosphate-guanine) island methylator phenotype type A (G-CIMP type A), 153
Glioma(s), 87, 111–112, 129, 318, 379
Gliomatosis cerebri, 130, 487–488
Gliomatous differentiation, 141
Glioneuronal tumors, 342
Gliosarcoma, 132, 141
Gliosis, 42
Globus pallidus, 9
Glomeruloid vessels, 131–132
Glossopharyngeal nerve, 37, 73
Glucagon-like peptide-1 (GLP-1), 304
Glucose transporter-1 (GLUT1), 201–202, 250, 440–441
Glycine-245-serine (Gly245Ser), 371

Glycoproteins, 45
Glypican-3, 428
GNAS complex locus (GNAS), 300
Goiter, 84
Golgi apparatus, 45
Gonadotrophic adenomas, 288
Gonadotrophic cell tumors, 305
Gorlin syndrome, 119
Graded potentials, 47, 49–52. See also Action potentials
  and action potentials, 54t
  passive processes, 52
Gradient echo sequence, 168
Granular cell(s), 136
  astrocytoma, 136
  glioblastoma, 136
  tumors, 318
Granzyme B, 358
Gray commissure, 26
Gray communicating rami, 37
Gray matter, 43, 45. See also White matter
Growth arrest and DNA-damage-inducible protein 45g (GAD-D45G), 292
Growth hormone-secreting adenomas (GH-secreting adenomas), 285
Growth hormone–regulating hormone (GHRH), 294
GTPase-activating protein (GAP), 197
Guanine nucleotide-binding protein subunit alpha (GNAQ), 337
Guanine–thymine mismatches (G-T mismatches), 380–381
Guanosine triphosphateases (GTPases), 133–134, 411–412
Gustatory cortex, 7
Gyriform rim of thick gray cortex, 134–135

H

H3 K27M-mutant, 483
HAART. See Highly active antiretroviral therapy (HAART)
Habenular nuclei, 11
Hair follicle receptors, 30
Hairy cell leukemia, 89–90
Hamartin, 115, 407–408
Hamartomas, 115
Hamartomatous intestinal polyps, 400–401
Haploinsufficiency, 209, 407–408
Headache, 286
Healthy life years lost per 1000 population per year (HeaLY), 94
Hemangioblast progenitor cells, 244–245
Hemangioblastoma, 134, 243–246, 389
  clinical manifestations, 245
  diagnosis, 245–246
  epidemiology, 243
  etiology and risk factors, 243–244
  pathology, 244–245
  prognosis, 246
  treatment, 246
Hemangioma, 90, 249–251
  clinical manifestations, 251
  diagnosis, 251

  epidemiology, 249
  etiology and risk factors, 249–250
  pathology, 250
  prognosis, 251
  treatment, 251
Hemangiopericytoma, 220, 246–249
  clinical manifestations, 248
  diagnosis, 248–249
  epidemiology, 246
  etiology and risk factors, 246
  pathology, 246–247
  prognosis, 249
  treatment, 249
Hematoxylin & eosin stain (H & E stain), 265–266, 303–304, 489
Hemicrania continua, 286–287
Hemihypophysectomy, 301
Hemiparesis, 115
Hemispheres, 12
Hemispheric lateralization, 8
Hemorrhage, 167
Hemosiderotic variants, 272
Hepatocyte growth factor, 390–391
Hepatoid variant, 426–427
HER2 protein, 456
HER2/neu. See Erb-b2 receptor tyrosine kinase 2 (ERBB2)
Herringbone pattern, 257
Heterodimeric complex, 311
Heterodimers, 380
Heterozygosity, 133
  loss of, 197–198
Hexaribonucleotide binding protein-3 (NeuN), 150
Hibernoma, 280
High-grade astrocytomas, 111–112
High-mobility group AT-hook 1 (HMGA1), 292
Higher mental functions, 8, 19–23
Highly active antiretroviral therapy (HAART), 259–260
Hindbrain, 66
Hippocampal formation, 14, 16–18, 18f
Hippocampus, 16–17, 22
Histiocyte-like cells, 272
Histiocytoid cells, 271
Histiocytoma, 272
Histochemistry, 325
Histologic appearance, 148
Histone 3 (H3), 137–138
Histone deacetylase 2 (HDAC2), 299
Histone deacetylase inhibitors, 483
Histone methylation on tail of histone H3 (H3K27me3), 481–482
Histopathological diagnosis, 135
HIV-1 Tat interactive protein 2 (HTATIP2), 308
Holocranial variants, 418
Homeostasis, 33
Homer-Wright rosettes, 330–332
Human chorionic gonadotropin (hCG), 419
Human herpesvirus 6 (HHV-6), 130
  HHV6A, 148–149
Human herpesvirus 8 (HHV8), 259

Human herpesviruses 6 and 8 (HHV6 and HHV8), 349
Hybrid nerve sheath tumors, 439–441
Hydrocephalus, 233–234
Hyperdiploidy, 234
Hyperglobulinemia, 222
Hypermethylated MGMT promoter, 153
Hypermutation, 351
Hyperpituitarism, 286
Hyperpolarization, 52
Hypertonic muscles, 28
Hypertrophic intracranial pachymeningitis, 270–271
Hypocortisolism. *See* Addison's disease
Hypodense central mass, 139–140
Hypodiploidy, 238
Hypogastric plexus, 36–37
Hypoglossal nerve, 73
Hypoglossal nucleus, 74
Hypomelanosis of Ito, 234
Hypomethylation, 418–419
Hypophosphorylated AKT, 399–400
Hypopituitarism, 286
Hypothalamic nuclei, 14
Hypothalamic sulcus, 75
Hypothalamohypophyseal axis, 419
Hypothalamus, 9–10, 67
  homeostatic activities, 11*t*
Hypotriploid tendency, 325
Hypoxia, 390
Hypoxia-inducible factor 1-alpha (HIF1A), 244, 390
Hypoxia-inducible factor 2-alpha (HIF2A), 244

I

Ikaros family zinc finger 1 (IKZF1), 292
Imatinib, 199
Immature teratomas, 423–424
Immunoblasts, 350
Immunodeficiency-associated central nervous system lymphomas, 356–357
Immunoexpression, 325
Immunoglobulin (IG), 351
Immunoglobulin heavy gene (IGH gene), 352
Immunohistochemistry, 149
Immunophenotype, 137–138
  immunophenotypic features, 314
Immunostain, 295
Immunotyping, 486
Importin 7 (IPO7), 308
Inborn reflex, 28
Incidence, 104, 106
  density rate, 104
  proportion, 105
  rate, 104–105
Incidentalomas, 288
Induced pluripotent stem cell-derived progenitor cells, 62
Infantile embryonal carcinoma, 426
Infantile hemangiomas, 249
Inferior colliculi, 12
Inferior mesenteric ganglion, 36
Inferior olivary nucleus, 74

Inferior sagittal sinus, 13
Inferior salivatory nucleus, 74
Infiltrative low-grade astrocytomas, 116–117
  clinical manifestations, 117
  diagnosis, 117
  epidemiology, 116
  etiology and risk factors, 116
  pathology, 117
  prognosis, 117
  treatment, 117
Inflammation, 137
  inflammation-rich meningioma, 222
  inflammatory myofibroblastic tumor, 270–271
Infratentorial ependymoblastomas, 173
Infratentorially glioblastoma, 130–131
Infundibulomas, 312
Infundibulum, 10
Inhibitory postsynaptic potential (IPSP), 59
Inhomogeneous enhancement, 443
Inner meningeal cranial dura, 23
Inositol polyphosphate 5-phosphatase1 (INPP5D), 351
Insula, 6
Insulin-like growth factor 2 (IGF2), 419
Insulin-like growth factor-1 receptor (IGF-1R), 276
Integrase interactor 1 (INI1), 441
Integration center, 8, 29
Interferon regulatory factor protein (IRF4), 350–351
Interleukin 8 (IL8), 137
Internal capsule, 8
Internal carotid arteries, 13
Internal jugular veins, 13
International Classification of Diseases (ICD), 94
International Cowden Consortium (ICC), 401–402
International League Against Epilepsy (ILAE), 409–411
Interneurons, 26, 46, 49
Internodes, 43
Interoceptors, 30, 48
Intertumoral variation, 137
Interventricular foramen, 76
Intrachromosomal inversion, 256–257
Intracortical lesion, 208
Intracranial calcifications, 209
Intradural tumors, 95–96, 471–474
  clinical manifestations, 472
  diagnosis, 472–473
  epidemiology, 471
  etiology and risk factors, 471
  pathology, 472
  prognosis, 474
  treatment, 473
Intradural-extramedullary tumors, 471
Intradural-intramedullary tumors, 471
Intralabyrinthine schwannomas, 179–180
Intralumbar injection, 460
Intralymphatic lymphomatosis, 353
Intramedullary tumors, 95–96, 465

Intraparenchymatous inflammatory fibrosarcoma, 258
Intratumoral hemorrhage, 139
Intratumoral variation, 137
Intravascular large B-cell lymphoma, 353–355
Intravascular lymphoma, 353
Intravascular lymphomatosis, 353
Intravascular papillary endothelial hyperplasia, 250
Intrinsic reflex. *See* Inborn reflex
Intumescences, 71
Ionic homeostasis, 43
Ipsilateral of body, 19–20
Irritative (radicular) syndrome, 467
Island of Reil. See Insula
Isochromosome 17q, 432
Isocitrate dehydrogenase (IDH), 132, 487
  IDH type 1 or 2 mutations, 147–148
  IDH1, 149
  isocitrate dehydrogenase-mutant, 141, 147–151
    clinical manifestations, 149
    diagnosis, 149–151
    epidemiology, 148
    etiology and risk factors, 148–149
    pathology, 149
    prognosis, 151
    treatment, 151
  isocitrate dehydrogenase-wild-type, 139–140
Isthmus, 75
IVA regimen, 266

J

Joint kinesthetic receptors, 30–31
Jumonji domain-containing 1C (JMJD1C), 419
Juvenile pilocytic astrocytoma (JPA). *See* Pilocytic astrocytoma
Juxtallocortex, 16–17
Juxtanuclear keratin aggresomes, 293

K

K-Rev interaction trapped 1 (CCM1), 249–250
Kaposi sarcoma, 258–260
Karnofsky performance status, 460–461
Karnofsky score, 151
KCNJ13. *See* Potassium inwardly rectifying channel, subfamily J, member 13 (KCNJ13)
KIAA1549L-BRAF fusion gene, 112
Killin (KLLN), 400
Kilobases, 369
Kinesin, 47
KIT receptor staining, 201–202
Klinefelter syndrome, 418
Knee-jerk reflex, 29
Kruppel-like factor 4 mutation, 222

L

Lactotrophic adenomas, 288
LAMB syndrome. *See* Carney complex

Lamellar corpuscles, 30–31, 31f
Lamina terminalis, 75
Laminas, 46
Language processing, 20–21
Large deletions, 207
Large tumor suppressor kinase 1 (LATS1), 370
Latency-associated nuclear antigen (LANA), 259
Latent membrane protein 1 (LMP1), 357
Lateral corticospinal tract, 27–28, 73
Lateral division, 27
Lateral funiculi, 27
Lateral rectus palsy, 286
Leak channels, 50
Learned reflex, 28
Legius syndrome, 203–204
Leiomyoma, 260, 275, 286
Leiomyosarcoma, 260–261, 424–425
Lens placode, 70–71
Lenticulostriate arteries, 9
Lenticulostriate arteries. See Posterior cerebral artery (PCA)
Lentiform nucleus, 9
Lentiginosis, 286
Leptomeningeal metastases, 89–90
Leptomeningeal spread, 149
Leptomeningeal tissues, 334
Leptomeninges, 200–201
Leptomyelolipomas. See Lumbosacral lipomas
LEU7, 187
Lhermitte–Duclos disease, 400
Life expectancy, 85–87
Life table, 85
Lifetime prevalence, 103–104
Li–Fraumeni syndrome, 139, 260, 367
   clinical cases, 373–376
   clinical manifestations, 371
   diagnosis, 371–372
   epidemiology, 367–368
   etiology and risk factors, 368–369
   pathology, 369–371
   prognosis, 373
   treatment, 372
Li–Fraumeni-like syndrome, 372
Ligands, 51
   ligand-gated ion channels, 51
LIM domain only 2 protein (LMO2 protein), 353
Limbic brain, 15
Limbic lobe, 14–15
Limbic midbrain areas, 15
Limbic system, 13–18, 18f
   functions, 15–16
Lin-28 homolog A (LIN28A), 428
Lipidization, 183
Lipidized cells, 136
Lipoblasts, 261
Lipofuscin, 45–46
Lipoma, 266–268
   clinical manifestations, 267
   diagnosis, 268
   epidemiology, 266
   etiology and risk factors, 266
   pathology, 266–267
   prognosis, 268
   treatment, 268
Liposarcoma, 261–262
Liposomal cytarabine, 339
Lobes, 5–6, 6f
Localized gigantism, 199
Localized intraneural neurofibroma, 199
Long ascending fibers, 27
Long descending fibers, 27
Long-term memories, 21
Longevity. See Life expectancy
Longitudinal fasciculi, 8
Longitudinal fissure, 5
Low-grade astrocytomas, 111–112, 318
Low-grade B-cell lymphomas, 355–356
Lower motor neurons, 27–28
Lumbar cistern, 33
Lumbar enlargements, 25–26
Lumbar nerve 1 (L_1), 25–26
Lumbar nerve 2 (L_2), 25–26
Lumbar puncture, 25–26
Lumbosacral enlargements, 25–26
Lumbosacral lipomas, 266
Lumbosacral schwannomas, 180–181
Luteinizing hormone (LH), 286
Lymphangioleiomyomas, 412–413
Lymphoepithelial lesions, 360
Lymphomagenesis, 357
Lymphomas, 134
Lymphomatoid granulomatosis, 357
Lymphomatosis cerebri, 353
Lymphoplasmacellular reaction, 421
Lymphoplasmacyte-rich meningioma, 222
Lymphoplasmacytoid cells, 355
Lynch syndrome, 380
Lysosomes, 45–46
LZTR1 mutations, 190

M

M133T. See Mutation at codon 133 (M133T)
Macroadenomas, 285
Macrocysts, 222
Macroglia, 41
Maffucci syndrome, 255
Magnetic resonance imaging (MRI), 89, 140, 164–165
Magnetic resonance spectroscopy (MRS), 134, 149–150, 351
Malcavernin (CCM2), 249–250
Malignant fibrous histiocytoma, 273–274
Malignant meningiomas, 215
Malignant perineuriomas, 449
Malignant peripheral nerve sheath tumors (MPNSTs), 187, 201, 442–449
   clinical manifestations, 443
   diagnosis, 443
   with divergent differentiation, 444–445
   epidemiology, 442
   etiology and risk factors, 442
   pathology, 442–443
   with perineurial differentiation, 449
   prognosis, 444
   treatment, 444
Malignant pineocytomas, 326
Malignant primary brain tumors, 87
Malignant teratoma, 425
Malignant transformation, 181–183
Malignant triton tumor, 444
MALT lymphoma of dura, 359–360
Mammalian target of rapamycin (mTOR), 199, 400, 419
Mammillary bodies, 10
Mammillary nuclei, 10
Mammillothalamic tract, 15
Mammosomatotroph adenomas, 291–292
Manifestations of TSC, 408
Matrix metallopeptidase 1 (MMP1), 300
Matrix metalloproteinase-2 (MMP2), 133–134
McCune–Albright syndrome, 292
Mechanically-gated ion channels, 51
Mechanistic target of rapamycin (mTOR), 115, 133–134, 299
Mechanoreceptors, 30
Meckel's cave, 339
Medial division, 27
Medulla. See Medulla oblongata
Medulla oblongata, 12, 71–73
Medulla spinalis, 71
Medullary pyramids, 73
Medulloblastoma, 134, 367–368, 380, 432–433
   clinical manifestations, 432
   diagnosis, 432–433
   epidemiology, 432
   etiology and risk factors, 432
   pathology, 432
   prognosis, 433
   treatment, 433
Medullomyoblastomas, 265–266
Megalencephaly, 401
Meissner's corpuscles. See Tactile corpuscles
Melan A. See Melanoma antigen recognized by T cells 1 (MART1)
Melanin granules, 45–46
Melanization, 234
Melanocytic neurilemmoma. See Melanotic schwannoma
Melanocytic tumors
   clinical cases, 344–346
   meningeal melanocytoma, 339–340
   meningeal melanocytosis, 342–344
   meningeal melanoma, 337–339
   meningeal melanomatosis, 340–342
Melanoma, 88–89
   melanoma-derived growth regulatory protein, 199
Melanoma antigen recognized by T cells 1 (MART1), 338
Melanophages, 337
Melanotic neurinoma. See Melanotic schwannoma
Melanotic schwannoma, 188–189
Melatonin, 11–12
Membrane potential, 49–56
   action potentials, 52
   graded potentials, 51–52
   resting membrane potential, 49–50

Memory, 21–22
　brain regions used in, 22
　consolidation, 21
Mendelian syndrome, 129–130
Meningeal layer, 23
Meningeal melanocytoma, 339–340
Meningeal melanocytosis, 342–344
Meningeal melanomatosis, 340–342
Meninges, 23–24
　arachnoid mater, 23
　dura mater and dural folds, 23
　pia mater, 23
Meningioangiomatosis, 208
Meningioma, 215–227, 318, 456–457
　anaplastic, 225–227
　angiomatous, 221–222
　atypical, 223–224
　chordoid, 222
　clear cell, 225
　clinical cases, 227–230
　clinical manifestations, 217
　diagnosis, 217–218
　epidemiology, 215
　etiology and risk factors, 215–216
　fibrous, 220
　lymphoplasmacyte-rich, 222
　meningothelial, 219–220
　metaplastic, 221
　microcystic, 222
　papillary, 227
　pathology, 216–217
　prognosis, 218–219
　psammomatous, 220
　rhabdoid, 227
　secretory, 222
　transitional, 220–221
　treatment, 218
Meningothelial meningioma, 219–220
Meningothelial proliferation, 208
Meninx primitive, 266
Merkel discs. See Tactile discs
Merlin, 179
Mesencephalon, 66–67, 75
Mesenchymal glioblastoma, 133
Mesenchymal neoplasm, 246
Mesenchymal tumors, 243
　angiolipoma, 268–269
　angiosarcoma, 253–255
　benign fibrous histiocytoma, 272–273
　chondroma, 276–278
　chondrosarcoma, 255–256
　clinical cases, 280–282
　desmoid-type fibromatosis, 269–270
　epithelioid hemangioendothelioma, 252–253
　Ewing sarcoma, 256–257
　fibrosarcoma, 257–258
　hemangioblastoma, 243–246
　hemangioma, 249–251
　hemangiopericytoma, 246–249
　hibernoma, 280
　inflammatory myofibroblastic tumor, 270–271
　Kaposi sarcoma, 258–260
　leiomyoma, 275
　leiomyosarcoma, 260–261
　lipoma, 266–268
　liposarcoma, 261–262
　malignant fibrous histiocytoma, 273–274
　myofibroblastoma, 271–272
　osteochondroma, 278–279
　osteoma, 279–280
　osteosarcoma, 262–264
　rhabdomyoma, 275–276
　rhabdomyosarcoma, 264–266
Mesenchymal-epithelial transition (MET), 484
Mesenteric ganglia, 34–36
Mesoblast, 426–427
Mesoderm, 65
Metachronous multiple tumors, 272
Metalloenzyme, 325
Metaplasia, 136
Metaplastic meningioma, 221
Metastatic pituitary tumors, 317
Metastatic RCC, 394
Metastatic tumors, 134, 455–461
　clinical cases, 461–463
　clinical manifestations, 458
　diagnosis, 458–460
　epidemiology, 455–456
　etiology and risk factors, 456
　pathology, 456–458
　prognosis, 460–461
　treatment, 460
Metencephalon, 67–68, 75
Methotrexate, 339
Methyl ethyl ketone 1 and 2 (MEK 1 and 2), 206
Methylation of MGMT, 133
Methylnitrosourea, 148–149
MIA gene, 199
Microadenomas, 285
Microcystic meningioma, 222
Microcysts, 114
Microduplication, 292
Microenvironmental factors, 133–134
Microfilaments, 45–46
Microgemistocytes, 150
Microglia, 43
MicroRNA-26a (miR-26a), 299
Microrosettes, 163
Microtubule-associated protein 2 (MAP2), 329, 411, 485
Microtubules, 45–46
Microvascular proliferation, 137
Midbrain, 12
Middle cerebral artery (MCA), 9, 13
Midkine, 197–198
Mifepristone, 301–302
Minigemistocytes, 150
Mismatch repair cancer syndrome. See Brain tumor-polyposis syndrome 1 (BTP1)
Mismatch repair genes, 380
Missense mutations, 207, 368–369, 389
Mitochondria, 45
Mitogen-activated protein kinase (MAPK), 117–118, 133–134, 299, 399–400, 490
Mitogen-activated protein kinase-kinase inhibitor (MEK inhibitor), 342
　MEK162, 342
Mitogens, 197–198
Mixed cranial nerves, 31
Mixed ependymoma/subependymoma, 167
Mixed germ cell tumors, 431
Mixed meningiomas. See Transitional meningioma
Mixed pineocytoma-pineoblastoma, 332
Mixed-lineage leukemia (MLL), 484
Moestrin, 411–412
Molecular cells, 12–13
Monoclonality, 358
Monosynaptic reflex, 29
Monro, 9
Mortality from brain and spinal tumors in United States, 96
Mortality rate, 94
Motor areas, 7
Motor cranial nerves, 31
Motor neurons, 29, 48–49
Motor nuclei, 31–33
Motor speech area. See Broca's area
Mouse double minute 2 (MDM2), 141, 489
Mucosa-associated lymphoid tissue lymphoma translocation protein 1 (MALT1), 351
Multinucleated giant cells, 136
Multinucleation, 380
Multiple endocrine neoplasia type 1 (MEN1), 286
Multiple hamartoma syndrome, 399
Multiple myeloma, 89–90
Multiple neurilemmomas, 190
Multiple neuroma, 202
Multiple schwannomas, 190
Multipolar neurons, 26, 47
Multipotency, 69–70
Multivacuolated cytoplasm, 261
Mural nodule, 489
Muscle sensing, 30
Muscle spindles, 30–31
Mutation at codon 133 (M133T), 369
Mutations, 147, 407
mutL homolog 1 (MLH1), 311, 380
mutS homolog 2 (MSH2), 380
mutS homolog 6 (MSH6), 311, 380
MutS protein homolog 4 (MSH4), 370
MutY homolog gene (MUTYH gene), 381
Myelencephalon, 68, 71–74
Myelin, 43, 56
　sheath, 43
Myelinated axon, 43
Myelination, 43
Myeloblastosis virus (MYB), 149
Myelocytomatosis/N-myc proto-oncogene protein (MYC/MYCN), 484
Myeloid differentiation primary response 88 (MYD88), 351
Myelotoxicity, 359
MYH glycosylase, 381
MYH-associated polyposis. See Autosomal recessive familial adenomatous polyposis

Myoepithelioma, 272
Myofibroblastoma, 271–272
Myogenin, 265–266
Myotomes, 467
Myxoid foci, 114
Myxopapillary ependymoma, 168–171
  clinical manifestations, 170
  diagnosis, 170
  epidemiology, 168–169
  etiology and risk factors, 169
  pathology, 169–170
  prognosis, 170–171
  treatment, 170

N

NAME syndrome. *See* Carney complex
NANOG, 422
Narrow posterior median sulcus, 26
National Comprehensive Cancer Network (NCCN), 372–373
National Institutes of Health (NIH), 206
Necrobiotic debris, 137
Necrosis, 137
Nelson's syndrome, 300
Neocortex, 6
Neoplasia, 379
Neoplasm, 233–234
Neoplastic syncytiotrophoblasts, 419
Neovascularity, 140
Nerve cell degeneration and regeneration, 61–62
Nerve fibers, 19–20, 46–47
Nervous system, 73
  ANS, 33–37
  cytology of
    glial cells, 41–44
    membrane potential, 49–56
    nerve cell degeneration and regeneration, 61–62
    neuronal integration of stimuli, 59–61
    neurons, 44–49
    synapses, 56–58
  PNS, 29–33
  subdivisions, 3, 4f
Nervous tissue, 41
Nervus erigentes, 37
Nestin, 137–138, 411
Neural cell adhesion molecule 1 (NCAM1), 167, 244, 329, 485
Neural crest, 69–70
  cells, 65
Neural fibrolipomas, 266
Neural fold, 65
Neural glioblastoma, 133
Neural groove, 65
Neural plate, 65
Neural responses to injuries, 44
Neural tube, 65–68
  bulges and flexures, 66–68
  differentiation and migration, 65–66
  sensory and motor areas, 66
Neuralgiform headache, 286–287
Neuraxial germ cell tumors, 418

Neurilemma, 43–44
Neurilemmomatosis, 190
Neurite outgrowth inhibitor A (NOGO-A), 150
Neuroblastoma rat sarcoma (NRAS), 337
Neurocristopathies, 69–70
Neurocutaneous melanosis, 338, 341
Neurodegenerative disease, 61
Neuroembryogenesis, 418
Neuroendoscopic procedures, 419–420
Neurofibrils, 45
Neurofibroma, 91, 197
  atypical neurofibroma, 200–202
  clinical cases, 211–213
  clinical manifestations, 199
  diagnosis, 199
  epidemiology, 197
  etiology and risk factors, 197–198
  NF1, 202–207
  NF2, 207–211
  pathology, 198
  plexiform neurofibroma, 201
  prognosis, 199–200
  treatment, 199
Neurofibromatosis type 1 (NF1), 111, 129–130, 139, 187, 197, 202–207, 419, 439, 465, 484
  clinical manifestations, 205
  diagnosis, 206
  epidemiology, 203
  etiology and risk factors, 203–204
  microdeletion syndrome, 203
  pathology, 204
  prognosis, 207
  treatment, 206
Neurofibromatosis type 2 (NF2), 179, 207–211, 370, 439
  clinical manifestations, 208–209
  diagnosis, 209
  epidemiology, 207
  etiology and risk factors, 207
  pathology, 207
  prognosis, 210–211
  treatment, 210
Neurofibromin, 112–113, 197
  neurofibromin 1, 133
Neurofibrosarcoma, 179
Neurofilament proteins (NFPs), 325, 338, 480–481
Neurofilaments, 45, 411
Neurogenesis, 62
Neuroglandular junction, 57
Neuroglia, 41
Neuroimaging, 149
Neurolemmocytes. *See* Schwann cells
Neuromatosis, 202
Neuromuscular choristomas, 276
Neuromuscular junction, 57
Neuronal differentiation 1 (NeuroD1), 300
Neuronal integration of stimuli, 59–61
  postsynaptic potentials, 59
  presynaptic regulation, 60
  rate of action potential generation, 60–61
Neuronal nuclear protein (NeuN), 485
Neuronal pathways, 27–29

Neurons, 6, 44–49, 66
  functional characteristics, 44
  functional classification, 48–49
  neuron-specific enolase, 167, 325
  processes, 46–47
  structural classification, 47
  structure, 45–47, 45f
  types, 48f
Neuropil, 46
Neuropores, 65
Neurotrophin receptor tyrosine kinase (NTRK), 485
  NTRK3, 489
Neurotubules, 45
Neurula, 65
Nevoid basal cell carcinoma syndrome (NBCCS), 432
NGFI-A-binding protein 2 (NAB2), 246
Nijmegen breakage syndrome, 432
Nissl bodies, 45
Nissl substance, 46–47
Nitrosoureas, 138
NK/T-cell lymphomas, 357–359
Nociceptors, 30
Nodes, 43
  of Ranvier, 43
Nodosal placode, 71
Nodular fasciitis, 270
Non-Hodgkin lymphoma, 89
Noncanonical planar cell polarity pathway, 382
Noncanonical Wnt/calcium pathway, 382
Noncommunicable disease, 84
Nondominant hemisphere, 19
Nonencapsulated nerve endings, 30
Nonfunctioning pituitary tumors, 305–307. *See also* Pituitary tumors
  clinical manifestations, 306
  diagnosis, 306–307
  epidemiology, 305
  etiology and risk factors, 305
  pathology, 305–306
  prognosis, 307
  treatment, 307
Nongenotoxic stimuli, 370
Nonhematopoietic CNS neoplasms, 456
Nonmelanotic plexiform schwannomas, 185
Nonneoplastic cells, 244–245
Nonrapid eye movement sleep (NREM sleep), 22
Nonsense mutations, 207, 369
NOS. *See* Not-otherwise-specified (NOS)
Not-otherwise-specified (NOS), 151
  anaplastic oligodendroglioma, 156
  oligodendroglioma, 149, 151–152
Notochord, 66
Nuclear envelope, 45
Nuclear pores, 45
Nuclear pseudopalisading, 132
Nuclear receptor coactivator 2 (NCOA2), 255
Nuclei, 46
Nucleolated nuclei, 411
Nucleolus, 45
Nucleophosmin 1 (NPM1), 360
Nucleotides, 369

Nucleus, 45
Nucleus accumbens, 9, 14
Nucleus ambiguous, 74
Nucleus basalis, 22
Nucleus of solitary tract. *See* Solitary nucleus
Nucleus pulposus, 66
Nucleus salivatorius inferior. *See* Inferior salivatory nucleus
Nucleus solitarius. *See* Solitary nucleus
Nucleus tractus solitarii. *See* Solitary nucleus
Null cell adenomas, 288
Nystagmus retractorius, 326−327

O

O-6-methylguanine-DNA methyltransferase (MGMT), 132−133, 147, 299
Obex, 73
Obstructive hydrocephalus, 327
Occipital lobe, 6
Occulta neural tube defect, 78
Octamer-binding transcription factor 4 (OCT4), 422, 428
Oculomotor nerve, 37
Oculomotor nucleus, 37
Olfactory cortex, 7, 14
Olfactory nerve, 31
Oligodendrocyte transcription factor 2 (OLIG2), 485
Oligodendrocytes, 41, 43
Oligodendroglial anaplastic glioma, 136
Oligodendroglioma, 147−152
    anaplastic, 153−156
    clinical cases, 157−158
    components, 136
    IDH-mutant and 1p/19q-codeleted oligodendroglioma, 147−151
    NOS, 151−152
    NOS anaplastic, 156
Oligodendrogliomatosis, 148−149
Oligosarcomas, 151, 155
Ollier disease, 255
OMG gene, 203
Optic chiasma, 286
Orthodenticle homeobox 2 (OTX2), 233
Osteoblastoma, 90
Osteochondroma, 90, 278−279
Osteoclastoma. *See* Giant cell tumor of bone
Osteoid osteoma, 90
Osteolipoma, 266
Osteoma, 279−280
Osteomata, 382−383
Osteosarcoma, 262−264
    clinical manifestations, 263
    diagnosis, 263
    epidemiology, 262
    etiology and risk factors, 262
    pathology, 263
    prognosis, 264
    treatment, 264
Osteosarcomatous differentiation, 442
Otic pit, 71
Otic placode, 71
Otic vesicle, 71
Oxysterol binding protein-like 10 (OSBPL10), 351
Oxytocin, 10

P

p-arm of chromosome 1 (1p deletion), 147
1p/19q co-deletion (1p19q codel), 136
    clinical manifestations, 149
    diagnosis, 149−151
    epidemiology, 148
    etiology and risk factors, 148−149
    oligodendroglioma, 147−151
    pathology, 149
    prognosis, 151
    treatment, 151
p53 protein, 367
Pachymeningeal fibrosis, 270−271
Pacinian corpuscles. *See* Lamellar corpuscles
Paget's disease, 262
Pagetoid bone, 253
Paired box 5 (PAX5), 350−351
Palatoglossus muscle, 73
Paleocortex, 6
Palmoplantar keratoses, 402
Papez circuit, 18, 19f
Papillary ependymoma, 171
Papillary form, 302
Papillary meningioma, 227
Papillary tumor of pineal region, 328−330
Papillomatous papules, 401−402
Parafalcine, 276−277
Paraparetic patient, 470−471
Parasellar masses, 318−319
Paraspinal tumors, 180
Parasympathetic neurons, 37
    and pathways, 38t
Paraventricular nuclei, 11
Parenchymal brain metastases, 88−89
Paresis, 28
Parietal lobe, 6
Parinaud syndrome, 325, 419
Passive ion channels, 50
Peau chagrin, 408
Pediatric-type oligodendroglioma, 149
Pegvisomant, 294
Pellagra, 84
Pelvic nerve, 37
Penetrance, 368
Perception, 30
Perforin, 358
Pericytes, 246
Perifocal zone of edema, 136−137
Periglandular pituitary infiltrates, 318
Perikaryal fibril, 408, 411
Perikaryon. *See* Cell body
Perineurioma, 201, 450−451
Perineuriomatous differentiation, 440
Period prevalence, 103−104
Periodic acid-Schiff reaction (PAS reaction), 300
Periosteal cranial dura, 23
Periosteal layer, 23
Peripheral GH-secreting tumors, 289
Peripheral nerve sheath tumors (PNSTs), 439. *See also* Malignant peripheral nerve sheath tumors (MPNSTs)
    clinical cases, 451−453
    epithelioid MPNSTs, 446−448
    hybrid nerve sheath tumors, 439−441
    malignant peripheral nerve sheath tumor with divergent differentiation, 444−445
    malignant peripheral nerve sheath tumor with perineurial differentiation, 449
    perineurioma, 450−451
Peripheral nervous system (PNS), 3, 29−33, 41, 42t, 65. *See also* Autonomic nervous system (ANS)
    cranial nerves, 31−33, 32f
    classifications, 32t
    sensory and motor nuclei, 31−33
    glial cells of, 43−44
    sensory receptors, 30−31
    spinal nerves, 33
Peripheral neuropathies, 209
Peripheral primitive neuroectodermal tumor. *See* Ewing sarcoma
Periphery, 66
Peritumoral edema, 139
Periventricular parenchyma (PP), 350
Person-time, 105
    incidence rate, 104
Petrosal placode, 71
Phacomatosis, 341
Phosphatase and tensin homolog (PTEN), 133, 139, 328, 399, 484
Phosphatidylinositol (3, 4, 5)-trisphosphate (PIP$_3$), 399−400
Phosphatidylinositol 3-kinase regulatory subunit alpha (PIK3R1), 484
Phosphatidylinositol 4,5-bisphosphate (PIP$_2$), 399−400
Phosphatidylinositol-4,5-bisphosphate 3-kinase (PI3K), 299, 399−400
Phosphatidylinositol-4,5-bisphosphate 3-kinase catalytic subunit alpha (PIK3CA), 400, 484
Phosphoinositide 3-kinase/protein kinase B pathway (PI3K/PKB pathway), 133
Photocoagulation therapy, 393
Photodynamic therapy, 199
Photoreceptors, 30
Photosensory differentiation, 325
Pia mater, 23
Pigmented schwannoma. *See* Melanotic schwannoma
Pilocytic astrocytoma, 111−114
    clinical manifestations, 113
    diagnosis, 114
    epidemiology, 112
    etiology and risk factors, 112
    pathology, 112−113
    prognosis, 114
    treatment, 114
Piloerection, 10
Piloid cells, 113
Pilomotor, 37
Pineal anlage tumors, 332

Pineal body. *See* Pineal gland
Pineal germinoma, 421*f*
Pineal gland, 11
Pineal parenchymal tumor with intermediate differentiation (PPTID), 326–328
Pineal parenchymal tumors
 clinical cases, 333–335
 papillary tumor of pineal region, 328–330
 pineoblastoma, 330–332
 pineocytoma, 323–326
 PPTID, 326–328
Pineal teratoma, 426*f*
Pinealocytes, 325
Pinealomas, 323
Pineoblastoma, 330–332
 with lobules, 326
Pineocytoma, 323–326
 with anaplasia, 326
 clinical manifestations, 325
 diagnosis, 325
 epidemiology, 323
 etiology and risk factors, 323
 pathology, 323–325
 prognosis, 325–326
 treatment, 325
Pineocytomatous rosettes, 323
Pirfenidone, 199
Pituicytes, 314
Pituicytoma, 312–314
Pituitary adenoma, 285–291
 clinical manifestations, 286–287
 diagnosis, 288–289
 epidemiology, 285
 etiology and risk factors, 286
 pathology, 286
 prognosis, 291
 screening tests, 289*t*
 treatment, 289–291
Pituitary apoplexy, 287
Pituitary carcinoma, 307–312
 clinical manifestations, 308–309
 diagnosis, 309–311
 epidemiology, 307
 etiology and risk factors, 307–308
 pathology, 308
 prognosis, 311–312
 treatment, 311
Pituitary choristomas, 318
Pituitary gland, 10
Pituitary tuberculoma, 289
Pituitary tumors, 285
 adrenocorticotropic hormone–secreting tumors, 299–302
 clinical cases, 319–321
 craniopharyngioma, 302–304
 nonfunctioning pituitary tumors, 305–307
 parasellar masses, 318–319
 pituicytoma, 312–314
 pituitary adenoma, 285–291
 pituitary carcinoma, 307–312
 prolactinoma, 295–297
 secondary tumors, 317–318
 somatotroph adenoma, 291–294
 spindle cell oncocytoma, 314–316

 thyrotropin-secreting tumors, 297–299
Pituitary-specific POU class homeodomain 1 (PIT1), 286, 297
Placental alkaline phosphatase (PLAP), 418, 428
Plasma cells, 355
Plastic ependymomas, 162
Platelet-derived growth factor (PDGF), 137
Platelet-derived growth factor receptor beta (PDGFRB), 270, 390
Platelet-derived growth factor receptor type A (PDGFRA), 133, 484
Pleomorphic cells, 135–138
Pleomorphic cytology, 326–327
Pleomorphic xanthoastrocytoma, 117–121, 136, 272, 488–491
 clinical manifestations, 119, 489
 diagnosis, 119, 489–490
 epidemiology, 117, 488
 etiology and risk factors, 117–118, 488
 pathology, 118, 488–489
 prognosis, 119, 491
 treatment, 119, 491
Plexiform neurilemomas, 185
Plexiform neurofibroma, 187, 199, 201
Plexiform schwannoma, 185–188
 clinical manifestations, 187
 diagnosis, 187
 epidemiology, 186
 etiology and risk factors, 187
 pathology, 187
 prognosis, 188
 treatment, 188
Plexus of Auerbach, 37
Plexus of Meissner, 37
Plurihormonal adenomas, 291
Pluripotent cells, 423
Pluripotent fetal cells, 481
PMS1 homolog 2 (PMS2), 311, 380
Pnemocystis jirovecii, 138
PNETs. *See* Primitive neuro-ectodermal tumors (PNETs)
Pneumogastric nerve, 73
Point prevalence, 103–104
Polycomb gene targets, 481–482
Polycystic kidney disease 1 gene, 407–408
Polycythemia, 245
Polyethylene glycol precipitation, 296
Polymerase chain reaction (PCR), 351
Polymorphic variant of Zulch, 155
Polyomaviruses, 349
Polyphenotypic differentiation, 480
Polypoid, 234
Polysynaptic reflex, 29
Polyubiquitination, 390
Polyvesicular vitelline patterns, 426–427
Pons, 12
Pontine flexure, 67–68
Population, 83, 103
Positron emission tomography (PET), 311
 scans, 134
Posterior cerebral artery (PCA), 9, 13
Posterior commissure, 11
Posterior funiculus, 26

Posterior horns, 26, 66
Posterior intermediate septum, 27
Posterior intermediate sulcus, 26
Posterior motor nucleus of vagus. *See* Dorsal nucleus of vagus nerve
Posterior pituitary astrocytomas, 312
Posterior root ganglion, 26
Posterior rootlets, 26
Posterolateral sulcus, 26
Postganglionic fibers, 49
Postganglionic sympathetic fibers, 37
Postsynaptic neuron, 56–57
Postsynaptic potentials, 59
 integrating, 59
Potassium inwardly rectifying channel, subfamily J, member 13 (KCNJ13), 233–234
Potential energy, 49
Potential spaces, 23
PR domain zinc finger protein 1 (PRDM1), 351
PR/SET domain 14 (PRDM14), 419
Prefrontal cortex, 8
Preganglionic fibers, 49
Preganglionic sympathetic fibers, 34–36
Premotor cortex, 7–8
Presynaptic neuron, 56–57
Presynaptic regulation, 60
Prethalamus. *See* Subthalamus
Prevalence, 103–104, 106
 of brain and spinal tumors, 106–107
Primary adrenal insufficiency. *See* Addison's disease
Primary CNS lymphomas, 90, 318
Primary diffuse leptomeningeal gliomatosis, 139–140
Primary glioblastomas, 129
Primary lymphoma of brain
 anaplastic large cell lymphoma, 360–361
 clinical cases, 361–364
 DLBCL, 349–353
 immunodeficiency-associated central nervous system lymphomas, 356–357
 intravascular large B-cell lymphoma, 353–355
 low-grade B-cell lymphomas, 355–356
 MALT lymphoma of dura, 359–360
 T-cell and NK/T-cell lymphomas, 357–359
Primary malignant and nonmalignant brain tumors, 88
Primary motor cortex, 7
Primary neurulation, 65
Primary NK/T-cell lymphoma, 357–358
Primary somatosensory cortex, 7
Primary spinal tumors, 90–91
Primary suprasellar malignant odontogenic tumor, 303–304
Primary T-cell lymphoma, 357–358
Primary vesicles, 66
Primitive neuro-ectodermal tumors (PNETs), 173, 493–494
Primitive neuronal cells, 135–136
Procarbazine-1-(2-chloroethyl)-3-cyclohexyl-1-nitrosourea (CCNU) (lomustine)-vincristine (PCV), 147

Procarbazine-lomustine-vincristine (CCNU), 138–139
Programmed cell death protein 10 (CCM3), 249–250
Projection fibers, 8
Projection neurons, 46
Prolactin (PRL), 286, 295
Prolactinoma, 295–297
　clinical manifestations, 295–296
　diagnosis, 296
　epidemiology, 295
　etiology and risk factors, 295
　pathology, 295
　prognosis, 297
　treatment, 296–297
Prolactinomas, 285
Proliferation, 137
　index, 139
Proneural alpha-internexin protein, 150
Proneural glioblastoma, 133
Propagation, 54–55
　speed, 55–56
Propranolol, 393
Proprioceptive fibers, 31
Proprioceptive neuronal columns, 71
Proprioceptors, 30, 48
Propriospinal fibers, 27
Propriospinal tract, 27
Prosencephalon, 66
Prosody, 21
Proteasomal degradation, 390
Protein capicua homolog (CIC), 256–257
Protein kinase C delta (PRKCD), 299
Protein kinase CAMP-dependent type I regulatory subunit alpha (PRKAR1A), 300
Protein phosphatase 1D (PPM1D), 484
Proteins, 389–390, 407
Proto-oncogene c-kit inhibitor, 199, 338
Proto-oncogene serine/threonine-protein kinase (PIM1), 351
Proton therapy, 289
Psammomatous meningioma, 220
Pseudarthrotic bone, 205
Pseudoangiomatous morphology, 328
Pseudodiploid tendency, 325
Pseudogliomatous growth, 457–458
Pseudohypoxia, 244–245
Pseudorosettes, 115, 161
Pseudotumoral lesion, 270–271
Pseudounipolar neurons. *See* Unipolar neurons
Pulmonary plexus, 37
Purkinje cells, 12–13
Pyramidal cells, 7
Pyramidal tract. *See* Lateral corticospinal tract

Q

Quality adjusted life year (QALY), 94
Quinagolide, 289

R

R-CHOP, 355
RAC-alpha serine/threonine-protein kinase (AKT1), 400
Radiation, 215–216
Radicular pain, 467
Radiosensitizers, 486
Radixin, 411–412
RAF mutations, 460
Ranibizumab, 199
Rap1GTPase-activating protein 1 (RAP1GAP), 407
Rapalogs, 412–413
Rapamycin, 403
Rapid eye movement sleep (REM sleep), 22
Rare brain tumors, 479–494
　anaplastic pleomorphic xanthoastrocytoma, 491–493
　AT-RT, 479–483
　clinical cases, 495–497
　diffuse midline gliomas, 483–487
　gliomatosis cerebri, 487–488
　pleomorphic xanthoastrocytoma, 488–491
　PNETs, 493–494
Ras association domain-containing protein 1 (RASSF1A), 292
Ras homolog family (Rho family), 411–412
　RhoA, 133–134
Ras homolog member H (RHOH), 351
Ras-related C3 botulinium toxin substrate (Rac), 133–134
RAS/MAPK pathway, 203–204
RASopathies, 203–204
Rathke's cyst, 318
Rathke's pouch, 302
RB transcriptional corepressor 1 (RB1), 419
Reactive astrocytes, 42
Receptive speech area, 20–21
Receptor, 28
Reflex center, 25–29
　components of reflex arc, 28–29
　deep tendon reflexes, 29t
　reflex activity, 28
　spinal reflexes, 29
Refractory period, 53–54
RELA fusion-positive ependymoma, 171–172
Relative refractory period, 53–54
Remak bundle, 197
Renal cell carcinoma (RCC), 389
Repolarization, 52
Resting membrane potential, 49–50
　membrane channel effects on, 50–51
Reticular activating system, 19
Reticular formation, 13, 19, 20f
Reticular pattern, 426–427
Reticulin, 338
　fibers, 246–247
Reticulospinal tracts, 19
Retiform patterns, 183
Retinoblastoma (RB), 133, 262
　RB1, 292, 484
Retrograde flow or transport, 47
Reverse transcription polymerase chain reaction (RT-PCR), 257
Rhabdoid meningioma, 227

Rhabdoid tumor predisposition syndrome 1, 190
Rhabdomyoblastic elements, 204
Rhabdomyoblasts, 332
Rhabdomyoma, 275–276
Rhabdomyosarcoma, 264–266
　clinical manifestations, 265
　diagnosis, 265–266
　epidemiology, 265
　etiology and risk factors, 265
　pathology, 265
　prognosis, 266
　treatment, 266
Rho family hydrolase enzymes, 133–134
Rhombencephalon, 66–68
Rhombic lips, 68–69
Rhomboid domain containing 3 (RHBDD3), 292
Ribosomal ribonucleic acid (rRNA), 45
RING-box protein 1, 389–390
Rolandic fissure. *See* Central sulcus
Rosai–Dorfman disease, 273, 360
Rosenthal fibers, 113
Rostral neuropore, 65
Rothmund–Thomson syndrome, 262
Rough endoplasmic reticulum (RER), 45
Rubinstein–Taybi syndrome, 428
Ruffini endings. *See* Bulbous corpuscles

S

S-100 protein, 218, 441, 485
S-arrestin, 325
S6 kinase beta-1, 408
Sacral spinal nerve 2 (S_2), 25–26
Sacral spinal nerve 3 (S_3), 25–26
Sal-like protein 4 (SALL4), 422, 428
Salivatory nuclei, 37
Saltatory propagation, 55
Sarcoma, 367
Sarcomatous differentiation, 141
Satellite cells, 44
Scatter factor, 390–391
Schiller–Duval bodies, 426–427
Schwann cells, 43–44
Schwannoma, 91, 179–189
　cellular, 184–185
　clinical cases, 192–195
　clinical manifestations, 180
　diagnosis, 180–183
　epidemiology, 179
　etiology and risk factors, 179
　melanotic, 188–189
　pathology, 179–180
　plexiform, 185–188
　prognosis, 183–184
　schwannoma/perineurioma hybrid tumors, 440–441
　schwannomatosis, 190–192
　treatment, 183
Schwannomatosis, 179, 190–192
　clinical manifestations, 191
　diagnosis, 191
　epidemiology, 190

Schwannomatosis (*Continued*)
  etiology and risk factors, 190
  pathology, 190–191
  prognosis, 192
  treatment, 191
Schwannosis, 208
SCL protein, 244–245
Secondary glioblastomas, 129, 133, 141
Secondary memories, 21
Secondary neurulation, 65
Secondary structures, 136–137
Secondary tumors, 317–318
Secondary vesicles, 66–68
Secretory meningioma, 222
Segmental overgrowth lipomatosis arteriovenous malformation epidermal nevus syndrome (SOLAMEN syndrome), 400
Sella expansion, 288
Selumetinib, 206
Seminomas, 419
Sensation, 30
Sense organs, 30
Sensory areas, 7
Sensory association areas, 7
Sensory neurons, 28, 48
Sensory nuclei, 31–33
Sensory receptors, 30–31
Septal nuclei, 14
Serotonin, 11–12
Serpiginous necrosis, 132
Shagreen patches, 408
Short-T1 inversion recovery (STIR), 202
Short-term memories, 21
Short-term synaptic depression. *See* Synaptic fatigue
Shorter propriospinal fibers, 27
Signal transducer and activator of transcription 6 (STAT6), 220, 246
Silent corticotroph adenomas, 300
Simian vacuolating virus 40 (SV40), 130
Sirolimus, 199, 403
Sjögren syndrome, 356–357
Skill memories, 21
Sleep
  importance, 23
  regulation, 22
  and sleep-wake cycles, 22–23
  types, 22
Small cells, 135
Small nuclear ribonucleoprotein polypeptide N gene (SNRPN gene), 418–419
Small-cell lung cancers, 89
SMARCB1 germline mutation, 190
SMILE protocol, 359
Smooth muscle actin (SMA), 247, 480–481
Smu region, 351
Sodium-potassium exchange pump, 50
  functions of, 54
Solitary nucleus, 73
Solute carrier family 2 member 11 (SLC2A11), 308
Soma. *See* Cell body
Somatic mosaicism, 207
Somatic motor neurons (SM neurons), 7, 48
Somatic nervous system (SNS), 48
Somatic reflexes, 29
Somatic sensory information (SS information), 7
Somatosensory association cortex, 7
Somatostatin receptor 5 (SSTR5), 299
Somatostatin receptor type 3 and 5 (SSTR3 and SSTR5), 297
Somatotrope adenomas. *See* Somatotroph adenoma
Somatotroph adenoma, 288, 291–294
  diagnosis, 294
  epidemiology, 292
  etiology and risk factors, 292
  pathology, 292–294
  prognosis, 294
  treatment, 294
Somatotrophinomas. *See* Somatotroph adenoma
Sonic hedgehog (SHH), 432
Sorafenib, 199
Southern blot technique, 392–393
SOX10, 187, 441
Sparsely granulated corticotroph adenomas, 300
Sparsely granulated somatotroph adenomas (SGSAs), 291
Spastic paralysis, 28
Spatial summation, 59
Special senses, 30
Spin-lattice effect, 119
Spinal accessory nerve, 73
Spinal accessory nucleus, 73
Spinal cord, 25–29, 65
  structure, 25*f*, 26–27
  white matter, 26–27
    ascending pathways, 27
    descending pathways, 27–28
    neuronal pathways, 27–29
    upper and lower motor neuron damage, 28*t*
Spinal cord limbic systems, 15
Spinal cord tumors, 95–96, 465–474
  clinical cases, 474–476
  clinical manifestations, 465–467
  diagnosis, 467–468
  epidemiology, 465
  etiology and risk factors, 465
  extradural tumors, 469–471
  intradural tumors, 471–474
  pathology, 465
  prognosis, 469
  treatment, 468–469
Spinal dura mater, 23
Spinal gray matter, 26
Spinal meninges, 23
Spinal nerves, 26, 33
Spinal reflexes, 29
Spinal schwannoma, 179
Spinal trigeminal nerve nuclei, 73–74
Spinal tumors, 93
Spindle cell oncocytoma, 314–316
Spindled cells, 440–441
Splanchnic nerves, 36
Splice-site mutations, 207, 369
Splicing factor 3B subunit 1 (SF3B1), 338
Sry-Box transcription factor 2 (SOX2), 429
SS division, 26
Stage one sleep, 22
Staghorn vessels, 246–247
Stanniocalcin-1 (STC1), 329
Startle reflex, 12
Stem cell(s), 163
  factor, 197–198
  rescue, 420
Stereotactic biopsy, 134
Stereotactic radiosurgery, 138, 183, 202, 218, 289–291
Stereotaxis needle biopsy specimens, 135
Steroidogenic factor 1 (SF-1), 286
Stevens–Johnson syndrome, 138
Stippled chromatin, 328
Straight sinus, 13
Stretch reflexes, 29
Stria medullaris thalami, 11
Striatum, 9
Stroma, 323–324, 423–424
Stromal cells, 243–245
Stromal desmoplasia, 421
Subarachnoid space, 23
Subcommissural organ, 328
Subdural space, 23
Subependymal giant cell astrocytoma (SEGA), 111, 115–121, 407–408
  clinical manifestations, 115
  diagnosis, 115
  epidemiology, 115
  etiology and risk factors, 115
  pathology, 115
  prognosis, 116
  treatment, 115
Subependymoma, 165–168, 166*f*
  clinical manifestations, 167–168
  diagnosis, 168
  epidemiology, 165
  etiology and risk factors, 166
  pathology, 166–167
  prognosis, 168
  treatment, 168
Subicular complex, 16–17
Subnucleus caudalis, 73–74
Subnucleus interpolaris, 73–74
Subnucleus oralis, 73–74
Substantia nigra, 9
Subthalamus, 9, 67
Succinate dehydrogenase genes (SDH genes), 400
Sulcus, 5
Sulcus limitans, 66
Summation, 59
  of EPSPs and IPSPs, 59
Superficial veins, 13
Superior cervical sympathetic ganglion, 37
Superior colliculi, 12
Superior mesenteric ganglia, 36
Superior sagittal sinus, 13
Supratentorial ependymoblastomas, 173

Supratentorial midline primitive neuroectodermal tumors, 330
Supratentorial primitive neuroectodermal tumor (sPNET), 425
Surveillance, Epidemiology, and End Results program (SEER program), 233, 307
Sushi domain containing 2 (SUSD2), 351
SV40. *See* Simian vacuolating virus 40 (SV40)
Sweat gland innervation, 37
SWI/SNF related matrix-associated actin-dependent regulator of chromatin subfamily B member 1 (SMARCB1), 209, 216, 299, 441, 479
Sympathetic neurons, 33–37
   adrenal medulla, 34–36
   and pathways, 38*t*
   postganglionic sympathetic fibers, 37
   preganglionic sympathetic fibers, 34–36
Synapses, 47, 56–58
   in central nervous system, 56*t*
   classification, 56–57
   structure, 56
Synapsins, 57
Synaptic delay, 58
Synaptic fatigue, 58
Synaptic junctions, 56
Synaptophysin, 135–136, 325, 480–481
Synchronous tumors, 272
Syncytial meningiomas, 220
Syncytiotrophoblastic giant cells, 417
Syringomyelic syndrome, 467

T

T-box factor, pituitary (TPIT), 299
T-box family member TBX19 (T-PIT), 286
T-cell lymphomas, 357–359
T1-weighted lesions, 117
T2-weighted lesions, 117
Tactile corpuscles, 30–31, 30*f*
Tactile discs, 30
Tactile sensation, 30
Tanycytic ependymoma, 171
Targeted therapy, 420
Tectal plate, 330–332, 419
Tecum, 75
Telencephalon, 9, 66, 75–76
Telodendria, 46–47
Telomerase reverse transcriptase (TERT), 139, 150–151, 338
Temozolomide (TMZ), 138, 294, 297
Temporal lobe, 6
Temporal summation, 59
Tendon organs, 30–31
Teneurin transmembrane protein 1 (TENM1), 308
Tensin, 399
Tentorium cerebelli, 6
Teratoid rhabdoid tumors, 92–93
Teratoma, 134, 423–426
   clinical manifestations, 425
   diagnosis, 425
   epidemiology, 423
   etiology and risk factors, 423
   pathology, 423–425
   prognosis, 426
   treatment, 425
Terminal arborizations, 47
Terminal branches, 47
Tertiary memories, 21
Tethered spinal cord syndrome, 266
Thalamus, 9–10, 67
   subportions, 9*f*
Thermoreceptors, 30
Thiotepa, 339
Thoracic spinal nerve 1 (T_1), 25–26
Threshold, 55
Thrombospondin 1 (THBS1), 292
Thyroid transcription factor 1 (TTF1), 312
Thyroid-regulating hormone (TRH), 298
Thyroid-stimulating hormone (TSH), 286
Thyrotrophic adenomas, 288
Thyrotropin-secreting tumors, 297–299
Thyrotropinomas (TSHomas), 297
Tight junctions, 41–43
TNF receptor associated factor 7 mutation, 222
Trabecular bone, 90
Tracts, 46
Transcobalamin C, 353
Transcription elongation factor B polypeptides 1 and 2, 389–390
Transcription elongation factor B1 (TECB1), 244–245
Transcription factor E3 (TFE3), 252
Transforming growth factor alpha (TGFA), 390
Transient receptor potential cation channel subfamily M member 3 (TRPM3), 233
Transitional meningioma, 220–221
Transporter molecules, 57
Transsphenoidal adenectomy, 289
Transverse fissure, 5
Transverse tracts, 26
Transversions, 369
Trevor's disease, 278
Trichilemmomas, 399
Trigeminal nerve, 12
Trigeminal placode, 71
Trilateral retinoblastoma syndrome, 330
Triple-negative, 456
Triventricular choroid plexus papilloma, 233–234
Truncating mutations, 368–369
Trunk ganglia, 34–36
Tuberectomy, 412–413
Tuberin, 115, 407–408
Tuberous sclerosis, 111
Tuberous sclerosis complex (TSC), 111, 407
   clinical cases, 413–415
   clinical manifestations, 408–409, 410*t*
   diagnosis, 409–412, 411*t*
   epidemiology, 407
   etiology and risk factors, 407–408
   pathology, 408
   prognosis, 413
   treatment, 412–413
   TSC1 gene, 407
   TSC2 gene, 407, 489
Tubers, 408
Tubonodular types, 267
Tumefactive multiple sclerosis, 134
Tumor necrosis, 137
Tumor necrosis factor (TNF), 137
Tumor protein 53 (TP53), 112–113, 133, 141, 367–369, 484
Tumor protein 73 (TP73), 292
Tumor susceptibility 101 (TSG101), 308
Tumors, 147, 179, 215
Turcot syndrome, 379–384
   BTP1, 379–381
   BTP2, 381–384
   clinical cases, 384–386
Turcot syndrome, 139
Two-hit hypothesis, 197–198
Type A fibers, 55–56
Type B fibers, 55–56
Type C fibers, 55–56
Typhoid, 84
Tyrosine kinase receptor inhibitor, 138, 199
Tyrosine-protein kinase (KIT)/rat sarcoma (RAS) signaling, 419

U

Ubiquitin carboxyl-terminal esterase L1 (UCHL1), 325
Ubiquitin-specific peptidase 8 (USP8), 299
Undifferentiated embryonic cell transcription factor 1 (UTF1), 422, 429
Unipolar neurons, 47
Unmyelinated axons, 43
Upper motor neurons, 28
Urokinase-type plasminogen activator (uPA), 133–134
Uveal tract, 337

V

VAC regimen, 266
Vagus nerve, 73
Vascular cooption, 457–458
Vascular endothelial growth factor (VEGF), 199
   VEGFA, 137, 390
Vascular meningiomas. *See* Angiomatous meningioma
Vasocentric arrangements, 340
Vasogenic edema, 131–132
Vasomotor, 37
Vasopressin, 10
Vasostimulatory factors, 137
VCB complex, 244–245
Venous dural sinuses, 13
Ventral (anterior) funiculi, 27
Ventral amygdalofugal pathway, 15
Ventral anterior/ventral lateral nucleus (VA/VL nucleus), 10
Ventral median fissure, 71
Ventral posteriomedial nucleus, 73–74
Ventral tegmental area, 15
Ventricles, 43
Ventricular system, 76, 162
Ventriculoperitoneal shunt, 115

Verbal autopsies, 94
Vermis, 12
Verocay bodies, 181–183, 439
Vertebral arteries, 13
Vertebral tumor, 95–96
Vestibular nuclei, 74
Vestibular schwannoma, 180
Vimentin, 170, 480–481
Virchow–Robin spaces, 340
Visceral motor (VM)
    division, 26
    functions, 19
    neurons, 49
Visceral reflexes. See Autonomic reflexes
Visceral sensory division (vs. division), 26
Visceral sensory neurons, 48
Visceroceptors, 30
Visual association area, 7
Visual cortex, 7
Vogt's triad, 407
Voltage-gated ion channels, 51
Von Hippel–Lindau disease (VHL disease), 243, 389–394
    clinical cases, 394–396
    clinical manifestations, 392
    diagnosis, 392
    epidemiology, 389
    etiology and risk factors, 389–391
    pathology, 391–392
    prognosis, 394
    treatment, 393

VHL tumor suppressor gene, 390f
Von Recklinghausen's disease, 202, 318
Von Willebrand factor, 244

W

Waardenburg-Shah syndrome, 69–70
Waldenström's macroglobulinemia, 355
Wallerian degeneration, 44
Weibel–Palade bodies, 254
Wernicke's area, 20
Wet keratin, 302
White communicating rami, 34–36, 36f
White epidermoid, 268
White matter, 3–5, 26–27, 43
    ascending pathways, 27
    descending pathways, 27–28
    neuronal pathways, 27–29
    upper and lower motor neuron damage, 28t
Wild-type p53, 370–371
Wildtype allele, 407–408
Wingless-interleukin (WNT)-betacatenin, 137
Wingless-interleukin (WNT) pathway, 382, 382f
Wiskott–Aldrich syndrome, 356–357
Withdrawal reflex, 28
Wood's lamp, 412
World Health Organization (WHO), 97
WW domain-containing transcription regulator protein 1 (WWTR1), 252

X

X-LAG, 286
X-ray spectroscopy, 460
Xanthogranulomas, 205
Xanthomatous cells, 118
Xenoestrogenic chemical, 295

Y

Years of life lived with disability (YLD), 93
Years of life lost (YLL), 93
Years of potential life lost (YPLL), 95
Yes-associated protein 1 (YAP1), 163–164, 252, 481–482
Yolk sac tumor, 426–428
    clinical manifestations, 428
    diagnosis, 428
    epidemiology, 426
    etiology and risk factors, 426
    pathology, 426–428
    prognosis, 428
    treatment, 428

Z

Zonula adherens, 330–332
Zonula occludens, 330–332
Zonulae adherentes, 163